Basic Analysis II

Basic Analysis II: A Modern Calculus in Many Variables

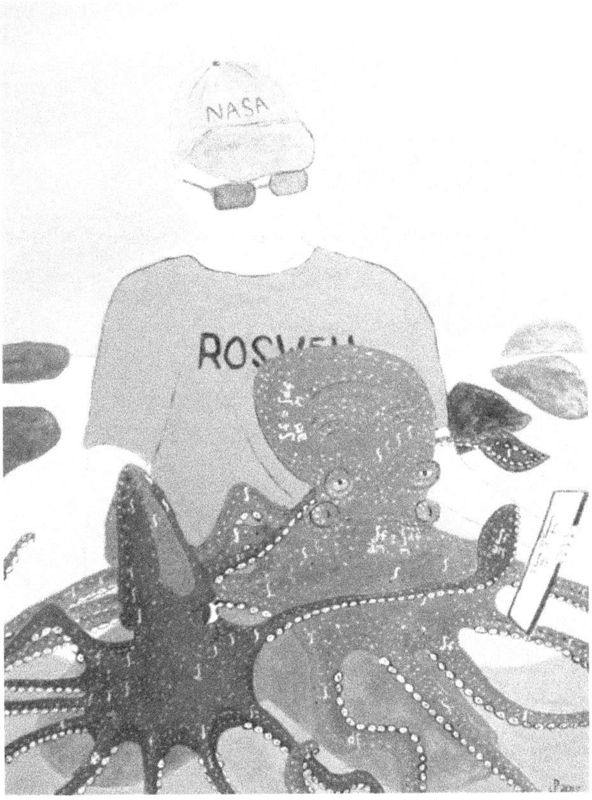

The cephalopods were eager to learn. Now both squid and octopi are coming to the lectures.

James K. Peterson
Department of Mathematical Sciences
Clemson University

CRC Press
Taylor & Francis Group
Boca Raton London New York

CRC Press is an imprint of the
Taylor & Francis Group, an **informa** business
A CHAPMAN & HALL BOOK

First edition published 2020
by CRC Press
6000 Broken Sound Parkway NW, Suite 300, Boca Raton, FL 33487-2742

and by CRC Press
2 Park Square, Milton Park, Abingdon, Oxon, OX14 4RN

CRC Press is an imprint of Taylor & Francis Group, LLC

ISBN: 978-1-138-05505-6 (hbk)
ISBN: 978-1-315-16621-6 (ebk)
LCCN: 2019059882

Dedication We dedicate this work to all of our students who have been learning these ideas of analysis through our courses. We have learned as much from them as we hope they have from us. We are a firm believer that all our students are capable of excellence and that the only path to excellence is through discipline and study. We have always been proud of our students for doing so well on this journey. We hope these notes in turn make you proud of our efforts.

Abstract This book introduces you to more ideas from multidimensional calculus. In addition to classical approaches, we will also discuss things like one forms, a little algebraic topology and a reasonable coverage of the multidimensional versions of the one variable Fundamental Theorem of Calculus ideas. We also continue your training in the *abstract* way of looking at the world. We feel that is a most important skill to have when your life's work will involve quantitative modeling to gain insight into the real world.

Acknowledgments I want to acknowledge the great debt I have to my wife, Pauli, for her patience in dealing with the long hours spent in typing and thinking. You are the love of my life.

The cover for this book is an original painting by me done in July 2017. It shows the moment when cephalopods first reached out to ask to be taught advanced mathematics. I was awed by their trust in picking us.

Table of Contents

Part I

Introduction

Chapter 1

Beginning Remarks

Our first book on basic analysis essentially discusses the abstract concepts underlying the study of calculus on the real line. A few higher dimensional concepts are touched on such as the development of rudimentary topology in $\Re^2$ and $\Re^3$, compactness and the tests for extrema for functions of two variables, but that is not a proper study of calculus concepts in two or more variables. A full discussion of the $\Re^n$ based calculus is quite complex and even our second volume on basic analysis cannot cover all the important things. The chosen focus here is on differentiation in $\Re^n$ and important concepts about mappings from $\Re^n$ to $\Re^m$ such as the inverse and implicit function theorem and change of variable formulae for multidimensional integration. These topics alone require much discussion and setup. These topics intersect nicely with many other important applied and theoretical areas which are no longer covered in mathematical science curricula. The knowledge here allows quantitatively inclined biologists and other scientists to more properly develop multivariable nonlinear ODE models for themselves and physicists to learn the proper background to study differential geometry and manifolds among many other applications. In addition, computer scientists who have not seen enough computationally slanted discussion can fill in gaps in their education. Also many mathematical sciences majors are not exposed to these ideas properly.

However, this course is just not taught at all anymore. It is material that students at all levels must figure out on their own. Most of the textbooks here are extremely terse and hard to follow as they assume a lot of abstract sophistication from their readers. This text is designed to be a self-study guide to this material. It is also designed to be taught from, but at my institution, it would be very difficult to find the requisite 10 students to register so that the course could be taught. Students who are coming in as master's students generally do not have the ability to allocate a semester course like this to learn this material. Instead, even if they have a solid introduction such as we try to give in our first book on analysis, they typically jump to the third volume on linear analysis which is an introduction to very abstract concepts such as metric spaces, normed linear spaces and inner products spaces along with many other needed deeper ideas. Such a transition is always problematic and the student is always trying to catch up on the holes in their background.

Also, many multivariable concepts and the associated theory are used in probability, operations research and optimization to name a few. In those courses, $\Re^n$ based ideas must be introduced and used despite the students not having a solid foundational course in such things. Hence, good students are reading about this themselves. This text we offer here is intended to give them a reasonable book to read and study from.

We also provide an nice introduction to line integrals and Green's Theorem in the plane and a re-branding of this material in terms of differential forms. New models of signals applied to a complex system use these sorts of ideas and it is a nice way to plant such a pointer to future things. We also have many examples of how we use these ideas in optimization and approximation algorithms.

Some key features of this approach to explaining this material are

- A very careful discussion of **some** of the many important concepts as we believe it is more important to teach **well** an important subset of this material rather than teaching a large quantity poorly. Hence, the coverage of the textbook is chosen to provide detailed coverage of an appropriate roadmap through this block of material.
- The discussion is also designed for self-study as greater mathematical maturity and training is desperately needed in the computer sciences, biological sciences and physical sciences at this critical junior and senior college level.
- Many pointers to ideas and concepts that are extensions of what is covered in the text so that students know this is just the first step of the journey.
- The emphasis is on learning how to think as no one can take all the courses they need in a tidy linear order. Hence, the ability to read and assess on their own is an essential skill to foster.
- An emphasis on learning how to understand the consequences of assumptions using a variety of tools to provide the proofs of propositions.
- An emphasis on learning to use abstraction as a key skill in solving problems.

We use Octave (Eaton et al. (4) 2020), which is an open source GPL licensed (Free Software Foundation (6) 2020) clone of MATLAB®, as a computational engine and we are not afraid to use it as an adjunct to all the theory we go over. Of course, you can use MATLAB® (MATLAB (18) 2018 - 2020) also if your university or workplace has a site license or if you have a personal license. Get used to it: theory, computation and science go hand in hand! Well, buckle up and let's get started!

1.1 Table of Contents

This text is based on quite a few years of teaching senior level courses in analysis and applied senior mathematics courses as well. It all began from handwritten notes starting roughly in the late 1990's but with material being added all the time. Along the way, these notes have been helped by the students we have taught and also by our own research interests as research informs teaching and teaching informs research in return.

In this text, we go over the following blocks of material.

Part One: These are our beginning remarks you are reading now which are in Chapter 1.

Part Two: Linear Mappings Here we are concerned with linear mappings.

- Chapter 2 goes over really basic things such as boundedness, infimums and so forth which are ideas from (Peterson (21) 2020) but we want to touch on them lightly here.
- Chapter 3 is a long chapter on vector spaces. Although you have probably seen this material before, we go to great pains to elaborate how the idea of vector spaces is used in practice in two and three dimensions and connect these ideas to more abstract finite dimensional vector spaces.
- Chapter 4 is about thinking carefully about linear mappings between finite dimensional vector spaces.
- Chapter 5 specializes the study of linear mappings to symmetric matrices. This includes coverage of rotating surfaces and systems of ordinary differential equations with complex roots.
- Chapter 6 is about extending the ideas of topology from two dimensional to n dimensions and includes discussions of compactness and functions of n variables.

- Chapter 7 works out the eigenvalue and eigenvector structure of a symmetric matrix using matrix norm ideas. We show such matrices always have an orthonormal basis of eigenvectors. This is almost the same proof we would use for the case of a self-adjoint linear operator in functional analysis (see (Peterson (22) 2020) and (Peterson (20) 2020)) but the tools we have here are set in $\Re^n$ so we cannot do that more general case.
- Chapter 8 works out a detailed example of how we use vector spaces and different coordinate systems to understand the basics of the movement of a satellite around the earth. This subject is called orbital mechanics.
- Chapter 9 proves the many properties of n dimensional determinants in detail. This is probably material you have not seen in other classes so it is good for you to see this.

Part Three: Calculus of Many Variables Here we are concerned with the differentiability of functions of n variables and the classical results called the inverse and implicit function theorems.

- Chapter 10 Here we present the theory of differentiation in n variables which includes tangent plane error.
- Chapter 11 We talk carefully about how to detect the presence of extremal values in a smooth function of n variables.
- Chapter 12 presents the important theorems called the inverse and implicit function theorems. These are complicated to prove and take some time.
- Chapter 13 is about using these ideas in some applications. First, we explain the use of tangent plane approximations in nonlinear ordinary systems of differential equations. Then we show how to use approximation ideas to use finite difference techniques to solve the diffusion equation numerically with code.

Part Four: Integration We are now ready to do n dimensional integration.

- Chapter 14 covers the full theory of n dimensional Riemann Integration. We must include a careful discussion of measure zero extended to $\Re^n$ and the idea of a multidimensional volume for a subset which is not a rectangle.
- Chapter 15 discusses how to change variables and Fubini's Theorem.
- Chapter 16 shows us how to build a version of the recapture theorem $\int_a^b f' = f(b) - f(a)$ in the two dimensional context. This uses ideas of conservative force fields, path independence of line integrals and the connection between integration over closed paths to integrals over the region enclosed by the path which is called Green's Theorem. We stop here and do not move into the higher dimensional versions which really require Differential Geometry and manifolds to do properly.
- Chapter 17 recasts the ideas of conservative force fields and path independence into the language of differential forms. This is just a taste of a conversation we will have once differential geometry is available to us.

Part Five: Applications We finish this text with two sophisticated applications.

- Chapter 18 shows you how to find the solution to the linear systems $x' = Ax$ in n dimensions using the e^{At} approach and the Jordan Canonical Form of the matrix A.
- Chapter 19 is about constrained optimization, which we can tackle reasonably well now that we have the implicit function theorem available for our use.

Part Six: Summary In Chapter 20 we talk about the things you have learned here and where you should go next to continue learning basic analysis.

There is much more we could have done, but these topics are a nice introduction into the further use of abstraction in your way of thinking about models.

1.2 Acknowledgments

Since I regained the ability to run in 2009, I am less grumpy than usual. Running on the local trails in the forest helps with the stress levels and counteracts my most important philosophical view: "*There is always room for a donut!*" Of course, now that I am a bit older, all those donuts, cookies and so forth have had to be cut back a bit. It turns out that as I age, my physiology slows to a crawl and even a cookie crumb expands into several pounds along the old waistline. However, I still look wistfully at all the sweets even though I have fewer samples!

My daily coffee for awhile was half-decaf but with all this typing, I have gone back to full strength. Strong, hot coffee helps to delay carpal tunnel, I hope. So, while holding a cup in my hands, I wish to thank all the students U have taught for helping me by listening to what I say in the lectures, finding my typographical errors (oh, so many) and other mistakes. I am always hopeful that my efforts help students to some extent and also impart some of my obvious enthusiasm for the subject. Of course, the reality is that I have been damaging students for years by forcing them to learn these abstract things. This is why I tell people at parties I am a roofer or electrician. If I am identified as a mathematician, it could go badly given the terrible things I inflict on the students in the classes. Who knows whom they have told about my tortuous methods. Hence, anonymity is best. Still, by writing these notes, I have gone public. Sigh. This could be bad. So before I am taken out with a very public hit, I, of course, also want to thank my family for all their help in many small and big ways. It is amazing to me that I have been teaching this material since our children were in grade school!

Jim Peterson
School of Mathematical and Statistical Sciences
Clemson University

Part II

Linear Mappings

Chapter 2

Preliminaries

We begin with some preliminaries. We do discuss these things in (Peterson (21) 2020) but a recap is always good and it is nice to keep stuff reasonably self-contained.

2.1 The Basics

Let S be a set of real numbers. We say S is bounded if there is a number $M > 0$ so that $x \leq M$ for all $x \in S$. For example, if $S = \{y : y = \tanh(x),\ x \in \Re\}$, then many numbers can play the role of a bound; e.g. $M = 4, 3, 2$ and 1, among others. Let's be more precise.

Definition 2.1.1 Bounded Sets

We say the set S of real numbers is bounded above if there is a number M so that $x \leq M$ for all $x \in S$. Any such number M is called an **upper bound** *of S. We usually abbreviate this by the letters* **u.b.** *or just* **ub**.

We say this same set S is bounded below if there is a number m so that $x \geq m$ for all $x \in S$. Any such number m is called a **lower bound** *of S; we abbreviate this a* **l.b.** *or simply* **lb**.

The next idea is more abstract; the idea of a **least upper bound** and **greatest lower bound**. We have to be careful to define this carefully.

Definition 2.1.2 Least Upper Bound and Greatest Lower Bound

The **least upper bound** *or* **lub** *or* **supremum** *or* **sup** *of a nonempty set S of real numbers is a number U satisfying*

1. *U itself is an upper bound of S*
2. *if M is any other upper bound of S, $U \leq M$.*

Similarly, the **greatest lower bound** *or* **glb** *or* **infimum** *or* **inf** *of the nonempty set S of real numbers is a number L satisfying*

1. *L itself is a lower bound of S*
2. *if m is any other lower bound of S, $m \leq L$.*

We often use the notation $\inf(S)$ and $\sup(S)$ to denote these numbers.

Note, $\inf(S)$ and $\sup(S)$ need not be in the set S itself!

We use the following conventions:

1. if S has no lower bound, we set $\inf(S) = -\infty$. If S has no upper bound, we set $\sup(S) = \infty$.
2. We can extend these definitions to an empty set which is sometimes nice to do. If $S = \emptyset$, since all positive numbers are not lower bounds, we can argue we should set $\inf(\emptyset) = \infty$. Similarly, we set $\sup(\emptyset) = -\infty$.

When the $inf(S)$ and $\sup(S)$ belong to S, this situation is special. It is worthy of a definition of a new set of terms.

Definition 2.1.3 The Minimum and Maximum of a Set

Let S be a set of real numbers. We say Q is a **maximum** *of S if $x \leq Q$ for all $x \in S$ and $Q \in S$. Then, we say q is a* **minimum** *of S when $x \geq q$ for all $x \in S$ and $q \in S$. We also say q is a* **minimal element** *of S and Q is a* **maximal element** *of S.*

Note if Q is a maximum of S, by definition, Q is an upper bound of S. Hence, $\sup(S) \leq M$. On the other hand, since $Q \in S$, we must also have $Q \leq \sup(S)$. We conclude $Q = \sup(S)$. So, when the supremum of a set is achieved by an element of the set, that element is the supremum. We can argue in a similar way to show $m = \inf(S)$.

Now, there are things we cannot prove about the set of real numbers $\Re$. Instead we must assume a certain property called the **Completeness Axiom.**

Axiom 1 The Completeness Axiom

Let S be a set of real numbers which is nonempty and bounded above. The $\sup(S)$ exists and is finite. If S is bounded below, then $\inf(S)$ exists and is finite.

So bounded sets of real numbers always have a finite infimum and a finite supremum. Of course, they need not possess a minimum or maximum! We can prove some useful things now.

Theorem 2.1.1 S has a Maximal Element if and only if sup(S) is in S

Let S be a nonempty set of real numbers which is bounded above. Then S has a **maximal element** *if and only if $\sup(S) \in S$.*

Proof 2.1.1

($\Longleftarrow$):

Assume $\sup(S) \in S$. By definition, $\sup(S)$ is an upper bound of S and so $x \leq \sup(S)$ for all $x \in S$. This shows $\sup(S)$ is the maximum.

($\Longrightarrow$):

Let Q denote the maximum of S. Then, by definition, $x \leq Q$ for all $x \in S$ which shows Q is an upper bound of S. Then, from the definition of the supremum of S, we must have $\sup(S) \leq Q$. However, $\sup(S)$ is also an upper bound of S and so we also have $Q \leq \sup(S)$. Combining these inequalities, we see $Q = \sup(S)$. ■

We can prove a similar theorem about the minimum.

Theorem 2.1.2 S has a Minimal Element if and only if inf(S) is in S

Let S be a nonempty set of real numbers which is bounded below. Then S has a **minimal element** *if and only if $\inf(S) \in S$.*

Proof 2.1.2
The argument is similar to the one we did for the maximum, so we'll leave this one to you. ■

Now the next two lemmas are very important in modern analysis.

Lemma 2.1.3 Supremum Tolerance Lemma

Let S be a nonempty set that is bounded above. Then we know $\sup(S)$ exists and is finite by the completeness axiom. Moreover, for all $\epsilon > 0$, there is an element $y \in S$ so that $\sup(S) - \epsilon < y \leq \sup(S)$.

Proof 2.1.3
We argue by contradiction. Suppose for a given $\epsilon > 0$, we cannot find such a y. Then we must have $y \leq \sup(S) - \epsilon$ for all $y \in S$. This tells us $\sup(S) - \epsilon$ is an upper bound of S and so by definition, $\sup(S) \leq \sup(S) - \epsilon$ which is not possible. Hence, our assumption is wrong and we can find such a y. ■

Lemma 2.1.4 Infimum Tolerance Lemma

Let S be a nonempty set that is bounded below. Then we know $\inf(S)$ exists and is finite by the completeness axiom. Moreover, for all $\epsilon > 0$, there is an element $y \in S$ so that $\inf(S) \leq y \leq \inf(S) + \epsilon$.

Proof 2.1.4
We argue by contradiction. Suppose for a given $\epsilon > 0$, we cannot find such a y. Then we must have $y \geq \inf(S) + \epsilon$ for all $y \in S$. This tells us $\inf(S) + \epsilon$ is a lower bound of S and so by definition, $\inf(S) \geq \inf(S) + \epsilon$ which is not possible. Hence, our assumption is wrong and we can find such a y. ■

We often abbreviate these lemmas as the **STL** and **ITL**, respectively.

Homework

Exercise 2.1.1 *If S is an infinite set bounded above, prove there is a sequence (x_n) in S with $\lim_{N\to\infty} x_n = \sup(S)$.*

Exercise 2.1.2 *If S is an infinite set bounded below, prove there is a sequence (x_n) in S with $\lim_{N\to\infty} x_n = \inf(S)$.*

Exercise 2.1.3 *Prove the nonempty set S has a minimal element m if and only if $m = \inf(S)$.*

Exercise 2.1.4 *Prove the nonempty set S has a maximal element M if and only if $M = \inf(S)$.*

There is also the notion of **compactness** which we have discussed thoroughly in (Peterson (21) 2020). However, let's go over it again as this is an important concept. We begin with **sequential compactness**. Recall

Definition 2.1.4 Sequential Compactness

Let S be a set of real numbers. We say S is **sequentially compact** *if every sequence (x_n) in S has at least one subsequence which converges to an element of S.*

This definition has many consequences. The first is a way to characterize the sequentially compact subsets of the real line.

Theorem 2.1.5 A Set S is Sequentially Compact if and only if It is Closed and Bounded

Let S be a set of real numbers. Then S is sequentially compact if and only if it is a bounded and closed set.

Proof 2.1.5
($\Longrightarrow$) If we assume S is sequentially compact, we must show S is closed and bounded.

a: *To show S is closed, we show S contains its limit points. Let x be a limit point of S. Then there is a sequence (x_n) in S so that $x_n \rightarrow x$. Since (x_n) is in S, the sequential compactness of S says there is a subsequence (x_n^1) of (x_n) and a $y \in S$ so that $x_n^1 \rightarrow y$. But the subsequential limits of a converging sequence must converge to the same limit. Hence, $y = x$. This tells us $x \in S$ and so the limit point x is in S. Since the choice of limit point x was arbitrary, this implies S is a closed set.*

b: *To show S is bounded, we assume it is not. Then there must be a sequence (x_n) in S with $|x_n| > n$ for all positive integers n. Now apply the sequential compactness of S again. This sequence must contain a convergent subsequence (x_n^1) which converges to an $x \in S$. But this implies the elements of the sequence (x_n^1) must be bounded as there is a theorem which tells us convergent sequences must be bounded. But this subsequence is from an unbounded sequence which is monotonic in absolute value. This is a contradiction. So our assumption that S is unbounded is wrong and we conclude S must be bounded.*

($\Longleftarrow$) We assume S is closed and bounded. We want to show S is sequentially compact. Let (x_n) be any sequence in S. Since S is a bounded set, by the Bolzano - Weierstrass Theorem, there is at least one subsequence (x_n^1) in (x_n) and a real number x with $x_n^1 \rightarrow x$. This says x is a limit point of S and since S is closed, we must have $x \in S$. Hence the sequence (x_n) in S has a subsequence which converges to a point in S. Since our choice of sequence in S is arbitrary, this shows S is sequentially compact. ■

Homework

Exercise 2.1.5 *Mimic the argument above to show a set S in $\Re^2$ is sequentially compact if and only it is closed and bounded. This uses the Bolzano - Weierstrass Theorem in 2D which you have seen in (Peterson (21) 2020). We will go over this again later, but you should be able to do this exercise now.*

Exercise 2.1.6 *Let $f : \Re \rightarrow \Re$ be given and let $\overline{S} = \{x \in \Re : f(x) = 0\}$. The set $\overline{S}$ is called the* **support** *of f. Do the C^∞ bump functions discussed in (Peterson (21) 2020) have compact support?*

Exercise 2.1.7 *Let $f : \Re^2 \rightarrow \Re$ be given and let $\overline{S} = \{\boldsymbol{x} \in \Re^2 : f(\boldsymbol{x}) = 0\}$. Again, the set $\overline{S}$ is called the* **support** *of f. Generalize the C^∞ bump functions discussed in (Peterson (21) 2020) to the 2D situation. Do these new C^∞ bump functions have compact support?*

Now add continuity to the mix. Here is a typical result.

Theorem 2.1.6 The Range of a Continuous Function on a Sequentially Compact Domain is Also Sequentially Compact

Let S be a nonempty sequentially compact set of real numbers. Assume $f : S \rightarrow \Re$ is continuous on S. Then the range $f(S)$ is sequentially compact.

Proof 2.1.6
We know $f(S) = \{f(x)|x \in S\}$. Let (y_n) be sequence in $f(S)$. Then there is an associated sequence

(x_n) in S so that $f(x_n) = y_n$. But S is sequentially compact, so there is a subsequence (x_n^1) which converges to a point $x \in S$. Since f is continuous at x, $f(x_n^1) \rightarrow f(x)$. Let y denote the number $f(x)$. We have shown there is a subsequence $(y_n^1) = f(x_n^1)$ with $y_n^1 \rightarrow y$ with $y \in f(S)$. Hence, we have shown $f(S)$ is sequentially compact. ■

This leads to our first **extremal theorem**.

Theorem 2.1.7 Continuous Functions on Sequentially Compact Domains Have a Minimum and Maximum Value

Let S be a sequentially compact set of real numbers and let $f : S \rightarrow \Re$ be continuous on S. Then there are two sequences x_n^m and x_n^M which converge to the points x_m and x_M respectively and $f(x) \geq f(x_m)$ and $f(x) \leq f(x_M)$ for all $x \in S$. Thus, $f(x_m)$ is the minimum value of f on S which is achieved at x_m. Moreover, $f(x_M)$ is the maximum value of f on S which is achieved at x_M.

We call the sequences (x_n^m) and (x_n^M) minimizing and maximizing sequences for f on S.

Proof 2.1.7

By Theorem 2.1.6, $f(S)$ is sequentially compact and nonempty. Hence, $f(S)$ is a closed and bounded set of real numbers by Theorem 2.1.5. We thus know $M = \sup(f(S))$ and $m = \inf(f(S))$ both exist and are finite. By the **ITL** *and STL, satisfying*

$$
\begin{aligned}
\exists\, x_1^M \in S \ni M - 1 &< f(x_1^M) \leq M \\
\exists\, x_2^M \in S \ni M - 1/2 &< f(x_2^M) \leq M \\
\exists\, x_3^M \in S \ni M - 1/3 &< f(x_3^M) \leq M \\
&\vdots \\
\exists\, x_n^M \in S \ni M - 1/n &< f(x_n^M) \leq M \\
&\vdots
\end{aligned}
$$

and

$$
\begin{aligned}
\exists\, x_1^m \in S \ni m &\leq f(x_1^m) < m + 1 \\
\exists\, x_2^m \in S \ni m &\leq f(x_2^m) < m + 1/2 \\
\exists\, x_3^m \subset S \supset m &\leq f(x_3^m) < m + 1/3 \\
&\vdots \\
\exists\, x_n^m \in S \ni m &\leq f(x_n^M) < m + 1/n \\
&\vdots
\end{aligned}
$$

We see $f(x_n^M) \rightarrow M$ and $f(x_n^m) \rightarrow m$. Hence $(f(x_n^M))$ and $(f(x_n^m))$ are convergent sequences in the closed set $f(S)$. So their limit values must be in $f(S)$. Hence, we know $m \in f(S)$ and $M \in f(S)$. Now apply Theorem 2.1.1 and Theorem 2.1.2 to see there are elements x_m and x_M in S so that $f(x_m) = m \leq f(x)$ and $f(x_M) = M \geq f(x)$ for all $x \in S$. ■

Homework

Exercise 2.1.8 *If $f : S \rightarrow \Re$ has a maximum M at $x_0 \in S$, prove there is a sequence (x_n) in S so that $f(x_n) \rightarrow M$.*

Exercise 2.1.9 *If $f : S \to \Re$ has a minimum m at $x_0 \in S$, prove there is a sequence (x_n) in S so that $f(x_n) \to m$.*

Exercise 2.1.10 *If f has an extremum value at x_0 in its domain, is it true x_0 is unique?*

Now add differentiation.

Theorem 2.1.8 Derivative of f at an Interior Point Local Extremum is Zero

> *Let $f : [a, b] \to \Re$ where $[a, b]$ is a finite interval. Assume f has a local extrema at an interior point $p \in [a, b]$. Then if f is differentiable at p, $f'(p) = 0$.*

Proof 2.1.8
Since p is an interior point, there is an $r_1 > 0$ so that $—x - p| < r_1 \Longrightarrow x \in [a, b]$. Also, since $f(p)$ is a local extreme value, there is another $r_2 > 0$ so that $f(p) \leq f(x)$ for $x \in B_{r_2}(p)$ with $B_{r_2}(p) \subset [a, b]$ if $f(p)$ is a local maximum or $f(p) \geq f(x)$ for $x \in B_{r_2}(p)$ with $B_{r_2}(p) \subset [a, b]$ if $f(p)$ is a local minimum. For convenience, we will assume $f(p)$ is a local maximum and let you argue the local minimum case for yourself.
Combining these facts, we see

$$|x - p| < r < \min(r_1, r_2) \Longrightarrow f(x) \quad \geq \quad f(p)$$

Thus, for $p < x < p + r$, $f(x) - f(p) \leq 0$ and $x - p > 0$ telling us the ratio $\frac{f(x)-f(p)}{x-p} \leq 0$. Hence, $f'(p^+) = \lim_{x \to p^+} \frac{f(x)-f(p)}{x-p} \leq 0$. Since f is differentiable at p, we know $f'(p^+) = f'(p)$. Thus, we have $f'(p) \leq 0$.
Next, for $p - r < x < p$, we have $f(x) - f(p) \leq 0$ and $x - p < 0$. Thus, the ratio $\frac{f(x)-f(p)}{x-p} \geq 0$. Hence, $f'(p^-) = \lim_{x \to p^-} \frac{f(x)-f(p)}{x-p} \geq 0$. Since f is differentiable at p, we know $f'(p^-) = f'(p)$. Thus, we have $f'(p) \geq 0$.
Combining, we see $f'(p) = 0$. You can do a similar argument for the case of a local minimum. ■

Homework

Exercise 2.1.11 *Give an example of a differentiable function f on a domain S with an extremum at x_0 in S but $f'(x_0) \neq 0$.*

Exercise 2.1.12 *Give an example of a differentiable function f on a domain S in $\Re^2$ with an extremum at $\boldsymbol{x_0}$ in S but $\nabla f(\boldsymbol{x_0}) \neq 0$. You have seen functions of two variables already and in your calculus class you have seen minimization/ maximization of a function of two variables subject to domain constraints. Use an example of that sort.*

Exercise 2.1.13 *If f has domain S in $\Re$ and possesses a finite infimum and a finite supremum, prove $\sup_{x \in S}(-f(x)) = -\inf_{x \in S}(f(x))$. Hint, look at $f(x) = 1 + x^2$.*

We will often be in situations where the classical limits fail to exist. However, even if this is the case, **limit inferiors** and **limit superiors** do exist! Let(a_n) be a sequence of real numbers, Let (a_n^1) be any subsequence of (a_n) which converges. If (a_n) is a bounded sequence, the Bolzano - Weierstrass Theorem guarantees there is at least one such subsequence. Consider the set

$$S \quad = \quad \{\alpha \in \Re | \exists (a_n^1) \subset (a_n) \ni a_n^1 \to \alpha\}$$

The set S is called the **set of subsequential limits** of the sequence (a_n). We define the limit inferior and limit superior of a sequence using S.

Definition 2.1.5 Limit Inferior and Limit Superior of a Sequence

The **limit inferior** *of the sequence* (a_n) *is denoted by* $\liminf(a_n) = \underline{\lim}(a_n)$ *and defined by* $\liminf(a_n) = \inf(S)$. *In a similar fashion, the* **limit superior** *of the sequence* (a_n) *is denoted by* $\limsup(a_n) = \overline{\lim}(a_n)$ *and defined by* $\limsup(a_n) = \sup(S)$.

Recall, it is possible for a sequence to converge to $\pm\infty$. For example, if $(a_n) = (n)$ for $n \geq 1$, we know the sequence is not bounded above and $\lim_n (a_n) = \lim_n (n) = \infty$. Similarly, if $(a_n) = (-n)$ for $n \geq 1$, we know the sequence is not bounded below and $\lim_n (a_n) = \lim_n (-n) = -\infty$. There are many other examples! Thus, it is possible to have sequences whose limits are not finite real numbers. The set of **extended real numbers** is denoted $\bar{\Re}$. and consists of all finite real numbers plus two additional symbols $+\infty$ and $-\infty$. Note, we use these symbols to connect with our intuition about *converging to* $\pm\infty$ but we could just as well used the symbols α and β. But for us, we have defined

$$\bar{\Re} = \Re \cup \{+\infty\} \cup \{-\infty\}$$

We use the following conventions: for addition we have

$$\begin{aligned} x + (+\infty) &= +\infty, \ \forall x \in \Re \\ x - (+\infty) &= -\infty, \ \forall x \in \Re \\ (+\infty) + x &= +\infty, \ \forall x \in \Re \\ (-\infty) + x &= -\infty, \ \forall x \in \Re \\ (-\infty) + (-\infty) &= -\infty, \\ (+\infty) + (+\infty) &= +\infty \end{aligned}$$

For multiplication,

$$\begin{aligned} (+\infty) \cdot x &= x \cdot (+\infty) = +\infty, \ \forall x > 0 \\ (+\infty) \cdot 0 &= 0 \cdot (+\infty) = 0, \\ (+\infty) \cdot x &= x \cdot (+\infty) = -\infty, \ \forall x < 0 \\ (-\infty) \cdot x &= x \cdot (-\infty) = -\infty, \ \forall x > 0 \\ (-\infty) \cdot 0 &= 0 \cdot (-\infty) = 0, \\ (-\infty) \cdot x &= x \cdot (-\infty) = +\infty, \ \forall x < 0 \\ (+\infty) \cdot (+\infty) &= +\infty \\ (-\infty) \cdot (-\infty) &= +\infty \\ (+\infty) \cdot (-\infty) &= -\infty \\ (-\infty) \cdot (+\infty) &= -\infty \end{aligned}$$

For division

$$\begin{aligned} \frac{x}{+\infty} &= 0, \ \forall x \in \Re \\ \frac{x}{-\infty} &= 0, \forall x \in \Re \\ \frac{+\infty}{x} &= +\infty, \ \forall x > 0 \\ \frac{+\infty}{x} &= -\infty, \ \forall x < 0 \\ \frac{-\infty}{x} &= -\infty, \ \forall x > 0 \end{aligned}$$

$$\frac{-\infty}{x} = \infty, \forall x < 0$$

There are several operations that are not defined. These are $0/0$, $(+\infty) + (-\infty)$, $(+\infty)/(-\infty)$, $(-\infty)/(+\infty)$, $(+\infty)/(+\infty)$ and $(-\infty)/(-\infty)$. However, with the above conventions, arithmetic in $\bar{\Re}$ is well-defined. With these definitions, we can extend limit inferiors and limit superiors to the extended reals. Let (a_n) be any sequence in $\bar{\Re}$. The set of subsequential limits now is

$$S = \{\alpha \in \bar{\Re} | \exists (a_n^1) \subset (a_n) \ni a_n^1 \to \alpha\}$$

where we now allow the possibility that $\pm\infty \in S$. We need to think carefully about limiting behavior here.

- If (a_n^1) is a subsequence which is in the reals, then $a_n^1 \to +\infty$ implies for sufficiently large n^1, $1/a_n^1 \to 0$. This means for any $\epsilon > 0$, there is an N so that $n^1 > N$ implies $|1/a_n^1| < \epsilon$ or $|a_n^1| > 1/\epsilon$. As $\epsilon \to 0^+$, $1/\epsilon \to +\infty$. So this is the same as saying for all $R > 0$, there is an N so that $|a_n^1| > R$ when $n^1 > N$.
- If the subsequence $(a_n^1) \to -\infty$, this is handled like we did above.
- If the subsequence has an infinite number of $+\infty$ values, then our tests for convergence, which require us to look at the difference $a_n^1 - \alpha$, where α is the limit, fail when we look at the term $\infty - \infty$ which is not a defined operation. In this case, we will simply posit that $\alpha = \infty$ or $\alpha = -\infty$ depending on the structure of the subsequence.

Here are some examples:

1. $(a_n) = (\tanh(n))$. Then $\underline{\lim}(a_n) = -1$ and $\overline{\lim}(a_n) = +1$.
2. $(a_n) = (n^2)$. Then $\underline{\lim}(a_n) = +\infty$ and $\overline{\lim}(a_n) = +\infty$.
3. $(a_n) = (\cos(n\pi/3))$ for $n \geq 0$. Then the range of this sequence is a set of repeating blocks

$$(a_n) = \{\textbf{Block}, \textbf{Block}, \ldots\}$$

where

$$\textbf{Block} = \{1(n = 0), 1/2(n = 1), -1/2(n = 2), -1(n = 3), -1/2(n = 4), 1/2(n = 5)\}$$

We see for integers k, we have subsequences that converge to each element in this block:

$$\begin{aligned}
\cos((6k + 0)\pi/3) &= 1, \Longrightarrow a_{6k+0} \to 1 \\
\cos((6k + 1)\pi/3) &= 1/2, \Longrightarrow a_{6k+1} \to 1/2 \\
\cos((6k + 2)\pi/3) &= -1/2, \Longrightarrow a_{6k+2} \to -1/2 \\
\cos((6k + 3)\pi/3) &= -1, \Longrightarrow a_{6k+3} \to -1 \\
\cos((6k + 4)\pi/3) &= -1/2, \Longrightarrow a_{6k+4} \to -1/2 \\
\cos((6k + 5)\pi/3) &= 1/2, \Longrightarrow a_{6k+5} \to 1/2
\end{aligned}$$

A little thought then allows us to conclude $S = \{-1. - 1/2, 1/2, 1\}$ and so $\underline{\lim}(a_n) = -1$ and $\overline{\lim}(a_n) = +1$.

We can extend these ideas to functional limits. Consider the function f defined on a finite interval $[a, b]$. Let $p \in [a, b]$. We say $\alpha \in \bar{\Re}$ is a **cluster point** of f as we approach p, if there is a sequence

(x_n) in $[a, b]$ with $x_n \rightarrow p$ and $f(x_n) \rightarrow \alpha$. We must now think of convergence in the sense of the extended reals, of course. We define the set of cluster points $S(p)$ then as

$$S(p) = \{\exists (x_n) \subset [a, b] \ni x_n \rightarrow p \text{ and } f(x_n) \rightarrow \alpha\}$$

and we define the limit inferior as x approaches p of f to be $\underline{\lim}_{x \rightarrow p} f(x) = \inf(S(p))$. Similarly, the limit superior as x approaches p of f is $\overline{\lim}_{x \rightarrow p} f(x) = \sup(S(p))$. In this version of these limits, we do not have an unbounded domain and so we cannot have a sequence $x_n \rightarrow \pm\infty$. The most general version has the domain of f simply being the set Ω and

$$S(p) = \{\exists (x_n) \subset \Omega \ni x_n \rightarrow p \text{ and } f(x_n) \rightarrow \alpha\}$$

In this case, we can have the extended real value limits $\pm\infty$ but the definitions of the limits above are still the same. We also use the notation $\underline{\lim}_{x \rightarrow p} f(x) = \liminf_{x \rightarrow} f(x)$ and $\overline{\lim}_{x \rightarrow p} f(x) = \limsup_{x \rightarrow p} f(x)$.

Example 2.1.1 *Find the cluster points of $f(x) = \cos(1/x)$ for $x \neq 0$.*

Solution *Here f is defined in $\Re$ not $\overline{\Re}$. Since $-1 \leq \cos(y) \leq 1$, given any $\alpha \in [-1, 1]$, the sequence (s_n) defined by $x_n = 1/(2n\pi + \cos^{-1}(\alpha))$ satisfies $x_n \rightarrow 0$ and $f(x_n) = \alpha$ for all n. Hence, $S(0) = [-1, 1]$ and $\underline{\lim}_{x \rightarrow 0} f(x) = \inf(S(0)) = -1$ and $\overline{\lim}_{x \rightarrow 0} f(x) = \sup(S(0)) = 1$. Because f is continuous at all $p \neq 0$, $S(p) = \{\cos(1/p)\}$ there and $\underline{\lim}_{x \rightarrow p} f(x) = \inf(S(p)) = \cos(1/p)$ and $\overline{\lim}_{x \rightarrow p} f(x) = \sup(S(p)) = \cos(1/p)$. Hence, $\lim_{x \rightarrow p} f(x) = f(p)$ at those points.*

We can prove an alternate definition of the limit inferior and limit superior of a sequence.

Theorem 2.1.9 Alternate Definition of the Limit Inferior and Limit Superior of a Sequence

Let (a_n) be a sequence of real numbers. Then

$$\underline{\lim}(a_n) = \sup_k \inf_{n \geq k} a_n \text{ and } \overline{\lim}(a_n) = \inf_k \sup_{n \geq k} a_n$$

To see this clearer, let

$$z_n = \inf\{a_k, a_{k+1}, \ldots\} = \inf_{n \geq k} a_n \text{ and } w_n = \sup\{a_k, a_{k+1}, \ldots\} = \sup_{n \geq k} a_n$$

Then,

$$z_1 \leq z_2 \leq \ldots \leq z_n \leq \ldots \leq w_n \leq w_{n-1} \leq \ldots \leq w_1$$

The (z_n) sequence is monotone increasing and the (w_n) sequence is monotone decreasing. Hence, $\lim_n z_n \uparrow \underline{\lim}(a_n)$ and $\lim_n w_n \downarrow \overline{\lim}(a_n)$.

Proof 2.1.9

Careful proofs of these ideas are in (Peterson (21) 2020). You should go back and review them. ■

Homework

Exercise 2.1.14 *Let $S_f(x_0)$ denote the set of cluster points of f at x_0. Assume both sets of cluster points are bounded sets. Is it true $S_f(x_0)\, S_g(x_0) \subset S_{fg}(x_0)$? If so, prove the result and if not, show* *a counterexample.*

Exercise 2.1.15 *Consider $f(x) = x$ and $g(x) = 1/x$ on the domain $(0,1)$. Calculate $S_f(0)$ $S_g(0)$ and $S_{fg}(0)$. Compare to the previous exercise. What is the difference?*

Exercise 2.1.16 *Consider $f(x) = x$ and $g(x) = \sin(1/x)$ on the domain $[-1,1]$. Calculate $S_f(0)$ $S_g(0)$ and $S_{fg}(0)$. Compare to the previous two exercises. What is the difference?*

Exercise 2.1.17 *Where does the proof of $S_{fg}(x_0) \subset S_f(x_0)\ S_g(x_0)$ break down?*

2.2 Some Topology in $\Re^2$

If you look at Figure 2.1, you can see the big difference between simple set topology in $\Re$ and in $\Re^2$.

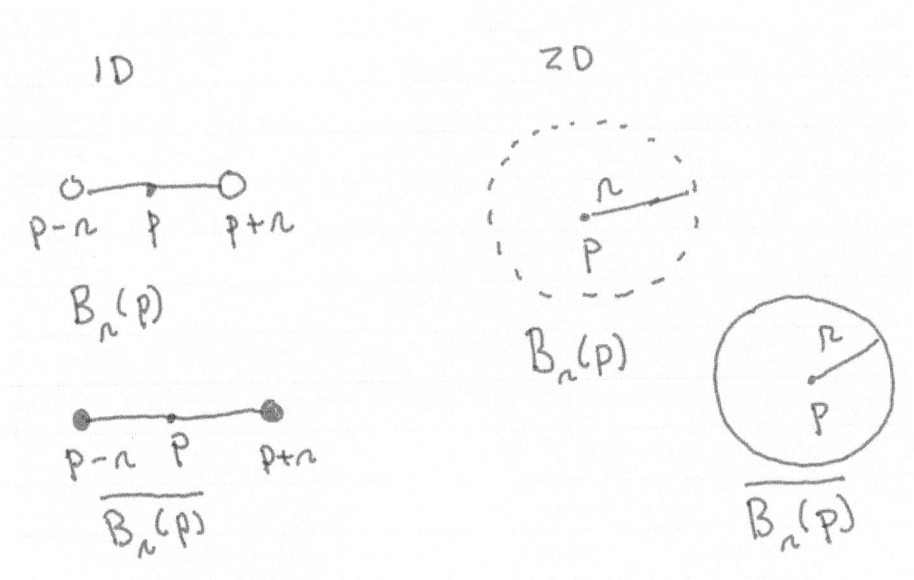

Figure 2.1: Open and closed balls in one and two dimensions.

The open and closed balls in $\Re$ are simply open intervals on the x-axis. You draw the interval centered at p and then it is an open ball if the endpoints are not included and a closed ball otherwise. We define this as $B_r(p) = \{x \in \Re : p - r \leq x \leq p + r\}$. In $\Re^2$, the open intervals become 2D circles centered at a point in $\Re^2$. Thus, $B_r(\boldsymbol{p}) = \{\boldsymbol{x} \in \Re^2 : \|\boldsymbol{x} - \boldsymbol{p} \leq r\}$ where $\boldsymbol{x}$ is the vector $\boldsymbol{x} = [x_1 // x_2]$ and the center is the vector $\boldsymbol{p} = [p_1 // p_2]$. Get familiar with this by drawing lots of pictures. From our discussions of topology in $\Re$ in (Peterson (21) 2020), the important concepts to move into $\Re^2$ are as follows:

1. If S is a subset of $\Re^2$, S^C is the **complement** of S and consists of all $\boldsymbol{x}$ not in S.
2. $\boldsymbol{p}$ is an **interior point** of the set S in $\Re^2$ if there is a positive r so that $B_r(\boldsymbol{p}) \subset S$. We define S to be an **open set** if all of its points are **interior points**. Note a curve in $\Re^2$ has no interior! Draw the graph of $y = x^2$ for $x \in [-1,1]$ and this arc has no interior points.
3. $\boldsymbol{p}$ is a **boundary point** of S if all balls $B_r(\boldsymbol{p})$ intersect both S and S^C in at least one point. In Figure 2.2, we see boundary point examples in both $\Re$ and $\Re^2$.
4. A set S is said to be **closed** if S^C is open.

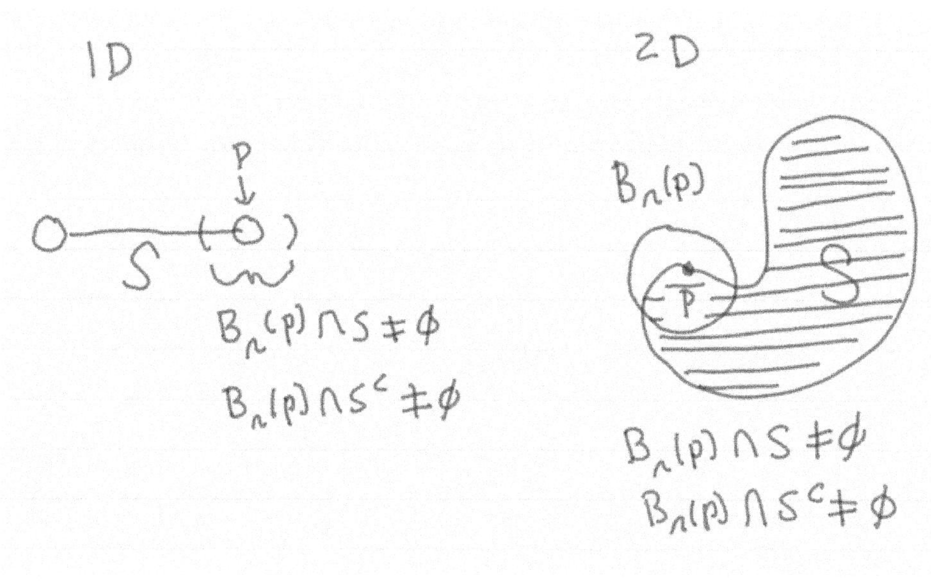

Figure 2.2: Boundary points in one and two dimensions.

A sequence in $\Re^2$ is a set of two dimensional vectors

$$\left(\begin{bmatrix} x_n \\ y_n \end{bmatrix}\right)$$

We say this sequence converges to the vector $\begin{bmatrix} x \\ y \end{bmatrix}$ if

$$\forall \epsilon > 0, \ \exists \, N \ni \|\boldsymbol{X_n} - \boldsymbol{X}\| < \epsilon, \ \forall n > N$$

where we let $\boldsymbol{X_n} = \begin{bmatrix} x_n \\ y_n \end{bmatrix}$ and $\boldsymbol{X} = \begin{bmatrix} x \\ y \end{bmatrix}$. Further, we use the norm

$$\|\boldsymbol{X_n} - \boldsymbol{X}\| = \sqrt{(x_n - x)^2 + (y_n - y)^2}$$

as our way to measuring the distance between vectors in $\Re^2$. Of course, other norms are possible but we will simply use the standard Euclidean norm for now. Now that we know what a sequence of vectors is and what we mean by the sequence converging, we can define some other special points.

1. A point $\boldsymbol{p}$ is an **accumulation point** of S if every ball $B_r(\boldsymbol{p})$ contains a point of S different from $\boldsymbol{p}$. This means there is a sequence $(\boldsymbol{x_n})$ from $B_{1/n}(\boldsymbol{p}) \setminus \{p\}$ which converges to $\boldsymbol{p}$.

2. A point $\boldsymbol{p}$ is a **cluster point** of S if there is a sequence $(\boldsymbol{x_n})$ from S with each $\boldsymbol{x_n} \neq \boldsymbol{p}$ so that $\boldsymbol{x_n} \to \boldsymbol{p}$. Note cluster points are accumulation points and accumulation points are cluster points.

3. A point $\boldsymbol{p}$ is a **limit point** of S if there is a sequence $(\boldsymbol{x_n})$ from S that converges to $\boldsymbol{p}$. Note an **isolated point** of a set is a point which is a positive distance from any other point in the set and an isolated point is a limit point but not an accumulation point.

4. S' is the set S and its limit points.

Note a set S need not be open or closed and $\Re^2$ is both open and closed. We can prove useful things essentially in the same way as we did in $\Re$.

Theorem 2.2.1 Basic Topology Results in $\Re^2$

Given S in $\Re^2$:

1. *S is closed if and only if $S = S'$; i.e. S is closed if and only if it contains its limit points.*
2. *S is closed if and only if S contains its boundary points.*
3. *Finite intersections of open sets are open.*
4. *Finite unions of open sets are open.*
5. *Countable unions of open sets are open. Countable intersections of open sets can be closed.*
6. *Finite intersections of closed sets are closed.*
7. *Finite unions of closed sets are closed.*
8. *Countable intersections of closed sets are closed. Countable unions of closed sets need not be open.*
9. *DeMorgan's Laws hold: i.e. $(\cup_n S_n)^C = \cap_n S_n^C$ and $(\cap_n S_n)^C = \cup_n S_n^C$*

Proof 2.2.1
You can mimic the one dimensional proofs fairly easily. ■

2.2.1 Homework

Exercise 2.2.1 *Prove finite unions of open sets are open.*

Exercise 2.2.2 *Prove finite intersections of open sets are open.*

Exercise 2.2.3 *Prove finite unions of closed sets are closed.*

Exercise 2.2.4 *Prove finite intersections of closed sets are closed.*

Exercise 2.2.5 *Prove a set S is closed if and only if $S' = S$.*

Exercise 2.2.6 *Prove DeMorgan's Laws.*

2.3 Bolzano - Weierstrass in $\Re^2$

We have discussed the Bolzano - Weierstrass theorem very carefully in $\Re$ and we mentioned casually how we would go about handling the proof in $\Re^n$ for $n \geq 2$ in (Peterson (21) 2020), but now we would like to show the details for the 2D proof. The proof is then very similar in 3D and so on and we just refer to this one below and say, "well, it is similar..."!

Theorem 2.3.1 Bolzano - Weierstrass Theorem in 2D

Every bounded sequence in $\Re^2$ has at least one convergent subsequence.

Proof 2.3.1
Let's assume the range of this sequence is infinite. If it were finite, there would be subsequences of it that converge to each of the values in the finite range. To make it easier to read this argument, we will use boldface font to indicate vectors in $\Re^2$ *and it is assumed they will have two components labeled with subscripts* 1 *and* 2*. So here the sequence is* $(\boldsymbol{a_n})$ *with* $\boldsymbol{a_n} = \begin{bmatrix} a_{n1} \\ a_{n2} \end{bmatrix}$*. We assume the sequences start at* $n = 1$ *for convenience and by assumption, there is a positive number* B *so that* $\|\boldsymbol{a_n}\| \leq B/2$ *for all* $n \geq 1$*. Hence, all the components of this sequence live in the square* $[-B/2, B/2] \times [-B/2, B/2] \subset \Re^2$*. Let's use a special notation for a* **box** *of this sort. Let* $J_0 = \boldsymbol{I_0^1} \times \boldsymbol{I_0^2}$ *where the interval* $\boldsymbol{I_0^1} = [\alpha_{01}, \beta_{01}]$ *and* $\boldsymbol{I_0^2} = [\alpha_{02}, \beta_{02}]$ *also. Here,* $\alpha_{0i} = -B/2$ *and* $\beta_{oi} = B/2$*. Note the area of* J_0*, denoted by* ℓ_0 *is* B^2*.*

Let $\mathcal{S}$ *be the range of the sequence which has infinitely many points and for convenience, we will let the phrase* infinitely many points *be abbreviated to* **IMPs**.

Step 1*:*
Bisect each axis interval of J_0 *into two pieces giving four subregions of* J_0 *all of which have area* $B^2/4$*. Now at least one of the subregions contains IMPs of* $\mathcal{S}$ *as otherwise each subregion has only finitely many points and that contradicts our assumption that* $\mathcal{S}$ *has IMPs. Now all may contain IMPs so select one such subregion containing IMPs and call it* J_1*. Then* $J_1 = \boldsymbol{I_1^1} \times \boldsymbol{I_1^2}$ *where the interval* $\boldsymbol{I_1^1} = [\alpha_{11}, \beta_{11}]$ *and* $\boldsymbol{I_1^2} = [\alpha_{12}, \beta_{12}]$ *also. Note the area of* J_1*, denoted by* ℓ_1 *is* $B^2/4$*.*
We see $J_1 \subset J_0$ *and*

$$-B/2 = \alpha_{0i} \leq \alpha_{1i} \leq \beta_{1i} \leq \beta_{0i} = B/2$$

Since J_1 *contains IMPs, we can select a sequence vector* $\boldsymbol{a_{n_1}}$ *from* J_1*.*

Step 2*:*
Now subdivide J_1 *into four subregions just as before. At least one of these subregions contain IMPs of* $\mathcal{S}$*. Choose one such subregion and call it* J_2*. Then* $J_2 = \boldsymbol{I_2^1} \times \boldsymbol{I_2^2}$ *where the interval* $\boldsymbol{I_2^1} = [\alpha_{21}, \beta_{21}]$ *and* $\boldsymbol{I_2^2} = [\alpha_{22}, \beta_{22}]$ *also. Note the area of* J_2*, denoted by* ℓ_1 *is* $B^2/(16)$*.*
We see $J_2 \subset J_1$ *and*

$$-B/2 = \alpha_{0i} \leq \alpha_{1i} \leq \alpha_{2i} \leq \beta_{2i} \leq \beta \leq 1i \leq \beta_{0i} = B/2$$

Since J_2 *contains IMPs, we can select a sequence vector* $\boldsymbol{a_{n_2}}$ *from* J_2*. It is easy to see this value can be chosen different from* $\boldsymbol{a_{n_1}}$*, our previous choice.*

You should be able to see that we can continue this argument using induction.

Proposition*:*
$\forall p \geq 1$*,* $\exists$ *an interval* $J_p = \boldsymbol{I_p^1} \times \boldsymbol{I_p^2}$ *with* $\boldsymbol{I_p^i} = [\alpha_{pi}, \beta_{pi}]$ *with the area of* J_p*,* $\ell_p = B^2/(2^{2p})$ *satisfying* $J_p \subseteq J_{p-1}$*,* J_p *contains IMPs of* $\mathcal{S}$ *and*

$$\alpha_{0i} \leq \ldots \leq \alpha_{p-1,i} \leq \alpha_{pi} \quad \leq \quad \beta_{pi} \leq \beta_{p-1,i} \leq \ldots \leq \beta_{0i}$$

Finally, there is a sequence vector $\boldsymbol{a_{n_p}}$ *in* J_p*, different from* $\boldsymbol{a_{n_1}}, \ldots, \boldsymbol{a_{n_{p-1}}}$*.*

Proof *We have already established the proposition is true for the basis step* J_1 *and indeed also for the next step* J_2*.*
Inductive: *We assume the interval* J_q *exists with all the desired properties. Since by assumption,* J_q *contains IMPs, bisect* J_q *into four subregions as we have done before. At least one of these subregions contains IMPs of* $\mathcal{S}$*. Choose one of the subregions and call it* J_{q+1} *and label* $J_{q+1} =$

$\boldsymbol{I}^{\mathbf{1}}_{\boldsymbol{q+1}} \times \boldsymbol{I}^{\mathbf{2}}_{\boldsymbol{q+1}}$ where $\boldsymbol{I}^{\boldsymbol{i}}_{\boldsymbol{q+1}} = [\alpha_{q+1,i}, \beta_{q+1,i}]$ for $i = 1, 2$. We see immediately $\ell_{q+1} = B^2/2^{2(q+1)}$ with $\alpha^i_q \leq \alpha^i_{q+1} \leq \beta^i_{q+1} \leq \beta^i_q$. This shows the nested inequality we want is satisfied. Finally, since J_{q+1} contains IMPs, we can choose $\boldsymbol{a}_{\boldsymbol{n}_{q+1}}$ distinct from the other $\boldsymbol{a}_{\boldsymbol{n}_i}$'s. So the inductive step is satisfied and by the POMI, the proposition is true for all n. □

From our proposition, we have proven the existence of sequences, (α_{pi}), (β_{pi}) and (ℓ_p) which have various properties. The sequence ℓ_p satisfies $\ell_p = (1/4)\ell_{p-1}$ for all $p \geq 1$. Since $\ell_0 = B^2$, this means $\ell_1 = B^2/4$, $\ell_2 = B^2/16$, $\ell_3 = B^2/(2^2)^3$ leading to $\ell_p = B^2/(2^2)^p$ for $p \geq 1$. Further, we have the inequality chain

$$\begin{aligned} -B/2 = \alpha_{0i} &\leq \alpha_{1i} \leq \alpha_{2i} \leq \ldots \leq \alpha_{pi} \\ &\leq \ldots \leq \\ \beta_{pi} &\leq \ldots \leq \beta_{2i} \leq \ldots \leq \beta_{0i} = B/2 \end{aligned}$$

The rest of this argument is almost identical to the one we did for the case of a bounded sequence in $\Re$. Note (α_{pi}) is bounded above by $B/2$ and $(\beta_{pi})_{p\geq 0}$ is bounded below by $-B/2$. Hence, by the completeness axiom, $\inf(\beta_{pi})$ *exists and equals the finite number β^i; also* $\sup(\alpha_{pi})$ *exists and is the finite number α^i.*

So if we fix p, it should be clear the number β_{pi} is an upper bound for all the α_{pi} values (look at our inequality chain again and think about this). Thus β_{pi} is an upper bound for (α_{pi}) and so by definition of a supremum, $\alpha^i \leq \beta_{pi}$ for all p. Of course, we also know since α^i is a supremum, that $\alpha_{pi} \leq \alpha^i$. Thus, $\alpha_{pi} \leq \alpha^i \leq \beta_{pi}$ for all p. A similar argument shows if we fix p, the number α^i_p is an lower bound for all the β_{pi} values and so by definition of an infimum, $\alpha_{pi} \leq \beta^i \leq \beta_{pi}$ for all the α_{pi} values. This tells us α^i and β^i are in $[\alpha_{pi}, \beta_{pi}]$ for all p. Next we show $\alpha{=}\beta^i$.
Let $\epsilon > 0$ be arbitrary. Since α^i and β^i are in an interval whose length is $\ell_p = (1/2^{2p})B^2$, we have $|\alpha^i - \beta^i| \leq (1/2^{2p})B^2$. Pick P so that $1/(2^{2P}B^2) < \epsilon$. Then $|\alpha{-}\beta^i| < \epsilon$. But $\epsilon > 0$ is arbitrary. Hence, $\alpha^i - \beta^i = 0$ implying $\alpha^i = \beta^i$. Finally, define the vector $\boldsymbol{\theta} = \begin{bmatrix} \alpha^0 \\ \alpha^1 \end{bmatrix}$.

We now must show $\boldsymbol{a}_{\boldsymbol{n}_k} \to \boldsymbol{\theta}$. This shows we have found a subsequence which converges to $\boldsymbol{\theta}$. We know $\alpha_{pi} \leq a^i_{n_p} \leq \beta_{pi}$ and $\alpha_{pi} \leq \alpha^i \leq \beta_{pi}$ for all p. Pick $\epsilon > 0$ arbitrarily. Given any p, we have

$$\begin{aligned} |a^i_{n_p,1} - \alpha^i| &= |a^i_{n_p,i} - \alpha_{pi} + \alpha_{pi} - \alpha^i|, && \textit{add and subtract trick} \\ &\leq |a^i_{n_p,i} - \alpha_{pi}| + |\alpha_{pi} - \alpha^i| && \textit{triangle inequality} \\ &\leq |\beta_{pi} - \alpha_{pi}| + |\alpha_{pi} - \beta_{pi}| && \textit{definition of length} \\ &= 2|\beta_{pi} - \alpha_{pi}| = 2\,(1/2^{2p})B^2. \end{aligned}$$

Choose P so that $(1/2^{2P})B^2 < \epsilon/2$. Then, $p > P$ implies $|a^i_{n_p,1} - \alpha^i| < 2\,\epsilon/2 = \epsilon$. Thus, $a_{n_k,i} \to \alpha^i$. This shows the subsequence converges to $\boldsymbol{\theta}$. ■

Note this argument is messy but quite similar to the one dimensional case. It should be easy for you to see how to extend this to $\Re^3$ and even $\Re^n$. It is more a problem of correct labeling than intellectual difficulty!

2.3.1 Homework

Exercise 2.3.1 *This exercise mimics the proof above. Prove every bounded sequence in $\Re^3$ has a least one subsequence which converges.*

Exercise 2.3.2 *This exercise mimics the proof above. Prove every bounded sequence in* $\Re^4$ *has a least one subsequence which converges.*

Exercise 2.3.3 *Prove every bounded infinite set in* $\Re^2$ *has at least one cluster point.*

Exercise 2.3.4 *Prove every bounded infinite set in* $\Re^3$ *has at least one cluster point.*

Exercise 2.3.5 *Given a bounded set* S *in* $\Re^2$ *such as* $(0,1) \times (0,1)$*, find its boundary points. How many paths are there in* $\Re^2$ *which terminate on a boundary point? Do these paths always have to be in* S^C*? Do they always have to be in* S*? Can they contain points in both* S *and* S^C*?*

Exercise 2.3.6 *Given a bounded set* S *in* $\Re^3$ *such as* $(0,1) \times (0,1) \times (0,1)$*, find its boundary points. How many paths are there in* $\Re^3$ *which terminate on a boundary point? Do these paths always have to be in* S^C*? Do they always have to be in* S*? Can they contain points in both* S *and* S^C*?*

Exercise 2.3.7 *Let* S *be the set of all subsets of* $\Re$*, define* $\oplus$ *by* $A \oplus B = A \cup B$ *and define* $\otimes$ *by* $A \odot B = A \cap B$*. Is* S *a group with respect to the operation* $\oplus$*? Is* S *a ring with respect to the two operations* $\oplus$ *and* $\otimes$*?*

Exercise 2.3.8 *Let* S *be the set of all subsets of* $\Re^2$*, define* $\oplus$ *by* $A \oplus B = A \cup B$ *and define* $\otimes$ *by* $A \odot B = A \cap B$*. Is* S *a group with respect to the operation* $\oplus$*? Is* S *a ring with respect to the two operations* $\oplus$ *and* $\otimes$*? What are the additive and multiplicative identities here?*

Chapter 3

Vector Spaces

We are now going to explore vector spaces in some generality. Let's go back and think about vectors in $\Re^2$. As you know, we think of these as arrows with a tail fixed at the origin of the two dimensional coordinate system we call the $x - y$ plane. They also have a length or magnitude and this arrow makes an angle with the positive x axis. Suppose we look at two such vectors, $\boldsymbol{E}$ and $\boldsymbol{F}$. Each vector has an x and a y component so that we can write

$$\boldsymbol{E} = \begin{bmatrix} a \\ b \end{bmatrix}, \ \boldsymbol{F} = \begin{bmatrix} c \\ d \end{bmatrix}$$

The cosine of the angle between them is proportional to the inner product $< \boldsymbol{E}, \boldsymbol{F} >= ac + bd$. If this angle is 0 or π, the two vectors lie along the same line. In any case, the angle associated with $\boldsymbol{E}$ is $\tan^{-1}(\frac{b}{a})$ and for $\boldsymbol{F}$, $\tan^{-1}(\frac{d}{c})$. Hence, if the two vectors lie on the same line, $\boldsymbol{E}$ must be a multiple of $\boldsymbol{F}$. This means there is a number β so that $\boldsymbol{E} = \beta\, \boldsymbol{F}$. We can rewrite this as $\begin{bmatrix} a \\ b \end{bmatrix} = \beta \begin{bmatrix} c \\ d \end{bmatrix}$. Now let the number 1 in front of $\boldsymbol{E}$ be called $-\alpha$. Then the fact that $\boldsymbol{E}$ and $\boldsymbol{F}$ lie on the same line implies there are 2 constants α and β, both not zero, so that $\alpha\, \boldsymbol{E} + \beta\, \boldsymbol{F} = \boldsymbol{0}$. where $\boldsymbol{0}$ is the zero vector. Note we could argue this way for vectors in $\Re^3$ and even in $\Re^n$. Of course, our ability to think of these things in terms of lying on the same line and so forth needs to be extended to situations we can no longer draw, but the idea is essentially the same. Instead of thinking of our two vectors as lying on the same line or not, we can *rethink* what is happening here and try to identify what is happening in a more abstract way. If our two vectors lie on the same line, they are not *independent* things in the sense one is a multiple of the other. As we saw above, this implies there was a linear equation connecting the two vectors which had to add up to 0. Hence, we might say the vectors were *not linearly independent* or simply, they are *linearly dependent*. Phrased this way, we are on to a way of stating this idea which can be used in many more situations. We want to be more general here: we are looking at objects of some *type* which we can add and scalar multiply. For the moment, we think of these as vectors in say $\Re^2$ but soon enough we will make this more abstract.

3.1 Vector Spaces over a Field

If we have a set of objects $\boldsymbol{u}$ with a way to add them to create new objects in the set and a way to *scale* them to make new objects, this is formally called a **Vector Space** with the set denoted by $\boldsymbol{V}$. For our purposes, we scale such objects with either real or complex numbers. If the scalars are real numbers, we say $\boldsymbol{V}$ is a vector space over the reals; otherwise, it is a vector space over the complex field. There are other fields, of course, but we will focus on these two.

Definition 3.1.1 Abstract Vector Space

Let $\boldsymbol{V}$ be a set of objects $\boldsymbol{u}$ with an additive operation $\oplus$ and a scaling method $\odot$. Formally, this means

VS1: *Given any $\boldsymbol{u}$ and $\boldsymbol{v}$, the operation of adding them together is written $\boldsymbol{u}\oplus\boldsymbol{v}$ and results in the creation of a new object in the vector space. This operation is* commutative *which means the order of the operation is not important; so $\boldsymbol{u}\oplus\boldsymbol{v}$ and $\boldsymbol{v}\oplus\boldsymbol{u}$ give the same result. Also, this operation is associative as we can group any two objects together first, perform this addition $\oplus$, then do the others and the order of the grouping does not matter.*

VS2: *Given any $\boldsymbol{u}$ and any number c (either real or complex, depending on the type of vector space we have), the operation $c\odot\boldsymbol{u}$ creates a new object. We call such numbers c* scalars.

VS3: *The scaling and additive operations are nicely compatible in the sense that order and grouping is not important. These are called the* distributive *laws for scaling and addition. They are*

$$\begin{aligned} c\odot(\boldsymbol{u}\oplus\boldsymbol{v}) &= (c\odot\boldsymbol{u})\oplus(c\odot\boldsymbol{v}) \\ (c+d)\odot\boldsymbol{u} &= (c\odot\boldsymbol{u})\oplus(d\odot\boldsymbol{u}). \end{aligned}$$

VS4: *There is a special object called $\boldsymbol{o}$ which functions as a* zero *so we always have $\boldsymbol{o}\oplus\boldsymbol{u}=\boldsymbol{u}\oplus\boldsymbol{o}=\boldsymbol{u}$.*

VS5: *There are* additive inverses, *which means to each $\boldsymbol{u}$ there is a unique object $\boldsymbol{u}^\dagger$ so that $\boldsymbol{u}\oplus\boldsymbol{u}^\dagger=\boldsymbol{o}$.*

Comment 3.1.1 *These laws imply*

$$(0+0)\odot\boldsymbol{u} = (0\odot\boldsymbol{u})\oplus(0\odot\boldsymbol{u})$$

which tells us $0\odot\boldsymbol{u}=0$. A little further thought then tells us that since

$$\mathbf{0} = (1-1)\odot\boldsymbol{u}=(1\odot\boldsymbol{u})\oplus(-1\odot\boldsymbol{u})$$

we have the additive inverse $\boldsymbol{u}^\dagger = -1\odot\boldsymbol{u}$.

Comment 3.1.2 *We usually say this much simpler. The set of objects $\boldsymbol{V}$ is a vector space over its scalar field if there are two operations which we denote by $\boldsymbol{u}+\boldsymbol{v}$ and $c\boldsymbol{u}$ which generate new objects in the vector space for any $\boldsymbol{u}$, $\boldsymbol{v}$ and scalar c. We then just add that these operations satisfy the usual commutative, associative and distributive laws and there are unique additive inverses.*

Comment 3.1.3 *The objects are often called* vectors *and sometimes we denote them by $\boldsymbol{u}$, although this notation is often too cumbersome.*

Comment 3.1.4 *To give examples of vector spaces, it is usually enough to specify how the additive and scaling operations are done.*

- *Vectors in $\Re^2$, $\Re^3$ and so forth are added and scaled by components.*
- *Matrices of the same size are added and scaled by components.*
- *A set of functions of similar characteristics uses as its additive operator, pointwise addition. The new function $(f\oplus g)$ is defined pointwise by $(f\oplus g)(t)=f(t)+g(t)$. Similarly, the new*

function $c \odot f$ *is defined as the function whose value at* t *is* $(cf)(t) = cf(t)$*. Classic examples are*

1. $C([a,b])$ *is the set of all functions whose domain is* $[a,b]$ *that are continuous on the domain.*
2. $C^1([a,b])$ *is the set of all functions whose domain is* $[a,b]$ *that are continuously differentiable on the domain.*
3. $RI([a,b])$ *is the set of all functions whose domain is* $[a,b]$ *that are Riemann integrable on the domain.*

There are many more, of course.

Vector spaces have two other important ideas associated with them. We have already talked about what we might mean by linearly independent objects in $\Re^2$ or $\Re^3$. In general, for any vector space, we would formalize that discussion and make the following definitions.

Definition 3.1.2 Two Linearly Independent Objects in a Vector Space

Let $\boldsymbol{E}$ *and* $\boldsymbol{F}$ *be two objects in a vector space* $\boldsymbol{V}$*. We say* $\boldsymbol{E}$ *and* $\boldsymbol{F}$ *are* **linearly dependent** *if we can find nonzero scalars* α *and* β *so that*

$$\alpha \odot \boldsymbol{E} + \beta \odot \boldsymbol{F} = \boldsymbol{0}.$$

Otherwise, we say they are **linearly independent.**

We can then easily extend this idea to any finite collection of such objects as follows.

Definition 3.1.3 Finitely Many Linearly Independent Objects

Let $\{\boldsymbol{E}_i : 1 \leq i \leq N\}$ *be* N *objects in a vector* $\boldsymbol{V}$*. We say* $\boldsymbol{E}$ *and* $\boldsymbol{F}$ *are* **linearly dependent** *if we can find nonzero constants* α_1 *to* α_N*, not all* 0*, so that*

$$\alpha_1 \odot \boldsymbol{E}_1 + \ldots + \alpha_N \odot \boldsymbol{E}_N = 0.$$

Note we have changed the way we define the constants. When there are more than two objects involved, we can't say, in general, that all *of the constants must be nonzero.*

As you know, the span of a set of vectors in $\Re^n$ is the set of all possible linear combinations of them. So if $\{\boldsymbol{A_1}, \ldots, \boldsymbol{A_p}\}$ is a set of p vectors in $\Re^n$, a vector in the span of these vectors has the form $\boldsymbol{V} = \sum_{i=1}^{p} a_i \boldsymbol{A_i}$ for some scalars a_i, $1 \leq i \leq p$.

Can we have three linearly independent vectors in $\Re^2$? The answer is no and the argument goes like this. Let's do this for three vectors that are linearly independent which means all are nonzero. Do Graham - Schmidt Orthogonalization (**GSO**) on the first two. This creates two vectors which are mutually orthogonal $\{\boldsymbol{E_1}, \boldsymbol{E_2}\}$ whose span is the same as the span of $\{\boldsymbol{A_1}, \boldsymbol{A_2}\}$. Let $\boldsymbol{W}$ be any vector in $\Re^2$. Then it is easy to see $\boldsymbol{W} =< \boldsymbol{W}, \boldsymbol{E_1} > \boldsymbol{E_1} + < \boldsymbol{W}, \boldsymbol{E_2} > \boldsymbol{E_2}$. This tells us the span of $\{\boldsymbol{A_1}, \boldsymbol{A_2}\}$ is $\Re^2$. Thus, the third vector $\boldsymbol{A_3}$ is a linear combination of $\boldsymbol{E_1}$ and $\boldsymbol{E_2}$, which also means it is a linear combination of $\boldsymbol{A_1}$ and $\boldsymbol{A_2}$. Clearly, this mean the set $\{\boldsymbol{A_1}, \ldots, \boldsymbol{A_3}\}$ is a linearly dependent set which violates the assumption they were linearly independent. So three vectors can't be linearly independent in $\Re^2$. It follows the maximum number of linearly independent vectors in $\Re^2$ is two. This is why we say the **dimension** of $\Re^2$ is two!

Homework

Exercise 3.1.1 *Given* $\boldsymbol{A} = \begin{bmatrix} 1 \\ 2 \end{bmatrix}$ *and* $\boldsymbol{B} = \begin{bmatrix} 4 \\ 5 \end{bmatrix}$, *prove the vectors are linearly independent.*

Exercise 3.1.2 *Given* $\boldsymbol{A} = \begin{bmatrix} 1 \\ 2 \end{bmatrix}$, $\boldsymbol{B} = \begin{bmatrix} 4 \\ 5 \end{bmatrix}$ *and* $\boldsymbol{C} = \begin{bmatrix} -3 \\ 7 \end{bmatrix}$ *prove the vectors are linearly dependent.*

Exercise 3.1.3 *Given* 5 *nonzero vectors in* $\Re^4$ *prove they form a dependent set.*

Exercise 3.1.4 *Prove the maximum number of linearly independent vectors in* $\Re^6$ *is six.*

Exercise 3.1.5 *Prove* $f(x) = x^2$ *and* $g(x) = \sin(x)$ *are linearly independent in* $C(\Re)$.

Exercise 3.1.6 *Prove* $f(x) = 2x$, $g(x) = x^3$ *and* $h(x) = \cos(x+1)$ *are linearly independent in* $C([0,1])$.

Exercise 3.1.7 *Prove the set of functions* $\{x^0, x^1, \ldots, x^p\}$ *is linearly independent in* $C(\Re)$ *for all* $p \geq 0$.

3.1.1 The Span of a Set of Vectors

The next concept is that of the **span** of a collection of vectors. We already know about this in $\Re^n$ but let's do it more generally now.

Definition 3.1.4 The Span of a Set of Vectors

> *Given a finite set of vectors in a vector space* $\boldsymbol{V}$, $\mathscr{W} = \{\boldsymbol{u}_1, \ldots, \boldsymbol{u}_N\}$ *for some positive integer* N, *the span of* $\mathscr{W}$ *is the collection of all new vectors of the form* $\sum_{i=1}^{N} c_i \boldsymbol{u}_i$ *for any choices of scalars* $c_1, \ldots, c_N$. *It is easy to see* $\mathscr{W}$ *is a vector space itself and since it is a subset of* $\mathscr{V}$, *we call it a* vector subspace. *The span of the set* $\mathscr{W}$ *is denoted by* $\boldsymbol{Sp}\boldsymbol{W}$. *If the set of vectors* $\mathscr{W}$ *is not finite, the definition is similar but we say the span of* $\mathscr{W}$ *is the set of all vectors which can be written as* $\sum_{i=1}^{N} c_i \boldsymbol{u}_i$ *for some finite set of vectors* $\boldsymbol{u}_1, \ldots \boldsymbol{u}_N$ *from* $\mathscr{W}$.

Let's talk about subspaces of a vector space in more detail. If we have a vector space $\boldsymbol{V}$ and we are given a finite collection of objects $\{\boldsymbol{V}_1, \ldots, \boldsymbol{V}_n\}$ in it, we have already noted the span of this collection satisfies the definition of a vector space itself. Since it is a subset of the original vector space, we distinguish it from the vector space $\boldsymbol{V}$ it is inside and call it a **subspace**. There are many examples and it is useful to look at some.

Example 3.1.1 *If we look at the vector space* $C([a,b])$ *of continuous functions on the interval* $[a,b]$, *the span of the function* $\mathbf{1}$, *i.e. the constant function* $x(t) = 1$ *on* $[a,b]$, *is the set of all real numbers. This is a subspace of* $C([a,b])$. *Note* $x(t) = t^0$.

The span of $\{\boldsymbol{t^0}, \boldsymbol{t^1}, \boldsymbol{t^2}, \ldots, \boldsymbol{t^n}\}$ *is all functions of the form* $p(t) = \sum_{i=0}^{n} a_i t^i$ *for any value of the real numbers* a_i. *This is the set of all polynomials of degree* n *over the real numbers. It is straightforward to show these polynomials are linearly independent in the vector space. The linear independence equation is*

$$c_0 \odot \mathbf{1} \oplus c_1 \odot \boldsymbol{t^1} \oplus \ldots \oplus c_n \odot \boldsymbol{t^n} \quad = \quad \mathbf{0}$$

which means

$$c_0 + c_1\, t^1 \; + \; \ldots + c_n\, t^n \quad = \quad 0, \quad a \leq t \leq b$$

Now take n derivatives of this equation, which will tell us $c_n = 0$, and then back substitute up through the n derivative equations to see the rest of the constants are zero also. Thus, this collection of functions is linearly dependent. Note every n^{th} degree polynomial is therefore uniquely expressed as a member of the span of $\{\boldsymbol{t^0}, \boldsymbol{t^1}, \boldsymbol{t^2}, \ldots, \boldsymbol{t^n}\}$. Let $\boldsymbol{P_n}$ denote the span of $\{\boldsymbol{t^0}, \boldsymbol{t^1}, \boldsymbol{t^2}, \ldots, \boldsymbol{t^n}\}$. Then $\{\boldsymbol{t^0}, \boldsymbol{t^1}, \boldsymbol{t^2}, \ldots, \boldsymbol{t^n}\}$ is a linearly independent set that span $\boldsymbol{P_n}$. For future reference, note it has $n+1$ elements.

Example 3.1.2 *The set of all solutions to the ordinary differential equation $x'' + 4x = 0$ is the span of the two linearly independent functions $\{\cos(2t), \sin(2t)\}$. Note this linearly independent spanning set consists of two functions.*

Comment 3.1.5 *Let $\boldsymbol{V}$ and $\boldsymbol{W}$ be two linearly independent vectors in a vector space. Is it possible to find another vector $\boldsymbol{Q}$ independent from $\boldsymbol{V}$ and $\boldsymbol{W}$ in the span of $\boldsymbol{V}$ and $\boldsymbol{W}$? Since $\boldsymbol{Q}$ is in the span of these two vectors, we know we can find constants a and b so that $\boldsymbol{Q} = a\boldsymbol{V} + \boldsymbol{W}$. But this says the set $\{\boldsymbol{V}, \boldsymbol{W}, \boldsymbol{Q}\}$ is a linearly dependent set. So it is not possible to find a third independent vector in the span.*

Comment 3.1.6 *The above comment can be extended to the case of n linearly independent vectors and their associated span. We see the maximum number of linearly independent vectors in the span is n.*

The comments above lead to the notation of a *basis* for a vector space.

Definition 3.1.5 A Basis for a Finite Dimensional Vector Space

> *If the set of vectors $\{\boldsymbol{V_1}, \ldots, \boldsymbol{V_n}\}$ is a linearly independent spanning set for the vector space $\mathscr{V}$, we say this set is a* **basis** *for $\mathscr{V}$ and that the* **dimension** *of $\mathscr{V}$ is n.*

Comment 3.1.7 *From our comments, we know the maximum number of linear independent objects in a vector space of dimension n is n.*

We can extend this to vector spaces that do not have a finite basis. First, we extend the idea of linear independence to sets that are not necessarily finite.

Definition 3.1.6 Linear Independence for Infinite Sets

> *Given a set of vectors in a vector space $\mathscr{V}$, $\mathscr{W}$, we say $\mathscr{W}$ is a linearly independent subset if every finite set of vectors from $\mathscr{W}$ is linearly independent in the usual manner.*

We have already defined the span of a set of vectors that is not finite. So as a final matter, we want to define what we mean by a **basis** for a vector space which does not have a finite basis.

Definition 3.1.7 A Basis for an Infinite Dimensional Vector Space

> *Given a set of vectors in an infinite dimensional vector space $\mathscr{V}$, $\mathscr{W}$, we say $\mathscr{W}$ is a* basis *for $\mathscr{V}$ if the span of $\mathscr{W}$ is all of $\mathscr{V}$ and if the vectors in $\mathscr{W}$ are linearly independent. Hence, a basis is a linearly independent spanning set for $\mathscr{V}$. Recall, this means any vector in $\mathscr{V}$ can be expressed as a finite linear combination of elements from $\mathscr{W}$ and any finite collection of objects from $\mathscr{W}$ is linearly independent in the usual sense for a finite dimensional vector space.*

Comment 3.1.8 *In a vector space like $\Re^n$, the maximum size of a set of linearly independent vectors is n, the dimension of the vector space.*

Comment 3.1.9 *Let's look at the vector space $C([0,1])$, the set of all continuous functions on $[0,1]$. Let $\mathscr{W}$ be the set of all powers of t, $\{1, t, t^2, t^3, \ldots\}$. We can use the derivative technique to show this set is linearly independent even though it is infinite in size. Take any finite subset from $\mathscr{W}$. Label* *the resulting powers as* $\{n_1, n_2, \ldots, n_p\}$. *Write down the linear dependence equation*

$$c_1 t^{n_1} + c_2 t^{n_2} + \ldots + c_p t^{n_p} \quad = \quad 0.$$

Take n_p *derivatives to find* $c_p = 0$ *and then* backtrack *to find the other constants are zero also. Hence* $C([0, 1])$ *is an infinite dimensional vector space. It is also clear that* $\mathscr{W}$ *does not span* $C([0, 1])$ *as if this was true, every continuous function on* $[0, 1]$ *would be a polynomial of some finite degree. This is not true as* $\sin(t)$, e^{-2t} *and many others are not finite degree polynomials.*

Homework

Exercise 3.1.8 *Prove the set of solutions of* $u'' + 5u' + 6u = 0$ *is a two dimensional vector space over* $\Re$. *Make sure you specify your choice of basis. You may want to review ODE material at this point. If you want to see differential equations and their numerical solution used in the study of various biological models, you can find those discussions in (J. Peterson (9) 2016) (one variable calculus and some two variable stuff) and (J. Peterson (10) 2016) (systems of two ODEs and more calculus) as well as many other texts.*

Exercise 3.1.9 *Prove the set of all* 2×2 *matrices over* $\Re$ *form a four dimensional vector space over* $\Re$.

Exercise 3.1.10 *Prove* $\mathbb{C}$ *is a two dimensional vector space over* $\Re$ *and a one dimensional vector space of* $\mathbb{C}$.

Exercise 3.1.11 *Prove the set of all* 2×2 *matrices over* $\mathbb{C}$ *form a four dimensional vector space over* $\mathbb{C}$.

Exercise 3.1.12 *Prove the set of all* 2×2 *matrices whose entries are functions in* $C([a, b])$ *is a vector space of* $\Re$.

Exercise 3.1.13 *Prove the set of all* 3×2 *matrices whose entries are functions in* $C([a, b])$ *is a vector space of* $\Re$.

Exercise 3.1.14 *Prove the set of all* $n \times m$ *matrices over* $\Re$ *form a* $n \times m$*dimensional vector space over* $\Re$.

3.2 Inner Products

Now there is an important idea that we use a lot in applied work. If we have an object $\boldsymbol{u}$ in a Vector Space $\boldsymbol{V}$, we often want to *approximate* $\boldsymbol{u}$ using an element from a given subspace $\mathscr{W}$ of the vector space. To do this, we need to add another property to the vector space. This is the notion of an **inner product**. A vector space with an inner product is called an **Inner Product Space**. We already know what an inner product is in a simple vector space like $\Re^2$. We usually denote such an inner product by $< \cdot, \cdot >$. For two vectors in $\Re^n$, with a given basis $\boldsymbol{E}$, we can find the representation of two vectors $\boldsymbol{x}$ and $\boldsymbol{y}$ with respect to $\boldsymbol{E}$ is $\boldsymbol{x} = \sum_{i=1}^{n} x_i \boldsymbol{E_i}$ and $\boldsymbol{y} = \sum_{i=1}^{n} y_i \boldsymbol{E_i}$, respectively and we define $< \boldsymbol{x}, \boldsymbol{y} >= \sum_{i=1}^{n} x_i y_i$. Many vector spaces can have such an inner product structure added easily. For example, in $C([a, b])$, since each object is continuous, each object is Riemann integrable. Hence, given two functions f and g from $C([a, b])$, the real number given by $\int_a^b f(s) g(s) ds$ is well-defined. It satisfies all the usual properties that the inner product for finite dimensional vectors in $\Re^n$ does also.

These properties are so common we will codify them into a definition for what an inner product for a vector space $\mathscr{V}$ should behave like.

Definition 3.2.1 Real Inner Product

Let $\mathscr{V}$ be a vector space with the reals as the scalar field. Then a mapping ω which assigns a pair of objects to a real number is called an inner product on $\mathscr{V}$ if

IP1: $\omega(\boldsymbol{u}, \boldsymbol{v}) = \omega(\boldsymbol{v}, \boldsymbol{u})$*; that is, the order is not important for any two objects.*

IP2: $\omega(c \odot \boldsymbol{u}, \boldsymbol{v}) = c\omega(\boldsymbol{u}, \boldsymbol{v})$*; that is, scalars in the* first slot *can be pulled out.*

IP3: $\omega(\boldsymbol{u} \oplus \boldsymbol{w}, \boldsymbol{v}) = \omega(\boldsymbol{u}, \boldsymbol{v}) + \omega(\boldsymbol{w}, \boldsymbol{v})$ *for any three objects.*

IP4: $\omega(\boldsymbol{u}, \boldsymbol{u}) \geq 0$ *and* $\omega(\boldsymbol{u}, \boldsymbol{u}) = 0$ *if and only if* $u = 0$.

These properties imply that $\omega(\boldsymbol{u}, c \odot \boldsymbol{v}) = c\omega(\boldsymbol{u}, \boldsymbol{v})$ *as well. A vector space $\mathscr{V}$ with an inner product is called an* **Inner Product Space**.

Comment 3.2.1 *The inner product is usually denoted with the symbol $<\ ,\ >$ instead of $\omega(\ ,\ )$. We will use this notation from now on.*

Comment 3.2.2 *When we have an inner product, we can* measure *the* size *or* magnitude *of an object, as follows. We define the analogue of the Euclidean norm of an object $\boldsymbol{u}$ using the usual $||\ ||$ symbol as*

$$||\boldsymbol{u}|| = \sqrt{< \boldsymbol{u}, \boldsymbol{u} >}.$$

This is called the norm induced by the inner product *of the object. In $C([a, b])$, with the inner product $< f, g >= \int_a^b f(s)g(s)ds$, the* norm *of a function f is thus $||f|| = \sqrt{\int_a^b f^2(s)ds}$. This has been discussed in the earlier volumes so you can go back and review there to get up to speed. This is called the L_2 norm of f and is denoted $||\cdot||_2$.*

It is possible to prove the Cauchy - Schwartz inequality in this more general setting also.

Theorem 3.2.1 General Cauchy - Schwartz Inequality

If $\mathscr{V}$ is an inner product space with inner product $<\ ,\ >$ and induced norm $||\ ||$, then

$$|< \boldsymbol{u}, \boldsymbol{v} >| \leq ||\boldsymbol{u}||\,||\boldsymbol{v}||$$

with equality occurring if and only if $\boldsymbol{u}$ and $\boldsymbol{v}$ are linearly dependent.

Proof 3.2.1
If $\boldsymbol{u} = t\boldsymbol{v}$, then

$$\begin{aligned} |< \boldsymbol{u}, \boldsymbol{v} >| &- |< t\boldsymbol{v}, \boldsymbol{v} >| = |t| < \boldsymbol{v}, \boldsymbol{v} >= |t|\boldsymbol{v}||^2| \\ &= (|t|||\boldsymbol{v}||)\,(||\boldsymbol{v}||) = ||\boldsymbol{u}||\,||\boldsymbol{v}|| \end{aligned}$$

Hence, the result holds if $\boldsymbol{u}$ and $\boldsymbol{v}$ are linearly dependent. In general, if $\boldsymbol{u}$ and $\boldsymbol{v}$ are linearly independent, look at the subspace spanned by $\boldsymbol{v}$. Any element in this subspace can be written as $a\boldsymbol{v}/||\boldsymbol{v}||$. Consider the function

$$f(a) = < \boldsymbol{u} - a\boldsymbol{v}||\boldsymbol{v}/||, \boldsymbol{u} - a\boldsymbol{v}/||\boldsymbol{v} >= ||\boldsymbol{u}||^2 - 2a < \boldsymbol{u}, \boldsymbol{v}/||\boldsymbol{v}|| > +a^2$$

Then

$$f'(a) = -2 < \boldsymbol{u}, \boldsymbol{v}/||\boldsymbol{v}|| > +2a$$

which implies the critical point is $a =< \boldsymbol{u}, \boldsymbol{v}/||\boldsymbol{v}|| >$. Since $f''(a) = 2 > 0$, the critical point is a *global minimum. Thus, the vector whose minimum norm distance to the subspace generated by* $\boldsymbol{v}$ *is* $\boldsymbol{P_{uv}} =< \boldsymbol{u}, \boldsymbol{v}/\|\boldsymbol{v}\| > \boldsymbol{v}\|/\boldsymbol{v}\|$ *of length* $| < \boldsymbol{u}, \boldsymbol{v}\|/\boldsymbol{v}\| > |$. *The vector* $\boldsymbol{P_{uv}}$ *is called the projection of* $\boldsymbol{u}$ *onto* $\boldsymbol{v}$. *We can then use GSO to find the orthonormal basis for the subspace spanned by* $\boldsymbol{u}$ *and* $\boldsymbol{v}$.

$$\begin{aligned} \boldsymbol{Q_1} &= \boldsymbol{v}/\|\boldsymbol{v}\| \\ \boldsymbol{W} &= \boldsymbol{u} - < \boldsymbol{u}, \boldsymbol{Q_1} > \boldsymbol{Q_1}, \quad \boldsymbol{Q_2} = \boldsymbol{W}/\|\boldsymbol{W}\| \end{aligned}$$

Thus we have the decomposition $\boldsymbol{u} =< \boldsymbol{u}, \boldsymbol{Q_1} > \boldsymbol{Q_1}+ < \boldsymbol{u}, \boldsymbol{Q_2} > \boldsymbol{Q_2}$. *Hence, if we assumed* $\boldsymbol{u}$ *and* $\boldsymbol{v}$ *were linearly independent with*

$$| < \boldsymbol{u}, \boldsymbol{v} >= \|\boldsymbol{u}\| \, \|\boldsymbol{v}\| \quad \Longrightarrow \|\boldsymbol{P_{uv}}\| = \left| \left\langle \boldsymbol{u}, \frac{\boldsymbol{v}}{\|\boldsymbol{v}\|} \right\rangle \right| = \|\boldsymbol{u}\|$$

$\boldsymbol{u}$ *is a multiple of* $\boldsymbol{v}$ *which is not possible as they are assumed independent. Hence, we conclude if* $< \boldsymbol{u}, \boldsymbol{v} >= \|\boldsymbol{u}\| \, \|\boldsymbol{v}\|$, $\boldsymbol{u}$ *and* $\boldsymbol{v}$ *must be dependent.*

However, if $\boldsymbol{u}$ *and* $\boldsymbol{v}$ *are linearly independent, we have the decomposition* $\boldsymbol{u} =< \boldsymbol{u}, \boldsymbol{Q_1} > \boldsymbol{Q_1}+ < \boldsymbol{u}, \boldsymbol{Q_2} > \boldsymbol{Q_2}$. *which implies* $\|\boldsymbol{u}\|^2 = | < \boldsymbol{u}, \boldsymbol{Q_1} > |^2 + | < \boldsymbol{u}, \boldsymbol{Q_2} > |^2$. *This immediately tells us*

$$| < \boldsymbol{u}, \boldsymbol{v}/\|\boldsymbol{v}\| > |^2 \leq \|\boldsymbol{u}\|^2 \Longrightarrow | < \boldsymbol{u}, \boldsymbol{v}/\|\boldsymbol{v}\| > | \leq \|\boldsymbol{u}\| \Longrightarrow | < \boldsymbol{u}, \boldsymbol{v} > | \leq \|\boldsymbol{u}\| \, /\|\boldsymbol{v}\|$$

which is the result we wanted to show. ■

Comment 3.2.3 *We can use the Cauchy - Schwartz inequality to define a notion of angle between objects exactly like we would do in* $\Re^2$. *We define the angle* θ *between* $\boldsymbol{u}$ *and* $\boldsymbol{v}$ *via its cosine as usual.*

$$\cos(\theta) = \frac{< \boldsymbol{u}, \boldsymbol{v} >}{||\boldsymbol{u}|| \, ||\boldsymbol{v}||}.$$

Hence, objects can be perpendicular or orthogonal even if we cannot interpret them as vectors in $\Re^2$. *We see two objects are orthogonal if their inner product is* 0.

Comment 3.2.4 *If* $\mathscr{W}$ *is a finite dimensional subspace, a basis for* $\mathscr{W}$ *is said to be an orthonormal basis if each object in the basis has* L_2 *norm* 1 *and all of the objects are mutually orthogonal. This means* $< \boldsymbol{u}_i, \boldsymbol{u}_j >$ *is* 1 *if* $i = j$ *and* 0 *otherwise. We typically let the Kronecker delta symbol* δ_{ij} *be defined by* $\delta_{ij} = 1$ *if* $i = j$ *and* 0 *otherwise so that we can say this more succinctly as* $< \boldsymbol{u}_i, \boldsymbol{u}_j >= \delta_{ij}$.

Let's go back to $\Re^2$. You have gone over these ideas quite a bit in your last few courses, so we will just refresh your mind here. Grab three pencils for the $\Re^2$ examples coming up and four pencils for the $\Re^3$ examples. We will try to look at these examples in a new light so you can build on earlier levels of understanding.

3.2.1 Homework

Exercise 3.2.1 *Prove* $< f, g >= \int_0^1 f(t)g(t)dt$ *defines an inner product on* $C([0, 1])$.

Exercise 3.2.2 *Find the angle between* $f(x) = x^2$ *and* $g(x) = x^3$ *on* $[-1, 2]$.

Exercise 3.2.3 *Do* **GSO** *of the functions* $f(x) = 2x$, $g(x) = 5x^2$ *and* $h(x) = 7x^3$ *on* $[-3, 2]$. *This is easiest to do using the MATLAB code of (Peterson (21) 2020).*

Exercise 3.2.4 *In the vector space of* 2×2 *matrices with entries that are continuous functions on* $[-1, 4]$, $M_{2\times 2}(C([-1,4]))$, *does*

$$< \boldsymbol{A}, \boldsymbol{B} > \;=\; \det \begin{bmatrix} \int_{-1}^{4} A_{11}(t)B_{11}(t)dt & \int_{-1}^{4} A_{12}(t)B_{12}(t)dt \\ \int_{-1}^{4} A_{21}(t)B_{21}(t)dt & \int_{-1}^{4} A_{22}(t)B_{22}(t)dt \end{bmatrix}$$

define an inner product?

3.3 Examples

Let's work through a number of important examples ranging from finite dimensional to infinite dimensional cases.

3.3.1 Two Dimensional Vectors in the Plane

Take two pencils and hold them in your hand. They don't have to be the same size. These are two vectors in the plane. It doesn't matter how you hold them relative to the room you are in either. These two pencils or vectors are linearly independent if they point in different directions. The only way they are linearly dependent is if they lay on top of one another or point directly opposite to one another. In the first case, the angle between the vectors in 0 radians and in the second the angle is π. Now these two pencils can be thought of as sitting in a sheet of paper which is our approximation of the plane determined by the two vectors represented by the pencils. Note as long as the vectors are linearly independent we can represent any other vector (now use that third pencil!) $\boldsymbol{W}$ in terms of the first two which we now call $\boldsymbol{E}$ and $\boldsymbol{F}$. We want to find the scalars a and b so that $a\boldsymbol{E} + b\boldsymbol{F} = \boldsymbol{W}$ where we have stopped representing scalar multiplication by $\odot$. Taking the usual representation of these vectors, we have

$$\begin{aligned} \boldsymbol{E} &= E_{11}\begin{bmatrix}1\\0\end{bmatrix} + E_{12}\begin{bmatrix}0\\1\end{bmatrix} \\ \boldsymbol{F} &= F_{11}\begin{bmatrix}1\\0\end{bmatrix} + F_{12}\begin{bmatrix}0\\1\end{bmatrix} \\ \boldsymbol{W} &= W_{11}\begin{bmatrix}1\\0\end{bmatrix} + W_{12}\begin{bmatrix}0\\1\end{bmatrix} \end{aligned}$$

We assume, of course, our vectors are nonzero. Recall we usually just write this as

$$\boldsymbol{E} = \begin{bmatrix}E_{11}\\E_{12}\end{bmatrix}, \quad \boldsymbol{F} = \begin{bmatrix}F_{11}\\F_{12}\end{bmatrix}, \quad \boldsymbol{W} = \begin{bmatrix}W_{11}\\W_{12}\end{bmatrix}$$

and it is very important to notice that we have **hidden** our way of representing the components of our vectors inside these column vectors. Using matrix and vector notation, we can write our problem of finding a and b as

$$\begin{bmatrix}E_{11} & F_{11}\\E_{12} & F_{12}\end{bmatrix}\begin{bmatrix}a\\b\end{bmatrix} = \begin{bmatrix}W_{11}\\W_{12}\end{bmatrix}$$

Call this 2×2 matrix $\boldsymbol{T}$. Now we all know how to take the determinant of a 2×2 matrix. Note if $det(\boldsymbol{T}) = E_{11}F_{12} - E_{12}F_{12}$. If this determinant was zero, then that would tell us $\frac{E_{11}}{E_{12}} = \frac{F_{11}}{F_{12}}$ in the case all the components are nonzero. But this tells us $\boldsymbol{E}$ and **F** point in the same direction. Hence, they can't be linearly independent. The argument going the other way is similar. So the $det(\boldsymbol{T}) = 0$ if and only if these two vectors are linearly dependent. Since we have assumed they are the opposite:

i.e. linearly independent, we must have $det(\boldsymbol{T}) \neq 0$. But then Cramer's rule tells us how to find a and b. Note each vector must have at least one nonzero component and it is easy to alter our argument above to handle the zeros we might see. We will let you do that yourself! Using Cramer's rule, we have

$$a = \frac{\begin{bmatrix} W_{11} & F_{11} \\ W_{12} & F_{12} \end{bmatrix}}{det(\boldsymbol{T})}, \quad b = \frac{\begin{bmatrix} E_{11} & W_{11} \\ E_{12} & W_{12} \end{bmatrix}}{det(\boldsymbol{T})}$$

Of course, if the vectors $\boldsymbol{E}$ and $\boldsymbol{F}$ were orthonormal, things are much simpler. First, note

$$a\boldsymbol{E} + b\boldsymbol{F} = \boldsymbol{W} \implies <\boldsymbol{E}, a\boldsymbol{E} + b\boldsymbol{F}> = <E, \boldsymbol{W}> == a <\boldsymbol{E}, \boldsymbol{E}> + b <\boldsymbol{E}, \boldsymbol{F}> = <E, \boldsymbol{W}>$$

But $<\boldsymbol{E}, \boldsymbol{F}> = 0$ and $<\boldsymbol{E}, \boldsymbol{E}> = 1$, so we have $a = <E, \boldsymbol{W}>$. In a similar way, we find

$$a\boldsymbol{E} + b\boldsymbol{F} = \boldsymbol{W} \implies <\boldsymbol{F}, a\boldsymbol{E} + b\boldsymbol{F}> = <F, \boldsymbol{W}> == a <\boldsymbol{F}, \boldsymbol{E}> + b <\boldsymbol{F}, \boldsymbol{F}> = <E, \boldsymbol{W}>$$

which tells us $b = <\boldsymbol{F}, \boldsymbol{W}>$. So we have found the unique solution is

$$\boldsymbol{W} = \begin{bmatrix} <\boldsymbol{W}, \boldsymbol{E}> \\ <\boldsymbol{W}, \boldsymbol{F}> \end{bmatrix} = \begin{bmatrix} E_{11}W_{11} + E_{12}W_{12} \\ F_{11}W_{11} + F_{12}W_{12} \end{bmatrix}$$

Now using standard notation, we can also write this as

$$\begin{bmatrix} a \\ b \end{bmatrix} = \begin{bmatrix} \boldsymbol{E}, \boldsymbol{F} \end{bmatrix} \boldsymbol{W}$$

It is easy to see, using the orthonormality of $\boldsymbol{E}$ and $\boldsymbol{F}$ that we also have

$$\begin{bmatrix} \boldsymbol{E}^T \\ \boldsymbol{F}^T \end{bmatrix} \begin{bmatrix} \boldsymbol{E}, \boldsymbol{F} \end{bmatrix} = \boldsymbol{I}$$

where $\boldsymbol{I}$ is the 2×2 identity matrix. Since we are in two dimensions, we note there is a nice relationship between the components of $\boldsymbol{E}$ and $\boldsymbol{F}$ because they are orthogonal. First, the slope of $\boldsymbol{F}$ is the negative reciprocal of the slope of $\boldsymbol{E}$ and hence we must have

$$\boldsymbol{E} = \begin{bmatrix} E_{11} \\ E_{12} \end{bmatrix} = \boldsymbol{F} = \begin{bmatrix} -E_{12} \\ E_{11} \end{bmatrix}$$

with $E_{11}^2 + E_{12}^2 = 1$. We can thus identify $E_{11} = \cos(\theta)$ and $E_{12} = \sin(\theta)$ for some angle θ. This shows us we can represent $\boldsymbol{E}$ and $\boldsymbol{F}$ as

$$\boldsymbol{E} = \begin{bmatrix} \cos(\theta) \\ \sin(\theta) \end{bmatrix} = \boldsymbol{F} = \begin{bmatrix} -\sin(\theta) \\ \cos(\theta) \end{bmatrix}$$

Thus, we see

$$\begin{aligned} \begin{bmatrix} \boldsymbol{E}, \boldsymbol{F} \end{bmatrix} \begin{bmatrix} \boldsymbol{E}^T \\ \boldsymbol{F}^T \end{bmatrix} &= \begin{bmatrix} \cos(\theta) & -\sin(\theta) \\ \sin(\theta) & \cos(\theta) \end{bmatrix} \begin{bmatrix} \cos(\theta) & \sin(\theta) \\ -\sin(\theta) & \cos(\theta) \end{bmatrix} \\ &= \begin{bmatrix} \cos^2(\theta) + \sin(\theta)^2 & \cos(\theta)\sin(\theta) - \cos(\theta)\sin(\theta) \\ \cos(\theta)\sin(\theta) - \cos(\theta)\sin(\theta) & \cos^2(\theta) + \sin(\theta)^2 \end{bmatrix} = \boldsymbol{I} \end{aligned}$$

You probably recognize this representation of $\boldsymbol{T}$ as the usual 2×2 matrix which denotes the rotation of a vector by the angle θ. We note the determinant here is always $+1$. From what we have just done, it is now clear that the matrix $\begin{bmatrix} \boldsymbol{E}^T, \boldsymbol{F}^T \end{bmatrix}$ is the inverse of the matrix $\boldsymbol{T}$. Indeed, it is clear the inverse is T^T; i.e. the inverse of T is its own transpose. Hence, any two dimensional vector in $\Re^2$ is a unique linear combination of E and F.

What we have shown here is that any vector W can be written as $aE + bF$ for a unique a and b. This kind of representation is what we have called a **linear combination of the vectors E and F**. Also note we can do this whether E and F form an orthonormal set. Clearly, the set of all possible linear combinations of E and F gives all of $\Re^2$; i.e. **the span of E and F is $\Re^2$**. Further, since E and F are linearly independent, we say **E and F form a linearly independent spanning set** or **basis** for $\Re^2$.

This method of attack will not be very useful for three dimensional vectors, of course as we will lose our intuition of dependent vectors being a simple scalar multiple of one another. Another way of showing $\begin{bmatrix} E & F \end{bmatrix}^T$ is the inverse of T is as follows. It does not use rotations at all. Let

$$A = [E, F] \begin{bmatrix} E^T \\ F^T \end{bmatrix}$$

We already know that any vector W can be expressed as the unique linear combination $aE + bF$ where

$$\begin{bmatrix} a \\ b \end{bmatrix} = [E, F]\; W$$

Thus

$$\begin{aligned} A\,W &= [E, F] \begin{bmatrix} E^T \\ F^T \end{bmatrix} (aE + bF) \\ &= a\,[E, F] \begin{bmatrix} E^T \\ F^T \end{bmatrix} E + b\,[E, F] \begin{bmatrix} E^T \\ F^T \end{bmatrix} F \\ &= a\,[E, F] \begin{bmatrix} E^T E \\ F^T E \end{bmatrix} + b\,[E, F] \begin{bmatrix} E^T F \\ F^T F \end{bmatrix} \\ &= a\,[E, F] \begin{bmatrix} 1 \\ 0 \end{bmatrix} + b\,[E, F] \begin{bmatrix} 0 \\ 1 \end{bmatrix} = aE + bF = W \end{aligned}$$

Since $AW = W$ for all W, this tells us $A = I$. Combining we have

$$\begin{bmatrix} E^T \\ F^T \end{bmatrix} [E, F] = [E, F] \begin{bmatrix} E^T \\ F^T \end{bmatrix} = I$$

So the inverse of T is T^T and this time we proved the result without resorting to rotation angles and slopes of vectors.

Of course, if E and F were not orthonormal, we could apply Graham - Schmidt Orthogonalization to create the orthonormal basis U_1, U_2 as usual:

$$\begin{aligned} U_1 &= \frac{E}{\|E\|} \\ V &= F - < F, U_1 > U_1 \\ U_2 &= \frac{V}{\|V\|} \end{aligned}$$

and $\{U_1, U_2\}$ would be an orthonormal basis for $\Re^2$.

3.3.1.1 Homework

Exercise 3.3.1 *Let* $\boldsymbol{E} = \begin{bmatrix} 1 \\ 3 \end{bmatrix}$, $\boldsymbol{F} = \begin{bmatrix} 4 \\ 5 \end{bmatrix}$ *and* $\boldsymbol{G} = \begin{bmatrix} -2 \\ 6 \end{bmatrix}$ *in* $\Re^2$. *Find* a *and* b *so that* $a\boldsymbol{E} + b\boldsymbol{F} = \boldsymbol{G}$.

Exercise 3.3.2 *Let* $\boldsymbol{E} = \begin{bmatrix} -2 \\ 9 \end{bmatrix}$, $\boldsymbol{F} = \begin{bmatrix} 8 \\ 1 \end{bmatrix}$ *and* $\boldsymbol{G} = \begin{bmatrix} 14 \\ 2 \end{bmatrix}$ *in* $\Re^2$. *Find* a *and* b *so that* $a\boldsymbol{E} + b\boldsymbol{F} = \boldsymbol{G}$.

Exercise 3.3.3 *Let* $\boldsymbol{E} = \begin{bmatrix} 1 \\ 3 \end{bmatrix}$, *and* $\boldsymbol{F} = \begin{bmatrix} 4 \\ 5 \end{bmatrix}$. *Use* **GSO** *to find an orthonormal basis* $\{\boldsymbol{U}, \boldsymbol{V}\}$ *for the span of* $\boldsymbol{E}$ *and* $\boldsymbol{F}$ *which is* $\Re^2$. *Show if* $\boldsymbol{A} = \begin{bmatrix} \boldsymbol{U} & \boldsymbol{V} \end{bmatrix}$, *then* $\boldsymbol{A}^T\boldsymbol{A} = \boldsymbol{A}\boldsymbol{A}^T = \boldsymbol{I}$ *so that* $\boldsymbol{A}^T = \boldsymbol{A}^{-1}$.

Exercise 3.3.4 *Let* $\boldsymbol{E} = \begin{bmatrix} -2 \\ 9 \end{bmatrix}$, *and* $\boldsymbol{F} = \begin{bmatrix} 8 \\ 1 \end{bmatrix}$. *Use* **GSO** *to find an orthonormal basis* $\{\boldsymbol{U}, \boldsymbol{V}\}$ *for the span of* $\boldsymbol{E}$ *and* $\boldsymbol{F}$ *which is* $\Re^2$. *Show if* $\boldsymbol{A} = \begin{bmatrix} \boldsymbol{U} & \boldsymbol{V} \end{bmatrix}$, *then* $\boldsymbol{A}^T\boldsymbol{A} = \boldsymbol{A}\boldsymbol{A}^T = \boldsymbol{I}$ *so that* $\boldsymbol{A}^T = \boldsymbol{A}^{-1}$. *Also, find* a *and* b *so that* $\boldsymbol{G} = a\boldsymbol{E} + b\boldsymbol{F}$ *and* c *and* d *so that* $\boldsymbol{G} = c\boldsymbol{U} + b\boldsymbol{V}$. *Let*

$$\boldsymbol{B} = \begin{bmatrix} <\boldsymbol{E}, \boldsymbol{U}> & <\boldsymbol{F}, \boldsymbol{U}> \\ <\boldsymbol{E}, \boldsymbol{V}> & <\boldsymbol{F}, \boldsymbol{V}> \end{bmatrix}$$

Calculate $\boldsymbol{B}\begin{bmatrix} a \\ b \end{bmatrix}$ *and compare to* $\begin{bmatrix} c \\ d \end{bmatrix}$. *What do you find?*

3.3.2 The Connection between Two Orthonormal Bases

Given the orthonormal basis $\{\boldsymbol{E_1}, \boldsymbol{E_2}\}$, we know from our remarks above that the matrix we form by using these vectors as columns, $\boldsymbol{T_E} = \begin{bmatrix} \boldsymbol{E_1}, \boldsymbol{E_2} \end{bmatrix}$, has an inverse which is its own transpose; i.e. $\boldsymbol{T_E}^{-1} = (\boldsymbol{T_E})^T$. For convenience, we now denote this orthonormal basis by $\boldsymbol{E}$. Hence, the orthonormal basis $\{\boldsymbol{F_1}, \boldsymbol{F_2}\}$ is denoted by $\boldsymbol{F}$. We now introduce notation we will use later when we think carefully about transformations from $\Re^n$ to $\Re^m$. Given a vector $\boldsymbol{A}$ it has a representation relative to any choice of basis. We need to recognize that this representation will depend on the choice of basis, so given that $\boldsymbol{A} = a\boldsymbol{E_1} + b\boldsymbol{E_2}$ for unique scalars a and b, the vector $\begin{bmatrix} a \\ b \end{bmatrix}$ is a convenient way to handle this representation. We use the subscript $\boldsymbol{E}$ to remind us these components depend on our choice of basis $\boldsymbol{E}$. When we write a two dimensional column vector, $\begin{bmatrix} a \\ b \end{bmatrix}$, have you ever thought about what the numbers a and b mean? You need to always remember that the idea of a two dimensional vector space is quite abstract. We can't really **see** these objects until we make a choice of basis. Thus we can't calculate inner products until we choose a basis also. The simplest basis is the usual one

$$\boldsymbol{i} = \begin{bmatrix} 1 \\ 0 \end{bmatrix}, \quad \boldsymbol{j} = \begin{bmatrix} 0 \\ 1 \end{bmatrix}$$

which is called the **standard basis**, which we will denote by $\boldsymbol{I}$. So all of our discussions above were using the standard basis $\boldsymbol{I}$. Now this is very important. With respect to the orthonormal basis $\boldsymbol{E}$, the representation of $\boldsymbol{E_1}$ is $\begin{bmatrix} E_{11} \\ E_{12} \end{bmatrix}_{\boldsymbol{I}}$ but with respect to the basis $\boldsymbol{E}$, its components would be $\begin{bmatrix} 1 \\ 0 \end{bmatrix}_{\boldsymbol{E}}$. This is what we are doing when we set up the cartesian coordinate system. We choose what we call the x and y axis oriented so they are 90° apart. The vectors $\boldsymbol{i}$ and $\boldsymbol{j}$ are then chosen along the positive x and y axis. However, rotating this choice of $\boldsymbol{i}$ and $\boldsymbol{j}$ by θ° gives us a new x' and y' axis still 90°

apart. So when we want to be very explicit, we write

$$[\boldsymbol{A}]_{\boldsymbol{E}} = \begin{bmatrix} a \\ b \end{bmatrix}_{\boldsymbol{E}}$$

to indicate this representation with respect to the basis $\boldsymbol{E}$. Now $\boldsymbol{A}$ also has a representation due to the basis $\boldsymbol{F}$.

$$[\boldsymbol{A}]_{\boldsymbol{F}} = \begin{bmatrix} c \\ d \end{bmatrix}_{\boldsymbol{F}}$$

From our first discussion, now explicitly realizing we were using representations relative to the standard basis, we should write

$$[\boldsymbol{A_I}]_{\boldsymbol{F_I}} = \begin{bmatrix} c \\ d \end{bmatrix}_{\boldsymbol{F_I}}$$

This is very cumbersome. The subscript $\boldsymbol{I}$ here reminds us that the components we see in the vectors are relative to the standard basis and that the components of $\boldsymbol{E}$ and $\boldsymbol{F}$ are also given with respect to the standard basis. If we are expressing everything in terms of the $\boldsymbol{E}$ basis remember $\boldsymbol{E_1}$ and $\boldsymbol{E_2}$ themselves have the usual standard basis components; i.e. we are finding components relative to a new x' and y' axis. So with all that said, given some component representation for the components of $\boldsymbol{E}$, we can find the components of the vectors in $\boldsymbol{F}$ with respect to the basis $\boldsymbol{E}$ as follows. Remember, the basis $\boldsymbol{F}$ corresponds to a new rotation to an x'' and y'' coordinate axis system. Using these ideas, we see we can also express $\boldsymbol{F}$ in terms of $\boldsymbol{E}$. We have

$$\begin{aligned} \boldsymbol{F_1} &= <\boldsymbol{F_1}, \boldsymbol{E_1}> \boldsymbol{E_1} + <\boldsymbol{F_1}, \boldsymbol{E_2}> \boldsymbol{E_2} = T_{11}\boldsymbol{E_1} + T_{12}\boldsymbol{E_2} \\ \boldsymbol{F_2} &= <\boldsymbol{F_2}, \boldsymbol{E_1}> \boldsymbol{E_1} + <\boldsymbol{F_2}, \boldsymbol{E_2}> \boldsymbol{E_2} = T_{21}\boldsymbol{E_1} + T_{22}\boldsymbol{E_2} \end{aligned}$$

Thus, using our two representations for $\boldsymbol{A}$, we have

$$\begin{aligned} \boldsymbol{A} &= c\boldsymbol{F_1} + d\boldsymbol{F_2} = c(T_{11}\boldsymbol{E_1} + T_{12}\boldsymbol{E_2}) + d(T_{21}\boldsymbol{E_1} + T_{22}\boldsymbol{E_2}) \\ &= (cT_{11} + dT_{21})\boldsymbol{E_1} + (cT_{12} + dT_{22})\boldsymbol{E_2} = a\boldsymbol{E_1} + b\boldsymbol{E_2} \end{aligned}$$

This tells us

$$\begin{aligned} a &= cT_{11} + dT_{21} \\ b &= cT_{12} + dT_{22} \end{aligned}$$

This implies

$$\begin{bmatrix} a \\ b \end{bmatrix} = \begin{bmatrix} T_{11} & T_{21} \\ T_{12} & T_{22} \end{bmatrix} \begin{bmatrix} c \\ d \end{bmatrix} \Longrightarrow [\boldsymbol{A}]_{\boldsymbol{E}} = \begin{bmatrix} T_{11} & T_{21} \\ T_{12} & T_{22} \end{bmatrix} [\boldsymbol{A}]_{\boldsymbol{F}}$$

or

$$[\boldsymbol{A}]_{\boldsymbol{E}} = \begin{bmatrix} <\boldsymbol{F_1}, \boldsymbol{E_1}> & <\boldsymbol{F_2}, \boldsymbol{E_1}> \\ <\boldsymbol{F_1}, \boldsymbol{E_2}> & <\boldsymbol{F_2}, \boldsymbol{E_2}> \end{bmatrix} [\boldsymbol{A}]_{\boldsymbol{F}}$$

We can do the same sort of analysis and interchange the role of $\boldsymbol{E}$ and $\boldsymbol{F}$ to get

$$[\boldsymbol{A}]_{\boldsymbol{F}} = \begin{bmatrix} <\boldsymbol{E_1}, \boldsymbol{F_1}> & <\boldsymbol{E_2}, \boldsymbol{F_1}> \\ <\boldsymbol{E_1}, \boldsymbol{F_2}> & <\boldsymbol{E_2}, \boldsymbol{F_2}> \end{bmatrix} [\boldsymbol{A}]_{\boldsymbol{E}}$$

The coefficients (T_{ij}) define a matrix which we will call $\boldsymbol{T_{FE}}$ as these coefficients are completely determined by the basis $\boldsymbol{E}$ and $\boldsymbol{F}$ and we are transforming $\boldsymbol{F}$ components into $\boldsymbol{E}$ components. Hence, we let

$$\boldsymbol{T_{FE}} = \begin{bmatrix} <\boldsymbol{F_1},\boldsymbol{E_1}> & <\boldsymbol{F_2},\boldsymbol{E_1}> \\ <\boldsymbol{F_1},\boldsymbol{E_2}> & <\boldsymbol{F_2},\boldsymbol{E_2}> \end{bmatrix} = \begin{bmatrix} T_{11} & T_{21} \\ T_{12} & T_{22} \end{bmatrix} = \boldsymbol{T_{EF}}^T$$

and

$$\boldsymbol{T_{EF}} = \begin{bmatrix} <\boldsymbol{E_1},\boldsymbol{F_1}> & <\boldsymbol{E_2},\boldsymbol{F_1}> \\ <\boldsymbol{E_1},\boldsymbol{F_2}> & <\boldsymbol{E_2},\boldsymbol{F_2}> \end{bmatrix} = \begin{bmatrix} T_{11} & T_{12} \\ T_{21} & T_{22} \end{bmatrix} = \boldsymbol{T_{FE}}^T$$

So $\boldsymbol{T_{FE}}$ and $\boldsymbol{T_{EF}}$ are transposes of each other. We can easily show $\boldsymbol{T_{FE}}^T$ is the inverse of $\boldsymbol{T_{FE}}$. We note for all vectors in $\Re^2$,

$$\boldsymbol{A_E} = \boldsymbol{T_{FE}}\,\boldsymbol{A_F} = \boldsymbol{T_{FE}}\,\boldsymbol{T_{EF}}\,\boldsymbol{A_E} = \boldsymbol{T_{FE}}\,\boldsymbol{T_{FE}}^T\,\boldsymbol{A_E}$$

and

$$\boldsymbol{A_F} = \boldsymbol{T_{EF}}\,\boldsymbol{A_E} = \boldsymbol{T_{FE}}^T\,\boldsymbol{T_{FE}}\,\boldsymbol{A_F} = \boldsymbol{T_{FE}}^T\,\boldsymbol{T_{FE}}\boldsymbol{A_E}$$

Hence $\boldsymbol{T_{FE}}\,\boldsymbol{T_{FE}}^T = \boldsymbol{T_{FE}}^T\,\boldsymbol{T_{FE}} = \boldsymbol{I}$ and $\boldsymbol{T_{FE}}^T$ is the inverse of $\boldsymbol{T_{FE}}$.

3.3.2.1 Homework

Exercise 3.3.5 *Let* $\boldsymbol{A} = \begin{bmatrix} 2 \\ -3 \end{bmatrix}$, $\boldsymbol{B} = \begin{bmatrix} 3 \\ 7 \end{bmatrix}$, $\boldsymbol{C} = \begin{bmatrix} -1 \\ 2 \end{bmatrix}$ *and* $\boldsymbol{D} = \begin{bmatrix} 4 \\ 1 \end{bmatrix}$ *in* $\Re^2$. *Use* **GSO** *on* $\boldsymbol{E} = \{\boldsymbol{A},\boldsymbol{B}\}$ *and* $\boldsymbol{F}\{\boldsymbol{C},\boldsymbol{D}\}$ *to find orthonormal bases* $\boldsymbol{G} = \{\boldsymbol{U},\boldsymbol{V}\}$ *and* $\boldsymbol{H} = \{\boldsymbol{P},\boldsymbol{Q}\}$, *respectively. Find* $\boldsymbol{T_{EG}}$ *and* $\boldsymbol{T_{FH}}$. *Feel free to do this in MATLAB. Verify all their properties.*

Exercise 3.3.6 *Let* $\boldsymbol{A} = \begin{bmatrix} 6 \\ -13 \end{bmatrix}$, $\boldsymbol{B} = \begin{bmatrix} 4 \\ 1 \end{bmatrix}$, $\boldsymbol{C} = \begin{bmatrix} -1 \\ -2 \end{bmatrix}$ *and* $\boldsymbol{D} = \begin{bmatrix} -3 \\ -4 \end{bmatrix}$ *in* $\Re^2$. *Use* **GSO** *on* $\boldsymbol{E} = \{\boldsymbol{A},\boldsymbol{B}\}$ *and* $\boldsymbol{F} = \{\boldsymbol{C},\boldsymbol{D}\}$ *to find orthonormal bases* $\boldsymbol{G} = \{\boldsymbol{U},\boldsymbol{V}\}$ *and* $\boldsymbol{H} = \{\boldsymbol{P},\boldsymbol{Q}\}$, *respectively. Find* $\boldsymbol{T_{EG}}$ *and* $\boldsymbol{T_{FH}}$. *Feel free to do this in MATLAB. Verify all their properties.*

3.3.3 The Invariance of the Inner Product

For two orthonormal bases $\boldsymbol{E}$ and $\boldsymbol{F}$, we now know the vector $\boldsymbol{A}$ transforms as follows:

$$\left[\boldsymbol{A}\right]_{\boldsymbol{F}} = \left[\boldsymbol{T_1},\boldsymbol{T_2}\right]^T \left[\boldsymbol{A}\right]_{\boldsymbol{E}}$$

Thus, for two vectors $\boldsymbol{A}$ and $\boldsymbol{B}$, we can calculate

$$< \left[\boldsymbol{A}\right]_{\boldsymbol{F}}, \left[\boldsymbol{B}\right]_{\boldsymbol{F}} > = < \left[\boldsymbol{T_1},\boldsymbol{T_2}\right]^T \left[\boldsymbol{A}\right]_{\boldsymbol{E}}, \left[\boldsymbol{T_1},\boldsymbol{T_2}\right]^T \left[\boldsymbol{B}\right]_{\boldsymbol{E}} >$$

It is easy to see that using vector representations, $<\boldsymbol{A},\boldsymbol{B}>= \boldsymbol{A}^T\boldsymbol{B} = \boldsymbol{B}^T\boldsymbol{A}$ where these are interpreted using the usual matrix multiplication routines we know. Thus,

$$\begin{aligned} < \left[\boldsymbol{A}\right]_{\boldsymbol{F}}, \left[\boldsymbol{B}\right]_{\boldsymbol{F}} > &= (< \left[\boldsymbol{T_1},\boldsymbol{T_2}\right]^T \left[\boldsymbol{A}\right]_{\boldsymbol{E}})^T \left[\boldsymbol{T_1},\boldsymbol{T_2}\right]^T \left[\boldsymbol{B}\right]_{\boldsymbol{E}} > \\ &= \left[\boldsymbol{A^T}\right]_{\boldsymbol{E}} (\left[\boldsymbol{T_1},\boldsymbol{T_2}\right]) \left[\boldsymbol{T_1},\boldsymbol{T_2}\right]^T \left[\boldsymbol{B}\right]_{\boldsymbol{E}} \end{aligned}$$

But $(\left[\boldsymbol{T_1},\boldsymbol{T_2}\right])^T$ is the inverse of $\left[\boldsymbol{T_1},\boldsymbol{T_2}\right]$. Thus, we find

$$< \left[\boldsymbol{A}\right]_{\boldsymbol{F}}, \left[\boldsymbol{B}\right]_{\boldsymbol{F}} > = \left[\boldsymbol{A^T}\right]_{\boldsymbol{E}} \left[\boldsymbol{B}\right]_{\boldsymbol{E}} =< \left[\boldsymbol{A}\right]_{\boldsymbol{E}}, \left[\boldsymbol{B}\right]_{\boldsymbol{E}} >$$

which tells us the inner product calculation is invariant under a transformation from one orthonormal basis to another. Of course, if one or both of these new basis are not orthonormal, the matrix $\boldsymbol{T}$ here is not its own inverse and the value of the inner product could change.
So in general, not writing down all the subscripts about components with respect to various bases, we find unique scalars using the $\{\boldsymbol{E_1}, \boldsymbol{E_2}\}$ basis so that

$$\boldsymbol{A} \;=\; a\boldsymbol{E_1} + b\boldsymbol{E_2}, \quad \boldsymbol{B} = c\boldsymbol{E_1} + d\boldsymbol{E_2}$$

Thus,

$$< \boldsymbol{A}, \boldsymbol{B} > \;=\; < a\boldsymbol{E_1} + b\boldsymbol{E_2}, c\boldsymbol{E_1} + d\boldsymbol{E_2} >= ac + bd$$

Using the other basis, we have

$$\begin{aligned} \boldsymbol{A} &= \alpha\boldsymbol{F_1} + \beta\boldsymbol{F_2} \\ \boldsymbol{B} &= \gamma\boldsymbol{F_1} + \delta\boldsymbol{F_2} \end{aligned}$$

and so

$$< \boldsymbol{A}, \boldsymbol{B} > \;=\; < \alpha\boldsymbol{F_1} + \beta\boldsymbol{F_2}, \gamma\boldsymbol{F_1} + \delta\boldsymbol{F_2} >= \alpha\gamma + \beta\delta$$

And we now know these two values will be the same!

3.3.3.1 Homework

Exercise 3.3.7 *Let* $\boldsymbol{A} = \begin{bmatrix} 2 \\ -3 \end{bmatrix}$, $\boldsymbol{B} = \begin{bmatrix} 3 \\ 7 \end{bmatrix}$, $\boldsymbol{C} = \begin{bmatrix} -1 \\ 2 \end{bmatrix}$ *and* $\boldsymbol{D} = \begin{bmatrix} 4 \\ 1 \end{bmatrix}$ *in* $\Re^2$. *Use* **GSO** *on* $\boldsymbol{E} = \{\boldsymbol{A}, \boldsymbol{B}\}$ *and* $\boldsymbol{F} = \{\boldsymbol{C}, \boldsymbol{D}\}$ *to find orthonormal bases* $\boldsymbol{G} = \{\boldsymbol{U}, \boldsymbol{V}\}$ *and* $\boldsymbol{H} = \{\boldsymbol{P}, \boldsymbol{Q}\}$, *respectively. Find* $\boldsymbol{T_{EG}}$ *and* $\boldsymbol{T_{FH}}$. *Feel free to do this in . Now compute inner products with respect to both bases. Let* $\boldsymbol{S} = \begin{bmatrix} 22 \\ 35 \end{bmatrix}$ *and* $\boldsymbol{T} = \begin{bmatrix} 41 \\ -4 \end{bmatrix}$. *Find* $[\boldsymbol{S}]_G$, $[\boldsymbol{T}]_G$ *and* $[\boldsymbol{S}]_H$, $[\boldsymbol{T}]_H$. *Then calculate* $< [\boldsymbol{S}]_G, [\boldsymbol{T}]_G >$ *and* $< [\boldsymbol{S}]_H, [\boldsymbol{T}]_H >$.

Exercise 3.3.8 *Let* $\boldsymbol{A} = \begin{bmatrix} 6 \\ -13 \end{bmatrix}$, $\boldsymbol{B} = \begin{bmatrix} 4 \\ 1 \end{bmatrix}$, $\boldsymbol{C} = \begin{bmatrix} -1 \\ -2 \end{bmatrix}$ *and* $\boldsymbol{D} = \begin{bmatrix} -3 \\ -4 \end{bmatrix}$ *in* $\Re^2$. *Use* **GSO** *on* $\boldsymbol{E} = \{\boldsymbol{A}, \boldsymbol{B}\}$ *and* $\boldsymbol{F} = \{\boldsymbol{C}, \boldsymbol{D}\}$ *to find orthonormal bases* $\boldsymbol{G} = \{\boldsymbol{U}, \boldsymbol{V}\}$ *and* $\boldsymbol{H} = \{\boldsymbol{P}, \boldsymbol{Q}\}$, *respectively. Find* $\boldsymbol{T_{EG}}$ *and* $\boldsymbol{T_{FH}}$. *Feel free to do this in MATLAB. Now compute inner products with respect to both bases. Let* $\boldsymbol{S} = \begin{bmatrix} 2 \\ 15 \end{bmatrix}$ *and* $\boldsymbol{T} = \begin{bmatrix} 4 \\ -19 \end{bmatrix}$. *Find* $[\boldsymbol{S}]_G$, $[\boldsymbol{T}]_G$ *and* $[\boldsymbol{S}]_H$, $[\boldsymbol{T}]_H$. *Then calculate* $< [\boldsymbol{S}]_G, [\boldsymbol{T}]_G >$ *and* $< [\boldsymbol{S}]_H, [\boldsymbol{T}]_H >$.

3.3.4 Two Dimensional Vectors as Functions

Let's review linear systems of first order ODE. You should have seen these before, but a quick review won't hurt. These have the form

$$\begin{aligned} x'(t) &= a\,x(t) \,+\, b\,y(t) \\ y'(t) &= c\,x(t) \,+\, d\,y(t) \\ x(0) &= x_0, \quad y(0) = y_0 \end{aligned}$$

for any numbers a, b, c and d and *initial conditions* x_0 and y_0. The full problem is called, as usual, an *Initial Value Problem* or **IVP** for short. The two initial conditions are just called the **IC**s for the problem to save writing.

- For example, we might be interested in the system

$$\begin{aligned} x'(t) &= -2\,x(t) + 3\,y(t) \\ y'(t) &= 4\,x(t) + 5\,y(t) \\ x(0) &= 5, \quad y(0) = -3 \end{aligned}$$

  Here the **IC**s are $x(0) = 5$ and $y(0) = -3$.

- Another sample problem might be the one below.

$$\begin{aligned} x'(t) &= 14\,x(t) + 5\,y(t) \\ y'(t) &= -4\,x(t) + 8\,y(t) \\ x(0) &= 2, \quad y(0) = 7 \end{aligned}$$

3.3.4.1 The Characteristic Equation

For linear first order problems like $u' = 3u$ and so forth, we find the solution has the form $u(t) = A\,e^{3t}$ for some number A. We then determine the value of A to use by looking at the initial condition. To find the solutions here, we begin by rewriting the model in matrix-vector notation.

$$\begin{bmatrix} x'(t) \\ y'(t) \end{bmatrix} = \begin{bmatrix} a & b \\ c & d \end{bmatrix} \begin{bmatrix} x(t) \\ y(t) \end{bmatrix}.$$

The 2×2 matrix above is called the **coefficient matrix** of this model and is usually denoted by $\boldsymbol{A}$. The initial conditions can then be redone in vector form as

$$\begin{bmatrix} x(0) \\ y(0) \end{bmatrix} = \begin{bmatrix} x_0 \\ y_0 \end{bmatrix}$$

Now it seems reasonable to believe that if a constant times e^{rt} solves a first order linear problem like $u' = ru$, perhaps a *vector* times e^{rt} will work here. Let's make this formal. So let's look at the problem below

$$\begin{aligned} x'(t) &= 3\,x(t) + 2\,y(t) \\ y'(t) &= -4\,x(t) + 5\,y(t) \\ x(0) &= 2 \\ y(0) &= -3 \end{aligned}$$

Assume the solution has the form $\boldsymbol{V}\,e^{rt}$. Let's denote the components of $\boldsymbol{V}$ as follows:

$$\boldsymbol{V} = \begin{bmatrix} V_1 \\ V_2 \end{bmatrix}.$$

We assume the solution is

$$\begin{bmatrix} x(t) \\ y(t) \end{bmatrix} = \boldsymbol{V}\,e^{rt}.$$

Then the derivative of $\boldsymbol{V}\,e^{rt}$ is

$$\left(\boldsymbol{V}\,e^{rt}\right)' = r\,\boldsymbol{V}\,e^{rt} \Longrightarrow \begin{bmatrix} x'(t) \\ y'(t) \end{bmatrix} = r\,\boldsymbol{V}\,e^{rt}.$$

When we plug these terms into the matrix-vector form of the problem, we find

$$r\,\boldsymbol{V}\,e^{rt} = \begin{bmatrix} 3 & 2 \\ -4 & 5 \end{bmatrix} \boldsymbol{V}\,e^{rt}$$

Rewrite using the identity matrix $\boldsymbol{I}$ as

$$\begin{aligned} r\,\boldsymbol{V}\,e^{rt} - \begin{bmatrix} 3 & 2 \\ -4 & 5 \end{bmatrix} \boldsymbol{V}\,e^{rt} &= r\boldsymbol{I}\,\boldsymbol{V}\,e^{rt} - \begin{bmatrix} 3 & 2 \\ -4 & 5 \end{bmatrix} \boldsymbol{V}\,e^{rt} \\ &= \left(r\boldsymbol{I} - \begin{bmatrix} 3 & 2 \\ -4 & 5 \end{bmatrix}\right) \boldsymbol{V}\,e^{rt} \\ &= \left(\begin{bmatrix} r & 0 \\ 0 & r \end{bmatrix} - \begin{bmatrix} 3 & 2 \\ -4 & 5 \end{bmatrix}\right) \boldsymbol{V}\,e^{rt} \\ &= \begin{bmatrix} r-3 & -2 \\ -(-4) & r-5 \end{bmatrix} \boldsymbol{V}\,e^{rt} \end{aligned}$$

Plugging this into our model, we find

$$\begin{bmatrix} r-3 & -2 \\ 4 & r-5 \end{bmatrix} \boldsymbol{V}\,e^{rt} = \begin{bmatrix} 0 \\ 0 \end{bmatrix}.$$

But e^{rt} is never 0, so we want r satisfying

$$\begin{bmatrix} r-3 & -2 \\ 4 & r-5 \end{bmatrix} \boldsymbol{V} = \begin{bmatrix} 0 \\ 0 \end{bmatrix}.$$

For each r, we get two equations in V_1 and V_2:

$$(r-3)V_1 - 2V_2 = 0, \quad 4V_1 + (r-5)V_2 = 0.$$

Let $\boldsymbol{A_r}$ be this matrix. Any r for which the det $\boldsymbol{A_r} \neq 0$ tells us these two lines have different slopes and so cross at the origin implying $V_1 = 0$ and $V_2 = 0$. Thus

$$\begin{bmatrix} x \\ y \end{bmatrix} = \begin{bmatrix} 0 \\ 0 \end{bmatrix} e^{rt} = \begin{bmatrix} 0 \\ 0 \end{bmatrix}$$

which will not satisfy **nonzero initial conditions**. So **reject** these r. Any value of r for which det $\boldsymbol{A_r} = 0$ gives an infinite number of solutions which allows us to pick one that matches the initial conditions we have. The equation

$$\det(r\boldsymbol{I} - \boldsymbol{A}) = \det \begin{bmatrix} r-3 & -2 \\ 4 & r-5 \end{bmatrix} = 0$$

is called the **characteristic equation** of this linear system. The **characteristic equation** is a quadratic, so there are three possibilities: two distinct roots, the real roots are the same and the roots are a complex conjugate pair.

Example 3.3.1 *Derive the characteristic equation for the system below*

$$\begin{aligned} x'(t) &= 8\,x(t) + 9\,y(t) \\ y'(t) &= 3\,x(t) - 2\,y(t) \\ x(0) &= 12 \\ y(0) &= 4 \end{aligned}$$

Solution *The matrix-vector form is*

$$\begin{bmatrix} x'(t) \\ y'(t) \end{bmatrix} = \begin{bmatrix} 8 & 9 \\ 3 & -2 \end{bmatrix} \begin{bmatrix} x(t) \\ y(t) \end{bmatrix}$$
$$\begin{bmatrix} x(0) \\ y(0) \end{bmatrix} = \begin{bmatrix} 12 \\ 4 \end{bmatrix}$$

The coefficient matrix $\boldsymbol{A}$ *is thus*

$$\boldsymbol{A} = \begin{bmatrix} 8 & 9 \\ 3 & -2 \end{bmatrix}$$

Assume the solution has the form $\boldsymbol{V}\ e^{rt}$ *Plug this into the system.*

$$r\,\boldsymbol{V}\ e^{rt} - \begin{bmatrix} 8 & 9 \\ 3 & -2 \end{bmatrix} \boldsymbol{V}\ e^{rt} = \begin{bmatrix} 0 \\ 0 \end{bmatrix}$$

Rewrite using the identity matrix I and factor

$$\left(r\,\boldsymbol{I} - \begin{bmatrix} 8 & 9 \\ 3 & -2 \end{bmatrix} \right) \boldsymbol{V}\ e^{rt} = \begin{bmatrix} 0 \\ 0 \end{bmatrix}$$

Since $e^{rt} \neq 0$ *ever, we find r and* $\boldsymbol{V}$ *satisfy*

$$\left(r\,\boldsymbol{I} - \begin{bmatrix} 8 & 9 \\ 3 & -2 \end{bmatrix} \right) \boldsymbol{V} = \begin{bmatrix} 0 \\ 0 \end{bmatrix}$$

If r is chosen so that det $(rI - A) \neq 0$, *the only solution to this system of two linear equations in the two unknowns* V_1 *and* V_2 *is* $V_1 = 0$ *and* $V_2 = 0$. *This leads to* $x(t) = 0$ *and* $y(t) = 0$ *always and this solution does not satisfy the initial conditions. Hence, we must find r which give det* $(rI - A) = 0$. *The* **characteristic equation** *is thus*

$$\det \begin{bmatrix} r-8 & -9 \\ -3 & r+2 \end{bmatrix} = (r-8)(r+2) - 27$$
$$= r^2 - 6r - 43 = 0$$

3.3.4.2 Finding the General Solution

The roots to the characteristic equation are, of course, the **eigenvalues** of the coefficient matrix $\boldsymbol{A}$. We usually organize the **eigenvalues** with the largest one first, although we don't have to. In fact, in the examples, we organize from small to large!

- Example: The eigenvalues are -2 and -1. So $r_1 = -1$ and $r_2 = -2$. Since e^{-2t} decays faster than e^{-t}, we say the root $r_1 = -1$ is the **dominant** part of the solution.

- Example: The eigenvalues are -2 and 3. So $r_1 = 3$ and $r_2 = -2$. Since e^{-2t} decays and e^{3t} grows, we say the root $r_1 = 3$ is the **dominant** part of the solution.

- Example: The eigenvalues are 2 and 3. So $r_1 = 3$ and $r_2 = 2$. Since e^{2t} grows slower than e^{3t}, we say the root $r_1 = 3$ is the **dominant** part of the solution.

For each eigenvalue r we want to find nonzero vectors $\boldsymbol{V}$ so that $(r\boldsymbol{I} - \boldsymbol{A})\boldsymbol{V} = \boldsymbol{0}$ where to help with our writing we let $\boldsymbol{0}$ be the two dimensional zero vector. The nonzero $\boldsymbol{V}$ are called the **eigenvectors** for **eigenvalue** r and satisfy $\boldsymbol{A}\boldsymbol{V} = r\boldsymbol{V}$.

For eigenvalue r_1, find $\boldsymbol{V}$ so that $(r_1\,\boldsymbol{I} - \boldsymbol{A})\,\boldsymbol{V} = \boldsymbol{0}$. There will be an infinite number of $\boldsymbol{V}$'s that solve this; we pick one and call it **eigenvector** $\boldsymbol{E_1}$.

For eigenvalue r_2, find $\boldsymbol{V}$ so that $(r_2\,\boldsymbol{I} - \boldsymbol{A})\,\boldsymbol{V} = \boldsymbol{0}$. There will again be an infinite number of $\boldsymbol{V}$'s that solve this; we pick one and call it **eigenvector** $\boldsymbol{E_2}$.

The general solution to our model will be

$$\begin{bmatrix} x(t) \\ y(t) \end{bmatrix} = A\boldsymbol{E_1}\, e^{r_1 t} + B\boldsymbol{E_2}\, e^{r_2 t}.$$

where A and B are arbitrary. We use the ICs to find A and B. It is best to show all this with some examples.

Example 3.3.2 *For the system below*

$$\begin{aligned} \begin{bmatrix} x'(t) \\ y'(t) \end{bmatrix} &= \begin{bmatrix} -20 & 12 \\ -13 & 5 \end{bmatrix} \begin{bmatrix} x(t) \\ y(t) \end{bmatrix} \\ \begin{bmatrix} x(0) \\ y(0) \end{bmatrix} &= \begin{bmatrix} -1 \\ 2 \end{bmatrix} \end{aligned}$$

- *Find the characteristic equation*
- *Find the general solution*
- *Solve the IVP*

Solution *The characteristic equation is*

$$\det\left(r\begin{bmatrix} 1 & 0 \\ 0 & 1 \end{bmatrix} - \begin{bmatrix} -20 & 12 \\ -13 & 5 \end{bmatrix}\right) = 0$$

$$\begin{aligned} 0 &= det\left(\begin{bmatrix} r+20 & -12 \\ 13 & r-5 \end{bmatrix}\right) \\ &= (r+20)(r-5) + 156 \\ &= r^2 + 15r + 56 \\ &= (r+8)(r+7) \end{aligned}$$

Hence, **eigenvalues** *or* **roots** *of the characteristic equation are $r_1 = -8$ and $r_2 = -7$. Note since this is just a calculation, we are not following our labeling scheme.*

For eigenvalue $r_1 = -8$, substitute the value into

$$\begin{bmatrix} r+20 & -12 \\ 13 & r-5 \end{bmatrix} \begin{bmatrix} V_1 \\ V_2 \end{bmatrix} \Rightarrow \begin{bmatrix} 12 & -12 \\ 13 & -13 \end{bmatrix} \begin{bmatrix} V_1 \\ V_2 \end{bmatrix}$$

This system of equations should be collinear: i.e. the rows should be multiples; i.e. both give rise to the same line. Our rows are multiples, so we can pick any row to find V_2 in terms of V_1. Picking the top row, we get $12V_1 - 12V_2 = 0$ implying $V_2 = V_1$. Letting $V_1 = a$, we find $V_1 = a$ and $V_2 = a$: so

$$\begin{bmatrix} V_1 \\ V_2 \end{bmatrix} = a \begin{bmatrix} 1 \\ 1 \end{bmatrix}$$

Choose $\boldsymbol{E_1}$:
The vector

$$\boldsymbol{E}_1 = \begin{bmatrix} 1 \\ 1 \end{bmatrix}$$

is our choice for an **eigenvector** *corresponding to eigenvalue $r_1 = -8$. So one of the solutions is*

$$\begin{bmatrix} x_1(t) \\ y_1(t) \end{bmatrix} = \boldsymbol{E_1}\, e^{-8t} = \begin{bmatrix} 1 \\ 1 \end{bmatrix} e^{-8t}.$$

For eigenvalue $r_2 = -7$, substitute the value into

$$\begin{bmatrix} r+20 & -12 \\ 13 & r-5 \end{bmatrix} \begin{bmatrix} V_1 \\ V_2 \end{bmatrix} \Rightarrow \begin{bmatrix} 13 & -12 \\ 13 & -12 \end{bmatrix} \begin{bmatrix} V_1 \\ V_2 \end{bmatrix}$$

This system of equations should be collinear: i.e. the rows should be multiples; i.e. both give rise to the same line. Our rows are multiples, so we can pick any row to find V_2 in terms of V_1. Picking the top row, we get $13V_1 - 12V_2 = 0$ implying $V_2 = (13/12)V_1$. Letting $V_1 = b$, we find $V_1 = b$ and $V_2 = (13/12)b$: so

$$\begin{bmatrix} V_1 \\ V_2 \end{bmatrix} = b \begin{bmatrix} 1 \\ \frac{13}{12} \end{bmatrix}$$

Choose $\boldsymbol{E_2}$ to be

$$\boldsymbol{E}_2 = \begin{bmatrix} 1 \\ \frac{13}{12} \end{bmatrix}$$

as our choice for an **eigenvector** *corresponding to eigenvalue $r_2 = -7$. So one of the solutions is*

$$\begin{bmatrix} x_2(t) \\ y_2(t) \end{bmatrix} = \boldsymbol{E_2}\, e^{-7t} = \begin{bmatrix} 1 \\ \frac{13}{12} \end{bmatrix} e^{-7t}.$$

Note, it is easy to see $\boldsymbol{E_1}$ and $\boldsymbol{E_2}$ are linearly independent vectors in $\Re^2$. The general solution is then

$$\begin{aligned} \begin{bmatrix} x(t) \\ y(t) \end{bmatrix} &= A\, \boldsymbol{E_1}\, e^{-8t} + B\, \boldsymbol{E_2}\, e^{-7t} \\ &= A \begin{bmatrix} 1 \\ 1 \end{bmatrix} e^{-8t} + B \begin{bmatrix} 1 \\ \frac{13}{12} \end{bmatrix} e^{-7t} \end{aligned}$$

It is also easy to see the functions $\boldsymbol{E_1}e^{-8t}$ and $\boldsymbol{E_2}e^{-7t}$ form a basis for a two dimensional vector space over $\Re$. Now let's solve the initial value problem: we find A and B.

$$\begin{aligned} \begin{bmatrix} x(0) \\ y(0) \end{bmatrix} &= \begin{bmatrix} -1 \\ 2 \end{bmatrix} = A \begin{bmatrix} 1 \\ 1 \end{bmatrix} e^0 + B \begin{bmatrix} 1 \\ \frac{13}{12} \end{bmatrix} e^0 \\ &= A \begin{bmatrix} 1 \\ 1 \end{bmatrix} + B \begin{bmatrix} 1 \\ \frac{13}{12} \end{bmatrix} \end{aligned}$$

So

$$\begin{aligned} A + B &= -1 \\ A + \frac{13}{12} B &= 2 \end{aligned}$$

Subtracting the bottom equation from the top equation, we get $-\frac{1}{12}B = -3$ *or* $B = 36$*. Thus,* $A = -1 - B = -37$*. So*

$$\begin{bmatrix} x(t) \\ y(t) \end{bmatrix} = -37 \begin{bmatrix} 1 \\ 1 \end{bmatrix} e^{-8t} + 36 \begin{bmatrix} 1 \\ \frac{13}{12} \end{bmatrix} e^{-7t}$$

We have only reviewed the case of distinct roots here. Of course, there are also the cases of repeated roots and complex roots which you can review for yourself.

Homework

Let's start with a review of single variable ODE.

Exercise 3.3.9 *Find the general solution to* $u'' + 7u' + 8u = 0$ *and show these solutions form a two dimensional vector space over* $\Re$*.*

Exercise 3.3.10 *Find the general solution to* $u'' + 2u' + u = 0$ *and show these solutions form a two dimensional vector space over* $\Re$*.*

Exercise 3.3.11 *Find the general complex solution to* $u'' + 6u' + 25u = 0$ *and show these solutions form a two dimensional vector space over* $\mathbb{C}$*. Then show the general real solutions form a two dimensional vector space over* $\Re$*. You will need to review the basics of the real and complex solutions here.*

Exercise 3.3.12 *Now review the differential operator point of view for these linear constant coefficient ODEs. Consider the ODE in factored form* $(D-1)(D-3)(D+4)u = 0$*. Find the general solution and show these solutions form a three dimensional vector space over* $\Re$*.*

Exercise 3.3.13 *Consider the ODE in factored form* $(D-1)(D^2+1)u = 0$*. Find the general solution and show the real solutions form a three dimensional vector space over* $\Re$*.*

Now let's do some systems. First, some work with eigenvalues and eigenvectors.

Exercise 3.3.14 *Find the general solution to*

$$\begin{aligned} x'(t) &= -4\,x(t) + y(t) \\ y'(t) &= -5\,x(t) + 2\,y(t) \end{aligned}$$

and show it forms a two dimensional vector space over $\Re$*.*

Exercise 3.3.15 *Find the general solution to*

$$\begin{aligned} x'(t) &= 5\,x(t) + y(t) \\ y'(t) &= -7\,x(t) - 3\,y(t) \end{aligned}$$

and show it forms a two dimensional vector space over $\Re$*.*

Now let's find the solutions to initial value systems.

Exercise 3.3.16 *Find the solution to*

$$\begin{aligned} x'(t) &= -4\,x(t) + y(t) \\ y'(t) &= -5\,x(t) + 2\,y(t) \\ x(0) &= 2, \quad y(0) = -3 \end{aligned}$$

You can use the general solution to this which you have already found in a previous problem.

Exercise 3.3.17 *Find the solution to*

$$\begin{aligned} x'(t) &= 5\,x(t) + \; y(t) \\ y'(t) &= -7\,x(t) - \; 3\,y(t) \\ x(0) &= 12, \quad y(0) = -33 \end{aligned}$$

Again, you can use the general solution to this which you have already found in a previous problem.

Let's look at vector spaces of solutions to the system

$$\begin{aligned} \begin{bmatrix} x'(t) \\ y'(t) \end{bmatrix} &= \begin{bmatrix} -20 & 12 \\ -13 & 5 \end{bmatrix} \begin{bmatrix} x(t) \\ y(t) \end{bmatrix} \\ \begin{bmatrix} x(0) \\ y(0) \end{bmatrix} &= \begin{bmatrix} -1 \\ 2 \end{bmatrix} \end{aligned}$$

again. For concreteness, let's do this on the interval $[0,10]$ although other intervals could be used. Let $C^1([0,10])$ be the set of all functions x on $[0,10]$ which are continuously differentiable. This is clearly a vector space with scalar multiplication $\odot$ defined by $c \odot y$ which is the function whose value at t is

$$(c \odot y)(t) \quad = \quad cx(t)$$

To define addition of vectors $\oplus$, we let $x \oplus y$ be the function whose value at t is

$$(x \odot y)(t) \quad = \quad x(t) + y(t)$$

With these operations $C^1([0,10])$ is a vector space over $\Re$. Now let $\boldsymbol{X} = C^1([0,10]) \times C^1([0,10])$ which is the collection of all ordered pairs of continuously differentiable functions over $\Re$. As usual, it is easy to see we make this into a vector space by defining $\odot$ and $\oplus$ to act on pairs of objects in the usual way. We can define an inner product on this space by

$$< \boldsymbol{f}, \boldsymbol{g} > \quad = \quad \int_0^{10} \left\langle \begin{bmatrix} f_1 \\ f_2 \end{bmatrix}, \begin{bmatrix} g_1 \\ g_2 \end{bmatrix} \right\rangle \, dt$$

where $\boldsymbol{f} = \begin{bmatrix} f_1 \\ f_2 \end{bmatrix}$ and $\boldsymbol{g} = \begin{bmatrix} g1 \\ g_2 \end{bmatrix}$ are arbitrary members of $\boldsymbol{X}$. The two solution $y_1(t) = \boldsymbol{E_1} e^{-8t}$ and $y_2(t) = \boldsymbol{E_2} e^{-7t}$ are members of $\boldsymbol{X}$ which are linearly independent as when we write down the linearly dependent check, we have

$$a \odot \boldsymbol{E_1} y_1 \oplus b \odot \boldsymbol{E_2} y_2 \quad = \quad \boldsymbol{0}$$

This means

$$a\boldsymbol{E_1} e^{-8t} + b\boldsymbol{E_2} e^{-7t} \quad = \quad 0, \; \forall\, t \geq 0$$

In particular at $t = 0$, we have

$$a\boldsymbol{E_1} + b\boldsymbol{E_2} \quad = \quad 0, \; \forall\, t \geq 0$$

which is easily seen to force $a = b = 0$ since the eigenvectors $\boldsymbol{E_1}$ and $\boldsymbol{E_2}$ are linearly independent vectors in $\Re^2$. Hence, these functions are linearly independent objects in this vector space. They play the same role as the two dimensional vectors in the plane $\Re^2$ in our first section called $\boldsymbol{E}$ and

$\boldsymbol{F}$. Their span $\boldsymbol{S}$ is all linear combinations of y_1 and y_2 which is the solution space of this ODE. Thus, $\{\boldsymbol{E}_1 y_1, \boldsymbol{E}_2 y_2\}$ is a basis for a two dimensional subspace inside $\boldsymbol{X}$. We can then use Graham - Schmidt Orthogonalization to create an orthonormal basis for this solution space.

Listing 3.1: **Computing the Orthonormal Basis**

```
%Convert the eigenvectors to unit vectors
A = [-1;2];
E_1 = A/norm(A)
B = [1; 13/12];
E_2 = B/norm(B);
% Now do GSO
G_1 = E_1;
V = E_2 - dot(E_2,G_1)*G_1;
G_2 = V/norm(V);
>> dot(E_1,E_2)
ans =   0.35389
>> dot(G_1,G_2)
ans = 0
```

We check in this code fragment that $\boldsymbol{E_1}$ and $\boldsymbol{E_2}$ are not orthogonal but the new basis $\boldsymbol{G}$ is. Then, the functions $\boldsymbol{u_1} = \boldsymbol{G_1} e^{-8t}$ and $\boldsymbol{u_2} = \boldsymbol{G_2} e^{-7t}$ also solve the ODE and we have

$$< \boldsymbol{u_1}, \boldsymbol{u_2} > \quad = \quad \int_0^{10} < \boldsymbol{G_1} e^{-8t}, \boldsymbol{G_2} e^{-7t} > dt = \int_0^{10} < \boldsymbol{G_1}, \boldsymbol{G_2} > e^{-15t}\, dt = 0$$

So we have found an orthonormal basis for the solution space.

Earlier, we made the connection that the set of solutions here is a two dimensional vector space over $\Re$. Now we have added an inner product and showed how to construct an orthonormal basis for the set of solutions. In fact, we can be even clearer with the mapping details. If we define the mapping $\phi : span(\{\boldsymbol{G_1}, \boldsymbol{G_2}\}) \rightarrow span(\{\boldsymbol{G_1} e^{-8t}, \boldsymbol{G_2} e^{-7t}\})$ by

$$\phi(\boldsymbol{E}) \quad = \quad \boldsymbol{G_1} e^{-8t}, \quad \phi(\boldsymbol{G_2}) = \boldsymbol{G_2} e^{-7t}$$

and extend linearly, then ϕ is a $1-1$ and onto mapping of the span of $\{\boldsymbol{G_1}, \boldsymbol{G_2}\}$ to the span of $\{\boldsymbol{G_1} e^{-8t}, \boldsymbol{G_2} e^{-7t}\}$. Hence, we can **identify** the solution space of this ODE with the two dimensional space $\Re^2$. We never set up this explicit mapping before because it is pretty obvious, but note that this kind of mapping is behind all of our identifications of these solution spaces with a two dimensional vector space.

3.3.4.3 Homework

Exercise 3.3.18 *Prove*

$$< \boldsymbol{f}, \boldsymbol{g} > \quad = \quad \int_a^b \left\langle \begin{bmatrix} f_1 \\ f_2 \end{bmatrix}, \begin{bmatrix} g_1 \\ g_2 \end{bmatrix} \right\rangle dt$$

where $\boldsymbol{f} = \begin{bmatrix} f_1 \\ f_2 \end{bmatrix}$ *and* $\boldsymbol{g} = \begin{bmatrix} g1 \\ g_2 \end{bmatrix}$ *are arbitrary members of* $C([a,b]) \times C([a,b])$ *defines an inner product.*

Exercise 3.3.19 *Does*

$$< \boldsymbol{f}, \boldsymbol{g} > \;=\; \int_a^b \left(\left\langle \begin{bmatrix} f_1 \\ f_2 \end{bmatrix}, \begin{bmatrix} g_1 \\ g_2 \end{bmatrix} \right\rangle + \left\langle \begin{bmatrix} f_1' \\ f_2' \end{bmatrix}, \begin{bmatrix} g_1' \\ g_2' \end{bmatrix} \right\rangle \right) dt$$

where $\boldsymbol{f} = \begin{bmatrix} f_1 \\ f_2 \end{bmatrix}$ *and* $\boldsymbol{g} = \begin{bmatrix} g1 \\ g_2 \end{bmatrix}$ *are arbitrary members of* $C^1([a,b]) \times C^1([a,b])$ *define an inner product?*

Exercise 3.3.20 *Does*

$$< \boldsymbol{f}, \boldsymbol{g} > \;=\; \int_a^b \left(\left\langle \begin{bmatrix} f_1 \\ f_2 \end{bmatrix}, \begin{bmatrix} g_1 \\ g_2 \end{bmatrix} \right\rangle + \left\langle \begin{bmatrix} f_1' \\ f_2' \end{bmatrix}, \begin{bmatrix} g_1' \\ g_2' \end{bmatrix} \right\rangle \right) dt$$

where $\boldsymbol{f} = \begin{bmatrix} f_1 \\ f_2 \end{bmatrix}$ *and* $\boldsymbol{g} = \begin{bmatrix} g1 \\ g_2 \end{bmatrix}$ *are arbitrary members of* $C^{1e}([a,b]) \times C^{1e}([a,b])$ *where* $C^{1e}([a,b])$ *is the set of all functions whose derivatives* **exist** *on* $[a,b]$ *but are not necessarily continuous define an inner product?*

We have done these problems already, but now we write down the explicit basis for each solution space.

Exercise 3.3.21 *Find an orthonormal basis for the solution space to*

$$\begin{aligned} x'(t) &= -4\,x(t) + y(t) \\ y'(t) &= -5\,x(t) + 2\,y(t) \end{aligned}$$

This uses MATLAB of course.

Exercise 3.3.22 *Find an orthonormal basis for the solution space to*

$$\begin{aligned} x'(t) &= 5\,x(t) + y(t) \\ y'(t) &= -7\,x(t) - 3\,y(t) \end{aligned}$$

This uses MATLAB of course.

3.3.5 Three Dimensional Vectors in Space

Now we will use four pencils to motivate what is happening in three dimensions. Pick any two and hold them in your hand. These two pencils determine a plane as long as the pencils are not collinear. This plane does not have to be oriented parallel to our standard planes. Based on what we have said already, the standard basis for $\Re^3$ is

$$\boldsymbol{e_1} = \begin{bmatrix} 1 \\ 0 \\ 0 \end{bmatrix}, \quad \boldsymbol{e_2} = \begin{bmatrix} 0 \\ 1 \\ 0 \end{bmatrix}, \quad \boldsymbol{e_3} = \begin{bmatrix} 0 \\ 0 \\ 1 \end{bmatrix},$$

where we use these labels for the standard basis instead of the $\boldsymbol{i}$, $\boldsymbol{j}$ and $\boldsymbol{k}$ you may have seen before. If we want to work in $\Re^4$ or higher dimensions, we need the $\boldsymbol{e_i}$ notation. Pick any axis to orient the $\boldsymbol{e_1}$ along. The starting point of the $\boldsymbol{e_1}$ vector determines the origin of the three dimensional coordinate system we are constructing. The direction of $\boldsymbol{e_1}$ determines the $+x$ axis. The $\boldsymbol{e_2}$ vector then starts at the same origin and is positioned so it is orthogonal to $\boldsymbol{e_2}$. The direction of $\boldsymbol{e_2}$ fixes the $+y$ axis. We then use the right-hand rule to determine the placement of e_3. This determines the $+z$ axis. We now have a standard $\Re^3$ coordinate system that uses the standard orthonormal basis. The first two pencils fit in this system. Their common origin doesn't have to match the origin of the $\Re^3$ system we have just constructed but we want our pencils to determine a subspace in $\Re^3$. So think of their common origin as the same as the one we use for $\Re^3$ as constructed. These two pencils may not have the same length and may not be orthogonal but they determine a plane. If the third pencil lies in this plane, then it is not independent from the first two. However, if it lies out of the plane, the three pencils are linearly independent vectors whose span is all three dimensional vectors. Call the first pencil $\boldsymbol{E_1}$, the second $\boldsymbol{E_2}$ and the third $\boldsymbol{E_3}$. Then if the fourth pencil is $\boldsymbol{W}$, we can find the unique numbers which give its linear combination in terms of the basis $\{\boldsymbol{E_1}, \boldsymbol{E_2}, \boldsymbol{E_3}\}$ by solving

$$a\boldsymbol{E_1} + b\boldsymbol{E_2} + c\boldsymbol{E_3} = \boldsymbol{W} \Longrightarrow \begin{bmatrix}\boldsymbol{E_1} & \boldsymbol{E_2} & \boldsymbol{E_3}\end{bmatrix}\begin{bmatrix}a\\b\\c\end{bmatrix} = \boldsymbol{W}$$

Since these three vectors are linearly independent,

$$a\boldsymbol{E_1} + b\boldsymbol{E_2} + c\boldsymbol{E_3} = \boldsymbol{0} \Longrightarrow \begin{bmatrix}\boldsymbol{E_1} & \boldsymbol{E_2} & \boldsymbol{E_3}\end{bmatrix}\begin{bmatrix}a\\b\\c\end{bmatrix} = \boldsymbol{0}$$

only has the solution $a = b = c = 0$. Hence, the kernel of this matrix $\boldsymbol{T} = \begin{bmatrix}\boldsymbol{E_1}, \boldsymbol{E_2}, \boldsymbol{E_3}\end{bmatrix}$ is $\boldsymbol{0}$ and so $\boldsymbol{T}$ is invertible. The vectors $\boldsymbol{E_1}$, $\boldsymbol{E_2}$ and $\boldsymbol{E_3}$ need not be orthogonal, so it is not so easy to find this inverse but what counts here is that the equation

$$a\boldsymbol{E_1} + b\boldsymbol{E_2} + c\boldsymbol{E_3} = \boldsymbol{W}$$

has a unique solution for each $\boldsymbol{W}$.

Linear dependence of the vectors $\boldsymbol{E_1}$, $\boldsymbol{E_2}$ and $\boldsymbol{E_3}$ is more interesting in $\Re^3$. If you pick any two vectors, linear dependence of the third on the other two means this:

- $\boldsymbol{E_3}$ is in the plane determined by $\boldsymbol{E_1}$ and $\boldsymbol{E_2}$
- $\boldsymbol{E_1}$ is in the plane determined by $\boldsymbol{E_2}$ and $\boldsymbol{E_3}$
- $\boldsymbol{E_2}$ is in the plane determined by $\boldsymbol{E_1}$ and $\boldsymbol{E_3}$

And if some of the vectors are collinear, the situation becomes degenerate giving these possibilities:

- $\boldsymbol{E_1}$ and $\boldsymbol{E_2}$ are collinear
- $\boldsymbol{E_2}$ and $\boldsymbol{E_3}$ are collinear
- $\boldsymbol{E_1}$ and $\boldsymbol{E_3}$ are collinear

So the span of $\boldsymbol{E_1}$, $\boldsymbol{E_2}$ and $\boldsymbol{E_3}$ can be one dimensional (all vectors are collinear), two dimensional (two of the three are linearly independent) and three dimensional (all three vectors are linearly independent). In the last case, the vectors form a linearly independent spanning set and so are a basis.

If the vectors are not mutually orthogonal, we can use Graham - Schmidt Orthogonalization to create an orthonormal basis as follows:

Listing 3.2: **Computing the 3D Orthonormal Basis**

```
E1 = [-1;2;3];
E2 = [4;5;6];
```

```
E3 = [-3;1;7];
% Now do GSO
G1 = E1/norm(E1);
V = E2 - dot(E2,G1)*G1;
G2 = V/norm(V);
V = E3-dot(E3,G1)*G1-dot(E3,G2)*G2;
G3 = V/norm(V);
```

If $\boldsymbol{E} = \{\boldsymbol{E_1}, \boldsymbol{E_2}, \boldsymbol{E_3}\}$ is an orthonormal basis and $\boldsymbol{F} = \{\boldsymbol{F_1}, \boldsymbol{F_2}, \boldsymbol{F_3}\}$ is another orthonormal basis, the representation of a vector $\boldsymbol{A}$ can be given with respect to the standard basis, the basis $\boldsymbol{E}$ or the basis $\boldsymbol{F}$. Hence, we have the unique representation

$$\boldsymbol{A_E} = a_1\boldsymbol{E_1} + a_2\boldsymbol{E_2} + a_3\boldsymbol{E_3} \Longrightarrow \begin{bmatrix}\boldsymbol{E_1} & \boldsymbol{E_2} & \boldsymbol{E_3}\end{bmatrix} \begin{bmatrix} a_1 \\ a_2 \\ a_3 \end{bmatrix} = \boldsymbol{A_E}$$

It is easy to see

$$\begin{bmatrix} \boldsymbol{E_1^T} \\ \boldsymbol{E_2^T} \\ \boldsymbol{E_3^T} \end{bmatrix} \begin{bmatrix}\boldsymbol{E_1} & \boldsymbol{E_2} & \boldsymbol{E_3}\end{bmatrix} = \boldsymbol{I}$$

Now let's consider

$$\begin{bmatrix}\boldsymbol{E_1} & \boldsymbol{E_2} & \boldsymbol{E_3}\end{bmatrix} \begin{bmatrix} \boldsymbol{E_1^T} \\ \boldsymbol{E_2^T} \\ \boldsymbol{E_3^T} \end{bmatrix} \boldsymbol{E_j} = \begin{bmatrix}\boldsymbol{E_1} & \boldsymbol{E_2} & \boldsymbol{E_3}\end{bmatrix} \boldsymbol{e_j} = \boldsymbol{E_j}$$

Thus, we see for any vector $a\boldsymbol{E_1} + b\boldsymbol{E_2} + c\boldsymbol{E_3}$, we have

$$\begin{bmatrix}\boldsymbol{E_1} & \boldsymbol{E_2} & \boldsymbol{E_3}\end{bmatrix} \begin{bmatrix} \boldsymbol{E_1^T} \\ \boldsymbol{E_2^T} \\ \boldsymbol{E_3^T} \end{bmatrix} (a\boldsymbol{E_1} + b\boldsymbol{E_2} + c\boldsymbol{E_3}) = a\boldsymbol{E_1} + b\boldsymbol{E_2} + c\boldsymbol{E_3}$$

Hence, the inverse of $\begin{bmatrix}\boldsymbol{E_1} & \boldsymbol{E_2} & \boldsymbol{E_3}\end{bmatrix}$ is its own transpose.
Finally, since $\boldsymbol{A}$ has a representation with respect to the basis $\boldsymbol{F}$ also, we have

$$\begin{aligned}
\boldsymbol{F_1} &= <\boldsymbol{F_1}, \boldsymbol{E_1}> \boldsymbol{E_1} + <\boldsymbol{F_1}, \boldsymbol{E_2}> \boldsymbol{E_2} + <\boldsymbol{F_1}, \boldsymbol{E_3}> \boldsymbol{E_3} = T_{11}\boldsymbol{E_1} + T_{12}\boldsymbol{E_2} + T_{13}\boldsymbol{E_3} \\
\boldsymbol{F_2} &= <\boldsymbol{F_2}, \boldsymbol{E_1}> \boldsymbol{E_1} + <\boldsymbol{F_2}, \boldsymbol{E_2}> \boldsymbol{E_2} + <\boldsymbol{F_2}, \boldsymbol{E_3}> \boldsymbol{E_3} = T_{21}\boldsymbol{E_1} + T_{22}\boldsymbol{E_2} + T_{23}\boldsymbol{E_3} \\
\boldsymbol{F_3} &= <\boldsymbol{F_3}, \boldsymbol{E_1}> \boldsymbol{E_1} + <\boldsymbol{F_3}, \boldsymbol{E_2}> \boldsymbol{E_2} + <\boldsymbol{F_3}, \boldsymbol{E_3}> \boldsymbol{E_3} = T_{31}\boldsymbol{E_1} + T_{32}\boldsymbol{E_2} + T_{33}\boldsymbol{E_3}
\end{aligned}$$

Thus, using our two representations for $\boldsymbol{A}$, we have

$$\begin{aligned}
\boldsymbol{A} &= b_1\boldsymbol{F_1} + b_2\boldsymbol{F_2} + b_3\boldsymbol{F_3} = b_1(T_{11}\boldsymbol{E_1} + T_{12}\boldsymbol{E_2} + T_{13}\boldsymbol{E_3}) + b_2(T_{21}\boldsymbol{E_1} + T_{22}\boldsymbol{E_2} + T_{23}\boldsymbol{E_3}) \\
&\quad + b_3(T_{31}\boldsymbol{E_1} + T_{32}\boldsymbol{E_2} + T_{33}\boldsymbol{E_3})
\end{aligned}$$

This tells us

$$\begin{aligned}
a_1 &= b_1T_{11} + b_2T_{21} + b_3T_{31} \\
a_2 &= b_1T_{12} + b_2T_{22} + b_3T_{32} \\
a_3 &= b_1T_{13} + b_2T_{23} + b_3T_{33}
\end{aligned}$$

This implies

$$\begin{bmatrix} a_1 \\ a_2 \\ a_e \end{bmatrix} = \begin{bmatrix} T_{11} & T_{21} & T_{31} \\ T_{12} & T_{22} & T_{32} \\ T_{13} & T_{23} & T_{33} \end{bmatrix} \begin{bmatrix} b_1 \\ b_2 \\ b_3 \end{bmatrix} \Longrightarrow \left[\boldsymbol{A}\right]_{\boldsymbol{E}} = \begin{bmatrix} T_{11} & T_{21} & T_{31} \\ T_{12} & T_{22} & T_{32} \\ T_{13} & T_{23} & T_{33} \end{bmatrix} \left[\boldsymbol{A}\right]_{\boldsymbol{F}}$$

or

$$\left[\boldsymbol{A}\right]_{\boldsymbol{E}} = \begin{bmatrix} <\boldsymbol{F_1}, \boldsymbol{E_1}> & <\boldsymbol{F_2}, \boldsymbol{E_1}> & \boldsymbol{F_3}, \boldsymbol{E_1} \\ <\boldsymbol{F_1}, \boldsymbol{E_2}> & <\boldsymbol{F_2}, \boldsymbol{E_2}> & \boldsymbol{F_2}, \boldsymbol{E_3} \\ <\boldsymbol{F_1}, \boldsymbol{E_3}> & <\boldsymbol{F_2}, \boldsymbol{E_3}> & \boldsymbol{F_3}, \boldsymbol{E_3} \end{bmatrix} \left[\boldsymbol{A}\right]_{\boldsymbol{F}}$$

We can do the same sort of analysis and interchange the role of $\boldsymbol{E}$ and $\boldsymbol{F}$ to get

$$\left[\boldsymbol{A}\right]_{\boldsymbol{F}} = \begin{bmatrix} <\boldsymbol{E_1}, \boldsymbol{F_1}> & <\boldsymbol{E_2}, \boldsymbol{F_1}> & \boldsymbol{E_3}, \boldsymbol{F_1} \\ <\boldsymbol{E_1}, \boldsymbol{F_2}> & <\boldsymbol{E_2}, \boldsymbol{F_2}> & \boldsymbol{E_3}, \boldsymbol{F_2} \\ <\boldsymbol{E_1}, \boldsymbol{F_3}> & <\boldsymbol{E_2}, \boldsymbol{F_3}> & \boldsymbol{E_3}, \boldsymbol{F_3} \end{bmatrix} \left[\boldsymbol{A}\right]_{\boldsymbol{E}}$$

As before, the coefficients (T_{ij}) define a matrix which we will call $\boldsymbol{T_{FE}}$ as these coefficients are completely determined by the basis $\boldsymbol{E}$ and $\boldsymbol{F}$ and we are transforming $\boldsymbol{F}$ components into $\boldsymbol{E}$ components. Hence, we let

$$\boldsymbol{T_{FE}} = \begin{bmatrix} <\boldsymbol{F_1}, \boldsymbol{E_1}> & <\boldsymbol{F_2}, \boldsymbol{E_1}> & \boldsymbol{F_3}, \boldsymbol{E_1} \\ <\boldsymbol{F_1}, \boldsymbol{E_2}> & <\boldsymbol{F_2}, \boldsymbol{E_2}> & \boldsymbol{F_2}, \boldsymbol{E_3} \\ <\boldsymbol{F_1}, \boldsymbol{E_3}> & <\boldsymbol{F_2}, \boldsymbol{E_3}> & \boldsymbol{F_3}, \boldsymbol{E_3} \end{bmatrix} = \begin{bmatrix} T_{11} & T_{21} & T_{31} \\ T_{12} & T_{22} & T_{23} \\ T_{13} & T_{23} & T_{33} \end{bmatrix} = \boldsymbol{T_{EF}}^T$$

and

$$\boldsymbol{T_{EF}} = \begin{bmatrix} <\boldsymbol{E_1}, \boldsymbol{F_1}> & <\boldsymbol{E_2}, \boldsymbol{F_1}> & \boldsymbol{E_3}, \boldsymbol{F_1} \\ <\boldsymbol{E_1}, \boldsymbol{F_2}> & <\boldsymbol{E_2}, \boldsymbol{F_2}> & \boldsymbol{E_3}, \boldsymbol{F_2} \\ <\boldsymbol{E_1}, \boldsymbol{F_3}> & <\boldsymbol{E_2}, \boldsymbol{F_3}> & \boldsymbol{E_3}, \boldsymbol{F_3} \end{bmatrix} = \begin{bmatrix} T_{11} & T_{21} & T_{31} \\ T_{12} & T_{22} & T_{23} \\ T_{13} & T_{23} & T_{33} \end{bmatrix} = \boldsymbol{T_{EF}}^T$$

So $\boldsymbol{T_{FE}}$ and $\boldsymbol{T_{EF}}$ are transposes of each other. We can again show $\boldsymbol{T_{FE}}^T$ is the inverse of $\boldsymbol{T_{FE}}$ by a calculation similar to the one we did for $\Re^2$. In fact, since the dimension of the underlying space is not shown, the argument is identical. We note for all vectors in $\Re^3$,

$$\boldsymbol{A_E} = \boldsymbol{T_{FE}}\,\boldsymbol{A_F} = \boldsymbol{T_{FE}}\,\boldsymbol{T_{EF}}\,\boldsymbol{A_E} = \boldsymbol{T_{FE}}\,\boldsymbol{T_{FE}}^T\,\boldsymbol{A_E}$$

and

$$\boldsymbol{A_F} = \boldsymbol{T_{EF}}\,\boldsymbol{A_E} = \boldsymbol{T_{FE}}^T\,\boldsymbol{T_{FE}}\,\boldsymbol{A_F} = \boldsymbol{T_{FE}}^T\,\boldsymbol{T_{FE}}\boldsymbol{A_E}$$

Hence $\boldsymbol{T_{FE}}\,\boldsymbol{T_{FE}}^T = \boldsymbol{T_{FE}}^T\,\boldsymbol{T_{FE}} = \boldsymbol{I}$ and $\boldsymbol{T_{FE}}^T$ is the inverse of $\boldsymbol{T_{FE}}$.

The calculation that an inner product in $\Re^3$ is invariant under a change of orthonormal basis is then the same. Also, the arguments we use here can be adapted with little change to the $\Re^n$ setting.

3.3.5.1 Homework

Exercise 3.3.23 *Let*

$$\boldsymbol{E_1} = \begin{bmatrix} 2 \\ -3 \\ 4 \end{bmatrix}, \boldsymbol{E_2} = \begin{bmatrix} 3 \\ 7 \\ 5 \end{bmatrix} \textit{ and } \boldsymbol{E_3} = \begin{bmatrix} -1 \\ 2 \\ -3 \end{bmatrix}$$

$$F_1 = \begin{bmatrix} -1 \\ 2 \\ 1 \end{bmatrix}, F_2 = \begin{bmatrix} 4 \\ 1 \\ 10 \end{bmatrix} \text{ and } F_3 = \begin{bmatrix} -6 \\ 9 \\ 1 \end{bmatrix}$$

in $\Re^3$. *Use* **GSO** *on* $\boldsymbol{E} = \{\boldsymbol{E_1}, \boldsymbol{E_2}, \boldsymbol{E_3}\}$ *and* $\boldsymbol{F} = \{\boldsymbol{F_1}, \boldsymbol{F_2}, \boldsymbol{F_3}\}$ *to find orthonormal bases* $\boldsymbol{G} = \{\boldsymbol{G_1}, \boldsymbol{G_2}, \boldsymbol{G_3}\}$ *and* $\boldsymbol{H} = \{\boldsymbol{H_1}, \boldsymbol{H_2}, \boldsymbol{H_3}\}$*, respectively. Find* $\boldsymbol{T_{EG}}$ *and* $\boldsymbol{T_{FH}}$*. Feel free to do this in MATLAB. Now compute inner products with respect to both bases.*

$$\text{Let } \boldsymbol{S} = \begin{bmatrix} 22 \\ 35 \\ 4 \end{bmatrix} \text{ and } \boldsymbol{T} = \begin{bmatrix} 41 \\ -4 \\ 1 \end{bmatrix}.$$

Find $[\boldsymbol{S}]_G$, $[\boldsymbol{T}]_G$ *and* $[\boldsymbol{S}]_H$, $[\boldsymbol{T}]_H$. *Then calculate* $< [\boldsymbol{S}]_G, [\boldsymbol{T}]_G >$ *and* $< [\boldsymbol{S}]_H, [\boldsymbol{T}]_H >$.

Exercise 3.3.24 *Let*

$$\boldsymbol{E_1} = \begin{bmatrix} -12 \\ -30 \\ 40 \end{bmatrix}, \boldsymbol{E_2} = \begin{bmatrix} 30 \\ 17 \\ -5 \end{bmatrix} \text{ and } \boldsymbol{E_3} = \begin{bmatrix} 1 \\ 21 \\ -35 \end{bmatrix}$$

$$\boldsymbol{F_1} = \begin{bmatrix} -1 \\ 20 \\ 1 \end{bmatrix}, \boldsymbol{F_2} = \begin{bmatrix} 4 \\ 11 \\ 3 \end{bmatrix} \text{ and } \boldsymbol{F_3} = \begin{bmatrix} -5 \\ 8 \\ 22 \end{bmatrix}$$

in $\Re^3$. *Use* **GSO** *on* $\boldsymbol{E} = \{\boldsymbol{E_1}, \boldsymbol{E_2}, \boldsymbol{E_3}\}$ *and* $\boldsymbol{F} = \{\boldsymbol{F_1}, \boldsymbol{F_2}, \boldsymbol{F_3}\}$ *to find orthonormal bases* $\boldsymbol{G} = \{\boldsymbol{G_1}, \boldsymbol{G_2}, \boldsymbol{G_3}\}$ *and* $\boldsymbol{H} = \{\boldsymbol{H_1}, \boldsymbol{H_2}, \boldsymbol{H_3}\}$*, respectively. Find* $\boldsymbol{T_{EG}}$ *and* $\boldsymbol{T_{FH}}$*. Feel free to do this in MATLAB. Now compute inner products with respect to both bases.*

$$\text{Let } \boldsymbol{S} = \begin{bmatrix} 22 \\ 35 \\ 4 \end{bmatrix} \text{ and } \boldsymbol{T} = \begin{bmatrix} 41 \\ -4 \\ 1 \end{bmatrix}.$$

Find $[\boldsymbol{S}]_G$, $[\boldsymbol{T}]_G$ *and* $[\boldsymbol{S}]_H$, $[\boldsymbol{T}]_H$. *Then calculate* $< [\boldsymbol{S}]_G, [\boldsymbol{T}]_G >$ *and* $< [\boldsymbol{S}]_H, [\boldsymbol{T}]_H >$.

3.3.6 The Solution Space of Higher Dimensional ODE Systems

Let's do a final example to show you how $\Re^6$ can be identified with the solution space of a linear system of ODEs. A general model of enhanced cytokine signaling production based on LRRK2 mutation is similar to the cancer model discussed in (J. Peterson (9) 2016) and (J. Peterson (10) 2016). The altered cytokine response starts with the activation of an allele called **A**, in a small compartment of cells. Initially, all cells have a correct version of the **LRRK2** gene. We will denote this by $\boldsymbol{A}^{-/-}$ where the superscript "$-/-$" indicates there are no LRRK2 mutations. One of the mutant alleles becomes activated at mutation rate u_1 to generate a cell type denoted by $\boldsymbol{A}^{+/-}$. The superscript $+/-$ tells us one allele is activated. The second allele becomes activated at rate $\hat{u}_2$ to become the cell type $\boldsymbol{A}^{+/+}$. In addition, $\boldsymbol{A}^{-/-}$ cells can also receive mutations that trigger **CIN**. This happens at the rate u_c resulting in the cell type $\boldsymbol{A}^{-/-\,CIN}$. This kind of a cell can activate the first allele of the LRRK2 gene with normal mutation rate u_1 to produce a cell with one activated allele (i.e. a $+/-$) which started from a CIN state. We denote these cells as $\boldsymbol{A}^{+/-\,CIN}$. We can also get a cell of type $\boldsymbol{A}^{+/-\,CIN}$ when a cell of type $\boldsymbol{A}^{+/-}$ receives a mutation which triggers **CIN**. We will assume this happens at the same rate u_c as before. The $\boldsymbol{A}^{+/-\,CIN}$ cell then rapidly undergoes **LOH** at rate $\hat{u}_3$ to produce cells having the second allele of LRRK2 which is of type $\boldsymbol{A}^{+/+\,CIN}$. Finally, $\boldsymbol{A}^{+/+}$ cells can experience **CIN** at rate u_c to generate $\boldsymbol{A}^{+/+\,CIN}$ cells. We show this information in Figure 3.1.

Let N be the population size within which the LRRK2 mutations occur. We will assume a typical value of N is 10^3 to 10^4. The first allele is activated by a point mutation. The rate at which this occurs is modeled by the rate u_1 as shown in Figure 3.1. We make the following assumptions:

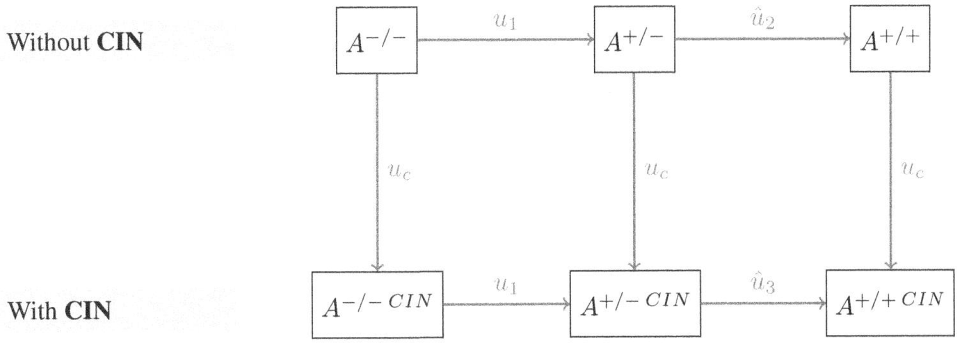

Figure 3.1: The pathways for the **LRRK2** allele gains.

- the mutations governed by the rates u_1 and u_c are **neutral**. This means that these rates do not depend on the size of the population N.
- The events governed by $\hat{u}_2$ and $\hat{u}_3$ give what is called **selective advantage**. This means that the size of the population size does matter.

Using these assumptions, we will model $\hat{u}_2$ and $\hat{u}_3$ as $\hat{u}_2 = N\,u_2$ and $\hat{u}_3 = N\,u_3$. where u_2 and u_3 are neutral rates. We can thus redraw our figure as Figure 3.2.

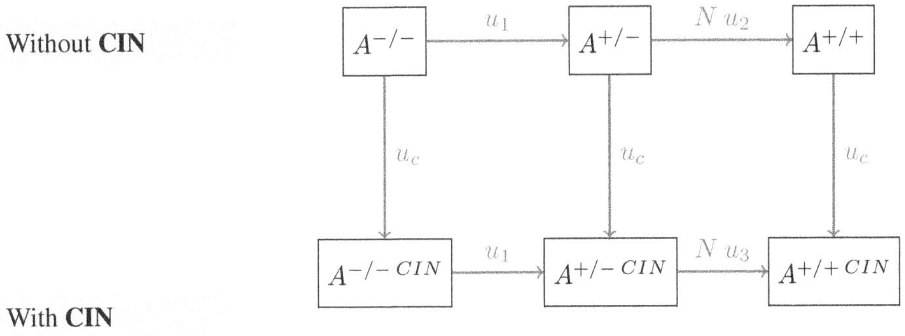

Figure 3.2: The pathways for the **LRRK2** allele gains rewritten using selective advantage.

The mathematical model is then set up as follows: Let

$X_0(t)$ is the probability a cell is in cell type $A^{-/-}$ at time t.

$X_1(t)$ is the probability a cell is in cell type $A^{+/-}$ at time t.

$X_2(t)$ is the probability a cell is in cell type $A^{+/+}$ at time t.

$Y_0(t)$ is the probability a cell is in cell type $A^{-/-\,CIN}$ at time t.

$Y_1(t)$ is the probability a cell is in cell type $A^{+/-\,CIN}$ at time t.

$Y_2(t)$ is the probability a cell is in cell type $A^{+/+\,CIN}$ at time t.

Looking at Figure 3.2, we can generate rate equations. First, let's rewrite Figure 3.2 using our variables as Figure 3.3. To generate the equations we need, note each box has arrows coming into

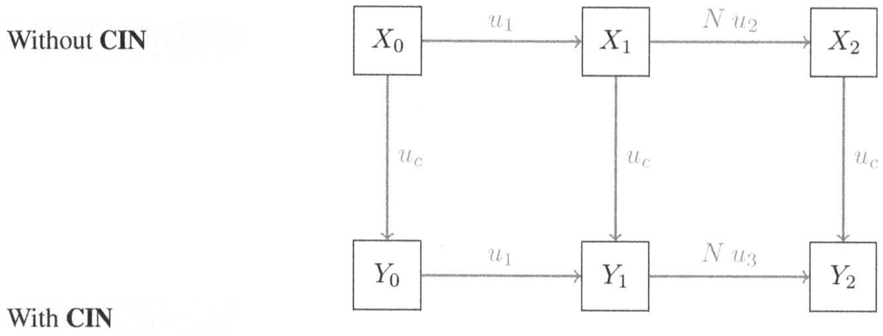

Figure 3.3: The pathways for the **LRRK2** allele gains rewritten using mathematical variables.

it and arrows coming out of it. The **arrows in** are **growth** terms for the net change of the variable in the box and the **arrows out** are the **decay or loss** terms. We model **growth** as **exponential growth** and **loss** as **exponential decay**. So X_0 only has arrows going out which tells us it only has **loss** terms. So we would say $(X_0')_{loss} = -u_1 X_0 - u_c X_0$ which implies $X_0' = -(u_1 + u_c)X_0$. Further, X_1 has arrows going in and out which tells us it has **growth** and **loss** terms. So we would say $(X_1')_{loss} = -Nu_2 X_1 - u_c X_1$ and $(X_1')_{growth} = u_1 X_0$ which implies $X_1' = u_1 X_0 - (Nu_2 + u_c)X1$. We can continue in this way to find all the model equations. We can then see the rate equations are

$$
\begin{aligned}
X_0' &= -(u_1 + u_c)\, X_0 && (3.1)\\
X_1' &= u_1\, X_0 - (u_c + N\, u_2)\, X_1 && (3.2)\\
X_2' &= N\, u_2\, X_1 - u_c\, X_2 && (3.3)\\
Y_0' &= u_c\, X_0 - u_1\, Y_0 && (3.4)\\
Y_1' &= u_c\, X_1 + u_1\, Y_0 - N\, u_3\, Y_1 && (3.5)\\
Y_2' &= N\, u_3\, Y_1 + u_c\, X_2 && (3.6)
\end{aligned}
$$

Initially, at time 0, all the cells are in the state X_0, so we have

$$
\begin{aligned}
X_0(0) &= 1,\ X_1(0) = 0,\ X_2(0) = 0 && (3.7)\\
Y_0(0) &= 0,\ Y_1(0) = 0,\ Y_2(0) = 0. && (3.8)
\end{aligned}
$$

Now under what circumstances is the CIN pathway to LRRK2 mutations the dominant one? In order to answer this, we need to analyze the trajectories of this model. We can rewrite this as a matrix-vector system:

$$
\begin{bmatrix} X_0' \\ X_1' \\ X_2' \\ Y_0' \\ Y_1' \\ Y_2' \end{bmatrix}
=
\begin{bmatrix}
-(u_1+u_c) & 0 & 0 & 0 & 0 & 0 \\
u_1 & -(u_c+Nu_2) & 0 & 0 & 0 & 0 \\
0 & Nu_2 & -u_c & 0 & 0 & 0 \\
u_c & 0 & 0 & -u_1 & 0 & 0 \\
0 & u_c & 0 & u_1 & -Nu_3 & 0 \\
0 & 0 & u_c & 0 & Nu_3 & 0
\end{bmatrix}
\begin{bmatrix} X_0 \\ X_1 \\ X_2 \\ Y_0 \\ Y_1 \\ Y_2 \end{bmatrix}
$$

The eigenvalues of this system are the roots of the polynomial $p(\lambda) = det(\lambda \boldsymbol{I} - \boldsymbol{A})$ where $\boldsymbol{A}$ is the 6×6 coefficient matrix above. We find

$$p(\lambda) = det \begin{bmatrix} \lambda + (u_1 + u_c) & 0 & 0 & 0 & 0 & 0 \\ -u_1 & \lambda + (u_c + Nu_2) & 0 & 0 & 0 & 0 \\ 0 & -Nu_2 & \lambda + u_c & 0 & 0 & 0 \\ -u_c & 0 & 0 & \lambda + u_1 & 0 & 0 \\ 0 & -u_c & 0 & -u_1 & \lambda + Nu_3 & 0 \\ 0 & 0 & -u_c & 0 & -Nu_3 & \lambda \end{bmatrix}$$

This is easily expanded using the properties of determinants (which we will discuss later!) to give

$$\begin{aligned} p(\lambda) &= (\lambda + (u_1 + u_c)) \begin{bmatrix} \lambda + (u_c + Nu_2) & 0 & 0 & 0 & 0 \\ -Nu_2 & \lambda + u_c & 0 & 0 & 0 \\ 0 & 0 & \lambda + u_1 & 0 & 0 \\ -u_c & 0 & -u_1 & \lambda + Nu_3 & 0 \\ 0 & -u_c & 0 & -Nu_3 & \lambda \end{bmatrix} \\ &= (\lambda + (u_1 + u_c))\,(\lambda + (u_c + Nu_2)) \begin{bmatrix} \lambda + u_c & 0 & 0 & 0 \\ 0 & \lambda + u_1 & 0 & 0 \\ 0 & -u_1 & \lambda + Nu_3 & 0 \\ -u_c & 0 & -Nu_3 & \lambda \end{bmatrix} \\ &= (\lambda + (u_1 + u_c))\,(\lambda + (u_c + Nu_2))\,(\lambda + u_c) \begin{bmatrix} \lambda + u_1 & 0 & 0 \\ -u_1 & \lambda + Nu_3 & 0 \\ 0 & -Nu_3 & \lambda \end{bmatrix} \\ &= (\lambda + (u_1 + u_c))\,(\lambda + (u_c + Nu_2))\,(\lambda + u_c)(\lambda + u_1)\,(\lambda + Nu_3)\,(\lambda) \end{aligned}$$

Hence, the eigenvalues are

$$\begin{bmatrix} \lambda_1 \\ \lambda_2 \\ \lambda_3 \\ \lambda_4 \\ \lambda_5 \\ \lambda_6 \end{bmatrix} = \begin{bmatrix} -(u_1 + u_c) \\ -(u_c + Nu_2) \\ -u_c \\ -u_1 \\ -Nu_3 \\ 0 \end{bmatrix}$$

Since this is a biological model, we get more insight into finding a solution with the critical parameters in this form rather than substituting numerical values and using a tool like MATLAB. It is possible to calculate the six needed eigenvectors by hand for this system which we enjoyed doing but we understand not all share this interest. We find the eigenvectors are

$$\boldsymbol{E_1} = \begin{bmatrix} Nu_2 - u_1 \\ u_1 \\ -Nu_2 \\ -(Nu_2 - u_1) \\ -u_1 \frac{Nu_2 - u_1 - u_c}{Nu_3 - u_1 - u_c} \\ \frac{Nu_3(Nu_2 - u_1) - Nu_2 u_c}{Nu_3 - u_1 - u_c} \end{bmatrix}, \quad \boldsymbol{E_2} = \begin{bmatrix} 0 \\ 1 \\ -1 \\ 0 \\ \frac{-u_c}{Nu_2 + u_c - Nu_3} \\ \frac{u_c}{Nu_2 + u_c - Nu_3} \end{bmatrix}, \quad \boldsymbol{E_3} = \begin{bmatrix} 0 \\ 0 \\ 1 \\ 0 \\ 0 \\ 0 \end{bmatrix},$$

$$E_4 = \begin{bmatrix} 0 \\ 0 \\ 0 \\ 1 \\ \frac{u_1}{Nu_3 - u_1} \\ \frac{Nu_3}{Nu_3 - u_1} \end{bmatrix}, \quad E_5 = \begin{bmatrix} 0 \\ 0 \\ 0 \\ 0 \\ 1 \\ -1 \end{bmatrix}, \quad E_6 = \begin{bmatrix} 0 \\ 0 \\ 0 \\ 0 \\ 0 \\ 1 \end{bmatrix},$$

The general solution is thus any linear combination of the form

$$c_1 \boldsymbol{E_1} e^{-(u_1+u_c)t} + c_2 \boldsymbol{E_2} e^{-(u_c+Nu_2)t} + c_3 \boldsymbol{E_3} e^{-u_c t} + c_4 \boldsymbol{E_4} e^{-u_1 t} + c_5 \boldsymbol{E_5} e^{-Nu_3 t} + c_6 \boldsymbol{E_6} \boldsymbol{1}$$

where $\boldsymbol{1}$ denotes the constant function $e^{0t} = 1$. If we let these six solutions be denoted by $\boldsymbol{y_i}(t) = \boldsymbol{E_i} e^{\lambda_i t}$, we see the general solution is a member of the span of $\{\boldsymbol{y_1}, \ldots, \boldsymbol{y_6}\}$ and since these functions are linearly independent in the space $\boldsymbol{X} = C^1([0,T]) \times C^1([0,T]) \times C^1([0,T]) \times C^1([0,T]) \times C^1([0,T]) \times C^1([0,T])$ for any appropriate T, we know this solution space is a six dimensional subspace of $\boldsymbol{X}$ with the basis $\{\boldsymbol{y_1}, \ldots, \boldsymbol{y_6}\}$. If $\boldsymbol{F}$ and $\boldsymbol{G}$ are two elements in $\boldsymbol{X}$, it is easy to show we can define an inner product on $\boldsymbol{X}$ to be

$$\int_0^T < \boldsymbol{F}(t), \boldsymbol{G}(t) > dt$$

We can construct an orthonormal basis for the solution space by applying GSO to the eigenvectors $\boldsymbol{E_i}$ to create the orthonormal basis $\boldsymbol{G}$. Then the new functions $\boldsymbol{w_i} = \boldsymbol{G_i} e^{\lambda_i t}$ are solutions to the ODE system which are mutually orthogonal. So it is straightforward to find an orthonormal basis here.

Note this solution space can be identified with any six dimensional subspace of any vector space by simply mapping one orthonormal basis to another and extending linearly.

It is not our intent here to pursue this problem further. It turns out looking at the solution space this way is not so helpful to understand what is going on. Finding approximate solutions using standard Taylor series expansions is better. You can read about this in other places if you are interested.

3.3.6.1 Homework

These are intense!

Exercise 3.3.25 *Verify the eigenvalues and eigenvectors here!*

Exercise 3.3.26 *For the initial conditions given, find the solution. This is indeed a paper and pencil exercise!!*

Exercise 3.3.27 *Once the previous exercise is done, estimate $Y_2(T)/X_2(T)$ where T is human lifetime. When is this ratio larger than one?*

3.4 Best Approximation in a Vector Space with Inner Product

An important problem is that of finding the best object in an subspace $\mathscr{W}$ that approximates a given object in the space. This is an easy theorem to prove.

Theorem 3.4.1 The Finite Dimensional Approximation Theorem

Let p be any object in the inner product space $\mathscr{V}$ with inner product $<,>$ and induced norm $\|\cdot\|$. Let $\mathscr{W}$ be a finite dimensional subspace with an orthonormal basis $\boldsymbol{E} = \{\boldsymbol{E}_1, \ldots \boldsymbol{E}_n\}$ where n is the dimension of the subspace. Then there is an unique object p^* in $\mathscr{W}$ which satisfies

$$||p - p^*|| = \min_{u \in \mathscr{W}} ||u - p||$$

with

$$p^* = \sum_{i=1}^{N} < \mathbf{p}, \boldsymbol{E}_i > \boldsymbol{E}_i.$$

Further, $p - p^*$ is orthogonal to the subspace $\mathscr{W}$; i.e. $< p^*, u >= 0$ for all $u \in \mathscr{W}$. The squared error made in this approximation is $E = \sum_{i=1}^{N} (< \mathbf{p}, \boldsymbol{E_i} >)^2$.

Proof 3.4.1

Any object in the subspace has the representation $\sum_{i=1}^{N} a_i \boldsymbol{E}_i$ for some scalars a_i. Consider the function of N variables

$$\begin{aligned} E(a_1, \ldots, a_N) &= \left\langle \mathbf{p} - \sum_{i=1}^{N} a_i \boldsymbol{E}_i, \mathbf{p} - \sum_{j=1}^{N} a_j \boldsymbol{E}_j \right\rangle \\ &= < \mathbf{p}, \mathbf{p} > -2 \sum_{i=1}^{N} a_i < \mathbf{p}, \boldsymbol{E}_i > \\ &\quad + \sum_{i=1}^{N} \sum_{j=1}^{N} a_i a_j < \boldsymbol{E}_i, \boldsymbol{E}_j > . \end{aligned}$$

Simplifying using the orthonormality of the basis, we find

$$E(a_1, \ldots, a_N) = < \mathbf{p}, \mathbf{p} > -2 \sum_{i=1}^{N} a_i < \mathbf{p}, \boldsymbol{E}_i > + \sum_{i=1}^{N} a_i^2.$$

This is a quadratic expression and setting the gradient of E to zero, we find the critical points $a_i =< \mathbf{p}, \boldsymbol{E}_i >$. This is a global minimum for the function E. Hence, the optimal p^* has the form

$$p^* = \sum_{i=1}^{N} < \mathbf{p}, \boldsymbol{E}_i > \boldsymbol{E}_i.$$

Finally, we see

$$\begin{aligned} < p - p^*, \boldsymbol{E_j} > &= < p, \boldsymbol{E_j} > - \sum_{k=1}^{N} < p, \boldsymbol{E_k} >< \boldsymbol{E_k}, \boldsymbol{E_j} > \\ &= < p, \boldsymbol{E_j} > - < p, \boldsymbol{E_j} > = 0, \end{aligned}$$

and hence, $p - p^*$ is orthogonal of $\mathscr{W}$.

The squared error term is easy to compute. ■

This is an extension of the Cauchy - Schwartz Theorem in a way. It was easy to see how to handle the best approximation idea when only two vectors were involved so we didn't need all the machinery above. But this is a powerful idea. We can put this into perspective by recalling some basic facts from Fourier Series.
Letting $u_n(x) = \sin(\frac{n\pi}{L}x)$ and using the standard inner product on $C([0,L])$, $< f, g >= \int_0^L f(t)\, g(t)\, dt$, we know

$$< u_i, u_j > \;=\; \begin{cases} \frac{L}{2}, & i = j \\ 0, & i \neq j \end{cases}$$

We define the new functions $\hat{u}_n$ by $\sqrt{\frac{2}{L}}u_n$. Then, $< \hat{u}_i, \hat{u}_j >= \delta_i^j$ and the sequence of functions $(\hat{u}_n)$ are all mutually orthogonal. It is clear $||\hat{u_n}||_2 = 1$ always. So the sequence of functions $(\hat{u}_n)$ are all mutually orthogonal and length one. Letting $v_n(x) = \cos(\frac{n\pi}{L}x)$, we also know

$$\begin{aligned} < v_0, v_0 > &= \frac{1}{L} \\ < v_i, v_j > &= \begin{cases} \frac{L}{2}, & i = j \\ 0, & i \neq j \end{cases} \end{aligned}$$

Hence, we can define the new functions

$$\begin{aligned} \hat{v}_0(x) &= \sqrt{\frac{1}{L}} \\ \hat{v}_n(x) &= \sqrt{\frac{2}{L}}\cos\left(\frac{n\pi}{L}x\right),\; n \geq 1, \end{aligned}$$

Then, $< \hat{v}_i, \hat{v}_j >= \delta_i^j$ and the sequence of functions $(\hat{v}_n)$ are all mutually orthogonal with length $||v_n|| = 1$.
If we start with a function f which is continuous on the interval $[0,L]$, we can define the trigonometric series associated with f as follows

$$\begin{aligned} S(x) &= \frac{1}{L} < f, 1 > \\ &+ \sum_{i=1}^{\infty}\left(\frac{2}{L}\left\langle f(x), \sin\left(\frac{i\pi}{L}x\right)\right\rangle \sin\left(\frac{i\pi}{L}x\right) + \frac{2}{L}\left\langle f(x), \cos\left(\frac{i\pi}{L}x\right)\right\rangle \cos\left(\frac{i\pi}{L}x\right)\right). \end{aligned}$$

In terms of the orthonormal families we have just defined, this can be rewritten as

$$S(x) = \lim_{N\to\infty}\left(< f, \hat{v}_0 1 > \hat{v}_0 + \sum_{n=1}^{N} < f(x), \hat{v}_n > \hat{v}_n + \sum_{n=1}^{N} < f(x), \hat{u}_n > \hat{u}_n\right)$$

Note the projection of the data function f, $P_N(f)$, onto the subspace spanned by $\{\hat{u}_1, \ldots, \hat{u}_N\}$ is the minimal norm solution to the problem $\inf_{u \in V_n} \|f - u\|_2$ where V_n is the span of $\{\hat{u}_1, \ldots, \hat{u}_N\}$.
Further, projection of the data function f, $Q_N(f)$, onto the subspace spanned by $\{1, \hat{v}_1, \ldots, \hat{v}_N\}$ is the minimal norm solution to the problem $\inf_{v \in W_n} \|f - v\|_2$ where W_n is the span of $\{\hat{v}_0, \ldots, \hat{v}_N\}$.
The Fourier Series of $f : [0, 2L]$ has partial sums that can be written as

$$S_N = P_N(f) + Q_N f = \sum_{n=1}^{N} < f, \hat{u}_n > \hat{u}_n + < f, \hat{v}_0 > \hat{v}_0 + \sum_{n=1}^{N} < f, \hat{v}_n > \hat{v}_n$$

and so the convergence of the Fourier Series is all about the conditions under which the projections of f to these subspaces converge pointwise to f.

3.4.1 Homework

Exercise 3.4.1 *Prove the squared error formula in the best approximation theorem.*

Exercise 3.4.2 *Let* $\boldsymbol{E} = \{x^0, x, x^2, x^3, x^4\}$ *in* $C([-1,1])$. *Use* **GSO** *to find an orthonormal basis* $\boldsymbol{F}$ *corresponding to* $\boldsymbol{E}$. *Then find the best approximation to* $f(x) = \sin(x)$ *and the squared error.*

Now find the Taylor polynomial of order four based at 0 *and estimate its error on this interval. How do these errors compare?*

Exercise 3.4.3 *Let* $\boldsymbol{E} = \{x^0, x, x^2, x^3, x^4\}$ *in* $C([-2,3])$. *Use* **GSO** *to find an orthonormal basis* $\boldsymbol{F}$ *corresponding to* $\boldsymbol{E}$. *Then find the best approximation to* $f(x) = \sin(x)$ *and the squared error.*

Also find the Taylor polynomial of order four based at 0 *and estimate its error on this interval. How do these errors compare?*

Exercise 3.4.4 *Let* $\boldsymbol{E} = \{x^0, x, x^2, x^3, x^4\}$ *in* $C([-1,1])$. *Use* **GSO** *to find an orthonormal basis* $\boldsymbol{F}$ *corresponding to* $\boldsymbol{E}$. *Then find the best approximation to* $f(x) = \cos(2x)$ *and the squared error.*

Now find the Taylor polynomial of order four based at 0 *and estimate its error on this interval. How do these errors compare?*

Exercise 3.4.5 *Let* $\boldsymbol{E} = \{x^0, x, x^2, x^3, x^4\}$ *in* $C([-2,3])$. *Use* **GSO** *to find an orthonormal basis* $\boldsymbol{F}$ *corresponding to* $\boldsymbol{E}$. *Then find the best approximation to* $f(x) = \cos(2x)$ *and the squared error.*

Next, find the Taylor polynomial of order four based at 0 *and estimate its error on this interval. How do these errors compare?*

Chapter 4

Linear Transformations

We are now going to study matrices and learn to interpret them in a more general way.

4.1 Organizing Point Cloud Data

The first place to start is what is called a **point cloud** of data. This is a collection in which each data point is itself a list of real numbers. So if x is such a data point, the list of numbers associated with it is $\{x_1, \ldots, x_n\}$ where n is the number of components in the data. For example, there is a complicated system of inferring what individual molecules are doing in fluid flow such as in an artery called flow cytometry. Roughly speaking, lots of molecules are separated as they *flow* down a column filled with liquid. The separation can be due to mass, charge and other things. Then laser light of various frequencies is shown through the liquid with the moving molecules and how the molecules absorb the light is collected and stored as lists of numbers. So a particular molecular complex that is of interest might have a list of 14 numbers associated with it and a list like this is obtained at regular time points. So if you collect data every millisecond, you have 1000 sets of these lists per molecular complex per second. A typical complex we might wish to track in the context of how signals are used for information processing in biology would be one of several cytokine molecules. A given flow cytometry experiment might collect such data on 10 different cytokines for 3 seconds. If we label the cytokines as $\boldsymbol{C}_1, \ldots, \boldsymbol{C}_{10}$, we have data of the form

$$\{\boldsymbol{X}_{i1}, \ldots, \boldsymbol{X}_{i,3000}\}$$

for each cytokine of type $\boldsymbol{C}_i$ for time points 1 millisecond to 3000 milliseconds. This vast collection of data is a set of lists with 14 components. Hence, we know $\boldsymbol{X}_{ij} \in \Re^{14}$, the collection of real numbers in 14 dimensions. Note, we can collect the data and store it as tables of numbers without referring to $\Re^{14}$ as a vector space with a specific orthonormal basis. The easiest way to organize this is to think of the initial data collection phase as generating vectors in $\Re^{14}$ using the standard orthonormal basis

$$\boldsymbol{e}_1 = \begin{bmatrix} 0, & i \neq 1 \\ 1, & i = 1 \end{bmatrix}, \quad \boldsymbol{e}_2 = \begin{bmatrix} 0, & i \neq 2 \\ 1, & i = 2 \end{bmatrix}, \ldots, \boldsymbol{e}_{14} = \begin{bmatrix} 0, & i \neq 14 \\ 1, & i = 14 \end{bmatrix},$$

To make sense of this collection of data, we want to find ways to **transform** the given **point cloud** into a new version of itself which we can use to understand the data better. We usually do this by applying what are called **Linear Transformations** which are mappings from $\Re^{14}$ into itself which satisfy a property called **linearity**. We find using a change of basis from one orthonormal basis to another and the use of the specialized orthonormal basis constructed from the eigenvectors to be very useful. So let's look at this transformation problem in general.

4.1.1 Homework

Let's look at the setups for different point cloud analysis problems.

Exercise 4.1.1 *So given a set of data in $\Re^{10}$ of size N, form the $10 \times N$ matrix $\boldsymbol{A}$ using the data as columns. Row reduce to find the linearly independent vectors that span this data. Then all the data lives in the subspace spanned by these vectors. So generate a set of 100 points in $\Re^{10}$ any way you like. Find a basis for the span of these vectors.*

Exercise 4.1.2 *Generate a set of data in $\Re^2$ any way you like and set up the optimization problem that finds the best straight line which minimizes the least squares error between the data and the line: the regression line. Note the regression line is a translated subspace of $\Re^2$ in general.*

Exercise 4.1.3 *Generate a set of data in $\Re^3$ any way you like and set up the optimization problem that finds the best straight line which minimizes the least squares error between the data and the line: the regression line. Note the regression line is a translated subspace of $\Re^3$ in general.*

Exercise 4.1.4 *Generate a set of data in $\Re^3$ any way you like and set up the optimization problem that finds the best plane which minimizes the least squares error between the data and the plane: the regression plane if you like. Note the regression plane is a translated subspace of $\Re^3$ in general.*

Exercise 4.1.5 *Generate a set of points in $\Re^2$ and set up an optimization problem which finds the best circle which minimizes the error between the data and the circle. This problem is based on conversations with Ben Merrit of the Boeing company about an optimization problem that arises in plane manufacturing.*

Exercise 4.1.6 *Generate a set of points in $\Re^3$ and set up an optimization problem which finds the best cylinder which minimizes the error between the data and the cylinder. This problem is based on conversations with Ben Merrit of the Boeing company about an optimization problem that arises in plane manufacturing.*

4.2 Linear Transformations

Given a basis $\boldsymbol{E}$ of $\Re^n$, $\boldsymbol{E} = \{\boldsymbol{e}_1, \ldots, \boldsymbol{e}_n\}$, we know any vector $\boldsymbol{x}$ in $\Re^n$ has a decomposition $\left[\boldsymbol{x}\right]_{\boldsymbol{E}}$ where the components of $\boldsymbol{X}$ with respect to the basis $\boldsymbol{E}$ are x_i with

$$\boldsymbol{x} = x_1 \boldsymbol{e}_1 + \ldots + x_n \boldsymbol{e}_n$$

We often wish to **measure** the magnitude or size of a vector $\boldsymbol{x}$. We have talked about doing this before and have used the term **norm** to denote the mapping that does the job. Let's be specific now.

Definition 4.2.1 Norm on a Vector Space

Let $\boldsymbol{V}$ be a vector space over the reals and let $\rho : \boldsymbol{V} \to \Re$ satisfy

N1: $\rho(\boldsymbol{x}) \geq 0$ *for all* $\boldsymbol{x} \in \boldsymbol{V}$.

N2: $\rho(\boldsymbol{x}) = 0$ *if and only if* $\boldsymbol{x} = \boldsymbol{0}$

N3: $\rho(\alpha \boldsymbol{x}) = |\alpha| \rho(\boldsymbol{x})$ *for all real numbers* α *and* $\boldsymbol{x} \in \boldsymbol{V}$

N4: $\rho(\boldsymbol{x} + \boldsymbol{y}) \leq \rho(\boldsymbol{x}) + \rho(\boldsymbol{y})$ *for all* $\boldsymbol{x}, \boldsymbol{y} \in \boldsymbol{V}$

Comment 4.2.1 *We usually denote the norm ρ by the symbol $\|\cdot\|$ and sometimes add a subscript to remind us where the norm comes from. We will see many examples of this soon.*

Comment 4.2.2 *As we have mentioned before, if the vector space* $\boldsymbol{V}$ *has an inner product, the mapping* $\rho(\boldsymbol{x}) =< \boldsymbol{x}, \boldsymbol{x} >$ *defines a norm on* $\boldsymbol{V}$. *The vector space* $\boldsymbol{V}$ *plus an inner product* $< \cdot >$ *is denoted as the pair* $(\boldsymbol{V}, < \cdot >)$ *and is called an* **inner product space**. *The vector space* $\boldsymbol{V}$ *plus a norm* ρ *is denoted as the pair* $(\boldsymbol{V}, \rho)$ *and is called a* **normed linear space**. *Hence, an inner product space induces a norm on the vector space. It is known that* **not all** *norms are induced by some inner product. Infinite dimensional vector spaces are known where this is not true, but that is a story for another time.*

Comment 4.2.3 *For the vector space* $C([a, b])$, *note:*

- *The inner product is* $< f, g >= \int_a^b f(t)g(t)dt$ *with induced norm* $\|f\|_2 = \sqrt{\int_a^b f^2(t)dt}$.
- *We can also use the norm* $\|f\|_1 = \int_a^b |f(t)|dt$ *which is not induced from an inner product.*
- *We can also use the norm* $\|f\|_\infty = \max_{a \le t \le b} |f(t)|$.

4.2.1 Homework

Exercise 4.2.1 *Prove* $\| \cdot \|_2$ *is a norm on* $C([a, b])$ *induced by the standard inner product* $< \cdot, \cdot >$.

Exercise 4.2.2 *Prove* $\| \cdot \|_1$ *is a norm on* $C([a, b])$.

Exercise 4.2.3 *Prove* $\| \cdot \|_\infty$ *is a norm on* $C([a, b])$

Exercise 4.2.4 *Define* $\| \cdot \|$ *on the set of all* $n \times n$ *matrices over* $\Re$ *by*

$$\|\boldsymbol{A}\| = \max_{1 \le j \le n} \begin{bmatrix} \sum_{i=1}^n |a_{i1}| \\ \vdots \\ \sum_{i=1}^n |a_{in}| \end{bmatrix}$$

Prove this is a norm.

Exercise 4.2.5 *Define* $\| \cdot \|$ *on the set of all* $n \times n$ *matrices over* $\Re$ *by*

$$\|\boldsymbol{A}\| = \max_{1 \le j \le n} \begin{bmatrix} \sum_{j=1}^n |a_{1j}| \\ \vdots \\ \sum_{j=1}^n |a_{nj}| \end{bmatrix}$$

Prove this is a norm.

Exercise 4.2.6 *Define* $\| \cdot \|$ *on the set of all* $n \times n$ *matrices over* $\Re$ *by* $\|\boldsymbol{A}\| = \sqrt{\sum_{i=1}^n \sum_{i=1}^n |a_{ij}|^2}$. *Prove this is a norm.*

Exercise 4.2.7 *Define* $\| \cdot \|$ *on the set of all* $n \times n$ *matrices over* $\Re$ *by* $\|\boldsymbol{A}\| = \sum_{i=1}^n \sum_{i=1}^n |a_{ij}|$. *Prove this is a norm.*

Exercise 4.2.8 *Define* $\| \cdot \|$ *on the set of all* $n \times n$ *matrices over* $\Re$ *by* $\|\boldsymbol{A}\| = \max_{1 \le i,j \le n} |a_{ij}|$. *Prove this is a norm.*

4.3 Sequence Spaces Revisited

Let's recall some ideas about sequence spaces. Given a sequence of real numbers (a_n), which we assume has indexing starting at $n = 1$ for convenience, the set of all sequences has a variety of interesting subsets which might even be subspaces. It is clear the set of all sequences is a vector space over the reals but whether or not a subset is a subspace depends on whether the subset is closed under scalar multiplication and vector addition. Now we say the positive numbers p and q are **conjugate exponents** if $p > 1$ and $1/p + 1/q = 1$. If $p = 1$, we define its conjugate exponent to be $q = \infty$. Conjugate exponents satisfy some fundamental identities. Clearly, if $p > 1$, $\frac{1}{p} + \frac{1}{q} \Longrightarrow 1 = \frac{p+q}{p\,q}$ and also $pq = p + q$ and $(p-1)\,(q-1) = 1$. We will use these identities quite a bit. We quote the following lemma whose proof is in many texts such as (Peterson (21) 2020).

Lemma 4.3.1 The $\alpha - \beta$ Lemma

Let α and β be positive real numbers and p and q be conjugate exponents. Then $\alpha\,\beta \leq \frac{\alpha^p}{p} + \frac{\beta^q}{q}$.

Proof 4.3.1
You should look at the proof in (Peterson (21) 2020) to refresh your memory of how this is done. ■

Important subsets of the set of all sequences can then be defined using conjugate exponents.

Definition 4.3.1 The ℓ^p Sequence Space

Let $p \geq 1$. The collection of all sequences, $(a_n)_{n=1}^\infty$ for which $\sum_{n=1}^\infty |a_n|^p$ converges is denoted by the symbol ℓ^p.
(1) $\ell^1 = \{(a_n)_{n=1}^\infty : \sum_{n=1}^\infty |a_n|$ converges.$\}$
(2) $\ell^2 = \{(a_n)_{n=1}^\infty : \sum_{n=1}^\infty |a_n|^2$ converges.$\}$
We also define $\ell^\infty = \{(a_n)_{n=1}^\infty : \sup_{n\geq 1} |a_n| < \infty\}$.

There is a fundamental inequality connecting sequences in ℓ^p and ℓ^q when p and q are conjugate exponents called **Hölder's Inequality**. Its proof is straightforward but has a few tricks.

Theorem 4.3.2 Hölder's Inequality

Let $p > 1$ and p and q be conjugate exponents. If $x \in \ell^p$ and $y \in \ell^q$, then

$$\sum_{n=1}^\infty |x_n\,y_n| \;\leq\; \left(\sum_{n=1}^\infty |x_n|^p\right)^{1/p} \left(\sum_{n=1}^\infty |y_n|^q\right)^{1/q}$$

where $x = (x_n)$ and $y = (y_n)$.

Proof 4.3.2
This inequality is clearly true if either of the two sequences x and y are the zero sequence. So we can assume both x and y have some nonzero terms in them. Then $x \in \ell^p$, we know

$$0 < u \;=\; \left(\sum_{n=1}^\infty |x_n|^p\right)^{1/p} < \infty, \quad 0 < v = \left(\sum_{n=1}^\infty |y_n|^q\right)^{1/q} < \infty$$

Now define new sequences, $\hat{x}$ and $\hat{y}$ by $\hat{x}_n = x_n/u$ and $\hat{y}_n = y_n/v$. Then, we have

$$\sum_{n=1}^\infty |\hat{x}_n|^p \;=\; \sum_{n=1}^\infty \frac{|x_n|^p}{u^p} = \frac{1}{u^p}\sum_{n=1}^\infty |x_n|^p = \frac{u^p}{u^p} = 1.$$
$$\sum_{n=1}^\infty |\hat{y}_n|^q \;=\; \sum_{n=1}^\infty \frac{|y_n|^q}{v^q} = \frac{1}{v^q}\sum_{n=1}^\infty |y_n|^q = \frac{v^q}{v^q} = 1.$$

Now apply the $\alpha - \beta$ Lemma to $\alpha = |\hat{x_n}|$ and $\beta = |\hat{y_n}|$ for any nonzero terms $\hat{x}_n$ and $\hat{y}_n$. Then

$|\hat{x_n}\,\hat{y_n}| \leq |\hat{x_n}|^p/p + |\hat{y_n}|^q/q$.
This is also true, of course, if either $\hat{x}_n$ *or* $\hat{y}_n$ *are zero although the* $\alpha - \beta$ *lemma does not apply! Now sum over* N *terms to get*

$$\sum_{n=1}^{N} |\hat{x_n}\,\hat{y_n}| \leq \frac{1}{p}\sum_{n=1}^{N} |\hat{x_n}|^p + \frac{1}{q}\sum_{n=1}^{N} |\hat{y_n}|^q$$

Since we know $x \in \ell^p$ *and* $y \in \ell^q$, *we know*

$$\begin{aligned} \sum_{n=1}^{N} |\hat{x_n}|^p \leq \sum_{n=1}^{\infty} |\hat{x_n}|^p &= 1 \\ \sum_{n=1}^{N} |\hat{y_n}|^q \leq \sum_{n=1}^{\infty} |\hat{y_n}|^q &= 1 \end{aligned}$$

So we have

$$\sum_{n=1}^{N} |\hat{x_n}\,\hat{y_n}| \leq \frac{1}{p} + \frac{1}{q} = 1$$

This is true for all N *so the partial sums* $\sum_{n=1}^{N} |\hat{x_n}\,\hat{y_n}|$ *are bounded above. Hence, the partial sums converge to this supremum which is denoted by* $\sum_{n=1}^{\infty} |\hat{x_n}\,\hat{y_n}|$. *We conclude* $\sum_{n=1}^{\infty} |\hat{x_n}\,\hat{y_n}| \leq 1$. *But* $\hat{x}_n\hat{y}_n = 1/(u\,v)\,x_n y_n$ *and so we have* $\frac{1}{u\,v}\sum_{n=1}^{\infty} |x_n\,y_n| \leq 1$ *which implies the result as*

$$\sum_{n=1}^{\infty} |x_n\,y_n| \leq u\,v = \left(\sum_{n=1}^{\infty} |x_n|^p\right)^{1/p} \left(\sum_{n=1}^{\infty} |y_n|^q\right)^{1/q}$$

■

We can also do this inequality for the case $p = 1$ and $q = \infty$.

Theorem 4.3.3 Hölder's Theorem for $p = 1$ **and** $q = \infty$

If $x \in \ell^1$ *and* $y \in \ell^\infty$, *then* $\sum_{n=1}^{\infty} |x_n y_n| \leq \left(\sum_{n=1}^{\infty} |x_n|\right) \sup_{n\geq 1} |y_n|$.

Proof 4.3.3
We know since $y \in \ell^\infty$, $|y_n| \leq \sup_{k\geq 1} |y_k|$. *Thus,*

$$\sum_{n=1}^{N} |x_n y_n| \leq \left(\sum_{n=1}^{\infty} |x_n|\right) \sup_{k\geq 1} |y_k|$$

Thus the sequence of partial sums $\sum_{n=1}^{N} |x_n y_n|$ *is bounded above by* $\left(\sum_{n=1}^{\infty} |x_n|\right) \sup_{k\geq 1} |y_k|$. *This gives us our result.* ■

Now we want to apply these ideas to $\Re^n$ which is a collection of numbers organized as vectors with n components. Here, we don't really care what the underlying orthonormal basis $\boldsymbol{E}$, which gives us these components, is chosen to be. Note the vector

$$\boldsymbol{x} = \begin{bmatrix} x_1 \\ \vdots \\ x_n \end{bmatrix}$$

can be identified with the sequence

$$(x_j) = \{x_1, \ldots, x_n, 0, \ldots\}$$

It is easy to see that since (x_j) has only n components, this sequence is in any ℓ^p. Hölder's Inequality specialized to $\Re^n$ gives:

Theorem 4.3.4 Hölder's Inequality in $\Re^n$

Let $p > 1$ and p and q be conjugate exponents. If $\boldsymbol{x} \in \Re^n$, then the associated sequence $(x_j) = \{x_1, \ldots, x_n, 0, \ldots\}$ is in ℓ^p for all p. We have for any two such sequences (x_j) and (y_j)

$$\sum_{i=1}^{n} |x_j\, y_j| \;\leq\; \left(\sum_{j=1}^{n} |x_j|^p\right)^{1/p} \left(\sum_{j=1}^{n} |y_j|^q\right)^{1/q}$$

and $\sum_{j=1}^{n} |x_j y_j| \leq \left(\sum_{j=1}^{n} |x_j|\right) \max_{1 \leq j \leq n} |y_j|$.

There is an associated inequality called Minkowski's Inequality which is also useful. The general version is then:

Theorem 4.3.5 Minkowski's Inequality

Let $p \geq 1$ and let x and y be in ℓ^p, Then,

$$\left(\sum_{n=1}^{\infty} |x_n + y_n|^p\right)^{\frac{1}{p}} \;\leq\; \left(\sum_{n=1}^{\infty} |x_n|^p\right)^{\frac{1}{p}} + \left(\sum_{n=1}^{\infty} |y_n|^p\right)^{\frac{1}{p}}$$

and for x and y in ℓ^∞,

$$\sup_{n\geq 1} |x_n + y_n| \;\leq\; \sup_{n\geq 1} |x_n| + \sup_{n\geq 1} |y_n|$$

Proof 4.3.4

(1): $p = \infty$

We know $|x_n + y_n| \leq |x_n| + |y_n|$ by the triangle inequality. So we have $|x_n + y_n| \leq \sup_{n\geq 1} |x_n| + \sup_{n\geq 1} |y_n|$. Thus, the right-hand side is an upper bound for all the terms of the left side. We then can say $\sup_{n\geq 1} |x_n + y_n| \leq \sup_{n\geq 1} |x_n| + \sup_{n\geq 1} |y_n|$ which is the result for $p = \infty$.

(2): $p = 1$

Again, we know $|x_n + y_n| \leq |x_n| + |y_n|$ by the triangle inequality. Sum the first N terms on both sides to get

$$\sum_{n=1}^{N} |x_n + y_n| \;\leq\; \sum_{n=1}^{N} |x_n| + \sum_{n=1}^{N} |y_n| \leq \sum_{n=1}^{\infty} |x_n| + \sum_{n=1}^{\infty} |y_n|$$

The right-hand side is an upper bound for the partial sums on the left. Hence, we have

$$\sum_{n=1}^{\infty} |x_n + y_n| \leq \sum_{n=1}^{\infty} |x_n| + \sum_{n=1}^{\infty} |y_n|$$

(3) $1 < p < \infty$
We have

$$\begin{aligned} |x_n + y_n|^p &= |x_n + y_n| \, |x_n + y_n|^{p-1} \leq |x_n| \, |x_n + y_n|^{p-1} + |y_n| \, |x_n + y_n|^{p-1} \\ \sum_{n=1}^{N} |x_n + y_n|^p &\leq \sum_{n=1}^{N} |x_n| \, |x_n + y_n|^{p-1} + \sum_{n=1}^{N} |y_n| \, |x_n + y_n|^{p-1} \end{aligned}$$

Let $a_n = |x_n|$, $b_n = |x_n + y_n|^{p-1}$, $c_n = |y_n|$ *and* $d_n = |x_n + y_n|^{p-1}$. *Hölder's Inequality applies just fine to finite sequences: i.e. sequences in* $\Re^N$. *So we have*

$$\sum_{n=1}^{N} a_n b_n \leq \left(\sum_{n=1}^{N} a_n^p\right)^{\frac{1}{p}} \left(\sum_{n=1}^{N} b_n^q\right)^{\frac{1}{q}}$$

But $b_n^q = |x_n + y_n|^{q(p-1)} = |x_n + y_n|^p$ *using the conjugate exponents identities we established. So we have found*

$$\sum_{n=1}^{N} |x_n| \, |x_n + y_n|^{p-1} \leq \left(\sum_{n=1}^{N} |x_n|^p\right)^{\frac{1}{p}} \left(\sum_{n=1}^{N} |x_n + y_n|^p\right)^{\frac{1}{q}}$$

We can apply the same reasoning to the terms c_n *and* d_n *to find*

$$\sum_{n=1}^{N} |y_n| \, |x_n + y_n|^{p-1} \leq \left(\sum_{n=1}^{N} |y_n|^p\right)^{\frac{1}{p}} \left(\sum_{n=1}^{N} |x_n + y_n|^p\right)^{\frac{1}{q}}$$

We can use the inequalities we just figured out to get the next estimate

$$\begin{aligned} \sum_{n=1}^{N} |x_n + y_n|^p \leq & \left(\sum_{n=1}^{N} |x_n|^p\right)^{\frac{1}{p}} \left(\sum_{n=1}^{N} |x_n + y_n|^p\right)^{\frac{1}{q}} \\ & + \left(\sum_{n=1}^{N} |y_n|^p\right)^{\frac{1}{p}} \left(\sum_{n=1}^{N} |x_n + y_n|^p\right)^{\frac{1}{q}} \end{aligned}$$

Now factor out the common term to get

$$\sum_{n=1}^{N} |x_n + y_n|^p \leq \left(\left(\sum_{n=1}^{N} |x_n|^p\right)^{\frac{1}{p}} + \left(\sum_{n=1}^{N} |y_n|^p\right)^{\frac{1}{p}} \right) \left(\sum_{n=1}^{N} |x_n + y_n|^p\right)^{\frac{1}{q}}$$

Rewrite again as

$$\left(\sum_{n=1}^{N} |x_n + y_n|^p\right)^{1-\frac{1}{q}} \leq \left(\sum_{n=1}^{N} |x_n|^p\right)^{\frac{1}{p}} + \left(\sum_{n=1}^{N} |y_n|^p\right)^{\frac{1}{p}}$$

But $1 - 1/q = 1/p$, *so we have* $\left(\sum_{n=1}^{N} |x_n + y_n|^p\right)^{\frac{1}{p}} \leq \left(\sum_{n=1}^{N} |x_n|^p\right)^{\frac{1}{p}} + \left(\sum_{n=1}^{N} |y_n|^p\right)^{\frac{1}{p}}$.

Now apply the final estimate to find $\left(\sum_{n=1}^{N} |x_n + y_n|^p\right)^{\frac{1}{p}} \leq \left(\sum_{n=1}^{\infty} |x_n|^p\right)^{\frac{1}{p}} + \left(\sum_{n=1}^{\infty} |y_n|^p\right)^{\frac{1}{p}}$. *This says the right-hand side is an upper bound for the partial sums on the left side. Hence, we know*

$$\left(\sum_{n=1}^{\infty} |x_n + y_n|^p\right)^{\frac{1}{p}} \leq \left(\sum_{n=1}^{\infty} |x_n|^p\right)^{\frac{1}{p}} + \left(\sum_{n=1}^{\infty} |y_n|^p\right)^{\frac{1}{p}}$$

■

Homework

Exercise 4.3.1 *Prove if* $(a_n) \in \ell^1$, *it is also in* ℓ^2.

Exercise 4.3.2 *Prove if* (a_n) *and* (b_n) *are in* ℓ^p *so is* $(a_n) + (b_n)$.

Exercise 4.3.3 *Prove if* f *and* g *are Riemann Integrable on* $[a, b]$, *then* $\int_a^b |f(t)g(t)|dt \leq \|f\|_\infty \|g\|_\infty$ $(b - a)$ *and* $\int_a^b |f(t)g(t)|dt \leq \|f\|_\infty \|g\|_1$.

Exercise 4.3.4 *If* f *is in* $C([a, b])$, *then* $\|f\|_p = (\int_a^b |f(t)|^p)^{1/p}$ *for* $1 \leq p < \infty$ *with* $\|f\|_\infty$ *defined as usual. Mimic the proof of the Hölder's and Minkowski's Inequality and prove them for* f *and* g *in* $C([a, b])$. *Can you prove these using this approach if we only know* f *and* g *are Riemann Integrable?*

When we specialize to $\Re^n$ we obtain the following theorem:

Theorem 4.3.6 Minkowski's Inequality in $\Re^n$

Let $p \geq 1$ *and let* $\boldsymbol{x}$ *and* $\boldsymbol{y}$ *be in* $\Re^n$ *with associated sequences* (x_j) *and* (y_j) *Then,*

$$\left(\sum_{j=1}^{n} |x_j + y_j|^p\right)^{\frac{1}{p}} \leq \left(\sum_{j=1}^{n} |x_j|^p\right)^{\frac{1}{p}} + \left(\sum_{j=1}^{n} |y_j|^p\right)^{\frac{1}{p}}$$

and for the case $p = \infty$, *we have*

$$\sup_{1 \leq j \leq n} |x_j + y_j| \leq \sup_{1 \leq j \leq n} |x_j| + \sup_{1 \leq j \leq n} |y_j|$$

In $\Re^n$, define

1. $\|\boldsymbol{x}\|_1 = \sum_{j=1}^{n} |x_j|$
2. $\|\boldsymbol{x}\|_2 = \sqrt{\sum_{j=1}^{n} |x_j|^2}$
3. $\|\boldsymbol{x}\|_\infty = \max_{1 \leq j \leq n} |x_j$

and in general $\|\boldsymbol{x}\|_p = (\sum_{j=1}^{n} |x_j|^p)^{1/p}$ for any $p \geq 1$. Minkowski's Inequality tells in all cases

$$\|\boldsymbol{x} + \boldsymbol{y}\|_p \leq \|\boldsymbol{x}\|_p + \|\boldsymbol{y}\|_p$$

This shows a number of things. Let $\boldsymbol{V}$ be the set of all sequences (x_j) identified with vectors from $\Re^n$. Then we can add them and scalar multiply them, so this is a vector space over $\Re$. If we look at $\|\boldsymbol{x}\|_p$, it clearly is a mapping from $\boldsymbol{V}$ to the reals which satisfies properties N1 to N3 for a norm on $\boldsymbol{V}$. Minkowski's Inequality then proves property N4 for a norm. Hence, each $\|\cdot\|_p$ is a norm on $\Re^n$ and can speak of the distinct normed linear spaces $(\Re^n, \|\cdot\|_1)$, $(\Re^n, \|\cdot\|_2)$, $(\Re^n, \|\cdot\|_\infty)$ and in general $(\Re^n, \|\cdot\|_p)$.

Convergence with respect to these norms is then defined in the usual way. We say $\boldsymbol{x_n}$ converges in $\|\cdot\|_p$ norm to $\boldsymbol{x}$ if

$$\forall\, \epsilon > 0,\ \exists\, N \ni n > N \quad \Longrightarrow \quad \|\boldsymbol{x_n} - \boldsymbol{x}\|_p < \epsilon$$

All of these normed linear spaces are complete. We proved this in the more general ℓ^p setting in (Peterson (21) 2020), but let's specialize these proofs to $\Re^n$ here. They are a bit simpler since we do not have to worry about the convergence of series.

Homework

Exercise 4.3.5 *If $(\boldsymbol{x_n})$ is a sequence ℓ^p and $\boldsymbol{x_n} \rightarrow \boldsymbol{x}$ for some $\boldsymbol{x} \in \ell^p$, prove $(\boldsymbol{x_n})$ is a Cauchy sequence with respect to $\|\cdot\|_p$.*

Exercise 4.3.6 *For any $\boldsymbol{x} \in \Re^n$, compute $\|\boldsymbol{x}\|_p$ for $p = 1, 2, 3, \ldots$. What do you think $\lim_{p\rightarrow\infty} \|\boldsymbol{x}\|_p$ is equal to? How would you prove it?*

Exercise 4.3.7 *Let*

$$f(x) \quad = \quad \begin{cases} 1 & x \in Q \cap [0,1] \\ -1 & x \in \mathbb{I} \cap [01,] \end{cases}$$

Prove f^2 is Riemann Integrable on $[0,1]$ but f is not. On the other hand, if f is Riemann integrable on $[0,1]$, prove f^2 is also Riemann Integrable.

Theorem 4.3.7 The Completeness of $\Re^n$ with the Sup Norm

$(\Re^n, \|\cdot\|_\infty)$ is complete; i.e. if $(\boldsymbol{x_k})$ is a Cauchy sequence in this normed linear space, there is an $\boldsymbol{x}$ in $(\Re^n, \|\cdot\|_\infty)$ so that $\boldsymbol{x_k}$ converges in $\|\cdot\|_\infty$ norm to $\boldsymbol{x}$.

Proof 4.3.5
Assume $(\boldsymbol{x_k})$ is a Cauchy sequence. The element $\boldsymbol{x_k}$ is a sequence itself which we denote by $(x_{k,j})$ for $1 \leq j \leq n$. Then for a given $\epsilon > 0$, there is an N so that

$$\max_{1\leq j\leq n} |x_{k,j} - x_{m,j}| \quad < \quad \epsilon/2 \text{ when } m > k > N_\epsilon$$

So for each fixed index j, we have

$$|x_{k,j} - x_{m,j}| \quad < \quad \epsilon/2 \text{ when } m > k > N_\epsilon$$

This says for fixed j, the sequence $(x_{k,j})$ is a Cauchy sequence of real numbers and hence must converge to a real number we will call a_k. This defines a new vector $\boldsymbol{a} \in \Re^n$. Does $\boldsymbol{x_n} \rightarrow \boldsymbol{a}$ in the $\|\cdot\|_\infty$ norm? We use the continuity of the function $|\cdot|$ to see for any $k > N_\epsilon$, we have

$$\lim_{m\rightarrow\infty} |x_{k,j} - x_{m,j}| \quad \leq \quad \epsilon/2 \Longrightarrow |x_{k,j} - \lim_{m\rightarrow\infty} x_{m,j}| \leq \epsilon/2$$

This argument works for all j and so $|x_{k,j} - a_j| \leq \epsilon/2$ when $k > N_\epsilon$ for all j which implies $\max_{1\leq j\leq n} |x_{k,j} - a_j| \leq \epsilon/2 < \epsilon$ or $\|\boldsymbol{x_k} - \boldsymbol{a}\|_\infty < \epsilon$ when $k > N_\epsilon$. So $\boldsymbol{x_n} \rightarrow \boldsymbol{a}$ in $\|\cdot\|_\infty$. ■

Next, let's look at vector convergence using the $\|\cdot\|_p$ norm for $p \geq 1$.

Theorem 4.3.8 $\Re^n$ with the p Norm is Complete

$(\Re^n, \|\cdot\|_p)$ is complete; i.e. if $(\boldsymbol{x_k})$ is a Cauchy sequence in this normed linear space, there is an $\boldsymbol{x}$ in $(\Re^n, \|\cdot\|_p)$ so that $\boldsymbol{x_k}$ converges in $\|\cdot\|_p$ norm to $\boldsymbol{x}$.

Proof 4.3.6

Let $(\boldsymbol{x_k})$ be a Cauchy sequence in $\|\cdot\|_p$. Then given $\epsilon > 0$, there is an N_ϵ so that $m > k > N_\epsilon$ implies

$$||\boldsymbol{x_k} - \boldsymbol{x_m}||_p = \left(\sum_{j=1}^{n} |x_{k,j} - x_{m,j}|^p\right)^{\frac{1}{p}} \quad < \quad \epsilon/2.$$

Thus, if $m > k > N_\epsilon$, $\sum_{j=1}^{n} |x_{k,j} - x_{m,j}|^p < (\epsilon/2)^p$. Since this is a sum on non-negative terms, each term must be less than $(\epsilon/2)^p$. So we must have $|x_{k,j} - x_{m,j}|^p < (\epsilon/2)^p$ or $|x_{k,j} - x_{m,j}| < (\epsilon/2)$ when $m > k > N_\epsilon$. This tells us immediately the sequence of real numbers $(x_{k,j})$ is a Cauchy sequence of real numbers and so must converge to a number we will call a_j. This defines the vector $\boldsymbol{a} \in \Re^n$. Since $|\cdot|^p$ is continuous, we can say

$$\begin{aligned}(\epsilon/2)^p \quad &\geq \quad \lim_{m\to\infty}\left(\sum_{j=1}^{n} |x_{k,j} - x_{m,j}|^p\right) = \left(\sum_{j=1}^{n} |x_{k,j} - \lim_{m\to\infty} x_{m,j}|^p\right) \\ &= \quad \sum_{j=1}^{n} |x_{k,j} - a_j|^p\end{aligned}$$

This says immediately that $||\boldsymbol{x_k} - \boldsymbol{a}||_p < \epsilon$ when $k > N_\epsilon$ so $\boldsymbol{x_k} \to \boldsymbol{a}$ in $\|\cdot\|_p$. ■

Comment 4.3.1 *It is easy to see $\Re^n$ has an inner product given by*

$$< \boldsymbol{x}, \boldsymbol{y} > \quad = \quad \sum_{j=1}^{n} x_j y_j$$

Hence, $(\Re^n, <,>)$ is an inner product space whose norm induces $\|\cdot\|_2$. Moreover Hölder's Inequality tells us

$$| < \boldsymbol{x}, \boldsymbol{y} > | \quad \leq \quad \sum_{j=1}^{n} |x_j y_j| \leq \|\boldsymbol{x}\|_2 \, \|\boldsymbol{x}\|_2$$

which is our standard Cauchy - Schwartz Inequality in $\Re^n$, which we use to define the angle between vectors in $\Re^n$ as usual.

4.3.1 Homework

Exercise 4.3.8 *In $\Re^2$, fix $\boldsymbol{x} \in \Re^2$ and define $f : \Re^2 \to \Re$ by $f(\boldsymbol{y}) =< \boldsymbol{y}, \boldsymbol{x} >$. Prove f is linear and*

$$\sup_{\boldsymbol{y}\neq\boldsymbol{0}} \frac{|f(\boldsymbol{y})}{\|\boldsymbol{y}\|_2} \quad = \quad \|\boldsymbol{x}\|_2$$

Exercise 4.3.9 *In $\Re^2$, fix $\boldsymbol{x} \in \Re^3$ and define $f : \Re^3 \to \Re$ by $f(\boldsymbol{y}) =< \boldsymbol{y}, \boldsymbol{x} >$. Prove f is linear and*

$$\sup_{\boldsymbol{y}\neq\boldsymbol{0}} \frac{|f(\boldsymbol{y})}{\|\boldsymbol{y}\|_2} \quad = \quad \|\boldsymbol{x}\|_2$$

Exercise 4.3.10 *In ℓ^2, fix $\boldsymbol{x} \in \ell^2$ and define $f : \ell^2 \rightarrow \Re$ by $f(\boldsymbol{y}) =< \boldsymbol{y}, \boldsymbol{x} >$. Recall the inner product on ℓ^2 is $\sum_{i=1}^{\infty} x_i y_i$ which is well-defined by Hölder's Inequality. Prove f is linear and*

$$\sup_{\boldsymbol{y} \neq \boldsymbol{0}} \frac{|f(\boldsymbol{y})}{\|\boldsymbol{y}\|_2} = \|\boldsymbol{x}\|_2$$

Exercise 4.3.11 *In ℓ^∞, fix $\boldsymbol{x} \in \ell^\infty$ and define $f : \ell^1 \rightarrow \Re$ by $f(\boldsymbol{y}) =< \boldsymbol{y}, \boldsymbol{x} >$. Prove f is linear and*

$$\sup_{\boldsymbol{y} \neq \boldsymbol{0}} \frac{|f(\boldsymbol{y})}{\|\boldsymbol{y}\|_2} \leq \|\boldsymbol{x}\|_\infty$$

4.4 Linear Transformations between Normed Linear Spaces

Let's start by defining linear transformations between vector spaces carefully

Definition 4.4.1 Linear Transformations between Vector Spaces

Let $\boldsymbol{X}$ and $\boldsymbol{Y}$ be two vector spaces over $\Re$. We say $\boldsymbol{T} : \boldsymbol{X} \rightarrow \boldsymbol{Y}$ is a linear transformation if

$$\boldsymbol{T}(\alpha \boldsymbol{x} + \beta \boldsymbol{y}) = \alpha \boldsymbol{T}(\boldsymbol{x}) + \beta \boldsymbol{T}(\boldsymbol{y})$$

To discuss linear transformations between finite dimensional vector spaces, we need to have a formal definition of matrices. We have used them quite a bit in our explanations, but it is time to be formal.

Definition 4.4.2 Real Matrices

Let $\boldsymbol{M}$ be a collection of real numbers organized in a table of m rows and n columns. The number M_{ij} refers to the entry in the ith row and j^{th} column of this table. The table is organized like this

$$\begin{matrix} M_{11} & \ldots & M_{1n} \\ M_{21} & \ldots & M_{2n} \\ \vdots & \vdots & \vdots \\ M_{m1} & \ldots & M_{mn} \end{matrix}$$

We identify the matrix $\boldsymbol{M}$ with this table and write

$$\boldsymbol{M} = \begin{bmatrix} M_{11} & \ldots & M_{1n} \\ \vdots & \vdots & \vdots \\ M_{m1} & \ldots & M_{mn} \end{bmatrix}$$

The set of all matrices with m rows and n columns is denoted by $\boldsymbol{M_{mn}}$. We also say $\boldsymbol{M}$ is an $m \times n$ matrix.

Comment 4.4.1 *The n columns of $\boldsymbol{M}$ are clearly vectors in $\Re^m$ and the m rows of $\boldsymbol{M}$ are vectors in $\Re^n$. Thus, we have some additional identifications:*

$$\boldsymbol{M} = \begin{bmatrix} M_{11} & \ldots & M_{1n} \\ \vdots & \vdots & \vdots \\ M_{m1} & \ldots & M_{mn} \end{bmatrix} = \begin{bmatrix} \boldsymbol{C_1} & \ldots & \boldsymbol{C_n} \end{bmatrix} = \begin{bmatrix} \boldsymbol{R_1} \\ \vdots \\ \boldsymbol{R_m} \end{bmatrix}$$

where column i is denoted by $\boldsymbol{C_j} \in \Re^m$ *and row j is denoted by* $\boldsymbol{R_j} \in \Re^n$. *Note the rows are vectors displayed as the transpose of the usual column vector notation we use.*

Comment 4.4.2 *So, for example, the data pairs of time and temperature we measure in a cooling model experiment might turn out to be the following:*

Listing 4.1: **Sample Data**

```
0 205
1 201
3 198
6 190
9 185
12 182
15 179
20 168
25 161
30 154
40 146
50 135
60 123
80 109
100 87
120 82
140 79
150 77
```

This table defines a matrix of 18 *rows and* 2 *columns and so defines a matrix in* $\boldsymbol{M_{18,2}}$.

4.4.0.1 Homework

Exercise 4.4.1 *Let* $\boldsymbol{A} = \begin{bmatrix} 0 & 0 \\ 1 & 2 \end{bmatrix}$. *Compute* $\boldsymbol{A}^j$ *for all* $j \geq 1$.

Exercise 4.4.2 *Let* $\boldsymbol{A} = \begin{bmatrix} 2 & 1 \\ 0 & 2 \end{bmatrix}$. *Compute* $\boldsymbol{A}^j$ *for all* $j \geq 1$.

Exercise 4.4.3 *Let*

$$\boldsymbol{A} = \begin{bmatrix} 2 & 1 & 0 \\ 0 & 2 & 0 \\ 0 & 0 & 3 \end{bmatrix}$$

Compute $\boldsymbol{A}^j$ *for all* $j \geq 1$.

Exercise 4.4.4 *Let*

$$\boldsymbol{A} = \begin{bmatrix} 2 & 1 & 1 & 0 \\ 0 & 2 & 1 & 0 \\ 0 & 0 & 2 & 0 \\ 0 & 0 & 0 & 3 \end{bmatrix}$$

Compute $\boldsymbol{A}^j$ *for all* $j \geq 1$.

Exercise 4.4.5 *Find* $p(\lambda) = \det(\boldsymbol{\lambda I} - \boldsymbol{A})$ *for any* $\boldsymbol{A}$ *in* $M_{2\times 2}$ *over* $\Re$. *Prove* $p(\boldsymbol{A}) = \boldsymbol{0}$. *This is the Cayley - Hamilton Theorem for the case of* 2×2 *matrices.*

4.4.1 Basic Properties

Consider a matrix $\boldsymbol{M}$ in $\boldsymbol{M_{mn}}$. We define the action of $\boldsymbol{M}$ on the vector space $\Re^n$ by

$$\boldsymbol{M}(\boldsymbol{x}) = \left(\begin{bmatrix}\boldsymbol{R_1} & \ldots & \boldsymbol{R_m}\end{bmatrix}\right)(\boldsymbol{x}) = \begin{bmatrix} <\boldsymbol{R_1}, \boldsymbol{x}> \\ \vdots \\ <\boldsymbol{R_m}, \boldsymbol{x}> \end{bmatrix}$$

We generally just write $\boldsymbol{Mx} = \boldsymbol{M}(\boldsymbol{x})$ to save a few parentheses. It is easy to see this definition of the action of $\boldsymbol{M}$ on $\Re^n$ defines a linear transformation from $\Re^n$ to $\Re^m$.
The set of $\boldsymbol{x}$ in $\Re^n$ with $\boldsymbol{Mx} = \boldsymbol{0} \in \Re^m$ is called the **kernel** or **nullspace** of $\boldsymbol{M}$. The span of the columns of $\boldsymbol{M}$ is called the column space of $\boldsymbol{M}$. We can prove a fundamental result.

Theorem 4.4.1 If M is an $m \times n$ Matrix, then dim(kernel) + dim(column space) = n

> *If $\boldsymbol{M}$ is a $m \times n$ matrix, if $p = dim(ker(\boldsymbol{M}))$ and q is the dimension of the span of the columns of $\boldsymbol{M}$, then $p + q = n$.*

Proof 4.4.1
It is easy to see that $ker(\boldsymbol{M})$ is a subspace of $\Re^n$ with dimension $p \leq n$. Let $\{\boldsymbol{K_1}, \ldots, \boldsymbol{K_p}\}$ be an orthonormal basis of $ker(\boldsymbol{M})$. We can use GSO to construct an additional $n - p$ vectors in $\Re^n$, $\{\boldsymbol{L_1}, \ldots, \boldsymbol{L_{n-p}}\}$ that are all orthogonal to $\ker(\boldsymbol{M})$ with length one. The subspace spanned by $\{\boldsymbol{L_1}, \ldots, \boldsymbol{L_{n-p}}\}$ is perpendicular to the subspace $ker(\boldsymbol{M})$ and is called $(\ker(\boldsymbol{M}))^\perp$ to denote it is what is called an orthogonal complement. We note $\Re^n = ker(\boldsymbol{M}) \oplus span\{\boldsymbol{L_1}, \ldots, \boldsymbol{L_{n-p}}\}$.
Thus, $\{\boldsymbol{K_1}, \ldots, \boldsymbol{K_p}, \boldsymbol{L_1}, \ldots, \boldsymbol{L_{n-p}}\}$ is an orthonormal basis for $\Re^n$ and hence $\boldsymbol{x} = a_1\boldsymbol{K_1} + \ldots + a_p\boldsymbol{K_p} + b_1\boldsymbol{L_1} + \ldots + b_{n-p}\boldsymbol{L_{n-p}}$. It follows using the linearity of $\boldsymbol{M}$ that

$$\boldsymbol{Mx} = b_1\boldsymbol{ML_1} + \ldots b_{n-p}\boldsymbol{ML_{n-p}}$$

Now if $b_1\boldsymbol{ML_1} + \ldots b_{n-p}\boldsymbol{ML_{n-p}} = 0$, this means $\boldsymbol{M}(b_1\boldsymbol{L_1} + \ldots + b_{n-p}\boldsymbol{L_{n-p}}) = 0$ too. This says $b_1\boldsymbol{L_1} + \ldots + b_{n-p}\boldsymbol{L_{n-p}}$ is in both $ker(\boldsymbol{M})$ and the $span\{\boldsymbol{L_1}, \ldots, \boldsymbol{L_{n-p}}\}$ which forces $b_1\boldsymbol{L_1} + \ldots + b_{n-p}\boldsymbol{L_{n-p}} = 0$. But $\{\boldsymbol{L_1}, \ldots, \boldsymbol{L_{n-p}}\}$ is a linearly independent set and so all coefficients $b_i = 0$. This says the set $\{\boldsymbol{ML_1}, \ldots, \boldsymbol{ML_{n-p}}\}$ is a linearly independent set in $\Re^m$.

Finally, note the column space of $\boldsymbol{M}$ is defined to be the $span\{\boldsymbol{C_1}, \ldots, \boldsymbol{C_n}\}$. A vector in this span has the look $\boldsymbol{y} = \sum_{i=1}^m a_i\boldsymbol{C_i}$. This is the same as $\boldsymbol{Ma}$ where $a \in \Re^n$. However, we know $\Re^n$ has the orthonormal basis $\{\boldsymbol{K_1}, \ldots, \boldsymbol{K_p}, \boldsymbol{L_1}, \ldots, \boldsymbol{L_{n-p}}\}$ and so $\boldsymbol{a} = \sum_{i=1}^p b_i\boldsymbol{K_i} + \sum_{i=1}^{n-p} c_i\boldsymbol{L_i}$. Applying $\boldsymbol{M}$ we find

$$\boldsymbol{Ma} = c_1\boldsymbol{ML_1} + \ldots + c_{n-p}\boldsymbol{ML_{n-p}}$$

Since $\{\boldsymbol{ML_1}, \ldots, \boldsymbol{ML_{n-p}}\}$ is linearly independent, we see the range of $\boldsymbol{M}$, the column space of $\boldsymbol{M}$, has dimension $n - p$. Thus, we have shown

$$dim(ker(\boldsymbol{M})) + dim(span(\{\boldsymbol{C_1}, \ldots, \boldsymbol{C_n}\})) = n$$

where, of course, we also have $n - p \leq m$. ■

Comment 4.4.3 *Thus, if $n - p = m$, the kernel of $\boldsymbol{M}$ must be $n - m$ dimensional.*

4.4.1.1 Homework

Exercise 4.4.6 *If*

$$\boldsymbol{A} = \begin{bmatrix} 1 & 2 & -3 \\ 0 & 4 & 5 \\ -1 & 2 & 8 \end{bmatrix}$$

Find an orthonormal basis for the kernel of $\boldsymbol{A}$ *and orthonormal basis for the span of the columns of* $\boldsymbol{A}$.

Exercise 4.4.7

$$\boldsymbol{A} = \begin{bmatrix} 1 & 2 & -3 \\ 0 & 4 & 5 \\ 0 & 0 & 0 \end{bmatrix}$$

Find an orthonormal basis for the kernel of $\boldsymbol{A}$ *and orthonormal basis for the span of the columns of* $\boldsymbol{A}$.

Exercise 4.4.8

$$\boldsymbol{A} = \begin{bmatrix} 1 & 2 & -3 & 6 \\ 0 & 4 & 5 & 7 \\ 0 & 0 & 0 & 0 \\ 0 & 0 & 0 & 0 \end{bmatrix}$$

Find an orthonormal basis for the kernel of $\boldsymbol{A}$ *and orthonormal basis for the span of the columns of* $\boldsymbol{A}$.

Exercise 4.4.9

$$\boldsymbol{A} = \begin{bmatrix} 1 & 2 & -3 & 6 \\ 0 & 4 & 5 & 7 \\ 0 & 0 & 9 & 1 \\ 0 & 0 & 0 & 0 \end{bmatrix}$$

Find an orthonormal basis for the kernel of $\boldsymbol{A}$ *and orthonormal basis for the span of the columns of* $\boldsymbol{A}$.

4.4.2 Mappings between Finite Dimensional Vector Spaces

Let $\boldsymbol{X}$ and $\boldsymbol{Y}$ be two finite dimensional vector spaces over $\Re$. Assume $\boldsymbol{X}$ has dimension n an $\boldsymbol{Y}$ has dimension m. Let $\boldsymbol{E}$ be a basis for $\boldsymbol{X}$ and $\boldsymbol{F}$, a basis for $\boldsymbol{Y}$. Before we go further, let's be clear about some of our notation.

- We use $\boldsymbol{x}$ to denote an object in the finite dimensional vector space $\boldsymbol{X}$.
- For a given basis $\boldsymbol{E}$, we can write $\boldsymbol{x} = x_1\boldsymbol{E_1} + \ldots + \boldsymbol{E_n}$ and the coefficients $\{x_1, \ldots, x_n\}$ are n real numbers and so give a vector in $\Re^n$. We let $\left[\boldsymbol{x}\right]_{\boldsymbol{E}}$ indicate this vector of real numbers. If **G** was another basis, there would be another set of n numbers given by $\boldsymbol{x} = u_1\boldsymbol{G_1} + \ldots + u_n\boldsymbol{G_n}$ and this vector would be denoted by $\left[\boldsymbol{x}\right]_{\boldsymbol{G}}$. Then, there will be a transformation law which converts $\left[\boldsymbol{x}\right]_{\boldsymbol{E}}$ into $\left[\boldsymbol{x}\right]_{\boldsymbol{G}}$ and we will talk about that soon. The point is $\boldsymbol{x}$ is the same object in $\boldsymbol{X}$ which we can represent in different ways with a mechanism to map one representation into another.

Theorem 4.4.2 The Change of Basis Mapping

Let $\boldsymbol{E}$ and $\boldsymbol{G}$ be bases for the finite dimensional vector space $\boldsymbol{X}$ which has dimension n. Then there is an invertible matrix $\boldsymbol{A_{EG}}$ so that $\left[\boldsymbol{x}\right]_{\boldsymbol{G}} = \boldsymbol{A_{EG}} \left[\boldsymbol{x}\right]_{\boldsymbol{E}}$. We read the matrix $\boldsymbol{A_{EG}}$ as mapping $\boldsymbol{E}$ into $\boldsymbol{G}$.

Proof 4.4.2
We know

$$\boldsymbol{x} = c_1\boldsymbol{E_1} + \ldots + c_n\boldsymbol{E_n}$$

We also know each $\boldsymbol{E_j}$ has an expansion in the $\boldsymbol{G}$ basis; i.e. $\boldsymbol{E_i} = \sum_{j=1}^{n} A_{ji}\boldsymbol{G_j}$. Using this, we find

$$\boldsymbol{x} = \sum_{i=1}^{n} c_i \left(\sum_{j=1}^{n} A_{ji}\boldsymbol{G_j}\right) = \sum_{j=1}^{n} \left(\sum_{i=1}^{n} c_i\, A_{ji}\right) \boldsymbol{G_j}$$

But $\boldsymbol{x}$ also has an expansion with respect to $\boldsymbol{G}$: $\boldsymbol{x} = \sum_{j=1}^{n} d_j\boldsymbol{G_j}$. Hence, equating coefficients, we have

$$d_j = \sum_{i=1}^{n} A_{ji}c_i$$

which can be organized into a familiar set of equations.

$$\begin{aligned} d_1 &= A_{11}c_1 + A_{12}c_2 + \ldots + A_{1n}c_n \\ &\vdots \\ d_j &= A_{j1}c_1 + A_{j2}c_2 + \ldots + A_{jn}c_n \\ &\vdots \\ d_n &= A_{n1}c_1 + A_{n2}c_2 + \ldots + A_{nn}c_n \end{aligned}$$

or

$$\begin{bmatrix} d_1 \\ \vdots \\ d_n \end{bmatrix} = \begin{bmatrix} A_{11} & \ldots & A_{1n} \\ \vdots & & \\ A_{n1} & \ldots & A_{nn} \end{bmatrix} \begin{bmatrix} c_1 \\ \vdots \\ c_n \end{bmatrix}$$

We then further rewrite as $\left[\boldsymbol{x}\right]_{\boldsymbol{G}} = \boldsymbol{A_{EG}} \left[\boldsymbol{x}\right]_{\boldsymbol{E}}$. A similar argument shows there is a matrix $\boldsymbol{B_{GE}}$ so that $\left[\boldsymbol{x}\right]_{\boldsymbol{E}} = \boldsymbol{B_{GE}} \left[\boldsymbol{x}\right]_{\boldsymbol{G}}$. Thus

$$\begin{aligned} \left[\boldsymbol{x}\right]_{\boldsymbol{G}} &= \boldsymbol{A_{EG}} \left[\boldsymbol{x}\right]_{\boldsymbol{E}} = \boldsymbol{A_{EG}}\boldsymbol{B_{GE}} \left[\boldsymbol{x}\right]_{\boldsymbol{G}} \\ \left[\boldsymbol{x}\right]_{\boldsymbol{E}} &= \boldsymbol{B_{GE}} \left[\boldsymbol{x}\right]_{\boldsymbol{G}} = \boldsymbol{B_{GE}}\boldsymbol{A_{EG}} \left[\boldsymbol{x}\right]_{\boldsymbol{E}} \end{aligned}$$

This tells us $\boldsymbol{A_{EG}}\boldsymbol{B_{GE}} = \boldsymbol{B_{GE}}\boldsymbol{A_{EG}} = \boldsymbol{I}$. Hence, $\boldsymbol{A_{EG}}^{-1} = \boldsymbol{B_{GE}}$. ■

Theorem 4.4.3 The Change of Basis Mapping in a Finite Dimensional Inner Product Space

Let $\boldsymbol{E}$ and $\boldsymbol{G}$ be two orthonormal bases for the finite dimensional vector space $\boldsymbol{X}$ which has dimension n. We assume $\boldsymbol{X}$ has an inner product $<,>$. Then there is an invertible matrix $\boldsymbol{A_{EG}}$ so that $\left[\boldsymbol{x}\right]_{\boldsymbol{G}} = \boldsymbol{A_{EG}}\left[\boldsymbol{x}\right]_{\boldsymbol{E}}$ which has the following form:

$$\boldsymbol{A_{EG}} = \begin{bmatrix} \boldsymbol{G_1}^T \\ \vdots \\ \boldsymbol{G_n}^T \end{bmatrix} \left[\boldsymbol{E_1}, \ldots, \boldsymbol{E_n}\right]$$

with

$$\boldsymbol{A_{EG}}^{-1} = \boldsymbol{A_{EG}}^T = \begin{bmatrix} \boldsymbol{E_1}^T \\ \vdots \\ \boldsymbol{E_n}^T \end{bmatrix} \left[\boldsymbol{G_1}, \ldots, \boldsymbol{G_n}\right]$$

Proof 4.4.3

Let $\boldsymbol{G}$ be another orthonormal basis for $\boldsymbol{X}$. Then since $\boldsymbol{x} = c_1\boldsymbol{E_1} + \ldots c_n\boldsymbol{E_n}$, we have

$$\begin{aligned} < \boldsymbol{x}, \boldsymbol{G_1} > &= c_1 < \boldsymbol{G_1}, \boldsymbol{E_1} > + \ldots + c_n < \boldsymbol{G_1}, \boldsymbol{E_n} > \\ \vdots &= \\ < \boldsymbol{x}, \boldsymbol{G_n} > &= c_1 < \boldsymbol{G_n}, \boldsymbol{E_1} > + \ldots + c_n < \boldsymbol{G_n}, \boldsymbol{E_n} > \end{aligned}$$

which in matrix-vector form is

$$\left[\boldsymbol{x}\right]_{\boldsymbol{G}} = \begin{bmatrix} \boldsymbol{G_1}^T \\ \vdots \\ \boldsymbol{G_n}^T \end{bmatrix} \left[\boldsymbol{E_1}, \ldots, \boldsymbol{E_n}\right] \left[\boldsymbol{x}\right]_{\boldsymbol{E}}$$

The orthonormality of $\boldsymbol{E}$ and $\boldsymbol{G}$ then tell us immediately

$$\boldsymbol{A_{EG}}^{-1} = \boldsymbol{A_{EG}}^T = \begin{bmatrix} \boldsymbol{E_1}^T \\ \vdots \\ \boldsymbol{E_n}^T \end{bmatrix} \left[\boldsymbol{G_1}, \ldots, \boldsymbol{G_n}\right]$$

■

Since there is an inner product on $\boldsymbol{X}$, we know its value does not depend on the choice of orthonormal basis for $\boldsymbol{X}$.

Homework

Let's assume for all these problems, we are working on the interval $[-1, 4]$, so the inner product is $\int_{-1}^{4} < F, G > dt$ where F and G come from the vector space in question.

Exercise 4.4.10 *Let $f(t) = \boldsymbol{E_1}e^{-2t}$ and $g(t) = \boldsymbol{E_2}e^{3t}$ where $\boldsymbol{E_1} = \begin{bmatrix} 1 \\ 2 \end{bmatrix}$ and $\boldsymbol{E_2} = \begin{bmatrix} -3 \\ 4 \end{bmatrix}$ are two linearly independent vectors in $\Re^2$. Note this set of functions is the typical set of solutions to a linear system of ODEs. These two functions form a two dimensional vector space over $\Re$. Let $\phi_1(t) = E_{11}e^{-2t} + E_{21}e^{3t}$ and $\phi_2(t) = E_{12}e^{-2t} + E_{22}e^{3t}$. Use* **GSO** *to find the orthonormal base $\boldsymbol{\psi} = \{\psi_1, \psi_2\}$ associated with $\boldsymbol{\phi} = \{\phi_1, \phi_2\}$. Compare these basis functions with the basis functions* *you obtain from using* **GSO** *to find the orthonormal base* $\boldsymbol{F}$ *associated with* $\boldsymbol{E} = \{\boldsymbol{E_1}, \boldsymbol{E_2}\}$. *Both of these sets of functions provide a basis for the span of* $\{\boldsymbol{E_1}e^{-2t}, \boldsymbol{E_2}e^{3t}\}$.

Exercise 4.4.11 *Let* $f(t) = \boldsymbol{E_1}e^{-2t}$ *and* $g(t) = \boldsymbol{E_2}e^{3t}$ *where* $\boldsymbol{E_1} = \begin{bmatrix} 1 \\ 2 \end{bmatrix}$ *and* $\boldsymbol{E_2} = \begin{bmatrix} -3 \\ 4 \end{bmatrix}$ *are two linearly independent vectors in* $\Re^2$. *These two functions form a two dimensional vector space over* $\Re$. *Use* **GSO** *to find the orthonormal base* $\boldsymbol{F}$ *associated with* $\boldsymbol{E} = \{\boldsymbol{E_1}, \boldsymbol{E_2}\}$. *Any function in the span of* $\{\boldsymbol{E_1}e^{-2t}, \boldsymbol{E_2}e^{3t}\}$ *is thus in the span of* $\{\boldsymbol{F_1}e^{-2t}, \boldsymbol{F_2}e^{3t}\}$. *Solve the equation* $a\boldsymbol{F_1}e^{-2t} + b\boldsymbol{F_2}e^{3t} = 2\boldsymbol{E_1}e^{-2t} + 3\boldsymbol{E_2}e^{3t}$.

Exercise 4.4.12 *Let* $f(t) = \boldsymbol{E_1}e^{-2t}$ *and* $g(t) = \boldsymbol{E_2}e^{3t}$ *where* $\boldsymbol{E_1} = \begin{bmatrix} 1 \\ 2 \end{bmatrix}$ *and* $\boldsymbol{E_2} = \begin{bmatrix} -3 \\ 4 \end{bmatrix}$ *are two linearly independent vectors in* $\Re^2$. *These two functions form a two dimensional vector space over* $\Re$. *Use* **GSO** *to find the orthonormal base* $\boldsymbol{F}$ *associated with* $\boldsymbol{E} = \{\boldsymbol{E_1}, \boldsymbol{E_2}\}$. *Any function in the span of* $\{\boldsymbol{E_1}e^{-2t}, \boldsymbol{E_2}e^{3t}\}$ *is thus in the span of* $\{\boldsymbol{F_1}e^{-2t}, \boldsymbol{F_2}e^{3t}\}$. *Find the transformation matrix* $\boldsymbol{A_{EG}}$ *that is relevant here that allows us to transform from representations with respect to* $\boldsymbol{E}$ *and* $\boldsymbol{G}$ *for this set of functions.*

Exercise 4.4.13 *Let* $f(t) = \boldsymbol{E_1}e^{4t}$ *and* $g(t) = \boldsymbol{E_2}e^{6t}$ *where* $\boldsymbol{E_1} = \begin{bmatrix} 4 \\ 2 \end{bmatrix}$ *and* $\boldsymbol{E_2} = \begin{bmatrix} 1 \\ 8 \end{bmatrix}$ *are two linearly independent vectors in* $\Re^2$. *These two functions form a two dimensional vector space over* $\Re$. *Use* **GSO** *to find the orthonormal base* $\boldsymbol{F}$ *associated with* $\boldsymbol{E} = \{\boldsymbol{E_1}, \boldsymbol{E_2}\}$. *Any function in the span of* $\{\boldsymbol{E_1}e^{4t}, \boldsymbol{E_2}e^{6t}\}$ *is thus in the span of* $\{\boldsymbol{F_1}e^{4t}, \boldsymbol{F_2}e^{6t}\}$. *Find the transformation matrix* $\boldsymbol{A_{EG}}$ *that is relevant here that allows us to transform from representations with respect to* $\boldsymbol{E}$ *and* $\boldsymbol{G}$ *for this set of functions.*

Exercise 4.4.14 *Let* $f_1(t) = \boldsymbol{E_1}e^{-3t}$, $f_2(t) = \boldsymbol{E_2}e^{-t}$ *and* $f_3(t) = \boldsymbol{E_3}e^{2t}$ *where* $\boldsymbol{E_1} = \begin{bmatrix} 1 \\ 2 \\ 4 \end{bmatrix}$, $\boldsymbol{E_2} = \begin{bmatrix} 2 \\ 1 \\ 5 \end{bmatrix}$ *and* $\boldsymbol{E_3} = \begin{bmatrix} 3 \\ 3 \\ 2 \end{bmatrix}$ *are linearly independent vectors in* $\Re^3$. *These three functions form a three dimensional vector space over* $\Re$. *Use* **GSO** *to find the orthonormal base* $\boldsymbol{F}$ *associated with* $\boldsymbol{E} = \{\boldsymbol{E_1}, \boldsymbol{E_2}, \boldsymbol{E_3}\}$. *Any function in the span of* $\{\boldsymbol{E_1}e^{-3t}, \boldsymbol{E_2}e^{-t}, \boldsymbol{E_3}e^{2t}\}$ *is thus in the span of* $\{\boldsymbol{F_1}e^{-3t}, \boldsymbol{F_2}e^{-2t}, \boldsymbol{F_3}e^{2t}\}$. *Solve the equation* $a\boldsymbol{F_1}e^{-3t} + b\boldsymbol{F_2}e^{-2t} + c\boldsymbol{F_3}e^{2t} = 2\boldsymbol{E_1}e^{-3t} + 3\boldsymbol{E_2}e^{-t} + 4\boldsymbol{E_3}e^{2t}$.

Exercise 4.4.15 *Let* $f_1(t) = \boldsymbol{E_1}e^{-3t}$, $f_2(t) = \boldsymbol{E_2}e^{-t}$ *and* $f_3(t) = \boldsymbol{E_3}e^{2t}$ *where* $\boldsymbol{E_1} = \begin{bmatrix} 1 \\ 2 \\ 4 \end{bmatrix}$, $\boldsymbol{E_2} = \begin{bmatrix} 2 \\ 1 \\ 5 \end{bmatrix}$ *and* $\boldsymbol{E_3} = \begin{bmatrix} 3 \\ 3 \\ 2 \end{bmatrix}$ *are linearly independent vectors in* $\Re^3$. *Use* **GSO** *to find the orthonormal base* $\boldsymbol{F}$ *associated with* $\boldsymbol{E} = \{\boldsymbol{E_1}, \boldsymbol{E_2}, \boldsymbol{E_3}\}$. *Any function in the span of* $\{\boldsymbol{E_1}e^{-3t}, \boldsymbol{E_2}e^{-t}, \boldsymbol{E_3}e^{2t}\}$ *is thus in the span of* $\{\boldsymbol{F_1}e^{-3t}, \boldsymbol{F_2}e^{-2t}, \boldsymbol{F_3}e^{2t}\}$. *Find the transformation matrix* $\boldsymbol{A_{EG}}$ *that is relevant here that allows us to transform from representations with respect to* $\boldsymbol{E}$ *and* $\boldsymbol{G}$ *for this set of functions.*

Theorem 4.4.4 The Inner Product on a Finite Dimensional Vector Space is Independent of Choice of Orthonormal Basis

> *Let* $\boldsymbol{X}$ *be an* n *dimensional vector space with inner product* $<,>$. *Then its value is independent of the choice of orthonormal basis* $\boldsymbol{E}$.

Proof 4.4.4
We know $\boldsymbol{x_E} = x_E^1 \boldsymbol{E_1} + \ldots x_E^n \boldsymbol{E_n}$ *and* $\boldsymbol{x_G} = x_G^1 \boldsymbol{G_1} + \ldots x_G^n \boldsymbol{G_n}$ *with a similar notation for the representations of* $\boldsymbol{y}$. *We have*

$$[\boldsymbol{x}]_{\boldsymbol{E}} = \begin{bmatrix} x_E^1 \\ \vdots \\ x_E^n \end{bmatrix}, \quad [\boldsymbol{x}]_{\boldsymbol{G}} = \begin{bmatrix} x_G^1 \\ \vdots \\ x_G^n \end{bmatrix}$$

and we know

$$[\boldsymbol{x}]_{\boldsymbol{G}} = \begin{bmatrix} \boldsymbol{G_1}^T \\ \vdots \\ \boldsymbol{G_n}^T \end{bmatrix} [\boldsymbol{E_1}, \ldots, \boldsymbol{E_n}] \, [\boldsymbol{x}]_{\boldsymbol{E}}$$

Thus, by orthonormality of $\boldsymbol{E}$ *and* $\boldsymbol{G}$, *we have*

$$\begin{aligned} < \boldsymbol{x_G}, \boldsymbol{y_G} > &= \sum_{i=1}^n x_G^i y_G^i = \langle [\boldsymbol{x}]_{\boldsymbol{G}}, [\boldsymbol{y}]_{\boldsymbol{G}} \rangle \\ &= \left\langle \begin{bmatrix} \boldsymbol{G_1}^T \\ \vdots \\ \boldsymbol{G_n}^T \end{bmatrix} [\boldsymbol{E_1}, \ldots, \boldsymbol{E_n}] \, [\boldsymbol{x}]_{\boldsymbol{E}}, \begin{bmatrix} \boldsymbol{G_1}^T \\ \vdots \\ \boldsymbol{G_n}^T \end{bmatrix} [\boldsymbol{E_1}, \ldots, \boldsymbol{E_n}] \, [\boldsymbol{y}]_{\boldsymbol{E}} \right\rangle \\ &= [\boldsymbol{x}]_{\boldsymbol{E}}^T \, [\boldsymbol{E_1}, \ldots, \boldsymbol{E_n}]^T \, [\boldsymbol{G_1}, \ldots, \boldsymbol{G_n}] \begin{bmatrix} \boldsymbol{G_1}^T \\ \vdots \\ \boldsymbol{G_n}^T \end{bmatrix} [\boldsymbol{E_1}, \ldots, \boldsymbol{E_n}] \, [\boldsymbol{y}]_{\boldsymbol{E}} \\ &= \langle [\boldsymbol{x}]_{\boldsymbol{E}}, [\boldsymbol{y}]_{\boldsymbol{E}} \rangle = \sum_{i=1}^n x_E^i y_E^i \end{aligned}$$

which shows the value is independent of the choice of orthonormal basis. ■

Homework
We use the same problems as the last homework but now we focus on inner products. Let's assume for all these problems, we are working on the interval $[-1, 4]$, so the inner product is $\int_{-1}^4 < F, G > dt$ where F and G come from the vector space in question.

Exercise 4.4.16 *Let* $f(t) = \boldsymbol{E_1} e^{-2t}$ *and* $g(t) = \boldsymbol{E_2} e^{3t}$ *where* $\boldsymbol{E_1} = \begin{bmatrix} 1 \\ 2 \end{bmatrix}$ *and* $\boldsymbol{E_2} = \begin{bmatrix} -3 \\ 4 \end{bmatrix}$ *are two linearly independent vectors in* $\Re^2$. *Use* **GSO** *to find the orthonormal base* $\boldsymbol{F}$ *associated with* $\boldsymbol{E} = \{\boldsymbol{E_1}, \boldsymbol{E_2}\}$.

Let $\phi(t) = 2f(t) + 3g(t)$ *and* $\psi(t) = -4f(t) + 6g(t)$. *Then* $\phi = \begin{bmatrix} 2 \\ 3 \end{bmatrix}$ *and* $\psi = \begin{bmatrix} -4 \\ 6 \end{bmatrix}$ *with respect to the basis for this space determined by* $\boldsymbol{E}$ *which we denote by* $[\phi_{\boldsymbol{E}}]$ *and* $[\psi_{\boldsymbol{E}}]$. *Find the representations of* ψ *and* ϕ *with respect to the basis determined by* $\boldsymbol{F}$. *Then compute* $< [\phi_{\boldsymbol{E}}], [\psi_{\boldsymbol{E}}] >$ *and* $< [\phi_{\boldsymbol{F}}], [\psi_{\boldsymbol{F}}] >$. *These should be the same!*

Exercise 4.4.17 *Let* $f(t) = \boldsymbol{E_1} e^{4t}$ *and* $g(t) = \boldsymbol{E_2} e^{6t}$ *where* $\boldsymbol{E_1} = \begin{bmatrix} 4 \\ 2 \end{bmatrix}$ *and* $\boldsymbol{E_2} = \begin{bmatrix} 1 \\ 8 \end{bmatrix}$ *are two linearly independent vectors in* $\Re^2$. *These two functions form a two dimensional vector space over* $\Re$. *Use* **GSO** *to find the orthonormal base* $\boldsymbol{F}$ *associated with* $\boldsymbol{E} = \{\boldsymbol{E_1}, \boldsymbol{E_2}\}$.

Let $\phi(t) = -3f(t) + 8g(t)$ *and* $\psi(t) = 14f(t) + 9g(t)$. *Then* $\phi = \begin{bmatrix} -3 \\ 8 \end{bmatrix}$ *and* $\psi = \begin{bmatrix} 14 \\ 9 \end{bmatrix}$ *with respect to the basis for this space determined by* $\boldsymbol{E}$ *which we denote by* $[\phi_{\boldsymbol{E}}]$ *and* $[\psi_{\boldsymbol{E}}]$. *Find the representations of* ψ *and* ϕ *with respect to the basis determined by* $\boldsymbol{F}$. *Then compute* $< [\phi_{\boldsymbol{E}}], [\psi_{\boldsymbol{E}}] >$ *and* $< [\phi_{\boldsymbol{F}}], [\psi_{\boldsymbol{F}}] >$.

Exercise 4.4.18 *Let* $f_1(t) = \boldsymbol{E_1}e^{-3t}$, $f_2(t) = \boldsymbol{E_2}e^{-t}$ *and* $f_3(t) = \boldsymbol{E_3}e^{2t}$ *where* $\boldsymbol{E_1} = \begin{bmatrix} 1 \\ 2 \\ 4 \end{bmatrix}$, $\boldsymbol{E_2} = \begin{bmatrix} 2 \\ 1 \\ 5 \end{bmatrix}$ *and* $\boldsymbol{E_3} = \begin{bmatrix} 3 \\ 3 \\ 2 \end{bmatrix}$ *are linearly independent vectors in* $\Re^3$. *Use* **GSO** *to find the orthonormal base* $\boldsymbol{F}$ *associated with* $\boldsymbol{E} = \{\boldsymbol{E_1}, \boldsymbol{E_2}, \boldsymbol{E_3}\}$.

Let $\phi(t) = -3f_1(t) + 8f_2(t) + 6f_3(t)$ *and* $\psi(t) = 14f_1(t) + 9f_2(t) - 4f_3(t)$. *Then* $\phi = \begin{bmatrix} -3 \\ 8 \\ 6 \end{bmatrix}$ *and* $\psi = \begin{bmatrix} 14 \\ 9 \\ 4 \end{bmatrix}$ *with respect to the basis for this space determined by* $\boldsymbol{E}$. *Find the representations of* ψ *and* ϕ *with respect to the basis determined by* $\boldsymbol{F}$. *Then compute* $< [\phi_{\boldsymbol{E}}], [\psi_{\boldsymbol{E}}] >$ *and* $< [\phi_{\boldsymbol{F}}], [\psi_{\boldsymbol{F}}] >$.

Exercise 4.4.19 *Let* $f_1(t) = \boldsymbol{E_1}e^{-3t}$, $f_2(t) = \boldsymbol{E_2}e^{-t}$ *and* $f_3(t) = \boldsymbol{E_3}e^{2t}$ *where* $\boldsymbol{E_1} = \begin{bmatrix} 1 \\ 2 \\ 4 \end{bmatrix}$, $\boldsymbol{E_2} = \begin{bmatrix} 2 \\ 1 \\ 5 \end{bmatrix}$ *and* $\boldsymbol{E_3} = \begin{bmatrix} 3 \\ 3 \\ 2 \end{bmatrix}$ *are linearly independent vectors in* $\Re^3$. *These three functions form a two dimensional vector space over* $\Re$. *Use* **GSO** *to find the orthonormal base* $\boldsymbol{F}$ *associated with* $\boldsymbol{E} = \{\boldsymbol{E_1}, \boldsymbol{E_2}, \boldsymbol{E_3}\}$.

Let $\phi(t) = 2f_1(t) + 5f_2(t) + 11f_3(t)$ *and* $\psi(t) = 4f_1(t) + 3f_2(t) + 18f_3(t)$. *Then* $\phi = \begin{bmatrix} 2 \\ 5 \\ 11 \end{bmatrix}$ *and* $\psi = \begin{bmatrix} 4 \\ 3 \\ 18 \end{bmatrix}$ *with respect to the basis for this space determined by* $\boldsymbol{E}$. *Find the representations of* ψ *and* ϕ *with respect to the basis determined by* $\boldsymbol{F}$. *Then compute* $< [\phi_{\boldsymbol{E}}], [\psi_{\boldsymbol{E}}] >$ *and* $< [\phi_{\boldsymbol{F}}], [\psi_{\boldsymbol{F}}] >$.

We define linear transformations the same really.

Theorem 4.4.5 Linear Transformations between Finite Dimensional Vector Spaces

Let $\boldsymbol{X}$ and $\boldsymbol{Y}$ be two finite dimensional vector spaces over $\Re$. Assume $\boldsymbol{X}$ has dimension n and $\boldsymbol{Y}$ has dimension m. Let $\boldsymbol{T}$ be a linear transformation between the spaces. Given a basis $\boldsymbol{E}$ for $\boldsymbol{X}$ and a basis $\boldsymbol{F}$ for $\boldsymbol{Y}$, we can identify $\Re^n$ with the span of $\boldsymbol{E}$ and $\Re^m$ with the span of $\boldsymbol{F}$. Then there is an $n \times m$ matrix $\boldsymbol{T_{EF}} : \Re^n \rightarrow \Re^m$ so that

$$\left[\boldsymbol{Tx}\right]_{\boldsymbol{F}} = \begin{bmatrix} T_{11} & \dots & T_{1n} \\ \vdots & \vdots & \vdots \\ T_{m1} & \dots & T_{mn} \end{bmatrix}_{\boldsymbol{EF}} \left[\boldsymbol{x}\right]_{\boldsymbol{E}}$$

Note $\boldsymbol{Tx}$ is the same object in $\boldsymbol{Y}$ no matter its representation via a choice of basis $\boldsymbol{F}$ for $\boldsymbol{Y}$.

Proof 4.4.5

Given $\boldsymbol{x}$ in $\boldsymbol{X}$, we can write

$$\boldsymbol{x} = x_1\boldsymbol{E_1} + \ldots + x_n\boldsymbol{E_n}$$

By the linearity of $\boldsymbol{T}$, we then have

$$\boldsymbol{Tx} = x_1\boldsymbol{TE_1} + \ldots + x_n\boldsymbol{TE_n}$$

But each $\boldsymbol{TE_i}$ is in $\boldsymbol{F}$, so we have the expansions

$$\begin{aligned} \boldsymbol{TE_1} &= \beta_{11}\boldsymbol{F_1} + \ldots + \beta_{m1}\boldsymbol{F_m} \\ \boldsymbol{TE_2} &= \beta_{12}\boldsymbol{F_1} + \ldots + \beta_{m2}\boldsymbol{F_m} \\ \vdots &= \vdots \\ \boldsymbol{TE_n} &= \beta_{1n}\boldsymbol{F_1} + \ldots + \beta_{mn}\boldsymbol{F_m} \end{aligned}$$

which implies since $\boldsymbol{Tx}$ is also in $\boldsymbol{Y}$ that

$$\boldsymbol{Tx} = x_1\left(\sum_{j=1}^{m} \beta_{j1}\boldsymbol{F_j}\right) + \ldots + x_n\left(\sum_{j=1}^{m} \beta_{jn}\boldsymbol{F_j}\right) = \sum_{j=1}^{m} \gamma_j\boldsymbol{F_j}$$

Now reorganize these sums to obtain

$$\sum_{j=1}^{m}\left(\sum_{i=1}^{n} \beta_{ji}x_i\right)\boldsymbol{F_j} = \sum_{j=1}^{m} \gamma_j\boldsymbol{F_j}$$

But expansions with respect to the $\boldsymbol{F}$ basis are unique, so we have

$$\begin{bmatrix} \gamma_1 \\ \vdots \\ \gamma_m \end{bmatrix}_{\boldsymbol{F}} = \begin{bmatrix} \beta_{11} & \beta_{12} & \beta_{13} & \dots & \beta_{1n} \\ \beta_{21} & \beta_{22} & \beta_{23} & \dots & \beta_{2n} \\ \vdots & \vdots & \vdots & \vdots & \vdots \\ \beta_{m1} & \beta_{m2} & \beta_{m3} & \dots & \beta_{mn} \end{bmatrix}_{\boldsymbol{EF}} \begin{bmatrix} x_1 \\ \vdots \\ x_n \end{bmatrix}_{\boldsymbol{E}}$$

Let $T_{ij} = \beta_{ij}$ and we have found the matrix $\boldsymbol{T}$ so that

$$\left[\boldsymbol{Tx}\right]_{\boldsymbol{F}} = \boldsymbol{T_{EF}}\left[\boldsymbol{x}\right]_{\boldsymbol{E}}$$

■

The change of a basis determines an equivalence relation on the set of all matrices.

Theorem 4.4.6 A Linear Transformation between Two Finite Dimensional Vector Spaces Determines an Equivalence Class of Matrices

Let $\boldsymbol{X}$ and $\boldsymbol{Y}$ be two finite dimensional vector spaces over $\Re$. Assume $\boldsymbol{X}$ has dimension n and $\boldsymbol{Y}$ has dimension m. Let $\boldsymbol{T}$ be a linear transformation between the spaces. Then $\boldsymbol{T}$ is associated with an equivalence class in the set of $m \times n$ matrices under the equivalence relation $\sim$ where $\boldsymbol{A} \sim \boldsymbol{B}$ if and only if there are invertible matrices $\boldsymbol{P}$ and $\boldsymbol{Q}$ so that $\boldsymbol{A} = \boldsymbol{P}^{-1}\boldsymbol{B}\boldsymbol{Q}$.

Proof 4.4.6

First, it is easy to see $\sim$ defines an equivalence relation on the set of all $m \times n$ matrices. We leave that to you. Any such linear transformation $\boldsymbol{T}$ is associated with an $m \times n$ matrix once the bases $\boldsymbol{E}$ and $\boldsymbol{F}$ are chosen. What happens if we change to a new basis $\boldsymbol{G}$ for $\boldsymbol{X}$ and a new basis $\boldsymbol{H}$ for $\boldsymbol{Y}$? We know from earlier calculations that there are invertible matrices $\boldsymbol{A_{EG}}$ and $\boldsymbol{B_{FH}}$ so that

$$\left[\boldsymbol{x}\right]_{\boldsymbol{E}} = \boldsymbol{A_{GE}} \left[\boldsymbol{x}\right]_{\boldsymbol{G}}, \quad \left[\boldsymbol{y}\right]_{\boldsymbol{F}} = \boldsymbol{B_{HF}} \left[\boldsymbol{y}\right]_{\boldsymbol{H}}$$

Using all this, we have

$$\left[\boldsymbol{Tx}\right]_{\boldsymbol{F}} = \boldsymbol{T_{EF}} \left[\boldsymbol{x}\right]_{\boldsymbol{E}} \Longrightarrow \boldsymbol{B_{HF}} \left[\boldsymbol{Tx}\right]_{\boldsymbol{H}} = \boldsymbol{T_{EF}}\, \boldsymbol{A_{GE}} \left[\boldsymbol{x}\right]_{\boldsymbol{G}}$$

Since, $\boldsymbol{B_{HF}}$ is invertible, this tells us

$$\left[\boldsymbol{Tx}\right]_{\boldsymbol{H}} = \boldsymbol{B_{HF}}^{-1}\, \boldsymbol{T_{EF}}\, \boldsymbol{A_{GE}} \left[\boldsymbol{x}\right]_{\boldsymbol{G}}$$

We also know $\left[\boldsymbol{Tx}\right]_{\boldsymbol{H}} = \boldsymbol{T_{GH}} \left[\boldsymbol{x}\right]_{\boldsymbol{G}}$ and thus we must have $\boldsymbol{T_{GH}} = \boldsymbol{B_{HF}}^{-1}\, \boldsymbol{T_{EF}}\, \boldsymbol{A_{GE}}$. This tells us $\boldsymbol{T_{GH}} \sim \boldsymbol{T_{EF}}$. ■

Comment 4.4.4 *Thus any linear transformation between two finite dimensional vector spaces is identified with an equivalence class of $m \times n$ matrices. When we pick basis $\boldsymbol{E}$ and $\boldsymbol{F}$ for $\boldsymbol{X}$ and $\boldsymbol{Y}$ respectively, we choose a particular representative from this equivalence class we denote by $\boldsymbol{T_{EF}}$. Note we can say more about the structure of $\boldsymbol{T_{EF}}$ if these bases are orthonormal.*

Homework

Exercise 4.4.20 *Prove $\sim$ is an equivalence relation.*

For the rest of these problems, let's assume we are working on the interval $[-1, 4]$ so that the inner product here is $\int_{-1}^{4} < F, G > dt$ where F and G come from the vector space in question.

Exercise 4.4.21 *Let $f_1(t) = \boldsymbol{F_1} e^{-2t}$ and $f_2(t) = \boldsymbol{F_2} e^{3t}$ where $\boldsymbol{F_1} = \begin{bmatrix} 1 \\ 2 \end{bmatrix}$ and $\boldsymbol{F_2} = \begin{bmatrix} -3 \\ 4 \end{bmatrix}$ are two linearly independent vectors in $\Re^2$. Use* **GSO** *to find the orthonormal base $\boldsymbol{U}$ associated with $\boldsymbol{F} = \{\boldsymbol{F_1}, \boldsymbol{F_2}\}$. Let this vector space be X and note we have found two bases for X.*

Let $g_1(t) = \boldsymbol{G_1} e^{-2t}$ and $g_2(t) = \boldsymbol{G_2} e^{3t}$ where $\boldsymbol{G_1} = \begin{bmatrix} 1 \\ 2 \end{bmatrix}$ and $\boldsymbol{G_2} = \begin{bmatrix} -3 \\ 4 \end{bmatrix}$ are two linearly independent vectors in $\Re^2$. Use **GSO** *to find the orthonormal base $\boldsymbol{V}$ associated with $\boldsymbol{G} = \{\boldsymbol{G_1}, \boldsymbol{G_2}\}$. Let this vector space be Y and note we have found two bases for Y.*

Define $M_{FG} : X \rightarrow Y$ *by* $M_{FG} = \begin{bmatrix} 2 & 3 \\ -3 & 5 \end{bmatrix}$. *Find* M_{UV}.

Exercise 4.4.22 *Let* $f_1(t) = \boldsymbol{F_1} e^{-3t}$, $f_2(t) = \boldsymbol{F_2} e^{-t}$ *and* $f_3(t) = \boldsymbol{F_3} e^{2t}$ *where* $\boldsymbol{F_1} = \begin{bmatrix} 1 \\ 2 \\ 4 \end{bmatrix}$, $\boldsymbol{F_2} = \begin{bmatrix} 2 \\ 1 \\ 5 \end{bmatrix}$ *and* $\boldsymbol{F_3} = \begin{bmatrix} 3 \\ 3 \\ 2 \end{bmatrix}$ *are linearly independent vectors in* $\Re^3$. *Use* **GSO** *to find the orthonormal base* $\boldsymbol{U}$ *associated with* $\boldsymbol{F} = \{\boldsymbol{F_1}, \boldsymbol{F_2}, \boldsymbol{F_3}\}$. *Let this vector space be* X *and note we have found two bases for* X.

Let $g_1(t) = \boldsymbol{G_1} e^{-3t}$, $g_2(t) = \boldsymbol{G_2} e^{-t}$ *and* $g3(t) = \boldsymbol{G_3} e^{2t}$ *where* $\boldsymbol{G_1} = \begin{bmatrix} 1 \\ 2 \\ 4 \end{bmatrix}$, $\boldsymbol{G_2} = \begin{bmatrix} 2 \\ 1 \\ 5 \end{bmatrix}$ *and* $\boldsymbol{G_3} = \begin{bmatrix} 3 \\ 3 \\ 2 \end{bmatrix}$ *are linearly independent vectors in* $\Re^3$. *Use* **GSO** *to find the orthonormal base* $\boldsymbol{V}$ *associated with* $\boldsymbol{G} = \{\boldsymbol{G_1}, \boldsymbol{G_2}, \boldsymbol{G_3}\}$. *Let this vector space be* Y *and note we have found two bases for* Y.

Define $M_{FG} : X \rightarrow Y$ *by* $M_{FG} = \begin{bmatrix} 2 & 4 \\ 5 & 3 \end{bmatrix}$. *Find* M_{UV}.

4.5 Magnitudes of Linear Transformations

For the linear transformation $\boldsymbol{T}$ mapping the n dimensional vector space $\boldsymbol{X}$ to the m dimensional vector space $\boldsymbol{Y}$, we know given a basis $\boldsymbol{E}$ for $\boldsymbol{X}$ and a basis $\boldsymbol{F}$ for $\boldsymbol{Y}$ $[\boldsymbol{Tx}]_{\boldsymbol{F}} = [\boldsymbol{T}]_{\boldsymbol{EF}} [\boldsymbol{x}]_{\boldsymbol{E}}$. Let's add a norm to the vector spaces $\Re^n$ and $\Re^m$. For now, let's assume we now use the normed linear spaces $(\Re^m, \|\cdot\|_2)$ and $(\Re^n, \|\cdot\|_2)$. Of course, we could use a different norm on the domain space and the range space, but we won't do that as it adds unnecessary confusion. Let the entries of $[\boldsymbol{T}]_{\boldsymbol{EF}}$ be called T_{ij} where we will not label these entries with an $\boldsymbol{EF}$ in order to avoid too much clutter. But of course, they **do** depend on $\boldsymbol{E}$ and $\boldsymbol{F}$; just keep that in the back of your mind. Note

$$\| [\boldsymbol{Tx}]_{\boldsymbol{F}} \|_2^2 = \sum_{j=1}^{m} (([\boldsymbol{Tx}]_{\boldsymbol{F}})_j)^2 = \sum_{i=1}^{m} \left(\sum_{j=1}^{n} T_{ij} x_j \right)$$

Apply the Cauchy - Schwartz Inequality here:

$$\left(\sum_{j=1}^{n} T_{ij} x_j \right)^2 \leq \left(\sum_{j=1}^{n} T_{ij}^2 \right) \left(\sum_{j=1}^{n} x_j^2 \right) = \left(\sum_{j=1}^{n} T_{ij}^2 \right) \|\boldsymbol{x_E}\|_2^2$$

Thus,

$$\| [\boldsymbol{Tx}]_{\boldsymbol{F}} \|_2^2 \leq \left(\sum_{i=1}^{m} \sum_{j=1}^{n} T_{ij}^2 \right) \|\boldsymbol{x_E}\|_2^2$$

which implies

$$\| \left[\boldsymbol{T}\boldsymbol{x} \right]_{\boldsymbol{F}} \|_2 \;\leq\; \sqrt{\sum_{i=1}^{m} \sum_{j=1}^{n} T_{ij}^2} \; \|\boldsymbol{x}_{\boldsymbol{E}}\|_2$$

This leads to the following definition:

Definition 4.5.1 The Frobenius Norm of a Linear Transformation

For the linear transformation $\boldsymbol{T}$ mapping the n dimensional vector space $\boldsymbol{X}$ to the m dimensional vector space $\boldsymbol{Y}$, we know given a basis $\boldsymbol{E}$ for $\boldsymbol{X}$ and a basis $\boldsymbol{F}$ for $\boldsymbol{Y}$ $\left[\boldsymbol{T}\boldsymbol{x}\right]_{\boldsymbol{F}} = \left[\boldsymbol{T}\right]_{\boldsymbol{EF}} \left[\boldsymbol{x}\right]_{\boldsymbol{E}}$. The **Frobenius Norm** *of $\boldsymbol{T}$ is denoted $\| \left[\boldsymbol{T}\right]_{\boldsymbol{EF}} \|_{\boldsymbol{Fr}}$ and is defined by*

$$\| \left[\boldsymbol{T}\right]_{\boldsymbol{EF}} \|_{\boldsymbol{Fr}} \;=\; \sqrt{\sum_{i=1}^{m} \sum_{j=1}^{n} T_{ij}^2}$$

where T_{ij} are the entries in the matrix $\left[\boldsymbol{T}\right]_{\boldsymbol{EF}}$.

We have just proven the following result:

Theorem 4.5.1 The Frobenius Norm Fundamental Inequality

For the linear transformation $\boldsymbol{T}$ mapping the n dimensional vector space $\boldsymbol{X}$ to the m dimensional vector space $\boldsymbol{Y}$, we know given a basis $\boldsymbol{E}$ for $\boldsymbol{X}$ and a basis $\boldsymbol{F}$ for $\boldsymbol{Y}$ $\left[\boldsymbol{T}\boldsymbol{x}\right]_{\boldsymbol{F}} = \left[\boldsymbol{T}\right]_{\boldsymbol{EF}} \left[\boldsymbol{x}\right]_{\boldsymbol{E}}$. The Frobenius Norm of this matrix satisfies

$$\| \left[\boldsymbol{T}\boldsymbol{x}\right]_{\boldsymbol{F}} \|_2 \;\leq\; \| \left[\boldsymbol{T}\right]_{\boldsymbol{EF}} \|_{\boldsymbol{Fr}} \; \|[\boldsymbol{x}_{\boldsymbol{E}}]\|_2$$

Proof 4.5.1
We have just gone over this argument. ■

4.5.0.1 Homework

Exercise 4.5.1 *Let* $\boldsymbol{A} = \begin{bmatrix} 4 & 3 \\ 2 & 1 \end{bmatrix}$. *Find* $\|\boldsymbol{A}\|_{\boldsymbol{Fr}}$.

Exercise 4.5.2 *Let* $\boldsymbol{A} = \begin{bmatrix} 9 & 1 & 8 \\ 7 & 6 & 5 \\ 4 & 3 & -1 \end{bmatrix}$. *Find* $\|\boldsymbol{A}\|_{\boldsymbol{Fr}}$.

Exercise 4.5.3 *Let $f_1(t) = \boldsymbol{F_1}e^{-2t}$ and $f_2(t) = \boldsymbol{F_2}e^{3t}$ where $\boldsymbol{F_1} = \begin{bmatrix} 1 \\ 2 \end{bmatrix}$ and $\boldsymbol{F_2} = \begin{bmatrix} -3 \\ 4 \end{bmatrix}$ are two linearly independent vectors in $\Re^2$. Use* **GSO** *to find the orthonormal base $\boldsymbol{U}$ associated with $\boldsymbol{F} = \{\boldsymbol{F_1}, \boldsymbol{F_2}\}$. Let this vector space be X and note we have found two bases for X.*

Let $g_1(t) = \boldsymbol{G_1}e^{-2t}$ and $g_2(t) = \boldsymbol{G_2}e^{3t}$ where $\boldsymbol{G_1} = \begin{bmatrix} 1 \\ 2 \end{bmatrix}$ and $\boldsymbol{G_2} = \begin{bmatrix} -3 \\ 4 \end{bmatrix}$ are two linearly independent vectors in $\Re^2$. Use **GSO** *to find the orthonormal base $\boldsymbol{V}$ associated with $\boldsymbol{G} = \{\boldsymbol{G_1}, \boldsymbol{G_2}\}$. Let this vector space be Y and note we have found two bases for Y.*

Define $M_{FG} : X \rightarrow Y$ *by* $M_{FG} = \begin{bmatrix} 2 & 3 \\ -3 & 5 \end{bmatrix}$. *Find* M_{UV}.

Verify

$$\|[M_{FG}\boldsymbol{x}]_{\boldsymbol{G}}\|_2 \;\leq\; \|[M_{FG}]\|_{\boldsymbol{Fr}} \;\|[\boldsymbol{x}_{\boldsymbol{F}}]\|_2$$

Verify

$$\|[M_{UV}\boldsymbol{x}]_{\boldsymbol{V}}\|_2 \;\leq\; \|[M_{UV}]\|_{\boldsymbol{Fr}} \;\|[\boldsymbol{x}_{\boldsymbol{U}}]\|_2$$

Exercise 4.5.4 *Let* $f_1(t) = \boldsymbol{F_1}e^{-3t}$, $f_2(t) = \boldsymbol{F_2}e^{-t}$ *and* $f_3(t) = \boldsymbol{F_3}e^{2t}$ *where* $\boldsymbol{F_1} = \begin{bmatrix} 1 \\ 2 \\ 4 \end{bmatrix}$, $\boldsymbol{F_2} = \begin{bmatrix} 2 \\ 1 \\ 5 \end{bmatrix}$ *and* $\boldsymbol{F_3} = \begin{bmatrix} 3 \\ 3 \\ 2 \end{bmatrix}$ *are linearly independent vectors in* $\Re^3$. *Use* **GSO** *to find the orthonormal base* $\boldsymbol{U}$ *associated with* $\boldsymbol{F} = \{\boldsymbol{F_1}, \boldsymbol{F_2}, \boldsymbol{F_3}\}$. *Let this vector space be* X *and note we have found two basis for* X.

Let $g_1(t) = \boldsymbol{G_1}e^{-3t}$, $g_2(t) = \boldsymbol{G_2}e^{-t}$ *and* $g3(t) = \boldsymbol{G_3}e^{2t}$ *where* $\boldsymbol{G_1} = \begin{bmatrix} 1 \\ 2 \\ 4 \end{bmatrix}$, $\boldsymbol{G_2} = \begin{bmatrix} 2 \\ 1 \\ 5 \end{bmatrix}$ *and* $\boldsymbol{G_3} = \begin{bmatrix} 3 \\ 3 \\ 2 \end{bmatrix}$ *are linearly independent vectors in* $\Re^3$. *Use* **GSO** *to find the orthonormal base* $\boldsymbol{V}$ *associated with* $\boldsymbol{G} = \{\boldsymbol{G_1}, \boldsymbol{G_2}, \boldsymbol{G_3}\}$. *Let this vector space be* Y *and note we have found two bases for* Y.

Define $M_{FG} : X \rightarrow Y$ *by* $M_{FG} = \begin{bmatrix} 2 & 4 \\ 5 & 3 \end{bmatrix}$. *Find* M_{UV}.

Verify

$$\|[M_{FG}\boldsymbol{x}]_{\boldsymbol{G}}\|_2 \;\leq\; \|[M_{FG}]\|_{\boldsymbol{Fr}} \;\|[\boldsymbol{x}_{\boldsymbol{F}}]\|_2$$

Verify

$$\|[M_{UV}\boldsymbol{x}]_{\boldsymbol{V}}\|_2 \;\leq\; \|[M_{UV}]\|_{\boldsymbol{Fr}} \;\|[\boldsymbol{x}_{\boldsymbol{U}}]\|_2$$

Chapter 5

Symmetric Matrices

Let's specialize to symmetric matrices now as they are of great interest to us in many applications in applied analysis.

5.1 The General Two by Two Symmetric Matrix

Let's start with a general 2×2 symmetric matrix $\boldsymbol{A}$ given by

$$\boldsymbol{A} = \begin{bmatrix} a & b \\ b & d \end{bmatrix}$$

where a, b and d are arbitrary nonzero numbers. The characteristic equation here is $r^2 - (a+d)r + ad - b^2$. Note that the term $ad - b^2$ is the determinant of $\boldsymbol{A}$. The roots are given by

$$\begin{aligned} r &= \frac{(a+d) \pm \sqrt{(a+d)^2 - 4(ad - b^2)}}{2} \\ &= \frac{(a+d) \pm \sqrt{a^2 + 2ad + d^2 - 4ad + 4b^2}}{2} \\ &= \frac{(a+d) \pm \sqrt{a^2 - 2ad + d^2 + 4b^2}}{2} \\ &= \frac{(a+d) \pm \sqrt{(a-d)^2 + 4b^2}}{2} \end{aligned}$$

It is easy to see the term in the square root here is always positive and so we have two real roots. Hence, for a general symmetric 2×2 matrix, the eigenvalues are always real. Note we can find the eigenvectors with a standard calculation. For eigenvalue $\lambda_1 = \frac{(a+d)+\sqrt{(a-d)^2+4b^2}}{2}$, we must find the vectors $\boldsymbol{V}$ so that

$$\begin{bmatrix} \frac{(a+d)+\sqrt{(a-d)^2+4b^2}}{2} - a & -b \\ -b & \frac{(a+d)+\sqrt{(a-d)^2+4b^2}}{2} - d \end{bmatrix} \begin{bmatrix} V_1 \\ V_2 \end{bmatrix} = \begin{bmatrix} 0 \\ 0 \end{bmatrix}$$

We can use the top equation to find the needed relationship between V_1 and V_2. We have

$$\left(\frac{(a+d) + \sqrt{(a-d)^2 + 4b^2}}{2} - a \right) V_1 - bV_2 = 0.$$

Thus, we have for $V_2 = \frac{d-a+\sqrt{(a-d)^2+4b^2}}{2}$, $V_1 = b$. Thus, the first eigenvector is

$$\boldsymbol{E_1} = \begin{bmatrix} b \\ \frac{d-a+\sqrt{(a-d)^2+4b^2}}{2} \end{bmatrix}$$

The second eigenvector is a similar calculation. We must find the vector $\boldsymbol{V}$ so that

$$\begin{bmatrix} \frac{(a+d)-\sqrt{(a-d)^2+4b^2}}{2} - a & -b \\ -b & \frac{(a+d)-\sqrt{(a-d)^2+4b^2}}{2} - d \end{bmatrix} \begin{bmatrix} V_1 \\ V_2 \end{bmatrix} = \begin{bmatrix} 0 \\ 0 \end{bmatrix}$$

We find

$$\left(\frac{(a+d) - \sqrt{(a-d)^2 + 4b^2}}{2} - a \right) V_1 - bV_2 = 0.$$

Thus, we have for $V_2 = d + \frac{(a+d)-\sqrt{(a-d)^2+4b^2}}{2}$, $V_1 = b$. Thus, the second eigenvector is

$$\boldsymbol{E_2} = \begin{bmatrix} b \\ \frac{d-a-\sqrt{(a-d)^2+4b^2}}{2} \end{bmatrix}$$

Note that $< \boldsymbol{E_1}, \boldsymbol{E_2} >$ is

$$\begin{aligned} < \boldsymbol{E_1}, \boldsymbol{E_2} > &= b^2 + \left(\frac{d-a+\sqrt{(a-d)^2+4b^2}}{2} \right) \left(\frac{d-a-\sqrt{(a-d)^2+4b^2}}{2} \right) \\ &= b^2 + \frac{(d-a)^2}{4} - \frac{(a-d)^2+4b^2}{4} b^2 + \frac{(d-a)^2-(a-d)^2}{4} - b^2 = 0. \end{aligned}$$

Hence, these two eigenvectors are **orthogonal** to each other. Note, the two eigenvalues are

$$\begin{aligned} \lambda_1 &= \frac{(a+d)+\sqrt{(a-d)^2+4b^2}}{2} \\ \lambda_2 &= \frac{(a+d)-\sqrt{(a-d)^2+4b^2}}{2} \end{aligned}$$

5.1.1 Examples

The only way both eigenvalues can be zero is if both $a + d = 0$ and $4(a + d)^2 + 4b^2 = 0$. That only happens if $a = b = d = 0$ which we explicitly ruled out at the beginning of our discussion because we said a, b and d were nonzero. However, both eigenvalues can be negative, both can be positive or they can be of mixed sign as our examples show.

Example 5.1.1 *For*

$$\boldsymbol{A} = \begin{bmatrix} 3 & 2 \\ 2 & 6 \end{bmatrix}$$

the eigenvalues are $(9 \pm 5)/2$ *and both are positive.*

Example 5.1.2 *For*

$$\boldsymbol{A} = \begin{bmatrix} 3 & 5 \\ 5 & 6 \end{bmatrix}$$

the eigenvalues are $(9 \pm \sqrt{9+100})/2$ *giving* $\lambda_1 = 9.72$ *and* $\lambda_2 = -5.22$ *and the eigenvalues have mixed sign.*

Example 5.1.3 *For*

$$A = \begin{bmatrix} 3 & 3\sqrt{2} \\ 3\sqrt{2} & 6 \end{bmatrix}$$

the eigenvalues are $(9 \pm 9)/2$ *giving* $\lambda_1 = 9$ *and* $\lambda_2 = 0$.

Example 5.1.4 *For*

$$A = \begin{bmatrix} 3 & 5 \\ 5 & 6 \end{bmatrix}$$

the eigenvalues are $(9 \pm \sqrt{9+100})/2$ *giving* $\lambda_1 = 9.72$ *and* $\lambda_2 = -5.22$ *and the eigenvalues have mixed sign.*

Example 5.1.5 *For*

$$A = \begin{bmatrix} -3 & 5 \\ 5 & -6 \end{bmatrix}$$

the eigenvalues are $(-9 \pm \sqrt{34})/2$ *giving* $\lambda_1 = -7.42$ *and* $\lambda_2 = -1.58$. *So here, both eigenvalues are negative.*

However, in all cases the eigenvectors are still orthogonal.

5.1.1.1 Homework

Exercise 5.1.1 *Find the eigenvalues and eigenvectors for* $A = \begin{bmatrix} 6 & 7 \\ 7 & 5 \end{bmatrix}$ *both by hand and using MATLAB. Show the eigenvectors are orthogonal.*

Exercise 5.1.2 *Find the eigenvalues and eigenvectors for* $A = \begin{bmatrix} -6 & 8 \\ -6 & 2 \end{bmatrix}$ *both by hand and using MATLAB. Show the eigenvectors are orthogonal.*

Exercise 5.1.3 *Find the eigenvalues and eigenvectors for* $A = \begin{bmatrix} 9 & 10 \\ 10 & -2 \end{bmatrix}$ *both by hand and using MATLAB. Show the eigenvectors are orthogonal.*

Exercise 5.1.4 *For each of the exercises above, imagine that the matrix* A *is the Hessian of a surface* $z = f(x, y)$ *at some critical point* (x_0, y_0). *Classify the extreme values using the standard second order two variable conditions we discussed in (Peterson (21) 2020).*

Exercise 5.1.5 *For each of the exercises above, form the matrix* $P = \begin{bmatrix} E_1 & E_2 \end{bmatrix}$ *where* $E = \{E_1, E_2\}$ *is the orthonormal basis obtained from the eigenvectors of* A. *Compute* $P^T A P$ *and note the result is a diagonal matrix. What are the entries on the diagonal?*

5.1.2 A Canonical Form

Now let's look at 2×2 symmetric matrices more abstractly. Don't worry, there is a payoff here in understanding! Let A be a general 2×2 symmetric matrix. Then it has two distinct eigenvalues λ_1 and another one λ_2. Consider the matrix P given by

$$P = \begin{bmatrix} E_1 & E_2 \end{bmatrix}$$

whose transpose is then

$$P^T = \begin{bmatrix} E_1^T \\ E_2^T \end{bmatrix}$$

It is easy to find that $P^T\, P = P\, P^T = I$. Hence, $P^{-1} = P^T$. Thus,

$$P^T\, A\, P = \begin{bmatrix} E_1^T \\ E_2^T \end{bmatrix} A \begin{bmatrix} E_1 & E_2 \end{bmatrix} = \begin{bmatrix} E_1^T \\ E_2^T \end{bmatrix} \begin{bmatrix} AE_1 & AE_2 \end{bmatrix} = \begin{bmatrix} E_1^T \\ E_2^T \end{bmatrix} \begin{bmatrix} \lambda_1 E_1 & \lambda_2 E_2 \end{bmatrix}$$

After we do the final multiplications, we have

$$P^T\, A\, P = \begin{bmatrix} \lambda_1 < E_1, E_1 > & \lambda_2 < E_1, E_2 > \\ \lambda_1 < E_2, E_1 > & \lambda_2 < E_2, E_2 > \end{bmatrix}$$

We know the eigenvectors are orthogonal, so we must have

$$P^T\, A\, P = \begin{bmatrix} \lambda_1 < E_1, E_1 > & 0 \\ 0 & \lambda_2 < E_2, E_2 > \end{bmatrix}$$

One last step and we are done! There is no reason, we can't choose as our eigenvectors, vectors of length one: here just replace E_1 by the new vector $E_1/||E_1||$ where $||E_1||$ is the usual Euclidean length of the vector. Similarly, replace E_2 by $E_2/||E_2||$. Assuming this is done, we have $< E_1, E_1 >= 1$ and $< E_2, E_2 >= 1$. We are left with the identity

$$P^T\, A\, P = \begin{bmatrix} \lambda_1 & 0 \\ 0 & \lambda_2 \end{bmatrix}$$

which can be rewritten as

$$A = P \begin{bmatrix} \lambda_1 & 0 \\ 0 & \lambda_2 \end{bmatrix} P^T$$

This is an important thing. We have shown the 2×2 matrix A can be decomposed into the product $A = P\, \Lambda\, P^T$ where Λ is the diagonal matrix whose entries are the eigenvalues of A which the most positive one in the $(1, 1)$ position.

It is now clear how we solve an equation like $A\, X = b$. We rewrite as $P\, \Lambda\, P^T X = b$ which leads to the solution

$$X = P\, \Lambda^{-1}\, P^T\, b$$

and we see the reciprocal eigenvalue sizes determine how large the solution can get. Another way to look at this is that the two eigenvectors can be used to find a representation of the data vector b and the solution vector X as follows:

$$\begin{aligned} b &= < b, E_1 > E_1 + < b, E_2 > E_2 = b_1 E_1 + b_2 E_2 \\ X &= < X, E_1 > E_1 + < X, E_2 > E_2 = X_1 E_1 + X_2 E_2 \end{aligned}$$

and so $\boldsymbol{A}\,\boldsymbol{X} = \boldsymbol{b}$ becomes

$$\begin{aligned} \boldsymbol{A}\Big(X_1\boldsymbol{E_1} + X_2\boldsymbol{E_2}\Big) &= b_1\boldsymbol{E_1} + b_2\boldsymbol{E_2} \\ \lambda_1 X_1\boldsymbol{E_1} + \lambda_2 X_2\boldsymbol{E_2} &= b_1\boldsymbol{E_1} + b_2\boldsymbol{E_2}. \end{aligned}$$

The only way this equation works is if the coefficients on the eigenvectors match. So we have $X_1 = \lambda_1^{-1} b_1$ and $X_2 = \lambda_2^{-1} b_2$. This shows very clearly how the solution depends on the size of the reciprocal eigenvalues. Thus, if our problem has a very small eigenvalue, we would expect our solution vector to be unstable. Also, if one of the eigenvalues is 0, we would have real problems! We can address this somewhat by finding a way to force all the eigenvalues to be positive.

The eigenvectors here are **independent vectors** in $\Re^2$ and since they **span** $\Re^2$, they form a **basis** which is an **orthonormal basis** because the vectors are orthogonal. Hence, any vector $\boldsymbol{V}$ in $\Re^2$ can be written as

$$\boldsymbol{V} = V_1\boldsymbol{E_1} + V_2\boldsymbol{E_2}$$

and the components V_1 and V_2 are known as the components of $\boldsymbol{V}$ relative to the basis $\{\boldsymbol{E_1}, \boldsymbol{E_2}\}$. We often just refer to this basis as $\boldsymbol{E}$. Hence, a vector $\boldsymbol{V}$ has many possible representations. To refresh your mind here, the one you are most used to is the one which uses the basis vectors

$$\boldsymbol{e_1} = \begin{bmatrix} 1 \\ 0 \end{bmatrix}, \text{ and } \boldsymbol{e_1} = \begin{bmatrix} 1 \\ 0 \end{bmatrix}$$

which is called the **standard basis**. When we write

$$\boldsymbol{V} = \begin{bmatrix} 3 \\ 5 \end{bmatrix}$$

unless it is otherwise stated, we assume these are the components of $\boldsymbol{V}$ with respect to the standard basis. Now let's go back to our general vector $\boldsymbol{V}$. Since the vectors $\boldsymbol{E_1}$ and $\boldsymbol{E_2}$ are orthogonal, we can take inner products on both sides of the representation of $\boldsymbol{V}$ with respect to the basis $\boldsymbol{E}$ to get

$$< \boldsymbol{V}, \boldsymbol{E_1} > \;=\; < V_1\boldsymbol{E_1}, \boldsymbol{E_1} > + < V_2\boldsymbol{E_2}, \boldsymbol{E_1} >= V_1 < \boldsymbol{E_1}, \boldsymbol{E_1} > + V_2 < \boldsymbol{E_2}, \boldsymbol{E_1} >$$

But $< \boldsymbol{E_2}, \boldsymbol{E_1} >= 0$ as the vectors are perpendicular and $< \boldsymbol{E_2}, \boldsymbol{E_1} >= E_{11}^2 + E_{12}^2$ where $E_1 1$ and E_{12} are the components of $\boldsymbol{E_1}$ in the standard basis. This is one though as we have chosen our eigenvectors to have length one. So we have $V_1 =< \boldsymbol{V}, \boldsymbol{E_1} >$. A similar calculation with inner products gives $V_2 =< \boldsymbol{V}, \boldsymbol{E_2} >$. So we could also have written

$$\boldsymbol{V} = < \boldsymbol{V}, \boldsymbol{E_1} > \boldsymbol{E_1} + < \boldsymbol{V}, \boldsymbol{E_2} > \boldsymbol{E_2}$$

From our discussions here, it should be easy to see that while we can do all of the needed calculations in this 2×2 case, we would have a lot of trouble if the symmetric matrix was 3×3, 4×4 or larger. Hence, let's explore another way. And yes it is more abstract and yes, you will probably be lying belly up on the floor with your legs and arms pointing straight up soon enough screaming "enough"! But if you keep at it, you'll see that when computation becomes too hard, the abstract approach is quite nice!

5.1.2.1 Homework

Exercise 5.1.6 *Find the eigenvalues and eigenvectors* $\boldsymbol{E} = \{\boldsymbol{E_1}, \boldsymbol{E_2}\}$ *for* $\boldsymbol{A} = \begin{bmatrix} -3 & 1 \\ 1 & -3 \end{bmatrix}$ *both by hand and using MATLAB. Let* $\boldsymbol{P} = \begin{bmatrix} \boldsymbol{E_1} & \boldsymbol{E_2} \end{bmatrix}$. *Let* $\boldsymbol{x} = \begin{bmatrix} 4 \\ 5 \end{bmatrix}$ *in the standard basis and convert to* $[\boldsymbol{x}]_{\boldsymbol{E}}$.

Exercise 5.1.7 *Find the eigenvalues and eigenvectors* $\boldsymbol{E} = \{\boldsymbol{E_1}, \boldsymbol{E_2}\}$ *for* $\boldsymbol{A} = \begin{bmatrix} 4 & 9 \\ 9 & 4 \end{bmatrix}$ *both by hand and using MATLAB. Let* $\boldsymbol{P} = \begin{bmatrix} \boldsymbol{E_1} & \boldsymbol{E_2} \end{bmatrix}$. *Let* $\boldsymbol{x} = \begin{bmatrix} -8 \\ 15 \end{bmatrix}$ *in the standard basis and convert to* $[\boldsymbol{x}]_{\boldsymbol{E}}$.

Exercise 5.1.8 *Find the eigenvalues and eigenvectors* $\boldsymbol{E} = \{\boldsymbol{E_1}, \boldsymbol{E_2}\}$ *for* $\boldsymbol{A} = \begin{bmatrix} -1 & 1 \\ 1 & -1 \end{bmatrix}$ *both by hand and using MATLAB. Let* $\boldsymbol{P} = \begin{bmatrix} \boldsymbol{E_1} & \boldsymbol{E_2} \end{bmatrix}$. *Let* $\boldsymbol{x} = \begin{bmatrix} 3 \\ -55 \end{bmatrix}$ *in the standard basis and convert to* $[\boldsymbol{x}]_{\boldsymbol{E}}$.

Exercise 5.1.9 *For each of the exercises above, imagine that the matrix* $\boldsymbol{A}$ *is the Hessian of a surface* $z = f(x, y)$ *at some critical point* (x_0, y_0). *Classify the extreme values using the standard second order two variable conditions we discussed in (Peterson (21) 2020).*

Exercise 5.1.10 *For each of the exercises above, as usual form the matrix* $\boldsymbol{P} = \begin{bmatrix} \boldsymbol{E_1} & \boldsymbol{E_2} \end{bmatrix}$ *and compute* $\boldsymbol{P^T A P} = \boldsymbol{D}$, *where* $\boldsymbol{D}$ *is diagonal matrix. Compare* $\boldsymbol{A}$ *and* $\boldsymbol{PDP^T}$.

5.1.3 Two Dimensional Rotations

Let's look closer at the matrices we have found using the eigenvectors of the symmetric matrix $\boldsymbol{A}$. We know the rows and columns of the matrix

$$\boldsymbol{P} = \begin{bmatrix} \boldsymbol{E_1} & \boldsymbol{E_2} \end{bmatrix}$$

are mutually orthogonal and the rows and columns are length one. Thus,

$$\boldsymbol{P} = \begin{bmatrix} E_{11} & E_{21} \\ E_{12} & E_{22} \end{bmatrix}$$

Since the eigenvectors are length one, we know $\sqrt{E_{11}^2 + E_{12}^2} = 1$ and $\sqrt{E_{21}^2 + E_{22}^2} = 1$. We also know $E_{11}E_{21} + E_{12}E_{22} = 0$ and $E_{11}E_{12} + E_{21}E_{22} = 0$. Hence, the first column defines an angle θ by $\cos(\theta) = E_{11}$ and $\sin(\theta) = E_{12}$. The second column defines the angle ψ by $\cos(\psi) = E_{21}$ and $\sin(\psi) = E_{22}$. Since $E_{11}E_{21} + E_{12}E_{22} = 0$ and $E_{11}E_{12} + E_{21}E_{22} = 0$ we also find θ and ψ are not independent as $\cos(\theta)\sin(\theta) + \cos(\psi)\sin(\psi) = 0$ and $\cos(\theta)\cos(\psi) + \sin(\theta)\sin(\psi) = 0$. The second equation is useful as it tells us $\cos(\theta - \psi) = 0$. This means $\psi = \theta \pm \pi/2, \pm 3\pi/2, \ldots$. For these choices of ψ, we have

- $\cos(\psi) = \cos(\theta + \pi/2) = -\sin(\theta)\sin(\pi/2) = -\sin(\theta)$ and $\sin(\psi) = \sin(\theta + \pi/2) = \cos(\theta)\sin(\pi/2) = \cos(\theta)$.

- $\cos(\psi) = \cos(\theta - \pi/2) = +\sin(\theta)\sin(\pi/2) = \sin(\theta)$ and $\sin(\psi) = \sin(\theta - \pi/2) = -\cos(\theta)\sin(\pi/2) = -\cos(\theta)$.

Both of these solutions work. You can check other choices such as $\psi = \theta \pm 3\pi/2$ do not give anything new. Hence, $\boldsymbol{P}$ can be written in two ways

$$\boldsymbol{P}_1 = \begin{bmatrix} \cos(\theta) & -\sin(\theta) \\ \sin(\theta) & \cos(\theta) \end{bmatrix} \text{ and } \boldsymbol{P}_2 = \begin{bmatrix} \cos(\theta) & \sin(\theta) \\ \sin(\theta) & -\cos(\theta) \end{bmatrix} = \begin{bmatrix} \cos(\theta) & -\sin(\theta) \\ \sin(\theta) & \cos(\theta) \end{bmatrix} \begin{bmatrix} 1 & 0 \\ 0 & -1 \end{bmatrix}$$

The matrix $\boldsymbol{P}_1$ is a classical rotation matrix. The standard basis vectors $\boldsymbol{i}$ and $\boldsymbol{j}$ when rotated in a counterclockwise direction of angle θ moves to new basis vectors

$$\boldsymbol{I}_1 = \begin{bmatrix} \cos(\theta) \\ \sin(\theta) \end{bmatrix} \text{ and } \boldsymbol{I}_2 = \begin{bmatrix} -\sin(\theta) \\ \cos(\theta) \end{bmatrix}$$

You should draw this in $\Re^2$! The matrix $\boldsymbol{P}_2$ is a rotation with a reflection. The basic vector $\boldsymbol{i}$ is rotated to the new basis vector $\boldsymbol{I}_1$, but instead of getting the new basis vector $\boldsymbol{I}_2$, we get the reflection of it through the x axis giving

$$\boldsymbol{I}_1^{reflection} = \boldsymbol{I}_1 = \begin{bmatrix} \cos(\theta) \\ \sin(\theta) \end{bmatrix} \text{ and } \boldsymbol{I}_2^{reflection} = -\boldsymbol{I}_2 = \begin{bmatrix} \sin(\theta) \\ -\cos(\theta) \end{bmatrix}$$

Note $det \boldsymbol{P}_1 = \cos(\theta)^2 + \sin(\theta)^2 = 1$ and $\det \boldsymbol{P}_2 = -\cos(\theta)^2 - \sin(\theta)^2 = -1$ for any θ. Clearly $\boldsymbol{P}_1{}^{-1} = \boldsymbol{P}_1{}^T$ and $\boldsymbol{P}_2{}^{-1} = \boldsymbol{P}_2{}^T$. You should draw this in $\Re^2$ also as it is instructive!
Thus, any 2×2 symmetric matrix $\boldsymbol{A}$ has the decomposition

$$\boldsymbol{A} = \boldsymbol{P}_1{}^T \begin{bmatrix} \lambda_1 & 0 \\ 0 & \lambda_2 \end{bmatrix} \boldsymbol{P}_1$$

or

$$\boldsymbol{A} = \boldsymbol{P}_2{}^T \begin{bmatrix} \lambda_1 & 0 \\ 0 & \lambda_2 \end{bmatrix} \boldsymbol{P}_2 = \begin{bmatrix} 1 & 0 \\ 0 & -1 \end{bmatrix}^T \boldsymbol{P}_1{}^T \begin{bmatrix} \lambda_1 & 0 \\ 0 & \lambda_2 \end{bmatrix} \boldsymbol{P}_1 \begin{bmatrix} 1 & 0 \\ 0 & -1 \end{bmatrix}$$

5.1.4 Homework

Exercise 5.1.11 *Show the set of all rotations is a group with respect to matrix multiplication as the operation of the group. Is this an abelian group?*

Exercise 5.1.12 *If $\boldsymbol{P}$ is reflection, what is* $\det \boldsymbol{P}$*? If $\boldsymbol{P}$ is rotation, what is* $\det \boldsymbol{P}$*?*

Exercise 5.1.13 *Pick a positive angle θ in quadrant one and draw $\boldsymbol{R}_\theta \boldsymbol{e}_1$ and $\boldsymbol{R}_\theta \boldsymbol{e}_2$. These two lines determine a new coordinate system we can call the (x', y') system with new orthonormal basis $\boldsymbol{e}_1' = \boldsymbol{R}_\theta \boldsymbol{e}_1$ and $\boldsymbol{e}_2' = \boldsymbol{R}_\theta \boldsymbol{e}_2$.*

Exercise 5.1.14 *Pick a positive angle θ in quadrant one and draw $\boldsymbol{R}_{-\theta} \boldsymbol{e}_1$ and $\boldsymbol{R}_{-\theta} \boldsymbol{e}_2$. These two lines also determine a new coordinate system we can call the (x', y') system with new orthonormal basis $\boldsymbol{e}_1' = \boldsymbol{R}_{-\theta} \boldsymbol{e}_1$ and $\boldsymbol{e}_2' = \boldsymbol{R}_{-\theta} \boldsymbol{e}_2$.*

Exercise 5.1.15 *Pick a positive angle θ in quadrant one and draw $\boldsymbol{R}_\theta \boldsymbol{e}_1$ and $\boldsymbol{R}_\theta \boldsymbol{e}_2$. These two lines determine a new coordinate system we can call the (x', y') system with new orthonormal basis $\boldsymbol{e}_1' = \boldsymbol{R}_\theta \boldsymbol{e}_1$ and $\boldsymbol{e}_2' = \boldsymbol{R}_\theta \boldsymbol{e}_2$. Now do it again and rotate by another positive angle ψ to construct a second coordinate system we can call the (x'', y'') system with new orthonormal basis $\boldsymbol{e}_1'' = \boldsymbol{R}_\psi \boldsymbol{e}_1'$ and $\boldsymbol{e}_2'' = \boldsymbol{R}_\psi \boldsymbol{e}_2'$. How does this compare to the new system we get using $\boldsymbol{R}_{\theta+\psi}$?*

Exercise 5.1.16 *Pick a positive angle θ in quadrant one and draw $\boldsymbol{R}_\theta \boldsymbol{e}_1$ and $\boldsymbol{R}_\theta \boldsymbol{e}_2$. These two lines determine a new coordinate system we can call the (x', y') system with new orthonormal basis $\boldsymbol{e}_1' = \boldsymbol{R}_\theta \boldsymbol{e}_1$ and $\boldsymbol{e}_2' = \boldsymbol{R}_\theta \boldsymbol{e}_2$. Now pick another positive angle in quadrant one and rotate by the negative angle $-\psi$ to construct a second coordinate system we can call the (x'', y'') system with new* *orthonormal basis* $\mathbf{e}_1'' = \boldsymbol{R}_{-\psi}\mathbf{e}_1'$ *and* $\mathbf{e}_2'' = \boldsymbol{R}_{-\psi}\mathbf{e}_2'$. *How does this compare to the new system we get using* $\boldsymbol{R}_{\boldsymbol{\theta}-\boldsymbol{\psi}}$*?*

Exercise 5.1.17 *Pick a positive angle* θ *in quadrant one and draw* $\boldsymbol{R}_{-\boldsymbol{\theta}}\boldsymbol{e}_1$ *and* $\boldsymbol{R}_{-\boldsymbol{\theta}}\boldsymbol{e}_2$. *These two lines also determine a new coordinate system we can call the* (x', y') *system with new orthonormal basis* $\mathbf{e}_1' = \boldsymbol{R}_{-\boldsymbol{\theta}}\boldsymbol{e}_1$ *and* $\mathbf{e}_2' = \boldsymbol{R}_{-\boldsymbol{\theta}}\boldsymbol{e}_2$.

5.2 Rotating Surfaces

Here is a nice example of these ideas. We will do this both by hand and using MATLAB so you can see how both approaches work. Consider the surface $z = 3x^2 + 2xy + 4y^2$ which is a rotated paraboloid with elliptical cross sections. Let's find the angle θ so that the change of variable $\boldsymbol{R}_{\boldsymbol{\theta}} \begin{bmatrix} x \\ y \end{bmatrix}$ to the new coordinates $\overline{x}$ and $\overline{y}$ leads to a surface equation with no cross terms $\overline{xy}$. First note

$$3x^2 + 2xy + 4y^2 = \begin{bmatrix} x \\ y \end{bmatrix}^T \begin{bmatrix} 3 & 1 \\ 1 & 4 \end{bmatrix} \begin{bmatrix} x \\ y \end{bmatrix}$$

where we let

$$\boldsymbol{H} = \begin{bmatrix} 3 & 1 \\ 1 & 4 \end{bmatrix}$$

The matrix $\boldsymbol{H}$ here is symmetric so using the eigenvectors and eigenvalues of $\boldsymbol{H}$, we can write

$$\begin{bmatrix} \boldsymbol{E}_1 & \boldsymbol{E}_2 \end{bmatrix}^T \begin{bmatrix} 3 & 1 \\ 1 & 4 \end{bmatrix} \begin{bmatrix} \boldsymbol{E}_1 & \boldsymbol{E}_2 \end{bmatrix} = \begin{bmatrix} \lambda_1 & 0 \\ 0 & \lambda_2 \end{bmatrix}$$

The orthonormal basis $\{\boldsymbol{E}_1, \boldsymbol{E}_2\}$ can be identified with either a rotation $\boldsymbol{R}_{\boldsymbol{\theta}}$ or a reflection $\boldsymbol{R}_{\boldsymbol{\theta}}^r$. We will choose the rotation. Then using the change of variables

$$\begin{bmatrix} \overline{x} \\ \overline{y} \end{bmatrix} = \boldsymbol{R}_{\boldsymbol{\theta}} \begin{bmatrix} x \\ y \end{bmatrix}$$

we find

$$z = \begin{bmatrix} x \\ y \end{bmatrix}^T \boldsymbol{R}_{\boldsymbol{\theta}}{}^T \begin{bmatrix} 3 & 1 \\ 1 & 4 \end{bmatrix} \boldsymbol{R}_{\boldsymbol{\theta}} \begin{bmatrix} x \\ y \end{bmatrix} = \lambda_1 \overline{x}^2 + \lambda_2 \overline{y}^2.$$

The characteristic equation for $\boldsymbol{H}$ is

$$det(\lambda \boldsymbol{I} - \boldsymbol{H}) = \lambda^2 - 7\lambda + 11 = 0$$

with roots

$$\lambda_1 = \frac{7+\sqrt{5}}{2} = 4.618, \quad \lambda_2 = \frac{7-\sqrt{5}}{2} = 2.382$$

It is straightforward to find the corresponding eigenvectors

$$\boldsymbol{W}_1 = \begin{bmatrix} 1 \\ \frac{1+\sqrt{5}}{2} \end{bmatrix}, \quad \boldsymbol{W}_1 = \begin{bmatrix} 1 \\ \frac{1-\sqrt{5}}{2} \end{bmatrix},$$

Then

$$\begin{aligned} \|\boldsymbol{W_1}\|_2^2 &= \sqrt{1+\left(\frac{1+\sqrt{5}}{2}\right)^2} \Longrightarrow \|\boldsymbol{W_1}\|_2 = \sqrt{3.618} = 1.902 \\ \|\boldsymbol{W_2}\|_2^2 &= \sqrt{1+\left(\frac{1-\sqrt{5}}{2}\right)^2} \Longrightarrow \|\boldsymbol{W_2}\|_2 = \sqrt{1.382} = 1.176 \end{aligned}$$

Then the orthonormal basis we need is

$$\begin{aligned} \boldsymbol{E_1} &= \frac{\boldsymbol{W_1}}{\|\boldsymbol{W_1}\|_2} = \begin{bmatrix} .526 \\ .8505 \end{bmatrix} \\ \boldsymbol{E_2} &= \pm\frac{\boldsymbol{W_2}}{\|\boldsymbol{W_2}\|_2} = \pm\begin{bmatrix} .8505 \\ -.526 \end{bmatrix} \end{aligned}$$

If we choose the minus choice, we get

$$\boldsymbol{E_2} = \begin{bmatrix} -.8505 \\ .526 \end{bmatrix}$$

and $\begin{bmatrix}\boldsymbol{E_1}, \boldsymbol{E_2}\end{bmatrix}$ is the rotation matrix $\boldsymbol{R_\theta}$. If we choose the plus choice, we get

$$\boldsymbol{E_2} = \begin{bmatrix} .8505 \\ -.526 \end{bmatrix}$$

and $\begin{bmatrix}\boldsymbol{E_1}, \boldsymbol{E_2}\end{bmatrix}$ is the reflected rotation matrix $\boldsymbol{R_\theta^r}$. Our rotation matrix is then

$$\boldsymbol{R_\theta} = \begin{bmatrix} .526 & -.8505 \\ .8505 & .526 \end{bmatrix} = \begin{bmatrix} \cos(1.017) & -\sin(1.017) \\ \sin(1.017) & \cos(1.017) \end{bmatrix}$$

where $\theta = 1.017$ radians. As a check,

$$\begin{aligned} \begin{bmatrix} .526 & .8505 \\ -.8505 & .526 \end{bmatrix} \begin{bmatrix} 3 & 1 \\ 1 & 4 \end{bmatrix} \begin{bmatrix} .526 & -.8505 \\ .8505 & .526 \end{bmatrix} &= \begin{bmatrix} .526 & .8505 \\ -.8505 & .526 \end{bmatrix} \begin{bmatrix} 2.4285 & -2.0255 \\ 3.928 & 1.2535 \end{bmatrix} \\ &= \begin{bmatrix} 4.618 & 0 \\ 0 & 2.382 \end{bmatrix} = \begin{bmatrix} \lambda_1 & 0 \\ 0 & \lambda_2 \end{bmatrix} \end{aligned}$$

The change of variables is

$$\begin{bmatrix} \overline{x} \\ \overline{y} \end{bmatrix} = \boldsymbol{R_\theta} \begin{bmatrix} x \\ y \end{bmatrix} = \begin{bmatrix} .526 & -.8505 \\ .8505 & .526 \end{bmatrix} \begin{bmatrix} x \\ y \end{bmatrix} = \begin{bmatrix} .526\,x - .8505\,y \\ .8505\,x + .526\,y \end{bmatrix}$$

Now let's do the same computations in MATLAB.

Listing 5.1: **Surface Rotation: Eigenvalues and Eigenvectors**

```
>> H = [3,1;1,4]
H =

     3     1
     1     4
```

```
>> [W,D] = eig(H)
W =

   -0.85065   0.52573
    0.52573   0.85065

D =

Diagonal Matrix

    2.3820        0
         0   4.6180
```

The command `eig` returns the unit eigenvectors as columns of V and the corresponding eigenvalues as the diagonal entries of D. You can see by looking at the matrix V the columns are not set up correctly to be a rotation matrix. We pick our orthonormal basis from V and hence our rotation matrix R as follows:

Listing 5.2: **Surface Rotation: Setting the Rotation Matrix**

```
>> E1 = W(:,2)
E1 =

    0.52573
    0.85065

>> E2 = W(:,1)
E2 =

   -0.85065
    0.52573

>> R = [E1,E2]
R =

    0.52573  -0.85065
    0.85065   0.52573
```

We then check the decomposition.

Listing 5.3: **Surface Rotation: Checking the Decomposition**

```
>> R'*H*R
ans =

    4.61803   0.00000
   -0.00000   2.38197
```

5.2.1 Homework

Exercise 5.2.1 *Analyze* $4x^2 + 6xy + 7y^2$.

Exercise 5.2.2 *Analyze* $-7x^2 + 2xy + 8y^2$.

Exercise 5.2.3 *Analyze* $2x^2 - 8xy + 6y^2$.

Exercise 5.2.4 *Analyze* $5x^2 + 12xy + 2y^2$.

5.3 A Complex ODE System Example

Let's look at a system of ODEs with complex roots to give a nontrivial example of how to use rotation matrices and these matrix decompositions to understand a solution. First, let's do a quick review as it has probably been a long time since you thought about this material. Let's begin with a theoretical analysis. If the real-valued matrix $\boldsymbol{A}$ has a complex eigenvalue $r = \alpha + i\beta$, then there is a nonzero vector $\boldsymbol{G}$ so that

$$\boldsymbol{A}\boldsymbol{G} = (\alpha + i\beta)\boldsymbol{G}.$$

Now take the complex conjugate of both sides to find

$$\overline{\boldsymbol{A}}\,\overline{\boldsymbol{G}} = \overline{(\alpha + i\beta)}\,\overline{\boldsymbol{G}}.$$

However, since $\boldsymbol{A}$ has real entries, its complex conjugate is simply $\boldsymbol{A}$ again. Thus, after taking complex conjugates, we find

$$\boldsymbol{A}\,\overline{\boldsymbol{G}} = (\alpha - i\beta)\,\overline{\boldsymbol{G}}$$

and we conclude that if $\alpha + i\beta$ is an eigenvalue of $\boldsymbol{A}$ with eigenvector $\boldsymbol{G}$, then the eigenvalue $\alpha - i\beta$ has eigenvector $\overline{\boldsymbol{G}}$. Hence, letting $\boldsymbol{E}$ be the real part of $\boldsymbol{G}$ and $\boldsymbol{F}$ be the imaginary part, we see $\boldsymbol{E} + i\boldsymbol{F}$ is the eigenvector for $\alpha + i\beta$ and $\boldsymbol{E} - i\boldsymbol{F}$ is the eigenvector for $\alpha - i\beta$.

5.3.1 The General Real and Complex Solution

We can write down the general complex solution immediately.

$$\begin{bmatrix} \phi(t) \\ \psi(t) \end{bmatrix} = c_1\,(\boldsymbol{E} + i\boldsymbol{F})\,e^{(\alpha+i\beta)t} + c_2\,(\boldsymbol{E} - i\boldsymbol{F})\,e^{(\alpha-i\beta)t}$$

for arbitrary complex numbers c_1 and c_2. We can reorganize this solution into a more convenient form as follows.

$$\begin{aligned} \begin{bmatrix} \phi(t) \\ \psi(t) \end{bmatrix} &= e^{\alpha t}\Big(c_1\,(\boldsymbol{E} + i\boldsymbol{F})\,e^{(i\beta)t} + c_2\,(\boldsymbol{E} - i\boldsymbol{F})\,e^{(-i\beta)t}\Big) \\ &= e^{\alpha t}\Big(\Big(c_1 e^{(i\beta)t} + c_2 e^{(-i\beta)t}\Big)\boldsymbol{E} + i\Big(c_1 e^{(i\beta)t} - c_2 e^{(-i\beta)t}\Big)\boldsymbol{F}\Big). \end{aligned}$$

The first real solution is found by choosing $c_1 = 1/2$ and $c_2 = 1/2$. This gives

$$\begin{bmatrix} x_1(t) \\ y_1(t) \end{bmatrix} = e^{\alpha t}\Big(\Big((1/2)\big(e^{(i\beta)t} + e^{(-i\beta)t}\big)\Big)\boldsymbol{E} + i\Big((1/2)\big(e^{(i\beta)t} - e^{(-i\beta)t}\big)\Big)\boldsymbol{F}\Big).$$

However, we know that $(1/2)\big(e^{(i\beta)t} + e^{(-i\beta)t}\big) = \cos(\beta t)$ and $(1/2)\big(e^{(i\beta)t} - e^{(-i\beta)t}\big) = i\,\sin(\beta t)$. Thus, we have

$$\begin{bmatrix} x_1(t) \\ y_1(t) \end{bmatrix} = e^{\alpha t}\Big(\boldsymbol{E}\,\cos(\beta t) - \boldsymbol{F}\,\sin(\beta t)\Big).$$

The second real solution is found by setting $c_1 = 1/2i$ and $c_2 = -1/2i$ which gives

$$\begin{aligned}\begin{bmatrix} x_2(t) \\ y_2(t) \end{bmatrix} &= e^{\alpha t}\Big(\Big((1/2i)\big(e^{(i\beta)t} - e^{(-i\beta)t}\big)\Big)\boldsymbol{E} + i\Big((1/2i)\big(e^{(i\beta)t} + e^{(-i\beta)t}\big)\Big)\boldsymbol{F}\Big) \\ &= e^{\alpha t}\Big(\boldsymbol{E}\sin(\beta t) + \boldsymbol{F}\cos(\beta t)\Big).\end{aligned}$$

The general real solution is therefore

$$\begin{bmatrix} x(t) \\ y(t) \end{bmatrix} = e^{\alpha t}\Big(a\,(\boldsymbol{E}\,\cos(\beta t) - \boldsymbol{F}\,\sin(\beta t)) + b\,(\boldsymbol{E}\sin(\beta t) + \boldsymbol{F}\cos(\beta t))\Big)$$

for arbitrary real numbers a and b.

Example 5.3.1

$$\begin{aligned} x'(t) &= 2\,x(t) + 5\,y(t) \\ y'(t) &= -x(t) + 4\,y(t) \\ x(0) &= 6 \\ y(0) &= -1 \end{aligned}$$

Solution *The characteristic is $r^2 - 6r + 13 = 0$ which gives the eigenvalues $3 \pm 2i$. The eigenvalue equation for the first root, $3 + 2i$ leads to this system to solve for nonzero $\boldsymbol{V}$.*

$$\begin{bmatrix} (3+2i) - 2 & -5 \\ 1 & (3+2i) - 4 \end{bmatrix}\begin{bmatrix} V_1 \\ V_2 \end{bmatrix} = \begin{bmatrix} 0 \\ 0 \end{bmatrix}$$

This reduces to the system

$$\begin{bmatrix} 1+2i & -5 \\ 1 & -1+2i \end{bmatrix}\begin{bmatrix} V_1 \\ V_2 \end{bmatrix} = \begin{bmatrix} 0 \\ 0 \end{bmatrix}$$

Although it is not immediately apparent, the second row is a multiple of row one. Multiply row one by $-1 - 2i$. This gives the row $[-1 - 2i, 5]$. Now multiple this new row by -1 to get $[1 + 2i, -5]$ which is row one. So even though it is harder to see, these two rows are equivalent and hence we only need to choose one to solve for V_2 in terms of V_1. The first row gives $(1 + 2i)V_1 + 5V_2 = 0$. Letting $V_1 = a$, we find $V_2 = (-1 - 2i)/5\ a$. Hence, the eigenvectors have the form

$$\boldsymbol{G} = a\begin{bmatrix} 1 \\ -\frac{1+2i}{5} \end{bmatrix} = \begin{bmatrix} 1 \\ -\frac{1}{5} \end{bmatrix} + i\begin{bmatrix} 0 \\ -\frac{2}{5} \end{bmatrix}$$

Hence,

$$\boldsymbol{E} = \begin{bmatrix} 1 \\ -\frac{1}{5} \end{bmatrix} \text{ and } \boldsymbol{F} = \begin{bmatrix} 0 \\ -\frac{2}{5} \end{bmatrix}$$

The general real solution is therefore

$$\begin{bmatrix} x(t) \\ y(t) \end{bmatrix} = e^{3t}\Big(a\,\Big(\boldsymbol{E}\,\cos(2t) - \boldsymbol{F}\,\sin(2t)\Big) + b\,\Big(\boldsymbol{E}\sin(2t) + \boldsymbol{F}\cos(2t)\Big)\Big)$$

$$= e^{3t}\left(a\left(\begin{bmatrix}1\\-\frac{1}{5}\end{bmatrix}\cos(2t)-\begin{bmatrix}0\\-\frac{2}{5}\end{bmatrix}\sin(2t)\right)+b\left(\begin{bmatrix}1\\-\frac{1}{5}\end{bmatrix}\sin(2t)+\begin{bmatrix}0\\-\frac{2}{5}\end{bmatrix}\cos(2t)\right)\right)$$

$$= e^{3t}\begin{bmatrix}a\cos(2t)+b\sin(2t)\\(-\frac{1}{5}\,a-\frac{2}{5}\,b)\,\cos(2t)(\frac{2}{5}\,a+\frac{1}{5}\,b)\,\sin(2t)\end{bmatrix}$$

Now apply the initial conditions to obtain

$$\begin{bmatrix}6\\-1\end{bmatrix} = e^{3t}\begin{bmatrix}a\\(-\frac{1}{5}\,a-\frac{2}{5}\,b)\end{bmatrix}$$

Thus, $a=6$ *and* $b=-1/2$. *The solution is therefore*

$$\begin{bmatrix}x(t)\\y(t)\end{bmatrix} = e^{3t}\begin{bmatrix}6\,\cos(2t)-\frac{1}{2}\,\sin(2t)\\-\cos(2t)-\frac{23}{10}\,\sin(2t)\end{bmatrix}$$

5.3.1.1 Homework

Exercise 5.3.1 *Show the real solutions to a complex eigenvalue linear systems ODE problem are a two dimensional vector space over* $\Re$.

Exercise 5.3.2 *Show the real solutions to a complex eigenvalue linear systems ODE are linearly independent over* $\Re$.

Exercise 5.3.3 *Show the complex solutions to a complex eigenvalue linear systems ODE are linearly independent over* $\mathbb{C}$.

Exercise 5.3.4 *For Example 5.3.1*

- *Plot* y *versus* x.
- *Plot* x *versus* t.
- *Plot* y *versus* t.

5.3.2 Rewriting the Real Solution

We know the real solution is written as

$$\begin{bmatrix}x(t)\\y(t)\end{bmatrix} = e^{\alpha t}\left(\left(a\,\boldsymbol{E}+b\,\boldsymbol{F}\right)\cos(\beta t)+\left(b\,\boldsymbol{E}-a\,\boldsymbol{F}\right)\sin(\beta t)\right)$$

Now rewrite again in terms of the components of $\boldsymbol{E}$ and $\boldsymbol{F}$ to obtain

$$\begin{bmatrix}x(t)\\y(t)\end{bmatrix} = e^{\alpha t}\left(\left(a\begin{bmatrix}E_1\\E_2\end{bmatrix}+b\begin{bmatrix}F_1\\F_2\end{bmatrix}\right)\cos(\beta t)+\left(b\begin{bmatrix}E_1\\E_2\end{bmatrix}-a\begin{bmatrix}F_1\\F_2\end{bmatrix}\right)\sin(\beta t)\right)$$

$$= e^{\alpha t}\begin{bmatrix}aE_1+bF_1 & bE_1-aF_1\\aE_2+bF_2 & bE_2-aF_2\end{bmatrix}\begin{bmatrix}\cos(\beta t)\\\sin(\beta t)\end{bmatrix}$$

Finally, we can move back to the vector form and write

$$\begin{bmatrix}x(t)\\y(t)\end{bmatrix} = e^{\alpha t}\left[a\boldsymbol{E}+b\boldsymbol{F},\;\; b\boldsymbol{E}-a\boldsymbol{F}\right]\begin{bmatrix}\cos(\beta t)\\\sin(\beta t)\end{bmatrix}.$$

Now we want nonzero solutions, so our initial conditions will give us a and b so that $\sqrt{a^2+b^2} \neq 0$. We have so far

$$\begin{aligned}
\begin{bmatrix} \frac{x(t)}{e^{\alpha t}} \\ \frac{y(t)}{e^{\alpha t}} \end{bmatrix} &= \begin{bmatrix} aE_1 + bF_1 & bE_1 - aF_1 \\ aE_2 + bF_2 & bE_2 - aF_2 \end{bmatrix} \begin{bmatrix} \cos(\beta t) \\ \sin(\beta t) \end{bmatrix} \\
&= \begin{bmatrix} \boldsymbol{E} & \boldsymbol{F} \end{bmatrix} \begin{bmatrix} a & b \\ b & -a \end{bmatrix} \begin{bmatrix} \cos(\beta t) \\ \sin(\beta t) \end{bmatrix} \\
&= \sqrt{a^2+b^2} \begin{bmatrix} \boldsymbol{E} & \boldsymbol{F} \end{bmatrix} \begin{bmatrix} \frac{a}{\sqrt{a^2+b^2}} & \frac{b}{\sqrt{a^2+b^2}} \\ \frac{b}{\sqrt{a^2+b^2}} & -\frac{a}{\sqrt{a^2+b^2}} \end{bmatrix} \begin{bmatrix} \cos(\beta t) \\ \sin(\beta t) \end{bmatrix}
\end{aligned}$$

Now define the angle θ by $\cos(\theta) = a/\sqrt{a^2+b^2}$ and so $\sin(\theta) = b/\sqrt{a^2+b^2}$. Further let $L = \sqrt{a^2+b^2}$. Then, we can rewrite the solution again as

$$\begin{bmatrix} \frac{x(t)}{Le^{\alpha t}} \\ \frac{y(t)}{Le^{\alpha t}} \end{bmatrix} = \begin{bmatrix} \boldsymbol{E} & \boldsymbol{F} \end{bmatrix} \begin{bmatrix} \cos(\theta) & \sin(\theta) \\ \sin(\theta) & -\cos(\theta) \end{bmatrix} \begin{bmatrix} \cos(\beta t) \\ \sin(\beta t) \end{bmatrix}$$

The matrix involving θ is a reflection matrix with angle θ which we will denote by $\mathbb{R}^r_\theta$. Then we have

$$\begin{bmatrix} \frac{x(t)}{Le^{\alpha t}} \\ \frac{y(t)}{Le^{\alpha t}} \end{bmatrix} = \begin{bmatrix} \boldsymbol{E} & \boldsymbol{F} \end{bmatrix} \mathbb{R}^r_\theta \begin{bmatrix} \cos(\beta t) \\ \sin(\beta t) \end{bmatrix}$$

which we can rewrite in terms of the rotation matrix $\boldsymbol{R_\theta}$ as

$$\begin{bmatrix} \frac{x(t)}{Le^{\alpha t}} \\ \frac{y(t)}{Le^{\alpha t}} \end{bmatrix} = \begin{bmatrix} \boldsymbol{E} & \boldsymbol{F} \end{bmatrix} \boldsymbol{R_\theta} \begin{bmatrix} 1 & 0 \\ 0 & -1 \end{bmatrix} \begin{bmatrix} \cos(\beta t) \\ \sin(\beta t) \end{bmatrix} = \begin{bmatrix} \boldsymbol{E} & \boldsymbol{F} \end{bmatrix} \boldsymbol{R_\theta} \begin{bmatrix} \cos(\beta t) \\ -\sin(\beta t) \end{bmatrix}$$

Let $u(t) = x(t)/(Le^{\alpha t})$ and $v(t) = y(t)/(Le^{\alpha t})$. Then

$$\begin{aligned}
u^2(t) + v^2(t) &= \left\langle \begin{bmatrix} \boldsymbol{E} & \boldsymbol{F} \end{bmatrix} \boldsymbol{R_\theta} \begin{bmatrix} \cos(\beta t) \\ -\sin(\beta t) \end{bmatrix}, \begin{bmatrix} \boldsymbol{E} & \boldsymbol{F} \end{bmatrix} \boldsymbol{R_\theta} \begin{bmatrix} \cos(\beta t) \\ -\sin(\beta t) \end{bmatrix} \right\rangle \\
&= \begin{bmatrix} \cos(\beta t) \\ -\sin(\beta t) \end{bmatrix}^T \boldsymbol{R_\theta}^T \begin{bmatrix} \boldsymbol{E} & \boldsymbol{F} \end{bmatrix}^T \begin{bmatrix} \boldsymbol{E} & \boldsymbol{F} \end{bmatrix}^T \boldsymbol{R_\theta} \begin{bmatrix} \cos(\beta t) \\ -\sin(\beta t) \end{bmatrix}
\end{aligned}$$

Now $\begin{bmatrix} \boldsymbol{E} & \boldsymbol{F} \end{bmatrix}^T \begin{bmatrix} \boldsymbol{E} & \boldsymbol{F} \end{bmatrix}$ is a 2×2 symmetric matrix. Hence, we have a representation of it in terms of its real eigenvalues λ_1 and λ_2 and associated eigenvectors $\boldsymbol{G_1}$ and $\boldsymbol{G_2}$.

$$\begin{bmatrix} \boldsymbol{E} & \boldsymbol{F} \end{bmatrix}^T \begin{bmatrix} \boldsymbol{E} & \boldsymbol{F} \end{bmatrix} = [\boldsymbol{G_1}, \boldsymbol{G_2}]^T \begin{bmatrix} \lambda_1 & 0 \\ 0 & \lambda_2 \end{bmatrix} [\boldsymbol{G_1}, \boldsymbol{G_2}]$$

We also know from our earlier discussions, that $[\boldsymbol{G_1}, \boldsymbol{G_2}]$ is either a rotation matrix $\boldsymbol{R_\psi}$ or a reflection $\boldsymbol{R_\psi J}$ where

$$\boldsymbol{J} = \begin{bmatrix} 1 & 0 \\ 0 & -1 \end{bmatrix}$$

Thus, we have, in the case it is a rotation matrix

$$u^2(t) + v^2(t) = \begin{bmatrix} \cos(\beta t) \\ -\sin(\beta t) \end{bmatrix}^T \boldsymbol{R_\theta}^T \boldsymbol{R_\psi}^T \begin{bmatrix} \lambda_1 & 0 \\ 0 & \lambda_2 \end{bmatrix} \boldsymbol{R_\psi} \boldsymbol{R_\theta} \begin{bmatrix} \cos(\beta t) \\ -\sin(\beta t) \end{bmatrix}$$

It is a straightforward calculation to show $\boldsymbol{R_\psi R_\theta} = \boldsymbol{R_{\psi+\theta}}$. Thus, for a rotation matrix

$$u^2(t) + v^2(t) \quad = \quad \begin{bmatrix} \cos(\beta t) \\ -\sin(\beta t) \end{bmatrix}^T \boldsymbol{R_{\psi+\theta}}^T \begin{bmatrix} \lambda_1 & 0 \\ 0 & \lambda_2 \end{bmatrix} \boldsymbol{R_{\psi+\theta}} \begin{bmatrix} \cos(\beta t) \\ -\sin(\beta t) \end{bmatrix}$$

Now note we can rewrite this in terms of new coordinates based on new unit vectors.

- The reflection

$$\boldsymbol{J} \begin{bmatrix} \cos(\beta t) \\ \sin(\beta t) \end{bmatrix} \quad = \quad \begin{bmatrix} \cos(\beta t) \\ -\sin(\beta t) \end{bmatrix}$$

  corresponds to $\boldsymbol{i} \to \boldsymbol{i}' = \boldsymbol{i}$ and $\boldsymbol{j} \to \boldsymbol{j}' = -\boldsymbol{j}$. Call this the $\boldsymbol{x}' - \boldsymbol{y}'$ coordinate system. Note these vectors live on the unit circle in the $\boldsymbol{x}' - \boldsymbol{y}'$ coordinate system.

- Now apply the rotation $\boldsymbol{R_{\psi+\theta}}$ which rotates the reflected system above counterclockwise an angle of $\psi+\theta$. The coordinates in this new system are $(\overline{x}, \overline{y})$. Call this the $\boldsymbol{x}'' - \boldsymbol{y}''$ coordinate system. Note these vectors live on the unit circle in the $\boldsymbol{x}'' - \boldsymbol{y}''$ coordinate system. The coordinates in this new system are $(\overline{\overline{x}}, \overline{\overline{y}})$.

So we have

$$u^2(t) + v^2(t) \quad = \quad \begin{bmatrix} \overline{\overline{x}} \\ \overline{\overline{y}} \end{bmatrix}^T \begin{bmatrix} \lambda_1 & 0 \\ 0 & \lambda_2 \end{bmatrix} \begin{bmatrix} \overline{\overline{x}} \\ \overline{\overline{y}} \end{bmatrix} = \lambda_1 (\overline{\overline{x}})^2 + \lambda_2 (\overline{\overline{y}})^2$$

which is an ellipse if both eigenvalues are positive, a circle if they are positive and equal and a hyperbola if they differ in algebraic sign. We get various degenerate possibilities if one of the eigenvalues is zero.

The analysis in the case of the reflected case is quite similar and the matrix $\boldsymbol{J}$ does not alter the result qualitatively. Rewriting again, we find

$$x^2(t) + y^2(t) \quad = \quad L^2 e^{2\alpha t} \left(\lambda_1 (\overline{\overline{x}})^2 + \lambda_2 (\overline{\overline{y}})^2 \right)$$

which shows we have spiral out, constant or spiral in trajectories depending on the sign of α.

5.3.2.1 Homework

Exercise 5.3.5 *Rewrite the real solution to*

$$\begin{aligned} x'(t) &= 2\,x(t) \;+\; 5\,y(t) \\ y'(t) &= -x(t) \;+\; 4\,y(t) \\ x(0) &= 4 \\ y(0) &= 3 \end{aligned}$$

as we discuss in this section. Plot $x^2(t) + y^2(t)$ *and* $u^2 t + v^2(t)$*. On the second plot, draw the* $(\overline{\overline{x}}, \overline{\overline{y}})$ *axes.*

Exercise 5.3.6 *Rewrite the real solution to*

$$\begin{aligned} x'(t) &= 2\,x(t) \;+\; 5\,y(t) \\ y'(t) &= -x(t) \;+\; 4\,y(t) \\ x(0) &= 3 \\ y(0) &= 8 \end{aligned}$$

as we discuss in this section. Plot $x^2(t)+y^2(t)$ and $u^2t+v^2(t)$. On the second plot, draw the $(\overline{\overline{x}},\overline{y})$ axes.

Exercise 5.3.7 *Rewrite the real solution to*

$$\begin{aligned} x'(t) &= 3\,x(t) \;-\; 2\,y(t) \\ y'(t) &= 2x(t) \;+\; 3\,y(t) \\ x(0) &= 4 \\ y(0) &= 3 \end{aligned}$$

as we discuss in this section. Plot $x^2(t)+y^2(t)$ and $u^2t+v^2(t)$. On the second plot, draw the $(\overline{\overline{x}},\overline{y})$ axes.

Exercise 5.3.8 *Rewrite the real solution to*

$$\begin{aligned} x'(t) &= 3\,x(t) \;-\; 2\,y(t) \\ y'(t) &= 2x(t) \;+\; 3\,y(t) \\ x(0) &= -5 \\ y(0) &= 7 \end{aligned}$$

as we discuss in this section. Plot $x^2(t)+y^2(t)$ and $u^2t+v^2(t)$. On the second plot, draw the $(\overline{\overline{x}},\overline{y})$ axes.

5.3.3 Signed Definite Matrices

A 2×2 matrix is said to be a **positive definite** matrix if $\boldsymbol{x}^T\boldsymbol{A}\boldsymbol{x}>0$ for all vectors $\boldsymbol{x}$. If we multiply this out, we find the inequality below

$$ax_1^2+2bx_1x_2+dx_2^2 \;>\; 0$$

If we complete the square, we find

$$a\left(\left(x_1+(b/a)x_2\right)^2+\left(\frac{ad-b^2}{a^2}x_2^2\right)\right) \;>\; 0$$

Now the leading term $a>0$, and if the determinant of $\boldsymbol{A}=ad-b^2>0$, we would have the quadratic $\left(x_1+(b/a)x_2\right)^2+\left(((ad-b^2)/a^2)x_2^2\right)$ is always positive. Note that since this determinant is positive, $ad>b^2$, which forces d to be positive as well. So in this case, a and d and $ad-b^2>0$. And the expression $\boldsymbol{x}^T\boldsymbol{A}\boldsymbol{x}>0$ in this case. Now recall what we found about the eigenvalues here. We had the eigenvalues were

$$r \;=\; \frac{(a+d)\pm\sqrt{(a-d)^2+4b^2}}{2}$$

Since $ad-b^2>0$, the term

$$(a-d)^2+4b^2 \;=\; a^2-2ad+d^2+4b^2<a^2-2ad+4ad+d^2=(a+d)^2.$$

Thus, the square root is smaller than $a+d$ as a and d are positive. The first root is always positive and the second root is too as $(a+d)-\sqrt{(a-d)^2+4b^2}>a+d-(a+d)=0$. So both eigenvalues are positive if a and d are positive and $ad-b^2>0$. Note the argument can go the other way. If we assume the matrix is positive definite, then we are forced to have $a > 0$ and $ad - b^2 > 0$ which gives the same result. We conclude our 2×2 symmetric matrix $\boldsymbol{A}$ is positive definite if and only if $a > 0$, $d > 0$ and the determinant of $\boldsymbol{A} > 0$ too. Note a positive definite matrix has positive eigenvalues.

A similar argument holds if we have determinant of $\boldsymbol{A} > 0$ but $a < 0$. The determinant condition will then force $d < 0$ too. We find that $\boldsymbol{x}^T \boldsymbol{A} \boldsymbol{x} < 0$. In this case, we say the matrix is **negative definite**. The eigenvalues are still

$$r = \frac{(a+d) \pm \sqrt{(a-d)^2 + 4b^2}}{2}.$$

But now, since $ad - b^2 > 0$, the term

$$(a-d)^2 + 4b^2 = a^2 - 2ad + d^2 + 4b^2 < a^2 - 2ad + 4ad + d^2 = |a+d|^2.$$

Since a and d are negative, $a + d < 0$ and so the second root is always negative. The first root's sign is determined by $(a+d) + |\sqrt{(a-d)^2 + 4b^2} < (a+d) + |a+d| = 0$. So both eigenvalues are negative. We have found the matrix $\boldsymbol{A}$ is negative definite if a and d are negative and the determinant of $\boldsymbol{A} > 0$. Note a negative definite matrix has negative eigenvalues.

5.3.4 Summarizing

Let's put all the information we have together. We have been studying a special type of 2×2 matrix which is symmetric. This matrix $\boldsymbol{A}$ has the form

$$\boldsymbol{A} = \begin{bmatrix} a & b \\ b & d \end{bmatrix}$$

where a, b and d are arbitrary nonzero numbers. The characteristic equation here is $r^2 - (a+d)r + ad - b^2$. Note that the term $ad - b^2$ is the determinant of $\boldsymbol{A}$. The roots are given by

$$r = \frac{(a+d) \pm \sqrt{(a-d)^2 + 4b^2}}{2}$$

- This matrix always has two distinct real eigenvalues and their respective eigenvectors $\boldsymbol{E}_1$ and $\boldsymbol{E}_2$ are mutually orthogonal. Hence, we usually normalize them so they form an orthonormal basis for $\Re^2$.
- We can use these eigenvectors to write $\boldsymbol{A}$ is a canonical form.

$$\boldsymbol{A} = \boldsymbol{P} \begin{bmatrix} \lambda_1 & 0 \\ 0 & \lambda_2 \end{bmatrix} \boldsymbol{P}^T$$

where

$$\boldsymbol{P} = \begin{bmatrix} \boldsymbol{E}_1 & \boldsymbol{E}_2 \end{bmatrix}$$

giving

$$\boldsymbol{P}^T = \begin{bmatrix} \boldsymbol{E}_1^T \\ \boldsymbol{E}_2^T \end{bmatrix}$$

- If we solve $\boldsymbol{A}\,\boldsymbol{X} = \boldsymbol{b}$, it is most illuminating to rewrite both the unknown $\boldsymbol{X}$ and the data $\boldsymbol{b}$ with respect to the orthonormal basis determined by the eigenvectors of $\boldsymbol{A}$. When we do this we find the solution is expressed quite nicely $X_1 = \lambda_1^{-1} b_1$ and $X_2 = \lambda_2^{-1} b_2$ where b_1 and b_2 are the components of $\boldsymbol{B}$ in the eigenvector basis. This shows very clearly how the solution depends on the size of the reciprocal eigenvalues.

- If $a > 0$ and the determinant of $\boldsymbol{A} = ad - b^2 > 0$, then $d > 0$ too and $\boldsymbol{A}$ is positive definite. In this case, both eigenvalues are positive. If we started by assuming $\boldsymbol{A}$ was positive definite, our chain of inference would lead us to the same conclusions about a, b and d.

- If $a < 0$ and the determinant of $\boldsymbol{A} = ad - b^2 > 0$, then $d < 0$ also and $\boldsymbol{A}$ is negative definite. In this case, both eigenvalues are negative. And assuming $\boldsymbol{A}$ is negative definite leads us backwards to $a < 0$, $d < -0$ and $ad - b^2 > 0$.

It seems obvious to ask if we can say similar things about symmetric matrices that are 3×3, 4×4 and so on. Clearly, we can't use our determinant route to find answers as analyzing cubics and quartics is quite impossible. So the next section tries another approach. The dreaded road we call **let's go abstract**.

5.3.4.1 Homework

Exercise 5.3.9 *For a 2×2 symmetric real matrix $\boldsymbol{A}$, we know $\boldsymbol{A} = \boldsymbol{P}\boldsymbol{D}\boldsymbol{P}^{\boldsymbol{T}}$ where $\boldsymbol{D}$ is the diagonal matrix of eigenvalues and $\boldsymbol{P}$ is a rotation or a reflection. Solve the equation $\boldsymbol{A}\boldsymbol{x_0} = \boldsymbol{b}$ and $\boldsymbol{A}\boldsymbol{x_1} = \boldsymbol{b} + \boldsymbol{h}$ for some $\boldsymbol{h}$. Find a formula for $\boldsymbol{x_1} - \boldsymbol{x_0} = \boldsymbol{\Delta}\boldsymbol{x_0}$ in terms of $\boldsymbol{h}$ in the case where the eigenvalues of $\boldsymbol{A}$ are nonzero. What is happening to $\boldsymbol{\Delta}\boldsymbol{x_0}$ if the absolute value of an eigenvalue is small?*

Exercise 5.3.10 *Using the previous exercise and the Frobenius norm of a 2×2 symmetric $\boldsymbol{A}$ real-valued matrix, prove the solutions to $\boldsymbol{A}\boldsymbol{x} = \boldsymbol{b}$ depend continuously on the data $\boldsymbol{b}$.*

Exercise 5.3.11 *For a 2×2 real matrix $\boldsymbol{A}$ which has an inverse $\boldsymbol{A}^{-1}$, prove the solutions to $\boldsymbol{A}\boldsymbol{x} = \boldsymbol{b}$ depend continuously on the data $\boldsymbol{b}$.*

Exercise 5.3.12 *For a 2×2 symmetric real matrix $\boldsymbol{A}$ with nonzero eigenvalues, find $\boldsymbol{A}^{-1}$.*

5.4 Symmetric Systems of ODEs

Here is a nice application of symmetric matrices to systems of two linear differential equations. Let's do some review first, as the ODE material seems to always slip away right when you need it. This was done quickly in Section 3.3.4.

Let's look at a specific symmetric ODE system and find its solution.

Example 5.4.1 *For the symmetric system below*

$$\begin{aligned} \begin{bmatrix} x'(t) \\ y'(t) \end{bmatrix} &= \begin{bmatrix} -20 & 12 \\ 12 & 5 \end{bmatrix} \begin{bmatrix} x(t) \\ y(t) \end{bmatrix} \\ \begin{bmatrix} x(0) \\ y(0) \end{bmatrix} &= \begin{bmatrix} -1 \\ 2 \end{bmatrix} \end{aligned}$$

- *Find the characteristic equation*

- *Find the general solution*

- *Solve the IVP*

Solution *The characteristic equation is*

$$\det\left(r \begin{bmatrix} 1 & 0 \\ 0 & 1 \end{bmatrix} - \begin{bmatrix} -20 & 12 \\ 12 & 5 \end{bmatrix} \right) = 0$$

Thus

$$\begin{aligned} 0 &= det\left(\begin{bmatrix} r+20 & -12 \\ -12 & r-5 \end{bmatrix} \right) \\ &= (r+20)(r-5) - 144 \\ &= r^2 + 15r - 244 \end{aligned}$$

Hence, **eigenvalues** *or* **roots** *of the characteristic equation are* $r_1 = 9.83$ *and* $r_2 = -24.83$.

For eigenvalue $r_1 = 9.83$*, substitute the value into*

$$\begin{bmatrix} r+20 & -12 \\ -12 & r-5 \end{bmatrix} \begin{bmatrix} V_1 \\ V_2 \end{bmatrix} \Rightarrow \begin{bmatrix} 29.83 & -12 \\ -12 & 4.83 \end{bmatrix} \begin{bmatrix} V_1 \\ V_2 \end{bmatrix}$$

Picking the top row, we get $29.83V_1 - 12V_2 = 0$ *implying* $V_2 = 6.18V_1$. *Letting* $V_1 = a$*, we find* $V_1 = a$ *and* $V_2 = 6.18a$*: so*

$$\boldsymbol{V} = \begin{bmatrix} V_1 \\ V_2 \end{bmatrix} = a \begin{bmatrix} 1 \\ 6.18 \end{bmatrix}$$

Let $a = 1/|| \begin{bmatrix} 1 & 6.18 \end{bmatrix}^T || = 1/6.26 = .16$ *and choose the unit eigenvector*

$$\boldsymbol{E_1} = 0.16 \begin{bmatrix} 1 \\ 6.18 \end{bmatrix} = \begin{bmatrix} .16 \\ .99 \end{bmatrix}$$

So one of the solutions is

$$\begin{bmatrix} x_1(t) \\ y_1(t) \end{bmatrix} = \boldsymbol{E_1}\, e^{9.83t} = \begin{bmatrix} .16 \\ .99 \end{bmatrix} e^{9.83t}.$$

For eigenvalue $r_2 = -24.83$*, substitute the value into*

$$\begin{bmatrix} r+20 & -12 \\ -12 & r-5 \end{bmatrix} \begin{bmatrix} V_1 \\ V_2 \end{bmatrix} \Rightarrow \begin{bmatrix} -4.83 & -12 \\ -12 & -29.83 \end{bmatrix} \begin{bmatrix} V_1 \\ V_2 \end{bmatrix}$$

Picking the top row, we get $-4.83V_1 - 29.83V_2 = 0$ *implying* $V_2 = -.16V_1$. *Letting* $V_1 = b$*, we find* $V_1 = b$ *and* $V_2 = \quad .16b$*: so*

$$\boldsymbol{V} = \begin{bmatrix} V_1 \\ V_2 \end{bmatrix} = b \begin{bmatrix} 1 \\ -.16 \end{bmatrix}$$

Let $b = 1/|| \begin{bmatrix} 1 & -.16 \end{bmatrix}^T || = 1/1.01 = .99$ *and choose the unit eigenvector*

$$\boldsymbol{E_2} = .99 \begin{bmatrix} 1 \\ -.16 \end{bmatrix} = \begin{bmatrix} .99 \\ -.16 \end{bmatrix}$$

Note $< \boldsymbol{E_1}, \boldsymbol{E_2} >= (.16)(.99) + (.99)(-.16) = 0$. *So the other solution is*

$$\begin{bmatrix} x_2(t) \\ y_2(t) \end{bmatrix} = \boldsymbol{E_2}\, e^{-29.83t} = \begin{bmatrix} .99 \\ -.16 \end{bmatrix} e^{-29.83t}.$$

The general solution

$$\begin{aligned} \begin{bmatrix} x(t) \\ y(t) \end{bmatrix} &= A\,\boldsymbol{E_1}\,e^{9.83t} + B\,\boldsymbol{E_2}\,e^{-29.83t} \\ &= A \begin{bmatrix} .16 \\ .99 \end{bmatrix} e^{9.83t} + B \begin{bmatrix} .99 \\ -.16 \end{bmatrix} e^{-29.83t} \end{aligned}$$

Finally, we find A and B using the initial conditions.

$$\begin{aligned} \begin{bmatrix} x(0) \\ y(0) \end{bmatrix} &= \begin{bmatrix} -1 \\ 2 \end{bmatrix} = A \begin{bmatrix} .16 \\ .99 \end{bmatrix} e^{0} + B \begin{bmatrix} .99 \\ -.16 \end{bmatrix} e^{0} \\ &= A \begin{bmatrix} .16 \\ .99 \end{bmatrix} + B \begin{bmatrix} .99 \\ -.16 \end{bmatrix} \end{aligned}$$

Thus,

$$\begin{bmatrix} \boldsymbol{E_1} & \boldsymbol{E_2} \end{bmatrix} \begin{bmatrix} A \\ B \end{bmatrix} = \begin{bmatrix} -1 \\ 2 \end{bmatrix} \Rightarrow \begin{bmatrix} \boldsymbol{E_1^T} & \boldsymbol{E_2^T} \end{bmatrix} \begin{bmatrix} \boldsymbol{E_1} & \boldsymbol{E_2} \end{bmatrix} \begin{bmatrix} A \\ B \end{bmatrix} = \begin{bmatrix} \boldsymbol{E_1^T} & \boldsymbol{E_2^T} \end{bmatrix} \begin{bmatrix} -1 \\ 2 \end{bmatrix}$$
$$\begin{bmatrix} A \\ B \end{bmatrix} = \begin{bmatrix} \boldsymbol{E_1^T} & \boldsymbol{E_2^T} \end{bmatrix} \begin{bmatrix} -1 \\ 2 \end{bmatrix}$$

Letting $\boldsymbol{D} = \begin{bmatrix} -1 & 2 \end{bmatrix}^T$, $A =< \boldsymbol{E_1}, \boldsymbol{D} >$ *and* $B =< \boldsymbol{E_2}, \boldsymbol{D} >$.

5.4.1 Writing the Solution Another Way

The coefficient matrix is

$$\boldsymbol{A} = \begin{bmatrix} -20 & 12 \\ 12 & 5 \end{bmatrix}$$

and the eigenvalues of $\boldsymbol{A}$ are $\lambda_1 = 9.83$ and $\lambda_2 = -29.83$. Then we know if

$$\boldsymbol{P} = \begin{bmatrix} \boldsymbol{E_1} & \boldsymbol{E_2} \end{bmatrix}$$

then

$$\boldsymbol{P^T}\,\boldsymbol{A}\,\boldsymbol{P} = \begin{bmatrix} \lambda_1 & 0 \\ 0 & \lambda_2 \end{bmatrix} \implies \boldsymbol{A} = \boldsymbol{P} \begin{bmatrix} \lambda_1 & 0 \\ 0 & \lambda_2 \end{bmatrix} \boldsymbol{P^T}$$

Let

$$\boldsymbol{\Lambda} = \begin{bmatrix} \lambda_1 & 0 \\ 0 & \lambda_2 \end{bmatrix} \implies \boldsymbol{A} = \boldsymbol{P \Lambda P^T}$$

Now let's calculate the powers of $\boldsymbol{A}$:

- $$\begin{aligned} \boldsymbol{A}^2 &= \left(\boldsymbol{P\Lambda P^T}\right)\left(\boldsymbol{P\Lambda P^T}\right) = \boldsymbol{P\Lambda}\left(\boldsymbol{P^T P}\right)\boldsymbol{\Lambda P^T} \\ &= \boldsymbol{P\Lambda^2 P^T} \\ \boldsymbol{A}^3 &= \left(\boldsymbol{P\Lambda P^T}\right)\left(\boldsymbol{A^2}\right) = \left(\boldsymbol{P\Lambda P^T}\right)\left(\boldsymbol{P\Lambda^2 P^T}\right) = \boldsymbol{P\Lambda}\left(\boldsymbol{P^T P}\right)\boldsymbol{\Lambda^2 P^T} \\ &= \boldsymbol{P\Lambda^3 P^T} \end{aligned}$$

- It is easy to see by POMI that $\boldsymbol{A^n} = \boldsymbol{P\Lambda^n P^T}$.

Recall for the scalar t, $\boldsymbol{A}t = t\boldsymbol{A}$ simply multiplies each entry of $\boldsymbol{A}$ by the number t. Hence, from the above we have $\boldsymbol{A^n}t = \boldsymbol{P\Lambda^n P^T}\, t$. Consider

$$\boldsymbol{A}t = \boldsymbol{P}\begin{bmatrix}\lambda_1 & 0\\ 0 & \lambda_2\end{bmatrix}\boldsymbol{P^T}\,t = \boldsymbol{P}\begin{bmatrix}\lambda_1 & 0\\ 0 & \lambda_2\end{bmatrix}\begin{bmatrix}E_{11}\,t & E_{12}\,t\\ E_{21}\,t & E_{22}\,t\end{bmatrix} = \boldsymbol{P}\begin{bmatrix}\lambda_1\,t & 0\\ 0 & \lambda_2\,t\end{bmatrix}\boldsymbol{P^T}$$

Thus,

$$\boldsymbol{A}t = \boldsymbol{P}\begin{bmatrix}\lambda_1\,t & 0\\ 0 & \lambda_2\,t\end{bmatrix}\begin{bmatrix}E_{11} & E_{12}\\ E_{21} & E_{22}\end{bmatrix}$$

A similar calculation shows

$$\boldsymbol{A^2}t^2 = \boldsymbol{P}\begin{bmatrix}\lambda_1^2\,t^2 & 0\\ 0 & \lambda_2^2\,t^2\end{bmatrix}\begin{bmatrix}E_{11} & E_{12}\\ E_{21} & E_{22}\end{bmatrix} = \boldsymbol{P}\begin{bmatrix}\lambda_1^2 t^2 & 0\\ 0 & \lambda_2^2 t^2\end{bmatrix}\boldsymbol{P^T}$$

And the function

$$\begin{aligned}W_3(t) &= \boldsymbol{I} + \boldsymbol{A}t + \boldsymbol{A^2}t^2/2! + \boldsymbol{A^3}t^3/3!\\ &= \boldsymbol{P}\begin{bmatrix}1+\lambda_1 t + \lambda_1^2 t^2/2 + \lambda_1^3 t^3/6 & 0\\ 0 & 1+\lambda_2 t + \lambda_2^2 t^2/2 + \lambda_2^3 t^3/6\end{bmatrix}\boldsymbol{P^T}\\ &= \boldsymbol{P}\begin{bmatrix}\sum_{k=0}^{3}(\lambda_1^k t^k)/k! & 0\\ 0 & \sum_{k=0}^{3}(\lambda_2^k t^k)/k!\end{bmatrix}\boldsymbol{P^T}\end{aligned}$$

In general

$$W_n(t) \quad = \quad \boldsymbol{P}\begin{bmatrix}\sum_{k=0}^{n}(\lambda_1^k t^k)/k! & 0\\ 0 & \sum_{k=0}^{n}(\lambda_2^k t^k)/k!\end{bmatrix}\boldsymbol{P^T}$$

We know $e^{\lambda_1 t} = \lim_{n\to\infty}\sum_{k=0}^{n}(\lambda_1^k t^k)/k!$ and $e^{\lambda_2 t} = \lim_{n\to\infty}\sum_{k=0}^{n}(\lambda_2^k t^k)/k!$ This suggests

$$\begin{aligned}\lim_{n\to\infty} W_n(t) &= \boldsymbol{P}\begin{bmatrix}\lim_{n\to\infty}\sum_{k=0}^{n}(\lambda_1^k t^k)/k! & 0\\ 0 & \lim_{n\to\infty}\sum_{k=0}^{n}(\lambda_2^k t^k)/k!\end{bmatrix}\boldsymbol{P^T}\\ &= \boldsymbol{P}\begin{bmatrix}e^{\lambda_1 t} & 0\\ 0 & e^{\lambda_2 t}\end{bmatrix}\boldsymbol{P^T}\end{aligned}$$

Define the matrix

$$e^{\Lambda t} \quad = \quad \begin{bmatrix}e^{\lambda_1 t} & 0\\ 0 & e^{\lambda_2 t}\end{bmatrix}$$

The system of ODEs $\boldsymbol{X'} = \boldsymbol{AX}$ can be written as $\boldsymbol{X'} = \boldsymbol{P}\Lambda\boldsymbol{P^T X}$. Let $\boldsymbol{Y} = \boldsymbol{P^T X}$. Then the system becomes $\boldsymbol{Y'} = \boldsymbol{\Lambda Y}$. with general solution

$$\boldsymbol{Y}(t) \quad = \quad \begin{bmatrix}\alpha e^{\lambda_1 t} & 0\\ 0 & \beta e^{\lambda_2 t}\end{bmatrix} = \begin{bmatrix}e^{\lambda_1 t} & 0\\ 0 & e^{\lambda_2 t}\end{bmatrix}\begin{bmatrix}\alpha & 0\\ 0 & \beta\end{bmatrix}$$

for arbitrary α and β. Define

$$\boldsymbol{\Theta} \quad = \quad \begin{bmatrix}\alpha & 0\\ 0 & \beta\end{bmatrix}$$

Then the general solution is $\boldsymbol{Y}(t) = \boldsymbol{P}^T\boldsymbol{X} = e^{\boldsymbol{\Lambda}t}\boldsymbol{\Theta}$. This gives $\boldsymbol{X}(t) = \boldsymbol{P}e^{\boldsymbol{\Lambda}t}\boldsymbol{\Theta}$. Now define the matrix $e^{\boldsymbol{A}t} = \boldsymbol{P}e^{\boldsymbol{\Lambda}t}\boldsymbol{P}^T$. Then $e^{\boldsymbol{A}t}\boldsymbol{P} = \boldsymbol{P}e^{\boldsymbol{\Lambda}t}$ and so $\boldsymbol{X}(t) = e^{\boldsymbol{A}t}\boldsymbol{P}\boldsymbol{\Theta}$. Note

$$\boldsymbol{P}\boldsymbol{\Theta} = \begin{bmatrix}\boldsymbol{E_1} & \boldsymbol{E_2}\end{bmatrix}\begin{bmatrix}\alpha & 0\\ 0 & \beta\end{bmatrix} = \begin{bmatrix}\alpha\boldsymbol{E_1} & \beta\boldsymbol{E_2}\end{bmatrix}$$

So for initial data, $\boldsymbol{X_0} = \begin{bmatrix}x_0^1 & x_0^2\end{bmatrix}^T$, we have

$$\begin{bmatrix}x_0^1\\ x_0^2\end{bmatrix} = \begin{bmatrix}\alpha\boldsymbol{E_1} & \beta\boldsymbol{E_2}\end{bmatrix}$$

implying $\alpha =< \boldsymbol{X_0}, \boldsymbol{E_1} >= x_0^1$ and $\beta =< \boldsymbol{X_0}, \boldsymbol{E_1} >= x_0^2$. Thus, the solution to the initial value problem is $\boldsymbol{X}(t) = e^{\boldsymbol{A}t}\,\boldsymbol{X_0}$ and the general solution to the dynamics has the form $\boldsymbol{X}(t) = e^{\boldsymbol{A}t}\,\boldsymbol{C}$ where $\boldsymbol{C}$ is an arbitrary vector.
The general solution to the scalar ODE $x' = \lambda x$ is $x(t) = e^{\lambda t}c$ and we now know we can write the general solution to the vector system $\boldsymbol{X}' = \boldsymbol{A}\boldsymbol{X}$ is $\boldsymbol{X}(t) = e^{\boldsymbol{A}t}\boldsymbol{C}$ also as long as we interpret the exponential matrix $e^{\boldsymbol{A}t}$ right. Our argument was for 2×2 symmetric matrices, but essentially the same argument is used in the general $n \times n$ case but the canonical form of $\boldsymbol{A}$ we need is called the **Jordan Canonical Form** which is discussed in more advanced classes.

We can do this from another point of view too. Formally define the matrix $E^{\boldsymbol{A}t} = \sum_{i=0}^{\infty} \frac{\boldsymbol{A}^i}{i!}\,t^i$. The special structure of $\boldsymbol{A}$ here since it is a symmetric matrix, allows us to sum this series as $\boldsymbol{A} = \boldsymbol{P}e^{\boldsymbol{\Lambda}t}\boldsymbol{P}^T$ as we showed using the partial sums $W_n(t)$ earlier. We look at this much more carefully in Chapter 18. Something to look forward to!

5.4.2 Homework

For these problems, which are not symmetric,

- Write the matrix-vector form.
- Derive the characteristic equation.
- Find the two eigenvalues. Label the largest one as r_1 and the other as r_2.
- Find the two associated eigenvectors as unit vectors.
- Define

$$\boldsymbol{P} = \begin{bmatrix}\boldsymbol{E_1} & \boldsymbol{E_2}\end{bmatrix}$$

- Compute $\boldsymbol{P}^{-1}$.
- Compute $\boldsymbol{P}^{-1}\boldsymbol{\Lambda}\boldsymbol{P}^T$ for $\boldsymbol{\Lambda}$ the diagonal matrix of eigenvalues of $\boldsymbol{A}$. Can you say $\boldsymbol{A} = \boldsymbol{P}\boldsymbol{\Lambda}\boldsymbol{P}^{-1}$?
- Write the general solution.
- Solve the IVP.

Exercise 5.4.1 *Solve*

$$\begin{aligned} x' &= 4\,x + 2y \\ y' &= -3\,x - 3y \\ x(0) &= 4 \\ y(0) &= -6. \end{aligned}$$

Exercise 5.4.2 *Solve*

$$\begin{aligned} x' &= x + 2\,y \\ y' &= 3\,x + 2\,y \\ x(0) &= 2 \\ y(0) &= -3. \end{aligned}$$

For these problems, which are symmetric,

- Write the matrix-vector form.
- Derive the characteristic equation.
- Find the two eigenvalues. Label the largest one as r_1 and the other as r_2.
- Find the two associated eigenvectors as unit vectors.
- Define

$$\boldsymbol{P} = \begin{bmatrix} \boldsymbol{E_1} & \boldsymbol{E_2} \end{bmatrix}$$

- Compute $\boldsymbol{P^T}\ \boldsymbol{A}\ \boldsymbol{P}$ where $\boldsymbol{A}$ is the coefficient matrix of the ODE system.
- Show $\boldsymbol{A} = \boldsymbol{P \Lambda P^T}$ for an appropriate $\boldsymbol{\Lambda}$.
- Write the general solution.
- Solve the IVP.

Exercise 5.4.3 *Solve*

$$\begin{aligned} x' &= x + 2y \\ y' &= 2\,x - 6y \\ x(0) &= 4 \\ y(0) &= -6. \end{aligned}$$

Exercise 5.4.4 *Solve*

$$\begin{aligned} x' &= 2x + 3\,y \\ y' &= 3\,x + 7\,y \\ x(0) &= 2 \\ y(0) &= -3. \end{aligned}$$

Exercise 5.4.5 *If $\boldsymbol{A} = \boldsymbol{P \Lambda P^T}$, prove that $\boldsymbol{A^n} = \boldsymbol{P \Lambda^n P^T}$ by POMI.*

Exercise 5.4.6 *If $\boldsymbol{A} = \boldsymbol{P \Lambda P^T}$, show via POMI*

$$\boldsymbol{A^n} t^n = \boldsymbol{P} \begin{bmatrix} \lambda_1^n t^n & 0 \\ 0 & \lambda_2^n t^n \end{bmatrix} \boldsymbol{P^T}$$

Exercise 5.4.7 *Calculate $e^{\boldsymbol{A}t}$ for*

$$\boldsymbol{A} = \begin{bmatrix} 2 & 3 \\ 3 & 5 \end{bmatrix}$$

Exercise 5.4.8 *Calculate* $e^{\boldsymbol{A}t}$ *for*

$$\boldsymbol{A} = \begin{bmatrix} -6 & 2 \\ 2 & 7 \end{bmatrix}$$

Chapter 6

Continuity and Topology

We went over some ideas about topology in $\Re^2$ in Section 2.2 but we need to do this in $\Re^n$ now.

6.1 Topology in n Dimensions

Most of this should be familiar, although it is set in $\Re^n$.

Definition 6.1.1 Balls in $\Re^n$

- *The ball about* $\mathbf{p}$ *in* $\Re^n$ *of radius* $r > 0$ *is*

$$B(\boldsymbol{p}; r) = \{\boldsymbol{x} \in \Re^n |\, \|\boldsymbol{x} - \boldsymbol{p}\| < r\}$$

where $\|\cdot\|$ *is any norm on* $\Re^n$*. This does not have to be the usual Euclidean norm.*

- *The closed ball about* $\mathbf{p}$ *in* $\Re^n$ *of radius* $r > 0$ *is*

$$\overline{B(\boldsymbol{p}; r)} = \{\boldsymbol{x} \in \Re^n |\, \|\boldsymbol{x} - \boldsymbol{p}\| \leq r\}$$

- *The punctured ball about* $\mathbf{p}$ *in* $\Re^n$ *of radius* $r > 0$ *is*

$$\hat{B}(\boldsymbol{p}; r) = \{\boldsymbol{x} \in \Re^n |\, 0 < \|\boldsymbol{x} - \boldsymbol{p}\| \leq r\}$$

We can draw these sets in $\Re$, $\Re^2$ and $\Re^3$ but, of course, we cannot do that for $n > 3$. Hence, we **must** begin to think about these ideas **abstractly**. There is also the idea of a boundary point.

Definition 6.1.2 Boundary Points of a Subset in $\Re^n$

Let D *be a subset of* $\Re^n$*. We say* $\mathbf{p}$ *is a* **boundary point** *of* D *if for all positive radii* r*,* $B(\boldsymbol{p}; r)$ *contains a point of* D *and a point of its complement* D^C*. The set of all boundary points of* D *is denoted by* ∂D*.*

Comment 6.1.1 *If* $D = \{\boldsymbol{x}_1, \boldsymbol{x}_2\}$*, then* D *consists of two vectors only. It is clear that every ball of radius* r *smaller about* $\boldsymbol{x}_1$ *that* $\|\boldsymbol{x}_1, \boldsymbol{x}_2\|$ *contains only* $\boldsymbol{x}_1$ *from* D*. We can say the same for* $\boldsymbol{x}_2$*. So each point in* D *is a boundary point. In fact, we would call these points* **isolated** *points as there is a ball about each which contains only them.*

Next, we define **open** and **closed** subsets.

Definition 6.1.3 Open and Closed Sets in $\Re^n$

Let D be a subset of $\Re^n$. If $\mathbf{p}$ is in D and there is a positive r so that $B(\mathbf{p}; r) \subset D$, we say $\mathbf{p}$ is an **interior point** *of D.*

- *If all $\mathbf{p}$ in D are interior points, we say D is an* **open** *subset of $\Re^n$.*
- *The set of points in $\Re^n$ not in D is called the* **complement** *of D. We say D is* **closed** *if its complement is open. We denote the complement of D by D^C.*

Note the complement of a closed set is open.

Now, as usual, we can talk about the convergence of sequences $\{\boldsymbol{x_n}\}$ of points in $\Re^n$.

Definition 6.1.4 Norm Convergence in n Dimensions

We say the sequence $\{\boldsymbol{x_n}\}$ in $\Re^n$ converges in norm $\|\cdot\|$ to $\boldsymbol{x}$ in $\Re^n$ if given $\epsilon > 0$, there is a positive integer N so that

$$n > N \implies \|\boldsymbol{x_n} - \boldsymbol{x}\| < \epsilon$$

We usually just write $\boldsymbol{x_n} \to \boldsymbol{x}$.

Comment 6.1.2 *All of the usual things about scalar convergence of sequences still hold for norm convergence in $\Re^n$, but we will not clutter up the narrative by going through these proofs.*

- *If a sequence converges, all of its subsequences converge to the same limit.*
- *The usual algebra of limits theorem except no products and quotients.*
- *Convergent sequences are bounded.*

There are more and we will leave them as exercises for you.

Homework

Exercise 6.1.1 *Prove if a sequence converges, all of its subsequences converge to the same limit.*

Exercise 6.1.2 *Prove the usual algebra of limits theorem. Note we do not have products and quotients though.*

Exercise 6.1.3 *Prove convergent sequences are bounded.*

There are then special points in a set D which can be accessed via sequences of points in D. We need to define several possibilities.

Definition 6.1.5 Limit Points, Cluster Points and Accumulation Points

Let D be a subset of $\Re^n$.

- *We say $\mathbf{p}$ is a* **limit point** *of D if there is a sequence $(\boldsymbol{x_n})$ in D so that $\boldsymbol{x_n} \rightarrow \mathbf{p}$. Note $\mathbf{p}$ need not be in D.*
- *We say $\mathbf{p}$ is a* **cluster point** *of D if there is a sequence $(\boldsymbol{x_n} \neq \mathbf{p})$ in D so that $\boldsymbol{x_n} \rightarrow \mathbf{p}$.*
- *We say $\mathbf{p}$ is an* **accumulation point** *of D, for any positive radius r, $\hat{B}(\mathbf{p}; r)$ contains a points $\boldsymbol{y}$ in D. Note this clearly implies there is a sequence $(\boldsymbol{x_n} \neq \mathbf{p})$ in D so that $\boldsymbol{x_n} \rightarrow \mathbf{p}$.*

Comment 6.1.3 *If $\mathbf{p}$ is an isolated point of D, then the constant sequence $\{\boldsymbol{x_n} = \mathbf{p}\}$ in D converges to $\mathbf{p}$ and so $\mathbf{p}$ is a limit point of D. However, an isolated point is* **not** *an accumulation point or cluster point.*

We can prove a nice theorem about this now.

Theorem 6.1.1 $D \subset \Re^n$ **is Closed if and only if It Contains All Its Limit Points.**

Let D be a subset of $\Re^n$. Let D' be the set of limit points of D. Then D is closed $\Longleftrightarrow$ $D' \subset D$.

Proof 6.1.1
($\Longrightarrow$):
Assume D is a closed set. Let $\mathbf{p}$ be a limit point of D which is not in D. Then there is a sequence $\{\boldsymbol{x_n}\}$ in D with $\boldsymbol{x_n} \rightarrow \mathbf{p}$. Since $\mathbf{p}$ is not in D, it is in D^C which is open. Thus $\mathbf{p}$ is an interior point of D^C and there is an $r > 0$ so $B(r, \mathbf{p}) \subset D^C$. However, for $\epsilon = r/2$, there is an N so that $n > N \Longrightarrow \|\boldsymbol{x_n} - \mathbf{p}\| < \epsilon = r/2$. Thus, for $n > N$, $\boldsymbol{x_n}$ is in D^C. This is a contradiction. Hence, $\mathbf{p}$ must be in D. This shows $D' \subset D$.

($\Longleftarrow$):
Assume D contains D' and consider D^C. If D^C is not open, then there is a $\mathbf{p}$ in D^C which is not an interior point of D^C. Hence for the sequence $\{r_n = 1/n\}$, there is a point $\boldsymbol{x_n}$ in $B(\mathbf{p}, 1/n)$ with $\boldsymbol{x_n}$ in D. Clearly, $\boldsymbol{x_n} \rightarrow \mathbf{p}$ and hence $\mathbf{p}$ is a limit point of D and so must be in D by assumption. But this is not possible by our assumption. So our assumption must be wrong and D^C must be open. This tells us D is closed. ■

This leads to the definition of the **closure** of a set.

Definition 6.1.6 The Closure of a Set

The closure of a set D in $\Re^n$ is denoted by $\overline{D} = D \cup D'$.

We can then prove a useful characterization of the closure of a set.

Theorem 6.1.2 A Set D **is Closed if and only if** $D = \overline{D}$

The set D is closed if and only if $D = \overline{D}$.

Proof 6.1.2
($\Longrightarrow$):

If D is closed, $D' \subset D$ and so $D \cup D' = D$. So $\overline{D} = D$.

($\Longleftarrow$):
If $D = \overline{D} = D \cup D'$, Thus, D' is in D, which tells us D is closed. ■

There is a lot more $\Re^n$ topology we could discuss but this is enough to get you started.

6.1.1 Homework

Exercise 6.1.4 *Prove finite unions of open sets in $\Re^n$ are open.*

Exercise 6.1.5 *Prove finite intersections of open sets in $\Re^n$ are open.*

Exercise 6.1.6 *Prove DeMorgan's Laws.*

Exercise 6.1.7 *Prove finite unions of closed sets in $\Re^n$ are closed.*

Exercise 6.1.8 *Prove finite intersections of closed sets in $\Re^n$ are closed.*

Exercise 6.1.9 *Boundary points in $\Re^2$ are much more interesting. Draw any rectangle you want and look at the edges of the rectangle. Make sure you understand how points on the edges are boundary points. Now do the same for a box in $\Re^3$ and focus on the faces and edges of the box. Make sure you understand which points are boundary points in this case.*

6.2 Cauchy Sequences

We have seen Cauchy sequences frequently in (Peterson (21) 2020) and the same idea is easy to move to the $\Re^n$ setting.

Definition 6.2.1 Cauchy Sequences in $\Re^n$

A sequence $\{\boldsymbol{x_n}\}$ in $\Re^n$ satisfies

$$\forall \epsilon > 0, \exists N \ni n, m > N \Longrightarrow \|\boldsymbol{x_n} - \boldsymbol{x_m}\| < \epsilon$$

We know from the $\Re$ setting that Cauchy sequences of real numbers must converge to a real number due to the **Completeness Axiom**. We also know from discussions in (Peterson (21) 2020) that Cauchy sequences in general normed spaces need not converge to an element in the space. Normed spaces in which Cauchy sequences converge to an element of the space are called **Complete Spaces**. We can easily prove $\Re^n$ is a complete space.

Theorem 6.2.1 $\Re^n$ is Complete

$\Re^n$ is a complete normed space.

Proof 6.2.1
Let $\{\boldsymbol{x_n}\}$ in $\Re^n$ be a Cauchy sequence in $\Re^n$. Pick $\epsilon > 0$. Then there is an N so that

$$n, m > N \Longrightarrow \|\boldsymbol{x_n} - \boldsymbol{x_m}\| < \epsilon/2$$

or

$$n, m > N \Longrightarrow \sum_{i=1}^{n} |x_{ni} - x_{mi}|^2 < \epsilon^2/4$$

Thus each piece of this sum satisfies

$$n, m > N \Longrightarrow |x_{ni} - x_{mi}| < \epsilon/2$$

This says $\{x_{ni}\}$ *is a Cauchy sequence in* $\Re$ *and so must converge to a number* x_i. *Let* $\boldsymbol{x}$ *be the vector whose components are* x_i. *Then, since* $|\cdot|$ *is continuous, for* $n, m > N$, *we have*

$$\lim_{m \to \infty} \left(\sum_{i=1}^{n} |x_{ni} - x_{mi}|^2 \right) \leq \epsilon^2/4$$

So

$$\| \boldsymbol{x_n} - \boldsymbol{x} \|^2 = \sum_{i=1}^{n} |x_{ni} - x_i|^2 \leq \epsilon^2/4$$

which implies $\|\boldsymbol{x_n} - \boldsymbol{x}\| < \epsilon$ *when* $n > N$. *This shows* $\Re^n$ *is complete.* ■

6.3 Compactness

Theorem 6.3.1 Bolzano - Weierstrass Theorem in $\Re^n$

Every bounded sequence in $\Re^n$ *has at least one convergent subsequence.*

Proof 6.3.1
Let's assume the range of this sequence is infinite. If it were finite, there would be subsequences of it that converge to each of the values in the finite range. We assume the sequences start at $n = 1$ *for convenience and by assumption, there is a positive number* B *so that* $\|\boldsymbol{a_n}\| \leq B/2$ *for all* $n \geq 1$. *Hence, all the components of this sequence live in what could be called a hyper rectangle of the form* $\mathscr{B} = [-B/2, B/2] \times \ldots [-B/2, B/2] \subset \Re^n$. *We can denote this by* $\mathscr{B} = \prod_{i=1^n} [-B/2, B/2]$. *Let* $J_0 = \prod_{i=1^n} \boldsymbol{I_0^i}$ *be the hyper rectangle* $[\alpha_{01}, \beta_{01}] \times \ldots [\alpha_{0n}, \beta_{0n}]$. *Here,* $\alpha_{0i} = -B/2$ *and* $\beta_{0i} = B/2$ *for all* i*; i.e. on each axis. The* n *dimensional area of* J_0, *denoted by* ℓ_0 *is* B^n.

Let S *be the range of the sequence which has infinitely many points and for convenience, we will let the phrase* infinitely many points *be abbreviated to* **IMPs**.

Step 1*:*
Bisect each axis interval of J_0 *into two pieces giving* 2^n *subregions of* J_0 *all of which have area* $B^2/4$. *Now at least one of the subregions contains IMPs of* S *as otherwise each subregion has only finitely many points and that contradicts our assumption that* S *has IMPs. Now all may contain IMPs so select one such subregion containing IMPs and call it* J_1. *Label the endpoints of the component intervals making up the cross product that define* J_1 *as* $\boldsymbol{I_1^i} = [\alpha_{1i}, \beta_{1i}]$. *hence,* $J_1 = \prod_{i=1}^{n} \boldsymbol{I_1^i}$ *with* $\ell_1 = B^2/4$. *We see* $J_1 \subset J_0$ *and on the* j^{th} *axis, the subinterval endpoints satisfy*

$$-B/2 = \alpha_{0i} \leq \alpha_{1i} \leq \beta_{1i} \leq \beta_{0i} = B/2$$

Since J_1 *contains IMPs, we can select a sequence vector* $\boldsymbol{a_{n_1}}$ *from* J_1.

Step 2*:*
Now subdivide J_1 *into* 2^n *subregions just as before. At least one of these subregions contains IMPs of* S. *Choose one such subregion and call it* J_2. *Then* J_2 *is the cross product of the intervals*

$\boldsymbol{I_2^i} = [\alpha_{2i}, \beta_{2i}]$. *The area of this subregion is now* $\ell_2 = B^2/16$. *We see* $J_2 \subseteq J_1 \subseteq J_0$ *and*

$$-B = \alpha_{0i} \leq \alpha_{1i} \leq \alpha_{2i} \leq \beta_{2i} \leq \beta_{1i} \leq \beta_{0i} = B$$

Since J_2 *contains IMPs, we can select a sequence vector* $\boldsymbol{a_{n_2}}$ *from* J_2. *It is easy to see this value can be chosen different from* $\boldsymbol{a_{n_1}}$, *our previous choice.*

You should be able to see that we can continue this argument using induction.

Proposition*:*
$\forall p \geq 1$, $\exists$ *an interval* $J_p = \prod_{i=1} n\, I_p^i$ *with* $\boldsymbol{I_p^i} = [\alpha_{pi}, \beta_{pi}]$ *with the area of* J_p, $\ell_p = B^2/(2^{2p})$ *satisfying* $J_p \subseteq J_{p-1}$, J_p *contains IMPs of* $\mathcal{S}$ *and*

$$-B/2 \leq \alpha_{0i} \leq \ldots \leq \alpha_{1,i} \leq \ldots \leq \alpha_{p,i} \quad \leq \quad \beta_{pi} \leq \cdots \leq \beta_{1i} \leq \beta_{0i} \leq B/2$$

Finally, there is a sequence vector $\boldsymbol{a_{n_p}}$ *in* J_p, *different from* $\boldsymbol{a_{n_1}}, \ldots, \boldsymbol{a_{n_{p-1}}}$.

Proof *We have already established the proposition is true for the basis steps* J_1 *and* J_2.
Inductive: *We assume the interval* J_q *exists with all the desired properties. Since by assumption,* J_q *contains IMPs, bisect* J_q *into* 2^n *subregions as we have done before. At least one of these subregions contains IMPs of* $\mathcal{S}$. *Choose one of the subregions and call it* J_{q+1} *and label* $J_{q+1} = \prod_{i=1}^n \boldsymbol{I_{q+1}^i}$ *where* $\boldsymbol{I_{q+1}^i} = [\alpha_{q+1,i}, \beta_{q+1,i}]$ *We see immediately* $\ell_{q+1} = B^2/2^{2(q+1)}$ *with*

$$\alpha_{q,i} \leq \alpha_{q+1,i} \quad \leq \quad \beta_{q+1,i} \leq \beta_{qi}$$

This shows the nested inequality we want is satisfied. Finally, since J_{q+1} *contains IMPs, we can choose* $\boldsymbol{a_{n_{q+1}}}$ *distinct from the other* $\boldsymbol{a_{n_i}}$*'s. So the inductive step is satisfied and by the POMI, the proposition is true for all* n. □

From our proposition, we have proven the existence of sequences on each axis: (α_{pi}), (β_{pi}) *and* (ℓ_p) *which have various properties. The sequence* ℓ_p *satisfies* $\ell_p = (1/4)\ell_{p-1}$ *for all* $p \geq 1$. *Since* $\ell_0 = B^2$, *this means* $\ell_1 = B^2/4$, $\ell_2 = B^2/16$, $\ell_3 = B^2/(2^2)^3$ *leading to* $\ell_p = B^2/(2^2)^p$ *for* $p \geq 1$. *Further, we have the inequality chain*

$$\begin{aligned} -B/2 = \alpha_{0i} & \leq \alpha_{1i} \leq \alpha_{2I} \leq \ldots \leq \alpha_{pi} \\ & \leq \ldots \leq \\ \beta_{pi} & \leq \ldots \leq \beta_{2i} \leq \ldots \leq \beta_{0I} = B/2 \end{aligned}$$

The rest of this argument is very familiar. Note (α_{pi}) *is bounded above by* $B/2$ *and* (β_{pi}) *is bounded below by* $-B/2$. *Hence, by the completeness axiom,* $\inf\,(\beta_{pi})$ *exists and equals the finite number* β^i; *also* $\sup\,(\alpha_{pi})$ *exists and is the finite number* α^i.

So if we fix p, it should be clear the number β_{pi} *is an upper bound for all the* α_{pi} *values (look at our inequality chain again and think about this). Thus* β_{pi} *is an upper bound for* (α_{pi}) *and so by definition of a supremum,* $\alpha^i \leq \beta_{pi}$ *for all p. Of course, we also know since* α^i *is a supremum, that* $\alpha_{pi} \leq \alpha^i$. *Thus,* $\alpha_{pi} \leq \alpha^i \leq \beta_{pi}$ *for all p. A similar argument shows if we fix p, the number* α_{pi} *is a lower bound for all the* β_{pi} *values and so by definition of an infimum,* $\alpha_{pi} \leq \beta^i \leq \beta_{pi}$ *for all the* α_{pi} *values. This tells us* α^i *and* β^i *are in* $[\alpha_{pi}, \beta_{pi}]$ *for all p. Next we show* $\alpha^i = \beta^i$.

Let $\epsilon > 0$ *be arbitrary. Since* α^i *and* β^i *are in an interval whose length is* $\ell_p = (1/2^{2p})B^2$, *we have* $|\alpha^i - \beta^i| \leq (1/2^{2p})B^2$. *Pick* P *so that* $1/(2^{2P}B^2) < \epsilon$. *Then* $|\alpha^- \beta^i| < \epsilon$. *But* $\epsilon > 0$ *is arbitrary.*

Hence, $\alpha^i - \beta^i = 0$ implying $\alpha^i = \beta^i$. Finally, define the vector

$$\boldsymbol{\theta} = \begin{bmatrix} \alpha^1 \\ \vdots \\ \alpha^n \end{bmatrix}$$

We now must show $\boldsymbol{a_{n_k}} \rightarrow \boldsymbol{\theta}$. This shows we have found a subsequence which converges to $\boldsymbol{\theta}$. We know $\alpha_{pi} \le a^i_{n_p} \le \beta_{pi}$ and $\alpha_{pi} \le \alpha^i \le \beta^i_p$ for all p. Pick $\epsilon > 0$ arbitrarily. Given any p, we have

$$\begin{aligned} |a^i_{n_p,1} - \alpha^i| &= |a^i_{n_p,i} - \alpha_{pi} + \alpha_{pi} - \alpha^i| \le |a^i_{n_p,i} - \alpha_{pi}| + |\alpha_{pi} - \alpha^i| \\ &\le |\beta_{pi} - \alpha_{pi}| + |\alpha_{pi} - \beta_{pi}| = 2|\beta_{pi} - \alpha_{pi}| = 2\,(1/2^{2p})B^2. \end{aligned}$$

Choose P so that $(1/2^{2P})B^2 < \epsilon/2$. Then, $p > P$ implies $|a^i_{n_p,1} - \alpha^i| < 2\,\epsilon/2 = \epsilon$. Thus, $a_{n_k,i} \rightarrow \alpha^i$. This shows the subsequence converges to $\boldsymbol{\theta}$. ■

Note this argument is messy but quite similar to the one and two dimensional cases. It is more a problem of correct labeling than intellectual difficulty!

There are several notions of **compactness** for a subset of $\Re^n$. There is **topological compactness** and **sequential compactness**.

Definition 6.3.1 Topologically Compact Subsets of $\Re^n$

Let D be a subset of $\Re^n$. We say the collection of open sets $\{U_\alpha\}$ where $\alpha \in I$ and I is any index set, whether finite, countable or uncountable, is an open cover of D if $D \subset \cup_\alpha U_\alpha$. We say this collection has a finite subcover if there are a finite number of sets in the collection, $\{U_{\beta_1}, \ldots, U_{\beta_p}\}$ so that $D \subset \cup_{i=1}^p U_{\beta_i}$. D is topologically compact if every open cover of D has a finite subcover.

Definition 6.3.2 Sequentially Compact Subsets of $\Re^n$

Let D be a subset of $\Re^n$. We say D is sequentially compact if every sequence $\{\boldsymbol{x_n}\}$ in D has at least one subsequence which converges to a point of D.

We need to prove their equivalence through a sequence of results. Compare how we prove these to their one dimensional analogues in (Peterson (21) 2020).

Theorem 6.3.2 D is Sequentially Compact if and only if it is Closed and Bounded

D in $\Re^n$ is sequentially compact if and only if it is bounded and closed.

Proof 6.3.3

($\Longrightarrow$)

We assume D is sequentially compact. Assume $(\boldsymbol{x_n})$ in D converges to $\boldsymbol{x}$. Since $(\boldsymbol{x_n})$ is a sequence in the sequentially compact set D, it has a subsequence $(\boldsymbol{x^1_n})$ which converges to $\boldsymbol{y}$ in D. Since limits of convergent sequences are unique, we must have $\boldsymbol{x} = \boldsymbol{y}$. So $\boldsymbol{x}$ is in D and hence D is closed.

To show D is bounded, we use contradiction. Assume it is not bounded. Then for each n, there is $\boldsymbol{x_n}$ in D so that $\|\boldsymbol{x_n}\| > n$. But since D is sequentially compact there is a subsequence $(\boldsymbol{x_{n_k}})$ which must converge to $\boldsymbol{x}$ in D. We know convergence sequences are bounded and so there is a positive number B so that $\|bsx_{n_k}\| < B$ for all elements of the subsequence. But eventually

$\|bsx_{n_k}\| > n_k > B$ which is the contradiction. We conclude D must be bounded.

($\Longleftarrow$)
Now we assume D is closed and bounded. We argue that D is sequentially compact using essentially the same procedure we used to prove the Bolzano - Weierstrass Theorem for $\Re^n$. First, if D is a finite set, it is sequentially compact. So we can assume D has infinitely many points. Since D is bounded, there is a positive number B so that $D \subset \prod_{i=1}^n [-B/2/, B/2]$. Let $D_0 = D$ and pick any x_0 in D_0. Now bisect each of the coordinate axis giving 2^n pieces. Since D is infinite, at least one piece contains infinitely many points of D. Call this piece D_1 and pick $\boldsymbol{x_1} \neq \boldsymbol{x_0} \in D_1$. Since both points are in D_0,

$$\|\boldsymbol{x_1} - \boldsymbol{x_0}\|^2 = \sum_{i=1}^n |x_{1i} - x_{0i}|^2 < nB^2 \quad \Longrightarrow \|\boldsymbol{x_1} - \boldsymbol{x_0}\| < \sqrt{n}B$$

Now do this again. We bisect D_1 on each axis to create 2^n pieces. Choose one piece containing infinitely many points of D and call it D_2. The lengths of the sides of this piece are now $B/4$. Pick $\boldsymbol{x_1} \in D_2$, not the same as the previous two. Then $\boldsymbol{x_2}$ and $\boldsymbol{x_1}$ are both in D_1 and so $\|\boldsymbol{x_2} - \boldsymbol{x_1}\| < \sqrt{n}(B/2)$. Continuing this process, we obtain a sequence $\boldsymbol{x_n}$ with $\boldsymbol{x_n} \in D_n$, The sides of D_n are length $B/2^n$ and both $\boldsymbol{x_{n-1}}$ and $\boldsymbol{x_n}$ are in D_{n-1}. So $\|\boldsymbol{x_{n-1}} - \boldsymbol{x_n}\| < \sqrt{n}(B/2^{n-1})$. We claim this sequence is a Cauchy sequence in $\Re^n$. For $k > m$,

$$\begin{aligned} \|\boldsymbol{x_k} - \boldsymbol{x_m}\| &\leq \sum_{j=m}^{k-1} \|\boldsymbol{x_{j+1}} - \boldsymbol{x_j}\| \leq \sum_{j=m}^{k-1} \sqrt{n}\frac{B}{2^j} \leq \sqrt{n}B\left(\sum_{j=0}^{k-1} \frac{1}{2^j} - \sum_{j=0}^{k-1} \frac{1}{2^j}\right) \\ &= \sqrt{n}B\left(\frac{1-1/2^k}{1-1/2} - \frac{1-1/2^m}{1-1/2}\right) \leq 2\sqrt{n}B\left(\frac{1}{2^m} - \frac{1}{2^k}\right) \leq \sqrt{n}B/2^{m-1} \end{aligned}$$

Then, given $\epsilon > 0$, there is an N so that $\sqrt{n}B/2^{m-1} < \epsilon$ if $m > N$. Thus, $\|\boldsymbol{x_k} - \boldsymbol{x_m}\| < \epsilon$ if $k > m > N$ which shows $(\boldsymbol{x_n})$ is a Cauchy sequence. Since $\Re^n$ is complete, there is $\boldsymbol{x}$ so that $\boldsymbol{x_n} \rightarrow \boldsymbol{x}$. Since D is closed, it follows $\boldsymbol{x}$ is in D. Now this same argument applies to any infinite sequence $(\boldsymbol{w_n})$ in D. This sequence will have a subsequence $(\boldsymbol{w_n^1})$ which converges to a point $\boldsymbol{w} \in D$. Hence, D is sequentially compact. ■

Theorem 6.3.3 $D \subset \Re^n$ **is Topologically Compact Implies it is Closed and Bounded**

$D \subset \Re^n$ is topologically compact implies it is closed and bounded.

Proof 6.3.4
The collection $\mathscr{O} = \{B(r, \boldsymbol{x}) | \boldsymbol{x} \in D\}$ is an open cover of D. Since D is topologically compact, $\mathscr{O}$ has a finite subcover $\mathscr{V} = \{B(r, \boldsymbol{x_i}) | 1 \leq i \leq N\}$ for some positive integer N. If $\boldsymbol{x} \in D$, there is an index i so that $\boldsymbol{x} \in B(r, \boldsymbol{x_i})$. Thus $\|\boldsymbol{x} - \boldsymbol{x_i}\| < r$ which implies $\|\boldsymbol{x}\| \leq r + \|\boldsymbol{x_i}\|$. Let $B = \max_{1 \leq i \leq N} \|\boldsymbol{x_i}\|$. Then, $\|\boldsymbol{x}\| \leq r + B$ which shows D is bounded.

Let's show D^C is open. Let $\boldsymbol{x} \in D^C$. For any $\boldsymbol{y}$ in D, let $d_{xy} = \|\boldsymbol{x} - \boldsymbol{y}\|$, which is not zero as $\boldsymbol{x}$ is in the complement. The collection $\mathscr{O} = \{B((1/2)d_{xy}, \boldsymbol{y}) | \boldsymbol{y} \in D\}$ is an open cover of D and so there is a finite subcover $\mathscr{V} = \{B((1/2)d_{x,y_i}, \boldsymbol{x_i}) | 1 \leq i \leq N\}$. Let $W = \cap B((1/2)d_{x,y_i}, \boldsymbol{x})$. By construction, each $B((1/2)d_{x,y_i}, \boldsymbol{x})$ is disjoint from $B((1/2)d_{x,y_i}, \boldsymbol{y_i})$. So if $\boldsymbol{z}$ is in W, $\boldsymbol{z}$ is not in all $B((1/2)d_{x,y_i}, \boldsymbol{y_i})$. But since $\mathscr{V}$ is an open cover for D, this says $\boldsymbol{z} \in D^C$. Hence, if $r = \min_{1 \leq i \leq N}\{(1/2)d_{x,y_i}\}$, then $B(r, \boldsymbol{x})$ is contained in D^C. This shows $\boldsymbol{x}$ is an interior point of D^C. So D^C is open which implies D is closed. ■

Theorem 6.3.4 A Hyper Rectangle is Topologically Compact

The hyper rectangle $D = \prod_{i=1}^{n}[a_i, b_i]$ is topologically compact when each segment $[a_i, b_i]$ is finite in length.

Proof 6.3.5

This proof is similar to how we proved the Bolzano - Weierstrass Theorem (BWT). So look at the similarities! Assume this is false and let $J_0 = \prod_{i=1}^{n}[a_i, b_i]$. Label this interval as $J_0 = [\alpha_0, \beta_0]$; i.e. $\alpha_0 = a$ and $\beta_0 = b$. The volume of this hyper rectangle is $\ell_0 = \prod_{i=1}^{n}(b_1 - a_i)$. Assume there is an open cover $\mathscr{U}$ which does not have a fsc.

Now divide J_0 into 2^n pieces by bisecting each segment $[a_i, b_i]$. At least one of these pieces cannot be covered by a fsc as otherwise all of D has a fsc, which we have assumed is not true. Call this piece J_1 and note the volume of J_1 is $\ell_1 = (1/2)\ell_0$. We can continue in this way by induction. Assume we have constructed the piece J_n in this fashion with volume $\ell_n = (1/2)\ell_{n-1} = \ell_0/2^n$ and there is no fsc of J_n. Now divide J_n into 2^n equal pieces as usual. At least one of them cannot be covered by a fsc. Call this interval J_{n+1} which has volume $\ell_{n+1} = (1/2)\ell_n$ like required.

We can see each $J_{n+1} \subset J_n$ by the way we have chosen the pieces and their volumes satisfy $\ell_n \to 0$. Using the same sort of arguments that we used in the proof of the BWT for sequences, we see there is a unique point $\boldsymbol{z}$ which lies in all the J_n; i.e. $\boldsymbol{z} \in \cap_{i=1}^{\infty} J_n$.

Since $\boldsymbol{z}$ is in J_1, there is a $\mathbb{O}_1$ in $\mathscr{U}$ with $\boldsymbol{z} \in \mathbb{O}_1$ because $\mathscr{U}$ covers D. Also, since $\mathbb{O}_1$ is open, there is a circle $B(r_1, \boldsymbol{z}) \subset \mathbb{O}_1$. Now choose K so that $\ell_K < \ell_1$. Then, J_k is contained in $\mathbb{O}_1$ for all $k > K$. This says $\mathbb{O}_1$ is a fsc of J_k which contradicts our construction process. Hence, our assumption that there is an open cover with no fsc is wrong and we have D is topologically compact.

■

Theorem 6.3.5 Closed Subsets of Hyper Rectangles are Topologically Compact

Let C be a closed subset of the hyper rectangle $D = \prod_{i=1}^{n}[a_i, b_i]$ when each segment $[a_i, b_i]$ is finite in length. Then D is topologically compact.

Proof 6.3.6

Let $\mathscr{U}$ be an open cover of B. Then the collection of open sets $\mathscr{O} = \mathscr{U} \cup B^C$ is an open cover of D and hence has a fsc $\mathscr{V} = \{\mathbb{O}_1, \ldots, \mathbb{O}_N, B^C\}$. for some N with each $\mathbb{O}_n \in \mathscr{U}$. Hence, $B \subset \cup_{i=1}^{N}\mathbb{O}_n$ and we have found a fsc of $\mathscr{U}$ which covers B. This shows B is topologically compact. ■

Theorem 6.3.6 *D* is Topologically Compact if and only if it is Closed and Bounded

D in $\Re^n$ is topologically compact if and only if it is closed and bounded

Proof 6.3.7

($\Longrightarrow$)

If D is topologically compact, by Theorem 6.3.3, it is closed and bounded.

($\Longleftarrow$)

If D is bounded, D is inside a hyper rectangle with finite edge lengths. Since D is a closed subset of the hyper rectangle, D is topologically compact by Theorem 6.3.5. ■

Theorem 6.3.7 D is Topologically Compact if and only if Sequentially Compact if and only if Closed and Bounded

Let D be a subset of $\Re^n$. Then D is topologically compact $\iff$ D is sequentially compact $\iff$ D is closed and bounded.

Proof 6.3.8
(Topologically Compact $\iff$ Closed and Bounded)
This is Theorem 6.3.6.

(Sequentially Compact $\iff$ Closed and Bounded)
This is Theorem 6.3.2.
■

Comment 6.3.1 *Since in these finite dimensional settings, the notions of* **sequential compactness, topological compactness** *and* **closed and bounded** *are equivalent, we often just say a set is* **compact** *and then use whichever characterization is useful for our purposes.*

6.3.1 Homework

Exercise 6.3.1 *Prove finite unions of compact subsets in $\Re^n$ are compact. Note you can do this three ways: use topological compactness, use sequential compactness and use the characterization of the compact sets as closed and bounded. Think through the various approaches.*

Exercise 6.3.2 *Prove finite intersections of compact subsets in $\Re^n$ are compact. Think through the various approaches here as well.*

Exercise 6.3.3 *If* $\left(\boldsymbol{x}_n = \begin{bmatrix} \sin(n) \\ \cos(n^2) \\ \sin(2n+5) \end{bmatrix} \right)$ *is a sequence in $\Re^3$, is it true it has at least one convergent subsequence?*

Exercise 6.3.4 *Let $f : [0,1] \times [0,1] \to [0,1] \times [0,1]$. Consider the sequence $(f^n(\boldsymbol{x_0}))$ where $f^n = f \circ \ldots \circ f$ is the composition of f with itself n times. Is it true it has at least one convergent subsequence?*

6.4 Functions of Many Variables

In (Peterson (21) 2020), we developed a number of simple MATLAB scripts and functions to help us visualize functions of two variables. Of course, these are of little use for functions of three or more variables. For example, if $y = f(x)$ we know how to *see* this graphically. We graph the pairs $(x, f(x))$ in the usual $x - y$ plane. If we had $z = f(x, y)$, we would graph the triples $(x, y, f(x, y))$ in three space. We would use the right-hand rule for the positioning of the $+z$ axis like usual. However, for $w = f(x, y, z)$, the visualization would require us to draw in $\Re^4$ which we cannot do. So much of what we want to do now must be done abstractly. Now let's look at this in the n dimensional setting. Let D be a subset of $\Re^n$ and assume $\boldsymbol{z} = f(\boldsymbol{x}) \in \Re^m$ is defined locally on $B(\boldsymbol{x_0}, r)$ for some positive r. This defines m **component** functions $f_1, \ldots, f_m$ by

$$f(\boldsymbol{x}) = f(x_1, \ldots, x_n) = (f_1(\boldsymbol{x}), \ldots, f_m(\boldsymbol{x}))$$

The set $D \subset \Re^n$ is at least an open set and we usually want it to be **path connected** as well. We will talk about this later, but a path connected subset of $\Re^n$ is one in which there is a **path** connecting any two points in U that lies entirely in D. We discuss this much more carefully in Chapter 16 but for now we will be casual about it.

Example 6.4.1 *Although $\boldsymbol{x}$ here is in $\Re^n$ and so is usually thought of as a column vector, we typically write it as a row vector. Here is a typical function $f : \Re^3 \rightarrow \Re^3$;*

$$\begin{aligned} f\left(\begin{bmatrix} x_1 \\ x_2 \\ x_3 \end{bmatrix}\right) = f(x_1, x_2, x_3) &= \begin{bmatrix} x_1^2 + 3x_3^2 \\ 2x_1 + 5x_2 - \sin(x_3^2) \\ 14x_2^4 x_3^2 + 20 \end{bmatrix} \\ &= \begin{bmatrix} x_1^2 + 3x_3^2, & 2x_1 + 5x_2 - \sin(x_3^2), & 14x_2^4 x_3^2 + 20 \end{bmatrix} \end{aligned}$$

These notational abuses are quite common. For example, it is very normal to use row vector notation for the domain of f and column vector notation for the range of f. Go figure. For this example, the component functions are

$$\begin{aligned} f_1(x_1, x_2, x_3) &= x_1^2 + 3x_3^2, \quad f_2(x_1, x_2, x_3) = 2x_1 + 5x_2 - \sin(x_3^2) \\ f_3(x_1, x_2, x_3) &= 14x_2^4 x_3^2 + 20 \end{aligned}$$

6.4.1 Limits and Continuity for Functions of Many Variables

Now let's look at continuity in this n dimensional setting. Let D be a subset of $\Re^n$. and assume $z = f(\boldsymbol{x})$ is defined locally on $B(\boldsymbol{x_0}, r)$ for some positive r. Here is the two dimensional extension of the idea of a limit of a function.

Definition 6.4.1 The $\Re^n$ to $\Re^m$ Limit

If $\lim_{\boldsymbol{x} \rightarrow \boldsymbol{x_0}} f(\boldsymbol{x})$ exists, this means there is a vector $\boldsymbol{L} \in \Re^m$ so that

$$\forall \epsilon > 0 \; \exists 0 < \delta < r \; \ni 0 < \|\boldsymbol{x} - \boldsymbol{x_0}\| < \delta \Rightarrow \|f(\boldsymbol{x}) - \boldsymbol{L}\| < \epsilon$$

We say $\lim_{\boldsymbol{x} \rightarrow \boldsymbol{x_0}} f(\boldsymbol{x}) = \boldsymbol{L}$. We use the same notation $\| \cdot \|$ for the norm in both the domain and the range. Of course, these norms are defined separately and we need to use the correct norms on both parts as needed.

We can now define new versions of limits and continuity for n variable functions.

Definition 6.4.2 $\Re^n$ to $\Re^m$ Continuity

If $\lim_{\boldsymbol{x} \rightarrow \boldsymbol{x_0}} f(\boldsymbol{x})$ exists and matches $f(\boldsymbol{x_0})$, we say f is continuous at $\boldsymbol{x_0}$. That is

$$\begin{aligned} &\forall \epsilon > 0 \; \exists 0 < \delta < r \; \ni \|\boldsymbol{x} - \boldsymbol{x_0}\| < \delta \\ &\Rightarrow \|f(\boldsymbol{x}) - f(\boldsymbol{x_0})\| < \epsilon \end{aligned}$$

We say $\lim_{\boldsymbol{x} \rightarrow \boldsymbol{x_0}} f(\boldsymbol{x}) = f(\boldsymbol{x_0})$.

Statements about limits and continuity of these functions are equivalent to statements about limits and continuity of the component functions.

Theorem 6.4.1 The Behavior of $f : D \subset \Re^n \rightarrow \Re^m$ is Equivalent to the Behavior of Its Component Functions

Let $f : D \subset \Re^n \rightarrow \Re^m$. Then

- *f has a limit at $\boldsymbol{x_0}$ of value $\boldsymbol{L}$ if and only if $f_i(\boldsymbol{x})$ has limit L_i*
- *f is continuous at $\boldsymbol{x_0}$ if and only if $f_i(\boldsymbol{x})$ is continuous at $\boldsymbol{x_0}$.*

Proof 6.4.1
We will leave this proof to you. ∎

Example 6.4.2 *Prove*

$$f(x,y) = \begin{bmatrix} 2x^2 + 3y^2 \\ 3x^2 + 5y^2 \end{bmatrix}$$

is continuous at $(2,3)$.

Solution *We find*

$$f(2,3) = \begin{bmatrix} 8 + 27 = 35 \\ 12 + 45 = 57 \end{bmatrix}$$

We will use the Euclidean norm on $\Re^2$ here. Note, if we choose $r < 1$, then $\|(x,y) - (2,3)\| < r$ implies $|x-2| < r$ and $|y-3| < r$. Then

$$\begin{aligned} \left\| \begin{bmatrix} 2x^2 + 3y^2 - 35 \\ 3x^2 + 5y^2 - 57 \end{bmatrix} \right\| &= \left\| \begin{bmatrix} 2(x-2+2)^2 + 3(y-3+3)^2 - 35 \\ 3(x-2+2)^2 + 5(y-3+3)^2 - 57 \end{bmatrix} \right\| \\ &= \left\| \begin{bmatrix} 2(x-2)^2 + 8(x-2) + 3(y-3)^2 + 6(y-3) \\ 3(x-2)^2 + 12(x-2) + 5(y-2)^2 + 30(y-3) \end{bmatrix} \right\| \\ &= \left\| \begin{bmatrix} 2(x-2)^2 \\ 3(x-2)^2 \end{bmatrix} + \begin{bmatrix} 8(x-2) \\ 12(x-2) \end{bmatrix} + \begin{bmatrix} 3(y-3)^2 \\ 5(y-2)^2 \end{bmatrix} + \begin{bmatrix} 6(y-3) \\ 30(y-3) \end{bmatrix} \right\| \end{aligned}$$

From the triangle inequality for $\|\cdot\|$, since this is the Euclidean norm, we have

$$\begin{aligned} \left\| \begin{bmatrix} 2x^2 + 3y^2 - 35 \\ 3x^2 + 5y^2 - 57 \end{bmatrix} \right\| &\le 3\sqrt{2}|x-2|^2 + 12\sqrt{2}|x-2| + 5\sqrt{2}|y-3|^2 + 30\sqrt{2}|y-3| \\ &\le 8\sqrt{2}r^2 + 42\sqrt{2}r < 50\sqrt{2}r \end{aligned}$$

Hence, given $\epsilon > 0$, if we choose $r < \min(1, \epsilon/(50\sqrt{2}))$ we have $\delta < r$ implies $\|f(x,y) - f(2,3)\| < \epsilon$ when $(x,y) \in B((2,3), r)$.

Homework

Exercise 6.4.1 *Prove Theorem 6.4.1:*
Let $f : D \subset \Re^n \rightarrow \Re^m$. Then

- *f has a limit at $\boldsymbol{x_0}$ of value $\boldsymbol{L}$ if and only if $f_i(\boldsymbol{x})$ has limit L_i*
- *f is continuous at $\boldsymbol{x_0}$ if and only if $f_i(\boldsymbol{x})$ is continuous at $\boldsymbol{x_0}$.*

Exercise 6.4.2 *Let*

$$f(x,y) = \begin{bmatrix} 6x^2 + 9y^2 \\ 3xy \end{bmatrix}$$

Prove f is continuous at $(0,0)$ both ways; i.e. as a function of two variables and with component functions.

Exercise 6.4.3 *Let*

$$f(x,y) = \begin{bmatrix} 6x^2 + 9y^2 \\ 3xy \end{bmatrix}$$

Prove f is continuous at $(2,5)$ both ways; i.e. as a function of two variables and with component functions.

Exercise 6.4.4 *Let*

$$f(x,y) = \begin{bmatrix} 2x^2 - 3y^2 \\ 7x + 8y \end{bmatrix}$$

Prove f is continuous at $(0,0)$ both ways; i.e. as a function of two variables and with component functions.

Exercise 6.4.5 *Let*

$$f(x,y) = \begin{bmatrix} 2x^2 - 3y^2 \\ 7x + 8y \end{bmatrix}$$

Prove f is continuous at $(-1,2)$ both ways; i.e. as a function of two variables and with component functions.

Note it is probably easier in practice to show each component is continuous separately. We get two different critical values of r and we just choose the minimum of them to prove continuity.

Discontinuities are much harder to analyze in multiple dimensions as a discontinuity need not be a single point. It could be a curve!

Example 6.4.3 *Show*

$$f(x,y) = \begin{cases} \begin{bmatrix} \frac{2x}{\sqrt{x^2+y^2}} \\ x^2 + 3y^2 \end{bmatrix}, & \text{if } (x,y) \neq (0,0) \\ 0, & \text{if } (x,y) = (0,0). \end{cases}$$

is not continuous at $(0,0)$.

Solution *First, although it is difficult to graph surfaces with discontinuities, it is possible. Let's modify the code* `DrawMesh` *in (Peterson (21) 2020). The function* `DrawMesh` *starts at the base point (x_0, y_0) and moves outward from there generating surface values that include the ones involving x_0 and/ or y_0. So to draw the surface, we have to add a conditional test as follows:*

Listing 6.1: **Adding Test to Avoid Discontinuity**

```
for i=-nx:nx
  if( i!= 0)
    u = [x0+i*delx];
    x = [x,u];
  end
end
```

The test to avoid `i = 0` *is all we need. The full code is below:*

Listing 6.2: **Draw a Surface with a Discontinuity**

```
function DrawDiscon(f,delx,nx,dely,ny,x0,y0)
% f is the function defining the surface
% delx is the size of the x step
% nx is the number of steps left and right from x0
% dely is the size of the y step
% ny is the number of steps left and right from y0
% (x0,y0) is the center of the rectangular domain
%
% set up x and y stuff
% want delx and dely to divide the domain evenly
% we avoid x0 and y0 which is where the problem is
x = [];
for i=-nx:nx
  if( i!= 0)
    u = [x0+i*delx];
    x = [x,u];
  end
end
%
y = [];
for i=-ny:ny
  if ( i!= 0)
    u = [y0+i*dely];
    y = [y,u];
  end
end
sx = 2*nx;
sy = 2*ny;
hold on

  % plot the surface
  % set up grid of x and y pairs (x(i),y(j))
  [X,Y] = meshgrid(x,y);
  % set up the surface
  Z = f(X,Y);
  mesh(X,Y,Z);
  xlabel('x');
  ylabel('y');
  zlabel('z');
  rotate3d on
  axis on
  grid on
hold off
end
```

If you look at the generated surface yourself, you can spin the plot around and really get a good feel for the surface discontinuity. At $y = 0$, the surface value is $+2$ if $x > 0$ and -2 if $x < 0$ and you can see the attempt the plotted image makes to capture that sudden discontinuity in the strip of unplotted points at $y = 0$ on the plot. We captured one version of the discontinuity in Figure 6.1.

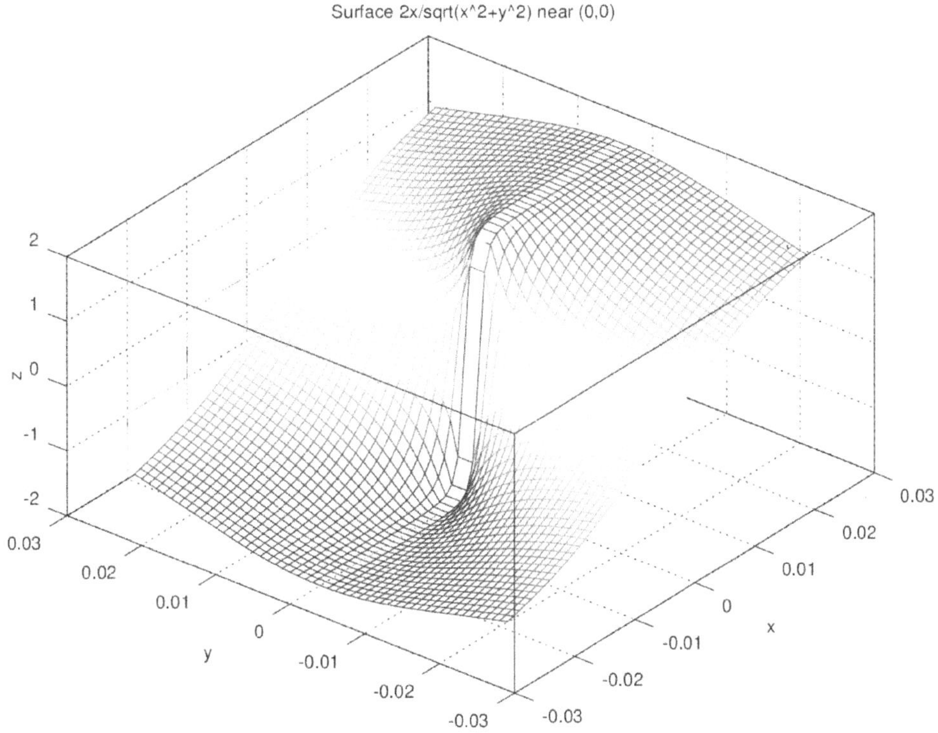

Figure 6.1: The surface $f(x, y) = 2x/\sqrt{x^2 + y^2}$ near the discontinuity at $(0, 0)$.

We know the component f_2 is continuous everywhere as the argument is similar to the last example. So we only have to show f_1 does not have a limit at $(0, 0)$ as this is enough to show f_1 is not continuous at $(0, 0)$ and so f is not continuous at $(0, 0)$. If this limit exists, we should get the same value for the limit no matter what path we take to reach $(0, 0)$.
Let the first path be given by $x(t) = t$ and $y(t) = 2t$. We have two paths really; one for $t > 0$ and one for $t < 0$. We find for $t > 0$, $f_1(t, 2t) = 2t/\sqrt{t^2 + 4t^2} = 2/\sqrt{5}$ and hence the limit along this path $2/\sqrt{5}$.

We find for $t < 0$, $f_1(t, 2t) = 2t/\sqrt{t^2 + 4t^2} = 2t/(|t|\ \sqrt{5}) = -2/\sqrt{5}$ and hence the limit along this path $-2/\sqrt{5}$. Since the limiting value differs on two paths, the limit can't exist. Hence, f_1 is not continuous at $(0, 0)$.

Comment 6.4.1 *The problem with visualization here is that we can never use this to help us with functions of more than two variables because we can't generate the plot. So we have to see it in our mind, so to speak.*

Homework

Exercise 6.4.6 *Let*

$$f(x, y) = \begin{bmatrix} \frac{3x}{\sqrt{(x+1)^2+2(y-2)^2}} \\ 4x^2 + 2y^2 \end{bmatrix}$$

Prove f is not continuous at $(-1,2)$. Draw the surface for f_1 here as well.

Exercise 6.4.7 *Let*

$$f(x,y) = \begin{bmatrix} \frac{4(x-3)}{\sqrt{(x-3)^2+2y^2}} \\ 5x^2+3y^2 \end{bmatrix}$$

Prove f is not continuous at $(3,0)$. Draw the surface for f_1 here as well.

Exercise 6.4.8 *Let*

$$f(x,y) = \begin{bmatrix} \frac{3(y-2)}{\sqrt{(x+1)^2+2(y-2)^2}} \\ \frac{3(x+1)(y-2)}{\sqrt{(x+1)^2+2(y-2)^2}} \end{bmatrix}$$

Prove f is not continuous at $(-1,2)$. Draw the surface for f_1 and f_2 here as well.

Next, we can examine what happens when a function f is continuous on a compact domain. We have carefully done this type of analysis in $\Re$ before, but now we are in a multidimensional situation.

Theorem 6.4.2 If f is Continuous on a Compact Domain, Its Range is Compact

Let $f : D \subset \Re^n \rightarrow \Re$ be continuous. If D is compact, then $f(D)$ is also compact.

Proof 6.4.2

If $f(D)$ were not bounded, there would be a sequence (y_n) in $f(D)$ so that $|y_n| > n$ for all n. But since $y_n = f(\boldsymbol{x_n})$ for some $\boldsymbol{x_n}$ in D, this means there is a sequence $(\boldsymbol{x_n})$ in D which has a convergent subsequence $(\boldsymbol{x_{n_k}})$ which converges to some $\boldsymbol{x}$ in D. Since convergent sequences are bounded, this means the sequence $(f(\boldsymbol{x_{n_k}}))$ must be bounded. Of course, it is not and this contradiction tells us our assumption $f(D)$ is not bounded is wrong. Hence $f(D)$ is bounded.

Next, let (y_n) be a convergent sequence in $f(D)$. So $y_n \rightarrow y$ for some y. There is then an associated sequence $(\boldsymbol{x_n})$ in D so that $y_n = f(\boldsymbol{x_n})$. Since D is compact, there is a subsequence $(\boldsymbol{x_{n_k}})$ which converges to some $\boldsymbol{x}$ in D. Since f is continuous on D, we have $f(\boldsymbol{x_{n_k}}) \rightarrow f(\boldsymbol{x}) \in f(D)$. But subsequences of convergent sequences converge to the same place, so we must have $y = f(\boldsymbol{x})$. Thus $y \in f(D)$ and $f(D)$ is closed.

Since $f(D)$ is closed and bounded, it is compact. ■

Theorem 6.4.3 If f is Continuous on a Compact Domain, It Has Global Extrema

Let $f : D \subset \Re^n \rightarrow \Re$ be continuous on the compact set D. Then, f achieves a global maximum and global minimum on D.

Proof 6.4.3

Since $f(D)$ is compact, it is bounded. Thus there is a positive B so that $|f(\boldsymbol{x})| \leq B$. Hence, $\alpha = \inf_{\boldsymbol{x} \in D} f(\boldsymbol{x})$ and $\beta = \sup_{\boldsymbol{x} \in D} f(\boldsymbol{x})$ both exist and are finite by the Completeness Axiom. Using both the Infimum and Supremum Tolerance Lemma, we can find sequences $(\boldsymbol{x_n})$ and $(\boldsymbol{y_n})$ so that $f(\boldsymbol{x_n}) \rightarrow \alpha$ and $f(\boldsymbol{y_n}) \rightarrow \beta$. Since D is compact, there exist subsequences $(\boldsymbol{x_n^1})$ and $(\boldsymbol{y_n^1})$ with $\boldsymbol{x_n^1} \rightarrow \boldsymbol{x_m} \in D$ and $\boldsymbol{y_n^1} \rightarrow \boldsymbol{x_M} \in D$. Then $f(\boldsymbol{x_n^1}) \rightarrow \alpha$ and $f(\boldsymbol{y_n^1}) \rightarrow \beta$ too. By continuity of f, this tells us $f(\boldsymbol{x_m}) = \alpha$ and $f(\boldsymbol{x_M}) = \beta$. Thus α is the minimum of f on D and β is the maximum of f on D. ■

Homework

Exercise 6.4.9 *Let* $I = [-1,1] \times [-1,1]$ *and let* $f(x,y,z) = \ln(x^2 + 3y^2)$ *on* $I \setminus \{0,0\}$.

- *What happens to the gradient of* f *on the edges and faces of* I*?*
- *Calculate the Laplacian of* f *on* I.

Exercise 6.4.10 *Let* $I = [-1,1] \times [-1,1] \times [-1,1]$ *and let* $f(x,y,z) = \ln(x^2 + y^2 + 2z^2)$ *on* $I \setminus \{0,0,0\}$.

- *What happens to the gradient of* f *on the edges and faces of* I*?*
- *Calculate the Laplacian of* f *on* I.

Exercise 6.4.11 *Let* $\boldsymbol{A} = \begin{bmatrix} 2 & 4 \\ 4 & 3 \end{bmatrix}$ *and define* $f(\boldsymbol{x}) = | < \boldsymbol{A}(\boldsymbol{x}), \boldsymbol{x} > |$ *for all* $\boldsymbol{x}$ *in* $\Re^2$. *Prove* f *has a maximum value on the set* $B = \{\boldsymbol{x} \mid \|\boldsymbol{x}\| = 1\}$ *which occurs at some* $\boldsymbol{x_0} \in B$.

Exercise 6.4.12 *Let* $\boldsymbol{A} = \begin{bmatrix} 2 & 4 & 3 \\ 4 & 3 & 2 \\ 3 & 2 & 7 \end{bmatrix}$ *and define* $f(\boldsymbol{x}) = | < \boldsymbol{A}(\boldsymbol{x}), \boldsymbol{x} > |$ *for all* $\boldsymbol{x}$ *in* $\Re^3$. *Prove* f *has a maximum value on the set* $B = \{\boldsymbol{x} \mid \|\boldsymbol{x}\| = 1\}$ *which occurs at some* $\boldsymbol{x_0} \in B$.

Exercise 6.4.13 *Let* $\boldsymbol{A} = \begin{bmatrix} 2 & 4 \\ 4 & 3 \end{bmatrix}$ *and define* $f(\boldsymbol{x}) =< \boldsymbol{A}(\boldsymbol{x}), \boldsymbol{A}(\boldsymbol{x}) >$ *for all* $\boldsymbol{x}$ *in* $\Re^2$. *Prove* f *has a maximum value on the set* $B = \{\boldsymbol{x} \mid \|\boldsymbol{x}\| = 1\}$ *which occurs at some* $\boldsymbol{x_0} \in B$.

Exercise 6.4.14 *Let* $\boldsymbol{A} = \begin{bmatrix} 2 & 4 & 3 \\ 4 & 3 & 2 \\ 3 & 2 & 7 \end{bmatrix}$ *and define* $f(\boldsymbol{x}) =< \boldsymbol{A}(\boldsymbol{x}), \boldsymbol{A}(\boldsymbol{x}) >$ *for all* $\boldsymbol{x}$ *in* $\Re^3$. *Prove* f *has a maximum value on the set* $B = \{\boldsymbol{x} \mid \|\boldsymbol{x}\| = 1\}$ *which occurs at some* $\boldsymbol{x_0} \in B$.

Chapter 7

Abstract Symmetric Matrices

Let's start with a few more ideas about matrices.

7.1 Input-Output Ratios for Matrices

The **size** of a matrix can be measured in terms of its rows and columns, but a better way is to think about how it acts on vectors. We can think of a matrix as an **engine** which transforms inputs into outputs and it is a natural thing to ask about the ratio of output to input. Such a ratio gives a measure of how the matrix alters the data. Since we can't divide by vectors, we typically use a measure of vector size for the output and input sides. Recall the **Euclidean norm** of a vector $\boldsymbol{x}$ is $||x|| = \sqrt{x_1^2 + \cdots + x_n^2}$ where x_1 through x_n are the components of $\boldsymbol{x}$ in $\Re^n$ with respect to the standard basis in $\Re^n$. This is also the $||\cdot||_2$ norm we have discussed before. Let $\boldsymbol{A}$ be an $n \times n$ matrix. Then $\boldsymbol{A}$ transforms vectors in $\Re^n$ to new vectors in $\Re^n$. We measure the ratio of output to input by calculating $||\boldsymbol{A}(\boldsymbol{x})||/||\boldsymbol{x}||$ and, of course, this doesn't make sense if $x = 0$. We want to see how big this ratio can get, so we are interested in the maximum value of $||\boldsymbol{A}(\boldsymbol{x})||/||\boldsymbol{x}||$. Now the matrix $\boldsymbol{A}$ is linear; i.e. $\boldsymbol{A}(c\boldsymbol{x}) = c\boldsymbol{A}(\boldsymbol{x})$. So in particular, if $y \neq 0$, we can say $\boldsymbol{A}(\boldsymbol{y}) = ||\boldsymbol{y}||\ \boldsymbol{A}(\boldsymbol{y}/||\boldsymbol{y}||)$. Thus,

$$\frac{||\boldsymbol{A}(\boldsymbol{y})||}{||\boldsymbol{y}||} = \frac{||\boldsymbol{y}||\boldsymbol{A}\Big(\boldsymbol{y}/||\boldsymbol{y}||\Big)||}{||\boldsymbol{y}||} = \frac{||\boldsymbol{A}\Big(\boldsymbol{y}/||\boldsymbol{y}||\Big)||}{||\Big(\boldsymbol{y}/||\boldsymbol{y}||\Big)}$$

Now let $\boldsymbol{x} = \boldsymbol{y}/||\boldsymbol{y}||$ and we have $\frac{||\boldsymbol{A}(\boldsymbol{y})||}{||\boldsymbol{y}||} = \frac{||\boldsymbol{A}(\boldsymbol{x})||}{||\boldsymbol{x}||}$ where $\boldsymbol{x}$ has norm 1. So

$$\max_{||\boldsymbol{y}|| \neq 0} \left(\frac{||\boldsymbol{A}(\boldsymbol{y})||}{||\boldsymbol{y}||} \right) = \max_{||\boldsymbol{x}||=1} ||\boldsymbol{A}(\boldsymbol{x})||$$

In order to understand this, first note that the numerator $||\boldsymbol{A}(\boldsymbol{x})|| = \sqrt{< \boldsymbol{A}(\boldsymbol{x}), \boldsymbol{A}(\boldsymbol{x}) >}$ and it is easy to see this is a continuous function of its arguments x_1 through x_n. We can't use just any arguments either; here we can only use those for which $x_1^2 + \ldots + x_n^2 = 1$. The set of such x_i's in $\Re^n$ is a bounded set which contains all of its boundary points. For example, in $\Re^2$, this set is the unit circle $x_1^2 + x_2^2 = 1$ and in $\Re^3$, it is the surface of the sphere $x_1^2 + x_2^2 + x_3^2 = 1$. These sets are bounded and contain their boundary points and are therefore compact subsets. We know any continuous function must have a global maximum and a global minimum over a compact domain. Hence, the problem of finding the maximum of $||\boldsymbol{A}(\boldsymbol{x})||$ over the closed and bounded set $||\boldsymbol{x}|| = 1$ has a solution. We define this maximum ratio to be the **norm** of the matrix $\boldsymbol{A}$. We denote the matrix norm as usual by $||\boldsymbol{A}||$ and so we have:

Definition 7.1.1 The Norm of a Symmetric Matrix

Let $\boldsymbol{A}$ *be an* $n \times n$ *symmetric matrix. The* **norm** *of* $\boldsymbol{A}$ *is*

$$||\boldsymbol{A}|| \quad = \quad \max_{||\boldsymbol{x}||=1} \left(\frac{||\boldsymbol{A}(\boldsymbol{x})||}{||\boldsymbol{x}||} \right)$$

7.1.1 Homework

Exercise 7.1.1 *Let* $A = \begin{bmatrix} 2 & 3 \\ 3 & 5 \end{bmatrix}$. *Estimate* $||\boldsymbol{A}||$.

Exercise 7.1.2 *Let* $A = \begin{bmatrix} -2 & 3 \\ 3 & 5 \end{bmatrix}$. *Prove* $||\boldsymbol{A}|| \leq ||\boldsymbol{P}|| \; ||\boldsymbol{D}|| \; \boldsymbol{P}^T$ *using the standard representation of* $\boldsymbol{A} = \boldsymbol{P}\boldsymbol{D}\boldsymbol{P}^T$ *we have often discussed.*

Exercise 7.1.3 *Prove* $| < \boldsymbol{A}(\boldsymbol{x}), \boldsymbol{x} > | \leq ||\boldsymbol{D}|| \; ||\boldsymbol{x}||^2$.

Exercise 7.1.4 *Write the eigenvalues of the symmetric* 2×2 *matrix* $\boldsymbol{A}$ *as* $\{\lambda_1, \lambda_2\}$. *Prove* $||\boldsymbol{A}|| = r_{max}$ *where* $r_{max} = \max\{|\lambda_1|, |\lambda_2|\}$

7.2 The Norm of a Symmetric Matrix

Now let's specialize to **symmetric** matrices. First, because of the symmetry, an easy calculation shows

$$< \boldsymbol{A}(\boldsymbol{x}), \boldsymbol{y} > \quad = \quad \boldsymbol{y}^T \, \boldsymbol{A} \, \boldsymbol{x} \; = \; \left(\boldsymbol{A} \, \boldsymbol{x} \right)^T \, \boldsymbol{y} \; = \; \boldsymbol{x}^T \, \boldsymbol{A}^T \, \boldsymbol{y}.$$

But this matrix is symmetric and so its transpose is the same as itself. We conclude

$$< \boldsymbol{A}(\boldsymbol{x}), \boldsymbol{y} > \quad = \quad \boldsymbol{x}^T \, \boldsymbol{A}^T \, \boldsymbol{y} \; = < \boldsymbol{x}, \boldsymbol{A}(\boldsymbol{y}) >$$

We will use this identity in a bit. Now, when the matrix is symmetric, we claim we also have

$$||\boldsymbol{A}|| \quad = \quad \max_{||\boldsymbol{x}||=1} \; | < \boldsymbol{A}(\boldsymbol{x}), \boldsymbol{x} > |$$

Let the maximum on the right-hand side be denoted by $\boldsymbol{J}$ for convenience. Now we know how inner products behave. We have for any vector $\boldsymbol{y}$

$$< \boldsymbol{A}(\boldsymbol{y}), \boldsymbol{y} > \quad \leq \quad ||\boldsymbol{A}(\boldsymbol{y})|| \; ||\boldsymbol{y}||$$

Next, from the way $||\boldsymbol{A}||$ is defined, for any particular nonzero vector $\boldsymbol{y}$, we have $||\boldsymbol{A}(\boldsymbol{y})|| \leq ||\boldsymbol{A}|| \, ||\boldsymbol{y}||$. Thus, combining these ideas, we have for any vector $\boldsymbol{x}$ with norm 1,

$$< \boldsymbol{A}(\boldsymbol{x}), \boldsymbol{x} > \quad \leq \quad ||\boldsymbol{A}(\boldsymbol{x})|| \; ||\boldsymbol{x}|| \; \leq \; ||\boldsymbol{A}|| \; ||\boldsymbol{x}|| \; ||\boldsymbol{x}|| \; = \; ||\boldsymbol{A}||.$$

Since this is true for all such vectors, we must have the maximum $\boldsymbol{J}$ satisfies $\boldsymbol{J} \leq ||\boldsymbol{A}||$. Next, we will show the reverse inequality and the two pieces together then show $\boldsymbol{J} = ||\boldsymbol{A}||$.

Now for any nonzero $\boldsymbol{y}$ we have

$$< \boldsymbol{A}\Big(\boldsymbol{y}/||\boldsymbol{y}||\Big), \Big(\boldsymbol{y}/||\boldsymbol{y}||\Big) > \quad \leq \quad \boldsymbol{J}$$

which implies $< \boldsymbol{A}(\boldsymbol{y}), \boldsymbol{y} > \leq \boldsymbol{J} \;/ ||\boldsymbol{y}||^2$. Now do the following calculations. These are just matrix multiplications and they are pretty straightforward, We have for any $\boldsymbol{x}$ and $\boldsymbol{y}$ that

$$< \boldsymbol{A}(\boldsymbol{x}+\boldsymbol{y}), \boldsymbol{x}+\boldsymbol{y} > \quad = \quad < \boldsymbol{A}(\boldsymbol{x}), \boldsymbol{x} > + < \boldsymbol{A}(\boldsymbol{y}), \boldsymbol{y} > +2 < \boldsymbol{A}(\boldsymbol{x}), \boldsymbol{y} > \leq \; \boldsymbol{J}||\boldsymbol{x}+\boldsymbol{y}||^2$$

because the matrix is symmetric. We also have

$$< \boldsymbol{A}(\boldsymbol{x}-\boldsymbol{y}), \boldsymbol{x}-\boldsymbol{y} > \quad = \quad < \boldsymbol{A}(\boldsymbol{x}), \boldsymbol{x} > + < \boldsymbol{A}(\boldsymbol{y}), \boldsymbol{y} > -2 < \boldsymbol{A}(\boldsymbol{x}), \boldsymbol{y} > \geq \; -\boldsymbol{J} \;||\boldsymbol{x}-\boldsymbol{y}||^2$$

Now subtract the second inequality from the first to get

$$\begin{aligned} 2 < \boldsymbol{A}(\boldsymbol{x}), \boldsymbol{y} > +2 < \boldsymbol{A}(\boldsymbol{y}), \boldsymbol{x} > \quad &\leq \quad |< \boldsymbol{A}(\boldsymbol{x}+\boldsymbol{y}), \boldsymbol{x}+\boldsymbol{y} >| + |< \boldsymbol{A}(\boldsymbol{x}-\boldsymbol{y}), \boldsymbol{x}-\boldsymbol{y} >| \\ &\leq \quad \boldsymbol{J}\Big(||\boldsymbol{x}+\boldsymbol{y}||^2 + ||\boldsymbol{x}-\boldsymbol{y}||^2\Big) = \boldsymbol{J}(2||\boldsymbol{x}||^2 + 2||\boldsymbol{y}||^2) \end{aligned}$$

Finally, we can calculate that

$$\Big(||\boldsymbol{x}+\boldsymbol{y}||^2 + ||\boldsymbol{x}+\boldsymbol{y}||^2\Big) \quad = \quad 2\Big(||\boldsymbol{x}||^2 + ||\boldsymbol{y}||^2\Big) \quad = \quad 4$$

as these vectors have length one. We conclude $< \boldsymbol{A}(\boldsymbol{x}), \boldsymbol{y} > \leq \boldsymbol{J}$. Now consider the case where $\boldsymbol{y} = \boldsymbol{A}(\boldsymbol{x})$. If $\boldsymbol{A}(\boldsymbol{x}) = 0$, we would have

$$||\boldsymbol{A}(\boldsymbol{x})||^2 \quad = \quad < \boldsymbol{A}(\boldsymbol{x}), \boldsymbol{A}(\boldsymbol{x}) > = < 0, 0 > = 0,$$

and so $|\boldsymbol{A}(\boldsymbol{x})|| \leq \boldsymbol{J}$. If $\boldsymbol{y} = \boldsymbol{A}(\boldsymbol{x}) \neq 0$, we can then say

$$< \boldsymbol{A}(\boldsymbol{x}), \Big(\boldsymbol{A}(\boldsymbol{x})/||\boldsymbol{A}(\boldsymbol{x})||\Big) > \leq \boldsymbol{J} \quad \leq \quad \boldsymbol{J}$$

which we can simplify to

$$\frac{< \boldsymbol{A}(\boldsymbol{x}), \boldsymbol{A}(\boldsymbol{x}) >}{||\boldsymbol{A}(\boldsymbol{x})||} \leq \boldsymbol{J}.$$

But $< \boldsymbol{A}(\boldsymbol{x}), \boldsymbol{A}(\boldsymbol{x}) > = ||\boldsymbol{A}(\boldsymbol{x})||^2$ and so dividing, we find $||\boldsymbol{A}(\boldsymbol{x})|| \leq \boldsymbol{J}$. This implies the maximum over all $||\boldsymbol{x}|| = 1$, also satisfies this inequality and so

$$||\boldsymbol{A}|| \quad = \quad \max_{||\boldsymbol{x}||=1} \; ||\boldsymbol{A}(\boldsymbol{x})|| \leq \boldsymbol{J}.$$

With the reverse inequality established, we have proven the result we want. Let's summarize our technical discussion. We have proven the following.

Theorem 7.2.1 Two Equivalent Ways to Calculate the Norm of a Symmetric Matrix

If $\boldsymbol{A}$ is a symmetric $n \times n$ matrix, then

$$\|\boldsymbol{A}\| \quad = \quad \max_{\|\boldsymbol{x}\|=1} \|\boldsymbol{A}(\boldsymbol{x})\| \;=\; \max_{\|\boldsymbol{x}\|=1} \,|<\boldsymbol{A}(\boldsymbol{x}), \boldsymbol{x}>| \;=\; \boldsymbol{J}.$$

Proof 7.2.1
See the discussions above. ∎

Now let's do some calculations with 2×2 matrices of real numbers. For these matrices, we know there are 2 nonzero eigenvalues λ_1 and λ_2 with orthogonal eigenvectors $\boldsymbol{E_1}$ and $\boldsymbol{E_2}$. Consider the computation

$$\begin{aligned} <\boldsymbol{A}(\boldsymbol{x}), \boldsymbol{x}> &= <\boldsymbol{A}(a\boldsymbol{E_1} + b\boldsymbol{E_2}), a\boldsymbol{E_1} + b\boldsymbol{E_2}> = < a\lambda_1 \boldsymbol{E_1} + b\lambda_2 \boldsymbol{E_2}, a\boldsymbol{E_1} + b\boldsymbol{E_2}> \\ &= a^2\lambda_1 + b^2\lambda_2 \end{aligned}$$

where for the arbitrary vector $\boldsymbol{x}$, we use its representation $\boldsymbol{x} = a\boldsymbol{E_1} + b\boldsymbol{E_2}$. If we want to find the maximum of $|<\boldsymbol{A}(\boldsymbol{x}), \boldsymbol{x}>|$ over $\|\boldsymbol{x}\| = 1$, we can first look at the maximum of $<\boldsymbol{A}(\boldsymbol{x}), \boldsymbol{x}>$ over these vectors. Replace a and b by $\cos(u)$ and $\sin(u)$ and let

$$f(u) \quad = \quad \cos^2(u)\lambda_1 + \sin^2(u)\lambda_2, \; 0 \leq u \leq 2\pi.$$

The critical points are when $f'(u) = 0$ or when $\sin(2u)(\lambda_2 - \lambda_1) = 0$. Note $f''(u) = 2\cos(2u)(\lambda_2 - \lambda_1)$.

- If $\lambda_1 > \lambda_2$, this implies $u = 0$ or $u = \pi/2$. If $u = 0$, $f''(0) = 2\,(\lambda_2 - \lambda_1) < 0$. Thus, the extremal vector is $\boldsymbol{x} = \boldsymbol{E_1}$, which gives a maximum of value λ_1.

  If $u = \pi/2$, $f''(\pi/2) = -2\,(\lambda_2 - \lambda_1) > 0$. The extremal vector is now $\boldsymbol{x} = \boldsymbol{E_2}$, which gives a minimum of value λ_2.

  A little thought then shows the maximum of $|<\boldsymbol{A}(\boldsymbol{x}), \boldsymbol{x}>|$ over $\|\boldsymbol{x}\| = 1$ is the larger of the two values $|\lambda_1|$ and $|\lambda_2|$.

- If $\lambda_1 < \lambda_2$, this also implies $u = 0$ or $u = \pi/2$. If $u = 0$, $f''(0) = 2\,(\lambda_2 - \lambda_1) > 0$. Then $\boldsymbol{x} = \boldsymbol{E_1}$ gives a minimum of value λ_1.

  If $u = \pi/2$, $f''(\pi/2) = -2\,(\lambda_2 - \lambda_1) < 0$. The extremal vector is now $\boldsymbol{x} = \boldsymbol{E_2}$, which gives a maximum of value λ_2.

  A little thought then again shows the maximum of $|<\boldsymbol{A}(\boldsymbol{x}), \boldsymbol{x}>|$ over $\|\boldsymbol{x}\| = 1$ is the larger of the two values $|\lambda_1|$ and $|\lambda_2|$.

- If the two eigenvalues are the same, $f(u) = \cos^2(u)\lambda_1 + \sin^2(u)\lambda_1 = \lambda_1$. Thus, the value of $f(u)$ is a constant and hence the maximum of $|<\boldsymbol{A}(\boldsymbol{x}), \boldsymbol{x}>|$ over $\|\boldsymbol{x}\| = 1$ is the maximum of the two values $|\lambda_1|$ and $|\lambda_2|$ as this is $|\lambda_1|$.

In the arguments above, we did not consider the case where one eigenvalue is 0 and other one nonzero. We will leave that to you. There is nothing in the argument above that uses the symmetry of $\boldsymbol{A}$.

Thus, if $\boldsymbol{A}$ was a matrix with two nonzero eigenvalues, which has a basis of unit eigenvectors $\boldsymbol{E_1}$ and $\boldsymbol{E_2}$, the only thing we are missing is the orthogonality of the two eigenvectors. It is easy to see

$| < \boldsymbol{A}(\boldsymbol{E_1}), \boldsymbol{A}(\boldsymbol{E_1}) > | = |\lambda_1^2|$ and $| < \boldsymbol{A}(\boldsymbol{E_2}), \boldsymbol{A}(\boldsymbol{E_2}) > | = |\lambda_2^2|$. Thus, we know the maximum of $< \boldsymbol{A}(\boldsymbol{x}), \boldsymbol{A}(\boldsymbol{x}) > \geq \max\{|\lambda_1^2|, |\lambda_2^2|\}$ over all vectors of length one. But it is much harder to pin down the equality here. One way to try to do this is to find an orthonormal basis. Apply GSO to these vectors to create an orthonormal basis $\{\boldsymbol{W_1}, \boldsymbol{W_2}\}$ where

$$\begin{aligned}
\boldsymbol{W_1} &= \boldsymbol{E_1} \\
\boldsymbol{F} &= \boldsymbol{E_2} - < \boldsymbol{E_2}, \boldsymbol{E_1} > \boldsymbol{E_1} \\
\boldsymbol{W_2} &= \boldsymbol{F}/\|\boldsymbol{F}\|
\end{aligned}$$

Let $\gamma =< \boldsymbol{E_1}, \boldsymbol{E_2} >$. Then

$$\begin{aligned}
< \boldsymbol{F}, \boldsymbol{F} > &= < \boldsymbol{E_2} - \gamma \boldsymbol{E_1}, \boldsymbol{E_2} - \gamma \boldsymbol{E_1} >= 1 + \gamma^2 \\
\boldsymbol{W_2} &= \frac{\boldsymbol{E_2} - \gamma \boldsymbol{E_1}}{\sqrt{1+\gamma^2}}
\end{aligned}$$

and

$$\begin{aligned}
\boldsymbol{A}(\boldsymbol{W_1}) &= \lambda_1 \boldsymbol{E_1} \\
\boldsymbol{A}(\boldsymbol{W_2}) &= \frac{1}{\sqrt{1+\gamma^2}} \boldsymbol{A}(\boldsymbol{E_2} - \gamma \boldsymbol{E_1}) = \frac{1}{\sqrt{1+\gamma^2}} (\lambda_2 \boldsymbol{E_2} - \gamma \lambda_1 \boldsymbol{E_1})
\end{aligned}$$

Then any vector $\boldsymbol{x}$ has representation $\boldsymbol{x} = a\boldsymbol{W_1} + b\boldsymbol{W_2}$. We have

$$\begin{aligned}
< \boldsymbol{A}(\boldsymbol{x}), A(\boldsymbol{x}) > &= < \boldsymbol{A}(a\boldsymbol{W_1} + b\boldsymbol{W_2}), \boldsymbol{A}(a\boldsymbol{W_1} + b\boldsymbol{W_2}) > \\
&= \left\langle a\lambda_1 \boldsymbol{E_1} + \frac{b}{\sqrt{1+\gamma^2}} (\lambda_2 \boldsymbol{E_2} - \gamma \lambda_1 \boldsymbol{E_1}), a\lambda_1 \boldsymbol{E_1} + \frac{b}{\sqrt{1+\gamma^2}} (\lambda_2 \boldsymbol{E_2} - \gamma \lambda_1 \boldsymbol{E_1}) \right\rangle \\
&= \left\langle \left(a - \frac{b\gamma}{\sqrt{1+\gamma^2}} \right) \lambda_1 \boldsymbol{E_1} + \frac{b\lambda_2}{\sqrt{1+\gamma^2}} \boldsymbol{E_2}, \right. \\
&\qquad \left. \left(a - \frac{b\gamma}{\sqrt{1+\gamma^2}} \right) \lambda_1 \boldsymbol{E_1} + \frac{b}{\sqrt{1+\gamma^2}} \lambda_2 \boldsymbol{E_2} \right\rangle
\end{aligned}$$

Thus,

$$< \boldsymbol{A}(\boldsymbol{x}), \boldsymbol{A}(\boldsymbol{x}) > = \left(a - \frac{b\gamma}{\sqrt{1+\gamma^2}} \right)^2 \lambda_1^2 + \frac{b^2}{1+\gamma^2} \lambda_2^2$$

Now $-1 < \gamma < 1$, so we can let $\phi = \frac{\gamma}{\sqrt{1+\gamma^2}}$ and $\psi = \frac{1}{\sqrt{1+\gamma^2}}$ with $\phi^2 + \psi^2 = 1$. Then, we have

$$\begin{aligned}
< \boldsymbol{A}(\boldsymbol{x}), \boldsymbol{A}(\boldsymbol{x}) > &= (a - b\psi)^2 \lambda_1^2 + b^2 \phi^2 \lambda_2^2 \\
&= a^2 \lambda_1^2 - 2ab\psi\lambda_1 + b^2\psi^2\lambda_1^2 + b^2\phi^2\lambda_2^2
\end{aligned}$$

and we need to extremize this over all choices of a and b where $a^2 + b^2 = 1$. A standard thing to do is to let $a = \cos(u)$ and $b = \sin(u)$, leading to

$$f(u) = \cos^2(u)\lambda_1^2 - 2\cos(u)\sin(u)\psi\lambda_1^2 + \sin^2(u)(\psi^2\lambda_1^2 + \phi^2\lambda_2^2)$$

The critical point is when $f'(u) = 0$ or

$$\begin{aligned}
0 &= \sin(2u)(\psi^2\lambda_1^2 + \phi^2\lambda_2^2 - \lambda_1^2) - \cos(2u)(2\psi\lambda_1^2) \\
&= \sin(2u)\phi^2(\lambda_2^2 - \lambda_1^2) - \cos(2u)(2\psi\lambda_1^2)
\end{aligned}$$

This implies

$$\tan(2u) = \frac{2\psi\lambda_1^2}{\phi^2(\lambda_2^2 - \lambda_1^2)}$$

Thus,

$$\cos(2u) = \frac{\phi^2(\lambda_2^2 - \lambda_1^2)}{D}, \quad \sin(2u) = \frac{2\psi\lambda_1^2}{D}$$

where $D^2 = 4\psi^2\lambda_1^4 + \phi^4(\lambda_2^2 - \lambda_1^2)^2$. Now rewriting $f(u)$ at this critical point, we have

$$\begin{aligned}
f(u) &= \frac{1+\cos(2u)}{2}\lambda_1^2 - \sin(2u)\psi\lambda_1^2 + \frac{1-\cos(2u)}{2}(\psi^2\lambda_1^2 + \phi^2\lambda_2^2) \\
&= \frac{\lambda_1^2 + \psi^2\lambda_1^2 + \phi^2\lambda_2^2}{2} + \frac{1}{2}\cos(2u)(\lambda_1^2 - \psi^2\lambda_1^2 - \phi^2\lambda_2^2) - \sin(2u)\psi\lambda_1^2 \\
&= \frac{\lambda_1^2 + \psi^2\lambda_1^2 + \phi^2\lambda_2^2}{2} + \frac{\phi^2(\lambda_2^2 - \lambda_1^2)}{2D}(\phi^2(\lambda_1^2 - \lambda_1^2)) - \frac{4\psi\lambda_1^2}{2D}\psi\lambda_1^2 \\
&= \frac{\lambda_1^2 + \psi^2\lambda_1^2 + \phi^2\lambda_2^2}{2} - \frac{\phi^4(\lambda_1^2 - \lambda_1^2)^2 + 4\psi^2\lambda_1^4}{2D} \\
&= \frac{\lambda_1^2 + \psi^2\lambda_1^2 + \phi^2\lambda_2^2}{2} - \frac{D}{2}
\end{aligned}$$

Hence, at the critical value, letting $r_{max} = \max\{|\lambda_1|, |\lambda_2|\}$

$$f(u) \leq \frac{\lambda_1^2 + \psi^2\lambda_1^2 + \phi^2\lambda_2^2}{2} \leq \frac{r_{max}^2 + \psi^2 r_{max}^2 + \phi^2 r_{max}^2}{2} = \frac{r_{max}^2 + r_{max}^2}{2} = r_{max}$$

Combining with our earlier observation, we see $\|\boldsymbol{A}\| = r_{max}$ just like it did in the symmetric matrix case. Of course, these things have been done only for 2×2 matrices either symmetric or possessing two linearly eigenvectors. We will leave it to you to examine other cases such as repeated roots, etc. It is clear the method of proof here for non-symmetric matrices is not at all clear. What we have done is show the matrix norm for some 2×2 matrices is the **spectral radius** of $\boldsymbol{A}$, which is always the largest absolute $|\lambda_i|$ where λ_i is the ith eigenvalue.

But the symmetric case does seem to have a good plan of attack. We just need to know the $n \times n$ symmetric real matrix always has n distinct eigenvalues and the corresponding eigenvectors form an orthonormal basis for $\Re^n$. This is what we do next.

Also, note in the symmetric case, we have

$$< \boldsymbol{A}(a\boldsymbol{E_1} + b\boldsymbol{E_2}), a\boldsymbol{E_1} + b\boldsymbol{E_2} > = < a\lambda_1\boldsymbol{E_1} + b\lambda_2\boldsymbol{E_2}, a\boldsymbol{E_1} + b\boldsymbol{E_2} >= a^2\lambda_1 + b^2\lambda_2$$

and we need to extremize this for all $a^2 + b^2 = 1$. This is best phrased as the problem of maximizing $f(u) = \lambda_1\cos^2(u) + \lambda_2\sin^2(u)$ over $0 \leq u \leq 2\pi$.

Homework

Exercise 7.2.1 *Let* $\boldsymbol{A} = \begin{bmatrix} 2 & 3 \\ 3 & 5 \end{bmatrix}$. *Compute* $\|\boldsymbol{A}\|$ *as a constrained optimization problem in* $\Re^2$ *using the* $\cos(u)$ *and* $\sin(u)$ *approach. Note we know the answer is* $\|\boldsymbol{A}\| = r_{max} = \max\{|\lambda_1|, |\lambda_2|\}$ *where* λ_i *are the eigenvalues of* $\boldsymbol{A}$.

Exercise 7.2.2 *Let* $\boldsymbol{A} = \begin{bmatrix} 2 & 3 \\ 3 & 5 \end{bmatrix}$. *Compute* $||\boldsymbol{A}||$ *by finding the eigenvalues and eigenvectors of* $\boldsymbol{A}$ *and mimicking the proof we did in the text.*

Exercise 7.2.3 *Let* $\boldsymbol{A} = \begin{bmatrix} -2 & 5 \\ 5 & 6 \end{bmatrix}$. *Compute* $||\boldsymbol{A}||$ *as a constrained optimization problem in* $\Re^2$ *using the* $\cos(u)$ *and* $\sin(u)$ *approach. Note we know the answer is* $||\boldsymbol{A}|| = r_{max} = \max\{|\lambda_1|, |\lambda_2|\}$ *where* λ_i *are the eigenvalues of* $\boldsymbol{A}$.

Exercise 7.2.4 *Let* $\boldsymbol{A} = \begin{bmatrix} -2 & 5 \\ 5 & 6 \end{bmatrix}$. *Compute* $||\boldsymbol{A}||$ *by finding the eigenvalues and eigenvectors of* $\boldsymbol{A}$ *and mimicking the proof we did in the text.*

Exercise 7.2.5 *This one is messy. Choose a non-symmetric matrix* $\boldsymbol{A}$ *with eigenvalues* 2 *and* 5. *Find the corresponding eigenvectors. Then mimic the argument we did in the text for this non-symmetric matrix to show* $||\boldsymbol{A}|| = 5$.

Exercise 7.2.6 *Let* $\boldsymbol{A} = \begin{bmatrix} 2 & 7 \\ 7 & 3 \end{bmatrix}$. *Prove* $||\boldsymbol{A}|| \leq ||\boldsymbol{P}|| \; ||\boldsymbol{D}|| \; \boldsymbol{P^T}$ *using the standard representation of* $\boldsymbol{A} = \boldsymbol{PDP^T}$ *we have often discussed.*

7.2.1 Constructing Eigenvalues

For the symmetric square matrix, we can construct its eigenvalues and eigenvectors using a procedure which also gives us useful estimates. Now let's find eigenvalues:

7.2.1.1 Eigenvalue One

Theorem 7.2.2 The First Eigenvalue

Either $||\boldsymbol{A}||$ *or* $-||\boldsymbol{A}||$ *is an eigenvalue for* $\boldsymbol{A}$. *Letting the eigenvalue be* λ_1, *then* $|\lambda_1| = ||\boldsymbol{A}||$ *with an associated eigenvector of norm* 1, $\boldsymbol{E_1}$.

Proof 7.2.2
We know that the maximum value of $| < \boldsymbol{A}(\boldsymbol{x}), \boldsymbol{x} > |$ *over all* $||\boldsymbol{x}|| = 1$ *occurs at some unit vector. Call this unit vector* $\boldsymbol{E_1}$. *For convenience, let* $\alpha = ||\boldsymbol{A}||$. *Then we know* $| < \boldsymbol{A}(\boldsymbol{E_1}), \boldsymbol{E_1} > | = \alpha$.
Case I: *We assume* $< \boldsymbol{A}(\boldsymbol{x}), \boldsymbol{x} >= -\alpha$. *Then,*

$$\begin{aligned} < \boldsymbol{A}(\boldsymbol{E_1}) - (-\alpha)\boldsymbol{E_1}, \boldsymbol{A}(\boldsymbol{E_1}) - (-\alpha)\boldsymbol{E_1} > &= < \boldsymbol{A}(\boldsymbol{E_1}) + \alpha\boldsymbol{E_1}, \boldsymbol{A}(\boldsymbol{E_1}) + \alpha\boldsymbol{E_1} > \\ &= < \boldsymbol{A}(\boldsymbol{E_1}), \boldsymbol{A}(\boldsymbol{E_1}) > + 2\alpha < \boldsymbol{A}(\boldsymbol{E_1}), \boldsymbol{E_1} > \\ &\quad + \alpha^2 < \boldsymbol{E_1}, \boldsymbol{E_1} > \\ &= ||\boldsymbol{A}(\boldsymbol{E_1})||^2 + 2\alpha < \boldsymbol{A}(\boldsymbol{E_1}), \boldsymbol{E_1} > + \alpha^2 \\ &= < \boldsymbol{A}(\boldsymbol{E_1}), \boldsymbol{A}(\boldsymbol{E_1}) > -2\alpha^2 + \alpha^2 \\ &= ||\boldsymbol{A}(\boldsymbol{E_1})||^2 - \alpha^2 \end{aligned}$$

since $||\boldsymbol{E_1}|| = 1$. *Then, overestimating, we have*

$$< \boldsymbol{A}(\boldsymbol{E_1}) - (-\alpha)\boldsymbol{E_1}, \boldsymbol{A}(\boldsymbol{E_1}) - (-\alpha)\boldsymbol{E_1} > \;\; \leq \;\; ||\boldsymbol{A}||^2 \, ||\boldsymbol{E_1}||^2 - \alpha^2.$$

But $\alpha = \|\boldsymbol{A}\|$, *so we have*

$$< \boldsymbol{A}(\boldsymbol{E}_1) - (-\alpha)\boldsymbol{E}_1, \boldsymbol{A}(\boldsymbol{E}_1) - (-\alpha)\boldsymbol{E}_1 > \;\leq\; \alpha^2 - \alpha^2 = 0.$$

We conclude $\|\boldsymbol{A}(\boldsymbol{E}_1) - (-\alpha)\boldsymbol{E}_1\| = 0$ *which tells us* $\boldsymbol{A}(\boldsymbol{E}_1) = -\alpha\boldsymbol{E}_1$. *Hence,* $-\alpha$ *is an eigenvalue of* $\boldsymbol{A}$ *and we can pick* $\boldsymbol{E}_1$ *as the associated eigenvector of norm* 1.
Case II: *We assume* $< \boldsymbol{A}(\boldsymbol{x}), \boldsymbol{x} >= \alpha$. *The argument is then quite similar. We conclude* $\|\boldsymbol{A}(\boldsymbol{E}_1) - \alpha\boldsymbol{E}_1\| = 0$ *which tells us* $\boldsymbol{A}(\boldsymbol{E}_1) = \alpha\boldsymbol{E}_1$. *Hence,* α *is an eigenvalue of* $\boldsymbol{A}$ *and we can pick* $\boldsymbol{E}_1$ *as the associated eigenvector of norm* 1. ■

Example 7.2.1 *Let* $\boldsymbol{A}$ *be defined to be*

$$\boldsymbol{A} = \begin{bmatrix} 1 & 3 \\ 3 & 2 \end{bmatrix}$$

Then, to find the first eigenvalue and eigenvector, we need to find a certain maximum. We let

$$f(x, y) = \begin{bmatrix} x \\ y \end{bmatrix}^T \begin{bmatrix} 1 & 3 \\ 3 & 2 \end{bmatrix} \begin{bmatrix} x \\ y \end{bmatrix} = x^2 + 6xy + 2y^2$$

and we want to find the maximum of $|f(x, y)|$ *subject to* $x^2 + y^2 = 1$. *To do this, let* $x = \cos(\theta)$ *and* $y = \sin(\theta)$. *Then our constrained optimization problem becomes maximize* $g(\theta)$ *where*

$$\begin{aligned} g(\theta) &= \cos^2(\theta) + 6\cos(\theta)\sin(\theta) + 2\sin^2(\theta) \\ &= 1 + 6\cos(\theta)\sin(\theta) + \sin^2(\theta) \\ &= 1 + 3\sin(2\theta) + (1/2)(1 - \cos(2\theta)) \\ &= (3/2) + 3\sin(2\theta) - (1/2)\cos(2\theta). \end{aligned}$$

over all $\theta \in [0, 2\pi]$. *The critical points here are at the two endpoint values* $\theta = 0$ *and* $\theta = 2\pi$, *and where* $g'(\theta) = 0$. *We have*

$$g'(\theta) = 6\cos(2\theta) + \sin(2\theta).$$

Hence, the derivative is zero when $6\cos(2\theta) + \sin(2\theta) = 0$ *or* $\tan(2\theta) = -6$. *Thus,* $2\theta = 1.7359$, 4.8776 *or* 8.0192 *which implies* $\theta = 0.8680$, 2.4388 *or* 4.0096. *This is a quadratic expression, so there is no need to look at second order tests. We'll just evaluate the function as all the critical points. We have* $g(0) = 1$, $g(2\pi) = 1$, $g(0.8680) = 4.5414$ *and* $g(2.4388) = 1.5414$. *The maximum occurs at* $\theta_1 = 0.8680$ *and has value* 4.5414. *This is our eigenvalue* λ_1. *The maximum occurs at*

$$\boldsymbol{E}_1 = \begin{bmatrix} \cos(0.8680) \\ \sin(0.8680) \end{bmatrix} = \begin{bmatrix} 0.64635 \\ 0.76304 \end{bmatrix}$$

To find the second eigenvalue, we construct the new function $h(x, y)$ *defined by*

$$h(x, y) = \begin{bmatrix} x \\ y \end{bmatrix}^T \left(\begin{bmatrix} 1 & 3 \\ 3 & 2 \end{bmatrix} - \lambda_1 \boldsymbol{E}_1 \boldsymbol{E}_1{}^T \right) \begin{bmatrix} x \\ y \end{bmatrix}$$

and maximize its absolute value over the vectors of length one. From our theoretical discussion, the extreme values here are found by maximizing $f(x, y)$ *over the vectors perpendicular to* $\boldsymbol{E}_1$. *These are the multiples of* $\boldsymbol{F}$ *given by*

$$\boldsymbol{F} = \begin{bmatrix} \sin(\theta_1) \\ -\cos(\theta_1) \end{bmatrix}$$

Hence, the extreme values occur at $\pm\boldsymbol{F}$. *The value of* $f(x, y)$ *at* $-\boldsymbol{F}$ *is* -1.5414 *which in absolute value is* 1.5414 *and so we choose* $\boldsymbol{E_2} = \boldsymbol{F}$. *These eigenvalues and their associated eigenvectors are exactly the same as the one we would find using the characteristic equation to find the eigenvalues and then solving the resulting eigenvector equations for a unit vector solution.*

You should do some of these constructions to get a feel for them.

Homework

Exercise 7.2.7 *Let* $\boldsymbol{A} = \begin{bmatrix} 2 & 3 \\ 3 & 5 \end{bmatrix}$. *As we have discussed in the previous section, we can compute the norm of* $\boldsymbol{A}$ *as an optimization problem. Set this up and solve it using the* $\cos(u)$ *and* $\sin(u)$ *approach. This is messy of course. This norm will be the* $r_{max} = \max\{|\lambda_1|, |\lambda_2|\}$ *and the minimizing vector will be the corresponding unit norm eigenvector.*

Compare your computed values to those obtained in MATLAB and to those computed by hand. They should be equivalent.

Exercise 7.2.8 *Let* $\boldsymbol{A} = \begin{bmatrix} 4 & 13 \\ 13 & 6 \end{bmatrix}$. *As we have discussed in the previous section, we can compute the norm of* $\boldsymbol{A}$ *as an optimization problem. Set this up and solve it using the* $\cos(u)$ *and* $\sin(u)$ *approach. This is messy of course. This norm will be the* $r_{max} = \max\{|\lambda_1|, |\lambda_2|\}$ *and the minimizing vector will be the corresponding unit norm eigenvector.*

Compare your computed values to those obtained in MATLAB and to those computed by hand. They should be equivalent.

Exercise 7.2.9 *Let* $\boldsymbol{A} = \begin{bmatrix} -2 & 1 & 3 \\ 1 & 4 & 2 \\ 3 & 2 & 5 \end{bmatrix}$. *As we have discussed in the previous section, we can compute the norm of* $\boldsymbol{A}$ *as an optimization problem. Set this up but don't solve it as this is much harder as it is an optimization over* $a^2 + b^2 + c^2 = 1$, *so it is now three dimensional.*

Use MATLAB to compute the eigenvalues and eigenvectors of $\boldsymbol{A}$. *This will tell you what* $||\boldsymbol{A}||$ *is. Verify one of* $||\boldsymbol{A}||$ *or* $||\boldsymbol{A}||$ *is an eigenvalue with an associated eigenvector. Then take that eigenvector and use is as the* $a^2 + b^2 + c^2 = 1$ *choice in the optimization. You should get the answer is the same* $||\boldsymbol{A}||$.

Exercise 7.2.10 *Let* $\boldsymbol{A} = \begin{bmatrix} 1 & 3 & 1 \\ 3 & 2 & 4 \\ 1 & 4 & 3 \end{bmatrix}$. *As we have discussed in the previous section, we can compute the norm of* $\boldsymbol{A}$ *as an optimization problem. Set this up but don't solve it as this is much harder as it is an optimization over* $a^2 + b^2 + c^2 = 1$, *so it is now three dimensional.*

Use MATLAB to compute the eigenvalues and eigenvectors of $\boldsymbol{A}$. *This will tell you what* $||\boldsymbol{A}||$ *is. Verify one of* $||\boldsymbol{A}||$ *or* $-||\boldsymbol{A}||$ *is an eigenvalue with an associated eigenvector. Then take that eigenvector and use it as the* $a^2 + b^2 + c^2 = 1$ *choice in the optimization. You should get the answer is the same* $||\boldsymbol{A}||$.

Exercise 7.2.11 *Let* $\boldsymbol{A} = \begin{bmatrix} -1 & 2 & 3 \\ 2 & 1 & -2 \\ 3 & -2 & 2 \end{bmatrix}$. *As we have discussed in the previous section, we can compute the norm of* $\boldsymbol{A}$ *as an optimization problem. Set this up but don't solve it as this is much harder as it is an optimization over* $a^2 + b^2 + c^2 = 1$, *so it is now three dimensional.*

Use MATLAB to compute the eigenvalues and eigenvectors of $\boldsymbol{A}$. *This will tell you what* $\|\boldsymbol{A}\|$ *is. Verify one of* $\|\boldsymbol{A}\|$ *or* $-\|\boldsymbol{A}\|$ *is an eigenvalue with an associated eigenvector. Then take that eigenvector and use it as the* $a^2 + b^2 + c^2 = 1$ *choice in the optimization. You should get the answer is the same* $\|\boldsymbol{A}\|$.

7.2.1.2 Eigenvalue Two

We can now find the next eigenvalue using a similar argument.

Theorem 7.2.3 The Second Eigenvalue

There is a second eigenvalue λ_2 *which satisfies* $|\lambda_2| \leq |\lambda_1|$. *Let the new symmetric matrix* $\boldsymbol{A_2}$ *be defined by*

$$\boldsymbol{A_2} = \boldsymbol{A} - \lambda_1 \, \boldsymbol{E_1} \, \boldsymbol{E_1^T},$$

Then, $|\lambda_2| = \|\boldsymbol{A}_2\|$ *and the associated eigenvector* $\boldsymbol{E_2}$ *is orthogonal to* $\boldsymbol{E_1}$, *that is,* $<\boldsymbol{E_1}, \boldsymbol{E_1}>= 0$.

Proof 7.2.3
The reasoning here is virtually identical to what we did for Theorem 7.2.2. First, note, if $\boldsymbol{x}$ *is a multiple of* $\boldsymbol{E_1}$, *we have* $\boldsymbol{x} = \mu \boldsymbol{E_1}$ *for some* μ *giving*

$$\begin{aligned} \boldsymbol{A}_2(\boldsymbol{x}) &= \boldsymbol{A}(\mu \boldsymbol{E_1}) \; - \; \lambda_1 \mu \boldsymbol{E_1} \, \boldsymbol{E_1^T} \boldsymbol{E_1} \\ &= \mu \lambda_1 \boldsymbol{E_1} \; - \; \mu \lambda_1 \boldsymbol{E_1} \; = \; 0 \end{aligned}$$

since $\boldsymbol{E_1}$ *is an eigenvector with eigenvalue* λ_1. *Hence, it is easy to see that* $\boldsymbol{A} = \boldsymbol{A_2}$ *on the orthogonal complement of* $\boldsymbol{E_1}$ *and so* $\|\boldsymbol{A}\| \geq \|\boldsymbol{A_2}\|$. *In fact,*

$$\max_{\|\boldsymbol{x}\|=1} |<\boldsymbol{A_2}(\boldsymbol{x}), \boldsymbol{x}>| \;\; = \;\; \max_{\|\boldsymbol{x}\|=1, \boldsymbol{x} \in \boldsymbol{W_1}} |<\boldsymbol{A_2}(\boldsymbol{x}), \boldsymbol{x}>|$$

where $\boldsymbol{W_1}$ *is the set of vectors that are orthogonal to eigenvector* $\boldsymbol{E_1}$. *The rest of the argument, applied to the new operator* $\boldsymbol{A}_2$, *is identical. Thus, we find*

$$|\lambda_2| \;\; = \;\; \|\boldsymbol{A_2}\| \;\; = \;\; \max_{\|\boldsymbol{x}\|=1, \boldsymbol{x} \in \boldsymbol{W_1}} |<\boldsymbol{A_2}(\boldsymbol{x}), \boldsymbol{x}>|.$$

Further, $|\lambda_2| \leq |\lambda_1|$ *and eigenvector* $\boldsymbol{E_2}$ *is orthogonal to* $\boldsymbol{E_1}$ *because it comes from* $\boldsymbol{W_1}$. ■

Homework

Exercise 7.2.12 *Let* $\boldsymbol{A} = \begin{bmatrix} -2 & 1 & 3 \\ 1 & 4 & 2 \\ 3 & 2 & 5 \end{bmatrix}$. *Use MATLAB to compute* $\boldsymbol{A}$*'s eigenvalues and eigenvectors. This will tell you* $\|\boldsymbol{A}\|$*'s value.*

Now calculate $\boldsymbol{A_2} = \boldsymbol{A} - \lambda_1 \boldsymbol{E_1}\boldsymbol{E_1}^T$. *Here* $\boldsymbol{E_1}$ *is the eigenvector associated with the dominant absolute eigenvalue. Now set up the optimization over* $a^2 + b^2 = 1$ *as usual to find* $\|\boldsymbol{A_2}\|$. *Solve this optimization as usual. Verify you get the next ranked eigenvalue and associated eigenvector by comparing the MATLAB results with your calculations.*

Exercise 7.2.13 *Let* $\boldsymbol{A} = \begin{bmatrix} 1 & 3 & 1 \\ 3 & 2 & 4 \\ 1 & 4 & 3 \end{bmatrix}$. *Use MATLAB to compute* $\boldsymbol{A}$*'s eigenvalues and eigenvectors. This will tell you* $\|\boldsymbol{A}\|$*'s value.*

Now calculate $\boldsymbol{A_2} = \boldsymbol{A} - \lambda_1 \boldsymbol{E_1}\boldsymbol{E_1}^T$. *Here* $\boldsymbol{E_1}$ *is the eigenvector associated with the dominant absolute eigenvalue. Now set up the optimization over* $a^2 + b^2 = 1$ *as usual to find* $\|\boldsymbol{A_2}\|$. *Solve this optimization as usual. Verify you get the next ranked eigenvalue and associated eigenvector by comparing the MATLAB results with your calculations.*

Exercise 7.2.14 *Let* $\boldsymbol{A} = \begin{bmatrix} -1 & 2 & 3 \\ 2 & 1 & -2 \\ 3 & -2 & 2 \end{bmatrix}$. *Use MATLAB to compute the eigenvalues and eigenvectors of* $\boldsymbol{A}$. *This will tell you* $\|\boldsymbol{A}\|$*'s value.*

Now calculate $\boldsymbol{A_2} = \boldsymbol{A} - \lambda_1 \boldsymbol{E_1}\boldsymbol{E_1}^T$. *Here* $\boldsymbol{E_1}$ *is the eigenvector associated with the dominant absolute eigenvalue. Now set up the optimization over* $a^2 + b^2 = 1$ *as usual to find* $\|\boldsymbol{A_2}\|$. *Solve this optimization as usual. Verify you get the next ranked eigenvalue and associated eigenvector by comparing the MATLAB results with your calculations.*

Exercise 7.2.15 *Let* $\boldsymbol{A} = \begin{bmatrix} -1 & 8 & 1 & 4 \\ 8 & -2 & 9 & -2 \\ 1 & 9 & 3 & 6 \\ 4- & 2 & 6 & 5 \end{bmatrix}$. *Use MATLAB to compute* $\boldsymbol{A}$*'s eigenvalues and eigenvectors.*

Now calculate $\boldsymbol{A_2} = \boldsymbol{A} - \lambda_1 \boldsymbol{E_1}\boldsymbol{E_1}^T$. *Here* $\boldsymbol{E_1}$ *is the eigenvector associated with the dominant absolute eigenvalue. Now set up the optimization over* $a^2 + b^2 + c^2 = 1$ *as usual to find* $\|\boldsymbol{A_2}\|$. *We can't easily solve this by hand! However, you can verify you get the next ranked eigenvalue and associated eigenvector by showing the next ranked eigenvalue with its associated eigenvector gives the maximizing vector* $a^2 + b^2 + c^2 = 1$.

Exercise 7.2.16 *Let* $\boldsymbol{A} = \begin{bmatrix} -2 & 4 & 1 & 2 \\ 4 & 3 & 5 & 8 \\ 2 & 5 & -1 & 9 \\ 4- & 8 & 9 & 7 \end{bmatrix}$. *Use MATLAB to compute the eigenvalues and eigenvectors of* $\boldsymbol{A}$.

Now calculate $\boldsymbol{A_2} = \boldsymbol{A} - \lambda_1 \boldsymbol{E_1}\boldsymbol{E_1}^T$. *Here* $\boldsymbol{E_1}$ *is the eigenvector associated with the dominant absolute eigenvalue. Now set up the optimization over* $a^2 + b^2 + c^2 = 1$ *as usual to find* $\|\boldsymbol{A_2}\|$. *We can't easily solve this by hand! However, you can verify you get the next ranked eigenvalue and associated eigenvector by showing the next ranked eigenvalue with its associated eigenvector gives the maximizing vector* $a^2 + b^2 + c^2 = 1$.

7.2.1.3 Eigenvalue Three

We can now find the next eigenvalue.

Theorem 7.2.4 The Third Eigenvalue

There is a third eigenvalue λ_3 which satisfies $|\lambda_3| \leq |\lambda_2|$. Let the new symmetric matrix $\boldsymbol{A_3}$ be defined by

$$\boldsymbol{A_3} = \boldsymbol{A} - \lambda_1 \boldsymbol{E_1} \boldsymbol{E_1^T} - \lambda_2 \boldsymbol{E_2} \boldsymbol{E_2^T},$$

Then, $|\lambda_3| = \|\boldsymbol{A_3}\|$ and the associated eigenvector $\boldsymbol{E_3}$ is orthogonal to $\boldsymbol{E_1}$ and $\boldsymbol{E_2}$.

Proof 7.2.4

The reasoning here is virtually identical what we did for the first two eigenvalues. First, note, if $\boldsymbol{x}$ is in the plane determined by $\boldsymbol{E_1}$ and $\boldsymbol{E_2}$, we have $\boldsymbol{x} = \mu_1\boldsymbol{E_1} + \mu_2\boldsymbol{E_2}$ for some constants μ_1 and μ_2 giving

$$\begin{aligned} \boldsymbol{A}_3(\boldsymbol{x}) &= \boldsymbol{A}(\mu_1\boldsymbol{E_1} + \mu_2\boldsymbol{E_2}) - \lambda_1\mu_1\boldsymbol{E_1}\,\boldsymbol{E_1^T}\boldsymbol{E_1} - \lambda_2\mu_2\boldsymbol{E_2}\,\boldsymbol{E_2^T}\boldsymbol{E_2} \\ &= \mu_1\lambda_1\boldsymbol{E_1} + \mu_2\lambda_2\boldsymbol{E_2} - \mu_1\lambda_1\boldsymbol{E_1} - \mu_2\lambda_2\boldsymbol{E_2} = 0. \end{aligned}$$

since $\boldsymbol{E_1}$ and $\boldsymbol{E_2}$ are eigenvectors. Further $\boldsymbol{A} = \boldsymbol{A_3}$ on the orthogonal complement of the span of $\boldsymbol{E_1}$ and $\boldsymbol{E_2}$ and $\boldsymbol{A_2} = \boldsymbol{A_3}$ on the orthogonal complement of the span of $\boldsymbol{E_1}$ and $\boldsymbol{E_2}$ as well. Hence, it is easy to see that $\|\boldsymbol{A}\| \geq \|\boldsymbol{A_2}\| \geq \|\boldsymbol{A_3}\|$ with

$$\max_{\|\boldsymbol{x}\|=1} |< \boldsymbol{A_3}(\boldsymbol{x}), \boldsymbol{x} >| = \max_{\|\boldsymbol{x}\|=1, \boldsymbol{x}\in\boldsymbol{W_2}} |< \boldsymbol{A_3}(\boldsymbol{x}), \boldsymbol{x} >|$$

where $\boldsymbol{W_2}$ is the set of vectors in that are orthogonal to the plane determined by the first two eigenvectors. Hence, we know $\max_{\|\boldsymbol{x}\|=1} |< \boldsymbol{A_2}(\boldsymbol{x}), \boldsymbol{x} >| \geq \max_{\|\boldsymbol{x}\|=1,\ \boldsymbol{x}\in\boldsymbol{W_2}} |< \boldsymbol{A_3}(\boldsymbol{x}), \boldsymbol{x} >|$. The rest of the argument, applied to the new operator $\boldsymbol{A_3}$ is identical. Thus, we find

$$|\lambda_2| = \|\boldsymbol{A_3}\| = \max_{\|\boldsymbol{x}\|=1, \boldsymbol{x}\in\boldsymbol{W_2}} |< \boldsymbol{A_3}(\boldsymbol{x}), \boldsymbol{x} >|.$$

Further, $|\lambda_3| \leq |\lambda_2|$ and eigenvector $\boldsymbol{E_3}$ is orthogonal to both $\boldsymbol{E_1}$ and $\boldsymbol{E_2}$ because it comes from $\boldsymbol{W_2}$. ■

Homework

Exercise 7.2.17 *Let* $\boldsymbol{A} = \begin{bmatrix} 4 & 2 & -1 \\ 2 & 6 & 7 \\ -1 & 7 & -3 \end{bmatrix}$. *Use MATLAB to compute the eigenvalues and eigenvectors of $\boldsymbol{A}$. This will tell you $\|\boldsymbol{A}\|$]s value.*

Now calculate $\boldsymbol{A_2} = \boldsymbol{A} - \lambda_1\boldsymbol{E_1}\boldsymbol{E_1}^T$. Here $\boldsymbol{E_1}$ is the eigenvector associated with the dominant absolute eigenvalue. Now set up the optimization over $a^2 + b^2 = 1$ as usual to find $\|\boldsymbol{A_2}\|$. Solve this optimization as usual. Verify you get the next ranked eigenvalue and associated eigenvector by comparing the MATLAB results with your calculations.

Now calculate $\boldsymbol{A_3} = \boldsymbol{A} - \lambda_1\boldsymbol{E_1}\boldsymbol{E_1}^T - \lambda_2\boldsymbol{E_2}\boldsymbol{E_2}^T$. Here $\boldsymbol{E_2}$ is the eigenvector associated with the next ranked absolute eigenvalue. Now set up the optimization over $a^2 = 1$ as usual to find $\|\boldsymbol{A_3}\|$. Note this optimization is trivial. Verify you get the next ranked eigenvalue and associated eigenvector by comparing the MATLAB results with your calculations.

Exercise 7.2.18 *Let* $\boldsymbol{A} = \begin{bmatrix} -9 & 4 & 6 \\ 4 & 1 & 2 \\ 6 & 2 & 8 \end{bmatrix}$. *Use MATLAB to compute the eigenvalues and eigenvectors of* $\boldsymbol{A}$*. This will tell you* $||\boldsymbol{A}||$*'s value.*

Now calculate $\boldsymbol{A_2} = \boldsymbol{A} - \lambda_1 \boldsymbol{E_1} \boldsymbol{E_1}^T$. *Here* $\boldsymbol{E_1}$ *is the eigenvector associated with the dominant absolute eigenvalue. Now set up the optimization over* $a^2 + b^2 = 1$ *as usual to find* $||\boldsymbol{A_2}||$*. Solve this optimization as usual. Verify you get the next ranked eigenvalue and associated eigenvector by comparing the MATLAB results with your calculations.*

Now calculate $\boldsymbol{A_3} = \boldsymbol{A} - \lambda_1 \boldsymbol{E_1} \boldsymbol{E_1}^T - \lambda_2 \boldsymbol{E_2} \boldsymbol{E_2}^T$. *Here* $\boldsymbol{E_2}$ *is the eigenvector associated with the next ranked absolute eigenvalue. Now set up the optimization over* $a^2 = 1$ *as usual to find* $||\boldsymbol{A_3}||$*. Note this optimization is trivial. Verify you get the next ranked eigenvalue and associated eigenvector by comparing the MATLAB results with your calculations.*

Exercise 7.2.19 *Let* $\boldsymbol{A} = \begin{bmatrix} -1 & 2 & 3 \\ 2 & 1 & -2 \\ 3 & -2 & 2 \end{bmatrix}$. *Use MATLAB to compute the eigenvalues and eigenvectors of* $\boldsymbol{A}$*. This will tell you* $||\boldsymbol{A}||$*'s value.*

Now calculate $\boldsymbol{A_2} = \boldsymbol{A} - \lambda_1 \boldsymbol{E_1} \boldsymbol{E_1}^T$. *Here* $\boldsymbol{E_1}$ *is the eigenvector associated with the dominant absolute eigenvalue. Now set up the optimization over* $a^2 + b^2 = 1$ *as usual to find* $||\boldsymbol{A_2}||$*. Solve this optimization as usual. Verify you get the next ranked eigenvalue and associated eigenvector by comparing the MATLAB results with your calculations.*

Now calculate $\boldsymbol{A_3} = \boldsymbol{A} - \lambda_1 \boldsymbol{E_1} \boldsymbol{E_1}^T - \lambda_2 \boldsymbol{E_2} \boldsymbol{E_2}^T$. *Here* $\boldsymbol{E_2}$ *is the eigenvector associated with the next ranked absolute eigenvalue. Now set up the optimization over* $a^2 = 1$ *as usual to find* $||\boldsymbol{A_3}||$*. Note this optimization is trivial. Verify you get the next ranked eigenvalue and associated eigenvector by comparing the MATLAB results with your calculations.*

Exercise 7.2.20 *Let* $\boldsymbol{A} = \begin{bmatrix} -1 & 8 & 1 & 4 \\ 8 & -2 & 9 & -2 \\ 1 & 9 & 3 & 6 \\ 4- & 2 & 6 & 5 \end{bmatrix}$. *Use MATLAB to compute the eigenvalues and eigenvectors of* $\boldsymbol{A}$*.*

Now calculate $\boldsymbol{A_2} = \boldsymbol{A} - \lambda_1 \boldsymbol{E_1} \boldsymbol{E_1}^T$. *Here* $\boldsymbol{E_1}$ *is the eigenvector associated with the dominant absolute eigenvalue. Now set up the optimization over* $a^2 + b^2 + c^2 = 1$ *as usual to find* $||\boldsymbol{A_2}||$*. We can't easily solve this by hand! However, you can verify you get the next ranked eigenvalue and associated eigenvector by showing the next ranked eigenvalue with its associated eigenvector gives the maximizing vector* $a^2 + b^2 + c^2 = 1$*.*

Now calculate $\boldsymbol{A_3} = \boldsymbol{A} - \lambda_1 \boldsymbol{E_1} \boldsymbol{E_1}^T - \lambda_2 \boldsymbol{E_2} \boldsymbol{E_2}^T$. *Here* $\boldsymbol{E_2}$ *is the eigenvector associated with the next ranked absolute eigenvalue. Now set up the optimization over* $a^2 + b^2 = 1$ *as usual to find* $||\boldsymbol{A_3}||$*. Note this optimization is similar to what we do for a* 2×2 *matrix. Verify you get the next ranked eigenvalue and associated eigenvector by comparing the MATLAB results with your calculations.*

Now calculate $\boldsymbol{A_4} = \boldsymbol{A} - \lambda_1 \boldsymbol{E_1} \boldsymbol{E_1}^T - \lambda_2 \boldsymbol{E_2} \boldsymbol{E_2}^T - \lambda_3 \boldsymbol{E_3} \boldsymbol{E_3}^T$. *Here* $\boldsymbol{E_3}$ *is the eigenvector associated with the next ranked absolute eigenvalue. Now set up the optimization over* $a^2 = 1$ *as usual to find* $||\boldsymbol{A_4}||$*. Note this optimization is trivial. Verify you get the next ranked eigenvalue and associated eigenvector by comparing the MATLAB results with your calculations.*

Exercise 7.2.21 *Let* $A = \begin{bmatrix} -2 & 4 & 1 & 2 \\ 4 & 3 & 5 & 8 \\ 2 & 5 & -1 & 9 \\ 4- & 8 & 9 & 7 \end{bmatrix}$. *Use MATLAB to compute the eigenvalues and eigenvectors of* A.

Now calculate $A_2 = A - \lambda_1 E_1 {E_1}^T$. *Note* A_2 *is a* 3×3 *symmetric matrix. Here* E_1 *is the eigenvector associated with the dominant absolute eigenvalue. Now set up the optimization over* $a^2 + b^2 + c^2 = 1$ *as usual to find* $\|A_2\|$. *We can't easily solve this by hand! However, you can verify you get the next ranked eigenvalue and associated eigenvector by showing the next ranked eigenvalue with its associated eigenvector gives the maximizing vector* $a^2 + b^2 + c^2 = 1$.

Now calculate $A_3 = A - \lambda_1 E_1 {E_1}^T - \lambda_2 E_2 {E_2}^T$. *Here* E_2 *is the eigenvector associated with the next ranked absolute eigenvalue. Now set up the optimization over* $a^2+b^2 = 1$ *as usual to find* $\|A_3\|$. *Note this optimization is similar to what we do for a* 2×2 *matrix. Verify you get the next ranked eigenvalue and associated eigenvector by comparing the MATLAB results with your calculations.*

Now calculate $A_4 = A - \lambda_1 E_1 {E_1}^T - \lambda_2 E_2 {E_2}^T - \lambda_3 E_3 {E_3}^T$. *Here* E_3 *is the eigenvector associated with the next ranked absolute eigenvalue. Now set up the optimization over* $a^2 = 1$ *as usual to find* $\|A_4\|$. *Note this optimization is trivial. Verify you get the next ranked eigenvalue and associated eigenvector by comparing the MATLAB results with your calculations.*

Exercise 7.2.22 *Let* $A = \begin{bmatrix} 11 & 3 & -6 & 2 \\ 3 & 6 & 1 & 8 \\ -6 & 1 & -2 & 7 \\ 2- & 8 & 7 & 9 \end{bmatrix}$. *Use MATLAB to compute the eigenvalues and eigenvectors of* A.

Now calculate $A_2 = A - \lambda_1 E_1 {E_1}^T$. *Note* A_2 *is a* 3×3 *symmetric matrix. Here* E_1 *is the eigenvector associated with the dominant absolute eigenvalue. Now set up the optimization over* $a^2 + b^2 + c^2 = 1$ *as usual to find* $\|A_2\|$. *We can't easily solve this by hand! However, you can verify you get the next ranked eigenvalue and associated eigenvector by showing the next ranked eigenvalue with its associated eigenvector gives the maximizing vector* $a^2 + b^2 + c^2 = 1$.

Now calculate $A_3 = A - \lambda_1 E_1 {E_1}^T - \lambda_2 E_2 {E_2}^T$. *Here* E_2 *is the eigenvector associated with the next ranked absolute eigenvalue. Now set up the optimization over* $a^2+b^2 = 1$ *as usual to find* $\|A_3\|$. *Note this optimization is similar to what we do for a* 2×2 *matrix. Verify you get the next ranked eigenvalue and associated eigenvector by comparing the MATLAB results with your calculations.*

Now calculate $A_4 = A - \lambda_1 E_1 {E_1}^T - \lambda_2 E_2 {E_2}^T - \lambda_3 E_3 {E_3}^T$. *Here* E_3 *is the eigenvector associated with the next ranked absolute eigenvalue. Now set up the optimization over* $a^2 = 1$ *as usual to find* $\|A_4\|$. *Note this optimization is trivial. Verify you get the next ranked eigenvalue and associated eigenvector by comparing the MATLAB results with your calculations.*

As you can see since A is a $n \times n$ matrix, this process terminates after n steps. We can now state the full result.

Theorem 7.2.5 All Eigenvalues

There is a sequence of eigenvalues λ_i which satisfies $|\lambda_1| \geq |\lambda_2| \geq \ldots \geq |\lambda_n|$ with associated eigenvectors $\boldsymbol{E_i}$ which form an orthonormal basis for $\Re^n$. Let the new matrix $\boldsymbol{A_i}$ be defined by

$$\boldsymbol{A_i} = \boldsymbol{A} - \sum_{j=1}^{i} \lambda_j \, \boldsymbol{E_j} \boldsymbol{E_j^T},$$

Then, $|\lambda_i| = \|\boldsymbol{A_i}\|$ and the associated eigenvector $\boldsymbol{E_i}$ is orthogonal to $\boldsymbol{E_j}$ for all $j \leq i$. Moreover, we can say

$$\boldsymbol{A} = \sum_{j=1}^{n} \lambda_j \, \boldsymbol{E_j} \boldsymbol{E_j^T},$$

7.3 What Does This Mean?

Let's put all of this together. Our symmetric $n \times n$ matrix always has n real eigenvalues and their respective eigenvectors $\{\boldsymbol{E_1}, \ldots \boldsymbol{E_n}\}$ are mutually orthogonal and so define an orthonormal basis for $\Re^n$. We can use these eigenvectors to write $\boldsymbol{A}$ in a canonical form just like we did for the 2×2 case. We define the two $n \times n$ matrices

$$\boldsymbol{P} = \begin{bmatrix} \boldsymbol{E_1} & \boldsymbol{E_2} & \ldots & \boldsymbol{E_n} \end{bmatrix}$$

which has transpose

$$\boldsymbol{P^T} = \begin{bmatrix} \boldsymbol{E_1^T} \\ \boldsymbol{E_2^T} \\ \ldots \\ \boldsymbol{E_n^T} \end{bmatrix}$$

We know the eigenvectors are mutually orthogonal, so we must have $\boldsymbol{P^T P} = \boldsymbol{I}$, $\boldsymbol{P P^T} = \boldsymbol{I}$ and

$$\begin{aligned} \boldsymbol{P^T} \, \boldsymbol{A} \, \boldsymbol{P} &= \begin{bmatrix} \lambda_1 < \boldsymbol{E_1}, \boldsymbol{E_1} > & 0 & 0 & \ldots & 0 \\ 0 & \lambda_2 < \boldsymbol{E_2}, \boldsymbol{E_2} > & 0 & \ldots & 0 \\ \vdots & \vdots & \ddots & \ldots & 0 \\ \vdots & \vdots & 0 & \ddots & 0 \\ 0 & 0 & 0 & 0 & \lambda_n < \boldsymbol{E_n}, \boldsymbol{E_n} > \end{bmatrix} \\ &= \begin{bmatrix} \lambda_1 & 0 & 0 & \ldots & 0 \\ 0 & \lambda_2 & 0 & \ldots & 0 \\ \vdots & \vdots & \ddots & \ldots & 0 \\ \vdots & \vdots & 0 & \ddots & 0 \\ 0 & 0 & 0 & 0 & \lambda_n \end{bmatrix} \end{aligned}$$

which can be rewritten as

$$A = P \begin{bmatrix} \lambda_1 & 0 & 0 & \dots & 0 \\ 0 & \lambda_2 & 0 & \dots & 0 \\ \vdots & \vdots & \ddots & \dots & 0 \\ \vdots & \vdots & 0 & \ddots & 0 \\ \vdots & \vdots & 0 & 0 & \lambda_n \end{bmatrix} P^T$$

And we have a nice representation for an arbitrary $n \times n$ symmetric matrix $\boldsymbol{A}$. Now at this point, we know all the eigenvalues are real and we can write them in descending order, but eigenvalues can be repeated and can even be zero. Also, we can represent the norm of the symmetric matrix $\boldsymbol{A}$ in terms of its eigenvalues.

Theorem 7.3.1 The Eigenvalue Representation of the norm of a symmetric matrix

Let $\boldsymbol{A}$ be an $n \times n$ symmetric matrix with eigenvalues $|\lambda_1| \geq |\lambda_2| \geq \ldots \geq |\lambda_n|$. Then

$$\|\boldsymbol{A}\| = |\lambda_1| = \max_{1 \leq i \leq n} |\lambda_i|$$

This also implies

$$\|\boldsymbol{A}\boldsymbol{x}\| \leq \|\boldsymbol{A}\| \, \|\boldsymbol{x}\|$$

for all $\boldsymbol{x}$.

Proof 7.3.1
We know $\|\boldsymbol{A}\| \geq | < \boldsymbol{A}\boldsymbol{E}_1, \boldsymbol{E}_1 > |$ where $\boldsymbol{E}_1$ is the unit norm eigenvector corresponding to eigenvalue λ_1. Thus,

$$\|\boldsymbol{A}\| \geq | < \lambda_1 \boldsymbol{E}_1, \boldsymbol{E}_1 > | = |\lambda_1|$$

Any $\boldsymbol{x}$ has representation $\sum_{i=1}^n c_i \boldsymbol{E}_i$ in terms on the orthonormal basis of eigenvectors of $\boldsymbol{A}$. If $\|\boldsymbol{x}\| = 1$, this implies $\sum_{i=1}^n c_i^2 = 1$. Then

$$\begin{aligned} | < \boldsymbol{A}\boldsymbol{x}, \boldsymbol{x} > &= \left| \sum_{i=1}^n \sum_{j=1}^n c_i c_j \lambda_i < \boldsymbol{E}_i, \boldsymbol{E}_j > \right| \\ &\leq \sum_{i=1}^n c_i^2 |\lambda_i| \leq (1)\, |\lambda_1| \end{aligned}$$

Thus, $\|\boldsymbol{A}\| = \max_{\|\boldsymbol{x}\|=1} | < \boldsymbol{A}\boldsymbol{x}, \boldsymbol{x} > | \leq |\lambda_1|$. Combining, we see $\|\boldsymbol{A}\| = \lambda_1|$. ■

7.4 Signed Definite Matrices Again

An $n \times n$ matrix is said to be a **positive definite** matrix if $\boldsymbol{x}^T \boldsymbol{A}\boldsymbol{x} > 0$ for all vectors $\boldsymbol{x}$. We can't do our usual algebraic tricks now (thank goodness!) so let's sneak up on this a different way. Rewrite $\boldsymbol{A}$ using our representation to get $\boldsymbol{x}^T \boldsymbol{P}\, \boldsymbol{D}\, \boldsymbol{P}^T \boldsymbol{x} > 0$ where $\boldsymbol{D}$ is the diagonal matrix having the eigenvalues λ_i along the main diagonal. Next, let $\boldsymbol{y} = \boldsymbol{P}^T \boldsymbol{x}$. Then the inequality can be rewritten as

$$(\boldsymbol{P^T x})^T \, \boldsymbol{D} \, \boldsymbol{P^T x} \quad = \quad \boldsymbol{y}^T \, \boldsymbol{D} \, \boldsymbol{y} > 0.$$

We can easily multiply this out to get $\lambda_1 y_1^2 + \lambda_2 y_2^2 + \ldots + \lambda_n y_n^2 > 0$. Since this is positive for all choices of y_i, set all but y_1 to zero. Then, we have $\lambda_1 y_1^2 > 0$ which implies $\lambda_1 > 0$. Similarly, setting all but y_2 to zero, we find $\lambda_2 > 0$. We can do this for all of the eigenvalues to conclude that $\boldsymbol{A}$ positive definite forces all the eigenvalues to be positive. Going the other direction, if all the eigenvalues were positive, that would force the matrix to be positive definite.
A similar argument shows that $\boldsymbol{A}$ is negative definite if and only if its eigenvalues are all negative. Note all, if we simply wanted $\boldsymbol{x}^T \boldsymbol{A} \boldsymbol{x} \geq 0$, we would say $\boldsymbol{A}$ is positive semidefinite and our arguments would say this happens if and only if all the eigenvalues are non-negative. And finally, if $\boldsymbol{x}^T \boldsymbol{A} \boldsymbol{x} \leq 0$, the matrix is negative semidefinite and all its eigenvalues are non-positive.

7.4.1 Homework

Exercise 7.4.1 *Let* $\boldsymbol{A} = \begin{bmatrix} 4 & 2 & -1 \\ 2 & 6 & 7 \\ -1 & 7 & -3 \end{bmatrix}$. *Use MATLAB to compute the eigenvalues and eigenvectors of* $\boldsymbol{A}$. *Determine if* $\boldsymbol{A}$ *is positive definite or not.*

Exercise 7.4.2 *Let* $\boldsymbol{A} = \begin{bmatrix} -9 & 4 & 6 \\ 4 & 1 & 2 \\ 6 & 2 & 8 \end{bmatrix}$. *Use MATLAB to compute the eigenvalues and eigenvectors of* $\boldsymbol{A}$. *Determine if* $\boldsymbol{A}$ *is positive definite or not.*

Exercise 7.4.3 *Let* $\boldsymbol{A} = \begin{bmatrix} -1 & 2 & 3 \\ 2 & 1 & -2 \\ 3 & -2 & 2 \end{bmatrix}$. *Use MATLAB to compute the eigenvalues and eigenvectors of* $\boldsymbol{A}$. *Determine if* $\boldsymbol{A}$ *is positive definite or not.*

Exercise 7.4.4 *Let* $\boldsymbol{A} = \begin{bmatrix} -1 & 8 & 1 & 4 \\ 8 & -2 & 9 & -2 \\ 1 & 9 & 3 & 6 \\ 4- & 2 & 6 & 5 \end{bmatrix}$. *Use MATLAB to compute the eigenvalues and eigenvectors of* $\boldsymbol{A}$. *Determine if* $\boldsymbol{A}$ *is positive definite or not.*

Exercise 7.4.5 *Let* $\boldsymbol{A} - \begin{bmatrix} -2 & 4 & 1 & 2 \\ 4 & 3 & 5 & 8 \\ 2 & 5 & -1 & 9 \\ 4- & 8 & 9 & 7 \end{bmatrix}$. *Use MATLAB to compute the eigenvalues and eigenvectors of* $\boldsymbol{A}$. *Determine if* $\boldsymbol{A}$ *is positive definite or not.*

Exercise 7.4.6 *Let* $\boldsymbol{A} = \begin{bmatrix} 11 & 3 & -6 & 2 \\ 3 & 6 & 1 & 8 \\ -6 & 1 & -2 & 7 \\ 2- & 8 & 7 & 9 \end{bmatrix}$. *Use MATLAB to compute the eigenvalues and eigenvectors of* $\boldsymbol{A}$. *Determine if* $\boldsymbol{A}$ *is positive definite or not.*

Chapter 8

Rotations and Orbital Mechanics

8.1 Introduction

It is easy to think that two dimensional vector spaces are somehow trivial and not worth a lot of our time. We think we can change your mind by looking at some problems in orbital mechanics. We are going to discuss how to write code to solve some problems in this field. This is a great example of how two dimensional vector spaces are used in practice. Our references for this material are

1. (R. Bate and D. Mueller and J. White (24) 1971): an old book on the basics behind interplanetary movement. It is a bit dated now.

2. (W. Thompson (30) 1961): a very nice book on the dynamics of space flight. It is old but most of this kind of information was worked out very early so don't be put off by the age of the text.

3. (G. Sutton and O. Biblarz (7) 2001): this is a standard senior level text in aerospace engineering undergraduate degree programs such as they have at the University of Illinois. It does not explain as much from the basics as (R. Bate and D. Mueller and J. White (24) 1971) and so it is a bit hard to read when you are starting out. But it has really interesting stuff on rocket design which is dated of course. However, the basic ways to do rocket propulsion and the technology to implement it have not changed all that much.

4. (A. E. Roy (1) 1982): this is a book we used when we worked at Aerospace Corporation back in the day. It is a nice complement to (R. Bate and D. Mueller and J. White (24) 1971).

5. (H. Curtis (8) 2014): a much newer book which replaces (R. Bate and D. Mueller and J. White (24) 1971).

There are a number of three dimensional coordinate systems we need to pay attention to.

- The **Heliocentric - Ecliptic** coordinate system. The origin here is the center of the sun. The earth rotates around the sun and its position at any time is a point on a plane. The basis vectors for this plane are denoted by $\boldsymbol{X_{he}}$ and $\boldsymbol{Y_{he}}$ with the $\boldsymbol{X_{he}}$ direction drawn from the center of the earth to the center of the sun on the first day of Spring. This is called the **vernal equinox** direction. The $\boldsymbol{Y_{he}}$ direction is then 90° rotated from $\boldsymbol{X_{he}}$ in the direction of the rotation of the earth and, of course, $\boldsymbol{Y_{he}}$ is in the orbital plane determined by the movement of the earth around the sun. The right-hand rule then determines the $\boldsymbol{Z_{he}}$ direction. Hence, we can write a vector $\boldsymbol{P}$ in the Heliocentric - Ecliptic system as

$$\boldsymbol{P} = x_{he}\boldsymbol{X_{he}} + y_{he}\boldsymbol{Y_{he}} + z_{he}\boldsymbol{Z_{he}}$$

  You can see this coordinate system in Figure 8.1.

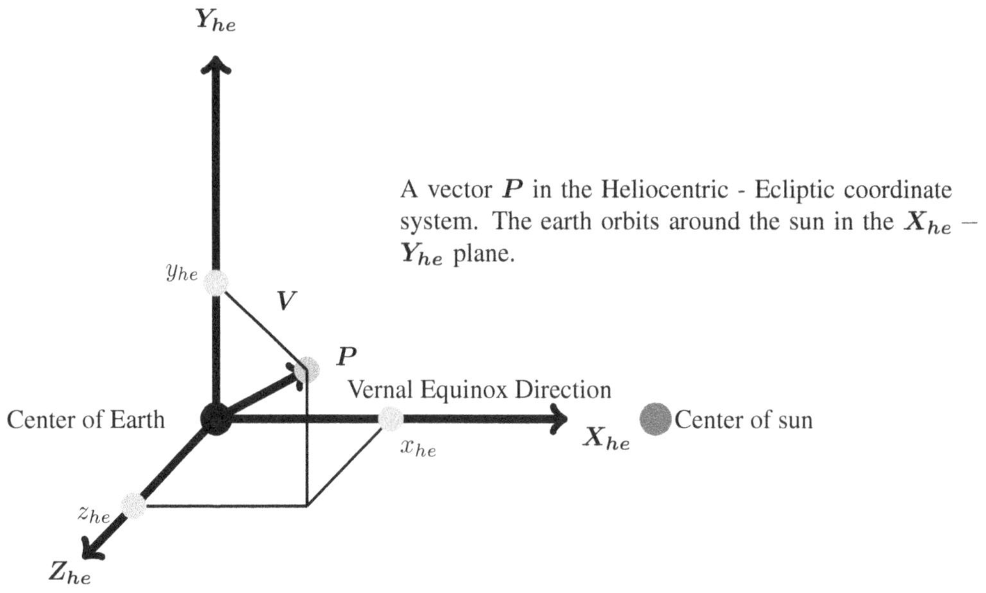

Figure 8.1: Heliocentric - Ecliptic coordinate system.

- The **Geocentric - Equatorial** coordinate system. The origin is now at the center of the earth. The equatorial circle determines a plane and we choose the unit vector for the $\boldsymbol{I}$ direction to be the vernal equinox direction we mentioned above. The $\boldsymbol{J}$ unit vector is then 90° from $\boldsymbol{I}$ and its direction is determined by the fact that the unit vector orthogonal to the equatorial plane is chosen to point toward the north pole. Hence, the right-hand rule determines the orientation of $\boldsymbol{J}$. We see $\boldsymbol{J}$ points east along the equator. A given vector $\boldsymbol{P}$ can then be written as

$$\boldsymbol{P} = x_{ge}\boldsymbol{I} + y_{ge}\boldsymbol{J} + z_{ge}\boldsymbol{K}$$

  You can see this coordinate system in Figure 8.2.

- The **Right Ascension - Declination** coordinate system. This is similar to a spherical coordinate system. For any point in three space, draw a line from the center of the earth to the point. Drop a perpendicular from that point to the equatorial plane or, if the point is below the equatorial plane, the perpendicular will point up until it hits the equatorial plane. This projection of the line connecting the center of the earth to the point will determine a line in the equatorial plane whose center is the center of the earth and whose terminal point is the point the vertical value of the point determines upon projection to the equatorial plane. The angle from the vernal equinox line, which is the $\boldsymbol{I}$ direction to this line, is the angle α, which is called the **right ascension**. Note this is exactly the same as the angle we call θ in the usual spherical coordinate system. We don't use the angle we call ϕ in the spherical coordinate system though. Recall ϕ is the angle from the positive z axis to the line connecting the center of the xyz coordinate system to the tip of the point $\boldsymbol{P}$ determines in three dimensional space. There is another angle, which is measured up from the $x - y$ plane to this line, which is $\pi/2 - \phi$. This angle is called the **declination** and is denoted by δ. To be clear, δ is measured from the line north to the $\boldsymbol{K}$ axis. So $\delta = \pi$ would be a point on the line going to the south pole. The length of the line from the center of the earth to the point is given by R. Then a given vector $\boldsymbol{P}$ can then be written as a ordered triple (R, α, δ). Note we are not writing $\boldsymbol{x}$ in terms of new unit vectors specialized

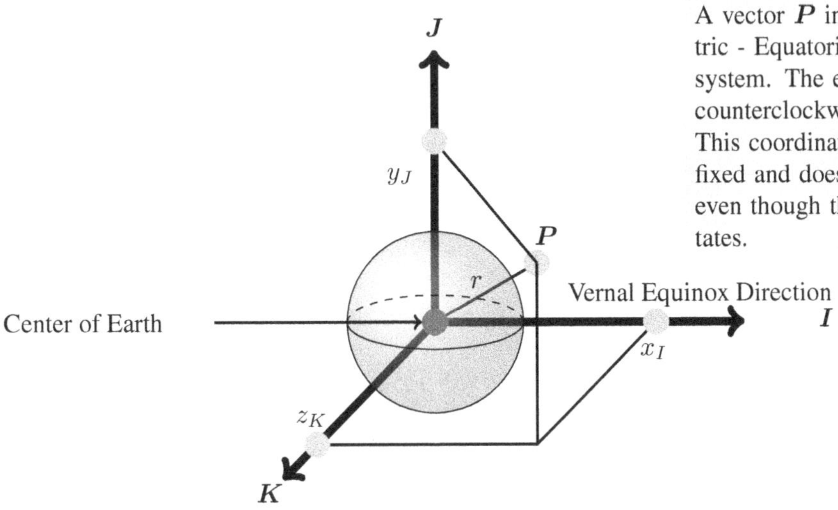

A vector $\boldsymbol{P}$ in the Geocentric - Equatorial coordinate system. The earth rotates counterclockwise $\boldsymbol{I}$ to $\boldsymbol{J}$. This coordinate system is fixed and does not rotate even though the earth rotates.

Figure 8.2: Geocentric - Equatorial coordinate system.

to the Right Ascension - Declination coordinate system. You can see this coordinate system in Figure 8.3.

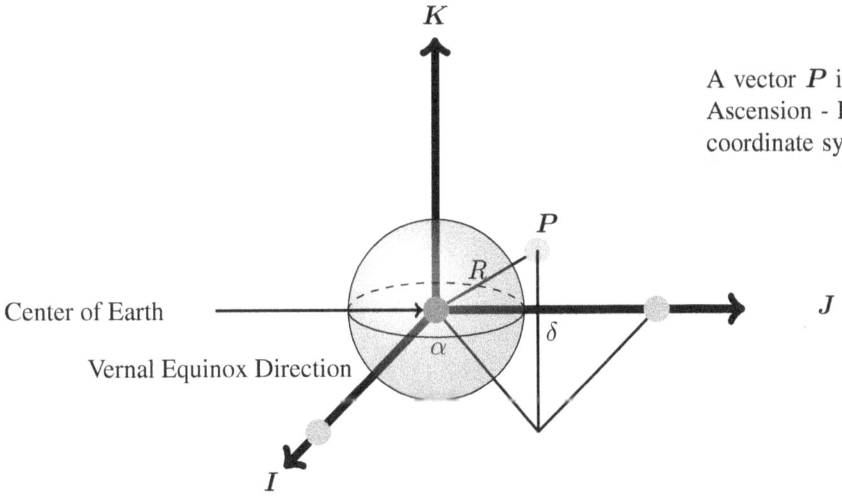

A vector $\boldsymbol{P}$ in the Right Ascension - Declination coordinate system.

Figure 8.3: Right Ascension - Declination coordinate system.

8.1.1 Homework

Exercise 8.1.1 *Let* $P = \begin{bmatrix} 200 \\ 300 \\ 150 \end{bmatrix}$ *be in* $\Re^3$. *We can interpret this as a vector in the* **Heliocentric - Ecliptic** *coordinate system easily. Draw it, carefully labeling all axes. Mark the vernal equinox direction and the orbital plane for the earth. The units here are in terms of the mean equatorial radius of the earth which is about* $3,443,934$ *nautical miles.*

Exercise 8.1.2 *Let* $\boldsymbol{P} = \begin{bmatrix} -200 \\ -300 \\ 150 \end{bmatrix}$ *be in* $\Re^3$. *be a vector in* $\Re^3$. *We can interpret this as a vector in the* **Heliocentric - Ecliptic** *coordinate system easily. Draw it carefully, labeling all axes. Mark the vernal equinox direction and the orbital plane for the earth. The units here are in terms of the mean equatorial radius of the earth which is about* $3,443,934$ *nautical miles.*

Exercise 8.1.3 *Let* $\boldsymbol{P} = \begin{bmatrix} 200 \\ 300 \\ 150 \end{bmatrix}$ *be in* $\Re^3$. *We can interpret this as a vector in the* **Geocentrix - Equatorial** *coordinate system easily. Draw it carefully, labeling all axes. Mark the vernal equinox direction and the orbital plane for the earth. The units here are in terms of the mean equatorial radius of the earth which is about* $3,443,934$ *nautical miles.*

Exercise 8.1.4 *Let* $\boldsymbol{P} = \begin{bmatrix} -200 \\ -300 \\ 150 \end{bmatrix}$ *be in* $\Re^3$. *We can interpret this as a vector in the* **Geocentrix - Equatorial** *coordinate system easily. Draw it carefully, labeling all axes. Mark the vernal equinox direction and the orbital plane for the earth. The units here are in terms of the mean equatorial radius of the earth which is about* $3,443,934$ *nautical miles.*

Exercise 8.1.5 *Let* $\boldsymbol{P} = \begin{bmatrix} -200 \\ -300 \\ 150 \end{bmatrix}$ *be in* $\Re^3$. *We can interpret this as a vector in the* **Right Ascension - Declination** *coordinate system easily. Draw it carefully, labeling all axes. Mark the vernal equinox direction and the orbital plane for the earth. The units here are in terms of the mean equatorial radius of the earth which is about* $3,443,934$ *nautical miles.*

Exercise 8.1.6 *Let* $\boldsymbol{P} = \begin{bmatrix} -200 \\ -300 \\ 150 \end{bmatrix}$ *be in* $\Re^3$. *We can interpret this as a vector in the* **Right Ascension - Declination** *coordinate system easily. Draw it carefully, labeling all axes. Mark the vernal equinox direction and the orbital plane for the earth. The units here are in terms of the mean equatorial radius of the earth which is about* $3,443,934$ *nautical miles.*

8.2 Orbital Planes

The above coordinate systems are not specialized to the movement of a satellite in its orbit. The **Perifocal** coordinate system describes the position of points in space in terms of a coordinate system based on the orbital motion of the satellite. Let M be the mass of the earth and m the mass of the satellite. We will think of position and velocity as numbers in $\Re^3$ and so each position and associated velocity defines vectors

$$\boldsymbol{r}(t) \quad = \quad \begin{bmatrix} x(t) \\ y(t) \\ z(t) \end{bmatrix}, \quad \boldsymbol{v}(t) = \boldsymbol{r}'(t) \quad = \quad \begin{bmatrix} x'(t) \\ y'(t) \\ z'(t) \end{bmatrix},$$

and we can decide to express this information in any choice of coordinate system we wish. The vector $\boldsymbol{r}$ starts at the origin of the coordinate system and ends at the position of the satellite. So there is an obvious unit vector we can use here, $\boldsymbol{E}(t) = \boldsymbol{r}t/\|\boldsymbol{r}\|_2$. The gravitational force between the earth and the satellite is modeled as if all the mass is concentrated at their individual centers and we can write down the usual gravitational force equations you know from basic physics. The force is proportional to the acceleration and is directed along the line between the earth and the satellite. The equation for the vector acceleration is then

$$\frac{d^2}{dt^2}\boldsymbol{r}(t) = -\frac{G(m+M)}{\|\boldsymbol{r}(t)\|_2^2}\,\boldsymbol{E}(t) = -\frac{G(m+M)}{\|\boldsymbol{r}(t)\|_2^3}\,\boldsymbol{r}(t)$$

where G is the universal gravitational constant. Since m is negligible compared to M, we approximate the vector acceleration by

$$\frac{d^2}{dt^2}\boldsymbol{r}(t) = -\frac{G(M)}{\|\boldsymbol{r}(t)\|_2^3}\,\boldsymbol{r}(t) = -\frac{\mu}{\|\boldsymbol{r}(t)\|_2^3}\,\boldsymbol{r}(t)$$

where μ is called the gravitational parameter. To make our arguments a bit more concise, let's start using a common notation for $\|\boldsymbol{r}\|_2$. This is often just written as $\|\boldsymbol{r}(t)\|$. For the rest of our arguments here, we will use this simpler notation. Hence, we have $\boldsymbol{r}''(t) = -(\mu/|\boldsymbol{r}(t)\|^3)\,\boldsymbol{r}(t)$. We will also drop the (t) now: just remember it is still there!

8.2.1 Orbital Constants

Now let's do some calculations: note $\boldsymbol{v} = \boldsymbol{r}'$ and $\boldsymbol{r}'' = \boldsymbol{v}'$.

$$< \boldsymbol{v}, \boldsymbol{v}' > = < \boldsymbol{r}', \boldsymbol{r}'' > = < \boldsymbol{r}', -(\mu/\|\boldsymbol{r}\|^3)\,\boldsymbol{r} > = < \boldsymbol{v}, -(\mu/\|\boldsymbol{r}\|^3)\,\boldsymbol{r} >$$

Thus

$$< \boldsymbol{v}, \boldsymbol{v}' > + < \boldsymbol{v}, (\mu/\|\boldsymbol{r}\|^3)\,\boldsymbol{r} > = 0$$

But we are using the $\|\cdot\|_2$ norm here which is the usual Euclidean norm in $\Re^3$. Now, it is easy to see

$$(\|\boldsymbol{v}\|)' = \frac{1}{2}\,\frac{2v_2v_1' + 2v_2v_2' + 2v_3v_3'}{\sqrt{v_1^2 + v_2^2 + v_3^2}} = \frac{< \boldsymbol{v}, \boldsymbol{v}' >}{\|\boldsymbol{v}\|}$$

which implies $\|\boldsymbol{v}\|\;\|\boldsymbol{v}\|' = < \boldsymbol{v}, \boldsymbol{v}' >$. It is usual to find a way to distinguish between the scalar $\|\boldsymbol{v}\|$ and the vector $\boldsymbol{v}$. We will let $\mathscr{V} = \|\boldsymbol{v}\|$. Hence, we have $\mathscr{V}\,\mathscr{V}' = < \boldsymbol{v}, \boldsymbol{v}' >$. We will also write $\mathscr{R} = \|\boldsymbol{r}\|$ and a similar calculation shows $\mathscr{R}\,\mathscr{R}' = < \boldsymbol{r}, \boldsymbol{r}' >$. Thus,

$$< \boldsymbol{v}, \boldsymbol{v}' > + < \boldsymbol{v}, (\mu/\|\boldsymbol{r}\|^3)\,\boldsymbol{r} > = 0 \implies \mathscr{V}\mathscr{V}' + (\mu/\mathscr{R}^2)\mathscr{V} = 0$$

But

$$(1/2)\frac{d}{dt}\left(\mathscr{V}^2\right)' = \mathscr{V}\mathscr{V}' \Longrightarrow \frac{d}{dt}\left(-\mu/\mathscr{R}\right) = \mu\frac{1}{\mathscr{R}^2}\mathscr{R}'$$

Hence, we conclude

$$\frac{d}{dt}\left((1/2)\mathscr{V}^2\right)' + \frac{d}{dt}\left(\frac{-\mu}{\mathscr{R}}\right) = 0$$

Our final equation is thus

$$\frac{d}{dt}\left(\frac{\mathscr{V}^2}{2} - \frac{\mu}{\mathscr{R}}\right) = 0$$

This says the term $\mathscr{V}/2 - \mu/\mathscr{R}$ is a constant which we call the **specific mechanical energy** and denote by $\mathcal{E}$.

Next, since

$$\frac{d^2}{dt^2}\boldsymbol{r} + \frac{\mu}{\|\boldsymbol{r}\|^3}\,\boldsymbol{r} = 0$$

we can find the cross product of this with $\boldsymbol{r}$. We find

$$\boldsymbol{r} \times \boldsymbol{r}'' + \frac{\mu}{\|\boldsymbol{r}\|^3}\,\boldsymbol{r} \times \boldsymbol{r} = \boldsymbol{0}$$

Since $\boldsymbol{r} \times \boldsymbol{r} = \boldsymbol{0}$, we have $\boldsymbol{r} \times \boldsymbol{r}'' = \boldsymbol{0}$. By direct calculation, we can check

$$\frac{d}{dt}(\boldsymbol{r} \times \boldsymbol{r}') = \boldsymbol{r}' \times \boldsymbol{r}' + \boldsymbol{r} \times \boldsymbol{r}''$$

But $\boldsymbol{r}' \times \boldsymbol{r}' = 0$, so

$$\boldsymbol{r} \times \boldsymbol{r}'' = \frac{d}{dt}(\boldsymbol{r} \times \boldsymbol{r}')$$

We have shown

$$\boldsymbol{0} = \boldsymbol{r} \times \boldsymbol{r}'' = \frac{d}{dt}(\boldsymbol{r} \times \boldsymbol{r}') = \frac{d}{dt}(\boldsymbol{r} \times \boldsymbol{v})$$

The angular momentum here is $\boldsymbol{h} = \boldsymbol{r} \times \boldsymbol{v}$. So we have shown the angular momentum $\boldsymbol{h}$ is constant. We let $\mathscr{H} = \|\boldsymbol{h}\|$. Thus, a satellite moves in an orbit around the center of the earth with constant angular momentum.

8.2.1.1 Homework

Exercise 8.2.1 *Let* $\boldsymbol{r} = \begin{bmatrix} 200 \\ 300 \\ 150 \end{bmatrix}$ *and* $\boldsymbol{v} = \begin{bmatrix} -20 \\ 30 \\ 15 \end{bmatrix}$ *be vectors in* $\Re^3$*. We can interpret* $\boldsymbol{r}$ *as a vector in the* **Heliocentric - Ecliptic** *coordinate system. The position units here are in terms of the mean equatorial radius of the earth which is about* $3,443,934$ *nautical miles and the velocity units are in the usual change in distance with respect to time units. Compute the angular momentum vector* $\boldsymbol{h}$ *and draw all three in the* **Heliocentric - Ecliptic** *coordinate system.*

Exercise 8.2.2 *Let* $\boldsymbol{r} = \begin{bmatrix} 200 \\ 300 \\ 150 \end{bmatrix}$ *and* $\boldsymbol{v} = \begin{bmatrix} -2 \\ 6 \\ 5 \end{bmatrix}$ *be vectors in* $\Re^3$*. We can interpret* $\boldsymbol{r}$ *as a vector in the* **Geocentric - Equatorial** *coordinate system. The position units here are in terms of the mean equatorial radius of the earth which is about* $3,443,934$ *nautical miles and the velocity units are in the usual change in distance with respect to time units. Compute the angular momentum vector* $\boldsymbol{h}$ *and draw all three in the* **Geocentric - Equatorial** *coordinate system.*

Exercise 8.2.3 *Let* $\boldsymbol{r} = \begin{bmatrix} -200 \\ 300 \\ -150 \end{bmatrix}$ *and* $\boldsymbol{v} = \begin{bmatrix} -12 \\ 16 \\ 5 \end{bmatrix}$ *be vectors in* $\Re^3$*. We can interpret* $\boldsymbol{r}$ *as a vector in the* **Right Ascension - Declination** *coordinate system. The position units here are in terms of the mean equatorial radius of the earth which is about* $3,443,934$ *nautical miles and the velocity units are in the usual change in distance with respect to time units. Compute the angular momentum vector* $\boldsymbol{h}$ *and draw all three in the* **Right Ascension - Declination** *coordinate system.*

8.2.2 The Orbital Motion

Since $\boldsymbol{r}$ and $\boldsymbol{v}$ determine the orbital plane, it is convenient to set up some conventions as shown in Figure 8.4.

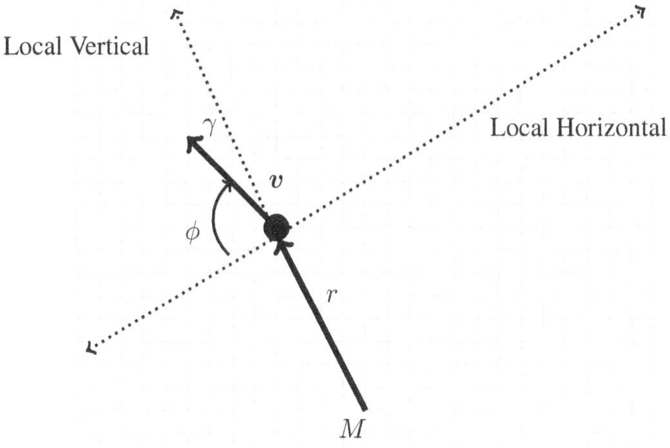

The radius vector from the center of the earth to the satellite determines a local vertical and local horizontal. The angle between $\boldsymbol{v}$ and the local vertical is γ and the angle between $\boldsymbol{v}$ and the local horizontal is ϕ.

Figure 8.4: The orbital plane.

The **flight path angle**, ϕ, is the angle between the local horizontal and $\boldsymbol{v}$. The **zenith** angle, γ, is the angle between $\boldsymbol{v}$ and the local vertical. We see $\gamma = \pi/2 - \phi$.

8.2.2.1 Homework

Exercise 8.2.4 *Let* $\boldsymbol{r} = \begin{bmatrix} 120 \\ 30 \\ 15 \end{bmatrix}$ *and* $\boldsymbol{v} = \begin{bmatrix} -20 \\ 30 \\ 25 \end{bmatrix}$. *Draw the* $\boldsymbol{r} - \boldsymbol{v}$ *orbital plane with orthogonal angular momentum vector indicating the* **flight path angle** ϕ *and the* **zenith** *angle.*

Exercise 8.2.5 *Let* $\boldsymbol{r} = \begin{bmatrix} 200 \\ 300 \\ 150 \end{bmatrix}$ *and* $\boldsymbol{v} = \begin{bmatrix} -2 \\ 6 \\ 5 \end{bmatrix}$. *Draw the* $\boldsymbol{r} - \boldsymbol{v}$ *orbital plane with orthogonal angular momentum vector indicating the* **flight path angle** ϕ *and the* **zenith** *angle.*

Exercise 8.2.6 *Let* $\boldsymbol{r} = \begin{bmatrix} -200 \\ 300 \\ -150 \end{bmatrix}$ *and* $\boldsymbol{v} = \begin{bmatrix} -12 \\ 16 \\ 5 \end{bmatrix}$. *Draw the* $\boldsymbol{r} - \boldsymbol{v}$ *orbital plane with orthogonal angular momentum vector indicating the* **flight path angle** ϕ *and the* **zenith** *angle.*

8.2.3 The Constant B Vector

We know the angular momentum $\boldsymbol{h}$ is a constant vector and so

$$\mathscr{H} = \|\boldsymbol{h}\| = \|\boldsymbol{r} \times \boldsymbol{v}\| = \mathscr{R}\mathscr{V}\sin(\gamma) = \mathscr{R}\mathscr{V}\cos(\phi)$$

Now we know $\boldsymbol{r}'' = -(\mu/\mathscr{R}^3)\boldsymbol{r}$. Thus,

$$\boldsymbol{r}'' \times \boldsymbol{h} \quad = \quad \frac{\mu}{\mathscr{R}^3}\boldsymbol{h} \times \boldsymbol{r}$$

Then, consider

$$(\boldsymbol{r}' \times \boldsymbol{h})' = \boldsymbol{r}'' \times \boldsymbol{h} + \boldsymbol{r}' \times \boldsymbol{h}' = \boldsymbol{r}'' \times \boldsymbol{h}$$

as $\boldsymbol{h}$ is a constant vector. Combining, we have

$$(\boldsymbol{r}' \times \boldsymbol{h})' = \frac{\mu}{\mathscr{R}^3} \boldsymbol{h} \times \boldsymbol{r}$$

Next, we look at

$$\frac{\mu}{\mathscr{R}^3} \boldsymbol{h} \times \boldsymbol{r} = \frac{\mu}{\mathscr{R}^3} \left((\boldsymbol{r} \times \boldsymbol{v}) \times \times \boldsymbol{r} \right)$$

You probably haven't worked much with all the identities associated with vector cross products, so let's go through this calculation.

$$\boldsymbol{r} \times \boldsymbol{v} = det \begin{bmatrix} \boldsymbol{i} & \boldsymbol{j} & \boldsymbol{k} \\ r_1 & r_2 & r_3 \\ v_1 & v_2 & v_3 \end{bmatrix} = (r_2 v_3 - v_2 r_3)\boldsymbol{i} + (r_3 v_1 - r_1 v_3)\boldsymbol{j} + (r_1 v_2 - v_1 r_2)\boldsymbol{k}$$

Thus, after some simplifying we leave to you

$$\begin{aligned} (\boldsymbol{r} \times \boldsymbol{v}) \times \boldsymbol{r} &= (-1) \cdot \begin{bmatrix} \boldsymbol{i} & \boldsymbol{j} & \boldsymbol{k} \\ r_1 & r_2 & r_3 \\ (r_2 v_3 - v_2 r_3) & (r_3 v_1 - r_1 v_3) & (r_1 v_2 - v_1 r_2) \end{bmatrix} \\ &= -\left(r_1 r_2 v_2 + r_1 r_3 v_3 - r_2^2 v_1 - r_3^2 v_1\right) \boldsymbol{i} + \left(-r_1 r_2 v_1 - r_2 r_3 v_3 + r_1^2 v_2 + r_3^2 v_2\right) \boldsymbol{j} \\ &\quad - \left(r_1 r_3 v_1 + r_2 r_3 v_2 - r_1^2 v_3 - r_2^2 v_3\right) \boldsymbol{k} \end{aligned}$$

We can rewrite this as

$$\begin{aligned} (\boldsymbol{r} \times \boldsymbol{v}) \times \boldsymbol{r} &= -\left(r_1(<\boldsymbol{r}, \boldsymbol{v}> - r_1 v_1) - v_1(<\boldsymbol{r}, \boldsymbol{r}> - r_1^2)\right) \boldsymbol{i} \\ &\quad + \left(-r_2(<\boldsymbol{r}, \boldsymbol{v}> - r_2 v_2) + v_2(<\boldsymbol{r}, \boldsymbol{r}> - r_2^2)\right) \boldsymbol{j} \\ &\quad - \left(r_3(<\boldsymbol{r}, \boldsymbol{v}> - r_3 v_3) - v_3(<\boldsymbol{r}, \boldsymbol{r}> - r_3^2)\right) \boldsymbol{k} \\ &= -(r_1 <\boldsymbol{r}, \boldsymbol{v}> - v_1(<\boldsymbol{r}, \boldsymbol{r})) \boldsymbol{i} + (-r_2 <\boldsymbol{r}, \boldsymbol{v}> + v_2 <\boldsymbol{r}, \boldsymbol{r}>) \boldsymbol{j} \\ &\quad - (r_3 <\boldsymbol{r}, \boldsymbol{v}> - v_3 <\boldsymbol{r}, \boldsymbol{r}>) \boldsymbol{k} \end{aligned}$$

We can reorganize the above to obtain our final formula:

$$(\boldsymbol{r} \times \boldsymbol{v}) \times \boldsymbol{r} = - <\boldsymbol{r}, \boldsymbol{v}> \boldsymbol{r} + <\boldsymbol{r}, \boldsymbol{r}> \boldsymbol{v}$$

Using this, we find

$$\begin{aligned} \frac{\mu}{\mathscr{R}^3} \boldsymbol{h} \times \boldsymbol{r} &= \frac{\mu}{\mathscr{R}^3} \left(- <\boldsymbol{r}, \boldsymbol{v}> \boldsymbol{r} + <\boldsymbol{r}, \boldsymbol{r}> \boldsymbol{v}\right) = \frac{\mu}{\mathscr{R}^3} \left(-\mathscr{R}\mathscr{R}' \boldsymbol{r} + \mathscr{R}^2 \boldsymbol{v}\right) \\ &= \frac{\mu}{\mathscr{R}} \boldsymbol{v} - \frac{\mu}{\mathscr{R}^2} \mathscr{R}' \boldsymbol{r} \end{aligned}$$

Recall, we know

$$(\boldsymbol{r}' \times \boldsymbol{h})' = \frac{\mu}{\mathscr{R}^3} \boldsymbol{h} \times \boldsymbol{r} = \frac{\mu}{\mathscr{R}} \boldsymbol{v} - \frac{\mu}{\mathscr{R}^2} \mathscr{R}' \boldsymbol{r}$$

But,

$$\mu\left(\frac{\boldsymbol{r}}{\mathscr{R}}\right)' = \mu\left(\frac{\boldsymbol{r}'}{\mathscr{R}} - \frac{\mathscr{R}'}{\mathscr{R}^2}\boldsymbol{r}\right) = \mu\left(\frac{\boldsymbol{v}}{\mathscr{R}} - \frac{\mathscr{R}'}{\mathscr{R}^2}\boldsymbol{r}\right)$$

This is what we had above! So we can say

$$(\boldsymbol{r}' \times \boldsymbol{h})' = \mu\left(\frac{\boldsymbol{r}}{\mathscr{R}}\right)'$$

We conclude the derivative of $\boldsymbol{r}' \times \boldsymbol{h} - \mu(\boldsymbol{r}/\mathscr{R})$ is a constant vector. Hence, there is a constant vector $\boldsymbol{B}$ so that

$$\boldsymbol{r}' \times \boldsymbol{h} = \mu\frac{\boldsymbol{r}}{\mathscr{R}} + \boldsymbol{B}$$

8.2.3.1 Homework

Exercise 8.2.7 *Let* $\boldsymbol{r} = \begin{bmatrix} 120 \\ 30 \\ 15 \end{bmatrix}$ *and* $\boldsymbol{v} = \begin{bmatrix} -20 \\ 30 \\ 25 \end{bmatrix}$. *Set* $\mu = 1$. *Calculate* $\boldsymbol{B}$.

Exercise 8.2.8 *Let* $\boldsymbol{r} = \begin{bmatrix} 200 \\ 300 \\ 150 \end{bmatrix}$ *and* $\boldsymbol{v} = \begin{bmatrix} -2 \\ 6 \\ 5 \end{bmatrix}$. *Set* $\mu = 1$. *Calculate* $\boldsymbol{B}$.

Exercise 8.2.9 *Let* $\boldsymbol{r} = \begin{bmatrix} -200 \\ 300 \\ -150 \end{bmatrix}$ *and* $\boldsymbol{v} = \begin{bmatrix} -12 \\ 16 \\ 5 \end{bmatrix}$. *Set* $\mu = 1$. *Calculate* $\boldsymbol{B}$.

8.2.4 The Orbital Conic

We can use the equation giving us the constant $\boldsymbol{B}$ vector to derive the equation of motion we associate with the satellite's motion. Consider

$$<\boldsymbol{r}, \boldsymbol{r}' \times \boldsymbol{h}> = \left\langle \boldsymbol{r}, \mu\frac{\boldsymbol{r}}{\mathscr{R}} + \boldsymbol{B} \right\rangle = \mu\frac{<\boldsymbol{r}, \boldsymbol{r}>}{\mathscr{R}} + <\boldsymbol{r}, \boldsymbol{B}> = \mu\mathscr{R} + <\boldsymbol{r}, \boldsymbol{B}>$$

Now, we need another vector computation you probably haven't seen. Let's look at $<\boldsymbol{A}, \boldsymbol{B} \times \boldsymbol{C}>$ for any vectors $\boldsymbol{A}$, $\boldsymbol{B}$ and $\boldsymbol{C}$.

$$\boldsymbol{B} \times \boldsymbol{C} = det \begin{bmatrix} \boldsymbol{i} & \boldsymbol{j} & \boldsymbol{k} \\ B_1 & B_2 & B_3 \\ C_1 & C_2 & C_3 \end{bmatrix} = (B_2C_3 - B_3C_2)\boldsymbol{i} + (B_3C_1 - B_1C_3)\boldsymbol{j} + (B_1C_2 - B_2C_1)\boldsymbol{k}$$

and so

$$\begin{aligned} <\boldsymbol{A}, \boldsymbol{B} \times \boldsymbol{C}> &= A_1(B_2C_3 - B_3C_2) + A_2(B_3C_1 - B_1C_3) + A_3(B_1C_2 - B_2C_1) \\ &= C_1(A_2B_3 - A_3B_2) + C_2(A_3B_1 - A_1B_3) + C_3(A_1B_2 - A_2B_1) \\ &= <\boldsymbol{A} \times \boldsymbol{B}, \boldsymbol{C}> \end{aligned}$$

So

$$<\boldsymbol{r}, \boldsymbol{r}' \times \boldsymbol{h}> = <\boldsymbol{r} \times \boldsymbol{r}', \boldsymbol{h}> = <\boldsymbol{h}, \boldsymbol{h}> = \mathscr{H}^2$$

This means

$$\mathscr{H}^2 \quad = \quad <\boldsymbol{r}, \boldsymbol{r}' \times \boldsymbol{h}> = \mu\mathscr{R} + <\boldsymbol{r}, \boldsymbol{B}> = \mu\mathscr{R} + \mathscr{R}\mathscr{B}\cos(\nu)$$

where ν is the angle between $\boldsymbol{r}$ and $\boldsymbol{B}$ and $\mathscr{B} = \|\boldsymbol{B}\|$ Rewriting, we have

$$\mathscr{R} \quad = \quad \frac{\left(\frac{\mathscr{H}^2}{\mu}\right)}{1 + \left(\frac{\mathscr{B}}{\mu}\right)\cos(\nu)}$$

The scalar $\mathscr{E} = \mathscr{B}/\mu$ is called the **eccentricity** of the orbit and so the vector $\boldsymbol{e} = \boldsymbol{B}/\mu$ is called the eccentricity vector. From what we know about the ellipse, the angle ν measures the angle between the vector pointing to the closest point on the ellipse to the current point. Hence, the constant vector $\boldsymbol{B}$ must point in the direction of periapsis. Therefore $\boldsymbol{e}$ points in the direction of periapsis.

We know $\boldsymbol{e} = \boldsymbol{B}/\mu = \boldsymbol{r}' \times \boldsymbol{h} - \mu\frac{\boldsymbol{r}}{\mathscr{R}}$. Thus,

$$\begin{aligned} \mu\,\boldsymbol{e} &= \boldsymbol{v} \times (\boldsymbol{r} \times \boldsymbol{v}) - \mu\boldsymbol{r}/\mathscr{R} & (8.1)\\ &= <\boldsymbol{v}, \boldsymbol{v}> \boldsymbol{r} - <\boldsymbol{r}, \boldsymbol{v}> \boldsymbol{v} - \mu\boldsymbol{r}/\mathscr{R} & (8.2)\\ &= (\mathscr{V}^2 - \mu/\mathscr{R})\boldsymbol{r} - \; <\boldsymbol{r}, \boldsymbol{v}> \boldsymbol{v} & (8.3)\end{aligned}$$

where we expand the triple cross product just like we did the previous one. This equation tells us the direction of periapsis for the conic!
You are probably more used to the Cartesian Coordinate version of an ellipse. Look at Figure 8.5.

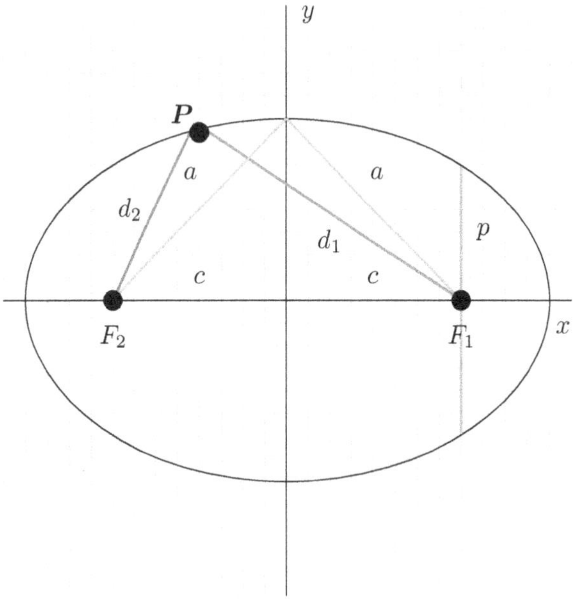

The defining parameters for an ellipse are $d_1 + d_2 = 2a$ and $a^2 = b^2 + c^2$. The eccentricity is $e = c/a$.

Figure 8.5: An ellipse.

There are two points called foci that are used to define an ellipse. These points are labeled F_1 and F_2 and are on the x axis with each a distance c from the origin. Hence, $F_1 = (c, 0)$ and $F_2 = (-c, 0)$. Let d_1 be the distance from a point on the ellipse to focus F_1 and d_2 be the distance from a point on the ellipse to focus F_2. The points on the ellipse must satisfy $d_1 + d_2 = 2a$ and we can use this to find certain relationships. At the top of the ellipse, we get $d_1 = d_2 = a$ which defined a right triangle as you can see in Figure 8.5. We let b be the distance from the center to the top of the ellipse: hence $A^2 = b^2 + c^2$. The ratio c/a is called the **eccentricity** and clearly for an ellipse, $0 < e < 1$. The length of the vertical line through the focus F_1 is $4p$ and half of this distance is called the latis rectum of the ellipse which is thus $2p$. The top part of the line defining the latis rectum gives us the equation $d_1 + 2p = 2a$ and the Pythagorean theorem for the triangle formed gives $d_1^2 = 4p^2 + 4c^2$. Thus $4(a-p)^2 = 4p^2 + 4c^2$. This implies $a^2 - 2ap = c^2$. Hence,

$$2ap = a^2 - c^2 = a^2(1-e^2) \Longrightarrow 2p = a(1-e^2)$$

With some work (and it is not so trivial!) we can derive the corresponding equation for an ellipse in polar coordinates. We will leave that to you. For the choice of using our labeled focus F_1 as the center, we find

$$\mathscr{R} = \frac{2p}{1 + e\cos(\nu)}$$

However, you can also use the Foci we label as F_2 as the center, which leads to the equation

$$\mathscr{R} = \frac{2p}{1 - e\cos(\nu)}$$

as the original angle ν now becomes $\pi - \nu$. We will use the first version in what follows.

Comparing the equation of motion we found for the satellite, we see $\mathscr{H}^2/\mu = 2p = a(1-e^2)$ and $\mathscr{B}/\mu = e$. Now in the orbital plane

$$\boldsymbol{r}(\nu) = \begin{bmatrix} \mathscr{R}\cos(\nu) \\ \mathscr{R}\sin(\nu) \end{bmatrix} = \begin{bmatrix} \frac{2p\cos(\nu)}{1+e\cos(\nu)} \\ \frac{2p\sin(\nu)}{1+e\cos(\nu)} \end{bmatrix}$$

We know the angular momentum is constant and $\mathscr{H} = \mathscr{R}\mathscr{V}\sin(\gamma) = \mathscr{R}\mathscr{V}\cos(\phi)$. Now $\mathscr{R}\cos(\phi)$ is the same as the tangential velocity $\mathscr{R}\nu'$ where ν' is the time derivative of ν. Hence, $\mathscr{R}\cos(\phi) = \mathscr{R}^2\nu'$. This tells us

$$\mathscr{H} = \mathscr{R}^2\nu'$$

We conclude

$$\boldsymbol{v} = (\boldsymbol{r})' = \boldsymbol{v}(\nu) = \frac{d\boldsymbol{r}}{d\nu}\nu' = \frac{d\boldsymbol{r}}{d\nu}\frac{\mathscr{H}}{\mathscr{R}^2(\nu)}$$

Now, a simple calculation shows

$$\begin{aligned} \boldsymbol{v}(\nu) &= \begin{bmatrix} \frac{-2p\sin(\nu)}{1+e\cos(\nu)} - \frac{2p\cos(\nu)}{(1+e\cos(\nu))^2}(-e\sin(\nu)) \\ \frac{2p\cos(\nu)}{1+e\cos(\nu)} - \frac{2p\sin(\nu)}{(1+e\cos(\nu))^2}(-e\sin(\nu)) \end{bmatrix} \\ &= \begin{bmatrix} \frac{-2p\sin(\nu)(1+e\cos(\nu))+2pe\sin(\nu)\cos(\nu)}{(1+e\cos(\nu))^2} \\ \frac{2p\cos(\nu)(1+e\cos(\nu))+2pe\sin^2(\nu)}{(1+e\cos(\nu))^2} \end{bmatrix} \end{aligned}$$

This can be simplified to

$$\boldsymbol{v}(\nu) = \begin{bmatrix} \mathscr{R}(\nu)(-\sin(\nu)) + (\mathscr{R}(\nu))^2\frac{e}{2p}\sin(\nu)\cos(\nu)) \\ \mathscr{R}(\nu)(\cos(\nu)) + (\mathscr{R}(\nu))^2\frac{e}{2p}\sin^2(\nu) \end{bmatrix}$$

Thus,

$$\boldsymbol{v} = \frac{\mathscr{H}}{2p\mathscr{R}(\nu)} \begin{bmatrix} -2p\sin(\nu) + \mathscr{R}(\nu)e\sin(\nu)\cos(\nu)) \\ 2p\cos(\nu) + \mathscr{R}(\nu)e\sin^2(\nu) \end{bmatrix}$$

We then have

$$|\boldsymbol{v}| = \frac{\mathscr{H}}{2p\mathscr{R}(\nu)} \sqrt{4p^2 + (\mathscr{R}(\nu))^2 e^2 \sin^2(\nu)}$$

We see then that

$$\begin{aligned} \boldsymbol{r}\cdot\boldsymbol{v} &= \frac{\mathscr{H}}{2p\mathscr{R}(\nu)} \begin{bmatrix} \mathscr{R}\cos(\nu) \\ \mathscr{R}\sin(\nu) \end{bmatrix} \cdot \begin{bmatrix} -2p\sin(\nu) + \mathscr{R}(\nu)e\sin(\nu)\cos(\nu)) \\ 2p\cos(\nu) + \mathscr{R}(\nu)e\sin^2(\nu) \end{bmatrix} \\ &= \mathscr{H} \begin{bmatrix} \cos(\nu) \\ \sin(\nu) \end{bmatrix} \cdot \begin{bmatrix} -2p\sin(\nu) + \mathscr{R}(\nu)e\sin(\nu)\cos(\nu)) \\ 2p\cos(\nu) + \mathscr{R}(\nu)e\sin^2(\nu) \end{bmatrix} \end{aligned}$$

Hence, at periapsis, $\nu = 0$ and

$$\boldsymbol{r}\cdot\boldsymbol{v} = \mathscr{H} \begin{bmatrix} 1 \\ 0 \end{bmatrix} \cdot \begin{bmatrix} 0 \\ 2p \end{bmatrix} = 0$$

Thus, at periapsis, the position vector and the velocity vector are perpendicular. A similar argument show this is true at apoapsis. Further, a simple calculation verifies that

$$\mathscr{R}(0)\mathscr{V}(0) = \mathscr{R}(\pi)\mathscr{V}(\pi) = \mathscr{H}$$

8.2.4.1 Homework

Exercise 8.2.10 *Look up the way we define a parabola in Cartesian coordinates. Using this definition, derive the polar form of the equation of a parabola. It is still* $\mathscr{R} = \frac{2p}{1+e\cos(\nu)}$ *but now* $e = 1$. *This is a nontrivial exercise but rewarding. We usually do this as part of our lectures on Orbital Mechanics in an applied mathematics modeling course.*

Exercise 8.2.11 *Look up the way we define a hyperbola in Cartesian coordinates. Using this definition, derive the polar form of the equation of a hyperbola. It is still* $\mathscr{R} = \frac{2p}{1+e\cos(\nu)}$ *but now* $e > 1$. *This is also a nontrivial exercise but rewarding. We usually do this as part of our lectures on Orbital Mechanics in an applied mathematics modeling course.*

Exercise 8.2.12 *On the same plot*

- *Graph* $\mathscr{R} = \frac{2p}{1+e\cos(\nu)}$ *for* $e = 0.25$ *and* $p = 4$.
- *Graph* $\mathscr{R} = \frac{2p}{1+e\cos(\nu)}$ *for* $e = 0.5$ *and* $p = 4$.
- *Graph* $\mathscr{R} = \frac{2p}{1+e\cos(\nu)}$ *for* $e = 0.75$ *and* $p = 4$.

Exercise 8.2.13 *On the same plot*

- *Graph* $\mathscr{R} = \frac{2p}{1+e\cos(\nu)}$ *for* $e = 1.25$ *and* $p = 4$.
- *Graph* $\mathscr{R} = \frac{2p}{1+e\cos(\nu)}$ *for* $e = 1.5$ *and* $p = 4$.
- *Graph* $\mathscr{R} = \frac{2p}{1+e\cos(\nu)}$ *for* $e = 1.75$ *and* $p = 4$.

Exercise 8.2.14 *On the same plot*

- *Graph* $\mathscr{R} = \frac{2p}{1+e\cos(\nu)}$ *for* $e = 1$ *and* $p = 0.5$.
- *Graph* $\mathscr{R} = \frac{2p}{1+e\cos(\nu)}$ *for* $e = 1$ *and* $p = 1$.
- *Graph* $\mathscr{R} = \frac{2p}{1+e\cos(\nu)}$ *for* $e = 1$ *and* $p = 2$.
- *Graph* $\mathscr{R} = \frac{2p}{1+e\cos(\nu)}$ *for* $e = 1$ *and* $p = 4$.

Exercise 8.2.15 *For these elliptical orbits, set* $\mu = 1$ *and compute* $\mathscr{B}$, $\mathscr{H}$, a *and* $\mathscr{H}$.

- $\mathscr{R} = \frac{2p}{1+e\cos(\nu)}$ *for* $e = 0.25$ *and* $p = 4$,
- $\mathscr{R} = \frac{2p}{1+e\cos(\nu)}$ *for* $e = 0.5$ *and* $p = 4$.
- $\mathscr{R} = \frac{2p}{1+e\cos(\nu)}$ *for* $e = 0.75$ *and* $p = 4$.

Exercise 8.2.16 *For these elliptical orbits, set* $\mu = 1$ *and compute* $\boldsymbol{r_p}$, $\boldsymbol{r_a}$ *and* $\mathscr{H}$.

- $\mathscr{R} = \frac{2p}{1+e\cos(\nu)}$ *for* $e = 0.125$ *and* $p = 1.5$,
- $\mathscr{R} = \frac{2p}{1+e\cos(\nu)}$ *for* $e = 0.55$ *and* $p = 3$.
- $\mathscr{R} = \frac{2p}{1+e\cos(\nu)}$ *for* $e = 0.85$ *and* $p = 4.5$.

8.3 Three Dimensional Rotations

If we have a 3×3 symmetric matrix $\boldsymbol{A}$, we know it has three mutually orthogonal eigenvectors corresponding to three real eigenvalues. Hence the unit eigenvectors form an orthonormal basis for $\Re^3$ and we can write

$$\begin{bmatrix} \boldsymbol{E_1}^T \\ \boldsymbol{E_2}^T \\ \boldsymbol{E_3}^T \end{bmatrix} \boldsymbol{A} \begin{bmatrix} \boldsymbol{E_1} & \boldsymbol{E_2} & \boldsymbol{E_3} \end{bmatrix} = \begin{bmatrix} \lambda_1 & 0 & 0 \\ 0 & \lambda_2 & 0 \\ 0 & 0 & \lambda_3 \end{bmatrix}$$

where $\{\boldsymbol{E_1}, \boldsymbol{E_2}, \boldsymbol{E_3}\}$ is an orthonormal basis for $\Re^3$ with corresponding eigenvalues λ_1, λ_2 and λ_3. We know the rows and columns of the matrix $\boldsymbol{P} = \begin{bmatrix} \boldsymbol{E_1} & \boldsymbol{E_2} & \boldsymbol{E_3} \end{bmatrix}$ are mutually orthogonal. Hence, like what happened in the two dimensional case, $\boldsymbol{P}$ is often a rotation matrix which transforms an orthonormal system of mutually orthogonal axis labeled x, y and z into a new orthogonal system labeled x', y' and z'. Let's study some of these rotation matrices. There are three simple building blocks:

- The axis of rotation is the z axis and we rotate the $x - y$ system about the z axis by θ degrees. The new system is thus x', y' and z.
- Now apply to this new system a rotation of ϕ degrees about the new y' axis. So the x' and z axis rotate into the new axis x'' and z'. This gives a new orthogonal system with axis x'', y' and z'.
- Now apply to this new system a rotation of δ degrees about the new x'' axis. This generates a new orthogonal system with axes x'', y'' and z''.

Of course, we could organize three rotations in different ways, but to explain this, it helps to have a concrete set of choices. We want to draw some of this to help you see it, so we use MATLAB. It is a bit complicated to draw things in any programming language. Let's begin with drawing a simple vector using `ThreeDToPlot.m` shown below.

Listing 8.1: **ThreeDToPlot.m**

```
function [x,y,z] = ThreeDToPlot(A)
%
% A is the vector to plot using plot3
% x,y,z is prepared vector that plot3 uses
% N is the number of points to plot
%
t = linspace(0,1,2);
VA = @(t) t*A;
for i=1:2
  C = VA(t(i));
  x(i) = C(1);
  y(i) = C(2);
  z(i) = C(3);
end
end
```

This looks a little strange, but the `plot3` MATLAB function draws the line between two vectors $\boldsymbol{A}$ and $\boldsymbol{B}$ like this. If

$$\boldsymbol{A} = \begin{bmatrix} A_1 \\ A_2 \\ A_3 \end{bmatrix}, \quad \boldsymbol{B} = \begin{bmatrix} B_1 \\ B_2 \\ B_3 \end{bmatrix}$$

then to draw the line between $\boldsymbol{A}$ and $\boldsymbol{B}$, we collect $\boldsymbol{x}$, $\boldsymbol{y}$ and $\boldsymbol{z}$ points as

$$\boldsymbol{x} = \begin{bmatrix} A_1 \\ B_1 \end{bmatrix}, \quad \boldsymbol{y} = \begin{bmatrix} A_2 \\ B_2 \end{bmatrix}, \quad \boldsymbol{z} = \begin{bmatrix} A_3 \\ B_3 \end{bmatrix}$$

and then `plot3` plots the line between the point $(x(1), y(1), z(1))$ and the point $(x(2), y(2), z(2))$. Since this is a pain to do for all the vectors we want to plot, we've rolled this into a convenience function so we don't have to type so much.

In addition, we need our rotation matrices functions which we call `RotateX.m`, `RotateY.m` and `RotateZ.m`. Here is the code. The function `RotateX` rotates about the x axis by positive $\theta°$ to give the new coordinate system (x, y', z'). This is a positive rotation which means your thumb points in the direction of the $+x$ axis and your fingers curl in the direction of the positive θ. Hence, this is a counterclockwise (ccw) rotation from the $+y$ axis giving the new coordinate system (x, y', z'). The rotation matrix here is $\boldsymbol{R}_{x,\theta}$ with

$$\boldsymbol{R}_{x,\theta} = \begin{bmatrix} 1 & 0 & 0 \\ 0 & \cos(\theta) & -\sin(\theta) \\ 0 & \sin(\theta) & \cos(\theta) \end{bmatrix} = \begin{bmatrix} 1 & \boldsymbol{0} \\ \boldsymbol{0} & \boldsymbol{R}_\theta \end{bmatrix}$$

where $\boldsymbol{0}$ here is whatever size of zero vector or its transpose we need and $\boldsymbol{R}_\theta$ is the usual 2×2 rotation matrix.

Listing 8.2: **RotateX.m**

```
function Rx = RotateX(theta)
% + theta means ccw which is what we usually want
% - theta means cw
```

```
ctheta = cos(theta);
stheta = sin(theta);
%
Rx = [1,0,0;0,ctheta, -stheta; 0,stheta, ctheta];
end
```

The function `RotateY` rotates about the y axis by positive θ° to give the new coordinate system (x', y, z'). The rotation matrix here is $\boldsymbol{R_{y,\theta}}$ with

$$\boldsymbol{R_{y,\theta}} = \begin{bmatrix} \cos(\theta) & 0 & -\sin(\theta) \\ 0 & 1 & 0 \\ \sin(\theta) & 0 & \cos(\theta) \end{bmatrix}$$

and you can clearly see $\boldsymbol{R_\theta}$ buried inside this matrix.

Listing 8.3: **RotateY.m**

```
function Ry = RotateY(theta)
% + theta means cw
% - theta means ccw which is what we usually want
ctheta = cos(theta);
stheta = sin(theta);
%
Ry = [ ctheta, 0, -stheta; 0, 1, 0; stheta, 0, ctheta];

end
```

The function `RotateZ` rotates about the z axis by positive θ° to give the new coordinate system (x', y', z). The rotation matrix here is $\boldsymbol{R_{z,\theta}}$ with

$$\boldsymbol{R_{z,\theta}} = \begin{bmatrix} \cos(\theta) & -\sin(\theta) & 0 \\ \sin(\theta) & \cos(\theta) & 0 \\ 0 & 0 & 1 \end{bmatrix}$$

and again you can clearly see $\boldsymbol{R_\theta}$ buried inside this matrix.

Listing 8.4: **RotateZ.m**

```
function Rz = RotateZ(theta)
% + theta means ccw which is what we usually want
% - theta means cw
ctheta = cos(theta);
stheta = sin(theta);
Rz = [ ctheta, -stheta,0; stheta, ctheta, 0; 0, 0, 1];

end
```

8.3.1 Homework

Exercise 8.3.1 *Compute the product $\boldsymbol{R}_{x,\theta}\, \boldsymbol{R}_{y,\phi}\, \boldsymbol{R}_{z,\psi}$ for general angles θ, ϕ and ψ. This gives you a 3×3 rotation matrix which is somewhat opaque.*

Exercise 8.3.2 *Compute the product $\boldsymbol{R}_{x,\theta}\, \boldsymbol{R}_{z,\phi}\, \boldsymbol{R}_{y,\psi}$ for general angles θ, ϕ and ψ. This gives you a 3×3 rotation matrix which is somewhat opaque.*

Exercise 8.3.3 *Given $\boldsymbol{P} = \begin{bmatrix} 10 \\ 5 \\ 7 \end{bmatrix}$, for $\theta = .5$, $\phi = .4$ and $\psi = .2$, calculate by hand*

- $\boldsymbol{R}_{x,\theta}\boldsymbol{P}$
- $\boldsymbol{R}_{y,\phi}\boldsymbol{P}$
- $\boldsymbol{R}_{z,\psi}\boldsymbol{P}$

Exercise 8.3.4 *Given $\boldsymbol{P} = \begin{bmatrix} 20 \\ 35 \\ 17 \end{bmatrix}$ for $\theta = .3$, $\phi = .5$ and $\psi = .15$, calculate by hand*

- $\boldsymbol{Q} = \boldsymbol{R}_{z,\psi}\, \boldsymbol{R}_{y,\phi}\, \boldsymbol{R}_{x,\theta}\boldsymbol{P}$
- $\boldsymbol{R}_{x,-\theta}\, \boldsymbol{R}_{y,-\phi}\, \boldsymbol{R}_{z,-\psi}\boldsymbol{Q}$

Exercise 8.3.5 *Given $\boldsymbol{P} = \begin{bmatrix} 10 \\ 5 \\ 7 \end{bmatrix}$, for $\theta = .5$, $\phi = .4$ and $\psi = .2$, calculate using MATLAB*

- $\boldsymbol{R}_{x,\theta}\boldsymbol{P}$
- $\boldsymbol{R}_{y,\phi}\boldsymbol{P}$
- $\boldsymbol{R}_{z,\psi}\boldsymbol{P}$

Exercise 8.3.6 *Given $\boldsymbol{P} = \begin{bmatrix} 20 \\ 35 \\ 17 \end{bmatrix}$ for $\theta = .3$, $\phi = .5$ and $\psi = .15$, calculate using MATLAB*

- $\boldsymbol{Q} = \boldsymbol{R}_{z,\psi}\, \boldsymbol{R}_{y,\phi}\, \boldsymbol{R}_{x,\theta}\boldsymbol{P}$
- $\boldsymbol{R}_{x,-\theta}\, \boldsymbol{R}_{y,-\phi}\, \boldsymbol{R}_{z,-\psi}\boldsymbol{Q}$

8.3.2 Drawing Rotations

The easiest way to understand the rotations is to look at some graphs. An even better way is to build a model out of cardboard: you should try it! We can draw the two sets of orthogonal axes we get from the application of rotations using the following code. This code is designed to work with drawing elliptical orbits so we set up the size of the axis system based on the size of the ellipse we will want to draw. Since the apoapsis of the ellipse is $R/(1-e)$ where e is the eccentricity and R is the scaling factor; we use that to set the axes dimensions. Then later, any ellipse we draw will fit into the coordinate system nicely. The rest of the code simply draws straight lines and determines the line style of the drawn line.

Listing 8.5: **DrawAxis.m**

```
function DrawAxis (R,e,N,E1,E2,E3,A,style ,color)
%
% Draw coordinate system
hold on;
Rmax = R/(1-e);
% Graph new z axis and new y axis
  t = linspace(-1.5*Rmax,1.5*Rmax,N);
  W1 = E1;
  vW1 = @(t) t*W1;
  for i=1:N
    s = t(i); D = vW1(s);
    x1(i) = D(1) + A(1);
    y1(i) = D(2) + A(2);
    z1(i) = D(3) + A(3);
  end
  %
  W2 = E2;
  vW2 = @(t) t*W2;
  for i=1:N
    s = t(i); D = vW2(t(i));
    x2(i) = D(1) + A(1);
    y2(i) = D(2) + A(2);
    z2(i) = D(3) + A(3);
  end
   %
  W3 = E3;
  vW3 = @(t) t*W3;
  for i=1:N
    s = t(i); D = vW3(t(i));
    x3(i) = D(1) + A(1);
    y3(i) = D(2) + A(2);
    z3(i) = D(3) + A(3);
  end
 if (style == 1)
  plot3(x1,y1,z1,'-','LineWidth',2,'Color',color);
  plot3(x2,y2,z2,'-','LineWidth',2,'Color',color);
  plot3(x3,y3,z3,'-','LineWidth',2,'Color',color);
 elseif (style == 2)
  if (style == 2)
  plot3(x1,y1,z1,'o','LineWidth',2,'Color',color);
  plot3(x2,y2,z2,'o','LineWidth',2,'Color',color);
  plot3(x3,y3,z3,'o','LineWidth',2,'Color',color);
 else
   disp("not one or two");
 end
hold off;
end
```

Let's look at a rotation about the $+z$ axis of angle θ. The code is simple.

Listing 8.6: **Rotation about the z Axis**

```
I = [1;0;0]; J = [0;1;0]; K = [0;0;1]; A = [0;0;0];
hold on;
```

```
DrawAxis (2,0.6,20,I,J,K,A,1,'black');
theta = pi/6;
Xp = RotateZ(theta)*I;
Yp = RotateZ(theta)*J;
Zp = RotateZ(theta)*K;
DrawAxis(2,0.6,20,Xp,Yp,Zp,A,1,'black!30');
hold off
Xpn = RotateZ(-theta)*I;
Ypn = RotateZ(-theta)*J;
Zpn = RotateZ(-theta)*K;
DrawAxis(2,0.6,20,Xpn,Ypn,Zpn,A,2,'red');
xlabel('x');
ylabel('y')
zlabel('z');
```

In Figure 8.6, we look down on the plot so that the $+z$ axis is pointing straight up and cannot be seen. The positive θ rotation moves the x axis counterclockwise. On the other hand, the negative rotation by $-\theta$ is shown with axis plotted by circles and it moves the $+x$ axis clockwise. So a positive rotation about the $+z$ axis means the 2×2 rotation matrix seen inside should be $\boldsymbol{R_\theta}$. The standard way to pick the kind of rotation we want is to define a positive rotation about a positive axis as follows: line the thumb of your right hand up along the positive axis. Then your fingers will curl in the direction of what we call a positive rotation angle. So positive rotations along the $+z$ axis give the usual embedded $\boldsymbol{R_\theta}$ part of the three dimensional rotation matrix.

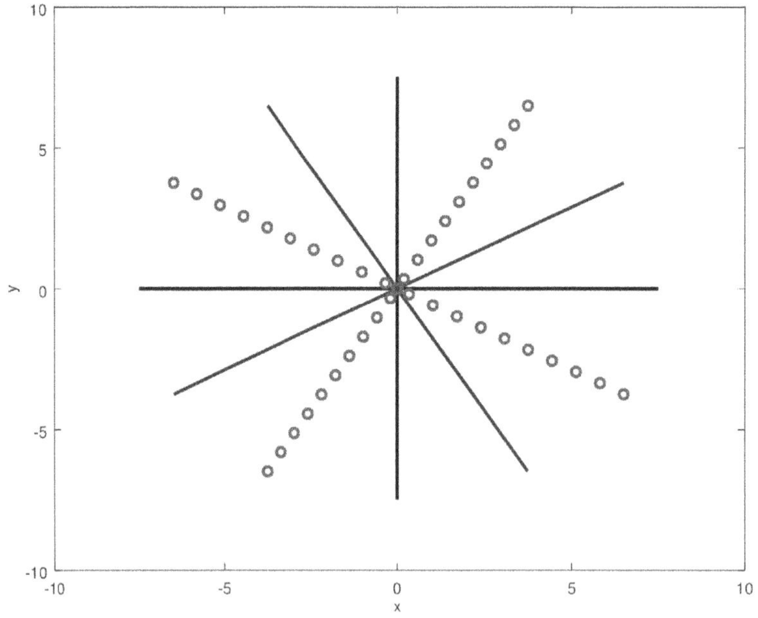

Figure 8.6: A positive and negative rotation about the z axis.

Homework

Exercise 8.3.7 *Given* $\boldsymbol{P} = \begin{bmatrix} 20 \\ 35 \\ 17 \end{bmatrix}$ *for* $\theta = .3$*, draw the rotation* $\boldsymbol{R_{z,\theta}P}$.

Exercise 8.3.8 *Given* $\boldsymbol{P} = \begin{bmatrix} -10 \\ 35 \\ 12 \end{bmatrix}$ *for* $\theta = -.3$*, draw the rotation* $\boldsymbol{R_{z,\theta}P}$.

Exercise 8.3.9 *Given* $\boldsymbol{P} = \begin{bmatrix} 2 \\ 5 \\ 7 \end{bmatrix}$ *for* $\theta = 1.2$*, draw the rotation* $\boldsymbol{R_{z,\theta}P}$.

Next, look at a rotation about the $+y$ axis of angle θ. The code is simple.

Listing 8.7: **Rotation about the y Axis**

```
I = [1;0;0]; J = [0;1;0]; K = [0;0;1]; A = [0;0;0];
hold on;
DrawAxis (2,0.6,20,I,J,K,A,1,'black');
theta = pi/6;
Xp = RotateY(theta)*I;
Yp = RotateY(theta)*J;
Zp = RotateY(theta)*K;
DrawAxis(2,0.6,20,Xp,Yp,Zp,A,1,'black!30');
hold off
Xpn = RotateY(-theta)*I;
Ypn = RotateY(-theta)*J;
Zpn = RotateY(-theta)*K;
DrawAxis(2,0.6,20,Xpn,Ypn,Zpn,A,2,'red');
xlabel('x');
ylabel('y')
zlabel('z');
```

In Figure 8.7, we look down on the plot so that the $+y$ axis is pointing straight up and cannot be seen. The positive θ rotation moves the z axis **counterclockwise** and the $+x$ axis **clockwise**. This corresponds to a rotation by $-\theta$. So a positive rotation about the $+y$ axis means the 2×2 rotation matrix seen inside should be $\boldsymbol{R_{-\theta}}$. Hence, if we want the $+x$ axis to rotate counterclockwise, we would apply $\boldsymbol{R_{\theta}}$.

Homework

Exercise 8.3.10 *Given* $\boldsymbol{P} = \begin{bmatrix} 20 \\ 35 \\ 17 \end{bmatrix}$ *for* $\theta = .35$*, draw the rotation* $\boldsymbol{R_{y,\theta}P}$.

Exercise 8.3.11 *Given* $\boldsymbol{P} = \begin{bmatrix} -10 \\ 35 \\ 12 \end{bmatrix}$ *for* $\theta = -.6$*, draw the rotation* $\boldsymbol{R_{y,\theta}P}$.

Exercise 8.3.12 *Given* $\boldsymbol{P} = \begin{bmatrix} 2 \\ 5 \\ 7 \end{bmatrix}$ *for* $\theta = 2.2$*, draw the rotation* $\boldsymbol{R_{y,\theta}P}$.

Finally, look at a rotation about the $+x$ axis of angle θ. The code is simple.

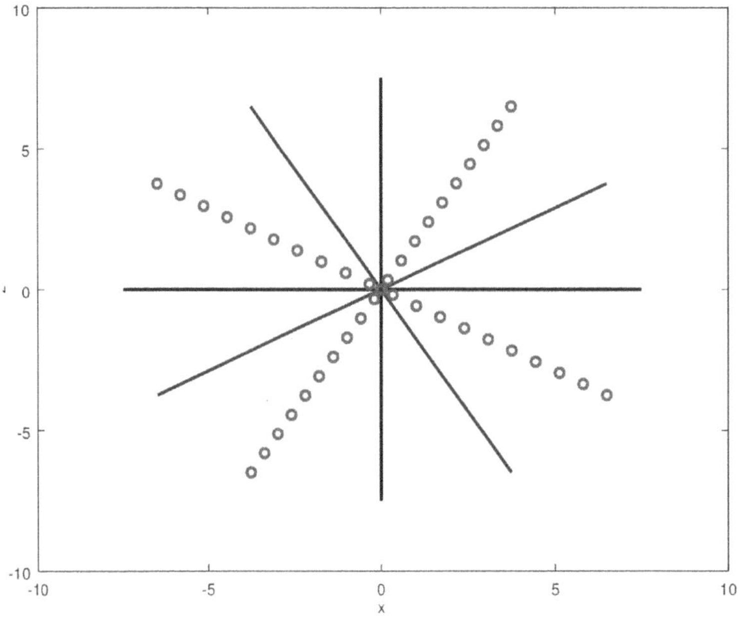

Figure 8.7: A positive and negative rotation about the y axis.

Listing 8.8: **Rotation about the x Axis**

```
I = [1;0;0]; J = [0;1;0]; K = [0;0;1]; A = [0;0;0];
hold on;
DrawAxis (2,0.6,20,I,J,K,A,1,'black');
theta = pi/6;
Xp = RotateX(theta)*I;
Yp = RotateX(theta)*J;
Zp = RotateX(theta)*K;
DrawAxis(2,0.6,20,Xp,Yp,Zp,A,1,'black!30');
hold off
Xpn = RotateX(-theta)*I;
Ypn = RotateX(-theta)*J;
Zpn = RotateX(-theta)*K;
DrawAxis(2,0.6,20,Xpn,Ypn,Zpn,A,2,'red');
xlabel('x');
ylabel('y');
zlabel('z');
```

In Figure 8.8, we look down on the plot so that the $+x$ axis is pointing up at almost a vertical angle. The positive θ rotation moves the z axis **counterclockwise** and the $+x$ axis **counterclockwise**. This corresponds to a rotation by θ. So a positive rotation about the $+x$ axis means the 2×2 rotation matrix seen inside should be $\boldsymbol{R_\theta}$.

Homework

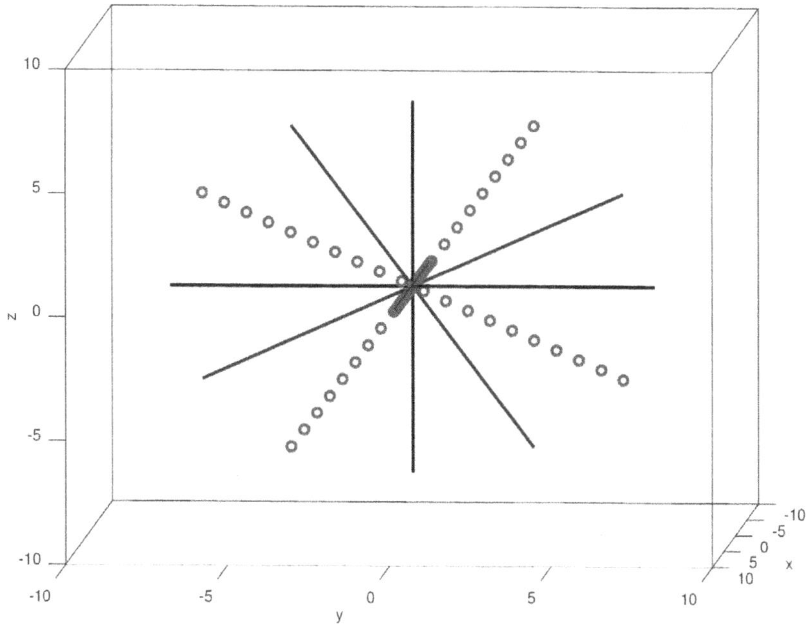

Figure 8.8: A positive and negative rotation about the x axis.

Exercise 8.3.13 *Given* $\boldsymbol{P} = \begin{bmatrix} 20 \\ 35 \\ 17 \end{bmatrix}$ *for* $\theta = .75$*, draw the rotation* $\boldsymbol{R}_{\boldsymbol{x},\boldsymbol{\theta}}\boldsymbol{P}$.

Exercise 8.3.14 *Given* $\boldsymbol{P} = \begin{bmatrix} -10 \\ 35 \\ 12 \end{bmatrix}$ *for* $\theta = -1.6$*, draw the rotation* $\boldsymbol{R}_{\boldsymbol{x},\boldsymbol{\theta}}\boldsymbol{P}$.

Exercise 8.3.15 *Given* $\boldsymbol{P} = \begin{bmatrix} 2 \\ 5 \\ 7 \end{bmatrix}$ *for* $\theta = 2.1$*, draw the rotation* $\boldsymbol{R}_{\boldsymbol{x},\boldsymbol{\theta}}\boldsymbol{P}$.

8.3.3 Rotated Ellipses

For a given ellipse in the $x-y$ plane, we can rotate it into a given orbital plane using various rotations. The general formula for an ellipse is $r = \rho/(1 + e\cos(t))$ and we use that to set up the plot. In this code, we make sure we move in the counterclockwise direction for the $+x$ axis when we do a rotation with respect to the y axis by explicitly using `RotateY(-phi)`.

Listing 8.9: **DrawEllipse.m**

```
function DrawEllipse(rho,e,theta,phi,delta)
% draw orbit in the new x-y coordinate system
Phi = linspace(0,2*pi,51);
R = @(t) rho/(1 + e*cos(t));
u = @(t) R(t)*cos(t);
v = @(t) R(t)*sin(t);
```

```
X = @(t) [u(t);v(t);0];
OP = @(t) RotateZ(theta)*X(t);
OPP = @(t) RotateY(-phi)*OP(t);
OPPP = @(t) RotateX(delta)*OPP(t);
x1 = zeros(3,51);
x2 = zeros(3,51);
x3 = zeros(3,51);
conic = zeros(3,51);
for i=1:51
  t = Phi(i);
  conic(:,i)= OPPP(t);
end
x1 = conic(1,:);
x2 = conic(2,:);
x3 = conic(3,:);
hold on;
plot3(x1,x2,x3,'-','LineWidth',3,'Color','black!60');
hold off;
end
```

Here is a typical rotated ellipse's graph session.

Listing 8.10: **A Typical Rotated Ellipse**

```
>> DrawEllipse(10,0.5,.5,.5,.4);
>> xlabel('x');
>> ylabel('y');
>> zlabel('z');
>> title('Rotated Ellipse: R = 10, e = 0.5, \theta=.5, \phi=0.5,\delta
   =0.4');
```

We show the resulting plot in Figure 8.9, but be warned, these plots are hard to see! We rotated it around until you can sort of see the rotations, but with three axis rotations, this gets hard fast.

8.3.3.1 Homework

Exercise 8.3.16 *Draw a rotated ellipse for* $R = 20$, $\theta = 0.8$, $\phi = 0.6$ *and* $\delta = .3$ *for* $e = 0.9$.

Exercise 8.3.17 *Draw a rotated ellipse for* $R = 30$, $\theta = 0.4$, $\phi = 1.9$ *and* $\delta = 1.3$ *for* $e = 0.4$.

Exercise 8.3.18 *Draw a rotated ellipse for* $R = 10$, $\theta = 1.1$, $\phi = 0.7$ *and* $\delta = .9$ *for* $e = 0.2$.

8.4 Drawing the Orbital Plane

Now let's put it all together and draw an orbital plane somewhat carefully. We do this by gluing together the various pieces of code we have written and we include a few functions we have not discussed. However, the code is available for viewing and it is easy enough to figure it out after all we have discussed so far. This code draws a snapshot of the orbital system using a mesh setup in a given 2D plane with the function `DrawAxisGrid(R,e,N)` where R is the radius vector in the polar orbit equation and e is the eccentricity. The mesh is then drawn using a $[-N, N] \times [-N \times N]$ grid using a $\Delta x = \Delta y = \frac{R_{max}}{N}$ where R_{max} is the maximum radius magnitude $R_{max} = \frac{R}{1-e}$.

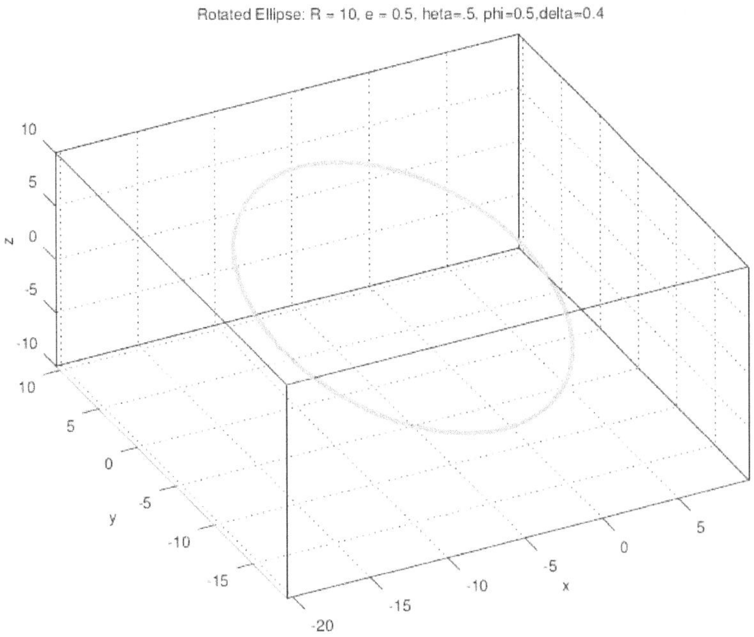

Figure 8.9: A rotated ellipse.

Listing 8.11: **DrawOrbit.m**

```
function DrawOrbit(R,e,N,theta,phi,delta,E1,E2,E3,A,style,color)
%
% Draw coordinate system
clf;
hold on;
% Graph the original axes
DrawAxisGrid(R,e,N);
DrawAxis(R,e,N,E1,E2,E3,A,1,color);
% Draw rotated system fixing z
Xp = RotateZ(theta)*E1;
Yp = RotateZ(theta)*E2;
Zp = E3;
%DrawAxis(Xp,Yp,Zp,A,'black!30');
Xpp = RotateY(-phi)*Xp;
Ypp = Yp;
Zpp = RotateY(-phi)*Zp;
%DrawAxis(Xpp,Ypp,Zpp,A,'black!60');
Xppp = Xpp;
Yppp = RotateX(delta)*Ypp;
Zppp = RotateX(delta)*Zpp;
DrawAxis(R,e,N,Xppp,Yppp,Zppp,A,1,'red');
DrawEllipse(R,e,N,theta,phi,delta);
DrawAxisGrid(R,e,N);
xlabel('x');
ylabel('y');
zlabel('z');
```

```
hold off;
end
```

The `DrawAxisGrid` code is as follows:

Listing 8.12: **DrawAxisGrid.m**

```
function DrawAxisGrid(R,e,N)
%
% draw a mesh in the original xy plane
hold on;
Rmax = R/(1-e);
delx = Rmax/(N);
dely = Rmax/(N);
x = [];
for i=-N:N
  u = [i*delx];
  x = [x,u];
end
%
y = [];
for i=-N:N
  u = [i*dely];
  y = [y,u];
end
%plot x, y grid
U = [-Rmax,Rmax];
W = [0,0];
for j=1:2*N+1
  V = [y(j),y(j)];
  plot3(U,V,W,'LineWidth',1);
end
V = [-Rmax,Rmax];
for j=1:2*N+1
  U = [x(j),x(j)];
  plot3(U,V,W,'LineWidth',1);
end
hold off;
end
```

To run this code, we use the command:

Listing 8.13: **Plotting an Orbital Plane with Rotated Axes**

```
I=[1;0;0]; J=[0;1;0]; K=[0;0;1]; A=[0;0;0];
DrawOrbit(2,0.6,20,pi/6,3*pi/4,3*pi/4,I,J,K,A,1,'black');
```

The plot we generate can be manipulated to get the image in Figure 8.10.
If you look closely, you can see the orbital plane crosses the equatorial plane and the details of this crossing are important to understanding this orbit. If you comment out the line in `DrawOrbit.m` which generates the rotated axes, you can see this a little more clearly as we show in Figure 8.11. The

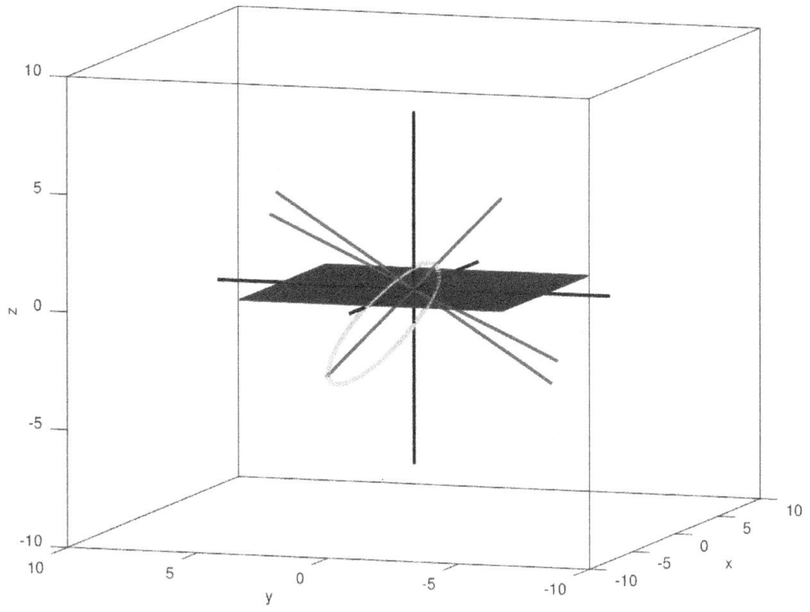

Figure 8.10: Drawing an orbital plane.

equatorial plane is drawn heavily crosshatched and the part of the orbital plane above the equatorial plane is reasonably easy to see.

Homework

Use the same $\boldsymbol{I}$, $\boldsymbol{J}$, $\boldsymbol{K}$ and $\boldsymbol{A}$ as in the example in these exercises.

Exercise 8.4.1 *Generate the orbital plane or* $R = 30$, $e = 0.7$, $N = 25$, $\theta = \pi/8$, $\phi = \pi/5$ *and* $\delta = 5\pi/6$. *Make sure you rotate the plot as needed to see the orbital plane.*

Exercise 8.4.2 *Generate the orbital plane or* $R = 40$, $e = 0.2$, $N = 35$, $\theta = \pi/4$, $\phi = \pi/3$ *and* $\delta = 7\pi/6$. *Make sure you rotate the plot as needed to see the orbital plane.*

Exercise 8.4.3 *Generate the orbital plane or* $R = 10$, $e = 0.6$, $N = 25$, $\theta = \pi/3$, $\phi = \pi/6$ *and* $\delta = 5\pi/8$. *Make sure you rotate the plot as needed to see the orbital plane.*

Exercise 8.4.4 *Generate the orbital plane or* $R = 40$, $e = 0.3$, $N = 55$, $\theta = \pi/6$, $\phi = \pi/3$ *and* $\delta = \pi/4$. *Make sure you rotate the plot as needed to see the orbital plane.*

8.4.1 The Perifocal Coordinate System

The position vector $\boldsymbol{r}$ and its velocity vector $\boldsymbol{v}$ determine a plane whose normal vector is $\boldsymbol{h}$. The closest point on the orbit to the center is called the **periapsis**, while the point farthest away is called the **apoapsis**. We let $\boldsymbol{x}_\omega$ denote the axis pointing towards periapsis and $\boldsymbol{y}_\omega$ is the axis rotated 90° in the direction of the orbital motion. Hence, the position of the satellite in the orbit is given by an ordered pair (u, v). We let $\boldsymbol{P}$ and $\boldsymbol{Q}$ be unit vectors along the $\boldsymbol{x}_\omega$ and $\boldsymbol{y}_\omega$ axes respectively. Of course, $\boldsymbol{P} = \boldsymbol{e}/\mathscr{E}$. Then, if the position of the satellite in the orbit is denoted by $\boldsymbol{X}$, then $\boldsymbol{X}$ has a unique representation in the two dimensional vector space determined by the orthonormal basis $\{\boldsymbol{P}, \boldsymbol{Q}\}$ given by $\boldsymbol{X} = x\boldsymbol{P} + y\boldsymbol{Q}$. The positive $\boldsymbol{z}_\omega$ axis is determined by the right-hand rule and the

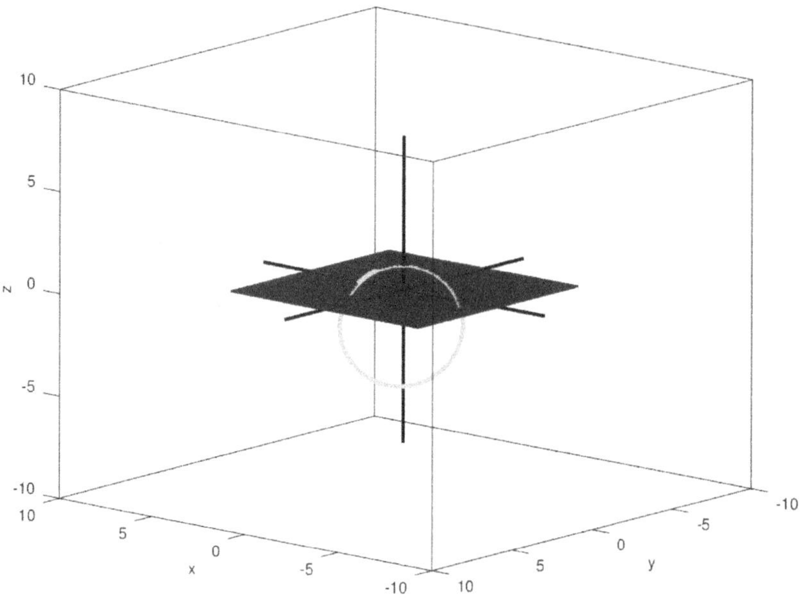

Figure 8.11: The orbital plane crosses the equatorial plane.

unit vector along $\boldsymbol{z}_\omega$ is called $\boldsymbol{W}$ or $\boldsymbol{R}$. Of course, a vector $a\boldsymbol{P} + b\boldsymbol{Q} + c\boldsymbol{W}$ in this system needs to be converted to coordinates in other systems. You can see this coordinate system in Figure 8.12.

8.4.1.1 Homework

Exercise 8.4.5 *Assume* $\boldsymbol{U} = \begin{bmatrix} 1 \\ -1 \\ 2 \end{bmatrix}$ *and* $\boldsymbol{V} = \begin{bmatrix} 2 \\ 4 \\ 1 \end{bmatrix}$. *Then set* $\boldsymbol{P} = \boldsymbol{U}/\|\boldsymbol{U}\|$ *and* $\boldsymbol{Q} = \boldsymbol{V}/\|\boldsymbol{V}\|$. *Let the position of the satellite be* $2\boldsymbol{P} + 3\boldsymbol{Q}$. *Draw the satellite in the* **Perifocal Coordinate System** *marking the* x_ω, y_ω *and* z_ω *axes. Draw the* $\boldsymbol{W}$ *vector too.*

Exercise 8.4.6 *Assume* $\boldsymbol{U} = \begin{bmatrix} 2 \\ 3 \\ -2 \end{bmatrix}$ *and* $\boldsymbol{V} = \begin{bmatrix} 2 \\ 4 \\ 8 \end{bmatrix}$. *Then set* $\boldsymbol{P} = \boldsymbol{U}/\|\boldsymbol{U}\|$ *and* $\boldsymbol{Q} = \boldsymbol{V}/\|\boldsymbol{V}\|$. *Let the position of the satellite be* $2\boldsymbol{P} + 3\boldsymbol{Q}$. *Draw the satellite in the* **Perifocal Coordinate System** *marking the* x_ω, y_ω *and* z_ω *axes. Draw the* $\boldsymbol{W}$ *vector too.*

Exercise 8.4.7 *Assume* $\boldsymbol{U} = \begin{bmatrix} 1 \\ -1 \\ -2 \end{bmatrix}$ *and* $\boldsymbol{V} = \begin{bmatrix} 4 \\ 8 \\ -2 \end{bmatrix}$. *Then set* $\boldsymbol{P} = \boldsymbol{U}/\|\boldsymbol{U}\|$ *and* $\boldsymbol{Q} = \boldsymbol{V}/\|\boldsymbol{V}\|$. *Let the position of the satellite be* $2\boldsymbol{P} + 3\boldsymbol{Q}$. *Draw the satellite in the* **Perifocal Coordinate System** *marking the* x_ω, y_ω *and* z_ω *axes. Draw the* $\boldsymbol{W}$ *vector too.*

Exercise 8.4.8 *Assume* $\boldsymbol{U} = \begin{bmatrix} 6 \\ -19 \\ 11 \end{bmatrix}$ *and* $\boldsymbol{V} = \begin{bmatrix} 20 \\ 4 \\ 4 \end{bmatrix}$. *Then set* $\boldsymbol{P} = \boldsymbol{U}/\|\boldsymbol{U}\|$ *and* $\boldsymbol{Q} = \boldsymbol{V}/\|\boldsymbol{V}\|$. *Let the position of the satellite be* $2\boldsymbol{P} + 3\boldsymbol{Q}$. *Draw the satellite in the* **Perifocal Coordinate System** *marking the* x_ω, y_ω *and* z_ω *axes. Draw the* $\boldsymbol{W}$ *vector too.*

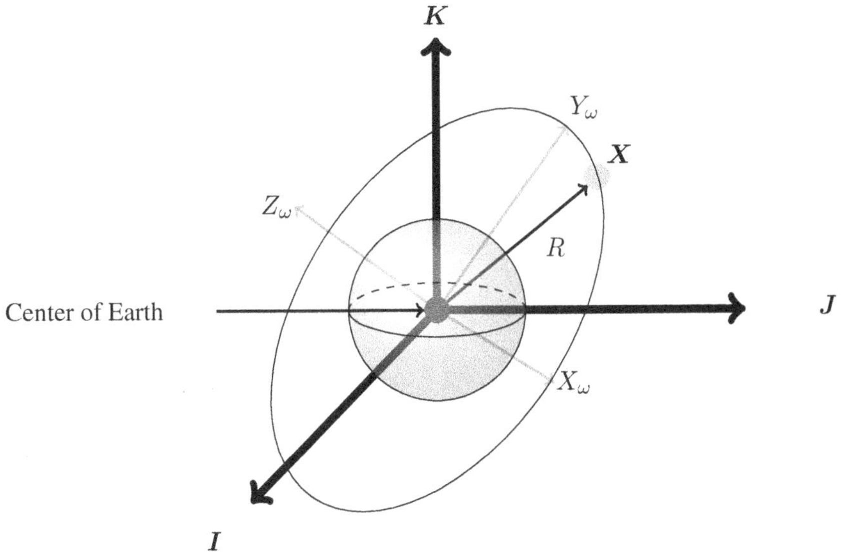

A vector $\boldsymbol{X}$ in the Perifocal coordinate system. $\boldsymbol{X}$ marks the position of the satellite in its orbit. The unit vector $\boldsymbol{P}$ is along the $\boldsymbol{X}_\omega$ axis, the unit vector $\boldsymbol{Q}$ is along the $\boldsymbol{Y}_\omega$ axis and the unit vector $\boldsymbol{W}$ lies on the $\boldsymbol{Z}_\omega$ axis. The $\boldsymbol{I} - \boldsymbol{J} - \boldsymbol{K}$ axis for the Geocentric - Equatorial and Right Ascension - Declination coordinate systems are shown for reference.

Figure 8.12: The Perifocal coordinate system.

8.4.2 Orbital Elements

The orbital elements of an orbit are enough to determine completely the look of the orbit. Let's describe them carefully.

- One half of the major axis of an ellipse is called a. You should look at Figure 8.5 to refresh your mind about this. From the origin of the ellipse at say $(0, 0)$, the point $(c, 0)$ is the focal point F_1. The point closest to a foci is $(a, 0)$ and so the distance from F_1 to the periapsis is $a - c$.

- The **eccentricity** of the elliptical orbit is called $\mathscr{E}$ and often just e. We use a boldface $\boldsymbol{e}$ to denote the eccentricity vector and the length of $\boldsymbol{e}$, which is traditionally just denoted by e, looks a lot like $\boldsymbol{e}$ here. So to be clear, we will let the eccentricity be called $\mathscr{E}$ although sometimes in code we just use e for convenience.

- The **inclination** angle is the angle between the unit vector $\boldsymbol{K}$ and the angular momentum vector $\boldsymbol{h}$. Hence, if we know $\boldsymbol{R} = R_1\boldsymbol{I} + R_2\boldsymbol{J} + R_3\boldsymbol{K}$ and $\boldsymbol{R} = R_1\boldsymbol{I} + R_2\boldsymbol{J} + R_3\boldsymbol{K}$, we can calculate

$$\boldsymbol{h} \;=\; \boldsymbol{r} \times \boldsymbol{V} = det \begin{bmatrix} \boldsymbol{I} & \boldsymbol{J} & \boldsymbol{K} \\ R_1 & R_2 & R_3 \\ V_1 & V_2 & V_3 \end{bmatrix}$$

Once $\boldsymbol{h}$ has been calculated, we can use it to find the **line of nodes** vector. The orbital plane crosses the equatorial plane except in degenerate cases such as an equatorial orbit. Now the normal to the orbital plane is $\boldsymbol{h}$. When the orbital plane crosses the equatorial plane, at those points, it is in the equatorial plane and so must be perpendicular to the equatorial plane too. So the line of intersection of the orbital plane and the equatorial plane is orthogonal to both $\boldsymbol{h}$ and $\boldsymbol{K}$. The cross product then determines a vector called the **node** vector which is the direction vector of the line of nodes. We let $\boldsymbol{n}$ denote this vector and we see $\boldsymbol{n} = \boldsymbol{h} \times \boldsymbol{K}$.

- Since the line of nodes is in the equatorial plane, we can measure the angle between the line of nodes vector and the I vector. This angle is called the **longitude of the ascending node**, Ω and we see $\Omega = \cos^{-1}(< \boldsymbol{n}, \boldsymbol{I} >)$.
- The **argument of periapsis** is the angle **in the orbital plane** measured from the line of nodes to the periapsis vector. We denote this by ω.
- The **true anomaly** is the angle **in the orbital plane** measured between the eccentricity vector or periapsis vector and the position vector. We denote this by ν_0.
- The **time of periapsis passage** is the time, T, when the satellite was at periapsis.

In Figure 8.13, we try to show you these in an easy to follow diagram. In the figure, we show the true anomaly as the angle between the periapsis vector and the position vector and it looks like these vectors extend to the center of the earth. Of course, this angle is measured using as center the focus of the ellipse that determines the orbit in the orbital plane. To keep the diagram just a bit less cluttered we did not draw in the focus. So just keep that in mind.

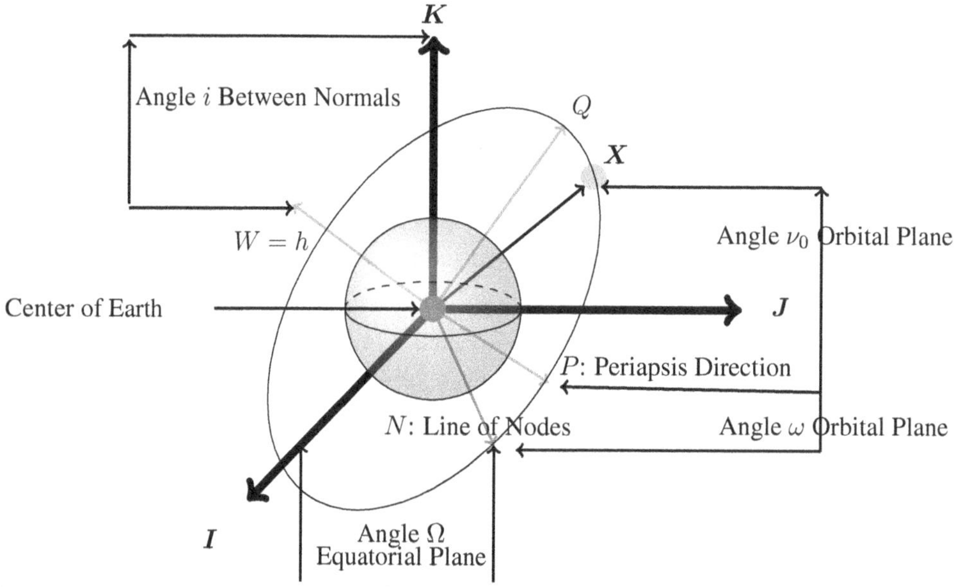

The orbital elements for a given orbit. All of these are straightforward to calculate given a measurement of $\boldsymbol{r}$ and $\boldsymbol{v}$ in the $\boldsymbol{IJK}$ system. Just remember ν_0 is actually measured from the foci of the orbit.

Figure 8.13: The orbital elements.

8.4.2.1 Homework

For these exercises, we need a value for $\boldsymbol{R}$ and $\boldsymbol{V}$. Set $\mu = 1$. Then we can calculate

- The eccentricity vector called $\boldsymbol{E}$ here as $\boldsymbol{E} = \frac{1}{\mu}(\|\boldsymbol{V}\|^2 - \frac{\mu}{\|\boldsymbol{R}\|})\boldsymbol{R} - < \boldsymbol{R}, \boldsymbol{V} > \boldsymbol{V}$.
- $\boldsymbol{h} = \boldsymbol{R} \times \boldsymbol{V}$.
- $\boldsymbol{n} = \boldsymbol{h} \times \boldsymbol{K}$.
- $\cos(\Omega) = \frac{<\boldsymbol{n},\boldsymbol{I}>}{\|\boldsymbol{n}\|}$.
- $\cos(\omega) = \frac{<\boldsymbol{n},\boldsymbol{E}>}{\|\boldsymbol{n}\|\,\|\boldsymbol{E}\|}$.

Exercise 8.4.9 *Find the orbital elements for* $\boldsymbol{R} = \begin{bmatrix} 10 \\ 20 \\ 5 \end{bmatrix}$ *and* $\boldsymbol{V} = \begin{bmatrix} -2 \\ 2 \\ 1 \end{bmatrix}$. *Make sure you find* Ω *and* ω. *Draw a sketch also.*

Exercise 8.4.10 *Find the orbital elements for* $\boldsymbol{R} = \begin{bmatrix} 10 \\ 20 \\ 5 \end{bmatrix}$ *and* $\boldsymbol{V} = \begin{bmatrix} -2 \\ 2 \\ 1 \end{bmatrix}$. *Make sure you find* Ω *and* ω. *Draw a sketch also.*

Exercise 8.4.11 *Find the orbital elements for* $\boldsymbol{R} = \begin{bmatrix} 10 \\ 20 \\ 5 \end{bmatrix}$ *and* $\boldsymbol{V} = \begin{bmatrix} -2 \\ 2 \\ 1 \end{bmatrix}$. *Make sure you find* Ω *and* ω. *Draw a sketch also.*

Exercise 8.4.12 *Find the orbital elements for* $\boldsymbol{R} = \begin{bmatrix} 10 \\ 20 \\ 5 \end{bmatrix}$ *and* $\boldsymbol{V} = \begin{bmatrix} -2 \\ 2 \\ 1 \end{bmatrix}$. *Make sure you find* Ω *and* ω. *Draw a sketch also.*

Exercise 8.4.13 *Find the orbital elements for* $\boldsymbol{R} = \begin{bmatrix} 10 \\ 20 \\ 5 \end{bmatrix}$ *and* $\boldsymbol{V} = \begin{bmatrix} -2 \\ 2 \\ 1 \end{bmatrix}$. *Make sure you find* Ω *and* ω. *Draw a sketch also.*

8.5 Drawing Orbital Planes Given Radius and Velocity Vectors

Now let's find and graph the orbital plane for a specific example and try to use our knowledge of orbital mechanics along the way. We will explicitly calculate the $\boldsymbol{P}$, $\boldsymbol{Q}$ and $\boldsymbol{W}$ orthonormal basis vectors of the perifocal coordinate system and the line of nodes vector $\boldsymbol{n}$. Once that is done, we can plot these vectors as lines in 3D space. So we will not use rotation matrices to do the plots here. We assume we are given the position vector $\boldsymbol{r}_0$ and velocity vector $\boldsymbol{v}_0$ of the satellite at the initial time 0. These are given as vectors in the equatorial plane coordinate system so each has an $\boldsymbol{I}$, $\boldsymbol{J}$ and $\boldsymbol{K}$ component. We need a cross product function, so for experience, let's build one.

Listing 8.14: **CrossProduct.m**

```
function C = CrossProduct(A,B)
% A is a 3D vector
% B is a 3D vector
% returns C, the cross product of A and B
%
% AxB = det(I,J,K;
C = [A(2)*B(3)-A(3)*B(2);A(3)*B(1)-A(1)*B(3);A(1)*B(2)-A(2)*B(1)];
end
```

First, we set up the standard equatorial frame unit vectors and find the angular momentum vector.

Listing 8.15: **Find Angular Momentum Vector**

```
% set standard I, J, K vectors
I = [1;0;0]; J = [0;1;0]; K = [0;0;1];
R0 = norm(r0);
V0 = norm(v0);
H = CrossProduct(r0,v0);
h = norm(H);
```

Now we can find $\mathscr{H}^2/\mu$, the eccentricity vector and its norm.

Listing 8.16: **Find the Eccentricity Vector**

```
p = h^2;
% find the eccentricity vector
E = (1/mu)*( V0^2 - mu/R0)*r0 - dot(r0,v0)*v0;
% find the eccentricity
e = norm(E);
```

Then find the inclination angle, i, the line of nodes vector $\boldsymbol{n}$ and the longitude of the ascending node, Ω.

Listing 8.17: **Find the Inclination, Line of Nodes and Ω**

```
% find the inclination
inc = acos( dot(H,K)/h );
% find the line of nodes vector
N = CrossProduct(K,H);
n = norm(N);
% find Omega
Omega = acos( dot(N,I)/n );
```

Then find the argument of periapsis, ω and the true anomaly, ν_0. The true anomaly is the angle between the periapsis direction $\boldsymbol{B}$ and e.

Listing 8.18: **Find the Argument of Periapsis and True Anomaly**

```
% find omega
omega = acos( dot(N,E)/(n*e) );
% find the true anomaly
nu0 = acos( dot(E,r0)/(e*R0) );
```

Now we can get the orbital plane.

Listing 8.19: **Find Perifocal Basis**

```
% unit normal to orbital plane
W = H/h;
% get perigee direction vector in orbital plane
P = E/e;
W = H/h;
Q = CrossProduct(W,P);
```

Now get the semimajor axis.

Listing 8.20: **Find the Semimajor Axis** a

```
rp = (h^2)/(1+e);
ra = (h^2)/(1-e);
a = (rp+ra)/2.0;
```

Next, plot the standard Geocentric - Equatorial and Perifocal coordinate axes. We start holding the plots now.

Listing 8.21: **Plot the Geocentric and Perifocal Axes**

```
hold on
% Draw IJK
[x,y,z] = ThreeDToPlot(ra*I);
plot3(x,y,z,'linewidth',2,'color','black');
[x,y,z] = ThreeDToPlot(ra*J);
plot3(x,y,z,'linewidth',2,'color','black');
[x,y,z] = ThreeDToPlot(ra*K);
plot3(x,y,z,'linewidth',2,'color','black');
% Draw PQW
[x,y,z] = ThreeDToPlot(ra*P);
plot3(x,y,z,'linewidth',2,'color','red');
[x,y,z] = ThreeDToPlot(-ra*P);
plot3(x,y,z,'linewidth',2,'color','red');
[x,y,z] = ThreeDToPlot(ra*Q);
plot3(x,y,z,'linewidth',2,'color','red');
[x,y,z] = ThreeDToPlot(-ra*Q);
plot3(x,y,z,'linewidth',2,'color','red');
[x,y,z] = ThreeDToPlot(ra*W);
plot3(x,y,z,'linewidth',2,'color','red');
```

Draw the line of nodes.

Listing 8.22: **Draw the Line of Nodes**

```
[x,y,z] = ThreeDToPlot(ra*N);
plot3(x,y,z,'linewidth',2,'color','black!60');
[x,y,z] = ThreeDToPlot(-ra*N);
```

Plot the orbit and stop the hold on plots.

Listing 8.23: **Plot the Orbit**

```
% Draw orbit
OrbitSize = 80;
rorbitplane=zeros(OrbitSize,3);
nu = linspace(0,2*pi,OrbitSize);
for i = 1:OrbitSize
  c = cos(nu(i)); s = sin(nu(i));
  rorbit = h^2/(1 + e*c);
  rorbitplane(i,:) = rorbit*c*P+rorbit*s*Q;
  T = rorbitplane(i,:);
  x(i) = T(1);
  y(i) = T(2);
  z(i) = T(3);
end
plot3(x,y,z,'linewidth',2,'color','black!30');
  %
  xlabel('x');
  ylabel('y');
  zlabel('z');
hold off
```

Print out the orbital elements.

Listing 8.24: **Print the Orbital Elements**

```
disp(sprintf('Omega: longitude of the ascending node = %8.4f',Omega
    *180/pi));
disp(sprintf('omega: argument of periapsis = %8.4f',omega*180/pi));
disp(sprintf('nu0: the true anomaly = %8.4f',nu0*180/pi));
disp(sprintf('inclination = %8.4f',inc*180/pi));
disp(sprintf('a: semimajor axis = %8.4f',a));
disp(sprintf('p = h^2/mu = %8.4f',p));
disp(sprintf('e: the eccentricity = %8.4f',e));
disp(sprintf('h: the angular momentum = %8.4f',h));
```

The full code is below. With some practice, you can read code documented this way pretty easily. Notice, we try to add comment lines that are short but informative before most of the things we try to do.

Listing 8.25: **DrawOrbitalPlane.m**

```
function [omega,Omega,nu0,a,p,e,E,inc,P,Q,W,h,H,N] = DrawOrbitalPlane(
    r0,v0)
%
% here mu is normalized to 1
% r0 is position vector at time 0
% v0 is velocity vector at time 0
% Omega is the longitude of the ascending node
%   which determines the line of nodes direction
% omega is the argument of periapsis
```

```
%nu0 is the true anomaly
% p = h^2/mu = h^2
% E is the eccentricity vector
% N is the line of nodes vector
% P, Q, W is the perifocal basis
% H is angular momentum vector
% N is the line of nodes unit vector
% inc is the inclination
%
% set standard I, J, K vectors
clf;
mu = 1;
I = [1;0;0]; J = [0;1;0]; K = [0;0;1];
% Find the angular momentum vector
R0 = norm(r0);
V0 = norm(v0);
H = CrossProduct(r0,v0);
h = norm(H);
p = h^2;
% find the eccentricity vector
E = (1/mu)*( V0^2 - mu/R0)*r0 - dot(r0,v0)*v0;
% find the eccentricity
e = norm(E);
% find the inclination
inc = acos( dot(H,K)/h );
% find the line of nodes vector
N = CrossProduct(K,H);
n = norm(N);
% find Omega
Omega = acos( dot(N,I)/n );
% find omega
omega = acos( dot(N,E)/(n*e) );
% find the true anomaly
nu0 = acos( dot(E,r0)/(e*R0) );
% to get orbital plane:
% unit normal to orbital plane
W = H/h;
% get perigee direction vector in orbital plane
P = E/e;
W = H/h;
Q = CrossProduct(W,P);
rp = (h^2)/(1+e);
ra = (h^2)/(1-e);
a = (rp+ra)/2.0;
hold on
% Draw IJK
[x,y,z] = ThreeDToPlot(ra*I);
plot3(x,y,z,'linewidth',2,'color','black');
[x,y,z] = ThreeDToPlot(ra*J);
plot3(x,y,z,'linewidth',2,'color','black');
[x,y,z] = ThreeDToPlot(ra*K);
plot3(x,y,z,'linewidth',2,'color','black');
% Draw PQW
[x,y,z] = ThreeDToPlot(ra*P);
plot3(x,y,z,'linewidth',2,'color','red');
[x,y,z] = ThreeDToPlot(-ra*P);
```

```
plot3(x,y,z,'linewidth',2,'color','red');
[x,y,z] = ThreeDToPlot(ra*Q);
plot3(x,y,z,'linewidth',2,'color','red');
[x,y,z] = ThreeDToPlot(-ra*Q);
plot3(x,y,z,'linewidth',2,'color','red');
[x,y,z] = ThreeDToPlot(ra*W);
plot3(x,y,z,'linewidth',2,'color','red');
% Draw line of nodes
[x,y,z] = ThreeDToPlot(ra*N);
plot3(x,y,z,'linewidth',2,'color','black!60');
[x,y,z] = ThreeDToPlot(-ra*N);
plot3(x,y,z,'linewidth',2,'color','black!60');
% Draw orbit
OrbitSize = 80;
rorbitplane=zeros(OrbitSize,3);
nu = linspace(0,2*pi,OrbitSize);
for i = 1:OrbitSize
  c = cos(nu(i)); s = sin(nu(i));
  rorbit = h^2/(1 + e*c);
  rorbitplane(i,:) = rorbit*c*P+rorbit*s*Q;
  T = rorbitplane(i,:);
  x(i) = T(1);
  y(i) = T(2);
  z(i) = T(3);
end
plot3(x,y,z,'linewidth',2,'color','black!30');
  %
  xlabel('x');
  ylabel('y');
  zlabel('z');
  axis on
  grid on
  box on;
hold off
disp(sprintf('Omega: longitude of the ascending node = %8.4f',Omega
   *180/pi));
disp(sprintf('omega: argument of periapsis = %8.4f',omega*180/pi));
disp(sprintf('nu0: the true anomaly = %8.4f',nu0*180/pi));
disp(sprintf('inclination = %8.4f',inc*180/pi));
disp(sprintf('a: semimajor axis = %8.4f',a));
disp(sprintf('p = h^2/mu = %8.4f',p));
disp(sprintf('e: the eccentricity = %8.4f',e));
disp(sprintf('h: the angular momentum = %8.4f',h));
end
```

Running this code generates a graph which you can grab with your mouse and rotate around to get it where you want it. Let's try it.

Listing 8.26: **Runtime**

```
r0 = [1.29;0.75;0.0];
   1.29900
   0.75000
   0.00000
```

```
v0 = [.35;0.61;0.707]
v0 =
   0.35000
   0.61000
   0.70700
[omega,Omega,nu0,a,p,e,E,inc,P,Q,W,h,H,N] = DrawOrbitalPlane(r0,v0);
Omega: longitude of the ascending node =  30.0007
omega: argument of periapsis =  94.9983
nu0: the true anomaly =  94.9983
inclination =  63.4500
a: semimajor axis =   2.9506
p = h^2/mu =   1.4054
e: the eccentricity =   0.7237
h: the angular momentum =   1.1855
```

You can see our chosen snapshot of this graph in Figure 8.14. The line of nodes is plotted and the $\boldsymbol{P}$ $\boldsymbol{Q}$ and $\boldsymbol{W}$ perifocal coordinate system and the Geocentric - Equatorial $\boldsymbol{I}$ $\boldsymbol{J}$ $\boldsymbol{K}$ coordinate system in different shades of black. It is hard to get a nice plot of this stuff, but as you can see with a bit of work, we can generate these automatically. The code actually plots these in colors but the printed output here in the text must be gray scale.

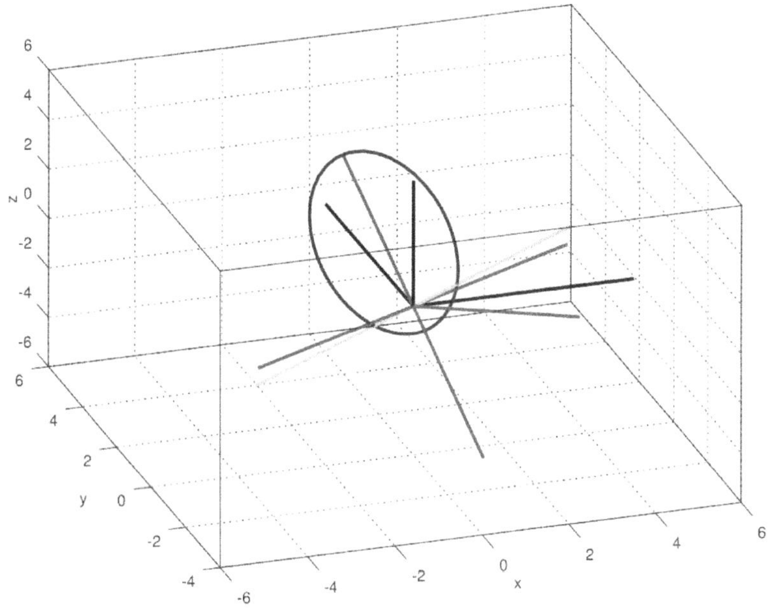

Figure 8.14: The orbit for $r_0 = [1.299; 0.75; 0.0]$, $v_0 = [0.35; 0.61; 0.707]$.

Here is another one:

Listing 8.27: **Runtime**

```
r0 = [1.29;0.75;1.2];
   1.29900
   0.75000
   1.20000
```

```
v0 = [.35;0.61;0.707]
v0 =
    0.35000
    0.61000
    0.70700
[omega,Omega,nu0,a,p,e,E,inc,P,Q,W,h,H,N] = DrawOrbitalPlane(r0,v0);
Omega: longitude of the ascending node =  22.2954
omega: argument of periapsis =  74.2970
nu0: the true anomaly = 135.9555
inclination =  45.4007
a: semimajor axis =  19.9894
p = h^2/mu =   0.5578
e: the eccentricity =   0.9859
h: the angular momentum =   0.7469
```

The resulting orbit is seen in Figure 8.15. However, you can't see where the orbit crosses the equatorial plane very well because $r_p = 0.2827$ with $r_1 = 39.698$ and so the orbit just looks like it starts at the origin.

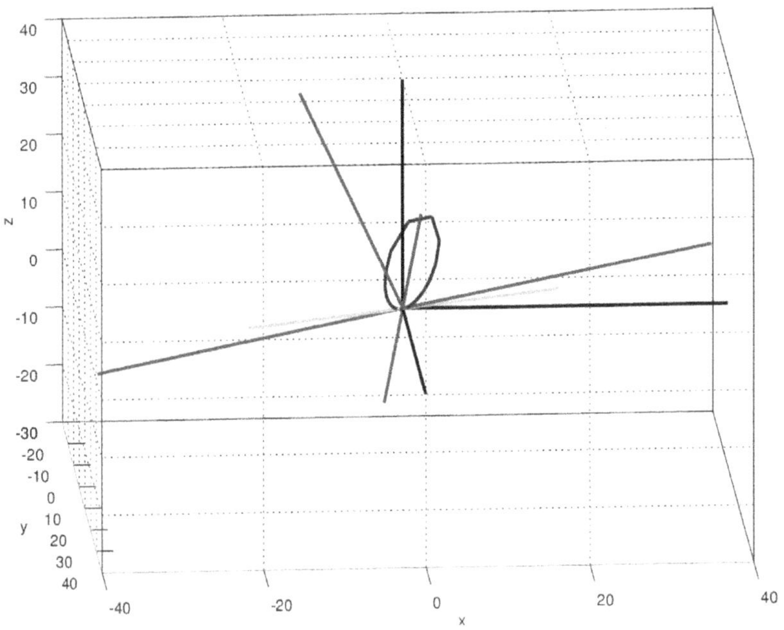

Figure 8.15: The orbit for $\boldsymbol{r}_0 = [1.299; 0.75; 1.2]$, $\boldsymbol{v}_0 = [0.35; 0.61; 0.707]$.

8.5.1 Homework

You can now generate orbital planes for yourself. Here you run the code `DrawOrbitalPlane` as we have done in the examples. The code fragment would be something like this:

Listing 8.28: **Runtime Fragment for the Homework**

```
r0 = <you enter the radius vector here>
```

```
v0 = <you enter the velocity vector here>

%you then call the function DrawOrbitPlane to both compute all
%the orbital elements and draw the 3D plot.

[omega,Omega,nu0,a,p,e,E,inc,P,Q,W,h,H,N] = DrawOrbitalPlane(r0,v0);
Omega: longitude of the ascending node = <you will get the value>
omega: argument of periapsis = <you will get the value>
nu0: the true anomaly = <you will get the value>
inclination = <you will get the value>
a: semimajor axis = <you will get the value>
p = h^2/mu = <you will get the value>
e: the eccentricity = <you will get the value>
h: the angular momentum = <you will get the value>
```

Of course, you should spin the resulting plot around until you are sure you understand the orbit. Also, our discussion was centered on elliptical orbits and we did not really go over parabolic and hyperbolic orbits. So if you make up radius and velocity vectors with components much larger than one, you will probably get hyperbolic orbits with $e > 1$ and our plot does not handle that well! If you wanted to do hyperbolic orbits, you would have to make appropriate code changes.

Exercise 8.5.1 *Use* $\boldsymbol{R} = \left[1.2; 1.5; 1.3\right]$ *and* $\boldsymbol{V} = \left[0.2; 0.45; 0.6\right]$ *and enter these as* `r_0` *and* `v_0` *in the function.*

Exercise 8.5.2 *Use* $\boldsymbol{R} = \left[1.16; 1.45; 2.11\right]$ *and* $\boldsymbol{V} = \left[0.4; -0.6; 0.3\right]$ *and enter these as* `r_0` *and* `v_0` *in the function.*

Exercise 8.5.3 *Use* $\boldsymbol{R} = \left[1.1; 1.24; 0.9\right]$ *and* $\boldsymbol{V} = \left[0.6; -0.55; 0.6\right]$ *and enter these as* `r_0` *and* `v_0` *in the function.*

Exercise 8.5.4 *Use* $\boldsymbol{R} = \left[1.33; 1; 24; 1.19\right]$ *and* $\boldsymbol{V} = \left[0.12; 0.5; 0.26\right]$ *and enter these as* `r_0` *and* `v_0` *in the function.*

Exercise 8.5.5 *Use* $\boldsymbol{R} = \left[.7; 1.2; 0.8\right]$ *and* $\boldsymbol{V} = \left[-0.2; 0.45; -0.6\right]$ *and enter these as* `r_0` *and* `v_0` *in the function.*

Exercise 8.5.6 *See if you can modify the code to handle hyperbolic orbits. Make sure you read up on these in one of the references!*

Chapter 9

Determinants and Matrix Manipulations

It is time for you to look at a familiar tool more carefully. Behold, determinants!

9.1 Determinants

We have already found the determinant of a 6×6 coefficient matrix that arose from a model of LRRK2 mutations. We know you have all done such calculations before, so there was nothing new there. But you probably have not discussed the determinant very carefully, which is what we will do now. We start with the determinant of a 3×3 matrix to set the stage. We let $\boldsymbol{A}$ be defined by

$$\boldsymbol{A} = \begin{bmatrix} V_{11} & V_{12} & V_{13} \\ V_{21} & V_{22} & V_{23} \\ V_{31} & V_{32} & V_{33} \end{bmatrix}$$

Now think about the matrix $\boldsymbol{A}$ as made up of the row vectors. Let these be denoted by $\boldsymbol{V_1}, \boldsymbol{V_2}$ and $\boldsymbol{V_3}$. We want a determinant function that acts on these three arguments and gives us back a real number. For the moment, call this function Φ. Then $\Phi(\boldsymbol{V_1}, \boldsymbol{V_2} < \boldsymbol{V_3})$ is a number and we want Φ to satisfy the properties our usual 2×2 determinant function does. Let $\boldsymbol{B}$ be a 2×2 matrix with row vectors $\boldsymbol{W_1}$ and $\boldsymbol{W_2}$ which also has the form

$$\boldsymbol{B} = \begin{bmatrix} W_{11} & W_{12} \\ W_{21} & W_{22} \end{bmatrix}$$

We know $\det(\boldsymbol{B}) = W_{11}W_{22} - W_{12}W_{21}$. So another way of saying this is

$$\det(\boldsymbol{W_1}, \boldsymbol{W_2}) = W_{11}W_{22} - W_{12}W_{21}$$

What basic properties does our familiar determinant have?

P1: $\det$ **is homogeneous: that is, *constants come out*:** What if we multiply a row by a constant c? Note

$$\det(\boldsymbol{W_1}, c\boldsymbol{W_2}) = W_{11}cW_{22} - W_{12}cW_{21} = c\det(\boldsymbol{W_1}, \boldsymbol{W_2}).$$

A similar calculation shows that $\det(c\boldsymbol{W_1}, \boldsymbol{W_2}) = c\det(\boldsymbol{W_1}, \boldsymbol{W_2})$.

P2: det **doesn't change if we add rows:** What is we add row two to row one? We have

$$\det(\boldsymbol{W_1} + \boldsymbol{W_2}, \boldsymbol{W_2}) = (W_{11} + W_{21})W_{22} - (W_{12} + W_{22})W_{21} = W_{11}cW_{22} - W_{12}cW_{21}$$
$$\det(\boldsymbol{W_1}, \boldsymbol{W_2}).$$

A similar computation shows that adding row one to row two doesn't change the determinant.

P3: The determinant of the identity is one: this is an easy computation; $\det(\boldsymbol{I}) = 1 \times 1 = 1$.

Let's assume we can find a function Φ defined on $n \times n$ matrices which we will think of as a function defined on n row vectors of size n that satisfies Properties **P1**, **P2** and **P3**. Let the row vectors now be $\boldsymbol{W_1}$ through $\boldsymbol{W_n}$. We assume Φ satisfies

P1: Φ is homogeneous: that is, *constants come out of* slots:

$$\Phi(\boldsymbol{W_1}, \boldsymbol{W_2}, \ldots, c\boldsymbol{W_i}, \ldots, \boldsymbol{W_n}) = c\Phi(\boldsymbol{W_1}, \boldsymbol{W_2}, \ldots, \boldsymbol{W_n}). \tag{9.1}$$

P2: Φ doesn't change if we add rows: As long as $j \neq i$,

$$\Phi(\boldsymbol{W_1}, \ldots, \boldsymbol{W_i} + \boldsymbol{W_j}, \ldots, \boldsymbol{W_n}) = \Phi(\boldsymbol{W_1}, \ldots, \boldsymbol{W_n}). \tag{9.2}$$

P3:

$$\Phi(\boldsymbol{I}) = 1. \tag{9.3}$$

9.1.1 Consequences One

Using these properties, we can figure out that others hold.

P4: If a row is zero, the Φ is zero: This follows from Property 9.1 by just setting $c = 0$.

If two rows are the same, the Φ is zero: This follows from Property 9.2. For convenience assume row one and row two are equal. We have

$$\begin{aligned}\Phi(\boldsymbol{W_1}, \boldsymbol{W_1}, \boldsymbol{W_3}, \ldots, \boldsymbol{W_n}) &= \Phi(\boldsymbol{W_1} - \boldsymbol{W_1}, \boldsymbol{W_1}, \boldsymbol{W_3}, \ldots, \boldsymbol{W_n}) \\ &= \Phi(\boldsymbol{0}, \boldsymbol{W_1}, \boldsymbol{W_3}, \ldots, \boldsymbol{W_n}) = 0\end{aligned}$$

P5: Replacing a row by adding a multiple of another row doesn't change the determinant: This uses our properties: the case $c = 0$ is obvious, so let's just assume $c \neq 0$: using Property 9.1, we have for $j \neq i$,

$$\Phi(\boldsymbol{W_1}, \ldots, \boldsymbol{W_i} + c\boldsymbol{W_j}, \ldots, \boldsymbol{W_n}) = (1/c)\, \Phi(\boldsymbol{W_1}, \ldots, \boldsymbol{W_i} + c\boldsymbol{W_j}, \ldots, c\boldsymbol{W_j}, \ldots, \boldsymbol{W_n})$$

But we now have a matrix whose j^{th} row is $c\boldsymbol{W_j}$. Hence, using Property 9.2, adding that row to row i does not change the value of Φ. So we have

$$\begin{aligned}\Phi(\boldsymbol{W_1}, \ldots, \boldsymbol{W_i} + c\boldsymbol{W_j}, \ldots, \boldsymbol{W_n}) &= (1/c)\, \Phi(\boldsymbol{W_1}, \ldots, \boldsymbol{W_i} + c\boldsymbol{W_j}, \ldots, c\boldsymbol{W_j}, \ldots, \boldsymbol{W_n}) \\ &= (1/c)\, \Phi(\boldsymbol{W_1}, \ldots, \boldsymbol{W_i}, \ldots, c\boldsymbol{W_j}, \ldots, \boldsymbol{W_n}) \\ &= \Phi(\boldsymbol{W_1}, \ldots, \boldsymbol{W_i}, \ldots, \boldsymbol{W_j}, \ldots, \boldsymbol{W_n})\end{aligned}$$

where we use Property 9.1 again.

P6: If the rows of B are linearly dependent, then Φ is zero: Let's do this for the case that $\boldsymbol{W_1}$ can be written in terms of the other $n-1$ vectors. You'll get the gist of the argument and then you'll be able to see in your mind how to handle the other cases. In this case, we can say $\boldsymbol{W_1} = \sum_{i=2}^{n} c_j \boldsymbol{W_j}$ for not all zero constants c_j. To save typing, we'll just write $\boldsymbol{\Lambda_n}$ for the arguments $\boldsymbol{W_2}$ through $\boldsymbol{W_n}$, $\boldsymbol{\Lambda_{n-1}}$ for the arguments $\boldsymbol{W_2}$ through $\boldsymbol{W_{n-1}}$, and so forth. So using our properties, we find

$$\begin{aligned}
\Phi(\boldsymbol{W_1}, \boldsymbol{\Lambda_n}) &= \Phi(\sum_{i=2}^{n} c_j \boldsymbol{W_j}, \boldsymbol{\Lambda_n}) \\
&= (1/c_n)\Phi(\sum_{i=2}^{n-1} c_j \boldsymbol{W_j} + c_n \boldsymbol{W_n}, \boldsymbol{\Lambda_{n-1}}, c_n \boldsymbol{W_n}) \\
&= \Phi(\sum_{i=2}^{n-1} c_j \boldsymbol{W_j}, \boldsymbol{\Lambda_{n-1}}, \boldsymbol{W_n})
\end{aligned}$$

We can do this over and over again, until we *peel* off all the terms of the summand in the first slot to obtain

$$\begin{aligned}
\Phi(\boldsymbol{W_1}, \boldsymbol{\Lambda}) &= \Phi(\boldsymbol{W_2}, \boldsymbol{W_2}, \ldots, \boldsymbol{W_n}) \\
&= \Phi(\boldsymbol{W_2} - \boldsymbol{W_2}, \boldsymbol{W_2}, \ldots, \boldsymbol{W_n}) \\
&= \Phi(\boldsymbol{0}, \boldsymbol{W_2}, \ldots, \boldsymbol{W_n}) = 0.
\end{aligned}$$

P7: Φ is linear in each argument: What if we have the sum of two vectors in row one, $\boldsymbol{\alpha} + \boldsymbol{\beta}$? We will use the same Λ_n notation as before to save typing. There are lots of cases here, so we'll just do one and let you think about the others. We will assume the vectors $\boldsymbol{W_2}$ through $\boldsymbol{W_n}$ are linearly independent. Add another vector, $\boldsymbol{\gamma}$, to them to make a basis and write down the expansions of both $\boldsymbol{\alpha}$ and $\boldsymbol{\beta}$ in them. $\boldsymbol{\alpha} = c\boldsymbol{\gamma} + \sum_{j=2}^{n} c_j \boldsymbol{W_j}$ and $\boldsymbol{\beta} = d\boldsymbol{\gamma} + \sum_{j=2}^{n} d_j \boldsymbol{W_j}$. Then using our properties,

$$\begin{aligned}
\Phi(\boldsymbol{\alpha} + \boldsymbol{\beta}, \boldsymbol{\Lambda_n}) &= \Phi\left((c\boldsymbol{\gamma} + \sum_{i=2}^{n} c_j \boldsymbol{W_j}) + (d\boldsymbol{\gamma} + \sum_{j=2}^{n} d_j \boldsymbol{W_j}), \boldsymbol{\Lambda_n}\right) \\
&- \Phi\left((c\boldsymbol{\gamma} + \sum_{i=2}^{n} c_j \boldsymbol{W_j}) - c_n \boldsymbol{W_n} + (d\boldsymbol{\gamma} + \sum_{j=2}^{n} d_j \boldsymbol{W_j}) - d_n \boldsymbol{W_n}, \boldsymbol{\Lambda_n}\right)
\end{aligned}$$

We then do this again and again; after doing this for $j = n-1$ to $j = 2$, we have

$$\Phi(\boldsymbol{\alpha} + \boldsymbol{\beta}, \boldsymbol{\Lambda_n}) = \Phi((c+d)\boldsymbol{\gamma}, \boldsymbol{\Lambda_n}) = (c+d)\, \Phi(\boldsymbol{\gamma}, \boldsymbol{\Lambda_n})$$

Now working the other way, we have

$$\begin{aligned}
\Phi(\boldsymbol{\alpha}, \boldsymbol{\Lambda_n}) + \Phi(\boldsymbol{\beta}, \boldsymbol{\Lambda_n}) &= \Phi\left(c\boldsymbol{\gamma} + \sum_{i=2}^{n} c_j \boldsymbol{W_j}, \boldsymbol{\Lambda_n}\right) + \Phi\left(d\boldsymbol{\gamma} + \sum_{j=2}^{n} d_j \boldsymbol{W_j}, \boldsymbol{\Lambda_n}\right) \\
&= \Phi(c\boldsymbol{\gamma}, \boldsymbol{\Lambda_n}) + \Phi(d\boldsymbol{\gamma}, \boldsymbol{\Lambda_n}) \\
&= \Phi((c+d)\boldsymbol{\gamma}, \boldsymbol{\Lambda_n}) = (c+d)\, \Phi(\boldsymbol{\gamma}, \boldsymbol{\Lambda_n})
\end{aligned}$$

These match and so we are done with this case. The other cases are similar. If $\boldsymbol{W_2}$ through $\boldsymbol{W_n}$ are dependent, the argument changes and so forth. We leave those details to your inquiring mind. We suggest a nice tall glass of stout will help with the pain.

P8: Interchanging two rows changes the sign of Φ: Let's do this for rows one and two to keep it simple. We have

$$\begin{aligned}
\Phi(\boldsymbol{W_1}, \boldsymbol{W_2}, \boldsymbol{\Lambda_{n-1}}) &= \Phi(\boldsymbol{W_1}, \boldsymbol{W_1} + \boldsymbol{W_2}, \boldsymbol{\Lambda_{n-1}}) \\
&= \Phi(\boldsymbol{W_1} - (\boldsymbol{W_1} + \boldsymbol{W_2}), \boldsymbol{W_1} + \boldsymbol{W_2}, \boldsymbol{\Lambda_{n-1}}) \\
&= \Phi(-\boldsymbol{W_2}, \boldsymbol{W_1} + \boldsymbol{W_2}, \boldsymbol{\Lambda_{n-1}}) \\
&= -\Phi(\boldsymbol{W_2}, \boldsymbol{W_1} + \boldsymbol{W_2}, \boldsymbol{\Lambda_{n-1}}) \;=\; -\,\Phi(\boldsymbol{W_2}, \boldsymbol{W_1}, \boldsymbol{\Lambda_{n-1}}).
\end{aligned}$$

Similar arguments, but messier notationally, work for the other rows. This implies a nice result. If we have a matrix whose rows are in the order $(j_1, j_2, \ldots, j_n)$, then if it takes an even number of row swaps to get back to the order $(1, 2, \ldots, n)$, then Φ doesn't change sign. But if it takes an odd number of interchanges, the sign flips.

9.1.2 Homework

Exercise 9.1.1 *Let* $\boldsymbol{A} = \begin{bmatrix} 1 & 2 & 3 \\ -1 & 12 & 6 \\ 4 & 5 & 7 \end{bmatrix}$. *Use Gaussian Elimination and* $\boldsymbol{P1}$ *to* $\boldsymbol{P8}$ *to reduce the determinant calculation to an upper triangular matrix.*

Exercise 9.1.2 *Let* $\boldsymbol{A} = \begin{bmatrix} 1 & 2 & 3 & 3 \\ -1 & 12 & 6 & -4 \\ 4 & 5 & 7 & 8 \\ 5 & 1 & 4 & 3 \end{bmatrix}$. *Use Gaussian Elimination and* $\boldsymbol{P1}$ *to* $\boldsymbol{P8}$ *to reduce the determinant calculation to an upper triangular matrix.*

Exercise 9.1.3 *Let*

$$\boldsymbol{A} = \begin{bmatrix} 1 & 2 & 3 & 3 & -10 \\ -1 & 12 & 6 & -4 & -3 \\ 1 & 5 & 7 & 8 & -2 \\ 5 & -1 & 4 & 3 & 6 \\ 11 & -3 & 4 & -5 & 1 \end{bmatrix}$$

Use Gaussian Elimination and $\boldsymbol{P1}$ *to* $\boldsymbol{P8}$ *to reduce the determinant calculation to an upper triangular matrix.*

Exercise 9.1.4 *Let*

$$\boldsymbol{A} = \begin{bmatrix} 1 & 2 & 3 & 3 & -10 & 7 \\ -1 & 12 & 6 & -4 & -3 & 2 \\ 1 & 5 & 7 & 8 & -2 & -5 \\ 5 & -1 & 4 & 3 & 6 & 1 \\ 11 & -3 & 4 & -5 & 1 & 10 \\ 1 & 2 & 1 & 2 & 1 & 3 \end{bmatrix}$$

Use Gaussian Elimination and $\boldsymbol{P1}$ *to* $\boldsymbol{P8}$ *to reduce the determinant calculation to an upper triangular matrix.*

9.1.3 Consequences Two

Our three fundamental assumptions about Φ have additional consequences. We still don't know such a function Φ exists, although we do know we have a nice candidate, det, that works on 2×2 matrices. Let's work out some more consequences.

P9: $\Phi(\boldsymbol{AB}) = \Phi(\boldsymbol{A})\,\Phi(\boldsymbol{B})$**:** This is a really useful thing. Let $\boldsymbol{A}$ have rows $\boldsymbol{V_1}$ through $\boldsymbol{V_n}$ and $\boldsymbol{B}$ have rows $\boldsymbol{W_1}$ through $\boldsymbol{W_n}$. Let $\boldsymbol{Y_i}$ be the columns of $\boldsymbol{B}$. The usual matrix multiplication gives

$$\boldsymbol{AB} = \begin{bmatrix} <\boldsymbol{V_1}, \boldsymbol{Y_1}> & \ldots & <\boldsymbol{V_1}, \boldsymbol{Y_n}> \\ \vdots & \vdots & \vdots \\ <\boldsymbol{V_n}, \boldsymbol{Y_1}> & \ldots & <\boldsymbol{V_n}, \boldsymbol{Y_n}> \end{bmatrix}$$

Each row of the product $\boldsymbol{AB}$ has a particular form. We'll show this for row one and you can figure out how to do it for the other rows from that. Do this on paper for a 3×3 and you'll see the pattern.

$$\begin{bmatrix} <\boldsymbol{V_1}, \boldsymbol{Y_1}> & \ldots & <\boldsymbol{V_1}, \boldsymbol{Y_n}> \end{bmatrix} = V_{11}\boldsymbol{W_1} + \ldots + V_{1n}\boldsymbol{W_n}.$$

Call this row vector $\boldsymbol{C_1}$. We can do a similar expansion for the other rows of the product matrix. Hence, we want to calculate $\Phi(\boldsymbol{C_1}, \ldots, \boldsymbol{C_n})$. We have, since Φ is additive,

$$\begin{aligned} \Phi(\boldsymbol{C_1}, \ldots, \boldsymbol{C_n}) &= \Phi(V_{11}\boldsymbol{W_1} + \ldots + V_{1n}\boldsymbol{W_n}, \boldsymbol{C_2}, \ldots, \boldsymbol{C_n}) \\ &= \sum_{j_1=1}^{n} V_{1,j_1}\, \Phi(\boldsymbol{W_{j_1}}, \boldsymbol{C_2}, \ldots, \boldsymbol{C_n}) \end{aligned}$$

Now do this again for $\boldsymbol{C_2}$. We have

$$\begin{aligned} \Phi(\boldsymbol{W_{j_1}}, \boldsymbol{C_2}, \ldots, \boldsymbol{C_n}) &= \Phi\left(\boldsymbol{W_{j_1}}, \sum_{j_2=1}^{n} V_{w,j_2}\, \boldsymbol{W_{j_2}}, \boldsymbol{C_3}, \ldots, \boldsymbol{C_n}\right) \\ &= \sum_{j_2=1}^{n} V_{2,j_2}\, \Phi(\boldsymbol{W_{j_1}}, \boldsymbol{W_{j_2}}, \boldsymbol{C_3}, \ldots, \boldsymbol{C_n}) \end{aligned}$$

So after two steps, we have

$$\Phi(\boldsymbol{C_1}, \ldots, \boldsymbol{C_n}) = \sum_{j_1=1}^{n} V_{1,j_1} \sum_{j_2=1}^{n} V_{2,j_2}\, \Phi(\boldsymbol{W_{j_1}}, \boldsymbol{W_{j_2}}, \boldsymbol{C_3}, \ldots, \boldsymbol{C_n})$$

Whew! Now finish up by handling the other rows. We obtain

$$\begin{aligned} \Phi(\boldsymbol{C_1}, \ldots, \boldsymbol{C_n}) &= \sum_{j_1=1}^{n} V_{1,j_1} \sum_{j_2=1}^{n} V_{2,j_2} \cdots \sum_{j_n=1}^{n} V_{n,j_n}\, \Phi(\boldsymbol{W_{j_1}}, \boldsymbol{W_{j_2}}, \ldots, \boldsymbol{W_{j_n}}) \\ &= \sum_{j_1=1}^{n} \cdots \sum_{j_n=1}^{n} V_{1,j_1} \cdots V_{n,j_n}\, \Phi(\boldsymbol{W_{j_1}}, \boldsymbol{W_{j_2}}, \ldots, \boldsymbol{W_{j_n}}). \end{aligned}$$

Now, many of these terms are zero as any time two rows match, Φ is zero. So we can say

$$\Phi(\boldsymbol{C_1},\ldots,\boldsymbol{C_n}) = \sum_{j_1,j_2,\ldots,j_n=1;\ j_1\neq j_2\ldots\neq j_n}^{n} V_{1,j_1}\ldots V_{n,j_n}\ \Phi(\boldsymbol{W_{j_1}},\boldsymbol{W_{j_2}},\ldots,\boldsymbol{W_{j_n}}).$$

Next, let $\epsilon_{j_1,\ldots,j_n}$ denote the -1 or $+1$ associated with the row swapping necessary to move $\{j_1,\ldots,j_n\}$ to $\{1,\ldots,n\}$. Then, we can rewrite again as

$$\Phi(\boldsymbol{C_1},\ldots,\boldsymbol{C_n}) = \sum_{j_1,j_2,\ldots,j_n=1;\ j_1\neq j_2\ldots\neq j_n}^{n} \epsilon_{j_1,\ldots,j_n}\ V_{1,j_1}\ldots V_{n,j_n}\ \Phi(\boldsymbol{W_1},\boldsymbol{W_2},\ldots,\boldsymbol{W_n}).$$

In particular, we can let $\boldsymbol{B}=\boldsymbol{I}$. Then we have $\boldsymbol{AB}=\boldsymbol{AI}=\boldsymbol{A}$ and

$$\begin{aligned}\Phi(\boldsymbol{V_1},\ldots,\boldsymbol{V_n}) &= \sum_{j_1,j_2,\ldots,j_n=1;\ j_1\neq j_2\ldots\neq j_n}^{n} \epsilon_{j_1,\ldots,j_n}\ V_{1,j_1}\ldots V_{n,j_n}\ \Phi(\boldsymbol{I_1},\boldsymbol{I_2},\ldots,\boldsymbol{I_n})\\ &= \sum_{j_1,j_2,\ldots,j_n=1;\ j_1\neq j_2\ldots\neq j_n}^{n} \epsilon_{j_1,\ldots,j_n}\ V_{1,j_1}\ldots V_{n,j_n}\end{aligned}$$

since $\Phi(\boldsymbol{I})=1$. We usually just refer to $\Phi(\boldsymbol{V_1},\ldots,\boldsymbol{V_n})$ as $\Phi(\boldsymbol{A})$, so we have the identity

$$\Phi(\boldsymbol{A}) = \sum_{j_1,j_2,\ldots,j_n=1;\ j_1\neq j_2\ldots\neq j_n}^{n} \epsilon_{j_1,\ldots,j_n}\ V_{1,j_1}\ldots V_{n,j_n}$$

Combining, we have shown $\Phi(\boldsymbol{AB})=\Phi(\boldsymbol{A})\,\Phi(\boldsymbol{B})$ since $\Phi(\boldsymbol{W_1},\boldsymbol{W_2},\ldots,\boldsymbol{W_n})=\Phi(\boldsymbol{B})$.

We have gotten pretty far in our discussions. We now know that if a function Φ satisfies Properties **P1**, **P2** and **P3**, it must have the value

$$\Phi(\boldsymbol{A}) = \sum_{j_1,j_2,\ldots,j_n=1;\ j_1\neq j_2\ldots\neq j_n}^{n} \epsilon_{j_1,\ldots,j_n}\ V_{1,j_1}\ldots V_{n,j_n}$$

This leads to our next consequence.

P10, $\Phi(\boldsymbol{A})=\Phi(\boldsymbol{A^T})$: Define a new function $\Psi(\boldsymbol{A})=\Phi(\boldsymbol{A^T})$. If we multiply a row in $\boldsymbol{A^T}$ by the constant c, this is the same as multiplying $\boldsymbol{A^T}$ by the matrix

$$\boldsymbol{R}(c) = \boldsymbol{D}(1,0,\ldots,c,1,\ldots,1)$$

where $\boldsymbol{D}$ is the diagonal matrix with ones on the main diagonal except in the (i,i) position which has the value c. We can use Property **P9** now.

$$\Psi(\boldsymbol{R}(c)\boldsymbol{A}) = \Phi(\boldsymbol{R}(c)\boldsymbol{A^T}) = \Phi(\boldsymbol{R}(c))\,\Phi(\boldsymbol{A^T}) = c\,\Phi(\boldsymbol{A^T}) = c\Psi(\boldsymbol{A})$$

and so Ψ satisfies **P1**. Next, adding row j to row i corresponds to multiplying $\boldsymbol{A^T}$ by

$$\boldsymbol{R} = \begin{bmatrix} 1 & 0 & \ldots & 0 & 0 & \ldots & 0 & 0 & \ldots & 0\\ 0 & 0 & \ldots & 1 \text{ position (i,i)} & 0 & \ldots & 0 & 1 \text{ position (i,j)} & \ldots & 0\\ 0 & 0 & \ldots & 0 & 1 & 0 & 0 & 0 & \ldots & 0\\ \vdots & \vdots & \vdots & \vdots & \vdots & \vdots & \vdots & \vdots & \vdots & \vdots\\ 0 & 0 & \ldots & 0 & 0 & \ldots & 0 & 0 & \ldots & 1\end{bmatrix}$$

Then

$$\Psi(\boldsymbol{RA}) \;=\; \Phi(\boldsymbol{RA}^T) = \Phi(\boldsymbol{R})\,\Phi(\boldsymbol{A}^T) \;=\; 1\,\Phi(\boldsymbol{A}^T) = \Psi(\boldsymbol{A})$$

and so Property **P2** holds. Finally, $\Psi(\boldsymbol{I}) = \Phi(\boldsymbol{I}^T) = \Phi(\boldsymbol{I}) = 1$ and so Property **P3** holds as well. Since all three properties hold, we must also have

$$\Psi(\boldsymbol{A}) \;=\; \sum_{j_1,j_2,\ldots,j_n=1;\; j_1\neq j_2\ldots\neq j_n}^{n} \epsilon_{j_1,\ldots,j_n}\, V_{1,j_1}\ldots V_{n,j_n} \;=\; \Phi(\boldsymbol{A}).$$

We conclude $\Phi(\boldsymbol{A}) = \Phi(\boldsymbol{A}^T)$. This also implies another really important result: **everything we have said about row manipulations applies equally to column manipulations**.

We could use the formula above as our definition of how to find Φ. However, there is an easier way. The proper way to do this would be to define how to calculate a **determinant** using what are called cofactors in a recursive fashion. Let's go through how this works for a 3×3. We have

$$\begin{aligned}\Phi(\boldsymbol{A}) &= \sum_{i,j,k=1;\; i\neq j\neq k}^{3} \epsilon_{i,j,k}\, V_{1i}V_{2j}V_{3k}\\ &= \epsilon_{123}V_{11}V_{22}V_{33} + \epsilon_{132}V_{11}V_{23}V_{32} + \epsilon_{213}V_{12}V_{21}V_{33}\\ &\quad +\epsilon_{231}V_{12}V_{23}V_{31} + \epsilon_{312}V_{13}V_{21}V_{32} + \epsilon_{321}V_{13}V_{22}V_{31}.\end{aligned}$$

Now, the ϵ terms refer to the sign (either 1 or -1) associated with the rows being in that order. Remember an odd number of row interchanges changes Φ by -1 and an even number of row interchanges leaves the sign unchanged.

$$\begin{aligned}\epsilon_{123} &= 1, \text{ because rows are already in proper order.}\\ \epsilon_{132} &= -1, \text{ because switching 3 with 2 gets 123 order.}\\ \epsilon_{213} &= -1, \text{ because switching 2 with 1 gets 123 order.}\\ \epsilon_{231} &= 1, \text{ because switching 3 with 1 gets 213 order; then 1 with 2 gives 123.}\\ \epsilon_{312} &= 1, \text{ because switching 3 with 1 gets 132 order; then 3 with 2 gives 123.}\\ \epsilon_{321} &= -1, \text{ because switching 3 with 1 gets 123 order.}\end{aligned}$$

Using these, we have

$$\begin{aligned}\Phi(\boldsymbol{A}) &= V_{11}V_{22}V_{33} - V_{11}V_{23}V_{32} - V_{12}V_{21}V_{33}\\ &\quad +V_{12}V_{23}V_{31} + V_{13}V_{21}V_{32} - V_{13}V_{22}V_{31}.\end{aligned}$$

Now choose any row or column you want: we will use column 3. We will organize these calculations around the entries of column 3. We write

$$\begin{aligned}\Phi(\boldsymbol{A}) &= V_{13}\,(V_{21}V_{32} - V_{22}V_{31}) + V_{23}\,(V_{12}V_{31} - V_{11}V_{32}) + V_{33}\,(V_{11}V_{22} - V_{12}V_{21})\\ &= V_{13}\,\det\left(\begin{bmatrix} V_{21} & V_{22}\\ V_{31} & V_{32}\end{bmatrix}\right) - V_{23}\,\det\left(\begin{bmatrix} V_{11} & V_{12}\\ V_{31} & V_{32}\end{bmatrix}\right) + V_{33}\,\det\left(\begin{bmatrix} V_{11} & V_{12}\\ V_{21} & V_{22}\end{bmatrix}\right)\\ &= (-1)^{1+3}\,V_{13}\,\det\left(\begin{bmatrix} V_{21} & V_{22}\\ V_{31} & V_{32}\end{bmatrix}\right) + (-1)^{2+3}\,V_{23}\,\det\left(\begin{bmatrix} V_{11} & V_{12}\\ V_{31} & V_{32}\end{bmatrix}\right)\\ &\quad +(-1)^{3+3}\,V_{33}\,\det\left(\begin{bmatrix} V_{11} & V_{12}\\ V_{21} & V_{22}\end{bmatrix}\right).\end{aligned}$$

Now look at the matrix $\boldsymbol{A}$ with column 3 singled out.

$$\boldsymbol{A} = \begin{bmatrix} V_{11} & V_{12} & \boldsymbol{V_{13}} \\ V_{21} & V_{22} & \boldsymbol{V_{23}} \\ V_{31} & V_{32} & \boldsymbol{V_{33}} \end{bmatrix}$$

Look at the entry $\boldsymbol{V_{13}}$ and mentally cross out row 1 and column 3 from the matrix. This leaves a 2×2 submatrix which is identical to the one after the V_{13} entry above. Call this the **cofactor** $\boldsymbol{C}_{13}$. Next, look at $\boldsymbol{V_{23}}$ and mentally cross out row 2 and column 3 from the matrix. This leaves a 2×2 submatrix which is identical to the one after the V_{23} entry above. Call this the **cofactor** $\boldsymbol{C}_{23}$. Finally, look at $\boldsymbol{V_{33}}$ and mentally cross out row 3 and column 3 from the matrix. This leaves a 2×2 submatrix which is identical to the one after the V_{33} entry above. Call this the **cofactor** $\boldsymbol{C}_{33}$. Hence, we can rewrite our computation as

$$\begin{aligned} \Phi(\boldsymbol{A}) &= (-1)^{1+3}\, V_{13} \det\left(\begin{bmatrix} V_{21} & V_{22} \\ V_{31} & V_{32} \end{bmatrix}\right) + (-1)^{2+3}\, V_{23} \det\left(\begin{bmatrix} V_{11} & V_{12} \\ V_{31} & V_{32} \end{bmatrix}\right) \\ &\quad + (-1)^{3+3}\, V_{33} \det\left(\begin{bmatrix} V_{11} & V_{12} \\ V_{21} & V_{22} \end{bmatrix}\right) \\ &= (-1)^{1+3}\, V_{13} \det(\boldsymbol{C}_{13}) + (-1)^{2+3}\, V_{23} \det(\boldsymbol{C}_{23}) + (-1)^{3+3}\, V_{33} \det(\boldsymbol{C}_{33}) \end{aligned}$$

What we have done here is to expand the computation of Φ using column 3 of the matrix. It is straightforward to choose a different row or column and do the same thing. The cofactors are defined similarly and we would find for a row expansion along row r

$$\Phi(\boldsymbol{A}) = (-1)^{r+1}\, V_{r1} \det(\boldsymbol{C}_{r1}) + (-1)^{r+2}\, V_{r2} \det(\boldsymbol{C}_{r2}) + (-1)^{r+3}\, V_{r3} \det(\boldsymbol{C}_{r3})$$

and for a column expansion along column c

$$\Phi(\boldsymbol{A}) = (-1)^{1+c}\, V_{1c} \det(\boldsymbol{C}_{1c}) + (-1)^{2+c}\, V_{2c} \det(\boldsymbol{C}_{2c}) + (-1)^{3+c}\, V_{3c} \det(\boldsymbol{C}_{3c})$$

This suggests Φ should be defined **recursively** starting with 3×3 matrices using the usual 2×2 determinant det. If $\boldsymbol{A}$ was 4×4, we would define the determinant of a 3×3 matrix as above and define the determinant $\det(\boldsymbol{A})$ as follows: for a row expansion along row r

$$\begin{aligned} \det(\boldsymbol{A}) &= (-1)^{r+1}\, V_{r1} \det(\boldsymbol{C}_{r1}) + (-1)^{r+2}\, V_{r2} \det(\boldsymbol{C}_{r2}) \\ &\quad + (-1)^{r+3}\, V_{r3} \det(\boldsymbol{C}_{r3}) + (-1)^{r+4}\, V_{r4} \det(\boldsymbol{C}_{r4}) \end{aligned}$$

where the dets are the 3×3 we defined in the previous step and the cofactors are now 3×3 submatrices. For a column expansion along column c, we have

$$\begin{aligned} \Phi(\boldsymbol{A}) &= (-1)^{1+c}\, V_{1c} \det(\boldsymbol{C}_{1c}) + (-1)^{2+c}\, V_{2c} \det(\boldsymbol{C}_{2c}) \\ &\quad + (-1)^{3+c}\, V_{3c} \det(\boldsymbol{C}_{3c}) + (-1)^{4+c}\, V_{4c} \det(\boldsymbol{C}_{4c}). \end{aligned}$$

It is clear we can continue to do this as the size of $\boldsymbol{A}$ increases. Hence, instead of using the computational formula $\Phi(\boldsymbol{A}) = \sum_{j_1, j_2, \ldots, j_n = 1;\; j_1 \neq j_2 \ldots \neq j_n}^{n} \epsilon_{j_1, \ldots, j_n} V_{1,j_1} \ldots V_{n,j_n}$ we define Φ recursively as the scheme above indicates. If you call the function defined by this recursive scheme Θ, we can prove it also satisfies Properties **P1**, **P2** and **P3** using what is called an induction argument. This argument works by assuming the stuff we have defined at step n (i.e. for $n \times n$ matrices) satisfies Properties **P1**, **P2** and **P3** and then showing the new version for matrices that are size $(n+1) \times (n+1)$ also satisfies the properties. Then Θ satisfies the usual Φ properties and so Θ and Φ must be the same function. As you can see, determinants of higher order matrices involve a lot of arithmetic! From this point on, we will simply call Φ by the name det!! Let's state what we know for posterity.

Theorem 9.1.1 The Determinant

There is a unique real-valued function det *defined on* $n \times n$ *matrices which satisfies the following properties:*

P1: *The value of the* det *is unchanged if a row or a column is multiplied by a constant c.*

P2: *The value of the* det *is unchanged if row a is replaced by the new row: row a plus row b as long as* $a \neq b$*. This is also true for columns.*

P3: *The determinant of the identity matrix is* 1.

Proof 9.1.1
We had a very long discussion about this. We showed that there is only one function satisfying these three properties. ■

These three properties imply many others which we have discussed. Here they are in a more compact form.

Theorem 9.1.2 The Determinant Properties

The real-valued function det *defined on* $n \times n$ *matrices which satisfies the following additional properties:*

P4: *If a row or column is zero, then* det *is zero.*

P5: *If row a is replaced by row* $a + c$ *row b,* det *is unchanged. The same is true this is done to columns.*

P6: *If the rows or columns are linearly dependent, then* det *is zero.*

P7: det *is linear in each row or column slot.*

P8: *Interchanging two rows or columns changes the sign of* det.

P9: $\det(\boldsymbol{AB}) = \det(\boldsymbol{A})\det(\boldsymbol{B})$.

P10: $\det(\boldsymbol{A}) = \det(\boldsymbol{A^T})$

Proof 9.1.2
We used the determinant's three properties to prove each of these additional ones in the arguments presented earlier. ■

Property **P9** is very important. From it, we derive the fundamental formula for the value of the det. Recall this is $\Phi(\boldsymbol{A}) = \sum_{j_1,j_2,\ldots,j_n=1;\; j_1 \neq j_2 \ldots \neq j_n}^{n} \epsilon_{j_1,\ldots,j_n} V_{1,j_1} \ldots V_{n,j_n}$ From this, we know immediately some very important things which we state as another theorem.

Theorem 9.1.3 The Determinant Smoothness

The real-valued function det *defined on* $n \times n$ *matrices is a continuous function of its* n^2 *parameters. Moreover, its partials of all orders exist and are continuous.*

Proof 9.1.3
The value of the determinant is just a large polynomial in n^2 *variables. Hence, the smoothness*

follows. ■

Finally, we have a recursive algorithm for calculating the determinant of a $n \times n$ matrix.

Theorem 9.1.4 The Determinant Algorithm

We can calculate the det *using a row expansion:*

$$\det(\boldsymbol{A}) = \sum_{j=1}^{n} (-1)^{i+j} A_{rj} \det(\boldsymbol{C}_{rj}).$$

or via a column expansion:

$$\Phi(\boldsymbol{A}) = \sum_{j=1}^{n} (-1)^{i+j} A_{jc} \det(\boldsymbol{C}_{jc}).$$

Proof 9.1.4
We showed this explicitly for a 4×4 *matrix earlier. This is just a generalization of that formula.* ■

Comment 9.1.1 *This calculation is expensive. A* 2×2 *takes* 2 *multiples and* 2 *adds. A* 3×3 *has* 3 2×2 *determinants in its computation. Hence, we have* 6 *multiplies and* 6 *adds for the determinants and then* 3 *more multiplies for the row or column entries and then* 3 *more for the* -1 *calculations. So, the total is* 12 *multiplies and* 6 *adds. What about a* 4×4 *matrix? We have* 4 *submatrix* 3×3 *determinants which gives* 48 *multiplies and* 24 *adds for them. Then the row or column entries give an addition* 8 *multiplies. The total is then* 56 *multiplies and* 24 *adds. So far, letting M denote multiplies and A denote adds:*

2×2 *: 2 M and 2 A which is larger than* $2!$

3×3 *: 12 M and 6 A which is larger than* $3!$.

4×4 *: 56 M and 24 A which is larger than* $4!$.

Let's look at the 5×5 *case. We have* 5 4×4 *submatrix determinants which gives* 280 *M and* 120 *A. We have an additional* 2×5 *multiplies for the row or column entries. The total is thus* 290 *M and* 120 *A which is larger than* $5!$. *Thus, as the size of the matrix increases, the number of arithmetic operations to calculate an* $n \times n$ *determinant is larger than* $n!$.

9.1.4 Homework

Exercise 9.1.5 *Let* $\boldsymbol{A} = \begin{bmatrix} 1 & 2 & 3 \\ -1 & 12 & 6 \\ 4 & 5 & 7 \end{bmatrix}$. *Use Gaussian Elimination and* $\boldsymbol{P1}$ *to* $\boldsymbol{P8}$ *to reduce the determinant calculation to an upper triangular matrix. Now calculate the determinant using Theorem 9.1.4.*

Exercise 9.1.6 *Let* $\boldsymbol{A} = \begin{bmatrix} 1 & 2 & 3 & 3 \\ -1 & 12 & 6 & -4 \\ 4 & 5 & 7 & 8 \\ 5 & 1 & 4 & 3 \end{bmatrix}$. *Use Gaussian Elimination and* $\boldsymbol{P1}$ *to* $\boldsymbol{P8}$ *to reduce the determinant calculation to an upper triangular matrix. Now calculate the determinant using Theorem 9.1.4.*

Exercise 9.1.7 *Let*

$$A = \begin{bmatrix} 1 & 2 & 3 & 3 & -10 \\ -1 & 12 & 6 & -4 & -3 \\ 1 & 5 & 7 & 8 & -2 \\ 5 & -1 & 4 & 3 & 6 \\ 11 & -3 & 4 & -5 & 1 \end{bmatrix}$$

Use Gaussian Elimination and $\boldsymbol{P1}$ to $\boldsymbol{P8}$ to reduce the determinant calculation to an upper triangular matrix. Now calculate the determinant using Theorem 9.1.4.

Exercise 9.1.8 *Let*

$$A = \begin{bmatrix} 1 & 2 & 3 & 3 & -10 & 7 \\ -1 & 12 & 6 & -4 & -3 & 2 \\ 1 & 5 & 7 & 8 & -2 & -5 \\ 5 & -1 & 4 & 3 & 6 & 1 \\ 11 & -3 & 4 & -5 & 1 & 10 \\ 1 & 2 & 1 & 2 & 1 & 3 \end{bmatrix}$$

Use Gaussian Elimination and $\boldsymbol{P1}$ to $\boldsymbol{P8}$ to reduce the determinant calculation to an upper triangular matrix. Now calculate the determinant using Theorem 9.1.4.

9.2 Manipulating Matrices

To understand our problem with finding conditions to check to see if a matrix is positive and negative definite, we need to dig deeper into how a matrix and another matrix which we have been altering by row combinations are related. To begin, we focus on what are called **elementary row operations**. You already know how to use these as you have been using Gaussian elimination in the homework sets. We are now going to do them carefully in code.

9.2.1 Elementary Row Operations and Determinants

Let's try to solve the system

$$\begin{bmatrix} 1 & -2 & 3 \\ -2 & 5 & 7 \\ 4 & 3 & -1 \end{bmatrix} \begin{bmatrix} x \\ y \\ z \end{bmatrix} = \begin{bmatrix} 2 \\ 5 \\ 8 \end{bmatrix}$$

We can manipulate this by adding multiples of one equation to another and by doing this sort of thing systematically, we can solve the problem. This is usually done much less formally when the code to perform an LU decomposition of the matrix $\boldsymbol{A}$ is discussed. But here we want to make some new points, so bear with us! These sorts of manipulations don't really need us to write down the x, y and z variables, so it is customary to make up a new matrix, by taking the coefficient matrix $\boldsymbol{A}$ above, and adding an extra column to it which consists of the data $\boldsymbol{b}$. This new matrix is called the **augmented matrix**. Let's call it $\boldsymbol{B}$. Then, we have

$$B = \left[\begin{bmatrix} 1 & -2 & 3 \\ -2 & 5 & 7 \\ 4 & 3 & -1 \end{bmatrix} \begin{bmatrix} 2 \\ 5 \\ 8 \end{bmatrix} \right]$$

Now if we multiple the original first equation by 2 and add it to the second equation, we get the new second equation

$$\begin{aligned} (2-2)x + (-4+5)y + (6+7)z &= 4+5 \\ y + 13z &= 9 \end{aligned}$$

This is the same as multiplying the coefficient matrix $\boldsymbol{A}$ by the matrix $\boldsymbol{R}_{21}(2)$

$$\boldsymbol{R}_{21}(2) = \begin{bmatrix} 0 & 1 & 0 \\ 2 & 1 & 0 \\ 0 & 0 & 1 \end{bmatrix}$$

which is read as replace row 2 of $\boldsymbol{A}$ by 2 times row 1 + row 2. We see

$$\boldsymbol{R}_{21}(2)\,\boldsymbol{A} = \begin{bmatrix} 1 & 0 & 0 \\ 2 & 1 & 0 \\ 0 & 0 & 1 \end{bmatrix} \begin{bmatrix} 1 & -2 & 3 \\ -2 & 5 & 7 \\ 4 & 3 & -1 \end{bmatrix} = \begin{bmatrix} 1 & -2 & 3 \\ 0 & 1 & 13 \\ 4 & 3 & -1 \end{bmatrix}$$

We do this to the data also giving

$$\boldsymbol{R}_{21}(2)\,\boldsymbol{b} = \begin{bmatrix} 1 & 0 & 0 \\ 2 & 1 & 0 \\ 0 & 0 & 1 \end{bmatrix} \begin{bmatrix} 2 \\ 5 \\ 8 \end{bmatrix} = \begin{bmatrix} 2 \\ 9 \\ 8 \end{bmatrix}$$

Hence, to save time, we can abuse notation and think of applying $\boldsymbol{R}_{21}(2)$ to the augmented matrix $\boldsymbol{B}$ (just apply $\boldsymbol{R}_{21}(2)$ to the fourth column separately!) giving

$$\boldsymbol{R}_{21}(2)\,\boldsymbol{B} = \begin{bmatrix} 1 & 0 & 0 \\ 2 & 1 & 0 \\ 0 & 0 & 1 \end{bmatrix} \left[\begin{bmatrix} 1 & -2 & 3 \\ -2 & 5 & 7 \\ 4 & 3 & -1 \end{bmatrix} \begin{bmatrix} 2 \\ 5 \\ 8 \end{bmatrix} \right] = \left[\begin{bmatrix} 1 & -2 & 3 \\ 0 & 1 & 13 \\ 4 & 3 & -1 \end{bmatrix} \begin{bmatrix} 2 \\ 9 \\ 8 \end{bmatrix} \right]$$

Now apply a new transformation, $\boldsymbol{R}_{31}(-4)$ to the new augmented matrix.

$$\boldsymbol{R}_{31}(-4)\,\boldsymbol{B} = \begin{bmatrix} 1 & 0 & 0 \\ 0 & 1 & 0 \\ -4 & 0 & 1 \end{bmatrix} \left[\begin{bmatrix} 1 & -2 & 3 \\ 0 & 1 & 13 \\ 4 & 3 & -1 \end{bmatrix} \begin{bmatrix} 2 \\ 9 \\ 8 \end{bmatrix} \right] = \left[\begin{bmatrix} 1 & -2 & 3 \\ 0 & 1 & 13 \\ 0 & 11 & -13 \end{bmatrix} \begin{bmatrix} 2 \\ 9 \\ 0 \end{bmatrix} \right]$$

Note we have *zeroed* out all of the entries below the A_{11} position. Now let's work on column two. Apply the transformation $\boldsymbol{R}_{32}(-11)$. We have

$$\boldsymbol{R}_{32}(-11)\,\boldsymbol{B} = \begin{bmatrix} 1 & 0 & 0 \\ 0 & 1 & 0 \\ 0 & -11 & 1 \end{bmatrix} \left[\begin{bmatrix} 1 & -2 & 3 \\ 0 & 1 & 13 \\ 0 & 11 & -13 \end{bmatrix} \begin{bmatrix} 2 \\ 9 \\ 0 \end{bmatrix} \right] = \left[\begin{bmatrix} 1 & -2 & 3 \\ 0 & 1 & 13 \\ 0 & 0 & -156 \end{bmatrix} \begin{bmatrix} 2 \\ 9 \\ -99 \end{bmatrix} \right]$$

Next, apply $\boldsymbol{R}_{12}(2)$. We find

$$\boldsymbol{R}_{12}(2)\,\boldsymbol{B} = \begin{bmatrix} 1 & 2 & 0 \\ 0 & 1 & 0 \\ 0 & 0 & 1 \end{bmatrix} \left[\begin{bmatrix} 1 & -2 & 3 \\ 0 & 1 & 13 \\ 0 & 0 & -156 \end{bmatrix} \begin{bmatrix} 2 \\ 9 \\ -99 \end{bmatrix} \right] = \left[\begin{bmatrix} 1 & 0 & 29 \\ 0 & 1 & 13 \\ 0 & 0 & -156 \end{bmatrix} \begin{bmatrix} 20 \\ 9 \\ -99 \end{bmatrix} \right]$$

Now we have *zeroed* out all the entries above and below the original A_{22}. We have found that if we applied a sequence of elementary row operations we could transform $\boldsymbol{A}$ into an upper triangular matrix. To summarize, we found constants c_{32}, c_{31} and c_{21} so that

$$R_{32}(c_{32})R_{31}(c_{31})R_{21}(c_{21})\,A = \begin{bmatrix} 1 & 0 & 29 \\ 0 & 1 & 13 \\ 0 & 0 & -156 \end{bmatrix}$$

Now each of the matrices $R_{32}(c_{32})$, $R_{31}(c_{31})$ and $R_{31}(c_{31})$ is formed by adding a multiple of some row of the identity to another row of the identity. Hence, all of these elementary row operation premultiplier matrices have determinant 1. Thus, we know

$$\begin{aligned} \det(A) &= det(R_{32}(c_{32}))\ \det(R_{31}(c_{31}))\ \det(R_{31}(c_{31}))\ \det(U) \\ &= \det(U) \end{aligned}$$

where U is the upper triangular matrix we have found by applying these transformations. The determinant of U is simple: applying our algorithm for the recursive computation of det here, we find $\det(U)$ is just the product of its diagonals (just expand using column 1). Hence, just 3 multiplies here. Now the elementary row operations don't need to be applied as real matrix multiplications. Instead, we just perform the alterations to rows as needed. So $R_{21}(c_{21})$ needs 3 M and 3 adds but since we know the first column will be zeroed out, we don't really have to do that one. So just 2 M and 2 A. The next one, $R_{31}(c_{31})$ only needs 1 M and 1 A as we don't really have to do the column two entry. Then we are done. We have found $\det(A)$ in just $2 + 1$ M and $2 + 1$ A to get U and then 3 M to get det. The total is 6 M and 3 A. The recursive algorithm requires 12 M and 6 A so already we have saved 50% on the operation count! It gets better and better as the size of A goes up. Hence, calculating det using the recursive algorithm is very expensive and we prefer to use elementary row operations instead.

We have also converted our original system into the new system

$$U \begin{bmatrix} x \\ y \\ z \end{bmatrix} = R_{32}(c_{32})R_{31}(c_{31})R_{21}(c_{21})b$$

which we can readily solve using backsubstitution.

9.2.1.1 Homework

These are problems we have done already but now we add a data vector and we want you to do all the steps like we have explained above using the appropriate $R_{ij}(c)$ matrices. These are *do by hand exercises*!

Exercise 9.2.1 *Let* $A = \begin{bmatrix} 1 & 2 & 3 \\ -1 & 12 & 6 \\ 4 & 5 & 7 \end{bmatrix}$ *and* $b = \begin{bmatrix} 1 \\ -2 \\ 8 \end{bmatrix}$. *Convert the augmented matrix to upper triangular form using explicit* $R_{ij}(c)$ *transformations.*

Exercise 9.2.2 *Let* $A = \begin{bmatrix} 1 & 2 & 3 & 3 \\ -1 & 12 & 6 & -4 \\ 4 & 5 & 7 & 8 \\ 5 & 1 & 4 & 3 \end{bmatrix}$ *and* $b = \begin{bmatrix} 1 \\ -2 \\ 8 \\ -4 \end{bmatrix}$. *Convert the augmented matrix to upper triangular form using explicit* $R_{ij}(c)$ *transformations.*

Exercise 9.2.3 *Let*

$$A = \begin{bmatrix} 1 & 2 & 3 & 3 & -10 \\ -1 & 12 & 6 & -4 & -3 \\ 1 & 5 & 7 & 8 & -2 \\ 5 & -1 & 4 & 3 & 6 \\ 11 & -3 & 4 & -5 & 1 \end{bmatrix}, \quad b = \begin{bmatrix} 14 \\ 2 \\ -10 \\ 35 \end{bmatrix}$$

Convert the augmented matrix to upper triangular form using explicit $R_{ij}(c)$ transformations.

Exercise 9.2.4 *Let*

$$A = \begin{bmatrix} 1 & 2 & 3 & 3 & -10 & 7 \\ -1 & 12 & 6 & -4 & -3 & 2 \\ 1 & 5 & 7 & 8 & -2 & -5 \\ 5 & -1 & 4 & 3 & 6 & 1 \\ 11 & -3 & 4 & -5 & 1 & 10 \\ 1 & 2 & 1 & 2 & 1 & 3 \end{bmatrix}, \quad b = \begin{bmatrix} 14 \\ 2 \\ -10 \\ 35 \\ 1 \end{bmatrix}$$

Convert the augmented matrix to upper triangular form using explicit $R_{ij}(c)$ transformations.

9.2.2 Code Implementations

Let's try making elementary operations on a matrix A to make it upper triangular. We are going to print out a lot of the intermediary calculations, so it will be a bit verbose! First, let's do column one.

Listing 9.1: **Doing Column 1**

```
>> A = [1,2,3;2,4,5;3,5,6];
% The original matrix
A =
    1   2   3
    2   4   5
    3   5   6
% we need to swap row 3 and row 1
W= [0,0,1;0,1,0;1,0,0]
W =
       0   0   1
       0   1   0
       1   0   0
% the swapped matrix is
A2 = W*A

    3   5   6
    2   4   5
    1   2   3
% now zero out the second entry in column one
% the row operation matrix is
S = [1,0,0;-0.66666667,1,0;0,0,1]
S =
    1.00000     0.00000     0.00000
   -0.66666667     1.00000     0.00000
    0.00000     0.00000     1.00000
% the new matrix is then
```

```
A3 = S*W*A
A3 =
     3.00000    5.00000    6.00000
     0.00000    0.66667    1.00000
     1.00000    2.00000    3.00000
% The combined row operation is then S*W
% which swaps column 1 and column 3
R = S*W
R =
     0.00000    0.00000    1.00000
     0.00000    1.00000   -0.66667
     1.00000    0.00000    0.00000
% now zero out the third entry in column one
% the new row operation matrix is
S2 = [1,0,0;-0,1,0;-0.3333333,0,1]
S2 =
     1.00000    0.00000    0.00000
     0.00000    1.00000    0.00000
    -0.33333    0.00000    1.00000
% the new matrix is
S2*R*A = S2*A3
 =
     3.00000    5.00000    6.00000
     0.00000    0.66667    1.00000
     0.00000    0.33333    1.00000
% the combined row operation matrix is
% the new one * the previous one giving
R2 = S2*S*W
=
     0.00000    0.00000    1.00000
     0.00000    1.00000   -0.66667
     1.00000    0.00000   -0.33333
```

Next, column two.

Listing 9.2: **Doing Column 2**

```
% We don't need to do another row swap, so
% the second swap matrix is just the identity
W2 = [1,0,0;0,1,0;0,0,1]
W2 =
       1    0    0
       0    1    0
       0    0    1
% The new A is unchanged as there is no swap
A4 = W2*A3
=
     3.00000    5.00000    6.00000
     0.00000    0.66667    1.00000
     0.00000    0.33333    1.00000
% now zero out the third entry in column 2
% the row operation matrix is
S3 = [1,0,0;0,1,0;0,-0.5,1]
S3 =
```

```
    1.00000    0.00000    0.00000
    0.00000    1.00000    0.00000
    0.00000   -0.50000    1.00000
the altered A is
S3*A4 =

    3.00000    5.00000    6.00000
    0.00000    0.66667    1.00000
    0.00000    0.00000    0.50000
% the combined row operation matrix
% is the new one times the previous one
R3 = S3*R2
=
    0.00000    0.00000    1.00000
    0.00000    1.00000   -0.66667
    1.00000   -0.50000    0.00000
%
P = R3
```

Note just as a check, `P*A` gives the result:

Listing 9.3: **Checking PA**

```
>> P*A
ans =

    3.00000    5.00000    6.00000
    0.00000    0.66667    1.00000
    0.00000    0.00000    0.50000
```

Our row operation matrix is given by

Listing 9.4: **The Multiplier P**

```
>> P
P =
    0.00000    0.00000    1.00000
    0.00000    1.00000   -0.66667
    1.00000   -0.50000    0.00000
```

Let's see if we can figure out how to find the multiplier matrices P using MATLAB. The calculations above were messy and it was hard to organize our labeling of the intermediate multiplier matrices. We can do better by making this code abstract. In our code, we return the multiplier P. First, let's introduce some needed utility functions. The first one, `LMult(i,j,c,n)` replaces row j in the matrix A by the new row j which is $c\,row(i) + row(j)$.

Listing 9.5: **The Fundamental Row Operation**

```
function R = LMult(i,j,c,n);
%
```

```
% R is the left multiplier which replaces
%    row j by c * row i + row j
%
R = eye(n,n);
R(j,i) = c;
end
```

We also need the code to do a row swap given by `SwapRows(i,j,n)` which swaps row i and row j of A.

Listing 9.6: **Swapping Rows**

```
function P = SwapRows(i,j,n)
%
row = n;
P = eye(row,row);
C = P(i,:);
D = P(j,:);
P(i,:) = D;
P(j,:) = C;
end
```

The code to find the multiplier formed by all the row operations is a modification of our old code `GePiv.m` so that we can keep track of all the multipliers. This is not very efficient, but it will help you see how it works.

Listing 9.7: **Finding the Multiplier**

```
function [P,piv] = GetP(A);
%
% A is nxn matrix
% piv is a nx1 integer premutation vector
%
[n,n] = size(A)
piv = 1:n;
R = eye(n,n);
RPrev = R;
Aorig = A;
%Mult = {};
W = {};
swapindex = 0;
A
for k=1:n-1
  % find maximum absolute value entry
  % in column k of rows k through n
  [maxc,r] = max(abs(A(k:n,k)));
  q = r+k-1;
  % this line swaps the k and q positions in piv
  piv([k q]) = piv([q k]);
  % then manually swap row k and q in A
  swapindex = swapindex+1;
  W{swapindex} = SwapRows(k,q,n);
```

```
  A = W{swapindex}*A;
    if A(k,k) ~=0
      % get multipliers
      c = -A(k+1:n,k)/A(k,k);
     % find multiplier matrix
      RPrev = W{swapindex}*RPrev;
      for u = k+1:n
        % get multiplier
        S = LMult(k,u,c(u-k),n);
        % reset A.  Note A has been possibly row swapped
        A = S*A;
        R = S*RPrev;
        RPrev = R;
      end
    end
  end
  P = R;
  end
```

Note the `GetP` returns `piv` which is the row permutation matrix. The upper triangular matrix $\boldsymbol{PA}$ thus has a determinant which must be multiplied by the sign of the permutation of `piv` to get the correct determinant value.

9.2.2.1 Homework

These are problems we have done already, but now we will mimic the careful line by line computations in MATLAB we have shown in our example and then after that we will use `GetP`.

Exercise 9.2.5 *Let* $\boldsymbol{A} = \begin{bmatrix} 1 & 2 & 3 \\ -1 & 12 & 6 \\ 4 & 5 & 7 \end{bmatrix}$.

- *Find* $\boldsymbol{P}$ *the line by line way.*
- *Find* $\boldsymbol{P}$ *using* `GetP`.

Exercise 9.2.6 *Let* $\boldsymbol{A} = \begin{bmatrix} 1 & 2 & 3 & 3 \\ -1 & 12 & 6 & -4 \\ 4 & 5 & 7 & 8 \\ 5 & 1 & 4 & 3 \end{bmatrix}$

- *Find* $\boldsymbol{P}$ *the line by line way.*
- *Find* $\boldsymbol{P}$ *using* `GetP`.

Exercise 9.2.7 *Let*

$$\boldsymbol{A} = \begin{bmatrix} 1 & 2 & 3 & 3 & -10 \\ -1 & 12 & 6 & -4 & -3 \\ 1 & 5 & 7 & 8 & -2 \\ 5 & -1 & 4 & 3 & 6 \\ 11 & -3 & 4 & -5 & 1 \end{bmatrix}$$

- *Find* $\boldsymbol{P}$ *the line by line way.*
- *Find* $\boldsymbol{P}$ *using* `GetP`.

Exercise 9.2.8 *Let*

$$A = \begin{bmatrix} 1 & 2 & 3 & 3 & -10 & 7 \\ -1 & 12 & 6 & -4 & -3 & 2 \\ 1 & 5 & 7 & 8 & -2 & -5 \\ 5 & -1 & 4 & 3 & 6 & 1 \\ 11 & -3 & 4 & -5 & 1 & 10 \\ 1 & 2 & 1 & 2 & 1 & 3 \end{bmatrix}$$

- *Find P the line by line way.*
- *Find P using* `GetP`.

9.2.3 Matrix Inverse Calculations

Let's go back to the problem

$$\begin{bmatrix} 1 & -2 & 3 \\ -2 & 5 & 7 \\ 4 & 3 & -1 \end{bmatrix} \begin{bmatrix} x \\ y \\ z \end{bmatrix} = \begin{bmatrix} 2 \\ 5 \\ 8 \end{bmatrix}$$

which had the augmented matrix

$$B = \left[\begin{bmatrix} 1 & -2 & 3 \\ -2 & 5 & 7 \\ 4 & 3 & -1 \end{bmatrix} \begin{bmatrix} 2 \\ 5 \\ 8 \end{bmatrix} \right]$$

After manipulation with elementary matrices we found

$$R_{32}(c_{32}) R_{31}(c_{31}) R_{21}(c_{21})\, A = \begin{bmatrix} 1 & 0 & 29 \\ 0 & 1 & 13 \\ 0 & 0 & -156 \end{bmatrix}$$

Now let's zero out the entries above the original A_{33} position with $R_{13}(c)$ and $R_{23}(d)$ for suitable c and d. Thus, applying $R_{13}(29/156)$ we have

$$\begin{aligned} R_{13}(29/156)\, B &= \begin{bmatrix} 1 & 0 & 29/156 \\ 0 & 1 & 0 \\ 0 & 0 & 1 \end{bmatrix} \left[\begin{bmatrix} 1 & 0 & 29 \\ 0 & 1 & 13 \\ 0 & 0 & -156 \end{bmatrix} \begin{bmatrix} 20 \\ 9 \\ -99 \end{bmatrix} \right] \\ &= \left[\begin{bmatrix} 1 & 0 & 0 \\ 0 & 1 & 13 \\ 0 & 0 & -156 \end{bmatrix} \begin{bmatrix} 1.5962 \\ 9 \\ -99 \end{bmatrix} \right] \end{aligned}$$

Finally, applying $R_{23}(13/156)$, we have

$$\begin{aligned} R_{23}(-1/13)\, B &= \begin{bmatrix} 1 & 0 & 0 \\ 0 & 1 & 13/156 \\ 0 & 0 & 1 \end{bmatrix} \left[\begin{bmatrix} 1 & 0 & 0 \\ 0 & 1 & 13 \\ 0 & 0 & -156 \end{bmatrix} \begin{bmatrix} 1.5962 \\ 9 \\ -99 \end{bmatrix} \right] \\ &= \left[\begin{bmatrix} 1 & 0 & 0 \\ 0 & 1 & 0 \\ 0 & 0 & -156 \end{bmatrix} \begin{bmatrix} 1.5962 \\ 0.75 \\ -99 \end{bmatrix} \right] \end{aligned}$$

The last row then tells us $z = 99/156 = 0.6346$. We had already found $x = 1.5962$ and $y = 0.75$ so we have solved the system. This is what we did in code with the LU decomposition of A and the use of upper and lower triangular solvers. You should go back and look at that again to get a better appreciation of what we have done here. Of course, not trusting our typing and arithmetic, we used our MATLAB code to double check all of these calculations above. All we can say is this has been intense! But let's see what we have found. We wanted to solve the problem $\boldsymbol{A}\boldsymbol{X} = \boldsymbol{b}$ where the data $\boldsymbol{b}$ is given and $\boldsymbol{X}$ denotes the triple $\begin{bmatrix} x & y & z \end{bmatrix}^T$. We found that if we applied a sequence of elementary row operations we could transform $\boldsymbol{A}$ into a diagonal matrix. To summarize, we found

$$\boldsymbol{R}_{23}(c_{23})\boldsymbol{R}_{13}(c_{13})\boldsymbol{R}_{12}(c_{12})\boldsymbol{R}_{32}(c_{32})\boldsymbol{R}_{31}(c_{31})\boldsymbol{R}_{21}(c_{21})\,\boldsymbol{A} = \begin{bmatrix} r_1 & 0 & 0 \\ 0 & r_2 & 0 \\ 0 & 0 & r_3 \end{bmatrix}$$

for appropriate constants c_{ij} and some nonzero constants r_1, r_2 and r_3. Once this was done it is easy to find the unique x, y and z that solve the problem: we simply divide the entries of the transformed $\boldsymbol{b}$ by the appropriate r_i. We can do this for any system of equations and, of course it fails if $\boldsymbol{A}$ has problems.

Note, we can think of this a different way. Assume you are trying to solve three separate problems simultaneously: $\boldsymbol{A}\boldsymbol{X} = \boldsymbol{e_1}$, $\boldsymbol{A}\boldsymbol{X} = \boldsymbol{e_2}$ and $\boldsymbol{A}\boldsymbol{X} = \boldsymbol{e_3}$ where $\boldsymbol{e_1} = \begin{bmatrix} 1 & 0 & 0 \end{bmatrix}^T$, $\boldsymbol{e_2} = \begin{bmatrix} 0 & 1 & 0 \end{bmatrix}^T$ and $\boldsymbol{e_3} = \begin{bmatrix} 0 & 0 & 1 \end{bmatrix}^T$. Form the new augmented matrix

$$\begin{bmatrix} \begin{bmatrix} A_{11} & A_{12} & A_{13} \\ A_{21} & A_{22} & A_{23} \\ A_{31} & A_{32} & A_{33} \end{bmatrix} & \begin{bmatrix} 1 & 0 & 0 \\ 0 & 1 & 0 \\ 0 & 0 & 0 \end{bmatrix} \end{bmatrix}$$

Then apply the elementary row operations like usual to convert $\boldsymbol{A}$ into an upper triangular matrix $\boldsymbol{U}$. If this system in solvable, we can always do this. Thus, we find

$$\begin{aligned} \boldsymbol{R}_{32}(c_{32})\boldsymbol{R}_{31}(c_{31})\boldsymbol{R}_{21}(c_{21})\,\boldsymbol{A} &= \boldsymbol{R}_{32}(c_{32})\boldsymbol{R}_{31}(c_{31})\boldsymbol{R}_{21}(c_{21})\,\boldsymbol{I} \\ \begin{bmatrix} U_{11} & U_{12} & U_{13} \\ 0 & U_{21} & U_{23} \\ 0 & 0 & U_{33} \end{bmatrix} &= \boldsymbol{R}_{32}(c_{32})\boldsymbol{R}_{31}(c_{31})\boldsymbol{R}_{21}(c_{21})\,\boldsymbol{I} \end{aligned}$$

Now apply the other three elementary row operations.

$$\begin{aligned} &\boldsymbol{R}_{23}(c_{23})\boldsymbol{R}_{13}(c_{13})\boldsymbol{R}_{12}(c_{12})\boldsymbol{R}_{32}(c_{32})\boldsymbol{R}_{31}(c_{31})\boldsymbol{R}_{21}(c_{21})\,\boldsymbol{A} \\ &\qquad = \boldsymbol{R}_{23}(c_{23})\boldsymbol{R}_{13}(c_{13})\boldsymbol{R}_{12}(c_{12})\boldsymbol{R}_{32}(c_{32})\boldsymbol{R}_{31}(c_{31})\boldsymbol{R}_{21}(c_{21})\,\boldsymbol{I} \\ &\begin{bmatrix} r_1 & 0 & 0 \\ 0 & r_2 & 0 \\ 0 & 0 & r_3 \end{bmatrix} = \boldsymbol{R}_{23}(c_{23})\boldsymbol{R}_{13}(c_{13})\boldsymbol{R}_{12}(c_{12})\boldsymbol{R}_{32}(c_{32})\boldsymbol{R}_{31}(c_{31})\boldsymbol{R}_{21}(c_{21})\,\boldsymbol{I} \end{aligned}$$

Now apply the diagonal matrix $\boldsymbol{D}$ whose main diagonal entries are $(1/r_1)$, $(1/r_2)$ and $(1/r_3)$ with every other entry zero. Let $\boldsymbol{B}$ be defined by

$$\boldsymbol{B} = \begin{bmatrix} 1/r_1 & 0 & 0 \\ 0 & 1/r_2 & 0 \\ 0 & 0 & 1/r_3 \end{bmatrix} \boldsymbol{R}_{23}(c_{23})\boldsymbol{R}_{13}(c_{13})\boldsymbol{R}_{12}(c_{12})\boldsymbol{R}_{32}(c_{32})\boldsymbol{R}_{31}(c_{31})\boldsymbol{R}_{21}(c_{21})$$

Then, we have

$$\boldsymbol{I} = \boldsymbol{B}\,\boldsymbol{A}$$

We have just found something cool! For convenience, let $\mathcal{R}$ be given by

$$\mathcal{R} = \boldsymbol{R}_{23}(c_{23})\boldsymbol{R}_{13}(c_{13})\boldsymbol{R}_{12}(c_{12})\boldsymbol{R}_{32}(c_{32})\boldsymbol{R}_{31}(c_{31})\boldsymbol{R}_{21}(c_{21})$$

Then note

$$\boldsymbol{B} = \boldsymbol{D}(1/r_1), (1/r_2), (1/r_3)\,\mathcal{R}$$

and $\boldsymbol{BA} = \boldsymbol{I}$ so that we have found the inverse of $\boldsymbol{A}$ using elementary row operations. It is

$$\boldsymbol{A}^{-1} = \boldsymbol{D}(1/r_1), (1/r_2), (1/r_3)\,\mathcal{R}!$$

This is a computationally efficient way to find a matrix inverse, if it exists.

Homework

For these problems, we can apply `GetP` to get the multiplier $\boldsymbol{P_1}$ which makes the augmented matrix for the inverse calculation upper triangular. Then you can figure out the next set of row operations to diagonalize. So the steps are

- $\boldsymbol{P_1}[\boldsymbol{A}|\boldsymbol{I}] = [\boldsymbol{U}|\boldsymbol{P_1}]$
- $\boldsymbol{P_2}[\boldsymbol{U}|\boldsymbol{P_1}] = [\boldsymbol{D}|\boldsymbol{P_2P}]$
- $\boldsymbol{D}^{-1}[\boldsymbol{D}|\boldsymbol{P_2P} = \boldsymbol{I}|\boldsymbol{D}^{-1}\boldsymbol{P_2P}]$

which tells us $\boldsymbol{A}^{-1} = \boldsymbol{D}^{-1}\boldsymbol{P_2P}$. So you will have to find $\boldsymbol{P_2}$ and $\boldsymbol{D}^{-1}$ by hand here. Have fun! This will help you understand at a gut level why code is nice.

Exercise 9.2.9 *Let* $\boldsymbol{A} = \begin{bmatrix} 1 & 2 & 3 \\ -1 & 12 & 6 \\ 4 & 5 & 7 \end{bmatrix}$.

- *Find* $\boldsymbol{P_1}$ *using* `GetP`.
- *Find* $\boldsymbol{P_2}$ *the line by line way.*
- *Find* $\boldsymbol{D}^{-1}$ *and finish the computation of the inverse.*

Exercise 9.2.10 *Let* $\boldsymbol{A} = \begin{bmatrix} 1 & 2 & 3 & 3 \\ -1 & 12 & 6 & -4 \\ 4 & 5 & 7 & 8 \\ 5 & 1 & 4 & 3 \end{bmatrix}$

- *Find* $\boldsymbol{P_1}$ *using* `GetP`.
- *Find* $\boldsymbol{P_2}$ *the line by line way.*
- *Find* $\boldsymbol{D}^{-1}$ *and finish the computation of the inverse.*

Exercise 9.2.11 *Let*

$$\boldsymbol{A} = \begin{bmatrix} 1 & 2 & 3 & 3 & -10 \\ -1 & 12 & 6 & -4 & -3 \\ 1 & 5 & 7 & 8 & -2 \\ 5 & -1 & 4 & 3 & 6 \\ 11 & -3 & 4 & -5 & 1 \end{bmatrix}$$

- *Find* $\boldsymbol{P_1}$ *using* `GetP`.
- *Find* $\boldsymbol{P_2}$ *the line by line way.*
- *Find* $\boldsymbol{D}^{-1}$ *and finish the computation of the inverse.*

Exercise 9.2.12 *Let*

$$\boldsymbol{A} = \begin{bmatrix} 1 & 2 & 3 & 3 & -10 & 7 \\ -1 & 12 & 6 & -4 & -3 & 2 \\ 1 & 5 & 7 & 8 & -2 & -5 \\ 5 & -1 & 4 & 3 & 6 & 1 \\ 11 & -3 & 4 & -5 & 1 & 10 \\ 1 & 2 & 1 & 2 & 1 & 3 \end{bmatrix}$$

- *Find* $\boldsymbol{P_1}$ *using* `GetP`.
- *Find* $\boldsymbol{P_2}$ *the line by line way.*
- *Find* $\boldsymbol{D}^{-1}$ *and finish the computation of the inverse.*

9.2.3.1 Code Implementations

We can implement these ideas using our row operation matrices explicitly as follows. Again, we will implement the code using a matrix name different from the $\boldsymbol{R}$'s above. For this code we will use Q and the $\boldsymbol{R}$ we want will end up being QP. Now, this is not so efficient, as normally, we would not store these matrices, but we wanted you to see how it works. We already know how to get the row operation matrix P that converts A into an upper triangular matrix. To finish the job, we don't have to worry about row swaps so the code is cleaner. We only apply this code to symmetric matrices as we know they can be diagonalized!

Consider the code for `GetQ2(A)` given below.

Listing 9.8: **GetQ2.m: Diagonalize A**

```
function Q = GetQ2(A);
%
% A is nxn matrix in upper triangular form
%
[n,n] = size(A);
R = eye(n,n);
for k=2:n
   if A(k,k) ~=0
      % get multipliers
      % look at rows k+1 to n and
      % divide them by the pivot value A(k,k)
      c = -A(1:k-1,k)/A(k,k);
     % find multiplier matrix
      for u = 1:k-1
         S = LMult(k,u,c(u),n);
         A = S*A;
         R = S*R;
      end
   end
end
```

```
Q = R;
end
```

This code finds the row operation matrices that convert PA into diagonal form. Let's do a 4×4 example.

Listing 9.9: **Upper Triangularizing A**

```
>> A = [4,1,3,2;1,5,6,3;3,6,10,5;2,3,5,1]
% original A
A =
      4    1    3    2
      1    5    6    3
      3    6   10    5
      2    3    5    1
>>  [P,piv] = GetP(A);
% PA is upper triangular
>> P*A
ans =
     4.00000    1.00000    3.00000    2.00000
     0.00000    5.25000    7.75000    3.50000
     0.00000    0.00000   -1.76190   -0.66667
     0.00000    0.00000    0.00000   -1.59459
% apply further row operations to diagonalize PA
>> Q = GetQ2(P*A);
>> Q
Q =

     1.00000   -0.19048    0.86486    0.47458
     0.00000    1.00000    4.39865    0.35593
     0.00000    0.00000    1.00000   -0.41808
     0.00000    0.00000    0.00000    1.00000
% note QP is the matrix which diagonalizes A
>> Q*P
ans =
     1.42373    0.81356   -1.15254    0.47458
     1.06780    4.36017   -3.11441    0.35593
     0.50767    1.04520   -0.74657   -0.41808
    -0.18919   -0.10811   -0.37838    1.00000
  % Q*P*A gives us the diagonal matrix D
Q*P*A
ans =
     4.00000   -0.00000    0.00000   -0.00000
    -0.00000    5.25000   -0.00000   -0.00000
     0.00000    0.00000   -1.76190    0.00000
     0.00000    0.00000    0.00000   -1.59459
```

Hence, we have $Q\,P\,A = D$ and so $D^{-1}\,Q\,P\,A = I$, which tells us $A^{-1} = D^{-1}\,Q\,P$! Now we can't just use this code on any matrix we make up. We have been cheating by plugging in our chosen matrix into MATLAB and calculating its determinant to see if it is invertible. But if we knew the matrix had n real eigenvalues and an associated orthonormal basis of eigenvectors, we would be guaranteed our procedure would work. Since this is true for symmetric matrices, we can finish with this nice piece of code.

Listing 9.10: **Simple Matrix Inverse Code**

```
function Ainv = GetInvSymm(A)
%
% A is an n x n symmetric matrix
%
n = size(A);
[P,piv] = GetP(A);
Q = GetQ2(P*A);
D = Q*P*A;
Dinv = D;
for k=1:n
  Dinv(k,k) = 1/D(k,k);
end
Ainv = Dinv*Q*P;
end
```

For our example, we find:

Listing 9.11: **Diagonalizing A**

```
>> Ainv = GetInvSymm(A);
% original matrix
A =

     4     1     3     2
     1     5     6     3
     3     6    10     5
     2     3     5     1
>> Ainv
Ainv =
    0.355932    0.203390   -0.288136    0.118644
    0.203390    0.830508   -0.593220    0.067797
   -0.288136   -0.593220    0.423729    0.237288
    0.118644    0.067797    0.237288   -0.627119
% Checks
>> A*Ainv
ans =
    1.0000e+00  -2.2482e-15   3.5527e-15  -2.2204e-16
   -1.7708e-14   1.0000e+00   1.8430e-14   4.2188e-15
   -1.9318e-14  -1.4599e-14   1.0000e+00   4.8850e-15
   -9.8949e-15  -7.8965e-15   9.5479e-15   1.0000e+00
>> Ainv*A
ans =
    1.00000  -0.00000   0.00000   0.00000
   -0.00000   1.00000   0.00000   0.00000
    0.00000   0.00000   1.00000   0.00000
    0.00000   0.00000  -0.00000   1.00000
```

We see the inverse checks out! Normally, we just do elementary row operations more efficiently. For example, our earlier *LU* decomposition works by using row operations.

Homework

Previously, we found the steps to find the inverse of a invertible $\boldsymbol{A}$ were as follows: So the steps are

- $P_1[A|I] = [U|P_1]$
- $P_2[U|P_1] = [D|P_2P]$
- $D^{-1}[D|P_2P = I|D^{-1}P_2P]$

which tells us $A^{-1} = D^{-1}P_2P$. Now we can do this in code and save our hands all that cramping! The basic plan is, thus, to do the following:

Listing 9.12: **Upper Triangularizing A**

```
>> A = <your matrix here>
>>  [P1,piv] = GetP(A);
% P1*A is upper triangular
>> P1*A
ans = < this is upper triangular
% apply further row operations to diagonalize P1 *A
>> P2 = GetQ2(P1*A);
>> P2 <you will get P2 now>
% note P2*P1 is the matrix which diagonalizes A
>> P2*P1*A < this gives a diagonal matrix D>
%now find DInverse then
%Ainverse = Dinverse*P2*P1
```

Exercise 9.2.13 *Let* $A = \begin{bmatrix} 1 & 2 & 3 \\ -1 & 12 & 6 \\ 4 & 5 & 7 \end{bmatrix}$. *Find the inverse of* $\boldsymbol{A}$ *using our code in the line by line fashion above.*

Exercise 9.2.14 *Let* $A = \begin{bmatrix} 1 & 2 & 3 & 3 \\ -1 & 12 & 6 & -4 \\ 4 & 5 & 7 & 8 \\ 5 & 1 & 4 & 3 \end{bmatrix}$ *Find the inverse of* $\boldsymbol{A}$ *using our code in the line by line fashion above.*

Exercise 9.2.15 *Let*

$$A = \begin{bmatrix} 1 & 2 & 3 & 3 & -10 \\ -1 & 12 & 6 & -4 & -3 \\ 1 & 5 & 7 & 8 & -2 \\ 5 & -1 & 4 & 3 & 6 \\ 11 & -3 & 4 & -5 & 1 \end{bmatrix}$$

Find the inverse of $\boldsymbol{A}$ *using our code in the line by line fashion above.*

Exercise 9.2.16 *Let*

$$A = \begin{bmatrix} 1 & 2 & 3 & 3 & -10 & 7 \\ -1 & 12 & 6 & -4 & -3 & 2 \\ 1 & 5 & 7 & 8 & -2 & -5 \\ 5 & -1 & 4 & 3 & 6 & 1 \\ 11 & -3 & 4 & -5 & 1 & 10 \\ 1 & 2 & 1 & 2 & 1 & 3 \end{bmatrix}$$

Find the inverse of A *using our code in the line by line fashion above.*

Exercise 9.2.17 *Make up any symmetric* 3×3 *matrix* A *and find its inverse using* `GetInvSymm`.

Exercise 9.2.18 *Make up any symmetric* 4×4 *matrix* A *and find its inverse using* `GetInvSymm`.

Exercise 9.2.19 *Make up any symmetric* 5×5 *matrix* A *and find its inverse using* `GetInvSymm`.

Exercise 9.2.20 *Make up any symmetric* 6×6 *matrix* A *and find its inverse using* `GetInvSymm`.

Exercise 9.2.21 *Make up any symmetric* 7×7 *matrix* A *and find its inverse using* `GetInvSymm`.

9.3 Back to Definite Matrices

Our interests are in symmetric matrices that are positive or negative definite. Recall, a symmetric matrix A can be written as

$$A = P \begin{bmatrix} \lambda_1 & 0 & 0 \\ 0 & \lambda_2 & 0 \\ 0 & 0 & \lambda_3 \end{bmatrix} P^T$$

where P is the matrix of A's eigenvectors. This P is different from the generic P we used in the upper triangularization code introduce above. So in this section, we have the matrix of eigenvectors P and new matrices of the form R which we will use to develop some ideas about minors. For our discussions here, we will ignore row swapping issues to keep our explanations from becoming too messy. However, you should be able to see adding the `piv` idea into the mixture is not terribly difficult. Note, letting Λ be the diagonal matrix $D(\lambda_1, \lambda_2, \lambda_3)$, we have

$$\begin{aligned} \det(A) &= \det(P)\ \det(\Lambda)\ \det(P^T) \\ &= \lambda_1\lambda_2\lambda_3 \end{aligned}$$

as $PP^T = I$ and so $\det(P)\det(P^T) = 1$. We can apply the ideas above to the matrix A to find a matrix $\mathcal{R}$ given by

$$\mathcal{R} = R_{32}(c_{32}) R_{31}(c_{31}) R_{21}(c_{21})$$

for appropriate constants c_{ij} so that

$$\mathcal{R}A = \begin{bmatrix} U_{11} & U_{12} & U_{13} \\ 0 & U_{22} & U_{23} \\ 0 & 0 & U_{33} \end{bmatrix}$$

Now let U denote the upper triangular matrix above. From our discussions above, we know $\det(U) = U_{11}U_{22}U_{33}$. However, this determinant is also $\lambda_1\lambda_2\lambda_3$. Hence, we know $\lambda_1\lambda_2\lambda_3 = U_{11}U_{22}U_{33}$. We will use this idea in a bit.

Part III

Calculus of Many Variables

Chapter 10

Differentiability

We now consider the differentiability properties of mappings $f : D \subset \Re^n \to \Re^m$ in earnest. In (Peterson (21) 2020), we worked through a good introduction to functions of two variables. Now we want to look at higher dimensional analogues.

10.1 Partial Derivatives

If f is locally defined on $B(\boldsymbol{x_0}, r)$ in $\Re^n$, we can look at the **trace** corresponding to a $\boldsymbol{x_0}$ in the direction of the unit vector $\boldsymbol{E}$; call this $f_{\boldsymbol{E}}(t) = f(\boldsymbol{x_0} + t\boldsymbol{E})$ for sufficiently small t so the function values are well-defined. This means

$$f_{\boldsymbol{E}}(t) \quad = \quad f(\boldsymbol{x_0} + t\boldsymbol{E}) = f(x_{01} + tE_1, \ldots, x_{0n} + tE_n)$$

If we calculate the difference between $f_{\boldsymbol{E}}(t)$ and $f_{\boldsymbol{E}}(0)$, we find for any $t \neq 0$

$$\frac{f_{\boldsymbol{E}}(t) - f_{\boldsymbol{E}}(0)}{t} = \frac{f(x_{01}, \ldots, x_{0,i-1}, \boldsymbol{x_{0i} + t}, \ldots, x_{0n}) - f(x_{01}, \ldots, x_{0,i-1}, \boldsymbol{x_{0i}}, \ldots, x_{0n})}{t}$$

where we have put in boldface the ith slot where all the action is taking place. It is useful to determine if $\lim_{t \to 0}(1/t)(f(\boldsymbol{x_0} + t\boldsymbol{E}) - f(\boldsymbol{x_0}))$ exists. In the case of two variables, this turns out to be the directional derivative of f in the direction of $\boldsymbol{E}$. But we need to do this carefully for n variables and we don't want to take advantage of two dimensional intuition. We start by looking at the standard basis $\{\boldsymbol{E_1}, \ldots, \boldsymbol{E_n}\}$ for $\Re^n$ where the components of $\boldsymbol{E_i} = \delta_{ij}$ as usual. Then, we can consider the limit

$$\lim_{t \to 0} \frac{f(\boldsymbol{x_0} + t\boldsymbol{E_i}) - f(\boldsymbol{x_0})}{t}$$

If this limit exists, it is called the **partial derivative** of f with respect to the ith coordinate slot. There are many notations for this.

Definition 10.1.1 Partial Derivatives

Let $z = f(\boldsymbol{x})$ be a function defined locally at $\boldsymbol{x}$ in $\Re^n$. The partial derivative of f with respect to its ith coordinate or slot at $\boldsymbol{x}$ is defined by the limit

$$\frac{\partial f}{\partial x}(\boldsymbol{x}) = \lim_{t \to 0} \frac{f(\boldsymbol{x} + t\boldsymbol{E}_i) - f(\boldsymbol{x})}{t} = \lim_{y \to x_i} \frac{f(\boldsymbol{x} + y\boldsymbol{E}_i) - f(\boldsymbol{x})}{y - x_i}$$

If this limit exists, it is denoted by a variety of symbols: for functions of two variables x and y, for the partial derivatives at the point (x_0, y_0), we often use $f_x(x_0, y_0)$, $z_x(x_0, y_0)$, $\frac{\partial f}{\partial x}(x_0, y_0)$ and $\frac{\partial z}{\partial x}(x_0, y_0)$ and $f_y(x_0, y_0)$, $z_y(x_0, y_0)$, $\frac{\partial f}{\partial y}(x_0, y_0)$ and $\frac{\partial z}{\partial y}(x_0, y_0)$. However, this notation is not very useful when $\boldsymbol{x}$ is in $\Re^n$. In the more general case, we use D_i to indicate the partial derivative with respect to the ith slot instead of $\frac{\partial f}{\partial x_i}$.

Comment 10.1.1 *It is easy to take partial derivatives. Just imagine the one variable held constant and take the derivative of the resulting function just like you did in your earlier calculus courses.*

Comment 10.1.2 *Notation can be really bad here. Consider the following partial we need to take in a problem with line integrals in Chapter 16. These calculations involve the chain rule which you have seen for two variables in (Peterson (21) 2020). We will discuss the chain rule for n variables carefully in a bit, but it is easy enough to use two variable ideas for this example. We need to find*

$$(\boldsymbol{A}\boldsymbol{G_1}_v + \boldsymbol{B}\boldsymbol{G_2}_v)_u$$

where $\boldsymbol{A}$ and $\boldsymbol{B}$ are functions of two variables and $\boldsymbol{G_1}$ and $\boldsymbol{G_2}$ are also. This is a change of variable problem with $x = \boldsymbol{G_1}(u, v)$ and $y = \boldsymbol{G_2}(u, v)$. In the chain rule, we need $\partial \boldsymbol{A}$ with respect to argument one and argument two. In the original $\boldsymbol{A}$ function, these would be $\boldsymbol{A}_x$ and $\boldsymbol{A}_y$ but that is really confusing here. That would give

$$\begin{aligned}(\boldsymbol{A}\boldsymbol{G_1}_v + \boldsymbol{B}\boldsymbol{G_2}_v)_u &= (\boldsymbol{A}_x\boldsymbol{G}_{1u} + \boldsymbol{A}_y\boldsymbol{G}_{2u})\boldsymbol{G}_{1v} + \boldsymbol{A}\boldsymbol{G}_{1vu} \\ &\quad +(\boldsymbol{B}_x\boldsymbol{G}_{1u} + \boldsymbol{B}_y\boldsymbol{G}_{2u})\boldsymbol{G}_{2v} + \boldsymbol{B}\boldsymbol{G}_{2vu}\end{aligned}$$

or we could use arg 1 and arg 2 which is ugly

$$\begin{aligned}(\boldsymbol{A}\boldsymbol{G_1}_v + \boldsymbol{B}\boldsymbol{G_2}_v)_u &= (\boldsymbol{A}_{arg\ 1}\boldsymbol{G}_{1u} + \boldsymbol{A}_{arg\ 2}\boldsymbol{G}_{2u})\boldsymbol{G}_{1v} + \boldsymbol{A}\boldsymbol{G}_{1vu} \\ &\quad +(\boldsymbol{B}_{arg\ 1}\boldsymbol{G}_{1u} + \boldsymbol{B}_{arg\ 2}\boldsymbol{G}_{2u})\boldsymbol{G}_{2v} + \boldsymbol{B}\boldsymbol{G}_{2vu}\end{aligned}$$

But it is cleaner to just use $\boldsymbol{A}_1$ and $\boldsymbol{A}_2$ and so forth to indicate these partials with respect to slot one and slot two.

$$\begin{aligned}(\boldsymbol{A}\boldsymbol{G_1}_v + \boldsymbol{B}\boldsymbol{G_2}_v)_u &= (\boldsymbol{A}_1\boldsymbol{G}_{1u} + \boldsymbol{A}_2\boldsymbol{G}_{2u})\boldsymbol{G}_{1v} + \boldsymbol{A}\boldsymbol{G}_{1vu} \\ &\quad +(\boldsymbol{B}_1\boldsymbol{G}_{1u} + \boldsymbol{B}_2\boldsymbol{G}_{2u})\boldsymbol{G}_{2v} + \boldsymbol{B}\boldsymbol{G}_{2vu}\end{aligned}$$

At any rate, don't be surprised if, in a complicated partial calculation, it all gets confusing!

10.1.1 Homework

Exercise 10.1.1 *Let $f(t, x, y) = \sqrt{x^2 + y^2 + t^2}$ and assume $x(t) = t^2 + 3$ and $y(t) = 2t^4 + 5t^2 + 8$. Find $\frac{df}{dt}$. Note how strange the notation for the derivative of f with respect to t and its partial with respect to t get!*

Exercise 10.1.2 *Assume $f_x(x_0, y_0) = f_x^0$ exists. Use an $\epsilon - \delta$ proof to show $\lim_{h \to 0} \frac{f(x_0+h, y_0) - f(x_0-h, y_0)}{2h} = f_x^0$.*

Exercise 10.1.3 *Let*

$$f(x,y) = \begin{cases} x^2+y^2, & x,y \in \mathbb{Q} \cap [-1,1] \\ -x^2-y^2, & x,y \in \mathbb{I} \cap [-1,1] \\ x^2+y^2, & x \in \mathbb{Q} \cap [-1,1], y \in \mathbb{I} \cap [-1,1] \\ -x^2-y^2, & x \in \mathbb{I} \cap [-1,1], y \in \mathbb{Q} \cap [-1,1] \end{cases}$$

Find $f_x(0,0)$ *and* $f_y(0,0)$. *Are* f_x *and* f_y *continuous at* $(0,0)$*?*

Exercise 10.1.4 *Let*

$$f(x,y) = \begin{cases} x+y, & x,y \in \mathbb{Q} \cap [-1,1] \\ -x-y, & x,y \in \mathbb{I} \cap [-1,1] \\ x+y, & x \in \mathbb{Q} \cap [-1,1], y \in \mathbb{I} \cap [-1,1] \\ -x-y, & x \in \mathbb{I} \cap [-1,1], y \in \mathbb{Q} \cap [-1,1] \end{cases}$$

Find $f_x(0,0)$ *and* $f_y(0,0)$.

Exercise 10.1.5 *Let*

$$f(x,y) = \begin{cases} x^2+y, & x,y \in \mathbb{Q} \cap [-1,1] \\ -x^2-y, & x,y \in \mathbb{I} \cap [-1,1] \\ x^2+y, & x \in \mathbb{Q} \cap [-1,1], y \in \mathbb{I} \cap [-1,1] \\ -x^2-y, & x \in \mathbb{I} \cap [-1,1], y \in \mathbb{Q} \cap [-1,1] \end{cases}$$

Do $f_x(0,0)$ *and* $f_y(0,0)$ *exist?*

10.2 Tangent Planes

It is very useful to have an analytical way to discuss tangent planes and their relationship to the real surface to which they are attached even if we can't draw this.

Definition 10.2.1 Planes in n Dimensions

A plane in $\Re^n$ through the point $\boldsymbol{x_0}$ is defined as the set of all vectors $\boldsymbol{x}$ so that the angle between the vectors $\boldsymbol{D}$ and $\boldsymbol{N}$ is 90° where $\boldsymbol{D}$ is the vector we get by connecting the point $\boldsymbol{x_0}$ to the point $\boldsymbol{x}$. Hence, for

$$\boldsymbol{D} = \boldsymbol{x} - \boldsymbol{x_0} = \begin{bmatrix} x_1 - x_{01} \\ \vdots \\ x_n - x_{0n} \end{bmatrix} \quad \text{and} \quad \boldsymbol{N} = \begin{bmatrix} N_1 \\ \vdots \\ N_n \end{bmatrix}$$

The plane is the set of points $\boldsymbol{x}$ so that $< \boldsymbol{D}, \boldsymbol{N} >= 0$. The vector $\boldsymbol{N}$ is called the **normal vector** *to the plane. If $n > 2$, we usually call this a* **hyperplane**. *Also note hyperplanes through the origin of $\Re^n$ are $n-1$ dimensional subspaces of $\Re^n$.*

If $\boldsymbol{N}$ is given, then the span of $\boldsymbol{N}$, $sp(\boldsymbol{N})$, is a one dimensional subspace of $\Re^n$ and the orthogonal complement of it which is

$$sp(\boldsymbol{N})^\perp = \{\boldsymbol{x} \in \Re^n \,|\, < \boldsymbol{x}, \boldsymbol{N} >= 0\}$$

is an $n-1$ dimensional subspace of $\Re^n$ and we can construct an orthonormal basis for it as follows:

- Pick any nonzero $\boldsymbol{w_1}$ not equal to a multiple of $\boldsymbol{N}$. Then $\boldsymbol{w_1}$ is not in $sp(\boldsymbol{N})$. Compute $\boldsymbol{F_1} = \boldsymbol{N}/\|\boldsymbol{N}\|$ and

$$\boldsymbol{v} = \boldsymbol{w_1} - < \boldsymbol{w_1}, \mathbf{F_1} > \boldsymbol{F_1}$$

  Then $\boldsymbol{v}$ is orthogonal to $\boldsymbol{F_1}$. Let $\boldsymbol{F_2} = \boldsymbol{v}/\|\boldsymbol{v}\|$.

- If the $sp(\boldsymbol{F_1}, \boldsymbol{F_2})$ is not $\Re^n$, we know there is a nonzero vector w in $(sp(\boldsymbol{F_1}, \boldsymbol{F_2}))^\perp$ from which we can construct a third vector $\boldsymbol{F_3}$ mutually orthogonal to $\boldsymbol{F_1}$ and $\boldsymbol{F_2}$ just we did before.

- If the $sp(\boldsymbol{F_1}, \boldsymbol{F_2}, \boldsymbol{F_3})$ is not $\Re^n$, we do this again. If not, we are done.

This constructs $n-1$ mutually orthonormal vectors $\boldsymbol{F_1}$ to $\boldsymbol{F_{n-1}}$ which are a basis for $sp(\boldsymbol{N})^\perp$. Note any $\boldsymbol{x}$ in this subspace satisfies $< \boldsymbol{x}, \boldsymbol{N} >= 0$.

On the other hand if $\boldsymbol{A_1}$ through $\boldsymbol{A_{n-1}}$ is a linearly independent set in $\Re^n$, the $span(\boldsymbol{A_1}, \ldots, \boldsymbol{A_{n-1}})$ is an $n-1$ dimensional subspace. To find the normal vector, note we are looking for a vector $\boldsymbol{N}$ so that

$$\begin{bmatrix} A_{11} & \ldots & A_{1,n} \\ A_{21} & \ldots & A_{2,n} \\ \vdots & \vdots & \vdots \\ A_{n-1,1} & \vdots & A_{n-1,n} \end{bmatrix} \begin{bmatrix} N_1 \\ \vdots \\ N_n \end{bmatrix} = \begin{bmatrix} 0 \\ \vdots \\ 0 \end{bmatrix}$$

This is the same as

$$N_1 \begin{bmatrix} A_{11} \\ A_{21} \\ \vdots \\ A_{n-1,1} \end{bmatrix} + N_2 \begin{bmatrix} A_{12} \\ A_{22} \\ \vdots \\ A_{n-1,2} \end{bmatrix} + \ldots + N_n \begin{bmatrix} A_{1n} \\ A_{2n} \\ \vdots \\ A_{n-1,n} \end{bmatrix} = \mathbf{0}.$$

Letting

$$\boldsymbol{W_i} = \begin{bmatrix} A_{1i} \\ A_{2i} \\ \vdots \\ A_{n-1,i} \end{bmatrix}$$

we see $\boldsymbol{W_1}, \ldots, \boldsymbol{W_n}$ is a set of n vectors in $\Re^{n-1}$ and so must be linearly dependent. Thus, the equation

$$N_1 \boldsymbol{W_1} + \ldots + N_n \boldsymbol{W_n} = \mathbf{0}$$

must have a nonzero solution $\boldsymbol{N}$ which will be the normal vector we seek.

Homework

Exercise 10.2.1 *Let* $\boldsymbol{N} = \begin{bmatrix} -1 \\ 2 \\ 4 \end{bmatrix} \in \Re^3$. *Find an orthonormal basis for* $\boldsymbol{N^p erp}$.

Exercise 10.2.2 *Let* $\boldsymbol{N} = \begin{bmatrix} 11 \\ -1 \\ 2 \\ 4 \end{bmatrix} \in \Re^4$. *Find an orthonormal basis for* $\boldsymbol{N^perp}$.

Exercise 10.2.3 *Let* $\boldsymbol{N} = \begin{bmatrix} 11 \\ -1 \\ 2 \\ 4 \\ -2 \end{bmatrix} \in \Re^5$. *Find an orthonormal basis for* $\boldsymbol{N^perp}$.

Exercise 10.2.4 *Let*

$$\boldsymbol{A} = \begin{bmatrix} 1 & 2 & 3 \\ 0 & 3 & 4 \\ 0 & 0 & 0 \end{bmatrix}$$

Find the $ker(\boldsymbol{A})$ *and an orthonormal basis for* $(ker(\boldsymbol{A}))^\perp$.

Exercise 10.2.5 *Let*

$$\boldsymbol{A} = \begin{bmatrix} 1 & 2 & 3 & -3 \\ 0 & 3 & 4 & 7 \\ 0 & 0 & 0 & 0 \\ 0 & 0 & 0 & 0 \end{bmatrix}$$

Find the $ker(\boldsymbol{A})$ *and an orthonormal basis for* $(ker(\boldsymbol{A}))^\perp$.

Now given our function of n variables, we calculate the n partial derivatives at a given point $\boldsymbol{x_0}$ giving $f_1(\boldsymbol{x_0})$ to $f_n(\boldsymbol{x_0})$. The vector

$$\boldsymbol{N} = \begin{bmatrix} f_1(\boldsymbol{x_0}) \\ \vdots \\ f_n(\boldsymbol{x_0}) \end{bmatrix}$$

therefore determines a hyperplane like we discussed above where the hyperplane is the set of all $\boldsymbol{x}$ in $\Re^n$ so that $< \boldsymbol{x}, \boldsymbol{N} >= 0$.

Now what does this have to do with what we have called tangent planes for functions of two variables? Let's look at that case for some motivation. Recall the tangent plane to a surface $z = f(x, y)$ at the point (x_0, y_0) was the plane determined by the tangent lines $T(x, y_0)$ and $T(x_0, y)$. The $T(x, y_0)$ line was determined by the vector

$$\boldsymbol{A} = \begin{bmatrix} 1 \\ 0 \\ \frac{\partial f}{\partial x}(x_0, y_0) \end{bmatrix} = \begin{bmatrix} 1 \\ 0 \\ 2x_0 \end{bmatrix}$$

and the $T(x_0, y)$ line was determined by the vector

$$\boldsymbol{B} = \begin{bmatrix} 0 \\ 1 \\ \frac{\partial f}{\partial y}(x_0, y_0) \end{bmatrix} = \begin{bmatrix} 0 \\ 1 \\ 2y_0 \end{bmatrix}$$

The vector perpendicular to both $\boldsymbol{A}$ and $\boldsymbol{B}$ is given by $\boldsymbol{A} \times \boldsymbol{B}$ but we can't use the cross product in the dimension $n > 3$. The cross product gives

$$\boldsymbol{N} = \begin{bmatrix} -f_x(x_0, y_0) \\ -f_y(x_0, y_0) \\ 1 \end{bmatrix}$$

The tangent plane to $z = f(x, y)$ at $(x_0, y_0, f(x_0, y_0))$ is defined to be the set of all (x, y, z) in $\Re^3$ so that

$$\left\langle \begin{bmatrix} -f_x(x_0, y_0) \\ -f_y(x_0, y_0) \\ 1 \end{bmatrix}, \begin{bmatrix} x - x_0 \\ y - y_0 \\ z - f(x_0, y_0) \end{bmatrix} \right\rangle = 0$$

or multiplying this out and rearranging

$$z = f(x_0, y_0) + f_x(x_0, y_0)(x - x_0) + f_y(x_0, y_0)(x - x_0)$$

We can use this as the motivation for defining the tangent vector to $z = f(\boldsymbol{x})$ at $\boldsymbol{x_0}$ as follows:

Definition 10.2.2 The Tangent Plane to $z = f(\boldsymbol{x})$ at $\boldsymbol{x_0}$ in $\Re^n$

The tangent plane in $\Re^n$ through the point $\boldsymbol{x_0}$ for the surface $z = f(\boldsymbol{x})$ is defined as the set of all vectors $\boldsymbol{x}$ so that

$$\left\langle \begin{bmatrix} -f_1(\boldsymbol{x_0}) \\ -f_2(\boldsymbol{x_0}) \\ \vdots \\ -f_n(\boldsymbol{x_0}) \\ 1 \end{bmatrix}, \begin{bmatrix} x_1 - x_{01} \\ x_2 - x_{02} \\ \vdots \\ x_n - x_{0n} \\ z - f(\boldsymbol{x_0}) \end{bmatrix} \right\rangle = 0$$

or multiplying this out and rearranging

$$z = f(\boldsymbol{x_0}) + f_1(\boldsymbol{x_0})(x_1 - x_{01}) + \ldots + f_n(\boldsymbol{x_0})(x_n - x_{0,n})$$

We can use another compact definition at this point. We can define the **gradient** of the function f to be the vector $\boldsymbol{\nabla} f$ which is defined as follows.

Definition 10.2.3 The Gradient

The gradient of the function $z = f(\boldsymbol{x})$ at the point $\boldsymbol{x_0}$ is defined to be the vector $\boldsymbol{\nabla} f$ where

$$\boldsymbol{\nabla} f(\boldsymbol{x_0}) = \begin{bmatrix} f_1(\boldsymbol{x_0}) \\ f_2(\boldsymbol{x_0}) \\ \vdots \\ f_n(\boldsymbol{x_0}) \end{bmatrix}$$

Note the gradient takes a scalar function argument and returns a vector answer.

Using the gradient, the tangent plane equation can be rewritten as

$$\begin{aligned} z &= f(\boldsymbol{x_0}) + < \boldsymbol{\nabla} f(\boldsymbol{x_0}), \boldsymbol{x} - \boldsymbol{x_0} > \\ &= f(\boldsymbol{x_0}) + (\boldsymbol{\nabla} f(\boldsymbol{x_0}))^T (\boldsymbol{x} - \boldsymbol{x_0}) \end{aligned}$$

It is also traditional to abbreviate $\boldsymbol{\nabla} f(\boldsymbol{x_0})$ by $\boldsymbol{\nabla} f^0$. Then, the tangent plane equation becomes

$$z = f(\boldsymbol{x_0}) + (\boldsymbol{\nabla} f^0)^T (\boldsymbol{x} - \boldsymbol{x_0})$$

The obvious question to ask now is how much of a discrepancy is there between the value $f(\boldsymbol{x})$ and the value of the tangent plane?

Example 10.2.1 *Find the gradient of $w = f(x, y, z) = x^2 + 4xyz + 9y^2 + 20z^4$ and the equation of the tangent plane to this surface at the point $(1, 2, 1)$.*

Solution

$$\boldsymbol{\nabla} f(x, y) = \begin{bmatrix} 2x + 4yz \\ 4xz + 18y \\ 4xy + 80z^3 \end{bmatrix} \Longrightarrow \boldsymbol{\nabla} f(1, 2, 1) = \begin{bmatrix} 2 + 8 = 10 \\ 4 + 36 = 40 \\ 4 + 80 = 84 \end{bmatrix}$$

The equation of the tangent plane at $(1, 2, 1)$ is then

$$\begin{aligned} z &= f(1, 2, 1) + \left\langle \begin{bmatrix} 10 \\ 40 \\ 84 \end{bmatrix}, \begin{bmatrix} x - 1 \\ y - 2 \\ z - 1 \end{bmatrix} \right\rangle \\ &= 65 + 10(x - 1) + 40(y - 2) + 84(z - 1) \end{aligned}$$

Homework

Exercise 10.2.6 *Find the tangent plane to $f(x_1, x_2, x_3, x_4) = 2x_1^2 + 3x_2^3 + 4x - 3^5 + 9x_1 x - 2x_3 x_4$ at $(1, 2, -1, 2)$.*

Exercise 10.2.7 *Find the tangent plane to $f(x_1, x_2, x_3, x_4, x_5) = 2x_1^2 + 3x_2^3 + 4x_3^5 + 9x_1 x_2 x_3 x_4 - 3x_1^2 x_5^4$ at $(1, 2, -1, 2, 1)$.*

Exercise 10.2.8 *Let $f(x, y) = 2x^2y^2$ and $g(x, y) = 4x^2 - 2xy_8y^2$. Find the tangent plane to f and g at $(0, 0)$.*

Exercise 10.2.9 *Let $f(x, y) = \frac{2+x+x^2}{1+x^2+4y^2}$. Find the gradient of f at $(-1, 1)$ and the tangent plane at $(2, 3)$.*

10.3 Derivatives for Scalar Functions of n Variables

Let $f : D \subset \Re^n \to \Re$ be a function of n variables. As long as $f(\boldsymbol{x})$ is defined locally at $\boldsymbol{x_0}$, the partials $f_1(\boldsymbol{x_0})$ through $f_n(\boldsymbol{x_0})$ exist if and only if there are error functions $\mathscr{E}_1$ through $\mathscr{E}_n$ so that

$$\begin{aligned} f_1(\boldsymbol{x_0} + h_1 \boldsymbol{E_1}) &= f(\boldsymbol{x_0}) + f_1(\boldsymbol{x_0})(h_1(x_1 - x_{01})) + \mathscr{E}_1(\boldsymbol{x_0}, h_1) \\ f_2(\boldsymbol{x_0} + h_2 \boldsymbol{E_2}) &= f(\boldsymbol{x_0}) + f_2(\boldsymbol{x_0})(h_2(x_1 - x_{02})) + \mathscr{E}_2(\boldsymbol{x_0}, h_2) \\ \vdots &= \vdots \\ f_n(\boldsymbol{x_0} + h_n \boldsymbol{E_n}) &= f(\boldsymbol{x_0}) + f_n(\boldsymbol{x_0})(h_n(x_n - x_{0n})) + \mathscr{E}_n(\boldsymbol{x_0}, h_n) \end{aligned}$$

with $\mathscr{E}_j \to 0$ and $\mathscr{E}_j / h_j \to 0$ as $h_j \to 0$ for all indices j. This suggests the correct definition for the differentiability of a function of n variables.

Definition 10.3.1 Error Form of Differentiability for Scalar Functions of n Variables

Let $f : D \subset \Re^n \rightarrow \Re$ be a function of n variables. If $f(\boldsymbol{x})$ is defined locally at $\boldsymbol{x_0}$, then f is differentiable at $\boldsymbol{x_0}$ if there is a vector $\boldsymbol{L}$ in $\Re^n$ so that the error function $\mathscr{E}(\boldsymbol{x_0}, \boldsymbol{x}) = f(\boldsymbol{x}) - f(\boldsymbol{x_0}) - < \boldsymbol{L}, \boldsymbol{x} - \boldsymbol{x_0} >$ satisfies $\lim_{\boldsymbol{x} \rightarrow \boldsymbol{x_0}} \mathscr{E}(\boldsymbol{x_0}, \boldsymbol{x}) = 0$ and $\lim_{\boldsymbol{x} \rightarrow \boldsymbol{x_0}} \mathscr{E}(\boldsymbol{x_0}, \boldsymbol{x}) / \|\boldsymbol{x} - \boldsymbol{x_0}\| = 0$. We define the derivative of f at $\boldsymbol{x_0}$ to be the linear map $\boldsymbol{L}$ whose value at $\boldsymbol{x}$ is $\boldsymbol{L}^T(\boldsymbol{x} - \boldsymbol{x_0})$. We denote this map by $\boldsymbol{D}f(\boldsymbol{x_0}) = \boldsymbol{L}^T$.

Note if f is differentiable at $\boldsymbol{x_0}$, f must be continuous at $\boldsymbol{x_0}$. The argument is simple:

$$f(\boldsymbol{x}) = f(\boldsymbol{x_0}) + < \boldsymbol{L}, \boldsymbol{x} - \boldsymbol{x_0} > + \mathscr{E}(\boldsymbol{x_0}, \boldsymbol{x})$$

and as $\boldsymbol{x} \rightarrow \boldsymbol{x_0}$, we have $f(\boldsymbol{x}) \rightarrow f(\boldsymbol{x_0})$, which is the definition of f being continuous at $(\boldsymbol{x_0})$. Hence, we can say

Theorem 10.3.1 Differentiable Implies Continuous: Scalar Function of n Variables

If f is differentiable at $\boldsymbol{x_0}$ then f is continuous at $\boldsymbol{x_0}$.

Proof 10.3.1
We have sketched the argument already. ■

Comment 10.3.1 *At each point $\boldsymbol{x}$ where $\boldsymbol{D}f(\boldsymbol{x})$ exists, we get a vector $\boldsymbol{L}(\boldsymbol{x})$. Hence, the derivative of a function of n variables defines a mapping from $\Re^n$ to $\Re^n$ by*

$$Df(x) = \boldsymbol{L}^T(\boldsymbol{x})$$

Comment 10.3.2 *If f is differentiable at $\boldsymbol{x_0}$, then there is a vector $\boldsymbol{L}$ so that*

$$f(\boldsymbol{x}) = f(\boldsymbol{x_0} + \boldsymbol{L}^T(\boldsymbol{x} - \boldsymbol{x_0}) + \mathscr{E})(\boldsymbol{x_0}, \boldsymbol{x})$$

If we have chosen a basis for $\Re^n$, then we have

$$f(\boldsymbol{x}) = f(\boldsymbol{x_0} + [\boldsymbol{L}]_{\boldsymbol{E}}^T([\boldsymbol{x} - \boldsymbol{x_0}]_{\boldsymbol{E}}) + \mathscr{E})([\boldsymbol{x_0}]_{\boldsymbol{E}}, [\boldsymbol{x}]_{\boldsymbol{E}})$$

From this definition, we can show if the scalar function f is differentiable at the point $\boldsymbol{x_0}$, then $L_i = f_i(\boldsymbol{x_0})$ for all indices i. The argument goes like this: since f is differentiable at $\boldsymbol{x_0}$, we can focus on what happens as we move from $\boldsymbol{x_0}$ to $\boldsymbol{x_0} + h_i\boldsymbol{E_i}$. Hence, all the components of $\boldsymbol{x_0} + h_i\boldsymbol{E_i}$ are the same as the ones in $\boldsymbol{x_0}$. From the definition of differentiability we then have

$$\lim_{h_i \rightarrow 0} \frac{f(\boldsymbol{x} + h_i\boldsymbol{E_i}) - f(\boldsymbol{x_0}) - L_i\, h_i}{|h_i|} = 0.$$

as $\|\boldsymbol{x_0} + h_i\boldsymbol{E_i} - \boldsymbol{x_0}\| = |h_i|$. For $h_i > 0$, we find

$$\lim_{h_i \rightarrow 0+} \frac{f(\boldsymbol{x} + h_i\boldsymbol{E_i}) - f(\boldsymbol{x_0}) - L_i\, h_i}{h_i} = 0.$$

Hence $(f_i(\boldsymbol{x_0}))^+ = L_i$. Similarly, if $h_i < 0$, we have

$$\lim_{h_i \rightarrow 0-} \frac{f(\boldsymbol{x} + h_i\boldsymbol{E_i}) - f(\boldsymbol{x_0}) - L_i\, h_i}{-h_i} = 0.$$

Hence $(f_i(\boldsymbol{x_0}))^- = L_i$ as well. Combining, we see $f_i(\boldsymbol{x_0}) = L_i$. The arguments for the other indices are much the same. Hence, we can say if the scalar function of n variables f is differentiable at $\boldsymbol{x_0}$ then $f_i(\boldsymbol{x_0})$ exist and

$$f(\boldsymbol{x}) \quad = \quad f(\boldsymbol{x_0}) + \sum_{i=1}^{n} f_i(\boldsymbol{x_0})(x_i - x_{0i}) + \mathscr{E}_f(\boldsymbol{x_0}, \boldsymbol{x})$$

where $\mathscr{E}_f(\boldsymbol{x_0}, \boldsymbol{x}) \to 0$ and $\mathscr{E}_f(\boldsymbol{x_0}, \boldsymbol{x})/||\boldsymbol{x} - \boldsymbol{x_0}|| \to 0$ as $\boldsymbol{x} \to \boldsymbol{x_0}$. Note this argument is a pointwise argument. **It only tells us that differentiability at a point implies the existence of the partial derivatives at that point**.

Theorem 10.3.2 The scalar function f is Differentiable implies $\boldsymbol{D}f = (\boldsymbol{\nabla}(f))^T$

If f is differentiable at $\boldsymbol{x_0}$ then $\boldsymbol{DF}(\boldsymbol{x_0}) = (\boldsymbol{\nabla}(f))^T(\boldsymbol{x_0})$.

Proof 10.3.2
We have just proven this. ■

Next, we look at the version of the chain rule for a scalar function of n variables.

10.3.1 The Chain Rule for Scalar Functions of n Variables

Let's review how you prove the chain rule. We assume there are two functions $u(x, y)$ and $v(x, y)$ defined locally about (x_0, y_0) and that there is a third function $f(u, v)$ which is defined locally around $(u_0 = u(x_0, y_0), v_0 = v(x_0, y_0))$. Now assume $f(u, v)$ is differentiable at (u_0, v_0) and $u(x, y)$ and $v(x, y)$ are differentiable at (x_0, y_0). Then we can say

$$\begin{aligned}
u(x, y) &= u(x_0, y_0) + u_x(x_0, y_0)(x - x_0) + u_y(x_0, y_0)(y - y_0) + \mathscr{E}_u(x_0, y_0, x, y) \\
v(x, y) &= v(x_0, y_0) + v_x(x_0, y_0)(x - x_0) + v_y(x_0, y_0)(y - y_0) + \mathscr{E}_v(x_0, y_0, x, y) \\
f(u, v) &= f(u_0, v_0) + f_u(u_0, v_0)(u - u_0) + f_v(u_0, v_0)(v - v_0) + \mathscr{E}_f(u_0, v_0, u, v)
\end{aligned}$$

where all the error terms behave as usual as $(x, y) \to (x_0, y_0)$ and $(u, v) \to (u_0, v_0)$. Note that as $(x, y) \to (x_0, y_0)$, $u(x, y) \to u_0 = u(x_0, y_0)$ and $v(x, y) \to v_0 = v(x_0, y_0)$ as u and v are continuous at the (u_0, v_0) since they are differentiable there. Let's consider the partial of f with respect to x. Let $\Delta u = u(x_0 + \Delta x, y_0) - u(x_0, y_0)$ and $\Delta v = v(x_0 + \Delta x, y_0) - v(x_0, y_0)$. Thus, $u_0 + \Delta u = u(x_0 + \Delta x, y_0)$ and $v_0 + \Delta v = v(x_0 + \Delta x, y_0)$. Hence,

$$\begin{aligned}
&\frac{f(u_0 + \Delta u, v_0 + \Delta v) - f(u_0, v_0)}{\Delta x} \\
&= \frac{f_u(u_0, v_0)(u - u_0) \ + \ f_v(u_0, v_0)(v - v_0) + \mathscr{E}_f(u_0, v_0, u, v)}{\Delta x} \\
&= f_u(u_0, v_0)\ \frac{u - u_0}{\Delta x} \ + \ f_v(u_0, v_0)\ \frac{v - v_0}{\Delta x} \ + \ \frac{\mathscr{E}_f(u_0, v_0, u, v)}{\Delta x}
\end{aligned}$$

Continuing

$$\begin{aligned}
\frac{f(u_0 + \Delta u, v_0 + \Delta v) - f(u_0, v_0)}{\Delta x} &= f_u(u_0, v_0)\ \frac{u_x(x_0, y_0)(x - x_0) + \mathscr{E}_u(x_0, y_0, x, y)}{\Delta x} \\
&+ f_v(u_0, v_0)\ \frac{v_x(x_0, y_0)(x - x_0) + \mathscr{E}_v(x_0.y_0, x, y)}{\Delta x} \ + \ \frac{\mathscr{E}_f(u_0, v_0, u, v)}{\Delta x} \\
&= f_u(u_0, v_0)\ u_x(x_0, y_0) \ + \ f_v(u_0, v_0)\ v_x(x_0, y_0) \\
&+ f_u(u_0, v_0)\ \frac{\mathscr{E}_u(x_0, y_0, x, y)}{\Delta x} \ + \ f_v(u_0, v_0)\ \frac{\mathscr{E}_v(x_0, y_0, x, y)}{\Delta x} \ + \ \frac{\mathscr{E}_f(u_0, v_0, u, v)}{\Delta x}.
\end{aligned}$$

If f was locally constant, then

$$\mathscr{E}_f(u_0, v_0, u, v) = f(u,v) - f(u_0, v_0) - f_u(u_0, v_0)(u - u_0) f_v((u_0, v_0)v - v_0) = 0$$

We know

$$\lim_{(a,b)\to(u_0,v_0)} \frac{\mathscr{E}_f(u_0, v_0, u, v)}{\|(a,b) - (u_0, v_0)\|} = 0$$

and this is true no matter what sequence (a_n, b_n) we choose that converges to $(u(x_0, y_0), v(u_0, y_0))$. If f is not locally constant, a little thought shows locally about $(u(x_0, y_0), v(u_0, y_0))$ there are sequences $(u_n, v_n) = (u(x_n, y_n), v(x_n, y_n)) \to (u_0, v_0) = (u(x_0, y_0), v(u_0, y_0))$ with $(u_n, v_n) \neq (u_0, v_0)$ for all n. Also, by continuity, as $\Delta x \to 0$, $(u_n, v_n) \to (u_0, v_0)$ too. For (u_n, v_n) from this sequence, we then have

$$\lim_{\Delta x \to 0} \frac{\mathscr{E}_f(u_0, v_0, u, v)}{\Delta x} = \left(\lim_{n\to\infty} \frac{\mathscr{E}_f(u_0, v_0, u, v)}{\|(u_n, v_n) - (u_0, v_0)\|}\right)\left(\lim_{\Delta x\to 0} \frac{\|(u_n, v_n) - (u_0, v_0)\|}{\Delta x}\right)$$

We know the first term goes to zero by the properties of $\mathscr{E}_f$. Expand the second term to get

$$\begin{aligned} &\lim_{\Delta x\to 0} \frac{\|(u_n, v_n) - (u_0, v_0)\|}{\Delta x} = \\ &\lim_{\Delta x\to 0} \frac{\|(u_x(x_0, y_0)\Delta x + \mathscr{E}_u(x_0, y_0, x, y), v_x(x_0, y_0)\Delta x + \mathscr{E}_v(x_0, y_0, x, y)\|}{\Delta x} \leq \\ &|u_x(x_0, y_0)| + |v_x(x_0, y_0)| \end{aligned}$$

Hence, the product limit is zero. The other error terms go to zero also as $(x, y) \to (x_0, y_0)$. Hence, we conclude

$$\frac{\partial f}{\partial x} = \frac{\partial f}{\partial u}\frac{\partial u}{\partial x} + \frac{\partial f}{\partial v}\frac{\partial v}{\partial x}.$$

A similar argument shows

$$\frac{\partial f}{\partial y} = \frac{\partial f}{\partial u}\frac{\partial u}{\partial y} + \frac{\partial f}{\partial v}\frac{\partial v}{\partial y}.$$

This result is known as the **Chain Rule** and we will find there is a better way to package this. Now assume there are more variables. We assume there are three functions $u_1(x_1, x_2, x_3)$, $u_2(x_1, x_2, x_3)$ and $u_3(x_1, x_2, x_3)$ defined locally about (x_{10}, x_{20}, x_{30}) and that there is another function $f(u_1, u_2, u_3)$, which are defined locally around $u_{i0} = u_i(x_{10}, x_{20}, x_{30})$. Now assume $f(u_1, u_2, u_3)$ is differentiable at (u_{10}, u_{20}, u_{30}) and $u_i(x_1, x_2, x_3)$ is differentiable at (x_{10}, x_{20}, x_{30}). Then we can say

$$\begin{aligned} u_i(x_1, x_2, x_3) &= u_i(x_{10}, x_{20}, x_{30}) + u_{i,x_1}(x_{10}, x_{20}, x_{30})(x_1 - x_{10}) \\ &\quad + u_{i.x_2}(x_{10}, x_{20}, x_{30})(x_2 - x_{20}) + u_{i,x_3}(x_{10}, x_{20}, x_{30})(x_3 - x_{30}) \\ &\quad + \mathscr{E}_{u_i}(x_{10}, x_{20}, x_{30}, x_1, x_2, x_3) \end{aligned}$$

where all the error terms behave as usual as $(x_1, x_2, x_3) \to (x_{10}, x_{20}, x_{30})$ and $(u_1, u_2, u_3) \to (u_{10}, u_{20}, u_{30})$. This is really messy to write down! To save space, let $\boldsymbol{x} = (x_1, x_2, x_3)$, $\boldsymbol{u} = (u_1, u_2, u_3)$ etc. Then we can rewrite the above as

$$\begin{aligned} u_i(\boldsymbol{x}) &= u_i(\boldsymbol{x}_0) + u_{i,x_1}(\boldsymbol{x}_0)(x_1 - x_{10}) + u_{i.x_2}(\boldsymbol{x}_0)(x_2 - x_{20}) + u_{i,x_3}(\boldsymbol{x}_0)(x_3 - x_{30}) \\ &\quad + \mathscr{E}_{u_i}(\boldsymbol{x}_0.\boldsymbol{x}) \end{aligned}$$

Again, note as $\boldsymbol{x} \rightarrow \boldsymbol{x}_0$, $u_i(\boldsymbol{x}) \rightarrow u_{i0} = u_i(\boldsymbol{x}_0)$ as u_i is continuous at $\boldsymbol{x}_0$ as it is differentiable there. Let's consider the partial of f with respect to x_j. Let $\Delta u_i = u_j(\boldsymbol{x}_0 + h_j \boldsymbol{E}_j) - u_j(\boldsymbol{x}_0)$. Thus, $u_{i0} + \Delta u_i = u_i(\boldsymbol{x}_0 + h_j \boldsymbol{E}_j)$. So we are using h_j to denote the changes in x_j which induce changes in u_i. We see

$$\frac{f(u_{10} + \Delta u_1, u_{20} + \Delta u_2, u_{30} + \Delta u_3) - f(\boldsymbol{u}_0)}{h_j} =$$
$$\frac{f_{u_1}(\boldsymbol{u}_0)(u_1 - u_{10}) + f_{u_2}(\boldsymbol{u}_0)(u_1 - u_{10}) + f_{u_3}(\boldsymbol{u}_0)(u_1 - u_{10})}{h_j} + \frac{\mathscr{E}_f(\boldsymbol{u}_0, \boldsymbol{u})}{h_j} =$$
$$\sum_{i=1}^{3} f_{u_i}(\boldsymbol{u}_0)\, \frac{(u_i - u_{i0})}{h_j} + \frac{\mathscr{E}_f(\boldsymbol{u}_0, \boldsymbol{u})}{h_j}$$

Now replace $u_i - u_{i0}$ by its expansion:

$$\begin{aligned} u_i - u_{i0} &= \sum_{k=1}^{3} u_{i,x_k}(\boldsymbol{x}_0)(x_k - x_{k0}) + \mathscr{E}_{u_i}(\boldsymbol{x}_0.\boldsymbol{x}) = u_{i,x_j}(\boldsymbol{x}_0)(x_j - x_{j0}) + \mathscr{E}_{u_i}(\boldsymbol{x}_0.\boldsymbol{x}) \\ &= u_{i,x_j}(\boldsymbol{x}_0)\, h_j + \mathscr{E}_{u_i}(\boldsymbol{x}_0.\boldsymbol{x}) \end{aligned}$$

This gives

$$\frac{f(u_{10} + \Delta u_1, u_{20} + \Delta u_2, u_{30} + \Delta u_3) - f(\boldsymbol{u}_0)}{h_j} =$$
$$\sum_{i=1}^{3} f_{u_i}(\boldsymbol{u}_0)\, u_{i,x_i}(\boldsymbol{x}_0) + \sum_{i=1}^{3} f_{u_i}(\boldsymbol{u}_0)\, \frac{\mathscr{E}_{u_i}(\boldsymbol{x}_0.\boldsymbol{x})}{h_j} + \frac{\mathscr{E}_f(\boldsymbol{u}_0, \boldsymbol{u})}{h_j}$$

As $h_j \rightarrow 0$, $\Delta \boldsymbol{u}_i \rightarrow 0$ and using the arguments we used in the two variable case, it is straightforward to show $\mathscr{E}_f(\boldsymbol{u}_0, \boldsymbol{u})/h_j \rightarrow 0$. The other two error terms go to zero also as $(x, y) \rightarrow (x_0, y_0)$. Hence, we conclude

$$\frac{\partial f}{\partial x_j} = \sum_{i=1}^{3} \frac{\partial f}{\partial u_i} \frac{\partial u_i}{\partial x_j}$$

It is straightforward to see how to extend this result to n variables. Just a bit messy! The notation here is quite messy and confusing. Using the $\frac{\partial}{\partial(\text{arg}:k)}$ doesn't work well as we have arg terms for both the x_j and the u_i variables. Also, in our argument above we had $\boldsymbol{u} : D_u \subset \Re^3 \rightarrow \Re^3$ with u defined locally about $\boldsymbol{x}_0$. Hence, $\boldsymbol{u}$ is a function of $(x_1, \ldots, x_3)$ with $\boldsymbol{u}(x_1, \ldots, x_3) = (u_1, \ldots, u_3) \in \Re^3$. Further, we assumed there is another function $f : D_f \subset \Re^3 \rightarrow \Re$ which is defined locally around $\boldsymbol{u}_0 = u(\boldsymbol{x}_0) \in \Re^3$. And we assume $f(\boldsymbol{u})$ is differentiable at $\boldsymbol{u}_0$ and $\boldsymbol{u}(x)$ is differentiable at $\boldsymbol{x}_0$. However, we could be more general: we could assume $\boldsymbol{u} : D_u \subset \Re^n \rightarrow \Re^m$ with u defined locally about $\boldsymbol{x}_0$. Hence, $\boldsymbol{u}$ is a function of $(x_1, \ldots, x_n)$ with $\boldsymbol{u}(x_1, \ldots, x_n) = (u_1, \ldots, u_m) \in \Re^m$. Further, assume there is another function $f : D_f \subset \Re^m \rightarrow \Re$ which is defined locally around $\boldsymbol{u}_0 = u(\boldsymbol{x}_0) \in \Re^n$. Further assume $f(\boldsymbol{u})$ is differentiable at $\boldsymbol{u}_0$ and $\boldsymbol{u}(x)$ is differentiable at $\boldsymbol{x}_0$. The arguments are quite similar and we find

$$\frac{\partial f}{\partial x_j} = \sum_{i=1}^{m} \frac{\partial f}{\partial u_i} \frac{\partial u_i}{\partial x_j}$$

This leads to our theorem.

Theorem 10.3.3 The Chain Rule for Scalar Functions of n Variables

Assume $\boldsymbol{u} : D_u \subset \Re^n \rightarrow \Re^m$ with u is defined locally about $\boldsymbol{x_0}$. Hence, $\boldsymbol{u}$ is a function of $(x_1, \ldots, x_n)$ with $\boldsymbol{u}(x_1, \ldots, x_n) = (u_1, \ldots, u_m) \in \Re^m$. Further, assume there is another function $f : D_f \subset \Re^m \rightarrow \Re$ which is defined locally around $\boldsymbol{u_0} = u(\boldsymbol{x_0}) \in \Re^n$. Further assume $f(\boldsymbol{u})$ is differentiable at $\boldsymbol{u_0}$ and $\boldsymbol{u}(\boldsymbol{x})$ is differentiable at $\boldsymbol{x_0}$. Then f_i exists at $\boldsymbol{x_0}$ and is given by

$$\frac{\partial f}{\partial x_j} = \sum_{i=1}^{m} \frac{\partial f}{\partial u_i} \frac{\partial u_i}{\partial x_j}$$

Proof 10.3.3
We have just gone over the argument. ■

Comment 10.3.3 *With a little bit of work, these arguments also prove the differentiability of the composition. In the two variable case, let $g(x, y) = f(u(x, y), v(x, y))$, i.e. $g = f \circ U$ where U is the vector function with components u and v, and define the error*

$$\mathscr{E}_g = f_u(u_0, v_0)\, \mathscr{E}_u(x_0, y_0, x, y) + f_v(u_0, v_0)\, \mathscr{E}_v(x_0, y_0, x, y) + \mathscr{E}_f(u_0, v_0, u, v).$$

It is clear from the assumptions that $\mathscr{E}_g(x_0, y_0, x, y) \rightarrow 0$ as $\Delta x \rightarrow 0$ and $\mathscr{E}_g(x_0, y_0, x, y) \rightarrow 0$ as $\Delta y \rightarrow 0$. The arguments above show $\frac{\mathscr{E}_g(x_0, y_0, x, y)}{\Delta x} \rightarrow 0$ as $\Delta x \rightarrow 0$ and $\frac{\mathscr{E}_g(x_0, y_0, x, y)}{\Delta y} \rightarrow 0$ as $\Delta y \rightarrow 0$. This shows $\mathscr{E}_g$ is the error function for the composition $g = f \circ U$. Hence $g = f \circ U$ is differentiable at (x_0, y_0) with

$$\begin{aligned}
\boldsymbol{Dg}(x_0, y_0) &= \begin{bmatrix} f_u(u_0, v_0) \\ f_v(u_0, v_0) \end{bmatrix}^T \begin{bmatrix} u_x(x_0, y_0) & v_x(x_0, y_0) \\ u_y(x_0, y_0) & v_y(x_0, y_0) \end{bmatrix} \\
&= \boldsymbol{Df}(u_0, v_0) \begin{bmatrix} u_x(x_0, y_0) & v_x(x_0, y_0) \\ u_y(x_0, y_0) & v_y(x_0, y_0) \end{bmatrix}
\end{aligned}$$

We will discuss this further.

Example 10.3.1 *Let $f(x, y, z) = x^2 + 2x + 5y^4 + z^2$. Then if $x = r\sin(\phi)\cos(\theta)$, $y = r\sin(\phi)\sin(\theta)$ and $z = r\cos(\phi)$. $y = r\sin(\theta)$, using the chain rule, we find*

$$\begin{aligned}
\frac{\partial f}{\partial r} &= \frac{\partial f}{\partial x}\frac{\partial x}{\partial r} + \frac{\partial f}{\partial y}\frac{\partial y}{\partial r} + \frac{\partial f}{\partial z}\frac{\partial z}{\partial r} \\
\frac{\partial f}{\partial \theta} &= \frac{\partial f}{\partial x}\frac{\partial x}{\partial \theta} + \frac{\partial f}{\partial y}\frac{\partial y}{\partial \theta} + \frac{\partial f}{\partial z}\frac{\partial z}{\partial \theta} \\
\frac{\partial f}{\partial \phi} &= \frac{\partial f}{\partial x}\frac{\partial x}{\partial \phi} + \frac{\partial f}{\partial y}\frac{\partial y}{\partial \phi} + \frac{\partial f}{\partial z}\frac{\partial z}{\partial \phi}
\end{aligned}$$

This becomes

$$\begin{aligned}
\frac{\partial f}{\partial r} &= \frac{\partial f}{\partial x}(\sin(\phi)\cos(\theta)) + \frac{\partial f}{\partial y}(-\sin(\phi)\sin(\theta)) + \frac{\partial f}{\partial z}(\cos(\phi)) \\
\frac{\partial f}{\partial \theta} &= \frac{\partial f}{\partial x}(-r\sin(\phi)\sin(\theta)) + \frac{\partial f}{\partial y}\frac{\partial y}{(}r\sin(\phi)\cos(\theta)) + \frac{\partial f}{\partial z}(0) \\
\frac{\partial f}{\partial \phi} &= \frac{\partial f}{\partial x}(r\cos(\phi)\cos(\theta)) + \frac{\partial f}{\partial y}(r\cos(\phi)\sin(\theta)) + \frac{\partial f}{\partial z}(-r\sin(\phi))
\end{aligned}$$

You can then substitute in for x and y to get the final answer in terms of r, θ and ϕ.

Example 10.3.2 *Let $u = x^2 + 2y^2 + z^2$ and $v = 4x^2 - 5y^2 + 7z$ and $f(u, v) = 10u^2v^4$. Then, by the chain rule*

$$\begin{aligned}\frac{\partial f}{\partial x} &= \frac{\partial f}{\partial u}\frac{\partial u}{\partial x} + \frac{\partial f}{\partial v}\frac{\partial v}{\partial x}\\ \frac{\partial f}{\partial y} &= \frac{\partial f}{\partial u}\frac{\partial u}{\partial y} + \frac{\partial f}{\partial v}\frac{\partial v}{\partial y}\\ \frac{\partial f}{\partial z} &= \frac{\partial f}{\partial u}\frac{\partial u}{\partial z} + \frac{\partial f}{\partial v}\frac{\partial v}{\partial z}\end{aligned}$$

This becomes

$$\begin{aligned}\frac{\partial f}{\partial x} &= \frac{\partial f}{\partial u}(2x) + \frac{\partial f}{\partial v}(8x)\\ \frac{\partial f}{\partial y} &= \frac{\partial f}{\partial u}(4y) + \frac{\partial f}{\partial v}(-10y)\\ \frac{\partial f}{\partial z} &= \frac{\partial f}{\partial u}(2z) + \frac{\partial f}{\partial v}(7)\end{aligned}$$

You can then substitute in for u and v to get the final answer in terms of x, y and z.

10.3.1.1 Homework

Exercise 10.3.1 *Let $x = \rho\sin(\phi)\cos(\phi)$, $y = \rho\sin(\phi)\sin(\phi)$ and $z = \rho\cos(\phi)$ be the usual equations which convert spherical coordinates to Cartesian coordinates. Given a function $f(x, y, z)$, express f_ρ, f_θ and f_ϕ in terms of the components f_x, f_y and f_z.*

- *Show this can be written as*

$$\begin{bmatrix} f_\rho \\ f_\theta \\ f_\phi \end{bmatrix} = \boldsymbol{A} \begin{bmatrix} f_x \\ f_y \\ f_z \end{bmatrix}$$

- *Find where $\boldsymbol{A}$ is invertible.*

- *Compute the inverse of $\boldsymbol{A}$. The easiest thing to do is to use old fashioned Gaussian Elimination by hand on the augmented matrix $[\boldsymbol{A}|\boldsymbol{I}]$. In your calculations above, you should find*

$$\begin{bmatrix} f_\rho \\ \frac{1}{\rho} f_\theta \\ \frac{1}{\rho} f_\phi \end{bmatrix} = \begin{bmatrix} \sin(\phi)\cos(\theta) & \sin(\phi)\sin(\theta) & \cos(\phi) \\ -\sin(\phi)\sin(\theta) & \sin(\phi)\cos(\theta) & 0 \\ \cos(\phi)\cos(\theta) & \cos(\phi)\sin(\theta) & -\sin(\phi) \end{bmatrix} \begin{bmatrix} f_x \\ f_y \\ f_z \end{bmatrix} = \hat{\boldsymbol{A}} \begin{bmatrix} f_x \\ f_y \\ f_z \end{bmatrix}$$

and after the laborious Gaussian Elimination you should find the inverse is

$$(\hat{\boldsymbol{A}})^{-1} = \begin{bmatrix} \sin(\phi)\cos(\theta) & -\frac{\sin(\theta)}{\sin(\phi)} & \cos(\theta)\cos(\phi) \\ \sin(\phi)\sin(\theta) & \frac{\cos(\theta)}{\sin(\phi)} & \sin(\theta)\cos(\phi) \\ \cos(\phi) & 0 & -\sin(\phi) \end{bmatrix} \Longrightarrow \begin{bmatrix} f_x \\ f_y \\ f_z \end{bmatrix} = (\hat{\boldsymbol{A}})^{-1} \begin{bmatrix} f_\rho \\ \frac{1}{\rho} f_\theta \\ \frac{1}{\rho} f_\phi \end{bmatrix}$$

Exercise 10.3.2 *Let $\rho = \sqrt{x^2 + y^2 + z^2}$, $\theta = \tan^{-1}(\frac{y}{x})$ and $\phi = \cos^{-1}\left(\frac{z}{\sqrt{x^2+y^2+z^2}}\right)$ be the usual equations which convert Cartesian coordinates into spherical coordinates. Given a function $f(\rho, \theta, \phi)$, express f_x, f_y and f_z in terms of the components f_ρ, f_θ and f_ϕ.*

- *Show this can be written as*

$$\begin{bmatrix} f_x \\ f_y \\ f_z \end{bmatrix} = \boldsymbol{B} \begin{bmatrix} f_\rho \\ f_\theta \\ f_\phi \end{bmatrix}$$

- *Find where $\boldsymbol{B}$ is invertible.*

Note this is another way to do the previous exercise!

Exercise 10.3.3 *Do the same two exercises above for cylindrical coordinate transformations.*

10.4 Partials and Differentiability

First, let's talk about some issues with smoothness for partials. We cover the following examples in (Peterson (21) 2020) also, but it is worth seeing them again. These examples are intense but you learn a lot by working through them.

10.4.1 Partials Can Exist but Not be Continuous

We start with the function f defined by

$$f(x,y) = \begin{cases} \frac{2xy}{\sqrt{x^2+y^2}}, & (x,y) \neq (0,0) \\ 0, & (x,y) = (0,0) \end{cases}$$

This function has a removeable discontinuity at $(0,0)$ because

$$\begin{aligned} \left| \frac{2xy}{\sqrt{x^2+y^2}} - 0 \right| &= 2\,\frac{|x|\;|y|}{\sqrt{x^2+y^2}} \leq 2\frac{\sqrt{x^2+y^2}\,\sqrt{x^2+y^2}}{\sqrt{x^2+y^2}} \\ &= 2\sqrt{x^2+y^2} \end{aligned}$$

because $|x| \leq \sqrt{x^2+y^2}$ and $|y| \leq \sqrt{x^2+y^2}$. Pick $\epsilon > 0$ arbitrarily. Then if $\delta < \epsilon/2$, we have

$$\left| \frac{2xy}{\sqrt{x^2+y^2}} - 0 \right| < 2\,\frac{\epsilon}{2} = \epsilon$$

which proves that the limit exists and equals 0 as $(x,y) \to (0,0)$. This tells us we can define $f(0,0)$ to match this value. Thus, as said, this is a *removeable discontinuity*.
The first order partials both exist at $(0,0)$ as is seen by an easy limit.

$$\begin{aligned} f_x(0,0) &= \lim_{x\to 0} \frac{f(x,0)-f(0,0)}{x} = \lim_{x\to 0} \frac{0-0}{x} = 0. \\ f_y(0,0) &= \lim_{y\to 0} \frac{f(0,y)-f(0,0)}{y} = \lim_{y\to 0} \frac{0-0}{y} = 0. \end{aligned}$$

We can also calculate the first order partials at any $(x,y) \neq (0,0)$.

$$\begin{aligned} f_x(x,y) &= \frac{2y(x^2+y^2)^{1/2} - 2xy\,(1/2)(x^2+y^2)^{-1/2}\,2x}{(x^2+y^2)^1} \\ &= \frac{2y(x^2+y^2) - 2x^2y}{(x^2+y^2)^{3/2}} = \frac{2y^3}{(x^2+y^2)^{3/2}} \end{aligned}$$

$$
\begin{aligned}
f_y(x,y) &= \frac{2x(x^2+y^2)^{1/2} - 2xy\,(1/2)(x^2+y^2)^{-1/2}\,2y}{(x^2+y^2)^1} \\
&= \frac{2x(x^2+y^2) - 2xy^2}{(x^2+y^2)^{3/2}} = \frac{2x^3}{(x^2+y^2)^{3/2}}
\end{aligned}
$$

We can show $\lim\,(x,y) \to (0,0)\ f_x(x,y)$ and $\lim\,(x,y) \to (0,0)\ f_y(x,y)$ do not exist as follows. We will find paths (x, mx) for $m \neq 0$ where we get different values for the limit depending on the value of m. We pick the path $(x, 5x)$ for $x > 0$. On this path, we find

$$
\lim_{x\to 0^+} f_x(x,5x) = \lim_{x\to 0^+} \frac{250x^3}{(26)^{3/2}|x^2|^{3/2}} = \lim_{x\to 0^+} \frac{250}{26\sqrt{26}}
$$

Then, for example, using $m = -5$ for $x < 0$, we get

$$
\lim_{x\to 0^+} f_x(x,-5x) = \lim_{x\to 0^+} \frac{-250x^3}{(26)^{3/2}|x^2|^{3/2}} = \lim_{x\to 0^+} \frac{-250}{26\sqrt{26}}
$$

Since these are not the same, this limit cannot exist. Hence f_x is not continuous at the origin. A similar argument shows $\lim\,(x,y) \to (0,0)\ f_y(x,y)$ does not exist either. So we have shown the first order partials exist locally around $(0,0)$ but f_x and f_y are not continuous at $(0,0)$.

Is f differentiable at $(0,0)$? If so, there a numbers L_1 and L_2 so that two things happen:

$$
\begin{aligned}
\lim_{(x,y)\to(0,0)} \left(\frac{2xy}{\sqrt{x^2+y^2}} - L_1 x - L_2 y \right) &= 0 \\
\lim_{(x,y)\to(0,0)} \frac{1}{\sqrt{x^2+y^2}} \left(\frac{2xy}{\sqrt{x^2+y^2}} - L_1 x - L_2 y \right) &= 0
\end{aligned}
$$

The first limit is 0 because

$$
\begin{aligned}
&\lim_{(x,y)\to(0,0)} \left| \frac{2xy}{\sqrt{x^2+y^2}} - L_1 x - L_2 y - 0 \right| \leq \\
&2\frac{|x||y|}{\sqrt{x^2+y^2}} + |L_1|\sqrt{x^2+y^2} + |L_2|\sqrt{x^2+y^2} \\
&\leq 2\frac{\sqrt{x^2+y^2}\sqrt{x^2+y^2}}{\sqrt{x^2+y^2}} + |L_1|\sqrt{x^2+y^2} + |L_2|\sqrt{x^2+y^2} \\
&\leq (2 + |L_1| + |L_2|)\sqrt{x^2+y^2}
\end{aligned}
$$

which goes to 0 as $(x,y) \to 0$. So that shows the first requirement is met.
However, the second requirement is not. Look at the paths (x, mx) and consider $\lim_{x\to 0^+}$ like we did before. We find, since $|x| = x$ here

$$
\begin{aligned}
&\lim_{x\to 0^+} \frac{1}{x\sqrt{1+m^2}} \left(\frac{2mx^2}{x\sqrt{1+m^2}} - L_1 x - mL_2 x \right) = \\
&\frac{1}{\sqrt{1+m^2}} \left(\frac{2mx}{x\sqrt{1+m^2}} - L_1 - L_2 m \right) = \frac{2m}{1+m^2} - \frac{L_1 + L_2 m}{\sqrt{1+m^2}}
\end{aligned}
$$

Now look at path $(x, -mx)$ for $x > 0$. This is the limit from the other side. We have

$$
\lim_{x\to 0^+} \frac{1}{x\sqrt{1+m^2}} \left(\frac{-2mx^2}{x\sqrt{1+m^2}} - L_1 x + mL_2 x \right) =
$$

$$\frac{1}{\sqrt{1+m^2}}\left(\frac{-2m}{\sqrt{1+m^2}} - L_1 + L_2 m\right) = \frac{-2m}{1+m^2} + \frac{-L_1 + L_2 m}{\sqrt{1+m^2}}$$

If these limits were equal we would have

$$\frac{2m}{1+m^2} + \frac{-L_1 - L_2 m}{\sqrt{1+m^2}} = \frac{-2m}{1+m^2} + \frac{-L_1 + L_2 m}{\sqrt{1+m^2}}$$

This implies $L_2 = 2/\sqrt{1+m^2}$ which tells us L_2 depends on m. This value should be independent of the choice of path and so this function cannot be differentiable at $(0, 0)$.

We see that just because the first order partials exist locally in some $B_r(0, 0)$ does not necessarily tell us f is differentiable at $(0, 0)$. So **differentiable** at (x_0, y_0) implies the first order partials exist and (x_0, y_0) and $L_1 = f_x(x_0, y_)$ and $L_2 = f_y(x_0, y_)$. **But the converse is not true in general.**
Homework

Exercise 10.4.1 *Let*

$$f(x,y) = \begin{cases} \frac{4xy}{\sqrt{x^2+y^2}}, & (x,y) \neq (0,0) \\ 0, & (x,y) = (0,0) \end{cases}$$

- *Show* $\lim_{(x,y)\to(0,0)} f(x,y) = 0$ *which is why it is ok to define the function as we have done as it has a removeable discontinuity at the origin. Thus,* f *is continuous on* $\Re^2$.
- *Show the gradient of* f *exists on* $\Re^2$*, but the first order partials are not continuous at* $(0, 0)$.

Exercise 10.4.2 *Let*

$$f(x,y) = \begin{cases} \frac{5xy}{\sqrt{2x^2+3y^2}}, & (x,y) \neq (0,0) \\ 0, & (x,y) = (0,0) \end{cases}$$

- *Show* $\lim_{(x,y)\to(0,0)} f(x,y) = 0$ *which is why it is ok to define the function as we have done as it has a removeable discontinuity at the origin. Thus,* f *is continuous on* $\Re^2$.
- *Show the gradient of* f *exists on* $\Re^2$*, but the first order partials are not continuous at* $(0, 0)$.

Exercise 10.4.3 *Let*

$$f(x,y) = \begin{cases} \frac{-2xy}{\sqrt{2x^2+3y^2}}, & (x,y) \neq (0,0) \\ 0, & (x,y) = (0,0) \end{cases}$$

- *Show* $\lim_{(x,y)\to(0,0)} f(x,y) = 0$ *which is why it is ok to define the function as we have done as it has a removeable discontinuity at the origin. Thus,* f *is continuous on* $\Re^2$.
- *Show the gradient of* f *exists on* $\Re^2$*, but the first order partials are not continuous at* $(0, 0)$.

We need to figure out the appropriate assumptions about the smoothness of f that will guarantee the existence of the derivative of f in this multivariable situation. We will do this for n dimensions as it is just as easy as doing the proof in $\Re^2$.

Theorem 10.4.1 Sufficient Conditions to Guarantee Differentiability

Let $f : D \subset \Re^n \to \Re$ *and assume* $\nabla(f)$ *exists and is continuous in a ball* $B_r(\mathbf{p})$ *of some radius* $r > 0$ *around the point* $\mathbf{p} = [p_1, \ldots, p_n]'$ *in* $\Re^n$. *Then* f *is differentiable at each point in* $B_r(\mathbf{p})$.

Proof 10.4.1
By definition of $\frac{\partial f}{\partial x_i}$ *at* $\mathbf{p}$,

$$f(\boldsymbol{p} + h\boldsymbol{E}_i) - f(\boldsymbol{p}) = \frac{\partial f}{\partial x_i}(\mathbf{p})h + \mathscr{E}_i(\mathbf{p}, \boldsymbol{p} + h\boldsymbol{E}_i)$$

Now if $\boldsymbol{x}$ *is in* $B(r, \mathbf{p})$, *all the first order partials of* f *are continuous. We have*

$$\boldsymbol{x} = \mathbf{p} + (x_1 - p_1)\boldsymbol{E}_1 + (x_2 - p_2)\boldsymbol{E}_2 + \ldots + (x_n - p_n)\boldsymbol{E}_n$$

Now let $r_i = x_i - p_i$. *Then we can write (this is the telegraphing sum trick!)*

$$\begin{aligned} f(\boldsymbol{x}) - f(\boldsymbol{p}) &= f(\boldsymbol{p} + \sum_{i=1}^{n} r_i\boldsymbol{E}_i) - f(\boldsymbol{p} + \sum_{i=1}^{n-1} r_i\boldsymbol{E}_i) + f(\boldsymbol{p} + \sum_{i=1}^{n-1} r_i\boldsymbol{E}_i) - f(\boldsymbol{p} + \sum_{i=1}^{n-2} r_i\boldsymbol{E}_i) \\ &\quad + f(\boldsymbol{p} + \sum_{i=1}^{n-2} r_i\boldsymbol{E}_i) - f(\boldsymbol{p} + \sum_{i=1}^{n-3} r_i\boldsymbol{E}_i) + \ldots \\ &\quad + f(\boldsymbol{p} + \sum_{i=1}^{2} r_i\boldsymbol{E}_i) - f(\boldsymbol{p} + \sum_{i=1}^{1} r_i\boldsymbol{E}_i) + f(\boldsymbol{p} + \sum_{i=1}^{1} r_i\boldsymbol{E}_i) - f(\boldsymbol{p}) \end{aligned}$$

To help you see this better, if $n = 3$, *this expansion would look like*

$$\begin{aligned} f(\boldsymbol{x}) - f(\boldsymbol{p}) &= f(\boldsymbol{p} + \sum_{i=1}^{3} r_i\boldsymbol{E}_i) - f(\boldsymbol{p} + \sum_{i=1}^{2} r_i\boldsymbol{E}_i) + f(\boldsymbol{p} + \sum_{i=1}^{2} r_i\boldsymbol{E}_i) - f(\boldsymbol{p} + \sum_{i=1}^{1} r_i\boldsymbol{E}_i) \\ &\quad + f(\boldsymbol{p} + \sum_{i=1}^{1} r_i\boldsymbol{E}_i) - f(\boldsymbol{p}) \end{aligned}$$

or

$$\begin{aligned} f(\boldsymbol{x}) - f(\boldsymbol{p}) &= f(p_1 + r_1, p_2 + r_2, p_3 + r_3) - f(p_1 + r_1, p_2 + r_2, p_3) \ \Delta \textit{ slot 3} \\ &+ f(p_1 + r_1, p_2 + r_2, p_3) - f(p_1 + r_1, p_2, p_3) \quad \Delta \textit{ slot 2} \\ &+ f(p_1 + r_1, p_2, p_3) - f(p_1, p_2, p_3) \quad \Delta \textit{ slot 1} \end{aligned}$$

So we have

$$f(\boldsymbol{x}) - f(\boldsymbol{p}) = \sum_{j=n}^{1} \left(f\left(\mathbf{p} + \sum_{i=1}^{j} r_i\boldsymbol{E}_i\right) - f\left(\mathbf{p} + \sum_{i=1}^{j-1} r_i\boldsymbol{E}_i\right)\right)$$

where we interpret $\sum_{j=1}^{0}$ *as* 0. *To see this done without all this notation, you can look at the argument for the two variable version of this theorem in (Peterson (21) 2020). That argument is messy too though, and in some ways, this more abstract form is better!*

Now apply the Mean Value Theorem in one variable to get

$$f(\boldsymbol{p} + \sum_{i=1}^{j} r_i\boldsymbol{E}_i) - f(\boldsymbol{p} + \sum_{i=1}^{j-1} r_i\boldsymbol{E}_i) = \frac{\partial f}{\partial x_i}(\boldsymbol{q}_j)\, r_j$$

where $\boldsymbol{q}_j$ *is between* $\mathbf{p} + \sum_{i=1}^{j-1} r_i \boldsymbol{E}_i$ *and* $\mathbf{p} + \sum_{i=1}^{j} r_i \boldsymbol{E}_i$*; i.e.* $\boldsymbol{q}_j = \mathbf{p} + \sum_{i=1}^{j-1} r_i \boldsymbol{E}_i + s_j \boldsymbol{E}_j$ *with* $|s_j| < |r_j| < r$*. So*

$$f(\boldsymbol{x}) - f(\mathbf{p}) \;=\; \sum_{j=n}^{1} \frac{\partial f}{\partial x_i}(\boldsymbol{q}_j)\, r_j$$

Since $\frac{\partial f}{\partial x_j}$ *is continuous on* $B_r(\mathbf{p})$*, given* $\epsilon > 0$ *arbitrary, there are* $\delta_j > 0$ *so that for* $1 \leq j \leq n$*,* $|\frac{\partial f}{\partial x_j}(\boldsymbol{x}) - \frac{\partial f}{\partial x_j}(\mathbf{p})| < \epsilon/n$*. Thus, if* $||\boldsymbol{x} - \mathbf{p}|| < \hat{\delta} = \min\{\delta_1, \ldots, \delta_n\}$*, since the points* $\boldsymbol{q}_j$ *are between* $\mathbf{p} + \sum_{i=1}^{j-1} r_i \boldsymbol{E}_i$ *and* $\mathbf{p} + \sum_{i=1}^{j} r_i \boldsymbol{E}_i$*, we have* $\boldsymbol{q}_j$ *is in* $B_r(\mathbf{p})$*.*

So we have $|\frac{\partial f}{\partial x_j}(\boldsymbol{q}_j) - \frac{\partial f}{\partial x_j}(\mathbf{p})| < \epsilon/n$*. Thus, we find*

$$\frac{\partial f}{\partial x_j}(\mathbf{p}) - \epsilon/n \;<\; \frac{\partial f}{\partial x_j}(\boldsymbol{q}_j) < \frac{\partial f}{\partial x_j}(\mathbf{p}) + \epsilon/n$$

Now $f(\boldsymbol{x}) - f(\mathbf{p}) = \sum_{j=n}^{1} \frac{\partial f}{\partial x_j}(\boldsymbol{q}_j)\, r_j$*. Let* I *be the set of indices where* $r_j > 0$ *and* J *be the other indices. Then* $\sum_{j=n}^{1} = \sum_{j \in I} + \sum_{j \in J}$*. For indices* $j \in I$*, multiplying the inequality* $r_j > 0$ *causes no changes:*

$$\left(\sum_{j\in I} \frac{\partial f}{\partial x_j}(\mathbf{p})\, r_j\right) - (\epsilon/n) \sum_{j\in I} r_j \;<\; \sum_{j\in I} \frac{\partial f}{\partial x_j}(\boldsymbol{q}_j)\, r_j < \left(\sum_{j\in I} \frac{\partial f}{\partial x_j}(\mathbf{p})\right) + (\epsilon/n) \sum_{j\in I} r_j$$

If $j \in J$*,* $r_j < 0$ *and multiplying changes the inequalities. We get*

$$\left(\sum_{j\in J} \frac{\partial f}{\partial x_j}(\mathbf{p})\, r_j\right) + (\epsilon/n) \sum_{j\in J} r_j \;<\; \sum_{j\in J} \frac{\partial f}{\partial x_j}(\boldsymbol{q}_j)\, r_j < \left(\sum_{j\in J} \frac{\partial f}{\partial x_j}(\mathbf{p})\right) - (\epsilon/n) \sum_{j\in J} r_j$$

We can combine these two forms by noting we can replace r_j *by* $sign(r_j) r_j$ *in each inequality. This gives*

$$\left(\sum_{j=n}^{1} \frac{\partial f}{\partial x_j}(\mathbf{p}) r_j\right) - \sum_{j=n}^{1} sign(r_j)\,(\epsilon/n)\, r_j < \sum_{j=n}^{1} \frac{\partial f}{\partial x_j}(\boldsymbol{q}_j) r_j$$
$$= f(\boldsymbol{x}) - f(\mathbf{p}) < \left(\sum_{j=n}^{1} \frac{\partial f}{\partial x_j}(\mathbf{p})\right) + \sum_{j=n}^{1} sign(r_j)(\epsilon/n)\, r_j$$

Thus,

$$|f(\boldsymbol{x}) - f(\mathbf{p}) - \sum_{j=n}^{1} \frac{\partial f}{\partial x_j}(\mathbf{p})\, r_j| < (\epsilon/n) \sum_{j=n}^{1} |r_j|$$

But $(\epsilon/n) \sum_{j=n}^{1} |r_j| \leq \sum_{j=n}^{1} (\epsilon/n)||\boldsymbol{x} - \mathbf{p}|| = \epsilon||\boldsymbol{x} - \mathbf{p}||$*. We therefore have the estimate*

$$|f(\boldsymbol{x}) - f(\mathbf{p}) - \sum_{j=n}^{1} \frac{\partial f}{\partial x_j}(\mathbf{p})\, r_j| < \epsilon||\boldsymbol{x} - \mathbf{p}||$$

The term inside the absolute values is the error function $E(\boldsymbol{x}, \mathbf{p})$ *and so we have* $E(\boldsymbol{x}, \mathbf{p})/||\boldsymbol{x} - \mathbf{p}|| < \epsilon$ *if* $||\boldsymbol{x} - \mathbf{p}|| < \hat{\delta}$*. This tells us both* $E(\boldsymbol{x}, \mathbf{p})/||\boldsymbol{x} - \mathbf{p}||$ *and* $E(\boldsymbol{x}, \mathbf{p})$ *go to* 0 *as* $\boldsymbol{x} \to \mathbf{p}$*. Hence,* f *is differentiable at* $\mathbf{p}$*.* ■

Homework
These are the same exercises we just did, but now we ask you to prove the functions are not differentiable.

Exercise 10.4.4 *Let*

$$f(x,y) = \begin{cases} \frac{4xy}{\sqrt{x^2+y^2}}, & (x,y) \neq (0,0) \\ 0, & (x,y) = (0,0) \end{cases}$$

- *Show* $\lim_{(x,y)\to(0,0)} f(x,y) = 0$ *which is why it is ok to define the function as we have done as it has a removeable discontinuity at the origin. Thus, f is continuous on* $\Re^2$.

- *Show the gradient of f exists on $\Re^2$, but the first order partials are not continuous at* $(0,0)$.

- *Show f is not differentiable at the origin.*

Exercise 10.4.5 *Let*

$$f(x,y) = \begin{cases} \frac{5xy}{\sqrt{2x^2+3y^2}}, & (x,y) \neq (0,0) \\ 0, & (x,y) = (0,0) \end{cases}$$

- *Show* $\lim_{(x,y)\to(0,0)} f(x,y) = 0$ *which is why it is ok to define the function as we have done as it has a removeable discontinuity at the origin. Thus, f is continuous on* $\Re^2$.

- *Show the gradient of f exists on $\Re^2$, but the first order partials are not continuous at* $(0,0)$.

- *Show f is not differentiable at the origin.*

Exercise 10.4.6 *Let*

$$f(x,y) = \begin{cases} \frac{-2xy}{\sqrt{2x^2+3y^2}}, & (x,y) \neq (0,0) \\ 0, & (x,y) = (0,0) \end{cases}$$

- *Show* $\lim_{(x,y)\to(0,0)} f(x,y) = 0$ *which is why it is ok to define the function as we have done as it has a removeable discontinuity at the origin. Thus, f is continuous on* $\Re^2$.

- *Show the gradient of f exists on $\Re^2$, but the first order partials are not continuous at* $(0,0)$.

- *Show f is not differentiable at the origin.*

Exercise 10.4.7 *Why is* $f(x,y) = \cos(x^2 + 2y^2)$ *differentiable on* $\Re^2$*?*

Exercise 10.4.8 *Why is* $f(x,y) = e^{(x^2+2y^2)}$ *differentiable on* $\Re^2$*?*

10.4.2 Higher Order Partials

We define the second order partials of f as follows.

Definition 10.4.1 Second Order Partials

If $f(\boldsymbol{x})$, $f_{x_i}(\boldsymbol{x})$ are defined locally at $\boldsymbol{x_0}$, we can attempt to find the following limits:

$$\lim_{h \to 0} \frac{f_{x_i}(\boldsymbol{x_0} + h\boldsymbol{E_j}) - f_{x_i}(\boldsymbol{x_0})}{h}$$

If the limit exists, it is called the partial of f_{x_i} with respect to x_j and is denoted by $f_{x_i\,x_j}$. Note, the order is potentially important as

$$\lim_{h \to 0} \frac{f_{x_j}(\boldsymbol{x_0} + h\boldsymbol{E_i}) - f_{x_i}(\boldsymbol{x_0})}{h}$$

if it exists, is the partial of f_{x_j} with respect to x_i which is denoted by $f_{x_j\,x_i}$. Of course, these do not have to match in general.

Comment 10.4.1 *When these second order partials exist at x_0, we use the following notations interchangeably: $f_{ij} = \partial_{x_j}(f_{x_i})$, $f_{ji} = \partial_{x_i}(f_{x_j})$, As usual, these notations can get confusing.*

The second order partials for a scalar function of n variables can be organized into a matrix called the **Hessian.**

Definition 10.4.2 The Hessian Matrix

If $f(x, y)$, f_x and f_y are defined locally at (x_0, y_0) and if the second order partials exist at (x_0, y_0), we define the Hessian, $\boldsymbol{H}(\boldsymbol{x_0})$ at $(\boldsymbol{x_0})$ to be the matrix

$$\boldsymbol{H}(\boldsymbol{x_0}) = \begin{bmatrix} f_{x_1\,x_1}(\boldsymbol{x_0}) & \dots & f_{x_1\,x_n}(\boldsymbol{x_0}) \\ \vdots & \vdots & \vdots \\ f_{x_n\,x_1}(\boldsymbol{x_0}) & \dots & f_{x_n\,x_n}(\boldsymbol{x_0}) \end{bmatrix} = \begin{bmatrix} f_{11}(\boldsymbol{x_0}) & \dots & f_{1n}(\boldsymbol{x_0}) \\ \vdots & \vdots & \vdots \\ f_{n1}(\boldsymbol{x_0}) & \dots & f_{nn}(\boldsymbol{x_0}) \end{bmatrix}$$

Hence for a scalar function of n variables, we have the gradient, $\boldsymbol{\nabla} f$ which is the derivative of f, Df, in some circumstances and is an $n \times 1$ column vector, and the Hessian of f which is a $n \times n$ matrix. If we have a vector function, it is messier. The analogue of the gradient for $f : D \subset \Re^n \to \Re^m$ is the Jacobian of f, $\boldsymbol{J_f}$ given by

$$\boldsymbol{J_f}(\boldsymbol{x_0}) = \begin{bmatrix} f_{1,x_1}(\boldsymbol{x_0}) & \dots & f_{1,\,x_n}(\boldsymbol{x_0}) \\ \vdots & \vdots & \vdots \\ f_{m,x_1}(\boldsymbol{x_0}) & \dots & f_{m\,x_n}(\boldsymbol{x_0}) \end{bmatrix} = \begin{bmatrix} f_{11}(\boldsymbol{x_0}) & \dots & f_{1n}(\boldsymbol{x_0}) \\ \vdots & \vdots & \vdots \\ f_{m1}(\boldsymbol{x_0}) & \dots & f_{mn}(\boldsymbol{x_0}) \end{bmatrix}$$

where f_i are the component functions of f. Note the Jacobian of f is the derivative of f, Df, in some circumstances and is an $n \times m$ matrix. You can see why the notation f_{ij} is so confusing. If f is a scalar function this has a different meaning than if f is a vector-valued function. So context is all. We can also define higher order derivatives such as $f_{x_i\,x_j\,x_k}$ which have more permutations. We will let you work through those notations as the need arises in your work.

Example 10.4.1 *Let $f(x, y, z) = 2x - 8xy + 4yz^2$. Find the first and second order partials of f and its Hessian.*

Solution *The partials are*

$$\begin{aligned} f_x(x, y, z) &= 2 - 8y, \quad f_y(x, y, z) = -8x + 4z^2, \quad f_z(x, y, z) = 8yz \\ f_{xx}(x, y, z) &= 0, \quad f_{xy}(x, y, z) = -8, \quad f_{xz}(x, y, z) = 0 \\ f_{yx}(x, y, z) &= -8, \quad f_{yy}(x, y, z) = 0, \quad f_{yz}(x, y, z) = 8z \end{aligned}$$

$$f_{zx}(x,y,z) \;=\; 0, \quad f_{zy}(x,y,z) = 8z, \quad f_{zz}(x,y,z) = 8y$$

and so the Hessian is

$$\boldsymbol{H}(x,y,z) \;=\; \begin{bmatrix} f_{xx}(x,y,z) & f_{xy}(x,y,z) & f_{xz}(x,y,z) \\ f_{yx}(x,y,z) & f_{yy}(x,y,z) & f_{yz}(x,y,z) \\ f_{zx}(x,y,z) & f_{zy}(x,y,z) & f_{zz}(x,y,z) \end{bmatrix} \;=\; \begin{bmatrix} 0 & -8 & 0 \\ -8 & 0 & 8z \\ 0 & 8z & 8y \end{bmatrix}$$

10.4.2.1 Homework

Exercise 10.4.9 *Let $f(x,y,z) = 2x^2y^2 + 4y^2z^4$. Find the Hessian of f.*

Exercise 10.4.10 *Let $f(x,y,z) = 2x^4y^2z^4$. Find the Hessian of f.*

Exercise 10.4.11 *Let $f(x,y,z) = \sin(2x^4y^2z^4)$. Find the Hessian of f.*

Exercise 10.4.12 *Let $f(x,y,z,w) = \cos(xyzw^2)$. Find the Hessian of f.*

10.4.3 When Do Mixed Partials Match?

Next, we want to know under what conditions the mixed order partials match. Here is a standard example with a function of two variables showing the two mixed partials don't agree at a point.

Example 10.4.2 *Let*

$$f(x,y) \;=\; \begin{cases} \frac{xy(x^2-y^2)}{x^2+y^2}, & (x,y) \neq (0,0) \\ 0, & (x,y) = (0,0) \end{cases}$$

Prove the mixed order partials do not match.

- *Prove $\lim_{(x,y)\to(0,0)} f(x,y) = 0$ thereby showing f has a removeable discontinuity at $(0,0)$ and so the definition for f above makes sense.*

  Solution

$$\begin{aligned} \lim_{(x,y)\to(0,0)} \left|\frac{xy(x^2-y^2)}{x^2+y^2}\right| &= \lim_{(x,y)\to(0,0)} \left|\frac{xy(x+y)}{x^2+y^2}\right| |x-y| = \lim_{(x,y)\to(0,0)} \left|\frac{x^2y+y^2x}{x^2+y^2}\right| |x-y| \\ &\leq \lim_{(x,y)\to(0,0)} \left(\left|\frac{x^2y}{x^2+y^2}\right| + \left|\frac{y^2x}{x^2+y^2}\right|\right)|x-y| \\ &\leq \lim_{(x,y)\to(0,0)} (|y|+|x|)|x-y| \leq 4(x^2+y^2) = 0 \end{aligned}$$

  Thus, f has a removeable discontinuity at $(0,0)$ of value 0.

- *Compute $f_x(0,0)$ and $f_y(0,0)$*

  Solution

$$\begin{aligned} f_x(0,0) &= \lim_{h\to 0} \frac{f(h,0)-f(0,0)}{h} = \frac{0}{h} = 0 \\ f_y(0,0) &= \lim_{h\to 0} \frac{f(0,h)-f(0,0)}{h} = \frac{0}{h} = 0 \end{aligned}$$

  Also, at any $(x,y) \neq (0,0)$, we have

$$f_x(x,y) \;=\; \frac{x^4y + 4x^2y^3 - y^5}{(x^2+y^2)^2}$$

$$f_y(x,y) = \frac{-xy^4 + x^5 - 4x^3y^2}{(x^2+y^2)^2}$$

Then

$$\begin{aligned}
&\lim_{(x,y)\to(0,0)} \left| \frac{yx^4 + 4x^2y^3 - y^5}{(x^2+y^2)^2} \right| \\
&\le \lim_{(x,y)\to(0,0)} \left(|y| \frac{x^4}{(x^2+y^2)^2} + |y| \frac{y^4}{(x^2+y^2)^2} + 4|y| \frac{x^2}{x^2+y^2} \frac{y^2}{x^2+y^2} \right) \\
&\le \lim_{(x,y)\to(0,0)} (2|y| + 4|y|) \le 5\sqrt{x^2+y^2} = 0
\end{aligned}$$

Hence f_x has a removeable discontinuity at $(0,0)$. A similar calculation shows f_y has a removeable discontinuity at $(0,0)$.

- *Compute $\partial_y(f_x)(0,0)$ and $\partial_x(f_y)(0,0)$. Are these equal?*

  Solution

$$\begin{aligned}
f_{xy}(0,0) &= \lim_{h\to 0} \frac{f_x(0,h) - f_x(0,0)}{h} = \frac{-h^5/h^4}{h} = -1 \\
f_{yx}(0,0) &= \lim_{h\to 0} \frac{f_y(0,h) - f_x(0,0)}{h} = \frac{h^5/h^4}{h} = 1
\end{aligned}$$

  Clearly, these mixed partials do not match.

- *Compute $\partial_y(f_x)(x,y)$ and $\partial_x(f_y)(x,y)$ for $(x,y) \neq (0,0)$.*

  Solution *At any $(x,y) \neq (0,0)$, we have*

$$\begin{aligned}
f_{xy}(x,y) &= \frac{x^6 - y^6 - 9y^4x^2 + 9y^2x^4}{(x^2+y^2)^3} \\
f_{yx}(x,y) &= \frac{x^6 - y^6 + 9y^4x^2 - 9y^2x^4}{(x^2+y^2)^3}
\end{aligned}$$

- *Determine if $\partial_y(f_x)(x,y)$ and $\partial_x(f_y)(x,y)$ are continuous at $(0,0)$.*

  Solution *If you look at the paths $(t, 2t)$ and $(t, 3t)$ for $t > 0$, you find you get different limiting values as $t \to 0^+$ for both f_{xy} and f_{yx}. Hence, these mixed partials are not continuous at $(0,0)$.*

We see the lack of continuity of the mixed partials in a ball about the origin is the problem.

This leads to our next theorem.

Theorem 10.4.2 Sufficient Conditions for the Mixed Order Partials to Match

Let $f : D \subset \Re^n \to \Re$ and fix $i \neq j$ for the set $\{1, \ldots, n\}$. Assume all the first order partials $\frac{\partial f}{\partial x_i}$ and $\frac{\partial f}{\partial x_j}$ and the second order partials $\partial_{x_j}\left(\frac{\partial f}{\partial x_i}\right)$ and $\partial_{x_i}\left(\frac{\partial f}{\partial x_j}\right)$ all exist in $B_r(\mathbf{p})$ for some $r > 0$. Also assume $\partial_{x_j}\left(\frac{\partial f}{\partial x_i}\right)$ and $\partial_{x_i}\left(\frac{\partial f}{\partial x_j}\right)$ are continuous in $B_r(\mathbf{p})$. Then $\partial_{x_j}\left(\frac{\partial f}{\partial x_i}(\mathbf{p})\right) = \partial_{x_i}\left(\frac{\partial f}{\partial x_j}\right)$ at $\mathbf{p}$.

Proof 10.4.2

Let's control this notational mess by noticing we are only looking at changes in x_i and x_j. Let $X_1 = x_i, X_2 = x_j$ and

$$\begin{aligned} F(X_1, X_2) &= f(x_1, \ldots, x_{i-1}, \boldsymbol{x_i} = X_1, x_{i+1}, \ldots, \\ &\quad x_{j-1}, \boldsymbol{x_j} = X_2, x_{j+1}, \ldots, x_n) \end{aligned}$$

Also, let $F_1 = \partial_{x_i} f$, $F_2 = \partial_{x_j} f$ and $F_{12} = \partial_{x_j}(\partial_{x_i} f)$ and $F_{21} = \partial_{x_i}(\partial_{x_j} f)$. Now with these variables defined, the argument to prove the mixed partials match is essentially the same as the one in (Peterson (21) 2020). So even though there are n variables involved, we just have to focus our attention on the two that matter. Fix h small enough for that $\mathbf{p} + h\boldsymbol{e_i} + h\boldsymbol{e_j}$ is in $B_r(\mathbf{p})$. Note $\mathbf{p} + h\boldsymbol{e_i} = X_1 + h$ and $\mathbf{p} + h\boldsymbol{e_j} = X_2 + h$.

Step One:
Define

$$\begin{aligned} G(h) &= F(X_1 + h, X_2 + h) - F(X_1 + h, X_2) - F(X_1, X_2 + h) + F(X_1, X_2) \\ \phi(x) &= F(x, X_2 + h) - F(x, X_2), \quad \psi(y) = \phi(X_1 + y) - \phi(X_1) \end{aligned}$$

Now apply the Mean Value Theorem to ϕ. Note $\phi'(x) = \partial_{X_1} F(x, X_2 + h) - \partial_{X_1} F(x, X_2) = f_{x_i}(\mathbf{p} + x\boldsymbol{e_i} + h\boldsymbol{e_j}) - f_{x_i}(\mathbf{p} + x\boldsymbol{e_i})$. Thus,

$$\begin{aligned} \phi(X_1 + h) - \phi(X_1) &= \phi'(X_1 + \theta_1\, h)\, h, \quad \theta_1 \text{ between } X_1 \text{ and } X_1 + h \\ &= (f_{x_i}(\mathbf{p} + \theta_1 h\boldsymbol{e_i} + h\boldsymbol{e_j}) - f_{x_i}(\mathbf{p} + \theta_1 h\boldsymbol{e_i}))h \\ &= (F_{X_1}(X_1 + \theta_1 h, X_2 + h) - F_{X_1}(X_1 + \theta_1 h, X_2))h \end{aligned}$$

where we use the definition of F. Now define for y between X_2 and $X_2 + h$, the function ξ by

$$\begin{aligned} \xi(z) &= F_{X_1}(X_1 + \theta_1 h, z) \\ \xi'(z) &= \partial_{X_2} F_{X_1}(X_1 + \theta_1 h, z) \end{aligned}$$

Apply the Mean Value Theorem to ξ: there is a θ_2 so that $X_2 + \theta_2 h$ is between X_2 and $X_2 + h$ with

$$\xi(X_2 + h) - \xi(X_2) = \partial_{X_2} F_{X_1}(X_1 + \theta_1 h, X_2 + \theta_2 h)\, h$$

Thus,

$$\begin{aligned} h\,(\xi(X_2 + h) - \xi(X_2)) &= h\,(F_{X_1}(X_1 + \theta_1 h, X_2 + h) - F_{X_1}(X_1 + \theta_1 h, X_2)) \\ &= \partial_{X_2} F_{X_1}(X_1 + \theta_1 h, X_2 + \theta_2 h)\, h^2 \end{aligned}$$

But,

$$\begin{aligned} \phi(X_1 + h) - \phi(X_1) &= (F(X_1 + h, X_2 + h) - F(X_1 + h, X_2)) \\ &\quad -(F(X_1, X_2 + h) - F(X_1, X_2)) \\ &= G(h) \\ &= (F_{X_1}(X_1 + \theta_1 h, X_2 + h) - F_{X_1}(X_1 + \theta_1 h, X_2))h \\ &= (\xi(X_2 + h) - \xi(X_2))h \\ &= \partial_{X_2} F_{X_1}(X_1 + \theta_1 h, X_2 + \theta_2 h)\, h^2 \end{aligned}$$

So

$$G(h) = \partial_{X_2} F_{X_1}(X_1 + \theta_1 h, X_2 + \theta_2 h)\, h^2$$

Step Two:
Now start again with $L(w) = F(X_1 + h, w) - F(X_1, w)$, noting $G(h) = L(X_2 + h) - L(X_2)$. Apply the Mean Value Theorem to L like before to find

$$\begin{aligned} G(h) &= L(X_2 + h) - L(X_2) = L'(X_2 + \theta_3 h)h \\ &= (\partial_{X_2} F(X_1 + h, X_2 + \theta_3 h) - \partial_{X_2} F(X_1, X_2 + \theta_3 h))\, h \end{aligned}$$

for an intermediate value of θ_3 with $X_2 + \theta_3 h$ between X_2 and $X_2 + h$.

Now define for z between X_2 and $X_2 + h$, the function γ by

$$\begin{aligned} \gamma(z) &= F_{X_2}(z, X_2 + \theta_3 h) \\ \gamma'(z) &= \partial_{X_1} F_{X_2}(z, X_2 + \theta_3 h) \end{aligned}$$

Now apply the Mean Value Theorem a fourth time: So there is a θ_4 with $X_1 + \theta_4 h$ between X_1 and $X_1 + h$ so that

$$\begin{aligned} \gamma(X_1 + h) - \gamma(X_1) &= \gamma'(X_1 + \theta_4 h)\, h \\ &= \partial_{X_1} F_{X_2}(X_1 + \theta_4 h, X_2 + \theta_3 h)\, h \end{aligned}$$

Using what we have found so far, we have

$$\begin{aligned} G(h) &= (\partial_{X_2} F(X_1 + h, X_2 + \theta_3 h) - \partial_{X_2} F(X_1, X_2 + \theta_3 h))\, h \\ &= (\gamma(X_1 + h) - \gamma(X_1))\, h = (\partial_{X_1} F_{X_2}(X_1 + \theta_4 h, X_2 + \theta_3 h))\, h^2 \end{aligned}$$

So we have two representations for $G(h)$:

$$G(h) = \partial_{X_2} F_{X_1}(X_1 + \theta_1 h, X_2 + \theta_2 h)\, h^2 = (\partial_{X_1} F_{X_2}(X_1 + \theta_4 h, X_2 + \theta_3 h))\, h^2$$

Canceling the common h^2, we have

$$\partial_{X_2} F_{X_1}(X_1 + \theta_1 h, X_2 + \theta_2 h) = (\partial_{X_1} F_{X_2}(X_1 + \theta_4 h, X_2 + \theta_3 h))$$

Now let $h \to 0$ and since these second order partials are continuous at $\mathbf{p}$, we have $\partial_{X_2} F_{X_1}(X_1, X_2) = \partial_{X_1} F_{X_2}(X_1, X_2)$. This shows the result we wanted. ■

10.4.3.1 Homework

Exercise 10.4.13

$$f(x, y) = \begin{cases} \frac{4xy(x^2-y^2)}{x^2+2y^2}, & (x, y) \neq (0, 0) \\ 0, & (x, y) = (0, 0) \end{cases}$$

- *Prove $\lim_{(x,y)\to(0,0)} f(x, y) = 0$ thereby showing f has a removeable discontinuity at $(0, 0)$ and so the definition for f above makes sense.*
- *Compute $f_x(0, 0)$ and $f_y(0, 0)$.*
- *Compute $f_x(x, y)$ and $f_y(x, y)$ for $(x, y) \neq (0, 0)$. Simplify your answers to the best possible form.*
- *Compute $\partial_y(f_x)(0, 0)$ and $\partial_x(f_y)(0, 0)$. Are these equal?*
- *Compute $\partial_y(f_x)(x, y)$ and $\partial_x(f_y)(x, y)$ for $(x, y) \neq (0, 0)$.*

- *Determine if* $\partial_y(f_x)(x,y)$ *and* $\partial_x(f_y)(x,y)$ *are continuous at* $(0,0)$.

Exercise 10.4.14

$$f(x,y) = \begin{cases} \frac{6xy(x^2-y^2)}{3x^2+2y^2}, & (x,y) \neq (0,0) \\ 0, & (x,y) = (0,0) \end{cases}$$

- *Prove* $\lim_{(x,y)\to(0,0)} f(x,y) = 0$ *thereby showing* f *has a removeable discontinuity at* $(0,0)$ *and so the definition for* f *above makes sense.*
- *Compute* $f_x(0,0)$ *and* $f_y(0,0)$.
- *Compute* $f_x(x,y)$ *and* $f_y(x,y)$ *for* $(x,y) \neq (0,0)$. *Simplify your answers to the best possible form.*
- *Compute* $\partial_y(f_x)(0,0)$ *and* $\partial_x(f_y)(0,0)$. *Are these equal?*
- *Compute* $\partial_y(f_x)(x,y)$ *and* $\partial_x(f_y)(x,y)$ *for* $(x,y) \neq (0,0)$.
- *Determine if* $\partial_y(f_x)(x,y)$ *and* $\partial_x(f_y)(x,y)$ *are continuous at* $(0,0)$.

10.5 Derivatives for Vector Functions of n Variables

If $f : D \subset \Re^n \to \Re^m$, then f determines m component functions $f_i : d \subset \Re^n \to \Re$; i.e. at any point $\boldsymbol{x_0}$ where f is locally defined, we have

$$f(\boldsymbol{x_0}) = \begin{bmatrix} f_1(\boldsymbol{x_0}) \\ \vdots \\ f_m(\boldsymbol{x_0}) \end{bmatrix}, \quad f(\boldsymbol{x_0}+\boldsymbol{\Delta}) = \begin{bmatrix} f_1(\boldsymbol{x_0}+\Delta_1\boldsymbol{E_1}) \\ \vdots \\ f_m(\boldsymbol{x_0}+\Delta_n\boldsymbol{E_m}) \end{bmatrix}$$

for suitable $\boldsymbol{\Delta}$. If we assume each f_i is differentiable at $\boldsymbol{x_0}$, we know

$$\begin{aligned} f_1(\boldsymbol{x_0}+\boldsymbol{\Delta}) &= f_1(\boldsymbol{x_0}) + \sum_{i=1}^{n} f_{1,x_i}(\boldsymbol{x_0})\,(x_{0i}-\Delta_i) + \mathscr{E}_1(\boldsymbol{x_0},\boldsymbol{\Delta}) \\ \vdots &= \vdots \\ f_m(\boldsymbol{x_0}+\boldsymbol{\Delta}) &= f_m(\boldsymbol{x_0}) + \sum_{i=1}^{n} f_{m,x_i}(\boldsymbol{x_0})\,(x_{0i}-\Delta_i) + \mathscr{E}_m(\boldsymbol{x_0},\boldsymbol{\Delta}) \end{aligned}$$

where

$$\lim_{\boldsymbol{\Delta}\to\boldsymbol{0}} \mathscr{E}_i(\boldsymbol{x_0},\boldsymbol{\Delta}) = 0, \quad \lim_{\boldsymbol{\Delta}\to\boldsymbol{0}} \frac{\mathscr{E}_i(\boldsymbol{x_0},\boldsymbol{\Delta})}{\|\boldsymbol{\Delta}\|} = 0$$

We can then write

$$f(\boldsymbol{x_0}+\boldsymbol{\Delta}) = \begin{bmatrix} f_1(\boldsymbol{x_0}) \\ \vdots \\ f_m(\boldsymbol{x_0}) \end{bmatrix} + \begin{bmatrix} f_{1,x_1}(\boldsymbol{x_0}) & \ldots & f_{1,x_n}(\boldsymbol{x_0}) \\ \vdots & \vdots & \\ f_{m,x_1}(\boldsymbol{x_0}) & \ldots & f_{m,x_n}(\boldsymbol{x_0}) \end{bmatrix} \begin{bmatrix} \Delta_1 \\ \vdots \\ \Delta_n \end{bmatrix} + \begin{bmatrix} \mathscr{E}_1(\boldsymbol{x_0},\boldsymbol{\Delta}) \\ \vdots \\ \mathscr{E}_m(\boldsymbol{x_0},\boldsymbol{\Delta}) \end{bmatrix}$$

The matrix of partials above occurs frequently and is called the **Jacobian** of f.

Definition 10.5.1 The Jacobian

Let $f : D \subset \Re^n \rightarrow \Re^m$ have partial derivatives at $\boldsymbol{x_0}$. The Jacobian of f is the $m \times n$ matrix

$$\mathbf{J_f}(\boldsymbol{x_0}) = \begin{bmatrix} f_{1,x_1}(\boldsymbol{x_0}) & \dots & f_{1,x_n}(\boldsymbol{x_0}) \\ \vdots & & \vdots \\ f_{m,x_1}(\boldsymbol{x_0}) & \dots & f_{m,x_n}(\boldsymbol{x_0}) \end{bmatrix}$$

If f is differentiable at $\boldsymbol{x_0}$, we can write

$$\boldsymbol{J_f}(\boldsymbol{x_0}) = \begin{bmatrix} (\boldsymbol{\nabla}(f_1))^T(\boldsymbol{x_0}) \\ \vdots \\ (\boldsymbol{\nabla}(f_m))^T(\boldsymbol{x_0}) \end{bmatrix} = \begin{bmatrix} \boldsymbol{D}f_1(\boldsymbol{x_0}) \\ \vdots \\ \boldsymbol{D}f_m(\boldsymbol{x_0}) \end{bmatrix}$$

Thus, we write

$$f(\boldsymbol{x_0} + \boldsymbol{\Delta}) = f(\boldsymbol{x_0}) + \boldsymbol{J_f}(\boldsymbol{x_0})\,\boldsymbol{\Delta} + \mathscr{E}(\boldsymbol{x_0}, \boldsymbol{\Delta})$$

where $\mathscr{E}(\boldsymbol{x_0}, \boldsymbol{\Delta})$ is the vector error

$$\mathscr{E}(\boldsymbol{x_0}, \boldsymbol{\Delta}) = \begin{bmatrix} \mathscr{E}_1(\boldsymbol{x_0}, \boldsymbol{\Delta}) \\ \vdots \\ \mathscr{E}_m(\boldsymbol{x_0}, \boldsymbol{\Delta}) \end{bmatrix}$$

and we see

$$\lim_{\boldsymbol{\Delta} \rightarrow \boldsymbol{0}} \|\mathscr{E}(\boldsymbol{x_0}, \boldsymbol{\Delta})\| = \lim_{\boldsymbol{\Delta} \rightarrow \boldsymbol{0}} \sqrt{(\mathscr{E}_1(\boldsymbol{x_0}, \boldsymbol{\Delta}))^2 + \ldots + (\mathscr{E}_m(\boldsymbol{x_0}, \boldsymbol{\Delta}))^2} = 0$$

$$\lim_{\boldsymbol{\Delta} \rightarrow \boldsymbol{0}} \left\| \frac{\mathscr{E}(\boldsymbol{x_0}, \boldsymbol{\Delta})}{\|\boldsymbol{\Delta}\|} \right\| = \lim_{\boldsymbol{\Delta} \rightarrow \boldsymbol{0}} \sqrt{\left(\frac{\mathscr{E}_1(\boldsymbol{x_0}, \boldsymbol{\Delta})}{\|\boldsymbol{\Delta}\|}\right)^2 + \ldots + \left(\frac{\mathscr{E}_m(\boldsymbol{x_0}, \boldsymbol{\Delta})}{\|\boldsymbol{\Delta}\|}\right)^2} = 0$$

The definition of differentiability for a vector function is thus:

Definition 10.5.2 The Differentiability of a Vector Function of n Variables

Let $f : D \subset \Re^n \rightarrow \Re^m$ be defined locally at $\boldsymbol{x_0}$. We say f is differentiable at $\boldsymbol{x_0}$ if there is a linear map $\boldsymbol{L}(\boldsymbol{x_0}) : \Re^n \rightarrow \Re^m$ and an error vector $\mathscr{E}(\boldsymbol{x_0}, \boldsymbol{\Delta})$ so that for given $\boldsymbol{\Delta}$,

$$f(\boldsymbol{x_0} + \boldsymbol{\Delta}) = f(\boldsymbol{x_0}) + \boldsymbol{L}(\boldsymbol{x_0})\,\boldsymbol{\Delta} + \mathscr{E}(\boldsymbol{x_0}, \boldsymbol{\Delta})$$

where

$$\lim_{\boldsymbol{\Delta} \rightarrow \boldsymbol{0}} \mathscr{E}(\boldsymbol{x_0}, \boldsymbol{\Delta}) = 0$$

$$\lim_{\boldsymbol{\Delta} \rightarrow \boldsymbol{0}} \frac{\mathscr{E}(\boldsymbol{x_0}, \boldsymbol{\Delta})}{\|\boldsymbol{\Delta}\|} = 0$$

where we use norm convergence with respect to the norm $\|\cdot\|$ used in $\Re^n$ and $\Re^m$. We then say the derivative of f at $\boldsymbol{x_0}$ is $\boldsymbol{L}(\boldsymbol{x_0})$. The derivative of f at $\boldsymbol{x_0}$ is denoted by $\boldsymbol{D}\boldsymbol{f}(\boldsymbol{x_0})$.

Comment 10.5.1 *For any given pair of orthonormal basis* $\boldsymbol{E}$ *of* $\Re^n$ *and* $\boldsymbol{F}$ *of* $\Re^m$*, the linear map* $\boldsymbol{L}(\boldsymbol{x_0})$ *has a representation* $\left[L(\boldsymbol{x_0})\right]_{\boldsymbol{EF}}$ *so that*

$$\left[f(\boldsymbol{x_0}+\boldsymbol{\Delta})\right]_{\boldsymbol{F}} = \left[f(\boldsymbol{x_0})\right]_{\boldsymbol{F}} + \left[L(\boldsymbol{x_0})\right]_{\boldsymbol{EF}} \left[\boldsymbol{\Delta}\right]_{\boldsymbol{E}} + \left[\mathscr{E}(\boldsymbol{x_0},\boldsymbol{\Delta})\right]_{\boldsymbol{F}}$$

Comment 10.5.2 *We usually are not so careful with the notation. It is understood that* $\Re^n$ *and* $\Re^m$ *have some choice of orthonormal basis* $\boldsymbol{E}$ *and* $\boldsymbol{F}$ *and so we* **identify**

$$\begin{aligned} \boldsymbol{L}(\boldsymbol{x_0}) &\equiv \left[L(\boldsymbol{x_0})\right]_{\boldsymbol{E},\boldsymbol{F}}, \quad f(\boldsymbol{x_0}+\boldsymbol{\Delta}) \equiv \left[f(\boldsymbol{x_0}+\boldsymbol{\Delta})\right]_{\boldsymbol{F}}, \quad \boldsymbol{\Delta} \equiv \left[\boldsymbol{\Delta}\right]_{\boldsymbol{E}} \\ \mathscr{E}(\boldsymbol{x_0},\boldsymbol{\Delta}) &\equiv \left[\mathscr{E}(\boldsymbol{x_0},\boldsymbol{\Delta})\right]_{\boldsymbol{F}} \end{aligned}$$

It is not so hard to mentally make the conversion back and forth and you should get used to it.

It is easy to prove the following result. Note we are using the identification of the linear mapping to its matrix representation here.

Theorem 10.5.1 If the Vector Function f **is Differentiable at** $\boldsymbol{x_0}$**,** $\boldsymbol{L}(\boldsymbol{x_0}) = \boldsymbol{J_f}(\boldsymbol{x_0})$

Let $f : D \subset \Re^n \rightarrow \Re^m$ *be defined locally at* $\boldsymbol{x_0}$*. If* f *is differentiable at* $\boldsymbol{x_0}$*, the* $(\boldsymbol{L}(\boldsymbol{x_0}))_{ij} = f_{i,x_j}(\boldsymbol{x_0})$*; i.e.* $\boldsymbol{L}(\boldsymbol{x_0}) = \boldsymbol{J_f}(\boldsymbol{x_0})$.

Proof 10.5.1
This argument is very similar to the one we did before. Just apply the argument to each component. We leave this to you. ■

10.5.0.1 Homework

Exercise 10.5.1 *Let*

$$f(x,y) = \begin{bmatrix} f_1(x,y) \\ f_2(x,y) \end{bmatrix}$$

be defined on $\Omega \subset \Re^2$.

- *Prove if* f_1 *and* f_2 *are differentiable on* Ω*, then* f *is differentiable on* Ω.
- *What conditions of* f_1 *and* f_2 *and therefore on* f *guarantee* f *to be differentiable on* Ω*?*

Exercise 10.5.2 *Let*

$$f(x,y,z) = \begin{bmatrix} f_1(x,y,z) \\ f_2(x,y,z) \end{bmatrix}$$

be defined on $\Omega \subset \Re^3$.

- *Prove if* f_1*,* f_2 *and* f_3 *are differentiable on* Ω*, then* f *is differentiable on* Ω.
- *What conditions of* f_1*,* f_2 *and* f_3 *and therefore on* f *guarantee* f *to be differentiable on* Ω*?*

Exercise 10.5.3 *Do the same as the previous exercise but this time in* $\Re^n$.

Exercise 10.5.4 *Let*

$$f(x,y) = \begin{bmatrix} f_1(x,y) \\ f_2(x,y) \end{bmatrix} = \begin{bmatrix} 2x^2+5y^2 \\ 4xy^4 \end{bmatrix}$$

- *Why are f_1 and f_2 differentiable on $\Re^2$?*
- *Find the Jacobian of f.*
- *From (Peterson (21) 2020), we know we can express the error in terms of the hessians of the component functions. Write down the error term here in terms of the two hessians.*

Exercise 10.5.5 *Let*

$$f(x,y,z) = \begin{bmatrix} f_1(x,y,z) \\ f_2(x,y,z) \\ f_3(x,y,z) \end{bmatrix} = \begin{bmatrix} 2x^2 + 5y^2 + 10z^4 \\ 4xy^4 + 20y^2z^5 \\ 2xyz^2 \end{bmatrix}$$

- *Why are f_1, f_2 and f_3 differentiable on $\Re^3$?*
- *Find the Jacobian of f.*
- *From (Peterson (21) 2020), we know we can express the error in terms of the Hessians of the component functions. Write down the error term here in terms of the three Hessians.*

10.5.1 The Chain Rule for Vector-Valued Functions

The next thing we want to do is to extend the chain rule to this setting. Note the proof here is different than the one we did in the scalar case. We are using matrix manipulations here instead of components.

Theorem 10.5.2 The Chain Rule for Vector Functions

Let $F_1 : D_1 \subset \Re^n \to \mathscr{R}(F_1) \subset \Re^m$ and $F_2 : D_2 \subset \Re^m \to \mathscr{R}(F_2) \subset \Re^r$. Here $\mathscr{F}$ denotes the range of F_i. We assume there is overlap: $\mathscr{F} \cap D_2 = S \neq \emptyset$ and let $E = F^{-1}(S)$. Further, assume F_1 is differentiable at $\boldsymbol{Q_0}$ in E and F_2 is differentiable at $\boldsymbol{P_0} = F_1(\boldsymbol{Q_0}) \in S$. Then, $F_3 = F_2 \circ F_1$ is differentiable at $\boldsymbol{Q_0}$ with matrix representation given by

$$\underset{r \times n}{\left[\boldsymbol{DF_3(Q_0)}\right]} = \underset{r \times m}{\left[\boldsymbol{DF_2(P_0)}\right]} \; \underset{m \times n}{\left[\boldsymbol{DF_1(Q_0)}\right]}$$

or in linear mapping terms

$$\boldsymbol{DF_3(Q_0)} = \boldsymbol{DF_2(P_0)} \circ \boldsymbol{DF_1(Q_0)}$$

Proof 10.5.2
You can see the rough flow of these mappings in Figure 10.1 which helps set the stage. We know

$$F_2(\boldsymbol{P}) = F_2(\boldsymbol{P_0}) + \boldsymbol{J_2}(\boldsymbol{P_0})(\boldsymbol{P} - \boldsymbol{P_0}) + \mathscr{E}_2(\boldsymbol{P_0}, \boldsymbol{P})$$

where $\boldsymbol{J_2}(\boldsymbol{P_0})$ is a linear map from $\Re^m$ to $\Re^r$ and $\mathscr{E}_2(\boldsymbol{P_0}, \boldsymbol{P}) \to \boldsymbol{0}$ as $\boldsymbol{P} \to \boldsymbol{P_0}$. Also,

$$F_1(\boldsymbol{Q}) = F_1(\boldsymbol{Q_0}) + \boldsymbol{J_1}(\boldsymbol{Q_0})(\boldsymbol{Q} - \boldsymbol{Q_0}) + \mathscr{E}_1(\boldsymbol{Q_0}, \boldsymbol{Q})$$

where $\boldsymbol{J_1}(\boldsymbol{Q_0})$ is a linear map from $\Re^n$ to $\Re^m$ and $\mathscr{E}_1(\boldsymbol{Q_0}, \boldsymbol{Q}) \to \boldsymbol{0}$ as $\boldsymbol{Q} \to \boldsymbol{Q_0}$. Let $\boldsymbol{P} = F_1(\boldsymbol{Q})$ so that $\boldsymbol{P_0} = F_1(\boldsymbol{Q_0})$. Then

$$\begin{aligned} F_3(\boldsymbol{Q}) - F_3(\boldsymbol{Q_0}) &= F_2(F_1(\boldsymbol{Q})) - F_2(F_1(\boldsymbol{Q_0})) \\ &= J_2(F_1(\boldsymbol{Q_0}))\,(F_1(\boldsymbol{Q}) - F_1(\boldsymbol{Q_0})) + \mathscr{E}_2(F_1(\boldsymbol{Q_0}), F_1(\boldsymbol{Q})) \end{aligned}$$

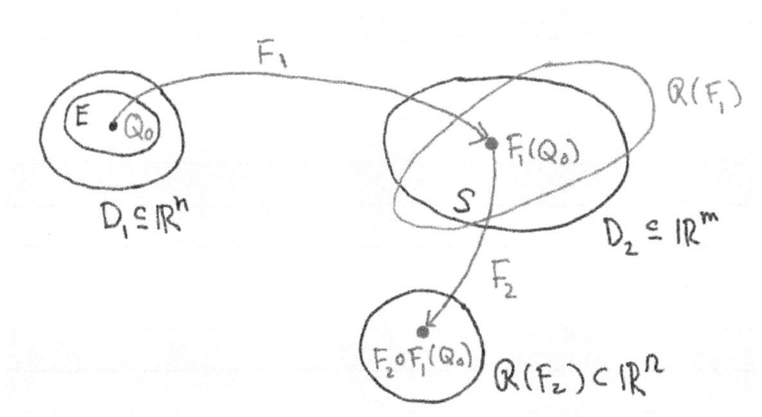

Figure 10.1: The vector chain rule.

$$\begin{aligned} &= J_2(F_1(\boldsymbol{Q_0}))\left\{J_1(\boldsymbol{Q_0})\,(\boldsymbol{Q}-\boldsymbol{Q_0})+\mathscr{E}_1(\boldsymbol{Q_0},\boldsymbol{Q})\right\} \\ &\quad +\mathscr{E}_2(F_1(\boldsymbol{Q_0}),F_1(\boldsymbol{Q})) \end{aligned}$$

Now reorganize:

$$\begin{aligned} F_3(\boldsymbol{Q})-F_3(\boldsymbol{Q_0}) &= J_2(F_1(\boldsymbol{Q_0}))\,J_1(\boldsymbol{Q_0})\,(\boldsymbol{Q}-\boldsymbol{Q_0})+J_2(F_1(\boldsymbol{Q_0}))\,\mathscr{E}_1(\boldsymbol{Q_0},\boldsymbol{Q}) \\ &\quad +\mathscr{E}_2(F_1(\boldsymbol{Q_0}),F_1(\boldsymbol{Q})) \end{aligned}$$

or

$$F_3(\boldsymbol{Q})-F_3(\boldsymbol{Q_0}) = J_2(F_1(\boldsymbol{Q_0}))\,J_1(\boldsymbol{Q_0})\,(\boldsymbol{Q}-\boldsymbol{Q_0})+\mathscr{E}_3(\boldsymbol{Q_0},\boldsymbol{Q})$$

where

$$\mathscr{E}_3(\boldsymbol{Q_0},\boldsymbol{Q}) = J_2(F_1(\boldsymbol{Q_0}))\,\mathscr{E}_1(\boldsymbol{Q_0},\boldsymbol{Q})+\mathscr{E}_2(F_1(\boldsymbol{Q_0}),F_1(\boldsymbol{Q}))$$

Since $\mathscr{E}_1(\boldsymbol{Q_0},\boldsymbol{Q})\to 0$ as $\boldsymbol{Q}\to\boldsymbol{Q_0}$, the first term in $\mathscr{E}_3$ goes to zero. For the second term, we know since F_1 is differentiable at $\boldsymbol{Q_0}$, F_1 is continuous at $\boldsymbol{Q_0}$. Thus,

$$\lim_{\boldsymbol{Q}\to\boldsymbol{Q_0}} \mathscr{E}_2(F_1(\boldsymbol{Q_0}),F_1(\boldsymbol{Q})) = \lim_{F_1(\boldsymbol{Q})\to F_1(\boldsymbol{Q_0})} \mathscr{E}_2(F_1(\boldsymbol{Q_0}),F_1(\boldsymbol{Q})) = \boldsymbol{0}$$

Next, we need to look at

$$\frac{\mathscr{E}_3(\boldsymbol{Q_0},\boldsymbol{Q})}{\|\boldsymbol{Q}-\boldsymbol{Q_0}\|} = \frac{J_2(F_1(\boldsymbol{Q_0}))\,\mathscr{E}_1(\boldsymbol{Q_0},\boldsymbol{Q})}{\|\boldsymbol{Q}-\boldsymbol{Q_0}\|}+\frac{\mathscr{E}_2(F_1(\boldsymbol{Q_0}),F_1(\boldsymbol{Q}))}{\|\boldsymbol{Q}-\boldsymbol{Q_0}\|}$$

Again, the first term goes to zero as $\boldsymbol{Q}\to\boldsymbol{Q_0}$ since that is a property of $\mathscr{E}_1$.

The second term is more interesting. Now if F_1 was locally constant at $\boldsymbol{Q_0}$, then $\boldsymbol{J_1}(\boldsymbol{Q_0})=\boldsymbol{0}$ and we know the error satisfies $\mathscr{E}_2(F_1(\boldsymbol{Q_0}),F_1(\boldsymbol{Q}))/\|\boldsymbol{Q}-\boldsymbol{Q_0}\|$ is identically zero and we have shown what we want. If F_1 is not locally constant, we can say

$$\frac{\mathscr{E}_2(F_1(\boldsymbol{Q_0}),F_1(\boldsymbol{Q}))}{\|\boldsymbol{Q}-\boldsymbol{Q_0}\|} = \begin{cases} 0, & \boldsymbol{Q}\neq\boldsymbol{Q_0},\ F_1(\boldsymbol{Q})=F_1(\boldsymbol{Q_0}) \\ \frac{\mathscr{E}_2(F_1(\boldsymbol{Q_0}),F_1(\boldsymbol{Q}))}{\|F_1(\boldsymbol{Q})-F_1(\boldsymbol{Q_0})\|}\,\frac{\|F_1(\boldsymbol{Q})-F_1(\boldsymbol{Q_0})\|}{\|\boldsymbol{Q}-\boldsymbol{Q_0}\|}, & \boldsymbol{Q}\neq\boldsymbol{Q_0},\ F_1(\boldsymbol{Q})\neq F_1(\boldsymbol{Q_0}) \end{cases}$$

Hence, for all $\boldsymbol{Q} \neq \boldsymbol{Q_0}$, $F_1(\boldsymbol{Q}) \neq F_1(\boldsymbol{Q_0})$ *and we have*

$$\begin{aligned}\frac{\mathscr{E}_2(F_1(\boldsymbol{Q_0}), F_1(\boldsymbol{Q}))}{\|\boldsymbol{Q} - \boldsymbol{Q_0}\|} &\leq \frac{\mathscr{E}_2(F_1(\boldsymbol{Q_0}), F_1(\boldsymbol{Q}))}{\|F_1(\boldsymbol{Q}) - F_1(\boldsymbol{Q_0})\|} \frac{\|F_1(\boldsymbol{Q}) - F_1(\boldsymbol{Q_0})\|}{\|\boldsymbol{Q} - \boldsymbol{Q_0}\|} \\ &= \frac{\mathscr{E}_2(F_1(\boldsymbol{Q_0}), F_1(\boldsymbol{Q}))}{\|F_1(\boldsymbol{Q}) - F_1(\boldsymbol{Q_0})\|} \frac{\|\boldsymbol{J_1}(\boldsymbol{Q_0})\|_{Fr} \, \|\boldsymbol{Q} - \boldsymbol{Q_0}\| + \|\mathscr{E}_1(\boldsymbol{Q_0}, \boldsymbol{Q})\|}{\|\boldsymbol{Q} - \boldsymbol{Q_0}\|}\end{aligned}$$

where we use the Frobenius norm inequalities from Chapter 4 in Theorem 4.5.1. So, simplifying a bit,

$$\frac{\mathscr{E}_2(F_1(\boldsymbol{Q_0}), F_1(\boldsymbol{Q}))}{\|\boldsymbol{Q} - \boldsymbol{Q_0}\|} \leq \underbrace{\frac{\mathscr{E}_2(F_1(\boldsymbol{Q_0}), F_1(\boldsymbol{Q}))}{\|F_1(\boldsymbol{Q}) - F_1(\boldsymbol{Q_0})\|}}_{\text{Term One}} \left(\|\boldsymbol{J_1}(\boldsymbol{Q_0})\|_{Fr} + \underbrace{\frac{\|\mathscr{E}_1(\boldsymbol{Q_0}, \boldsymbol{Q})\|}{\|\boldsymbol{Q} - \boldsymbol{Q_0}\|}}_{\text{Term Two}} \right)$$

As $\boldsymbol{Q} \rightarrow \boldsymbol{Q_0}$, $F_1(\boldsymbol{Q}) \rightarrow F_1(\boldsymbol{Q_0})$ *also by continuity. The case where all* $F_1(\boldsymbol{Q}) = F_1(\boldsymbol{Q_0})$ *has an error bounded above by the case with* $F_1(\boldsymbol{Q}) \neq F_1(\boldsymbol{Q_0})$*. Hence, from the estimates above, term one goes to zero and the second piece goes to* $\boldsymbol{J_1}(\boldsymbol{Q_0})$ *as term two goes to zero. Hence, the limit is zero and we have shown the second part of the condition for* $\mathscr{E}_3$*. We conclude* F_3 *is differentiable at* $\boldsymbol{Q_0}$ *and*

$$\boldsymbol{DF_3}(\boldsymbol{Q_0}) = \boldsymbol{DF_2}(\boldsymbol{P_0})\,\boldsymbol{DF_1}(\boldsymbol{Q_0})$$

■

10.5.1.1 Homework

These are complicated notational messes and you should work some out in detail so you understand fully.

Exercise 10.5.6 *Using matrix notation, write down the chain rule for* $F_1 : D_1 \subset \Re^2 \rightarrow \Re^3$, $F_2 : F_1(D_1) \subset \Re^3 \rightarrow \Re^4$.

Exercise 10.5.7 *Using matrix notation, write down the chain rule for* $F_1 : D_1 \subset \Re^2 \rightarrow \Re^2$, $F_2 : F_1(D_1) \subset \Re^2 \rightarrow \Re^2$*. For this one, draw nice pictures!*

Exercise 10.5.8 *Using matrix notation, write down the chain rule for* $F_1 : D_1 \subset \Re^1 \rightarrow \Re^2$, $F_2 : F_1(D_1) \subset \Re^2 \rightarrow \Re^3$.

10.6 Tangent Plane Error

Now that we have the chain rule, we can quickly develop other results such as how much error we make when we approximate our surface $f(x, y)$ using a tangent plane at a point (x_0, y_0). We approximate nonlinear mappings for many purposes. We go through some examples in Chapter 13 and they are well worth studying. The first thing we need is to know when a scalar function of n variables is differentiable. Just because its partials exist at a point is not enough to guarantee that! But we know if the partials are continuous around that point, then the derivative does exist. And that means we can write the function in terms of its tangent plane plus an error. The arguments to do this are not terribly hard, but they are a bit involved. For now, let's go back to the old idea of a tangent plane to a surface. For the surface $z = f(x, y)$ if its partials are continuous functions (they usually are for our work!) then f is differentiable and hence we know that

$$f(x, y) = f(x_0, y_0) + f_x(x_0, y_0)(x - x_0) + f_y(x_0, y_0)(y - y_0) + E(x, y, x_0, y_0)$$

where $\mathscr{E}(x, y, x_0, y_0)/\sqrt{(x - x_0)^2 + (y - y_0)^2} \to 0$ and $E(x, y, x_0, y_0)$ go to 0 as $(x, y) \to (x_0, y_0)$. We can characterize the error much better if we have access to the second order partial derivatives of f. We have shown that if the second order partials are continuous locally near (x_0, y_0) then the mixed order partials f_{xy} and f_{yx} must match at the point (x_0, y_0). Hence, for these **smooth** surfaces, the Hessian for a scalar function is a symmetric matrix!

10.6.1 The Mean Value Theorem

We can approximate a scalar function of many variables using their gradients. This is the first step in developing Taylor polynomial approximations of arbitrary order, but the first step uses gradient information only. The approximations using second order partial information use both the gradient and the Hessian and will be covered next. The following theorem is called the Mean Value Theorem.

Theorem 10.6.1 The Mean Value Theorem

Let $f : D \subset \Re^n \to \Re$ be differentiable locally at $\boldsymbol{x_0}$; i.e. f is differentiable in $B(r, \boldsymbol{x_0})$ for some $r > 0$. Then if $\boldsymbol{x}$ and $\boldsymbol{y}$ are in $B(r, \boldsymbol{x_0})$, there is a point $\boldsymbol{u}$ on the line connecting $\boldsymbol{x}$ and $\boldsymbol{y}$ so that

$$f(\boldsymbol{x}) - f(\boldsymbol{y}) = (\boldsymbol{\nabla}(f))^T(\boldsymbol{u})\,(\boldsymbol{x} - \boldsymbol{y}) = \boldsymbol{Df}(\boldsymbol{u})\,(\boldsymbol{x} - \boldsymbol{y})$$

The line connecting $\boldsymbol{x}$ and $\boldsymbol{y}$ is often denoted by $[\boldsymbol{x}, \boldsymbol{y}]$ even though $\boldsymbol{x}$ and $\boldsymbol{y}$ are vectors.

Proof 10.6.1

Let $\boldsymbol{v} = \boldsymbol{x} - \boldsymbol{y}$ and define $P : \Re \to \Re^n$ by $P(t) = \boldsymbol{y} + t(\boldsymbol{x} - \boldsymbol{y})$. Then

$$P'(t) = \lim_{h \to 0} \frac{P(t + h) - P(t)}{h} = \boldsymbol{v}$$

Let $F : \Re \to \Re$ be defined by $F(t) = f(P(t)) = f(\boldsymbol{y} + t\boldsymbol{v})$. By the Mean Value Theorem for functions of one variable, there is a c between 0 and 1 so that

$$F(1) - F(0) = F'(c)\,(1 - 0)$$

Thus, by the chain rule

$$F'(c) = \boldsymbol{Df}(p(c))\,P'(c) = (\boldsymbol{\nabla}(f))^T(p(c))\,\boldsymbol{v} = \sum_{i=1}^{n} f_{x_i}(p(c))\,v_i$$

Now $F(1) = f(P(1)) = f(\boldsymbol{x})$ and $F(0) = f(P(0)) = f(\boldsymbol{y})$. Also, let $P(c) = \boldsymbol{y} + c(\boldsymbol{x} - \boldsymbol{y}) = \boldsymbol{u}$ which is a point on $[\boldsymbol{x}, \boldsymbol{y}]$. So we have shown

$$f(\boldsymbol{x}) - f(\boldsymbol{y}) = \boldsymbol{Df}(\boldsymbol{u})\,(\boldsymbol{x} - \boldsymbol{y})$$

■

10.6.1.1 Homework

Exercise 10.6.1 *Let $f(x, y) = x^2 + 3y^2$, $\boldsymbol{A} = (x_0, y_0) = (1, 2)$ and $\boldsymbol{B} = (x_1, y_1) = (3, 4)$. Find (u, v) on the line joining these two points so that $f(x_1, y_1) - f(x_0, y_0) =< \nabla f(u, v), \boldsymbol{B} - \boldsymbol{A} >$.*

Exercise 10.6.2 *Let* $f(x,y,z) = x^2 + 3y^2 + 4z^2$, $\boldsymbol{A} = (x_0, y_0, z_0) = (1,2,-1)$ *and* $\boldsymbol{B} = (x_1, y-1, z_1) = (3, 4.1)$. *Find* (u,v,w) *on the line joining these two points so that* $f(x_1,y_1,z_1) - f(x_0,y_0,z_0) =< \nabla f(u,v,w), \boldsymbol{B} - \boldsymbol{A} >$.

10.6.2 Hessian Approximations

We can now explain the most common approximation result for tangent planes.

Two Variables:
Let $h(t) = f(x_0 + t\Delta x, y_0 + t\Delta y)$. Then we know we can write $h(t) = h(0) + h'(0)t + h''(c)\frac{t^2}{2}$. Using the chain rule, we find

$$h'(t) = f_x(x_0 + t\Delta x, y_0 + t\Delta y)\Delta x + f_y(x_0 + t\Delta x, y_0 + t\Delta y)\Delta y$$

and

$$\begin{aligned} &h''(t) \\ &= \partial_x\Big(f_x(x_0 + t\Delta x, y_0 + t\Delta y)\Delta x + f_y(x_0 + t\Delta x, y_0 + t\Delta y)\Delta y \Big)\Delta x \\ &+ \partial_y\Big(f_x(x_0 + t\Delta x, y_0 + t\Delta y)\Delta x + f_y(x_0 + t\Delta x, y_0 + t\Delta y)\Delta y \Big)\Delta y \\ &= f_{xx}(x_0 + t\Delta x, y_0 + t\Delta y)(\Delta x)^2 + f_{yx}(x_0 + t\Delta x, y_0 + t\Delta y)(\Delta y)(\Delta x) \\ &+ f_{xy}(x_0 + t\Delta x, y_0 + t\Delta y)(\Delta x)(\Delta y) + f_{yy}(x_0 + t\Delta x, y_0 + t\Delta y)(\Delta y)^2 \end{aligned}$$

We can rewrite this in matrix-vector form as

$$h''(t) = \begin{bmatrix}\Delta x & \Delta y\end{bmatrix} \begin{bmatrix} f_{xx}(x_0 + t\Delta x, y_0 + t\Delta y) & f_{yx}(x_0 + t\Delta x, y_0 + t\Delta y) \\ f_{xy}(x_0 + t\Delta x, y_0 + t\Delta y) & f_{yy}(x_0 + t\Delta x, y_0 + t\Delta y) \end{bmatrix} \begin{bmatrix}\Delta x \\ \Delta y\end{bmatrix}$$

Of course, using the definition of $\boldsymbol{H}$, this can be rewritten as

$$h''(t) = \begin{bmatrix}\Delta x \\ \Delta y\end{bmatrix}^T \boldsymbol{H}(x_0 + t\Delta x, y_0 + t\Delta y) \begin{bmatrix}\Delta x \\ \Delta y\end{bmatrix}$$

Thus, our tangent plane approximation can be written as

$$h(1) = h(0) + h'(0)(1-0) + h''(c)\frac{1}{2}$$

for some c between 0 and 1. Substituting for the h terms, we find

$$\begin{aligned} f(x_0 + \Delta x, y_0 + \Delta y) &= f(x_0, y_0) + f_x(x_0, y_0)\Delta x + f_y(x_0, y_0)\Delta y \\ &+ \frac{1}{2}\begin{bmatrix}\Delta x \\ \Delta y\end{bmatrix}^T \boldsymbol{H}(x_0 + c\Delta x, y_0 + c\Delta y) \begin{bmatrix}\Delta x \\ \Delta y\end{bmatrix} \end{aligned}$$

Clearly, we have shown how to express the error in terms of second order partials. There is a point c between 0 and 1 so that

$$\mathscr{E}(x_0, y_0, \Delta x, \Delta y) = \frac{1}{2}\begin{bmatrix}\Delta x \\ \Delta y\end{bmatrix}^T \boldsymbol{H}(x_0 + c\Delta x, y_0 + c\Delta y) \begin{bmatrix}\Delta x \\ \Delta y\end{bmatrix}$$

Note, the error is a quadratic expression in terms of the Δx and Δy. We also now know for a function of two variables, $f(x,y)$, we can estimate the error made in approximating using the gradient at the given point (x_0, y_0) as follows: We have

$$f(x,y) = f(x_0,y_0) + \ < \nabla(f)(x_0,y_0), [x-x_0, y-y_0]^T > \\ + (1/2)[x-x_0, y-y_0] H(x_0 + c(x-x_0), y_0 + c(y-y_0))[x-x_0, y-y_0]^T$$

Three Variables:

Let's shift to using (x_1, x_2, x_3) instead of (x, y, z). Define let $h(t) = f(x_{01} + t\Delta_1, y_{01} + t\Delta_2, z_{01} + t\Delta_3)$. For convenience, let

$$\boldsymbol{x} \ = \ \begin{bmatrix} x_1 \\ x_2 \\ x_3 \end{bmatrix}, \quad \boldsymbol{x_0} = \begin{bmatrix} x_{01} \\ x_{02} \\ x_{03} \end{bmatrix}, \quad \boldsymbol{\Delta} = \begin{bmatrix} \Delta_1 \\ \Delta_2 \\ \Delta_3 \end{bmatrix}$$

Then, $h(t) = f(\boldsymbol{x_0} + t\boldsymbol{\Delta})$ and we know $h(t) = h(0) + h'(0)t + h''(c)\frac{t^2}{2}$. Using the chain rule, we find

$$h'(t) \ = \ \sum_{i=1}^{3} f_{x_i}(\boldsymbol{x_0} + t\boldsymbol{\Delta})\Delta_i$$

and

$$h''(t) \ = \ \sum_{i=1}^{3}\Big(\sum_{j=1}^{3} f_{x_i,x_j}(\boldsymbol{x_0} + t\boldsymbol{\Delta})\Delta_i\,\Delta_j\Big)$$

We can rewrite this in matrix-vector form as

$$h''(t) = \begin{bmatrix} \Delta_1 \\ \Delta 2 \\ \Delta_3 \end{bmatrix}^T \begin{bmatrix} f_{x_1,x_1}(\boldsymbol{x_0} + t\boldsymbol{\Delta}) & f_{x_1,x_2}(\boldsymbol{x_0} + t\boldsymbol{\Delta}) & f_{x_1,x_3}(\boldsymbol{x_0} + t\boldsymbol{\Delta}) \\ f_{x_2,x_1}(\boldsymbol{x_0} + t\boldsymbol{\Delta}) & f_{x_2,x_2}(\boldsymbol{x_0} + t\boldsymbol{\Delta}) & f_{x_2,x_3}(\boldsymbol{x_0} + t\boldsymbol{\Delta}) \\ f_{x_3,x_1}(\boldsymbol{x_0} + t\boldsymbol{\Delta}) & f_{x_3,x_2}(\boldsymbol{x_0} + t\boldsymbol{\Delta}) & f_{x_3,x_3}(\boldsymbol{x_0} + t\boldsymbol{\Delta}) \end{bmatrix} \begin{bmatrix} \Delta_1 \\ \Delta_2 \\ \Delta_3 \end{bmatrix}$$

Of course, using the definition of $\boldsymbol{H}$, this can be rewritten as

$$\begin{bmatrix} \Delta_1 \\ \Delta 2 \\ \Delta_3 \end{bmatrix}^T \quad \boldsymbol{H}(\boldsymbol{x_0} + t\boldsymbol{\Delta}) \begin{bmatrix} \Delta_1 \\ \Delta_2 \\ \Delta_3 \end{bmatrix} = \boldsymbol{\Delta}^T\,\boldsymbol{H}(\boldsymbol{x_0} + t\boldsymbol{\Delta})\,\boldsymbol{\Delta}$$

Thus, our tangent plane approximation can be written as

$$h(1) \ = \ h(0) + h'(0)(1-0) + h''(c)\frac{1}{2}$$

for some c between 0 and 1. Substituting for the h terms, we find

$$f(\boldsymbol{x_0} + \boldsymbol{\Delta}) \ = \ f(\boldsymbol{x_0}) + \sum_{i=1}^{3} f_{x_i}(\boldsymbol{x_0})\Delta_i + \frac{1}{2}\boldsymbol{\Delta}^T\boldsymbol{H}(\boldsymbol{x_0} + c\boldsymbol{\Delta})\,\boldsymbol{\Delta}$$

Clearly, we have shown how to express the error in terms of second order partials. There is a point c between 0 and 1 so that

$$\mathscr{E}(\boldsymbol{x_0}, \boldsymbol{\Delta}) \ = \ \boldsymbol{\Delta}^T\boldsymbol{H}(\boldsymbol{x_0} + c\boldsymbol{\Delta})\,\boldsymbol{\Delta}$$

You can see how the analysis does not change much if we did the case of n variables. We'll leave that to you!

Vector-Valued Functions:
We are now really doing a linear approximation to a vector-valued function and this is not a tangent plane, but the basic idea is the same. We now have a vector function $f : D \subset \Re^n \rightarrow \Re^m$, f would have m component functions. Now $\boldsymbol{\Delta}$ and $\boldsymbol{x_0}$ have n components. A similar analysis uses

$$h(t) = h(t) = f(\boldsymbol{x_0} + t\boldsymbol{\Delta}) = \begin{bmatrix} f_1(\boldsymbol{x_0} + t\boldsymbol{\Delta}) \\ \vdots \\ f_m(\boldsymbol{x_0} + t\boldsymbol{\Delta}) \end{bmatrix}$$

Using the chain rule, we find

$$h'(t) = \begin{bmatrix} \sum_{i=1}^n f_{1,x_i}(\boldsymbol{x_0} + t\boldsymbol{\Delta})\Delta_i \\ \vdots \\ \sum_{i=1}^n f_{m,x_i}(\boldsymbol{x_0} + t\boldsymbol{\Delta})\Delta_i \end{bmatrix} = \boldsymbol{J_f}(\boldsymbol{x_0} + t\boldsymbol{\Delta})\,\boldsymbol{\Delta}$$

and

$$h''(t) = \begin{bmatrix} \sum_{i=1}^n \sum_{j=1}^n f_{1,x_i,x_j}(\boldsymbol{x_0} + t\boldsymbol{\Delta})\Delta_i\,\Delta_j \\ \vdots \\ \sum_{i=1}^n \sum_{j=1}^n f_{1,x_i,x_j}(\boldsymbol{x_0} + t\boldsymbol{\Delta})\Delta_i\,\Delta_j \end{bmatrix}$$

Each component in the vector above can be written in a matrix-vector form as

$$h''(t) = \begin{bmatrix} \boldsymbol{\Delta}^T \boldsymbol{H_1}(\boldsymbol{x_0} + t\boldsymbol{\Delta})\,\boldsymbol{\Delta} \\ \vdots \\ \boldsymbol{\Delta}^T \boldsymbol{H_n}(\boldsymbol{x_0} + t\boldsymbol{\Delta})\,\boldsymbol{\Delta} \end{bmatrix}$$

where $\boldsymbol{H_i}$ is the Hessian for f_i. Thus, our linear approximation is

$$h(1) = h(0) + h'(0)(1-0) + h''(c)\frac{1}{2}$$

for some c between 0 and 1. Substituting for the h terms, we find

$$f(\boldsymbol{x_0} + \boldsymbol{\Delta}) = f(\boldsymbol{x_0}) + \boldsymbol{J_f}(\boldsymbol{x_0})\,\boldsymbol{\Delta} + \frac{1}{2}\begin{bmatrix} \boldsymbol{\Delta}^T \boldsymbol{H_1}(\boldsymbol{x_0} + c\boldsymbol{\Delta})\,\boldsymbol{\Delta} \\ \vdots \\ \boldsymbol{\Delta}^T \boldsymbol{H_n}(\boldsymbol{x_0} + c\boldsymbol{\Delta})\,\boldsymbol{\Delta} \end{bmatrix}$$

Each of these entries corresponds to the error $\mathscr{E}_i$ for component f_i and the vector error is

$$\mathscr{E} = \begin{bmatrix} \mathscr{E}_1 \\ \vdots \\ \mathscr{E}_n \end{bmatrix} \Longrightarrow \|\mathscr{E}\| = \sqrt{(\mathscr{E}_1)^2 + \ldots + (\mathscr{E}_n)^2}$$

Example 10.6.1 *Let*

$$f(x, y, z) = \begin{bmatrix} x^2y^4 + 2x + 3yz^2 + 10 \\ 4x^2 + 5z^3y^2 \end{bmatrix}$$

Estimate the linear approximation error about $(0, 0, 0)$.

Solution *We have*

$$\begin{aligned}
f_{1,x} &= 2xy^4+2, \quad f_{1,y}=4x^2y^3+3z^2, \quad f_{1,z}=6yz\\
f_{1,xx} &= 2y^4, \quad f_{1,xy}=8xy^3, \quad f_{1,xz}=0\\
f_{1,yx} &= 8xy^3, \quad f_{1,yy}=12x^2y^2, \quad f_{1,yz}=6z,\\
f_{1,zx} &= 0, \quad f_{1,zy}=6z, \quad f_{1,zz}=6y,\\
f_{2,x} &= 8x, \quad f_{2,y}=10z^3, \quad f_{2,z}=15y^2z^2\\
f_{2,xx} &= 8, \quad f_{2,xy}=0, \quad f_{2,xz}=0\\
f_{2,yx} &= 0, \quad f_{2,yy}=0, \quad f_{2,yz}=30z^2,\\
f_{2,zx} &= 0, \quad f_{2,zy}=30yz^2, \quad f_{2,zz}=30y^2z,
\end{aligned}$$

Hence,

$$\begin{aligned}
\boldsymbol{J}_f(x,y,z) &= \begin{bmatrix} 2xy^4+2 & 4x^2y^3+3z^2 & 6yz\\ 8x & 10z^3 & 15y^2z^2\end{bmatrix}\\
\boldsymbol{H}_1(x,y,z) &= \begin{bmatrix} 2y^4 & 8xy^3 & 0\\ 8xy^3 & 12x^2y^2 & 6z\\ 0 & 6z & 6y\end{bmatrix}, \quad \boldsymbol{H}_2(x,y,z)=\begin{bmatrix} 8 & 0 & 0\\ 0 & 0 & 30z^2\\ 0 & 20yz^2 & 30z^2\end{bmatrix}
\end{aligned}$$

This gives the errors

$$\begin{aligned}
\mathscr{E}_1(0,0,\boldsymbol{\Delta}) &= \frac{1}{2}\boldsymbol{\Delta}^T\boldsymbol{H_1}(c\Delta)\boldsymbol{\Delta}\\
\mathscr{E}_2(0,0,\boldsymbol{\Delta}) &= \frac{1}{2}\boldsymbol{\Delta}^T\boldsymbol{H_2}(c\Delta)\boldsymbol{\Delta}
\end{aligned}$$

for a c between 0 *and* 1*. To find how much error is made for at* $\boldsymbol{x_0}+\boldsymbol{\Delta}$*, we would approximate each error separately, for example, as we detail in (Peterson (21) 2020) for functions of two variables. All of these calculations are terribly messy! So at* $(0,0,0)$*, letting* (x^*,y^*,z^*) *denote the in between point* $\boldsymbol{x_0}+c\boldsymbol{\Delta}$*, we have*

$$\mathscr{E}_1(\boldsymbol{x_0},\boldsymbol{\Delta}) = \begin{bmatrix}\Delta_1 & \Delta_2 & \Delta_3\end{bmatrix}\begin{bmatrix} 2(y^*)^4 & 8x^*(y^*)^3 & 0\\ 8x^*(y^*)^3 & 12(x^*)^2(y^*)^2 & 6z^*\\ 0 & 6z^* & 6y^*\end{bmatrix}\begin{bmatrix}\Delta_1\\ \Delta_2\\ \Delta_3\end{bmatrix}$$

If we restrict our attention to points in $B(r,(0,0,0))$*, we see all the terms in* $\boldsymbol{\Delta}$ *satisfy* $\Delta_i| < r$ *and* $|x^*|$*,* $|y^*|$ *and* $|z^*|$ *are all less than* r*. Thus*

$$\begin{aligned}
\mathscr{E}_1(\boldsymbol{x_0},\boldsymbol{\Delta}) &= \begin{bmatrix}\Delta_1 & \Delta_2 & \Delta_3\end{bmatrix}\begin{bmatrix} 2(y^*)^4\Delta_1+8x^*(y^*)^3\Delta_2\\ 8x^*(y^*)^3\Delta_1+12(x^*)^2(y^*)^2\Delta_2+6z^*\Delta_3\\ 6z^*\Delta_2+6y^*\Delta_3\end{bmatrix}\\
&= (2(y^*)^4\Delta_1+8x^*(y^*)^3\Delta_2)\Delta_1\\
&\quad +(8x^*(y^*)^3\Delta_1+12(x^*)^2(y^*)^2\Delta_2+6z^*\Delta_3)\Delta_2\\
&\quad +(6z^*\Delta_2+6y^*\Delta_3)\Delta_3
\end{aligned}$$

Thus,

$$\begin{aligned}
|\mathscr{E}_1(\boldsymbol{x_0},\boldsymbol{\Delta})| &\leq (2r^5+8r^5)r+(8r^5+12r^5+6r^2)r+(6r^2+6r^2)r\\
&= 30r^6+18r^3
\end{aligned}$$

If we further restrict our attention to $r < 1$, *then* $r^6 < r^2$ *and* $r^3 < r^2$. *Thus, our overestimate of the error is* $|\mathscr{E}_1(\boldsymbol{x_0}, \boldsymbol{\Delta})| < 48r^2$. *A similar analysis will show an overestimate* $|\mathscr{E}_2(\boldsymbol{x_0}, \boldsymbol{\Delta})| < Kr^2$ *for some integer* K. *Hence, the total overestimate of the error is*

$$\|\mathscr{E}\| = \sqrt{(\mathscr{E}_1(\boldsymbol{x_0}, \boldsymbol{\Delta}))^2 + (\mathscr{E}_2(\boldsymbol{x_0}, \boldsymbol{\Delta}))^2} \leq \sqrt{(48+K)r^2} = \sqrt{(48+K)}\, r$$

We can therefore determine the value of r *which will guarantee the error in making our linear approximation is as small as we like.*

Comment 10.6.1 *These calculations are horrible if the point involved is different from the origin. We do some for the two variable case in (Peterson (21) 2020) so, in principle, we know what to do. But really intense!*

10.6.2.1 Homework

Exercise 10.6.3 *For the following function find its gradient and Hessian and the tangent plane approximation plus error at the point* $(0, 0, 0)$.

$$f(x, y, z) = x^2 + x^4y^3z^2 + 14x^2y^3z^4$$

Find the approximate error as a function of r.

Exercise 10.6.4 *Find the linear approximation of*

$$f(x, y, z) = \begin{bmatrix} 2x^2 + 3y^4 + 5z^4 \\ -2x + 5y - 8z^2 \end{bmatrix}$$

near $(0, 0, 0)$. *Find the approximate error* $\mathscr{E}_1$ *as a function of* r.

Exercise 10.6.5 *Find the linear approximation of*

$$f(x, y) = \begin{bmatrix} 2x^2 + 3y^4 \\ -2x^5y^2 + 5y^6 \end{bmatrix}$$

near $(0, 0)$. *Find the approximate errors* $\mathscr{E}_1$, $\mathscr{E}_2$ *and the total error* $\mathscr{E}$ *as a function of* r.

10.7 A Specific Coordinate Transformation

Given a function $g(x, y)$, the Laplacian of g is

$$\boldsymbol{\nabla}^2 g(x, y) = \frac{\partial^2 g}{\partial x^2}(x, y) + \frac{\partial^2 g}{\partial y^2}(x, y) = g_{xx}(x, y) + g_{yy}(x, y)$$

We usually leave off the (x, y) as it is understood and just write $\boldsymbol{\nabla}^2 g = g_{xx} + g_{yy}$. Let's convert this to polar coordinates. We assume g has continuous first and second order partials and so the mixed order partials match. Define $T : \Re^2 \to \Re^2$ by

$$T\left(\begin{bmatrix} r \\ \theta \end{bmatrix}\right) = \begin{bmatrix} T_1(r, \theta) \\ T_2(r, \theta) \end{bmatrix} = \begin{bmatrix} r\cos(\theta) \\ r\sin(\theta) \end{bmatrix} = \begin{bmatrix} x \\ y \end{bmatrix}$$

We casually abuse the notation here all the time. We usually write the column vector $\begin{bmatrix} r \\ \theta \end{bmatrix}$ as the ordered pair (r, θ) which is a lot like identifying the column vector with its transpose $\begin{bmatrix} r & \theta \end{bmatrix}$. There are similar confusions built into the notation for (x, y). This is because the vector space $\Re^2$ can be interpreted as ordered pairs, column vectors or row vectors and we generally just make these transformations without thinking about it. But, of course, they are always there. So we would normally write the statements above as

$$T(r,\theta) = \begin{bmatrix} T_1(r,\theta) \\ T_2(r,\theta) \end{bmatrix} = \begin{bmatrix} r\cos(\theta) \\ r\sin(\theta) \end{bmatrix} = (x,y)$$

Now define $f : \Re^2 \rightarrow \Re$ by

$$f(r,\theta) = (g \circ T)(r,\theta) = g(T_1(r,\theta), T_2(r,\theta)) = g(r\cos(\theta), r\sin(\theta)) = g(x,y)$$

We know

$$\begin{aligned} \boldsymbol{Df}(r,\theta) &= (\boldsymbol{\nabla}(f))^T(r,\theta) = \begin{bmatrix} f_r(r,\theta) & f_\theta(r,\theta) \end{bmatrix} \\ \boldsymbol{Dg}(x,y) &= (\boldsymbol{\nabla}(g))^T(x,y) = \begin{bmatrix} g_x(x,y) & g_y(x,y) \end{bmatrix} \end{aligned}$$

and

$$\boldsymbol{DT}(r,\theta) = \boldsymbol{J_T}(r,\theta) = \begin{bmatrix} T_{1,r}(r,\theta) & T_{1,\theta}(r,\theta) \\ T_{2,r}(r,\theta) & T_{2,\theta}(r,\theta) \end{bmatrix} = \begin{bmatrix} x_r(r,\theta) & x_\theta(r,\theta) \\ y_r(r,\theta) & y_\theta(r,\theta) \end{bmatrix}$$

Hence, by the chain rule

$$\boldsymbol{Df}(r,\theta) = \boldsymbol{Dg}(x,y)\,\boldsymbol{DT}(r,\theta)$$

or

$$\begin{bmatrix} f_r(r,\theta) & f_\theta(r,\theta) \end{bmatrix} = \begin{bmatrix} g_x(x,y) & g_y(x,y) \end{bmatrix} \begin{bmatrix} x_r(r,\theta) & x_\theta(r,\theta) \\ y_r(r,\theta) & y_\theta(r,\theta) \end{bmatrix}$$

Let's drop all of the (x,y) and (r,θ) terms as it just adds clutter. We have

$$\begin{bmatrix} f_r & f_\theta \end{bmatrix} = \begin{bmatrix} g_x & g_y \end{bmatrix} \begin{bmatrix} x_r & x_\theta \\ y_r & y_\theta \end{bmatrix}$$

which when multiplied out, gives our familiar expansions

$$\begin{aligned} f_r &= g_x\,x_r + g_y\,y_r \\ f_\theta &= g_x\,x_\theta + g_y\,y_\theta \end{aligned}$$

Thus, we have

$$\begin{aligned} f_r &= \cos(\theta)g_x + \sin(\theta)g_y \\ f_\theta &= -r\sin(\theta)g_x + r\cos(\theta)g_y \end{aligned}$$

This can be rewritten as

$$\begin{bmatrix} f_r \\ f_\theta \end{bmatrix} = \begin{bmatrix} \cos(\theta) & \sin(\theta) \\ -r\sin(\theta) & r\cos(\theta) \end{bmatrix} \begin{bmatrix} g_x \\ g_y \end{bmatrix}$$

Now we apply the chain rule again:

$$f_{\theta\theta} = \frac{\partial}{\partial\theta}(-r\sin(\theta))g_x + r\cos(\theta)g_y$$

$$
\begin{aligned}
&= (-r\sin(\theta)(g_{xx}x_\theta + g_{xy}y_\theta) + \frac{\partial}{\partial\theta}(-r\sin(\theta))g_x \\
&\quad +(r\cos(\theta))(g_{yx}x_\theta + g_{yy}y_\theta)) + \frac{\partial}{\partial\theta}(r\cos(\theta))g_y
\end{aligned}
$$

Thus,

$$
\begin{aligned}
f_{\theta\theta} &= (-r\sin(\theta))(-g_{xx}r\sin(\theta) + g_{xy}r\cos(\theta)) - r\cos(\theta)g_x \\
&\quad +(r\cos(\theta))(g_{yx}(-r\sin(\theta)) + g_{yy}(r\cos(\theta))) - r\sin(\theta)g_y \\
&= r^2(\sin^2(\theta)g_{xx} - \sin(\theta)\cos(\theta)g_{xy}) + r^2(-\sin(\theta)\cos(\theta)g_{yx} + \cos^2(\theta)g_{yy}) \\
&\quad -r(\cos(\theta)g_x + \sin(\theta)g_y)
\end{aligned}
$$

So, we now know

$$
\begin{aligned}
\frac{1}{r^2}f_{\theta\theta} &= \sin^2(\theta)g_{xx} - 2\sin(\theta)\cos(\theta)g_{xy} + \cos^2(\theta)g_{yy} - \frac{1}{r}(\cos(\theta)g_x + \sin(\theta)g_y) \\
&= \sin^2(\theta)g_{xx} - 2\sin(\theta)\cos(\theta)g_{xy} + \cos^2(\theta)g_{yy} - \frac{1}{r}f_r
\end{aligned}
$$

Hence, we have shown

$$
\frac{1}{r^2}f_{\theta\theta} + \frac{1}{r}f_r = \sin^2(\theta)g_{xx} - 2\sin(\theta)\cos(\theta)g_{xy} + \cos^2(\theta)g_{yy}
$$

Next, we need

$$
\begin{aligned}
f_{rr} &= \frac{\partial}{\partial r}(\cos(\theta)g_x + \sin(\theta)g_y) \\
&= \cos(\theta)(g_{xx}x_r + g_{xy}y_r + \frac{\partial}{\partial r}(\cos(\theta))g_x + \sin(\theta)(g_{yx}x_r + g_{yy}y_r)) \\
&\quad +\frac{\partial}{\partial r}(\sin(\theta))g_y \\
&= \cos(\theta)(g_{xx}\cos(\theta) + g_{xy}\sin(\theta)) + \sin(\theta)(g_{yx}\cos(\theta) + g_{yy}\sin(\theta))
\end{aligned}
$$

Thus,

$$
f_{rr} = \cos^2(\theta)g_{xx} + 2\cos(\theta)\sin(\theta)g_{xy} + \sin^2(\theta)g_{yy}
$$

Combining

$$
\begin{aligned}
f_{rr} + \frac{1}{r^2}f_{\theta\theta} + \frac{1}{r}f_r &= \sin^2(\theta)g_{xx} - 2\sin(\theta)\cos(\theta)g_{xy} + \cos^2(\theta)g_{yy} \\
&\quad + \cos^2(\theta)g_{xx} + 2\cos(\theta)\sin(\theta)g_{xy} + \sin^2(\theta)g_{yy} \\
&= g_{xx} + g_{yy}
\end{aligned}
$$

So the Laplacian in polar coordinates is

$$
\boldsymbol{\nabla}^2_{r\theta}f = f_{rr} + \frac{1}{r^2}f_{\theta\theta} + \frac{1}{r}f_r = \boldsymbol{\nabla}^2_{xy}g
$$

where $f(r,\theta) = g(T(r,\theta))$.

10.7.1 Homework

Converting partial differential equations in Cartesian coordinates into their equivalent forms under a nonlinear coordinate transformation (these are called curvilinear transformations) is a really useful idea. We will leave much of the detail of the following problems to you as they are nicely detailed and intense. Remember, going through this sort of thing on your own is a very useful exercise and helps to build your personal skill sets. As a reference, any book on old-fashioned advanced calculus and mathematical physics goes over this stuff. Stay away from the newer books and pick up an old text. A good choice is (R. Wrede and M. Spiegel (26) 2010).

Exercise 10.7.1 *Find the Laplacian in cylindrical coordinates.*

Exercise 10.7.2 *Find the Laplacian in spherical coordinates.*

Exercise 10.7.3 *Find the Laplacian in paraboloidal coordinates.*

Exercise 10.7.4 *Find the Laplacian in ellipsoidal coordinates.*

Chapter 11

Multivariable Extremal Theory

Now let's look at the extremes of functions of many variables.

11.1 Differentiability and Extremals

We are now ready to connect the value of partial derivatives at an extremum such as a minimum or maximum of a scalar function of n variables. We know continuous functions on compact domains possess such extreme values, but we do not really know how to find them. Here is a start.

Theorem 11.1.1 At an Interior Extreme Point, the Partials Vanish

Let $f : D \subset \Re^n \rightarrow \Re$ have an extreme value at the interior point $\boldsymbol{x_0}$ in D. Then if all the partials $f_{x_i}(\boldsymbol{x_0})$ exist, they must be zero.

Proof 11.1.1

Let $\boldsymbol{x_0}$ be an extreme value for f which is an interior point. For convenience of exposition, let's assume the extremum is a maximum locally. So there is a radius $r > 0$ so that if $\boldsymbol{y} \in B(r, \boldsymbol{x_0})$, $f(\boldsymbol{y}) \leq f(\boldsymbol{x_0})$. Then if each $f_{x_i}(\boldsymbol{x_0})$ exists, for small enough $h > 0$, we must have

$$f(\boldsymbol{x_0} + h\boldsymbol{E_i}) \quad \leq \quad f(\boldsymbol{x_0})$$

Thus,

$$\frac{f(\boldsymbol{x_0} + h\boldsymbol{E_i}) - f(\boldsymbol{x_0})}{h} \leq 0 \implies \lim_{h \rightarrow 0^+} \frac{f(\boldsymbol{x_0} + h\boldsymbol{E_i}) - f(\boldsymbol{x_0})}{h} \leq 0$$

This tells us $(f_{x_i}(\boldsymbol{x_0}))^+ \leq 0$. On the other hand, for sufficiently small $h < 0$, we have

$$f(\boldsymbol{x_0} + h\boldsymbol{E_i}) \quad \leq \quad f(\boldsymbol{x_0})$$

Thus, because h is negative

$$\frac{f(\boldsymbol{x_0} + h\boldsymbol{E_i}) - f(\boldsymbol{x_0})}{h} \geq 0 \implies \lim_{h \rightarrow 0^-} \frac{f(\boldsymbol{x_0} + h\boldsymbol{E_i}) - f(\boldsymbol{x_0})}{h} \geq 0$$

This tells us $(f_{x_i}(\boldsymbol{x_0}))^- \geq 0$. Since we assume $f_{x_i}(\boldsymbol{x_0})$ exists, we must have $(f_{x_i}(\boldsymbol{x_0}))^- = (f_{x_i}(\boldsymbol{x_0}))^+ = f_{x_i}(\boldsymbol{x_0})$. Thus $0 \leq f_{x_i}(\boldsymbol{x_0}) \leq 0$ which tells us $f_{x_i}(\boldsymbol{x_0}) = 0$. We can do this for each index i. Thus, we have shown that if $\boldsymbol{x_0}$ is an interior point which corresponds to a local max-

imum, then if the partials $f_{x_i}(\boldsymbol{x_0})$ exist they must be zero; i.e. $\boldsymbol{\nabla}(f) = \boldsymbol{0}$. We can do a similar argument for a local minimum. ■

Comment 11.1.1 *Now if $f : D \subset \Re^n \rightarrow \Re$ is differentiable, then $\boldsymbol{Df} = (\boldsymbol{\nabla}(f))^T$ and so the result above tells that if f is differentiable at an extreme point which is an interior point $\boldsymbol{x_0}$ then $\boldsymbol{Df}(\boldsymbol{x_0}) = \boldsymbol{0}$.*

Points where the gradient is zero are thus important. Points where extreme behavior occurs are called **critical points** of f. Once we have found them, the next question is to classify them as minima, maxima or something else. For completeness, we state a formal definition.

Definition 11.1.1 Critical Points

Let $f : D \subset \Re^n \rightarrow \Re$. The point $\boldsymbol{x_0}$ is called a **critical point** *of f if $\boldsymbol{\nabla}(f)(\boldsymbol{x_0}) = \boldsymbol{0}$, the $\boldsymbol{\nabla}(f)(\boldsymbol{x_0})$ fails to exist or $\boldsymbol{x_0}$ is a boundary point of D. Recall this means $B(r, \boldsymbol{x_0})$ has points of D and D^C for all choices of $r > 0$.*

We have already discussed conditions for classifying critical points in the two variable case in (Peterson (21) 2020). For completeness, we will go over this again.

11.2 Second Order Extremal Conditions

At a place where the extremum of a function of n variables might occur with $\boldsymbol{\nabla}(f)(\boldsymbol{x_0}) = \boldsymbol{0}$, it is clear the tangent plane will be **flat** as all of the partials will be zero. Of course, we can only see the flatness of this plane in three space but we will use that idea to help our intuition in higher dimensional spaces. From our discussion of differentiability, we know

$$f(\boldsymbol{x_0} + \boldsymbol{\Delta}) = f(\boldsymbol{x_0}) + \sum_{i=1}^{n} f_{x_i}(\boldsymbol{x_0})\Delta_i + \frac{1}{2}\boldsymbol{\Delta}^T \boldsymbol{H}(\boldsymbol{x_0} + c\boldsymbol{\Delta})\,\boldsymbol{\Delta}$$

as long as all the second order partials of f all exist locally at $\boldsymbol{x_0}$. Here,

$$\frac{1}{2}\boldsymbol{\Delta}^T \boldsymbol{H}(\boldsymbol{x_0} + c\boldsymbol{\Delta})\,\boldsymbol{\Delta} = \frac{1}{2}\sum_{i=1}^{n}\sum_{j=1}^{n} f_{x_i x_j}(\boldsymbol{x_0} + c\boldsymbol{\Delta})\Delta_i\Delta_j$$

Now since $\boldsymbol{x_0}$ is a critical point with $\boldsymbol{\nabla}(f)(\boldsymbol{x_0}) = \boldsymbol{0}$, we can write

$$\begin{aligned} f(\boldsymbol{x_0} + \boldsymbol{\Delta}) &= f(\boldsymbol{x_0}) + \frac{1}{2}\sum_{i=1}^{n}\sum_{j=1}^{n} f_{x_i x_j}(\boldsymbol{x_0} + c\boldsymbol{\Delta})\Delta_i\Delta_j \\ &= f(\boldsymbol{x_0}) + \frac{1}{2}\boldsymbol{\Delta}^T \boldsymbol{H}(\boldsymbol{x_0} + c\boldsymbol{\Delta})\,\boldsymbol{\Delta} \end{aligned}$$

We can decide if we have a minimum or a maximum easily. The matrix $\boldsymbol{H}(\boldsymbol{x_0} + c\boldsymbol{\Delta})$ is an $n \times n$ matrix. If we assume all the second order partials are continuous locally at $\boldsymbol{x_0}$, then $det\,\boldsymbol{H}(\boldsymbol{x_0} + c\boldsymbol{\Delta})$ is also continuous locally at $\boldsymbol{x_0}$. Hence, we can say there is a $B(r, \boldsymbol{x_0})$ where both $\boldsymbol{H}$ and $det\boldsymbol{H}$ are continuous. We also know if a continuous function is positive at $\boldsymbol{x_0}$, it is strictly positive locally at $\boldsymbol{x_0}$. Similarly, if a continuous function is negative at $\boldsymbol{x_0}$, it is strictly negative locally at $\boldsymbol{x_0}$. Thus, there is a $B(r_1, \boldsymbol{x_0})$ where $r_1 < r$ so that

- If $det\boldsymbol{H}(\boldsymbol{x_0}) > 0$, then $det\boldsymbol{H}(\boldsymbol{y}) > 0$ if $\boldsymbol{y} \in B(r_1, \boldsymbol{x_0})$.
- If $det\boldsymbol{H}(\boldsymbol{x_0}) < 0$, then $det\boldsymbol{H}(\boldsymbol{y}) < 0$ if $\boldsymbol{y} \in B(r_1, \boldsymbol{x_0})$.

So if we start with $\|\boldsymbol{\Delta}\| < r_1$, the c we obtain above will give us $\boldsymbol{x_0} + c\boldsymbol{\Delta} \in B(r_1, \boldsymbol{x_0})$. Thus, we will be able to say

- If $det\boldsymbol{H}(\boldsymbol{x_0}) > 0$, then $det\boldsymbol{H}(\boldsymbol{x_0} + c\boldsymbol{\Delta}) > 0$
- If $det\boldsymbol{H}(\boldsymbol{x_0}) < 0$, then $det\boldsymbol{H}(\boldsymbol{x_0} + c\boldsymbol{\Delta}) < 0$

Since all the second order partials are continuous locally, we know the mixed order partials must be the same. Hence the Hessian here is a symmetric matrix.

11.2.1 Positive and Negative Definite Hessians

We can get the algebraic sign we need for $det\ \boldsymbol{H}(\boldsymbol{x_0})$ using the notion of positive and negative definite matrices. The conditions above become:

- If $\boldsymbol{H}(\boldsymbol{x_0})$ is positive definite, then $\boldsymbol{\Delta}^T\boldsymbol{H}(\boldsymbol{x_0})\ \boldsymbol{\Delta} > 0$ for any $\boldsymbol{\Delta}$. Since $\boldsymbol{\Delta}^T\boldsymbol{H}(\boldsymbol{x_0})\ \boldsymbol{\Delta}$ is continuous at $\boldsymbol{x_0}$ and $\boldsymbol{\Delta}^T\boldsymbol{H}(\boldsymbol{x_0})\ \boldsymbol{\Delta} > 0$, this means there is a radius $r_2 < r_1$ so that $\boldsymbol{\Delta}^T\boldsymbol{H}(\boldsymbol{y})\ \boldsymbol{\Delta} > 0$ in $B(r_2, \boldsymbol{x_0})$. By choosing $\|\boldsymbol{\Delta}\| < r_2$, we guarantee c is small enough so that $\boldsymbol{x_0} + c\boldsymbol{\Delta} \in B(r_2, \boldsymbol{x_0})$. Hence, $\boldsymbol{\Delta}^T\boldsymbol{H}(\boldsymbol{x_0} + c\boldsymbol{\Delta})\ \boldsymbol{\Delta} > 0$. Thus, we have

  $$f(\boldsymbol{x_0} + \boldsymbol{\Delta}) \quad = \quad f(\boldsymbol{x_0}) + \textbf{ a positive number}$$

  implying $f(\boldsymbol{x_0} + \boldsymbol{\Delta}) > f(\boldsymbol{x_0})$ locally. Hence, f has a local minimum at $\boldsymbol{x_0}$. We also know since $\boldsymbol{H}(\boldsymbol{x_0})$ is positive definite, all of its eigenvalues are positive.

- If $\boldsymbol{H}(\boldsymbol{x_o})$ is negative definite, then we can analyze just as we did above. We have $\boldsymbol{\Delta}^T\boldsymbol{H}(\boldsymbol{x_0})$ $\boldsymbol{\Delta} < 0$ for any $\boldsymbol{\Delta}$. Since $\boldsymbol{\Delta}^T\boldsymbol{H}(\boldsymbol{x_0})\boldsymbol{\Delta}$ is continuous at $\boldsymbol{x_0}$ and $\boldsymbol{\Delta}^T\boldsymbol{H}(\boldsymbol{x_0})\boldsymbol{\Delta} < 0$, this means there is a radius $r_2 < r_1$ so that $\boldsymbol{\Delta}^T\boldsymbol{H}(\boldsymbol{y})\ \boldsymbol{\Delta} < 0$ in $B(r_2, \boldsymbol{x_0})$. By choosing $\|\boldsymbol{\Delta}\| < r_2$, we guarantee c is small enough so that $\boldsymbol{x_0} + c\boldsymbol{\Delta} \in B(r_2, \boldsymbol{x_0})$. Hence, $\boldsymbol{\Delta}^T\boldsymbol{H}(\boldsymbol{x_0} + c\boldsymbol{\Delta})\ \boldsymbol{\Delta} > 0$. Thus, we have

  $$f(\boldsymbol{x_0} + \boldsymbol{\Delta}) \quad = \quad f(\boldsymbol{x_0}) + \textbf{ a negative number}$$

  implying $f(\boldsymbol{x_0} + \boldsymbol{\Delta}) > f(\boldsymbol{x_0})$ locally. Hence, f has a local maximum at $\boldsymbol{x_0}$. We also know since $\boldsymbol{H}(\boldsymbol{x_0})$ is negative definite, all of its eigenvalues are negative.

It is time we mentioned the idea of the **Principle Minors** of a matrix. The principle minors of a $n \times n$ matrix $\boldsymbol{A}$ are defined as follows:

$$\begin{aligned}
\boldsymbol{M_1} &= A_{11}, \quad (\boldsymbol{PM})_{\boldsymbol{1}} = \det(M_1) = A_{11} \\
\boldsymbol{M_2} &= \begin{bmatrix} A_{11} & A_{12} \\ A_{12} & A_{22} \end{bmatrix}, \quad (\boldsymbol{PM})_{\boldsymbol{2}} = \det(M_2) \\
\boldsymbol{M_3} &= \begin{bmatrix} A_{11} & A_{12} & A_{13} \\ A_{12} & A_{22} & A_{23} \\ A_{13} & A_{23} & A_{33} \end{bmatrix}, \quad (\boldsymbol{PM})_{\boldsymbol{3}} = \det(M_3)
\end{aligned}$$

and so forth. Now our interests are in symmetric $n \times n$ matrices $\boldsymbol{A}$. Recall our symmetric matrix is positive definite if $\boldsymbol{x^T A x} > 0$ for all nonzero $\boldsymbol{x} \in \Re^n$. Let $\boldsymbol{E} = \{\boldsymbol{E_1}, \ldots, \boldsymbol{E_n}\}$ be the standard orthonormal basis for $\Re^n$. Now $\boldsymbol{A} = \boldsymbol{PDP^T}$ where $\boldsymbol{P}$ is the matrix whose columns are the eigenvectors of $\boldsymbol{A}$ normalized to have length one. We know these eigenvectors then form an orthonormal basis of $\Re^n$ also. The matrix $\boldsymbol{D}$ is a diagonal matrix whose entries are the eigenvalues of $\boldsymbol{A}$ ranked as $|\lambda_1| \geq |\lambda_2| \geq \ldots \geq |\lambda_n|$.

- Let $\boldsymbol{x} = \begin{bmatrix} x_1 \\ 0 \\ \vdots \\ 0 \end{bmatrix}$ be nonzero. Then

$$\boldsymbol{x}^T \boldsymbol{A} \boldsymbol{x} > 0 \Longrightarrow A_{11} > 0$$

This says the minor $\boldsymbol{M_1}$ is positive definite and so it has positive eigenvalues. Of course this is a silly way to say it as it is just a scalar!

- Let $\boldsymbol{x} = \begin{bmatrix} x_1 \\ x_2 \\ 0 \\ \vdots \\ 0 \end{bmatrix}$ be nonzero. Then upon multiplication we find

$$.\boldsymbol{x}^T \boldsymbol{A} \boldsymbol{x} > 0 \Longrightarrow \begin{bmatrix} x_1 \\ x_2 \end{bmatrix}^T \begin{bmatrix} A_{11} & A_{12} \\ A_{21} & A_{22} \end{bmatrix} \begin{bmatrix} x_1 \\ x_2 \end{bmatrix} > 0$$

This says the minor $\boldsymbol{M_2}$ is positive definite and so it has positive eigenvalues. Hence $(\boldsymbol{PM})_2 > 0$.

- Let $\boldsymbol{x} = \begin{bmatrix} x_1 \\ x_2 \\ x_3 \\ 0 \\ \vdots \\ 0 \end{bmatrix}$ be nonzero. Then upon multiplication we find

$$.\boldsymbol{x}^T \boldsymbol{A} \boldsymbol{x} > 0 \Longrightarrow \begin{bmatrix} x_1 \\ x_2 \\ x_3 \end{bmatrix}^T \begin{bmatrix} A_{11} & A_{12} & A_{13} \\ A_{21} & A_{22} & A_{23} \\ A_{31} & A_{32} & A_{33} \end{bmatrix} \begin{bmatrix} x_1 \\ x_2 \\ x_3 \end{bmatrix} > 0$$

This says the minor $\boldsymbol{M_3}$ is positive definite and so it has positive eigenvalues. Hence $(\boldsymbol{PM})_3 > 0$.

- We can clearly continue this process and we find all the minors $\boldsymbol{M_i}$ are positive definite with $(\boldsymbol{PM})_i > 0$.

- We also know the original matrix $\boldsymbol{A}$ is equivalent to the diagonal matrix D and $\det(\boldsymbol{A}) = (\lambda_1)(\lambda_2)\cdots(\lambda_n)$ and since all these eigenvalues are positive, the algebraic signs of the determinants of the minors of $\boldsymbol{D}$ **match** the algebraic signs of the determinants of the minors of $\boldsymbol{A}$.

We can conclude the algebraic sign pattern of the determinants of the minors of $\boldsymbol{A}$ match the sign pattern of the determinants of the minors of $\boldsymbol{D}$.

We can do a similar analysis of the case where $\boldsymbol{A}$ is negative definite.

- Let $\boldsymbol{x} = \begin{bmatrix} x_1 \\ 0 \\ \vdots \\ 0 \end{bmatrix}$ be nonzero. Then

$$\boldsymbol{x}^{\boldsymbol{T}} \boldsymbol{A} \boldsymbol{x} < 0 \Longrightarrow A_{11} < 0$$

  This says the minor $\boldsymbol{M}_{\mathbf{1}}$ is negative definite and so it has negative eigenvalues. Again, this is a silly way to say it as it is just a scalar!

- Let $\boldsymbol{x} = \begin{bmatrix} x_1 \\ x_2 \\ 0 \\ \vdots \\ 0 \end{bmatrix}$ be nonzero. Then upon multiplication we find

$$.\boldsymbol{x}^{\boldsymbol{T}} \boldsymbol{A} \boldsymbol{x} < 0 \Longrightarrow \begin{bmatrix} x_1 \\ x_2 \end{bmatrix}^T \begin{bmatrix} A_{11} & A_{12} \\ A_{21} & A_{22} \end{bmatrix} \begin{bmatrix} x_1 \\ x_2 \end{bmatrix} < 0$$

  This says the minor $\boldsymbol{M}_{\mathbf{2}}$ is negative definite and so it has two negative eigenvalues. Hence $(\boldsymbol{PM})_{\mathbf{2}}$ is the product of these eigenvalues and so it is positive.

- Let $\boldsymbol{x} = \begin{bmatrix} x_1 \\ x_2 \\ x_3 \\ 0 \\ \vdots \\ 0 \end{bmatrix}$ be nonzero. Then upon multiplication we find

$$.\boldsymbol{x}^{\boldsymbol{T}} \boldsymbol{A} \boldsymbol{x} < 0 \Longrightarrow \begin{bmatrix} x_1 \\ x_2 \\ x_3 \end{bmatrix}^T \begin{bmatrix} A_{11} & A_{12} & A_{13} \\ A_{21} & A_{22} & A_{23} \\ A_{31} & A_{32} & A_{33} \end{bmatrix} \begin{bmatrix} x_1 \\ x_2 \\ x_3 \end{bmatrix} < 0$$

  This says the minor $\boldsymbol{M}_{\mathbf{3}}$ is negative definite and so it has three negative eigenvalues. Hence $\boldsymbol{PM}_{\mathbf{3}}$ is the product of these eigenvalues and so it is negative

- We can clearly continue this process and we find all the minors $\boldsymbol{M}_{\boldsymbol{i}}$ are negative definite with $(\boldsymbol{PM})_{\boldsymbol{i}} = (-1)^i$.

- We also know the original matrix $\boldsymbol{A}$ is equivalent to the diagonal matrix D and $\det(\boldsymbol{A}) = (\lambda_1)(\lambda_2)\cdots(\lambda_n)$ and since all these eigenvalues are negative, the algebraic signs of the determinants of the minors of $\boldsymbol{D}$ **match** the algebraic signs of the determinants of the minors of $\boldsymbol{A}$.

Note we have a local minimum if the algebraic sign of the minors is $+ + + \ldots +$. We can easily find the eigenvalues in MATLAB and determine this. For $n > 2$, these minor calculations are intense. Note we are not phrasing the conditions for the local minimum in terms of the second order partials at $\boldsymbol{x_0}$ although we can.

Also, we have a local maximum if the algebraic sign of the minors is $- + - \ldots$. Again, for larger matrices, finding the eigenvalues of $\boldsymbol{A}$ using MATLAB is probably the best strategy.

Since the Hessian $\boldsymbol{H}(\boldsymbol{y})$ is symmetric, its eigenvectors determine an orthonormal basis for $\Re^n$, $\{\boldsymbol{E}_1(\boldsymbol{y}), \ldots, \boldsymbol{E}_n(\boldsymbol{y})\}$. Defining the matrix

$$\boldsymbol{P}(\boldsymbol{y}) = \begin{bmatrix}\boldsymbol{E}_1(\boldsymbol{y}) & \ldots & \boldsymbol{E}_n(\boldsymbol{y})\end{bmatrix}$$

whose i^{th} column is $\boldsymbol{E}_i(\boldsymbol{y})$, we have

$$\begin{aligned}\boldsymbol{H}(\boldsymbol{x}_0) &= \boldsymbol{P}(\boldsymbol{x}_0)\boldsymbol{D}(\boldsymbol{x}_0)\boldsymbol{P}^T(\boldsymbol{x}_0)\\ \boldsymbol{H}(\boldsymbol{x}_0 + c\boldsymbol{\Delta}) &= \boldsymbol{P}(\boldsymbol{x}_0 + c\boldsymbol{\Delta})\boldsymbol{D}(\boldsymbol{x}_0 + c\boldsymbol{\Delta})\boldsymbol{P}^T(\boldsymbol{x}_0 + c\boldsymbol{\Delta})\end{aligned}$$

where $\boldsymbol{D}(\boldsymbol{y})$ is the diagonal matrix of the eigenvalues of $\boldsymbol{H}(\boldsymbol{y})$.

Now if $\boldsymbol{H}(\boldsymbol{x}_0)$ was not positive or negative definite at a critical point, then there must be two unit vectors $\boldsymbol{U}$ and $\boldsymbol{V}$ so that $\boldsymbol{U}^T\boldsymbol{H}(\boldsymbol{x}_0)\boldsymbol{U} < 0$ and $\boldsymbol{V}^T\boldsymbol{H}(\boldsymbol{x}_0)\boldsymbol{V} > 0$. A little thought shows $\boldsymbol{U}$ and $\boldsymbol{V}$ must be linearly independent. Hence $\boldsymbol{H}(\boldsymbol{x}_0)$ is positive definite on the line determined by $\boldsymbol{V}$ passing through $\boldsymbol{x}_0$ and it is negative definite on the line determined by $\boldsymbol{U}$ passing through $\boldsymbol{x}_0$. Further, by the same arguments presented earlier, we can see for sufficiently small c, there are constants d_1 and d_2 so

$$\begin{aligned}f(\boldsymbol{x}_0 + c\boldsymbol{U}) &= f(\boldsymbol{x}_0) + \frac{1}{2}c\boldsymbol{U}^T\boldsymbol{H}(\boldsymbol{x}_0 + d_1c\boldsymbol{U})\, c\boldsymbol{U} = f(\boldsymbol{x}_0) - \textbf{A Positive Number}\\ f(\boldsymbol{x}_0 + c\boldsymbol{V}) &= f(\boldsymbol{x}_0) + \frac{1}{2}c\boldsymbol{V}^T\boldsymbol{H}(\boldsymbol{x}_0 + d_2c\boldsymbol{V})\, c\boldsymbol{V} = f(\boldsymbol{x}_0) + \textbf{A Positive Number}\end{aligned}$$

Thus, in the direction $\boldsymbol{U}$, f goes down and so locally $f(\boldsymbol{x}_0)$ is a maximum. However, in the direction $\boldsymbol{V}$, f goes up and so locally $f(\boldsymbol{x}_0)$ is a minimum. This, of course, is what we call saddle behavior at the critical point.

Here is another way to look at these things. If $\boldsymbol{H}(\boldsymbol{x}_0)$ is positive or negative definite, its minors are either strictly positive or negative and hence, because $det\ \boldsymbol{H}(\boldsymbol{x}_0)$ is continuous at $\boldsymbol{x}_0$, there is a radius $r_2 < r_1$, where all the minors have the same algebraic sign as long as $\boldsymbol{x}_0 + c\boldsymbol{\Delta} \in B(r_3, \boldsymbol{x}_0)$. By choosing $\|\boldsymbol{\Delta}\|$ sufficiently small we can guarantee this. Hence, the remarks above concerning the minors and eigenvalues still hold. We have

$$\begin{aligned}f(\boldsymbol{x}_0 + \boldsymbol{\Delta}) &= f(\boldsymbol{x}_0) + \frac{1}{2}\boldsymbol{\Delta}^T\boldsymbol{H}(\boldsymbol{x}_0 + c\boldsymbol{\Delta})\,\boldsymbol{\Delta}\\ &= f(\boldsymbol{x}_0) + \frac{1}{2}\boldsymbol{\Delta}^T\boldsymbol{P}(\boldsymbol{x}_0 + c\boldsymbol{\Delta})\boldsymbol{D}(\boldsymbol{x}_0 + c\boldsymbol{\Delta})\boldsymbol{P}^T(\boldsymbol{x}_0 + c\boldsymbol{\Delta})\boldsymbol{\Delta}\\ &= f(\boldsymbol{x}_0) + \frac{1}{2}\Big(\boldsymbol{P}^T(\boldsymbol{x}_0 + c\boldsymbol{\Delta})\boldsymbol{\Delta}\Big)^T\boldsymbol{D}(\boldsymbol{x}_0 + c\boldsymbol{\Delta})\Big(\boldsymbol{P}^T(\boldsymbol{x}_0 + c\boldsymbol{\Delta})\boldsymbol{\Delta}\Big)\end{aligned}$$

Let $\boldsymbol{\xi}_0^c = \boldsymbol{P}^T(\boldsymbol{x}_0 + c\boldsymbol{\Delta})\boldsymbol{\Delta}$. Then we have

$$f(\boldsymbol{x}_0 + \boldsymbol{\Delta}) = f(\boldsymbol{x}_0) + \frac{1}{2}\boldsymbol{\xi}_0^{cT}\boldsymbol{D}(\boldsymbol{x}_0 + c\boldsymbol{\Delta})\boldsymbol{\xi}_0^c$$

Let the eigenvalues of $\boldsymbol{H}(\boldsymbol{x}_0 + c\boldsymbol{\Delta})$ be denoted by λ_{i0}^c for convenience. Then we can rewrite as

$$f(\boldsymbol{x}_0 + \boldsymbol{\Delta}) = f(\boldsymbol{x}_0) + \frac{1}{2}\left(\lambda_0 1^c(\xi_{10}^c)^2 + \ldots + \lambda_{n0}^c(\xi_{n0}^c)^2\right)$$

and the signs of these eigenvalues are the same as the signs of the eigenvalues for $\boldsymbol{H}(\boldsymbol{x}_0)$ by our choice of $\|\boldsymbol{\Delta}\| < r_3$. Note the following: the equation above makes it very clear that

- If $\boldsymbol{H}(\boldsymbol{x_0})$ is positive definite, so is $\boldsymbol{H}(\boldsymbol{x_0} + c\boldsymbol{\Delta})$ and so $\lambda_{10}^c(\xi_{10}^c)^2 + \ldots + \lambda_{n0}^c(\xi_{n0}^c)^2 > 0$ gives us a local minimum.
- If $\boldsymbol{H}(\boldsymbol{x_0})$ is negative definite, so is $\boldsymbol{H}(\boldsymbol{x_0} + c\boldsymbol{\Delta})$ and so $\lambda_{10}^c(\xi_{10}^c)^2 + \ldots + \lambda_{n0}^c(\xi_{n0}^c)^2 < 0$ gives us a local maximum.

Also, since

$$f(\boldsymbol{x_0} + \boldsymbol{\Delta}) = f(\boldsymbol{x_0}) + \frac{1}{2}\left(\lambda_{10}^c(\xi_{10}^c)^2 + \ldots + \lambda_{n0}^c(\xi_{n0}^c)^2 \right)$$

if the eigenvalues are not all the same sign, we can have what is called a saddle structure. Suppose λ_{i0}^c and λ_{j0}^c had different signs for some $i \neq j$. For concreteness, suppose $\lambda_{i0}^c > 0$ and let $\lambda_{i0}^c = \alpha > 0$. We also have $\lambda_{j0}^c < 0$. Write $\lambda_{j0}^c = -\beta$ where $\beta > 0$. Then on the corresponding surface trace for the ith eigenvector

$$f(\boldsymbol{x_0} + \boldsymbol{\Delta}) = f(\boldsymbol{x_0}) + \frac{1}{2}\,\alpha(\xi_{i0}^c)^2$$

Hence, on this trace, f has a local minimum at $\boldsymbol{x_0}$. On the additional trace for the jth eigenvector

$$f(\boldsymbol{x_0} + \boldsymbol{\Delta}) = f(\boldsymbol{x_0}) - \frac{1}{2}\beta(\xi_j^c)^2$$

which shows f has a local maximum at $\boldsymbol{x_0}$. This type of behavior, where f behaves like it has a minimum along some directions and a maximum along others is the traditional **saddle** behavior. Hence, by several arguments, we conclude if $\boldsymbol{H}(\boldsymbol{x_0})$ is not positive or negative definite, it will have eigenvalues of different sign and therefore have a saddle at the critical point.

Of course, if $det\ \boldsymbol{H}(\boldsymbol{x_0}) = 0$, we don't know anything. It might have a local minimum, a local maximum, a saddle or none of those things.

11.2.2 Expressing Conditions in Terms of Partials

The minor conditions can be expressed in terms of the partials. We have

$$\boldsymbol{PM_1} = det\ \left[f_{x_1x_1}(\boldsymbol{x_0})\right] = f_{x_1x_1}(\boldsymbol{x_0})$$

Then the second minor is

$$\boldsymbol{PM_2} = det\ \begin{bmatrix} f_{x_1x_1}(\boldsymbol{x_0}) & f_{x_1x_2}(\boldsymbol{x_0}) \\ f_{x_2x_1}(\boldsymbol{x_0}) & f_{x_2x_2}(\boldsymbol{x_0}) \end{bmatrix} = f_{x_1x_1}(\boldsymbol{x_0})f_{x_2x_2}(\boldsymbol{x_0}) - (f_{x_1x_2}(\boldsymbol{x_0}))^2$$

The third minor requires a 3×3 determinant

$$\boldsymbol{PM_3} = det\ \begin{bmatrix} f_{x_1x_1}(\boldsymbol{x_0}) & f_{x_1x_2}(\boldsymbol{x_0}) & f_{x_1x_3}(\boldsymbol{x_0}) \\ f_{x_2x_1}(\boldsymbol{x_0}) & f_{x_2x_2}(\boldsymbol{x_0}) & f_{x_2x_3}(\boldsymbol{x_0}) \\ f_{x_3x_1}(\boldsymbol{x_0}) & f_{x_3x_2}(\boldsymbol{x_0}) & f_{x_3x_3}(\boldsymbol{x_0}) \end{bmatrix}$$

which is much more messy to write out and so we will not do that. We discuss the two variable case in (Peterson (21) 2020) using different strategies of proof that involve a completing the square. Clearly that technique is limited to the two variable case and the extra theory we use here allows us to get a better grasp on the extremal structure of f. In the two variable case, we find

- f has a local minimum at $\boldsymbol{x_0}$ with $\boldsymbol{\nabla}(f)(\boldsymbol{x_0}) = \boldsymbol{0}$ when
  1. $f_{x_1x_1}(\boldsymbol{x_0}) > 0$ (first minor condition).

2. $f_{x_1x_1}(\boldsymbol{x_0})f_{x_2x_2}(\boldsymbol{x_0}) - (f_{x_1x_2}(\boldsymbol{x_0}))^2 > 0$ (second minor condition)

- f has a local maximum at $\boldsymbol{x_0}$ with $\boldsymbol{\nabla}(f)(\boldsymbol{x_0}) = \boldsymbol{0}$ when
  1. $f_{x_1x_1}(\boldsymbol{x_0}) < 0$ (first minor condition).
  2. $f_{x_1x_1}(\boldsymbol{x_0})f_{x_2x_2}(\boldsymbol{x_0}) - (f_{x_1x_2}(\boldsymbol{x_0}))^2 > 0$ (second minor condition)
- f has a saddle at $\boldsymbol{x_0}$ if the two eigenvalues of $\boldsymbol{H}(\boldsymbol{x_0})$ have different signs. This implies $f_{x_1x_1}(\boldsymbol{x_0})f_{x_2x_2}(\boldsymbol{x_0}) - (f_{x_1x_2}(\boldsymbol{x_0}))^2 < 0$

The standard two variable theorem is thus:

Theorem 11.2.1 Extreme Test

If the partials of f are zero at the point (x_0, y_0), we can determine if that point is a local minimum or local maximum of f using a second order test. We must assume the second order partials are continuous at the point (x_0, y_0).

- *If $f^0_{xx} > 0$ and $f^0_{xx}\, f^0_{yy} - (f^0_{xy})^2 > 0$ then $f(x_0, y_0)$ is a local minimum.*
- *If $f^0_{xx} < 0$ and $f^0_{xx}\, f^0_{yy} - (f^0_{xy})^2 > 0$ then $f(x_0, y_0)$ is a local maximum.*
- *If $f^0_{xx}\, f^0_{yy} - (f^0_{xy})^2 < 0$ then $f(x_0, y_0)$ is a local saddle.*

We just don't know anything if the test $f^0_{xx}\, f^0_{yy} - (f^0_{xy})^2 = 0$.

Example 11.2.1 *Use our tests to show $f(x, y) = x^2 + 3y^2$ has a minimum at $(0, 0)$.*

Solution *The partials here are $f_x = 2x$ and $f_y = 6y$. These are zero at $x = 0$ and $y = 0$. The Hessian at this critical point is*

$$H(x, y) = \begin{bmatrix} 2 & 0 \\ 0 & 6 \end{bmatrix} = H(0, 0)$$

as H is constant here. Our second order test says the point $(0, 0)$ corresponds to a minimum because $f_{xx}(0, 0) = 2 > 0$ and $f_{xx}(0, 0)\, f_{yy}(0, 0) - (f_{xy}(0, 0))^2 = 12 > 0$.

Example 11.2.2 *Use our tests to show $f(x, y) = x^2 + 6xy + 3y^2$ has a saddle at $(0, 0)$.*

Solution *The partials here are $f_x = 2x + 6y$ and $f_y = 6x + 6y$. These are zero when $2x + 6y = 0$ and $6x + 6y = 0$ which has solution $x = 0$ and $y = 0$. The Hessian at this critical point is*

$$H(x, y) = \begin{bmatrix} 2 & 6 \\ 6 & 6 \end{bmatrix} = H(0, 0)$$

as H is again constant here. Our second order test says the point $(0, 0)$ corresponds to a saddle because $f_{xx}(0, 0) = 2 > 0$ and $f_{xx}(0, 0)\, f_{yy}(0, 0) - (f_{xy}(0, 0))^2 = 12 - 36 < 0$.

Example 11.2.3 *Show our tests fail on $f(x, y) = 2x^4 + 4y^6$ even though we know there is a minimum value at $(0, 0)$.*

Solution *For $f(x, y) = 2x^4 + 4y^6$, you find that the critical point is $(0, 0)$ and all the second order partials are 0 there. So all the tests fail. Of course, a little common sense tells you $(0, 0)$ is indeed the place where this function has a minimum value. Just think about how the surface looks. But the tests just fail. This is much like the curve $f(x) = x^4$ which has a minimum at $x = 0$ but all the tests fail on it also.*

Example 11.2.4 *Show our tests fail on $f(x, y) = 2x^2 + 4y^3$ and the surface does not have a minimum or maximum at the critical point $(0, 0)$.*

Solution *For $f(x, y) = 2x^2 + 4y^3$, the critical point is again $(0,0)$ and $f_{xx}(0,0) = 4$, $f_{yy}(0,0) = 0$ and $f_{xy}(0,0) = f_{yx}(0,0) = 0$. So $f_{xx}(0,0)\, f_{yy}(0,0) - (f_{xy}(0,0))^2 = 0$ so the test fails. Note the $x = 0$ trace is $4y^3$ which is a cubic and so is negative below $y = 0$ and positive above $y = 0$. Not much like a minimum or maximum behavior on this trace! But the trace for $y = 0$ is $2x^2$ which is a nice parabola which does reach its minimum at $x = 0$. So the behavior of the surface around $(0,0)$ is not a maximum or a minimum.*

Homework

Exercise 11.2.1 *Assume you have found a critical point for the smooth function $f(x, y)$ at (x_0, y_0) and at that point $\boldsymbol{H}(\boldsymbol{x_0}) = \begin{bmatrix} 1 & 2 \\ 2 & 3 \end{bmatrix}$. Find the principle minors and their determinants and classify this critical point. Do this in MATLAB also to make sure you understand how to do it in this simple case.*

Exercise 11.2.2 *Assume you have found a critical point for the smooth function $f(x, y)$ at (x_0, y_0) and at that point $\boldsymbol{H}(\boldsymbol{x_0}) = \begin{bmatrix} 1 & 2 \\ 2 & 5 \end{bmatrix}$. Find the principle minors and their determinants and classify this critical point.*

Exercise 11.2.3 *Assume you have found a critical point for the smooth function $f(x, y)$ at (x_0, y_0) and at that point $\boldsymbol{H}(\boldsymbol{x_0}) = \begin{bmatrix} -1 & -2 \\ -2 & 1 \end{bmatrix}$. Find the principle minors and their determinants and classify this critical point.*

The theorem for n variables is this:

Theorem 11.2.2 Definiteness Extrema Test for n Variables

We assume the second order partials are continuous at the point $\boldsymbol{x_0}$ and $\boldsymbol{\nabla}(f)(\boldsymbol{x_0}) = \boldsymbol{0}$.

- *If the Hessian $\boldsymbol{H}(\boldsymbol{x_0})$ is positive definite, the critical point is a local minimum.*
- *If the Hessian $\boldsymbol{H}(\boldsymbol{x_0})$ is negative definite, the critical point is a local maximum.*
- *If the local Hessian $\boldsymbol{H}(\boldsymbol{x_0})$ has eigenvalues of different sign, the critical point is a saddle.*

Moreover, the Hessian $\boldsymbol{H}(\boldsymbol{x_0})$ is positive definite if its principle minors are all positive and have positive eigenvalues. Further, $\boldsymbol{H}(x_0, y_0)$ is negative definite if its principle minors alternate in sign starting with $-$ and all eigenvalues are negative.

Proof 11.2.1
We have gone over the proof of this result in our earlier discussions. ■

Example 11.2.5 *Let $f(x_1, x_2, x_3, x_4, x_5) = 2x_1^2 + 4x_2^2 + 9x_3^2 + 5x_4^2 + 8x_5^2$. Classify the critical points.*

Solution *The only critical point is $(0,0,0,0,0)$. The Hessian is*

$$\boldsymbol{H} = \begin{bmatrix} 4 & 0 & 0 & 0 & 0 \\ 0 & 8 & 0 & 0 & 0 \\ 0 & 0 & 18 & 0 & 0 \\ 0 & 0 & 0 & 10 & 0 \\ 0 & 0 & 0 & 0 & 16 \end{bmatrix}$$

$\boldsymbol{H}$ is clearly positive definite with eigenvalues $4, 8, 18, 10, 6$ and the eigenvectors are the standard basic vectors. The minors are $4, 32, 576, 5760, 92160$ and are all positive. Hence, this critical point is a minimum.

Example 11.2.6 *Let $f(x_1, x_2, x_3, x_4, x_5) = 2x_1^2 - 4x_2^2 + 9x_3^2 + 5x_4^2 + 8x_5^2$. Classify the critical points.*

Solution *The only critical point is $(0, 0, 0, 0, 0)$. The Hessian is*

$$\boldsymbol{H} = \begin{bmatrix} 4 & 0 & 0 & 0 & 0 \\ 0 & -8 & 0 & 0 & 0 \\ 0 & 0 & 18 & 0 & 0 \\ 0 & 0 & 0 & 10 & 0 \\ 0 & 0 & 0 & 0 & 16 \end{bmatrix}$$

$\boldsymbol{H}$ is clearly not positive or negative definite with eigenvalues $4, -8, 18, 10, 6$ and the eigenvectors are the standard basic vectors. The minors are $4, -32, -576, 5760, -92160$. Hence, this critical point is a saddle.

Example 11.2.7 *Let $f(x_1, x_2, x_3, x_4, x_5) = 100 - 2x_1^2 - 4x_2^2 - 9x_3^2 - 5x_4^2 - 8x_5^2$. Classify the critical points.*

Solution *The only critical point is $(0, 0, 0, 0, 0)$. The Hessian is*

$$\boldsymbol{H} = \begin{bmatrix} -4 & 0 & 0 & 0 & 0 \\ 0 & -8 & 0 & 0 & 0 \\ 0 & 0 & -18 & 0 & 0 \\ 0 & 0 & 0 & -10 & 0 \\ 0 & 0 & 0 & 0 & -16 \end{bmatrix}$$

$\boldsymbol{H}$ is clearly not positive or negative definite with eigenvalues $-4, -8, -18, -10, -6$ and the eigenvectors are the standard basic vectors. The minors are $-4, 32, -576, 5760, -92160$. Hence, this critical point is a maximum.

Homework

The following 2×2 matrices give the Hessian $\boldsymbol{H}^0$ at the critical point $(1, 1)$ of an extremal value problem.

- Determine if $\boldsymbol{H}^0$ is positive or negative definite.
- Determine if the critical point is a maximum or a minimum.
- Find the two eigenvalues of $\boldsymbol{H}^0$. Label the largest one as r_1 and the other as r_2.
- Find the two associated eigenvectors as unit vectors.
- Define

$$\boldsymbol{P} = \begin{bmatrix} \boldsymbol{E}_1 & \boldsymbol{E}_2 \end{bmatrix}$$

  Compute $\boldsymbol{P}^T \, \boldsymbol{H}^0 \, \boldsymbol{P}$ and show $\boldsymbol{H}^0 = \boldsymbol{P}\boldsymbol{\Lambda}\boldsymbol{P}^T$.
- Find all the minors for an appropriate $\boldsymbol{\Lambda}$.

Exercise 11.2.4

$$\boldsymbol{H}^0 = \begin{bmatrix} 1 & -3 \\ -3 & 12 \end{bmatrix}$$

Exercise 11.2.5

$$H^0 = \begin{bmatrix} -2 & 7 \\ 7 & -40 \end{bmatrix}$$

The following 3×3 matrices give the Hessian H^0 at the critical point $(1, 1, 2)$ of an extremal value problem.

- Determine if H^0 is positive or negative definite.
- Classify the critical point.
- Find the three eigenvalues of H^0. Label the largest one as r_1 and the smallest as r_3.
- Find the three associated eigenvectors as unit vectors.
- Define the usual P, compute $P^T\ H^0\ P$ and show $H^0 = P \Lambda P^T$ for an appropriate Λ.
- Find all the minors.

Exercise 11.2.6

$$H^0 = \begin{bmatrix} 1 & -3 & 4 \\ -3 & 12 & 5 \\ 4 & 5 & 4 \end{bmatrix}$$

Exercise 11.2.7

$$H^0 = \begin{bmatrix} -2 & 7 & -1 \\ 7 & -4 & 5 \\ -1 & 5 & 12 \end{bmatrix}$$

The following 4×4 matrices give the Hessian H^0 at the critical point $(1, 1, 2, -3)$ of an extremal value problem.

- Determine if H^0 is positive or negative definite.
- Classify the critical point.
- Find the four eigenvalues of H^0. Label the largest one as r_1 and the smallest as r_4.
- Find the four associated eigenvectors as unit vectors.
- Define the usual P, compute $P^T\ H^0\ P$ and show $H^0 = P \Lambda P^T$ for an appropriate Λ.
- Find all the minors.

Exercise 11.2.8

$$H^0 = \begin{bmatrix} 1 & -3 & 4 & 8 \\ -3 & 12 & 5 & 2 \\ 4 & 5 & 4 & -1 \\ 8 & 2 & -1 & 8 \end{bmatrix}$$

Exercise 11.2.9

$$H^0 = \begin{bmatrix} -2 & 7 & -1 & 3 \\ 7 & -4 & 5 & 10 \\ -1 & 5 & 12 & 5 \\ 3 & 10 & 5 & 9 \end{bmatrix}$$

Chapter 12

The Inverse and Implicit Function Theorems

Let's look at mappings that transform one set of coordinates into another set.

12.1 Mappings

First, look at linear maps. This particular one could be replaced by any other with a nonzero determinant. So you can make up plenty of examples.

Example 12.1.1 *Define* $T : \Re^2 \rightarrow \Re^2$ *by* $T(x, y) = (u, v)$ *where* $u = x - 3y$ *and* $v = 2x + 5y$. *We can write this in matrix/ vector form*

$$T\left(\begin{bmatrix} x \\ y \end{bmatrix}\right) = \begin{bmatrix} T_1(x,y) \\ T_2(x,y) \end{bmatrix} = \begin{bmatrix} u \\ v \end{bmatrix} = \begin{bmatrix} x - 3y \\ 2x + 5y \end{bmatrix} = \begin{bmatrix} 1 & -3 \\ 2 & 5 \end{bmatrix} \begin{bmatrix} x \\ y \end{bmatrix}$$

So there are many ways to choose to express this mapping. The matrix representation of T *with respect to an orthonormal basis* $\boldsymbol{E}$ *is*

$$\left[T\right]_{\boldsymbol{E},\boldsymbol{E}} = \begin{bmatrix} 1 & -3 \\ 2 & 5 \end{bmatrix}$$

We could also write the mapping simply as $\boldsymbol{U} = \left[T\right]_{\boldsymbol{E},\boldsymbol{E}} \boldsymbol{X}$ *where we let* $\boldsymbol{U}$ *be the vector with components* u *and* v *and* $\boldsymbol{X}$, *the vector with components* x *and* y. *However, this is too cluttered and it is easy to identify* T *and its matrix representation* $\left[T\right]_{\boldsymbol{E},\boldsymbol{E}} \boldsymbol{X}$ *which we will do from now on. By direct calculation,* $det(T) = 11 \neq 0$, *so* T^{-1} *exists. Hence if* $T(x_1, y_1) = T(x_2, y_2)$, *we would have*

$$\begin{bmatrix} 1 & -3 \\ 2 & 5 \end{bmatrix} \begin{bmatrix} x_1 \\ y_1 \end{bmatrix} = \begin{bmatrix} 1 & -3 \\ 2 & 5 \end{bmatrix} \begin{bmatrix} x_2 \\ y_2 \end{bmatrix} \Longrightarrow \begin{bmatrix} 1 & -3 \\ 2 & 5 \end{bmatrix} \begin{bmatrix} x_1 - x_2 \\ y_1 - y_2 \end{bmatrix} = \begin{bmatrix} 0 \\ 0 \end{bmatrix} \Longrightarrow x_1 = x_2,\ y_1 = y_2$$

Thus, T *is a* $1 - 1$ *map. Further, if we chose a vector* $\boldsymbol{B}$ *in* $\Re^2$, *then*

$$\begin{bmatrix} 1 & -3 \\ 2 & 5 \end{bmatrix} \begin{bmatrix} x \\ y \end{bmatrix} = \begin{bmatrix} b_1 \\ b_2 \end{bmatrix} \Longrightarrow \begin{bmatrix} x \\ y \end{bmatrix} = \left(\begin{bmatrix} 1 & -3 \\ 2 & 5 \end{bmatrix}\right)^{-1} \begin{bmatrix} b_1 \\ b_2 \end{bmatrix}$$

and so we also know T is an onto map. Further, it is easy to see $\boldsymbol{J_T} = T$, so we have since T is a linear map

$$\boldsymbol{DT} = \boldsymbol{J_T} = \begin{bmatrix} 1 & -3 \\ 2 & 5 \end{bmatrix}$$

Therefore, in this case, $det(\boldsymbol{DT}(x_0, y_0)) = det(T) \neq 0$ on $\Re^2$.

Now let's do a nonlinear mapping.

Example 12.1.2 *We are now going to study $T : \Re^2 \to \Re^2$ defined by $T(x,y) = (x^2 + y^2, 2xy)$. This mapping T is not linear and not $1-1$ because $T(x,y) = T(-x,-y)$. What is the range of T? If (u,v) is in the range, then $u = x^2 + y^2$ implying $u \geq 0$ always. Further, $v = 2xy$. Note*

$$\begin{aligned} u+v &= x^2 + 2xy + y^2 = (x+y)^2 \geq 0 \\ u-v &= x^2 - 2xy + y^2 = (x-y)^2 \geq 0 \end{aligned}$$

Now $u+v$ divides the $u-v$ plane into three parts as seen in Figure 12.1(a): where $u+v > 0$, $u+v = 0$ and $u-v < 0$. Also, $u-v$ divides the plane into three parts also as seen in Figure 12.1(b). Hence, to be in the range, both $u+v$ and $u-v$ must be non-negative and we already know

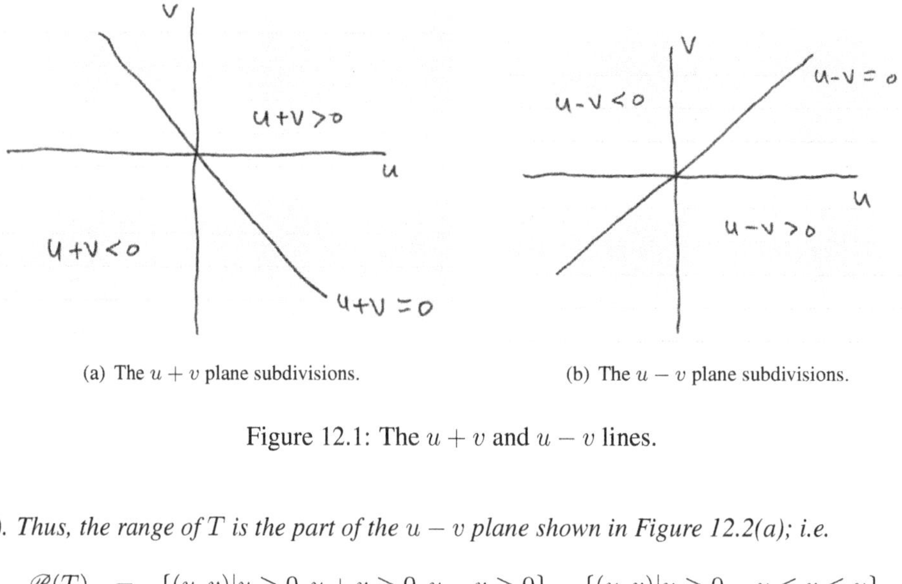

(a) The $u+v$ plane subdivisions. (b) The $u-v$ plane subdivisions.

Figure 12.1: The $u+v$ and $u-v$ lines.

$u \geq 0$. Thus, the range of T is the part of the $u-v$ plane shown in Figure 12.2(a); i.e.

$$\mathscr{R}(T) = \{(u,v) | u \geq 0, u+v \geq 0, u-v \geq 0\} = \{(u,v) | u \geq 0, -u \leq v \leq u\}$$

The domain of T is $\Re^2$ and it can be divided into four regions as shown in Figure 12.2(b). Thus, the four indicated regions are the disjoint and open sets

$$\begin{aligned} S_1 &= \{(x,y) | x > 0, -x < y < x\}, \quad S_2 = \{(x,y) | y > 0, -y < x < y\} \\ S_3 &= \{(x,y) | x < 0, -|x| < y < |x|\}, \quad S_4 = \{(x,y) | y < 0, -|y| < x < |y|\} \end{aligned}$$

Proposition 12.1.1 Finding Where $(x^2 + y^2, 2xy)$ is a Bijection

The mapping $T : \Re^2 \to \Re^2$ defined by $T(x,y) = (x^2 + y^2, 2xy)$. is invertible on $\overline{S_2}$ and $T|_{\overline{S_2}}$ is onto $\mathscr{R}(T)$.

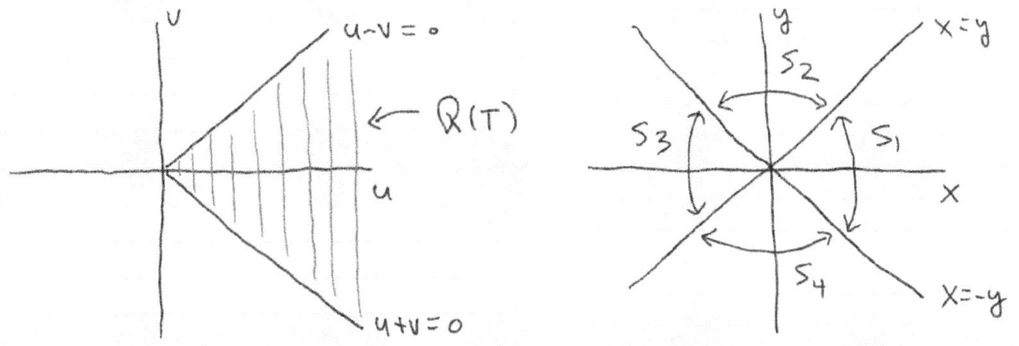

(a) The range of the mapping T.

(b) The subdivided domain of T.

Figure 12.2: The range and domain of T.

Proof 12.1.1

Assume $T(x_0, y_0) = T(x_1, y_1)$. Since (x_0, y_0) and (x_1, y_1) are in S_2, we have $y_0 > 0, -y_0 < x_0 < y_0$ and $y_1 > 0, -y_1 < x_1 < y_1$. Now if $T(x_0, y_0) = (u_0, v_0)$ and $T(x_1, y_1) = (u_1, v_1)$,

$$\begin{aligned} u_0 + v_0 &= (x_0 + y_0)^2, \quad u_0 - v_0 = (x_0 - y_0)^2 \\ u_1 + v_1 &= (x_1 + y_1)^2, \quad u_1 - v_1 = (x_1 - y_1)^2 \end{aligned}$$

Since $T(x_0, y_0) = T(x_1, y_1)$, $(x_0 + y_0)^2 = (x_1 + y_1)^2$ and $(x_0 - y_0)^2 = (x_1 - y_1)^2$. Now in S_2, $x_0 + y_0 > 0$ and $x_1 + y_1 > 0$. Thus, we can take the positive square root to find $x_0 + y_0 = x_1 + y_1$. Also, in S_2, $x_0 - y_0 > 0$ and $x_1 - y_1 > 0$. Hence, in this case, we take the negative square root to get $y_0 - x_0 = y_1 - x_1$. We now know

$$\begin{aligned} x_0 + y_0 &= x_1 + y_1 \\ -x_0 + y_0 &= -x_1 + y_1 \end{aligned}$$

Adding, we have $y_0 = y_1$ which then implies $x_0 = x_1$ as well. Hence, T is $1 - 1$ on S_2.

In fact if $x \geq 0$ and $y = x$, $T(x, y) = T(x, x) = (2x^2, 2x^2)$ which is on the line $y = x$ in the range of T. Moreover, if $x \geq 0$ and $y = -x$, $T(x, y) = T(x, -x) = (2x^2, -2x^2)$ which is on the line $y = -x$ in the range of T. It is then easy to see this shows T is $1 - 1$ on these lines too. So we can say T is $1 - 1$ on $\overline{S_2}$. Hence, T^{-1} exists on $\overline{S_2}$.

To show $T : \overline{S_2} \to \mathscr{R}(T)$ is onto is next. We'll do this by constructing the inverse. If $T(x_0, y_0) = (u_0, v_0)$ with $(x_0, y_0) \in \overline{S_2}$, then

$$u_0 + v_0 = (x_0 + y_0)^2 \geq 0, \quad u_0 - v_0 = (x_0 - y_0)^2 \geq 0$$

In S_2, $x_0 + y_0 > 0$ and $x_0 - y_0 < 0$. Taking square roots, we have

$$x_0 + y_0 = \sqrt{u_0 + v_0}, \quad x_0 - y_0 = -\sqrt{u_0 - v_0}$$

which implies

$$x_0 = \frac{1}{2}(\sqrt{u_0 + v_0} - \sqrt{u_0 - v_0})$$

$$y_0 = \frac{1}{2}(\sqrt{u_0 + v_0} + \sqrt{u_0 - v_0})$$

Hence,

$$T^{-1}(u_0, v_0) = (\frac{1}{2}(\sqrt{u_0 + v_0} - \sqrt{u_0 - v_0}). \frac{1}{2}(\sqrt{u_0 + v_0} + \sqrt{u_0 - v_0}))$$

On the line $u = v$, *if* $(x, y) \in \overline{S_2}$, $y \geq 0$, $x + y \geq 0$ *and* $x - y \leq 0$. *Thus*

$$u + v = (x + y)^2, \;\; u - v = (x - y)^2 \implies 2u = (x + y)^2, \;\; 0 = (x - y)^2 \implies \frac{1}{2}\sqrt{2u} = x, y = x$$

Thus, if $u = v$, $T^{-1}(u, u) = (\frac{1}{2}\sqrt{2u}, \frac{1}{2}\sqrt{2u})$.
If $v = -u$, *the argument is similar: we have*

$$\begin{aligned} u + v = (x + y)^2, \quad u - v = (x - y)^2 &\implies 0 = (x + y)^2, \quad 2u = (x - y)^2 \\ &\implies y = -x, y > 0, -\frac{1}{2}\sqrt{2u} = x \end{aligned}$$

So if $u = -v$, $T^{-1}(u, -u) = (-\frac{1}{2}\sqrt{2u}, \frac{1}{2}\sqrt{2u})$. *This completes the argument that* $T : \overline{S_2} \rightarrow \mathscr{R}(T)$ *is onto.* ∎

The map T *is differentiable with*

$$\boldsymbol{DT}(x_0, y_0) = \boldsymbol{J_T}(x_0, y_0) = \begin{bmatrix} 2x_0 & 2y_0 \\ 2y_0 & 2x_0 \end{bmatrix}$$

with $\det(\boldsymbol{DT}(x_0, y_0)) = 4(x_0^2 - x_1^2)$. *Hence, if* $(x_0, y_0) \in S_2$, $\det(\boldsymbol{DT}(x_0, y_0)) \neq 0$ *and if* (x_0, y_0) *is on the boundary of* S_2, $\det(\boldsymbol{DT}(x_0, y_0)) = 0$. *Thus, the determinant vanishes on the lines* $y = \pm x$ *in* s_2. *Finally, in the interior of* S_2, $x + y > 0$ *and* $x - y < 0$ *and so* $\det(\boldsymbol{DT}(x_0, y_0)) = 4(x_0 - x_1)(x_0 + x_1) < 0$.

These two examples give us some insight.

- If $T : \Re^2 \rightarrow \Re^2$ has $det(\boldsymbol{DT}(x_0, y_0)) \neq 0$ does not necessarily force T^{-1} to exist on all of $\Re^2$. Our second example shows us we may need to restrict T to a suitable domain for the inverse to exist. This also shows us T^{-1} can exist even when $det(\boldsymbol{DT}(x_0, y_0)) = 0$ as we found on the lines $y = \pm x$.
- Given $T : \Re^2 \rightarrow \Re^2$ which is linear, we have a mapping **too nice** to be interesting!

Homework

Exercise 12.1.1 *Analyze* $T(x, y) = (2x - 4y, 5x + 6y)$ *as we have discussed above.*

Exercise 12.1.2 *Analyze* $T(x, y) = (-x + 3yy, 2x - 8y)$ *as we have discussed above.*

Exercise 12.1.3 *Analyze* $T(x, y) = (4x - 2y, x + 6y)$ *as we have discussed above.*

Exercise 12.1.4 *Analyze* $T(x, y) = (x^2 + y^2, 3xy)$ *as we have discussed above.*

Exercise 12.1.5 *Analyze* $T(x, y) = (2x^2 + 4y^2, 5xy)$ *as we have discussed above.*

We have been using the determinant of the Jacobian of a mapping and we know $\boldsymbol{DF} = \boldsymbol{J_F}$ for our vector-valued differentiable functions. In the literature, the determinant of the Jacobian is very useful and some texts use the term **Jacobian** for the **determinant of the Jacobian**. We won't do that here and instead will always use $det\boldsymbol{J_F}$ to denote this computation. Let's look at our first **invertibility** theorem.

Theorem 12.1.2 The First Inverse Function Theorem

Let $f : D \subset \Re^n \to \Re^n$ with D an open set. Assume f_{i,x_j} are all continuous on D and assume $\det(\boldsymbol{J_f}(\boldsymbol{x_0})) \neq 0$ at $\boldsymbol{x_0} \in D$. Then there is a radius $r > 0$ so that T is $1-1$ on $B(r, \boldsymbol{x_0})$; i.e. F^{-1} exists locally.

Proof 12.1.2

If $n = 1$, this is a simple scalar function of one variable. We have no partial derivatives and $\boldsymbol{Df}(x_0) = f'(x_0)$. If $f'(x_0) \neq 0$, then given $\epsilon = |f'(x_0)|/2$, there is a $\delta > 0$ so that $|f'(x)| > |f'(x_0)|/2$ if $x \in B(\delta, x_0)$. Pick any two distinct points p and q in $B(\delta, x_0)$ and let $[p, q]$ be the line segment between p and q. By the Mean Value Theorem, there is point c in $[p, q]$ so that $f(p) - f(q) = f'(c)(q-p)$. Now if $f(p) = f(q)$, this would mean $0 = f'(c)(p-q)$. But $f'(c) \neq 0$ and $p \neq q$ and so this is not possible. Hence in $B(\delta, x_0)$, f must be $1-1$ and so is invertible.

If $n > 1$, the argument gets more interesting. Before we do the general one, let $n = 2$. Define the function $M : \Re^2 \times \Re^2 \to \Re$ by

$$M(\boldsymbol{x}, \boldsymbol{y}) = \begin{bmatrix} f_{1,x_1}(\boldsymbol{x}) & f_{1,x_2}(\boldsymbol{x}) \\ f_{2,x_1}(\boldsymbol{y}) & f_{2,x_2}(\boldsymbol{y}) \end{bmatrix}$$

Then

$$detM(\boldsymbol{x}, \boldsymbol{y}) = f_{1,x_1}(\boldsymbol{x})\, f_{2,x_2}(\boldsymbol{y}) - f_{1,x_2}(\boldsymbol{x})\, f_{2,x_1}(\boldsymbol{y})$$

Since f_{1,x_1}, f_{1,x_2}, f_{2,x_1} and f_{2,x_2} are continuous in D, so are the functions $M(\boldsymbol{x}, \boldsymbol{y})$ and $detM(\boldsymbol{x}, \boldsymbol{y})$. We assume $det\boldsymbol{J_f}(\boldsymbol{x_0}) = M(\boldsymbol{x_0}, \boldsymbol{x_0}) \neq 0$ and hence, for $\epsilon = |detM(\boldsymbol{x_0}, \boldsymbol{x_0})|/2$, there is a positive δ so that if $\boldsymbol{x}$ and $\boldsymbol{y}$ are in $B(\delta, \boldsymbol{x_0})$

$$|detM(\boldsymbol{x}, \boldsymbol{y})| > |detM(\boldsymbol{x_0}, \boldsymbol{x_0})|/2 > 0$$

Now choose any two distinct points $\boldsymbol{p}$ and $\boldsymbol{q}$ in $B(\delta, \boldsymbol{x_0})$. By the Mean Value Theorem, there are points $\boldsymbol{c_1}$ and $\boldsymbol{c_2}$ on the line segment $[\boldsymbol{p}, \boldsymbol{q}]$ between $\boldsymbol{p}$ and $\boldsymbol{q}$ so that

$$\begin{aligned} f_1(\boldsymbol{p}) - f_1(\boldsymbol{q}) &= < (\boldsymbol{\nabla}(f_1))^T(\boldsymbol{c_1}), \boldsymbol{p} - \boldsymbol{q} > \\ f_2(\boldsymbol{p}) - f_2(\boldsymbol{q}) &= < (\boldsymbol{\nabla}(f_2))^T(\boldsymbol{c_2}), \boldsymbol{p} - \boldsymbol{q} > \end{aligned}$$

But if $f(\boldsymbol{p}) = f(\boldsymbol{q})$, then

$$\begin{aligned} < (\boldsymbol{\nabla}(f_1))^T(\boldsymbol{c_1}), \boldsymbol{p} - \boldsymbol{q} > &= 0 \\ < (\boldsymbol{\nabla}(f_2))^T(\boldsymbol{c_2}), \boldsymbol{p} - \boldsymbol{q} > &= 0 \end{aligned}$$

But we know $detM(\boldsymbol{c_1}, \boldsymbol{c_2}) \neq 0$ and so the only solution is $\boldsymbol{p} = \boldsymbol{q}$ which is not possible. So the assumption that $f(\boldsymbol{p}) = f(\boldsymbol{q})$ is not correct. We conclude f is $1-1$ on $B(\delta, \boldsymbol{x_0})$ and is invertible there.

To do the general theorem, we use the material on determinants developed in Chapter 9. By Theorem 9.1.3,

$$M(\boldsymbol{x_1},\ldots,\boldsymbol{x_n}) \quad = \quad det \begin{bmatrix} (\boldsymbol{\nabla}(f_1))^T(\boldsymbol{x_1}) \\ \vdots \\ (\boldsymbol{\nabla}(f_n))^T(\boldsymbol{x_n}) \end{bmatrix}$$

is a continuous function of the variables $\boldsymbol{x_1},\ldots,\boldsymbol{x_n}$. *Since* $det(\boldsymbol{J_f}(\boldsymbol{x_0})) = det M(\boldsymbol{x_0},\ldots,\boldsymbol{x_0}) \neq 0$, *for* $\epsilon = |det M(\boldsymbol{x_0},\boldsymbol{x_0})|/2$, *there is a positive* δ *so that if* $\boldsymbol{x_1}$ *through* $\boldsymbol{x_n}$ *are in* $B(\delta,\boldsymbol{x_0})$

$$|det M(\boldsymbol{x_1},\ldots,\boldsymbol{x_n})| \quad > \quad |det M(\boldsymbol{x_0},\boldsymbol{x_0})|/2 > 0$$

Now choose any two distinct points $\mathbf{p}$ *and* $\boldsymbol{q}$ *in* $B(\delta,\boldsymbol{x_0})$. *By the Mean Value Theorem, there are points* $\mathbf{c_1}$ *through* $\mathbf{c_n}$ *on the line segment* $[\mathbf{p},\boldsymbol{q}]$ *between* $\mathbf{p}$ *and* $\boldsymbol{q}$ *so that*

$$\begin{aligned} f_1(\mathbf{p}) - f_1(\boldsymbol{q}) &= \; < (\boldsymbol{\nabla}(f_1))^T(\boldsymbol{c_1}), \boldsymbol{p}-\boldsymbol{q} > \\ \vdots \;\; &= \;\; \vdots \\ f_n(\mathbf{p}) - f_n(\boldsymbol{q}) &= \; < (\boldsymbol{\nabla}(f_n))^T(\boldsymbol{c_n}), \boldsymbol{p}-\boldsymbol{q} > \end{aligned}$$

But $f(\mathbf{p}) = f(\boldsymbol{q})$ *and so*

$$\begin{aligned} < (\boldsymbol{\nabla}(f_1))^T(\boldsymbol{c_1}), \boldsymbol{p}-\boldsymbol{q} > &= 0 \\ \vdots \;\; &= \;\; \vdots \\ < (\boldsymbol{\nabla}(f_n))^T(\boldsymbol{c_n}), \boldsymbol{p}-\boldsymbol{q} > &= 0 \end{aligned}$$

The determinant is zero if its rows are linearly independent and it is nonzero if its rows are linearly independent. Hence, the only solution to the above equation is $\boldsymbol{p} = \boldsymbol{q} = \boldsymbol{0}$. *But this is not possible and hence* f *must be* $1-1$ *on* $B(\delta,\boldsymbol{x_0})$ *and it invertible there.* ■

Homework

Exercise 12.1.6 *Let* $f(x,y) = \begin{bmatrix} x^2+y^2 \\ 2xy \end{bmatrix}$.

- *Find* $\boldsymbol{J_f}$.
- *Find* $\det(\boldsymbol{J_f})$.
- *Is* f *invertible locally at* $(1,2)$?
- *Is* f *invertible locally at* $(2,2)$?
- *Is* f *invertible at any point in* $\Re^2$?

Exercise 12.1.7 *Let* $f(x,y) = \begin{bmatrix} x^2+y^2 \\ 2xz \\ y^2+z^4 \end{bmatrix}$.

- *Find* $\boldsymbol{J_f}$.
- *Find* $\det(\boldsymbol{J_f})$.
- *Is* f *invertible locally at* $(1,2,1)$?

- *Is f invertible at any point in $\Re^3$?*

Exercise 12.1.8 *Let* $f(x,y) = \begin{bmatrix} 2x^2+4y^2 \\ 5xz \\ 3y^2+z^4 \end{bmatrix}$.

- *Find $\boldsymbol{J_f}$.*
- *Find* $\det(\boldsymbol{J_f})$.
- *Is f invertible locally at $(-1,2,2)$?*
- *Is f invertible at any point in $\Re^3$?*

Exercise 12.1.9 *Let* $f(x,y) = \begin{bmatrix} 2+3xyz \\ 2x^2+4y^2 \\ 5xz \\ 3y^2+z^4 \end{bmatrix}$.

- *Find $\boldsymbol{J_f}$.*
- *Find* $\det(\boldsymbol{J_f})$.
- *Is f invertible locally at $(-1,2,2,-1)$?*
- *Is f invertible at any point in $\Re^4$?*

12.2 Invertibility Results

We begin with a technical result.

Theorem 12.2.1 Interior Points of the Range of f

Let $D \subset \Re^n$ be open and $f : D \subset \Re^n \rightarrow \Re^n$ with continuous partial derivatives on D. Assume $\boldsymbol{x_0}$ is in D and there is a positive r so that $\overline{B(r,\boldsymbol{x_0})} \subset D$ with det $\boldsymbol{J_f}(\boldsymbol{x}) \neq 0$ for all $\boldsymbol{x}$ in $\overline{B(r,\boldsymbol{x_0})}$. Further, assume f is $1-1$ on $\overline{B(r,\boldsymbol{x_0})}$. Then $f(\boldsymbol{x_0})$ is an interior point of $f(B(r,\boldsymbol{x_0}))$.

Proof 12.2.1
For $\boldsymbol{x}$ in $\overline{B(r,\boldsymbol{x_0})}$, let $g(\boldsymbol{x}) = \|f(\boldsymbol{x}) - f(\boldsymbol{x_0})\|$. The g is a non-negative real-valued function whose domain in compact. We know the map $\|\cdot\| : \Re^n \rightarrow \Re$ is continuous and f is continuous since it is differentiable. Note the continuous partials imply its differentiability. Hence, the composition $h(\boldsymbol{x}) = \|f(\boldsymbol{x}) - f(\boldsymbol{x_0})\|$ is continuous on the compact set $\overline{B(r,\boldsymbol{x_0})}$. We see $g(\boldsymbol{x}) = 0$ implies $f(\boldsymbol{x}) = f(\boldsymbol{x_0})$. Since we assume f is $1-1$ on this domain, this means $\boldsymbol{x} = \boldsymbol{x_0}$. Hence h is $1-1$ on $\overline{B(r,\boldsymbol{x_0})}$. Finally, note the boundary $S = \{\boldsymbol{x} | \|\boldsymbol{x} - \boldsymbol{x_0}\| = r\}$ is a compact set also. Since f is $1-1$ on this set, g cannot be zero on S; so $g(\boldsymbol{y}) > 0$ if $\boldsymbol{y} \in S$.
Since g is continuous on the compact set S, g attains a global minimum of value $m > 0$ for some $\boldsymbol{y} \in S$.

Show $B(m/2, f(\boldsymbol{x_0})) \subset f(B(r,\boldsymbol{x_0}))$: *Let $\boldsymbol{z}$ be in $\underline{B(m/2, f(\boldsymbol{x_0}))}$. We must find a $\mathbf{p}$ in $B(r,\boldsymbol{x_0})$ with $z = f(\mathbf{p})$. Let ϕ be the real-valued function on $\overline{B(r,\boldsymbol{x_0})}$ defined by*

$$\phi(\boldsymbol{x}) \quad = \quad \|f(\boldsymbol{x}) - \boldsymbol{z}\| = \sqrt{\sum_{i=1}^{n} (f_i(\boldsymbol{x}) - z_i)^2}$$

We see ϕ is continuous on the compact set $\overline{B(r, \boldsymbol{x_0})}$ and so attains a minimum value α at some point $\boldsymbol{y_0}$ in $\overline{B(r, \boldsymbol{x_0})}$. Since $\boldsymbol{z}$ is in $B(m/2, f(\boldsymbol{x_0}))$, $\|f(\boldsymbol{x_0}) - \boldsymbol{z}\| < m/2$ and so $\phi(\boldsymbol{x_0}) = \|f(\boldsymbol{x_0}) - \boldsymbol{z}\| < m/2$. This tells us

$$\alpha = \min_{\boldsymbol{x} \in \overline{B(r, \boldsymbol{x_0})}} \phi(\boldsymbol{x}) = \min_{\boldsymbol{x} \in \overline{B(r, \boldsymbol{x_0})}} \|f(\boldsymbol{x}) - \boldsymbol{z}\| = \phi(\boldsymbol{y_0}) \le \phi(\boldsymbol{x_0}) < m/2.$$

Is it possible for $\boldsymbol{y_0} \in S$? If so,

$$\begin{aligned} \phi(\boldsymbol{y_0}) &= \|f(\boldsymbol{y_0}) - \boldsymbol{z}\| = \|f(\boldsymbol{y_0}) - f(\boldsymbol{x_0}) + f(\boldsymbol{x_0}) - \boldsymbol{z}\| \\ &\ge \|f(\boldsymbol{y_0}) - f(\boldsymbol{x_0})\| - \|f(\boldsymbol{x_0}) - \boldsymbol{z}\| \end{aligned}$$
$$= g(\boldsymbol{y_0}) - \|f(\boldsymbol{x_0}) - \boldsymbol{z}\|$$

But $\|f(\boldsymbol{x_0}) - \boldsymbol{z}\| < m/2$ and $g(\boldsymbol{y_0}) \ge m > 0$ if $\boldsymbol{y_0}$ is in S. This implies

$$\phi(\boldsymbol{y_0}) \ge m - m/2 = m/2 > 0$$

But we know

$$\min_{\boldsymbol{x} \in \overline{B(r, \boldsymbol{x_0})}} \phi(\boldsymbol{x}) = \alpha < m/2$$

This means $\phi(\boldsymbol{y_0}) \ge m/2$ is not possible. We conclude $\boldsymbol{y_0}$ is not in S. So the point where ϕ attains its minimum on $\overline{B(r, \boldsymbol{x_0})}$ is not on the boundary S; so it is an interior point and $\boldsymbol{y_0} \in B(r, \boldsymbol{x_0})$.

Since ϕ has a minimum at $\boldsymbol{y_0}$, so does ϕ^2. Since the minimum is an interior point, we must have $(\phi^2)_{x_i}(\boldsymbol{y_0}) = 0$ for all i. Now

$$\phi^2(\boldsymbol{x}) = \sum_{i=1}^{n} (f_i(\boldsymbol{x}) - z_i)^2 \Longrightarrow (\phi^2)_{x_i} = 2 \sum_{i=1}^{n} (f_i(\boldsymbol{x}) - z_i)\, f_{i,x_i}$$

Hence at $\boldsymbol{y_0}$,

$$(\phi^2)_{x_i}(\boldsymbol{x_0}) = 2 \sum_{i=1}^{n} (f_i(\boldsymbol{x_0}) - z_i)\, f_{i,x_i}(\boldsymbol{x_0}) = 0$$

This is the same as $\boldsymbol{J_f}(\boldsymbol{y_0})\,(f(\boldsymbol{y_0}) - \boldsymbol{z}) = \boldsymbol{0}$. Since det $\boldsymbol{J_f}(\boldsymbol{y_0}) \neq 0$, this tells the unique solution is $f(\boldsymbol{y_0}) = \boldsymbol{z}$. We conclude $\boldsymbol{z} \in B(m/2, f(\boldsymbol{x_0}))$ or $f(\boldsymbol{y_0}) \in f(B(r, \boldsymbol{x_0}))$ as $\boldsymbol{y_0} \in B(r, \boldsymbol{x_0})$. Hence, $B(m/2, f(\boldsymbol{x_0})) \subset f(B(r, \boldsymbol{x_0}))$.

This, of course, says $f(\boldsymbol{x_0})$ is an interior point of $f(B(r, \boldsymbol{x_0}))$ and completes the proof. ■

We are now ready to tackle the **Inverse Function Theorem**. First, a result about inverse images of open sets.

Theorem 12.2.2 Inverse Images of Open Sets Under Continuous Maps are Open

Let $f : D \subset \Re^n \to \Re^m$ be a continuous mapping of the open set D. If V is open in the range of f, then $f^{-1}(V)$ is open in D.

Proof 12.2.2
Let $\boldsymbol{x_0} \in f^{-1}(V)$. Then $f(\boldsymbol{x_0}) \in V$. Since V is open, $f(\boldsymbol{x_0})$ is an interior point of V and so there is a radius $r > 0$ so that $B(r, f(\boldsymbol{x_0})) \subset V$. Since f is continuous on D, for $\epsilon = r$, there is a $\delta > 0$

so that $\|f(\boldsymbol{x}) - f(\boldsymbol{x_0})\| < r$ *is* $\|\boldsymbol{x} - \boldsymbol{x_0}\|, \delta$. *Thus,* $f(\boldsymbol{x}) \in B(r, f(\boldsymbol{x_0}))$ *is* $\boldsymbol{x} \in B(\delta, \boldsymbol{x_0}) \cap D$. *Since* D *is open,* $B(\delta, \boldsymbol{x_0}) \cap D$ *is open and so there is a* $\delta_1 > 0$ *with* $B(\delta_1, \boldsymbol{x_0}) \subset B(\delta, \boldsymbol{x_0}) \cap D$. *This tells us* $\boldsymbol{x} \in B(\delta_1, \boldsymbol{x_0})$ *implies* $f(\boldsymbol{x}) \in B(r, f(\boldsymbol{x_0})) \subset V$. *We conclude* $B(\delta_1, \boldsymbol{x_0}) \subset f^{-1}(V)$ *and so* $\boldsymbol{x_0}$ *is an interior point of* $f^{-1}(V)$. *Since* $\boldsymbol{x_0}$ *is arbitrary, this shows* $f^{-1}(V)$ *is open.* ■

Theorem 12.2.3 The Inverse Function Theorem

Let $D \subset \Re^n$ *be an open set and assume* $f : D \subset \Re^n \rightarrow \Re^n$ *has continuous first partials derivatives,* f_{i,x_j} *in* D. *Let* $\boldsymbol{x_0} \in D$ *with* $\det \boldsymbol{J_f}(\boldsymbol{x_0}) \neq 0$. *Then there is an open set* U *containing* $\boldsymbol{x_0}$ *and an open set* $V \subset f(U)$ *and a map* $g : V \subset \Re^n \rightarrow U \subset \Re^n$ *so that*

- $f : U \subset \Re^n$ *is* $1-1$ *and onto* V.
- $g : V \subset \Re^n \rightarrow$ *is* $1-1$ *and onto* U.
- $g \circ f(\boldsymbol{x}) = \boldsymbol{x}$ *on* U.
- g_{i,x_j} *is continuous on* V *implying* $\boldsymbol{Dg}$ *exists on* V.

Comment 12.2.1 *If* $\boldsymbol{y} \in V$, $g(\boldsymbol{y}) \in U$. *So* $f \circ g(\boldsymbol{y}) = f(g(\boldsymbol{y})) = \boldsymbol{z} \in V$. *But there is an* $\boldsymbol{x} \in U$ *so that* $f(\boldsymbol{x}) = \boldsymbol{y}$. *Hence,* $g(\boldsymbol{y}) = g(f(\boldsymbol{x})) = \boldsymbol{x}$ *by this theorem. We conclude* $f \circ g(\boldsymbol{y}) = f(\boldsymbol{x}) = \boldsymbol{y}$. *So we see both* $f \circ g(\boldsymbol{x}) = \boldsymbol{x}$ *and* $g \circ f(\boldsymbol{y}) = \boldsymbol{y}$ *on* U *and* V *respectively. So* $g = f^{-1}$ *and* $f = g^{-1}$.

Proof 12.2.3
Since $\det \boldsymbol{J_f}(\boldsymbol{x_0}) \neq 0$ *and* f_{i,x_j} *is continuous in* D, *the continuity of* $\det \boldsymbol{J_f}(\boldsymbol{x_0})$ *implies we can find a positive* r_0 *so that* $B(r_0, \boldsymbol{x_0}) \subset D$ *and by Theorem 12.1.2, there is an* $r_1 < r_0 0$, *so that* f *is* $1-1$ *on* $B(r_1, \boldsymbol{x_0})$. *Choose any* $r < r_1$. *Then*

$$\overline{B(r, \boldsymbol{x_0})} \subset B(r_1, \boldsymbol{x_0}) \implies f1 : -1 \text{ on } \overline{B(r_1, \boldsymbol{x_0})}$$

Now by Theorem 12.2.1, $f(\boldsymbol{x_0})$ *is an interior point of* $f(B(r, \boldsymbol{x_0}))$. *This tells us there is an* $r_2 > 0$ *so that* $B(r_2, f(\boldsymbol{x_0})) \subset f(B(r, \boldsymbol{x_0}))$. *Let* $V = B(r_2, f(\boldsymbol{x_0}))$ *and* $U = f^{-1}(V)$. *Note* $f^{-1}(V)$ *is a subset of* $B(r, \boldsymbol{x_0})$. *You can see these sets in Figure 12.3. Since* V *is open,* $f^{-1}(V)$ *is open also since*

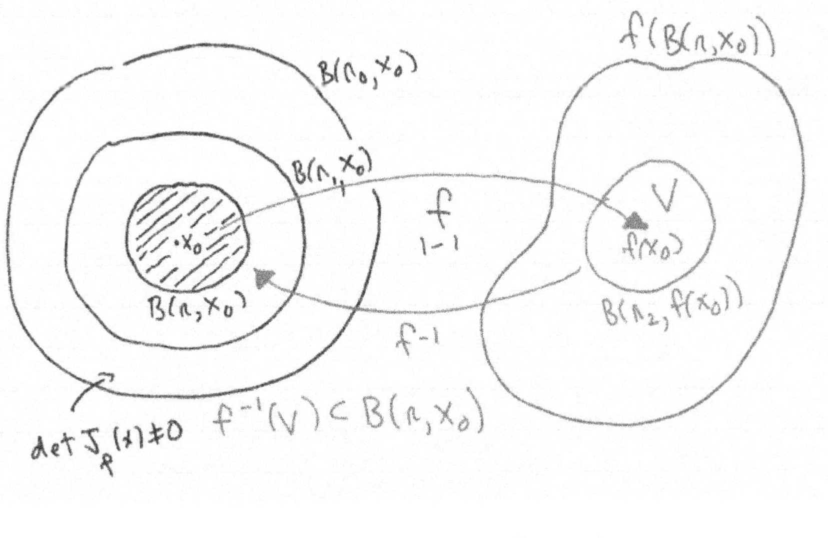

Figure 12.3: Inverse function mappings and sets.

f is continuous. We now know f is $1-1$ on $B(r, \boldsymbol{x_0})$ implying f is $1-1$ on U with $U \subset B(r, \boldsymbol{x_0})$. Thus, $g = f^{-1}$ does exist on $f(U) = V$ and $g \circ f(\boldsymbol{x}) = \boldsymbol{x}$ on U. This is part of what we need to prove. There is a lot more to do.

f **is onto** V:

Proof *Let $\boldsymbol{y} \in V$. Then $\boldsymbol{y} \in B(r_2, f(\boldsymbol{x_0})) \subset f(B(r, \boldsymbol{x_0}))$. Hence, there is $\boldsymbol{x} \in B(r, \boldsymbol{x_0})$ with $\boldsymbol{y} = f(\boldsymbol{x})$ and so $\boldsymbol{x} \in f^{-1}(V) = U$. So f is onto V.* □

g **is onto** U:

Proof *If $\boldsymbol{x} \in U = f^{-1}(V)$, then there is $\boldsymbol{y} \in V$ so that $f(\boldsymbol{x}) = \boldsymbol{y}$ because f is onto V. Thus, $g(f(\boldsymbol{x})) = g(\boldsymbol{y})$. However, we know g is f^{-1} here, so $g(\boldsymbol{y}) = \boldsymbol{x}$ and we have shown g is onto U.* □

g^{-1} **is continuous on** U:

Proof *We know f is continuous and $1-1$ on the compact set $\overline{B(r, \boldsymbol{x_0})}$. Hence, $g = f^{-1}$ is continuous on $f(\overline{B(r, \boldsymbol{x_0})})$. This implies g is continuous on V. What about g^{-1}'s continuity? Let $(\boldsymbol{y_n})$ in V converge to $\boldsymbol{y} \in V$. We know f is $1-1$ on $\overline{B(r, \boldsymbol{x_0})}$ to $f(\overline{B(r, \boldsymbol{x_0})})$. So there is a sequence $(\boldsymbol{x_n})$ in $\overline{B(r, \boldsymbol{x_0})}$ with $f(\boldsymbol{x_n}) = \boldsymbol{y_n}$. since $\overline{B(r, \boldsymbol{x_0})}$ is compact, there is a subsequence $(\boldsymbol{x^1_n})$ which converges to a point $\boldsymbol{x}$ in $\overline{B(r, \boldsymbol{x_0})}$. Since f is continuous there, we must have $f(\boldsymbol{x^1_n}) \to f(\boldsymbol{x})$. But $f(\boldsymbol{x^1_n}) = \boldsymbol{y^1_n}$ and since $\boldsymbol{y_n} \to \boldsymbol{y}$, we have $\boldsymbol{y^1_n} \to \boldsymbol{y}$. Thus, $f(\boldsymbol{x}) = \boldsymbol{y}$.*

If there was another subsequence $(\boldsymbol{x^2_n})$ which converged to $\boldsymbol{x}^*$ in $\overline{B(r, \boldsymbol{x_0})}$, by the continuity of f, we have $f(\boldsymbol{x^2_n}) \to f(\boldsymbol{x}^*)$. But $f(\boldsymbol{x^2_n}) = \boldsymbol{y^2_n}$ and so $\boldsymbol{y^2_n} \to \boldsymbol{y}$. Thus, $f(\boldsymbol{x}^*) = \boldsymbol{y}$.

But $f(\boldsymbol{x}) = f(\boldsymbol{x}^*)$ means $\boldsymbol{x} = \boldsymbol{x}^*$ as f is $1-1$. We have shown any convergent subsequence of $(\boldsymbol{x_n})$ must convergent to the same point $\boldsymbol{x}$. These vector sequences have n component sequences. We have shown the set of subsequential limits for each component sequence consists of one number. Hence, each component sequence of x_i converges. Thus, the set of subsequential limits of the sequence is $\boldsymbol{x}$ and so $\boldsymbol{x_n} \to \boldsymbol{x}$. But this says $g^{-1}(\boldsymbol{y_n}) \to g^{-1}(\boldsymbol{y})$. □

g_{x_i} **exists and is continuous on** V:

Proof *Fix $\boldsymbol{y} \in V$ and choose indices j and k from $\{1, \ldots, n\}$. We need to show*

$$\lim_{h \to 0} \frac{g_j(\boldsymbol{y} + h\boldsymbol{E_k}) - g_j(\boldsymbol{y})}{h}$$

exists and that g_{y_j} is continuous. For small enough h, the line segment $[\boldsymbol{y}, \boldsymbol{y} + h\boldsymbol{E_k}]$ is in V. Let

$$\begin{aligned} \boldsymbol{x} &= g(\boldsymbol{y}) = f^{-1}(\boldsymbol{y}) \in U \\ \boldsymbol{z} &= g(\boldsymbol{y} + h\boldsymbol{E_k}) = f^{-1}(\boldsymbol{y} + h\boldsymbol{E_k}) \in U \end{aligned}$$

Then, $\boldsymbol{x}$ and $\boldsymbol{z}$ are distinct as f^{-1} is $1-1$. Note the line segment $[\boldsymbol{x}, \boldsymbol{z}]$ is in $B(r, \boldsymbol{x_0})$. Since g is continuous on V, $\lim_{h \to 0} g(\boldsymbol{y} + h\boldsymbol{E_k}) = g(\boldsymbol{y})$. But this says $\lim_{h \to 0} \boldsymbol{z} = g(\boldsymbol{y}) = \boldsymbol{x}$

Now $f(\boldsymbol{z}) - f(\boldsymbol{x}) = \boldsymbol{y} + h\boldsymbol{E_k} - \boldsymbol{y} = h\boldsymbol{E_k}$, *so*

$$\begin{bmatrix} f_1(\boldsymbol{z}) - f_1(\boldsymbol{x}) \\ \vdots \\ f_{k-1}(\boldsymbol{z}) - f_{k-1}(\boldsymbol{x}) \\ f_k(\boldsymbol{z}) - f_k(\boldsymbol{x}) \\ f_{k+1}(\boldsymbol{z}) - f_k(\boldsymbol{x}) \\ \vdots \\ f_n(\boldsymbol{z}) - f_n(\boldsymbol{x}) \end{bmatrix} = \begin{bmatrix} 0 \\ \vdots \\ 0 \\ h, \text{ component } k \\ 0 \\ \vdots \\ 0 \end{bmatrix}$$

Thus, $f_i(\boldsymbol{z}) = f_i(\boldsymbol{x})$ *if* $i \neq k$ *and* $f_k(\boldsymbol{z}) - f_k(\boldsymbol{x}) = h$. *Now apply the Mean Value Theorem to each* f_i. *Then, there is a* $\boldsymbol{w_i} \in [\boldsymbol{z}, \boldsymbol{x}]$ *with*

$$f_i(\boldsymbol{z}) - f(\boldsymbol{x}) = (\boldsymbol{\nabla}(f_i))^T(\boldsymbol{z} - \boldsymbol{x}) = (\boldsymbol{\nabla}(f_i))^T(\boldsymbol{w_i})(g_i(\boldsymbol{y} + h\boldsymbol{E_k}) - g(\boldsymbol{y}))$$

This implies

$$\begin{bmatrix} (\boldsymbol{\nabla}(f_1))^T(\boldsymbol{w_1}) \\ \vdots \\ (\boldsymbol{\nabla}(f_{k-1}))^T(\boldsymbol{w_{k-1}}) \\ (\boldsymbol{\nabla}(f_k))^T(\boldsymbol{w_k}) \\ (\boldsymbol{\nabla}(f_{k+1}))^T(\boldsymbol{w_{k+1}}) \\ \vdots \\ (\boldsymbol{\nabla}(f_n))^T(\boldsymbol{w_n}) \end{bmatrix} \begin{bmatrix} (g_1(\boldsymbol{y} + h\boldsymbol{E_k}) - g_1(\boldsymbol{y}))/h \\ \vdots \\ (g_{k-1}(\boldsymbol{y} + h\boldsymbol{E_k}) - g_{k-1}(\boldsymbol{y}))/h \\ (g_k(\boldsymbol{y} + h\boldsymbol{E_k}) - g_k(\boldsymbol{y}))/h \\ (g_{k+1}(\boldsymbol{y} + h\boldsymbol{E_k}) - g_{k+1}(\boldsymbol{y}))/h \\ \vdots \\ (g_n(\boldsymbol{y} + h\boldsymbol{E_k}) - g_n(\boldsymbol{y}))/h \end{bmatrix} \begin{bmatrix} 0 \\ \vdots \\ 0 \\ 1, \text{ component } k \\ 0 \\ \vdots \\ 0 \end{bmatrix}$$

Since each $\boldsymbol{w_i}$ *is in the line segment* $[\boldsymbol{z}, \boldsymbol{x}]$ *which is in* $B(r, \boldsymbol{x_0}) \subset B(r_0, \boldsymbol{x_0})$, *we know for*

$$\Delta(\boldsymbol{w_1}, \ldots, \boldsymbol{w_n}) = \begin{bmatrix} (\boldsymbol{\nabla}(f_1))^T(\boldsymbol{w_1}) \\ \vdots \\ (\boldsymbol{\nabla}(f_{k-1}))^T(\boldsymbol{w_{k-1}}) \\ (\boldsymbol{\nabla}(f_k))^T(\boldsymbol{w_k}) \\ (\boldsymbol{\nabla}(f_{k+1}))^T(\boldsymbol{w_{k+1}}) \\ \vdots \\ (\boldsymbol{\nabla}(f_n))^T(\boldsymbol{w_n}) \end{bmatrix}$$

det $\Delta(\boldsymbol{w_1}, \ldots, \boldsymbol{w_n}) \neq 0$ *and there is a unique solution for each* $(g_k(\boldsymbol{y} + h\boldsymbol{E_k}) - g_k(\boldsymbol{y}))/h$. *We can find this solution by Cramer's rule. Let* Δ_i *be the submatrix we get by replacing column* i *with the data vector* $\boldsymbol{E_k}$.

$$\Delta_i(\boldsymbol{w_1}, \ldots, \boldsymbol{w_n}) =$$
$$\begin{bmatrix} f_{1,x_1}(\boldsymbol{w_1}) & \ldots & f_{1,x_{i-1}}(\boldsymbol{w_1}) & 0 & f_{1,x_{i+1}}(\boldsymbol{w_1}) & \ldots & f_{1,x_n}(\boldsymbol{w_1}) \\ \vdots & \vdots & \vdots & 0 & \vdots & \vdots & \vdots \\ f_{k-1,x_1}(\boldsymbol{w_{k-1}}) & \ldots & f_{k-1,x_{i-1}}(\boldsymbol{w_{k-1}}) & 0 & f_{k-1,x_{i+1}}(\boldsymbol{w_{k-1}}) & \ldots & f_{k-1,x_n}(\boldsymbol{w_{k-1}}) \\ f_{k,x_1}(\boldsymbol{w_k}) & \ldots & f_{k,x_{i-1}}(\boldsymbol{w_k}) & 1 & f_{k,x_{i+1}}(\boldsymbol{w_k}) & \ldots & f_{k,x_n}(\boldsymbol{w_k}) \\ f_{k+1,x_1}(\boldsymbol{w_{k+1}}) & \ldots & f_{k+1,x_{i-1}}(\boldsymbol{w_{k+1}}) & 0 & f_{k+1,x_{i+1}}(\boldsymbol{w_{k+1}}) & \ldots & f_{k+1,x_n}(\boldsymbol{w_{k+1}}) \\ \vdots & \vdots & \vdots & 0 & \vdots & \vdots & \vdots \\ f_{n,x_1}(\boldsymbol{w_n}) & \ldots & f_{n,x_{i-1}}(\boldsymbol{w_n}) & 0 & f_{n,x_{i+1}}(\boldsymbol{w_n}) & \ldots & f_{n,x_n}(\boldsymbol{w_n}) \end{bmatrix}$$

From Cramer's rule, we know

$$(g_i(\boldsymbol{y} + h\boldsymbol{E_k}) - g_i(\boldsymbol{y}))/h = \frac{det\ \Delta_i(\boldsymbol{w_1}, \ldots, \boldsymbol{w_n})}{det\ \Delta(\boldsymbol{w_1}, \ldots, \boldsymbol{w_n})}$$

We assume the continuity of the partials on D and so det $\Delta_i(\boldsymbol{w_1}, \ldots, \boldsymbol{w_n})$ *and det* $\Delta(\boldsymbol{w_1}, \ldots, \boldsymbol{w_n})$ *are continuous functions of* $\boldsymbol{w_1}, \ldots, \boldsymbol{w_n}$*. Therefore we know*

$$\lim_{h \to 0} \frac{det\ \Delta_i(\boldsymbol{w_1}, \ldots, \boldsymbol{w_n})}{det\ \Delta(\boldsymbol{w_1}, \ldots, \boldsymbol{w_n})} = \frac{det\ \Delta_i(\boldsymbol{z_1}, \ldots, \boldsymbol{z_n})}{det\ \Delta(\boldsymbol{z_1}, \ldots, \boldsymbol{z_n})}$$

Thus, $g_{i,y_k}(\boldsymbol{y}) = \lim_{h \to 0}(g_i(\boldsymbol{y} + h\boldsymbol{E_k}) - g_i(\boldsymbol{y}))/h$ *exists. If* $\boldsymbol{Y_p}$ *were a sequence converging to* $\boldsymbol{y}$*, then we would have*

$$g_{i,y_k}(\boldsymbol{Y_p}) = \frac{det\ \Delta_i(\boldsymbol{z_1^p}, \ldots, \boldsymbol{z_n^p})}{det\ \Delta(\boldsymbol{z_1^p}, \ldots, \boldsymbol{z_n^p})}$$

where the points obtained by the Mean Value Theorem now depend on p and so are labeled $\boldsymbol{z_i^p}$*. The determinants are still continuous on D and so*

$$\lim_{p \to \infty} g_{i,y_k}(\boldsymbol{Y_p}) = \lim_{p \to \infty} \frac{det\ \Delta_i(\boldsymbol{z_1^p}, \ldots, \boldsymbol{z_n^p})}{det\ \Delta(\boldsymbol{z_1^p}, \ldots, \boldsymbol{z_n^p})}$$

When we apply the Mean Value Theorem, we find $\boldsymbol{w_i^p} \in [g(\boldsymbol{Y_p}), g(\boldsymbol{Y_p} + h\boldsymbol{E_k})]$*. As* $p \to \infty$*,* $\boldsymbol{w_i^p} \to \boldsymbol{w_i} \in [g(\boldsymbol{y}), g(\boldsymbol{y} + h\boldsymbol{E_k})]$ *since g is continuous. Hence*

$$\lim_{p \to \infty} \frac{det\ \Delta_i(\boldsymbol{z_1^p}, \ldots, \boldsymbol{z_n^p})}{det\ \Delta(\boldsymbol{z_1^p}, \ldots, \boldsymbol{z_n^p})} = \frac{det\ \Delta_i(\boldsymbol{z_1}, \ldots, \boldsymbol{z_n})}{det\ \Delta(\boldsymbol{z_1}, \ldots, \boldsymbol{z_n})} = g_{i,y_k}(\boldsymbol{y})$$

which tells us g_{i,y_k} *is continuous.* **Note how useful Cramer's Rule is here. Cramer's rule is not very good computationally for** $n > 2$ **but it is a powerful theoretical tool for the study of smoothness of mappings.** □

This completes the proof. ■

Example 12.2.1 *For* $(x_1, x_2) = \boldsymbol{x} \in \Re^2$*, let* $\boldsymbol{y} = (y_1, y_2) = (x_1^2 - x_2^2, 2x_1, x_2)$*. This defined a map F. Then*

$$\boldsymbol{J_F}(x_1, x_2) = \begin{bmatrix} 2x_1 & -2x_2 \\ 2x_2 & 2x_1 \end{bmatrix}$$

and det $\boldsymbol{J_f}(x_1, x_2) = 4x_1^2 + 4x_2^2$*. This is positive unless* $(x_1, x_2) = (0, 0)$*. Hence, at any* $\mathbf{p} = (p_1, p_2) \neq \mathbf{0}$*, we can invoke the Inverse Function Theorem to find an open set U containing* $\mathbf{p}$ *and an open set V containing* $F(\mathbf{p})$ *so that* $f : U \to V$ *is* $1-1$ *and onto and* $f^{-1} : V \to U$ *is also* $1-1$ *and onto. Further* f^{-1} *has continuous partials on V. If you look back at Example 12.1.2, we figured out all of this for a slightly different map* $G(x_1, x_2) = (x_1^2 + x_2^2, 2x_1x_2)$ *in great detail. We could have done that here. Of course, these kinds of details are really needed if you want to know the explicit functional form of these mappings.*

12.2.1 Homework

These questions are quite similar to the ones in the last section. The difference is that the Inverse Function Theorem tells us the inverse is a bijection locally! Answer these questions like we do in the example above. Note in general it is impossible to explicitly determine g!

Exercise 12.2.1 *Let* $f(x, y) = \begin{bmatrix} x^2 + y^2 \\ 2xy \end{bmatrix}$.

- *Find* $\boldsymbol{J_f}$.
- *Find* $\det(\boldsymbol{J_f})$.
- *Explain where we can apply the Inverse Function Theorem in* $\Re^2$.
- *For each point* $\boldsymbol{x_0} \in \Re^2$, *write down what the Inverse Function Theorem says about the inverse* g.
- *Is* g *differentiable?*

Exercise 12.2.2 *Let* $f(x, y) = \begin{bmatrix} x^2 + y^2 \\ 2xz \\ y^2 + z^4 \end{bmatrix}$.

- *Find* $\boldsymbol{J_f}$.
- *Find* $\det(\boldsymbol{J_f})$.
- *Explain where we can apply the Inverse Function Theorem in* $\Re^2$.
- *For each point* $\boldsymbol{x_0} \in \Re^2$, *write down what the Inverse Function Theorem says about the inverse* g.
- *Is* g *differentiable?*

Exercise 12.2.3 *Let* $f(x, y) = \begin{bmatrix} 2x^2 + 4y^2 \\ 5xz \\ 3y^2 + z^4 \end{bmatrix}$.

- *Find* $\boldsymbol{J_f}$.
- *Find* $\det(\boldsymbol{J_f})$.
- *Explain where we can apply the Inverse Function Theorem in* $\Re^2$.
- *For each point* $\boldsymbol{x_0} \in \Re^2$, *write down what the Inverse Function Theorem says about the inverse* g.
- *Is* g *differentiable?*

Exercise 12.2.4 *Let* $f(x, y) = \begin{bmatrix} 2 + 3xyz \\ 2x^2 + 4y^2 \\ 5xz \\ 3y^2 + z^4 \end{bmatrix}$.

- *Find* $\boldsymbol{J_f}$.
- *Find* $\det(\boldsymbol{J_f})$.
- *Explain where we can apply the Inverse Function Theorem in* $\Re^2$.
- *For each point* $\boldsymbol{x_0} \in \Re^2$, *write down what the Inverse Function Theorem says about the inverse* g.
- *Is* g *differentiable?*

12.3 Implicit Function Results

To motivate the implicit function theorem, we want to go over a long example which will illustrate the general approach to the problem. We often want to handle this kind of problem.

Example 12.3.1 *We know $x^2 + y^2 + z^2 = 10$. Can we solve for z in terms of x and y? Let $F(x, y, z) = x^2 + y^2 + z^2 - 10$. Then $f(x, y, z) = 0$. We know $z = \pm\sqrt{10 - x^2 - y^2}$ and for real solutions, we must have $\sqrt{x^2 + y^2} \leq \sqrt{10}$. $F_z(x, y, z) = 2z$ which is not zero as long as $z \neq 0$. Since $x^2 + y^2 = 10 - z_0^2$, we can't pick just any $z_0 \neq 0$. We must choose $-\sqrt{10} \leq z_0 \leq \sqrt{10}$. So if $0 < z_0 < \sqrt{10}$, then the set S of viable choices of (s, y) is $S = \{(x, y)|x^2+y^2 = 10-z_0^2\}$ which is a circle about the origin in $\Re^2$ of radius $\sqrt{10 - z_0^2}$. For concreteness, pick a point (x_0, y_0) in Quadrant One with $x_0^2+y_0^2+z_0^2-10 = 0$. Then in the open set $V_0 = \{(x, y) : x^2+y^2 < 10-z_0^2\}$, we can solve for z in terms of (x, y) as $z = \sqrt{10 - x^2 - y^2}$. This defines a mapping $G(x, y) = \sqrt{10 - x^2 - y^2}$ which has continuous partials and $F(x, y, G(x, y)) = x^2 + y^2 + 10 - x^2 - y^2 - 10 = 0$. Finally, we have $G(x_0, y_0) = \sqrt{1 - -x_0^2 - y_0^2} = z_0$. This is an example of finding a function defined implicitly by a constraint surface. Note at $z = 0$, $F_z(x_0, y_0, 0) = 0$ always and the constraint surface reduces to $x^2 = y^2 - 10 = 0$. For that constraint, there is no way to find z in terms of (x, y). So it seems a restriction on $F_z \neq 0$ is important.*

So this is the problem of given $x^2 + y^2 + z^2 = 10$, solve for z locally in terms of (x, y). Note we could also ask to solve for x in terms of (y, z) or y in terms of (x, z). We can do this explicitly here.

Example 12.3.2 *The constraint surface is now $F(x, y, z) = sin(x^2 + 2y^4z^2) + 5z^4 - 25 = 0$. This tells us $(24/5)^{0.25} < z < (26/5)^{0.25}$ or $1.48 < z < 1.51$. However there are no constraints on (x, y) at all. It is really hard to see how to solve for z in terms of (x, y). But notice $F_z = 8y^4z\cos(x^2 + 2y^4z^2) + 20z^3$ which is not zero at many (x, y, z). It turns out we can prove there is a way to solve for z in terms of (x, y) locally around a point (x_0, y_0) where $F(x_0, y_0, z_0) = 0$ with $F_z(x_0, y_0, z_0) \neq 0$. This result is called the implicit function theorem and it is our next task.*

Example 12.3.3 *Consider the following situation. We are given $U \subset \Re^3$ which is an open set and a mapping $F(x, y, z)$ with domain U so that F has continuous first order partials in U. Assume there is a point (x_0, y_0, z_0) in U with $F(x_0, y_0, z_0) = 0$ with $F_z(x_0, y_0, z_0) \neq 0$. We claim we can find an open set V_0 containing (x_0, y_0) in $\Re^2$ and a mapping G with continuous partials on V_0 so that $G(x_0, y_0) = z_0$ and $F(x, y, G(x, y)) = 0$ for all (x, y) in V_0. This means the* **constraint** *surface $F(x, y, z) = 0$ can be used to provide a local solution for z in terms of x and y. This is something we do all the time as you can see from Example 12.3.1 and Example 12.3.2.*

For any $(x, y, z) \in U$ define the projections $f_1(x, y, z) = x$, $f_2(x, y, z) = y$ and define $f_3(x, y, z) = F(x, y, z)$. Then letting $f = (f_1, f_2, f_2)$, we have

$$\boldsymbol{J_f}(x, y, z) = \begin{bmatrix} f_{1x} & f_{1y} & f_{1z} \\ f_{2x} & f_{2y} & f_{2z} \\ f_{3x} & f_{3y} & f_{3z} \end{bmatrix} = \begin{bmatrix} 1 & 0 & 0 \\ 0 & 1 & 0 \\ F_x & F_y & F_z \end{bmatrix}$$

Thus $\det \boldsymbol{J_F}(x, y, z) = F_z(x, y, z)$ and by our assumption $\det \boldsymbol{J_F}(x_0, y_0, z_0) = F_z(x_0, y_0, z_0) \neq 0$. Apply the Inverse Function Theorem to $f : \Re^3 \to \Re^3$. Then there is an open set $U_1 \subset U$ containing (x_0, y_0, z_0), an open set $V_1 \subset f(U_1)$ containing $f(x_0, y_0, z_0)$ and $g = f^{-1}$ so that

$$f : U_1 \to V_1, 1-1, \text{ onto}$$
$$g : V_1 \to U_1, 1-1, \text{ onto}$$

Let the components of $g = f^{-1}$ be denoted by (g_1, g_2, h). Then, if $(X_1, X_2, X_3) \in V_1$,

$$g(X_1, Y_1, Z_1) \quad = \quad (g_1(X_1, Y_1, Z_1), g_2(X_1, Y_1, Z_1), h(X_1, Y_1, Z_1))$$

Since f maps U_1 to V_1 $1-1$ and onto, there exist a unique $(x, y, z) \in U_1$ with $f(x, y, z) = (X_1, Y_1, Z_1)$. But using the definition of f, this says $(x, y, F(x, y, z)) = (X_1, Y_1, Z_1)$. Thus, since $g = f^{-1}$,

$$\begin{aligned} (x, y, z) &= (g \circ f)(x, y, z) = g(f(x, y, z)) = g(X_1, Y_1, Z_1) \\ &= (g_1(X_1, Y_1, Z_1), g_2(X_1, Y_1, Z_1), h(X_1, Y_1, Z_1)) \\ &= (g_1(x_1, y_1, Z_1), g_2(x_1, y_1, Z_1), h(x_1, y_1, Z_1)) \end{aligned}$$

We see

$$x \quad = \quad g_1(x, y, Z_1), \quad y = g_2(x, y, Z_1), \quad z = h(x, y, Z_1)$$

and recall (x, y, z) is the unique point satisfying $f(x, y, z) = (X_1, Y_1, Z_1)$. Since $x = X_1$ and $y = Y_1$, this can be rewritten as $f(x, y, z) = (x, y, Z_1)$.

Let

$$V_0 \quad = \quad \{(x, y)\Re^2 | (x, y, 0) \in V_1\}$$

We claim V_0 is an open set in $\Re^2$ since V_1 is open in $\Re^3$. Let $(x, y) \in V_0$. Then $(x, y, 0) \in V_1$ and since V_1 is open, $(x, y, 0)$ is an interior point. So there is a radius $r > 0$ so that $B(r, (x, y, 0)) \subset V_1$. If $(X, Y) \in B(r, x, y)$, then $\sqrt{(X-x)^2 + (Y-y)^2} < r$, which implies $(X, Y, 0) \in B(r, x, y, 0) \subset V_1$. Thus, $(X, Y, 0) \in V_1$ telling us $(X, Y) \in V_0$. We conclude $B(r, (x, y)) \subset V_0$ which says (x, y) is an interior point of V_0. Thus, V_0 is open in $\Re^2$.

All of this works for any choice of (X_1, Y_1, Z_1). In particular, if we have $Z_1 = 0$, we get the unique point z so that $f(x, y, z) = (x, y, 0)$. But that means $z = h(x, y, 0)$. Define the mapping G by $G(x, y) = h(x, y, 0) = z$. Now $g = f^{-1}$ has continuous partials in V_1 and since h is a component of g, h also has continuous partials in V_1. We conclude G also have continuous partials on V_0 as $V_0 \subset V_1$.

Next, consider $G(x_0, y_0) = h(x_0, y_0, 0)$. By construction, $h(x_0, y_0, 0) = w$ where w is the unique point satisfying $f(x_0, y_0, w) = (x_0, y_0, 0)$. But $f(x_0, y_0, w) = (x_0, y_0, F(x_0, y_0, w))$, so we see $F(x_0, y_0, w) = 0$. There is only one point which has this property, so w must be z_0. Finally, if $(x, y) \in V_0$, then $(x, y, 0) \in V_1$ and $h(x, y, 0) = z$ where (x, y, z) is the unique point with $f(x, y, z) - (x, y, 0)$. So

$$(x, y, z) \quad = \quad g \circ f(x, y, z) = g(f(x, y, z)) = g(x, y, 0) = (g_1(x, y, 0), g_2(x, y, 0), h(x, y, 0))$$

So $x = g_1(x, y, 0)$, $y = g_2(x, y, 0)$ and $z = h(x, y, 0)$. Thus, $g(x, y, 0) = (x, y, h(x, y, 0)) = (x, y, G(x, y))$. Note $G(x, y) = z$ is the unique point so that $(x, y, z) \in U_1$ and $f(x, y, z) = (x, y, 0)$. We also know that $f(x, y, G(x, y)) = (x, y, F(x, y, G(x, y))) = (x, y, 0)$. Thus, $F(x, y, G(x, y)) = 0$ for $(x, y) \in U_0$.

So you can see how we might go about proving a more general theorem. We will do two cases: first, we extend this argument of a scalar function of $n+1$ variables and solve for the $(n+1)^{st}$ variable in terms of the others. This is Theorem 12.3.1. Then, we extend to vector-valued functions. We think of the domain as $\Re^{n+m}$ and we solve for the variables $x_{n+1}, \ldots, x_{n+m}$ in terms of $x_1, \ldots, x_n$.

This is Theorem 12.3.2. We have been trying to work through this slowly so that you can follow the arguments below. It is like the ones we have already done, but a bit more abstract.

Theorem 12.3.1 Implicit Function Theorem

Let $U \subset \Re^{n+`1}$ be an open set. Let $\boldsymbol{u} \in U$ be written as $\boldsymbol{u} = (\mathbb{X}, y)$ where $\mathbb{X} \in \Re^n$. Assume $F : U \rightarrow \Re$ has continuous first order partials in U and there is a point $(\mathbb{X}_0, y_0) \in U$ satisfying $F(\mathbb{X}_0, y_0) = 0$ with $F_y(\mathbb{X}_0, y_0) \neq 0$. Then there is an open set V_0 containing $\mathbb{X}_0$ and a function $G : V_0 \rightarrow \Re$ with continuous first order partials so that $G(\mathbb{X}_0) = y_0$ and $F(\mathbb{X}, G(\mathbb{X})) = 0$ on V_0.

Proof 12.3.1

The argument is very similar to the one we used in the example at the start of this section. Basically, we replace the use of (x_1, x_2) with $\mathbb{X}$ and make some obvious notational changes. Let's get started.

Here we are using $\mathbb{X}$ to denote the components $(x_1, x_2, \ldots, x_n)$ and we are setting the last component x_{n+1} to be the variable y. For any $(\mathbb{X}, y) \in U$ define the projections $f_i(\mathbb{X}, y) = x_i$ and define $f_{n+1}(\mathbb{X}, y) = F(\mathbb{X}, y)$. Then letting $f = (f_1, \ldots, f_n, F)$, we have

$$\boldsymbol{J_f}(\mathbb{X}, y) = \begin{bmatrix} f_{1,x_1} & \ldots & f_{1,x_n} & f_{1,y} \\ f_{2,x_1} & \ldots & f_{2,x_n} & f_{2,y} \\ \vdots & \vdots & \vdots & \vdots \\ f_{n,x_1} & \ldots & f_{n,x_n} & f_{n,y} \\ F_{x_1} & \ldots & F_{x_n} & F_y \end{bmatrix} = \begin{bmatrix} 1 & 0 & 0 & \ldots & 0 & 0 & 0 \\ 0 & 1 & 0 & \ldots & 0 & 0 & 0 \\ \vdots & \vdots & \vdots & \vdots & 0 & 0 & 0 \\ 0 & 0 & 0 & \ldots & 0 & 1 & 0 \\ F_{x_1} & F_{x_2} & F_{x_3} & \ldots & F_{x_n} & F_y & \end{bmatrix}$$

Thus det $\boldsymbol{J_F}(\mathbb{X}, y) = F_y(\mathbb{X}, y)$ and by our assumption det $\boldsymbol{J_F}(\mathbb{X}_0, y_0) = F_y(\mathbb{X}_0, y_0) \neq 0$. Apply the Inverse Function Theorem to $f : \Re^{n+1} \rightarrow \Re^{n+1}$. Then there is an open set $U_1 \subset U$ containing $(\mathbb{X}_0, y_0)$, an open set $V_1 \subset f(U_1)$ containing $f(\mathbb{X}_0, y_0)$ and $g = f^{-1}$ so that

$$f : U_1 \rightarrow V_1, 1-1, \text{ onto}$$
$$g : V_1 \rightarrow U_1, 1-1, \text{ onto}$$

Let the components of $g = f^{-1}$ be denoted by $(g_1, \ldots, g_n, h)$. Then, if $(\mathbb{Z}, u) \in V_1$,

$$g(\mathbb{Z}, u) = (g_1(\mathbb{Z}, u), \ldots, g_n(\mathbb{Z}, u), h(\mathbb{Z}, u))$$

Since f maps U_1 to V_1 $1-1$ and onto, there exist a unique $(\mathbb{X}, y) \in U_1$ with $f(\mathbb{X}, y) = (\mathbb{Z}, u)$. But using the definition of f, this says $(\mathbb{X}, F(\mathbb{X}, y)) = (\mathbb{Z}, u)$. Thus, since $g = f^{-1}$, $\mathbb{X} = \mathbb{Z}$ and $F(\mathbb{X}, y) = u$. Further,

$$\begin{aligned} (\mathbb{X}, y) &= (g \circ f)(\mathbb{X}, y) = g(f(\mathbb{X}, y)) = g(\mathbb{Z}, u) = g(\mathbb{X}, u) \\ &= (g_1(\mathbb{X}, u), \ldots, g_n(\mathbb{X}, u), h(\mathbb{X}, u)) \end{aligned}$$

We see $x_i = g_i(\mathbb{X}, u)$. Now recall $(\mathbb{X}, y)$ is the unique point satisfying $f(\mathbb{X}, y) = (\mathbb{Z}, u)$. Since $\mathbb{X} = \mathbb{Z}$, this can be rewritten as $f(\mathbb{X}, y) = (\mathbb{X}, u)$.

Let

$$V_0 = \{\mathbb{X} \in \Re^n | (\mathbb{X}, 0) \in V_1\}$$

Using an argument like before, it is straightforward to show V_0 is an open set in $\Re^n$ since V_1 is open in $\Re^{n+1}$.

All of this works for any choice of $(\mathbb{X}, y)$. *In particular, if we have* $y = 0$, *we get the unique point* u *so that* $f(\mathbb{X}, u) = (\mathbb{X}, 0)$. *But that means* $y = h(\mathbb{X}, 0)$. *Define the mapping* G *by* $G(\mathbb{X}, y) = h(\mathbb{X}, 0) = y$. *Now* $g = f^{-1}$ *has continuous partials in* V_1 *and since* h *is a component of* g, h *also has continuous partials in* V_1. *We conclude* G *also have continuous partials on* V_0 *as* $V_0 \subset V_1$.

Next, consider $G(\mathbb{X}_0) = h(\mathbb{X}_0, 0)$. *By construction,* $h(\mathbb{X}_0, 0) = w$ *where* w *is the unique point satisfying* $f(\mathbb{X}_0, w) = (\mathbb{X}_0, 0)$. *But* $f(\mathbb{X}_0, w) = (\mathbb{X}_0, F(\mathbb{X}_0, w))$, *so we see* $F(\mathbb{X}_0, w) = 0$. *There is only one point which has this property so* w *must be* y_0. *Finally, if* $(\mathbb{X}) \in V_0$, *then* $(\mathbb{X}, 0) \in V_1$ *and* $h(\mathbb{X}, 0) = y$ *where* $(\mathbb{X}, y)$ *is the unique point with* $f(\mathbb{X}, y) = (\mathbb{X}, 0)$. *So*

$$(\mathbb{X}, y) = g \circ f(\mathbb{X}, y) = g(f(\mathbb{X}, y)) = g(\mathbb{X}, 0) = (g_1(\mathbb{X}, 0), \ldots, g_n(\mathbb{X}, 0), h(\mathbb{X}, 0))$$

So $x_i = g_i(\mathbb{X}, 0)$ *and* $y = h(\mathbb{X}, 0)$. *Thus,* $g(\mathbb{X}, 0) = (\mathbb{X}, h(\mathbb{X}, 0)) = (\mathbb{X}, G(\mathbb{X}))$. *Note* $G(\mathbb{X}) = y$ *is the unique point so that* $(\mathbb{X}, y) \in U_1$ *and* $f(\mathbb{X}, y) = (\mathbb{X}, 0)$. *We also know that* $f(\mathbb{X}, G(\mathbb{X})) = (\mathbb{X}, F(\mathbb{X}, G(\mathbb{X}))) = (\mathbb{X}, 0)$. *Thus,* $F(\mathbb{X}, G(\mathbb{X})) = 0$ *for* $\mathbb{X} \in U_0$. ■

Example 12.3.4 *Let* $f(x, y, z) = \tanh(xyz) - xyz - 10 = 0$. *Letting* $u = xyz$, *this is equivalent to* $\tanh(u) - u - 10 = 0$ *or* $\tanh(u) = u + 10$. *It is easy to see there is a solution by graphing* $\tanh(u)$ *and* $u + 10$ *and finding the intersection. So there is a value* u^* *where* $\tanh(u^*) = u^* + 10$. *In particular, for* $z_0 = 1$, *there is a value* $(x_0$ *and* y_0 *so that* $\tanh(x_0 y_0) - x_0 y_0 - 10 = 0$; *i.e.* $f(x_0, y_0, z_0, 1) = 0$. *Note*

$$F_z(x, y, z) = sech^2(xyz)xy - xy = xy(sech^2(xyz) - 1)$$

From the graph where we found the intersection (you should sketch this yourself!), we can see $x_0 y_0 \neq 0$ *so* $sech^2(x_0 y_0) - 1 \neq 0$. *We conclude* $F_x(x_0, y_0, 1) \neq 0$.

Applying the Implicit Function Theorem, there is an open set $U \in \Re^2$ *containing* $(x_0, y_0, 1)$ *and a mapping* G *with continuous partials in* U *so that* $G(x_0, y_0) = 1$ *and*

$$\tanh(xy\, G(x, y)) - xy\, G(x, y) - 10 = 0$$

for all $(x, y) \in U$.

Theorem 12.3.2 An Extension of the Implicit Function Theorem

Let $U \subset \Re^{n+m}$ *be an open set. Let* $\boldsymbol{u} \in U$ *be written as* $\boldsymbol{u} = (\mathbb{X}, \mathbb{Y})$ *where* $\mathbb{X} \in \Re^n$ *and* $\mathbb{Y} \in \Re^m$. *Assume* $F : U \to \Re^m$ *has continuous first order partials in* U *and there is a point* $(\mathbb{X}_0, \mathbb{Y}_0) \in U$ *satisfying* $F(\mathbb{X}_0, \mathbb{Y}_0) = \mathbb{O}$ *(here* $\mathbb{O}$ *is the zero vector in* $\Re^m$ *)with*

$$det \begin{bmatrix} F_{n+1,y_1}(\mathbb{X}_0, \mathbb{Y}_0) & \ldots & F_{n+1,y_m}(\mathbb{X}_0, \mathbb{Y}_0) \\ \vdots & \vdots & \vdots \\ F_{n+m,y_1}(\mathbb{X}_0, \mathbb{Y}_0) & \ldots & F_{n+m,y_m}(\mathbb{X}_0, \mathbb{Y}_0) \end{bmatrix} \neq 0$$

where the matrix above is a submatrix of the full $\boldsymbol{J_F}$. *Then there is an open set* V_0 *containing* $\mathbb{X}_0 \in \Re^n$ *and* $G : V_0 \to \Re^m$ *with continuous partials so that* $G(\mathbb{X}_0) = \mathbb{Y}_0$ *and* $F(\mathbb{X}, G(\mathbb{X})) = \mathbb{O}$ *on* V_0.

Proof 12.3.2

The argument is again very similar to the one we just used. This time instead of a single scalar variable y, *we have a vector* $\mathbb{Y}$. *It is mostly a matter of notation. If you compare these proofs* *carefully, we mostly replace y by $\mathbb{Y}$, u by $\mathbb{U}$ and so forth as part played by the variable x_{n+1} has now been expanded to m variables and hence we must use vector notation. However, the structure of the argument is very much the same.*

Here we are using $\mathbb{X}$ to denote the components $(x_1, x_2, \ldots, x_n)$ and we are setting the components $x_{n+1}, \ldots, x_{n+m}$ to be the variable $\mathbb{Y}$ for the new variables $y_1, \ldots, y_m$. For any $(\mathbb{X}, \mathbb{Y}) \in U$, define $f(\mathbb{X}, \mathbb{Y}) = (\mathbb{X}, F(\mathbb{X}, \mathbb{Y}))$. Thus, the first n components are simply the usual projections. Then, since F has m components F_1 through F_m, we have

$$\boldsymbol{J_f}(\mathbb{X}, \mathbb{Y}) = \begin{bmatrix} 1 & 0 & 0 & 0 & 0 & 0 & 0 \\ 0 & 1 & 0 & 0 & 0 & 0 & 0 \\ \vdots & \vdots & \vdots & 0 & 0 & 0 & 0 \\ 0 & 0 & 1 & 0 & 0 & 0 & 0 \\ F_{1,x_1} & \ldots & F_{1,x_n} & F_{1,y_1} & F_{1,y_2} & \ldots & F_{1,y_m} \\ \vdots & \vdots & \vdots & \vdots & \vdots & \vdots & \vdots \\ F_{m,x_1} & \ldots & F_{m,x_n} & F_{m,y_1} & F_{m,y_2} & \ldots & F_{m,y_m} \end{bmatrix}$$

where all of the partials are evaluated at $(\mathbb{X}, \mathbb{Y})$ although we have not put that evaluation point in so we save space. Thus , at $(\mathbb{X}_0, \mathbb{Y}_0)$

$$det\, \boldsymbol{J_F}(\mathbb{X}_0, \mathbb{Y}_0) = det \begin{bmatrix} F_{1,y_1}(\mathbb{X}, \mathbb{Y}_0) & \ldots & F_{1,y_m}(\mathbb{X}, \mathbb{Y}_0) \\ \vdots & \vdots & \vdots \\ F_{m,y_1}(\mathbb{X}, \mathbb{Y}_0) & \ldots & F_{m,y_m}(\mathbb{X}, \mathbb{Y}_0) \end{bmatrix} \neq 0$$

by our assumption. Apply the Inverse Function Theorem to $f : \Re^{n+m} \rightarrow \Re^{n+m}$. Then there is an open set $U_1 \subset U$ containing $(\mathbb{X}_0, \mathbb{Y}_0)$, an open set $V_1 \subset f(U_1)$ containing $f(\mathbb{X}_0, \mathbb{Y}_0)$ and $g = f^{-1}$ so that

$$\begin{aligned} f &: U_1 \rightarrow V_1, 1-1, \ onto \\ g &: V_1 \rightarrow U_1, 1-1, \ onto \end{aligned}$$

Let the components of $g = f^{-1}$ be denoted by $(g_1, \ldots, g_n, h_1, \ldots, h_m)$. Then, if $(\mathbb{Z}, \mathbb{U}) \in V_1$,

$$\begin{aligned} g(\mathbb{Z}, \mathbb{U}) &= (g_1(\mathbb{Z}, \mathbb{U}), \ldots, g_n(\mathbb{Z}, \mathbb{U}), \\ & \quad h_1(\mathbb{Z}, \mathbb{U}), \ldots, h_1(\mathbb{Z}, \mathbb{U})) \end{aligned}$$

Since f maps U_1 to V_1 $1-1$ and onto, there exist a unique $(\mathbb{X}, \mathbb{Y}) \in U_1$ with $f(\mathbb{X}, \mathbb{Y}) = (\mathbb{Z}, \mathbb{U})$. But using the definition of f, this says $(\mathbb{X}, F(\mathbb{X}, \mathbb{Y})) = (\mathbb{Z}, \mathbb{U})$. Thus, since $g = f^{-1}$, $\mathbb{X} = \mathbb{Z}$ and $F(\mathbb{X}, \mathbb{Y}) = \mathbb{U}$. Further,

$$\begin{aligned} (\mathbb{X}, \mathbb{Y}) &= (g \circ f)(\mathbb{X}, \mathbb{Y}) = g(f(\mathbb{X}, \mathbb{Y})) = g(\mathbb{Z}, \mathbb{U}) = g(\mathbb{X}, \mathbb{U}) \\ &= (g_1(\mathbb{X}, \mathbb{U}), \ldots, g_n(\mathbb{X}, \mathbb{U})) \\ & \quad (h_1(\mathbb{X}, \mathbb{U}), \ldots, h_m(\mathbb{X}, \mathbb{U})) \end{aligned}$$

We see $x_i = g_i(\mathbb{X}, \mathbb{U})$. Now recall $(\mathbb{X}, \mathbb{Y})$ is the unique point satisfying $f(\mathbb{X}, \mathbb{Y}) = (\mathbb{Z}, \mathbb{U})$. Since $\mathbb{X} = \mathbb{Z}$, this can be rewritten as $f(\mathbb{X}, \mathbb{Y}) = (\mathbb{X}, \mathbb{U})$.

Let

$$V_0 = \{\mathbb{X} \in \Re^n | (\mathbb{X}, \mathbb{O}) \in V_1\}$$

Using an argument like before, it is straightforward to show V_0 is an open set in $\Re^n$ since V_1 is open *in* $\Re^{n+m}$.

All of this works for any choice of $(\mathbb{X}, \mathbb{Y})$. *In particular, if we have* $\mathbb{Y} = \mathbb{O}$, *we get the unique point* $(\mathbb{U})$ *so that* $f(\mathbb{X}, \mathbb{U}) = (\mathbb{X}, \mathbb{O})$. *But that means* $\mathbb{Y} = (h_1(\mathbb{X}, \mathbb{O}), \ldots, h_m(\mathbb{X}, \mathbb{O}))$. *For convenience, let*

$$h(\mathbb{X}, \mathbb{Y}) = (h_1(\mathbb{X}, \mathbb{O}), \ldots, h_m(\mathbb{X}, \mathbb{O}))$$

Then define the mapping G *by* $G(\mathbb{X}, \mathbb{Y}) = h(\mathbb{X}, \mathbb{O}) = \mathbb{Y}$. *Now* $g = f^{-1}$ *has continuous partials in* V_1 *and since* h_i *are components of* g, h *also has continuous partials in* V_1. *We conclude* G *also have continuous partials on* V_0 *as* $V_0 \subset V_1$.

Next, consider $G(\mathbb{X}_0) = h(\mathbb{X}_0, \mathbb{O})$. *By construction,* $h(\mathbb{X}_0, \mathbb{O}) = \mathbb{W}$ *where* $\mathbb{W}$ *is the unique point satisfying* $f(\mathbb{X}_0, \mathbb{W}) = (\mathbb{X}_0, \mathbb{O})$. *But* $f(\mathbb{X}_0, \mathbb{W}) = (\mathbb{X}_0, F(\mathbb{X}_0, \mathbb{W}))$, *so we see* $F(\mathbb{X}_0, \mathbb{W}) = \mathbb{O}$. *There is only one point which has this property, so* $\mathbb{W}$ *must be* $\mathbb{Y}_0$. *Finally, if* $(\mathbb{X}) \in V_0$, *then* $(\mathbb{X}, \mathbb{O}) \in V_1$ *and* $h(\mathbb{X}, \mathbb{O}) = \mathbb{Y}$ *where* $(\mathbb{X}, \mathbb{Y})$ *is the unique point with* $f(\mathbb{X}, \mathbb{Y}) = (\mathbb{X}, \mathbb{O})$. *So*

$$(\mathbb{X}, \mathbb{Y}) = g \circ f(\mathbb{X}, \mathbb{Y}) = g(f(\mathbb{X}, \mathbb{Y})) = g(\mathbb{X}, \mathbb{O}) = (g_1(\mathbb{X}, \mathbb{O}), \ldots, g_n(\mathbb{X}, \mathbb{O}), h(\mathbb{X}, \mathbb{O}))$$

So $x_i = g_i(\mathbb{X}, \mathbb{O})$ *and* $\mathbb{Y} = h(\mathbb{X}, \mathbb{O})$. *Thus,* $g(\mathbb{X}, \mathbb{O}) = (\mathbb{X}, h(\mathbb{X}, \mathbb{O})) = (\mathbb{X}, G(\mathbb{X}))$. *Note* $G(\mathbb{X}) = \mathbb{Y}$ *is the unique point so that* $(\mathbb{X}, \mathbb{Y}) \in U_1$ *and* $f(\mathbb{X}, \mathbb{Y}) = (\mathbb{X}, \mathbb{O})$. *We also know that* $f(\mathbb{X}, G(\mathbb{X})) = (\mathbb{X}, F(\mathbb{X}, G(\mathbb{X}))) = (\mathbb{X}, \mathbb{O})$. *Thus,* $F(\mathbb{X}, G(\mathbb{X})) = \mathbb{O}$ *for* $\mathbb{X} \in U_0$. ■

Comment 12.3.1 *It is not important that this happens at* $F(\mathbb{X}_0, \mathbb{Y}_0) = \mathbb{O}$. *If we had* $F(\mathbb{X}_0, \mathbb{Y}_0) = \mathbf{c} \in \Re^m$, *we just apply the theorem to* $H(\mathbb{X}_0, \mathbb{Y}_0) = F(\mathbb{X}_0, \mathbb{Y}_0) - \mathbf{c}$ *as we do in the example below. We would then find* $F(\mathbb{X}, G(\mathbb{X})) = \mathbf{c}$ *on* V_0.

Example 12.3.5 *Let* $F(x, y, u, v) = (u^2 + 2xu - v^2 - 7y, u^3 + 3\sin(u) + 15x^2 e^y + \cos(v) + 10)$. *Here* F *maps* $\Re^4$ *to* $\Re^2$ *so n = m = 2. Then*

$$\boldsymbol{J_F}(x, y, u, v) = \begin{bmatrix} 2u & -7 & 2u + 2x & -2v \\ 30xe^y & 15x^2 e^y & 3u^2 + 3\cos(u) & -\sin(v) \end{bmatrix}$$

At $(0, 0, 0, 0)$,

$$\boldsymbol{J_F}(0, 0, 0, 0) = \begin{bmatrix} 0 & -7 & 0 & 0 \\ 0 & 0 & 3 & -0 \end{bmatrix}$$

and $F(0, 0, 0, 0) = (0, 11)$. *The submatrix*

$$A = \begin{bmatrix} -7 & 0 \\ 0 & 3 \end{bmatrix}$$

has a nonzero determinant. Hence, the Implicit Function Theorem, Theorem 12.3.2, tells us we can solve for (y, u) *in terms of* (x, v) *to find* $G(x, v)$ *so that by relabeling the order of the variables,* $F(x, v, G(x, v)) = 11$ *in an open set* U *containing* $(0, 0)$ *with* $F(0, 0, 0, 0) = (0, 11)$.

Another way to say this, is we write $F(x, v, y(x, v), u(x, v)) = 11$ *uniquely locally. Note although our original order was* (x, y, u, v) *we find it convenient to reorganize in the order* (x, v, y, u) *to match what the implicit function theorem says. We could have written* $F(x, y(x, v), u(x, v), v) = 11$ *uniquely locally instead. So you have to get used to these mental relabelings!*

At the point $(1, 1, 2, 3)$, *we have*

$$\boldsymbol{J_F}(1,1,2,3) \quad = \quad \begin{bmatrix} 4 & -7 & 6 & -6 \\ 30e & 15e & 12+3\cos(2) & -\sin(3) \end{bmatrix}$$

and $F(1, 1, 2, 3) = (-8, 18 + 3\sin(2) + 15e + \cos(3))$. *We can then solve for any two variables in terms of the others locally near this point. Just apply the implicit function theorem to any pair of variables in* $\boldsymbol{J_F}$ *with nonzero determinant.*

Example 12.3.6 *Consider* $f(x, y) = x^2 + y^2$. *Then* $\boldsymbol{J_F} = \begin{bmatrix} 2x & 2y \end{bmatrix}$. *We have* f *maps* $\Re^2$ *to* $\Re$ *so* $n = m = 1$. *At any point where* $f_x \neq 0$ *we can solve locally in a bijective way for* $y(x)$ *and at any point where* $f_y \neq 0$, *we can solve locally for* $x(y)$.

- *At* $(1, 1)$, $f(1, 1) = 2$ *and we can solve for* $y = \sqrt{2 - x^2}$ *or* $y = -\sqrt{2 - x^2}$.
- *At* $(1, 1)$, $f(1, 1) = 2$ *and we can solve for* $x = \sqrt{2 - y^2}$ *or* $x = -\sqrt{2 - y^2}$.

Now in this example, we can actually find the bijections which is rare. The implicit function theorem just tells us we can do this locally.

Example 12.3.7 *Consider* $f(x, y) = x^2 + y^2 + z^2$. *Then* $\boldsymbol{J_F} = \begin{bmatrix} 2x & 2y & 2z \end{bmatrix}$. *We have* f *maps* $\Re^3$ *to* $\Re$ *so* $n = 2$ *and* $m = 1$. *At any point where* $f_x \neq 0$ *we can solve locally in a bijective way for* $(y(x), z(x))$, *at any point where* $f_y \neq 0$, *we can solve locally for* $(x(y), z(y))$ *and at any point where* $f_z \neq 0$, *we can find* $(x(z), y(z))$. *In fact, we can construct these bijections as follows:*

- *At* $(1, 1, 1)$, $f(1, 1, 1) = 3$ *and so we must have* $x^2 + y^2 + z^2 = 3$. *Hence for any* x, *we can let* $y = \cos(x)\sqrt{3 - x^2}$ *and* $z = \sin(x)\sqrt{3 - x^2}$. *Then we get* $x^2 + (3 - x^2)\cos^2(x) + (3 - x^2)\sin^2(x) = 3$ *as desired. There are many more such mapping choices of course and you should find more of them to make sure you see what is going on. Hence, locally we have* $x^2 + (y(x))^2 + (z(x))^2 = 3$.
- *At* $(1, 1, 1)$, $f(1, 1, 1) = 3$ *and so we must have* $x^2 + y^2 + z^2 = 3$. *Hence for any* y, *we can let* $x = \cos(y)\sqrt{3 - y^2}$ *and* $z = \sin(y)\sqrt{3 - y^2}$. *Then we get* $(x(y))^2 + y^2 + (z(y))^2 = 3$.
- *You can do a similar thing for the* z *case. We leave that to you.*

Note again in this example, we can actually find the bijections which is rare. The implicit function theorem just tells us we can do this locally via an existence result.

12.3.1 Homework

Exercise 12.3.1 *Let* $F(x, y, u, v) = (u^2 + 3xu - v^2 - 7y, u^3 + 3u + 15x^2y^2 + v^3 + 10)$. *This is a map from* $\Re^4$ *to* $\Re^2$. *So* $n = 2$ *and* $m = 2$ *in the Implicit Function Theorem.*

- *Find* $\boldsymbol{J_f}$. *use the order* (x, y, u, v).
- *Which pairs of variables can you solve for locally in terms of the other two at the point* $(1, 1, 1, 1)$? *What do you know about the resulting mapping G if it exists?*
- *Which pairs of variables can you solve for locally in terms of the other two at the point* $(0, 0, 0, 0)$? *What do you know about the resulting mapping G if it exists?*
- *Which pairs of variables can you solve for locally in terms of the other two at the point* $(1, 0, 1, 0)$? *What do you know about the resulting mapping G if it exists?*
- *Which pairs of variables can you solve for locally in terms of the other two at the point* $(0, 1, 0, 1)$? *What do you know about the resulting mapping G if it exists?*

Exercise 12.3.2 *Let* $F(x, y, u, v) = (u^2 + 3xu - v^2 - 7y^2, u^3 + 15x^2y^4 + xv^3)$. *This is a map from* $\Re^4$ *to* $\Re^2$. *So* $n = 2$ *and* $m = 2$ *in the Implicit Function Theorem.*

- *Find* $\boldsymbol{J_f}$. *use the order* (x, y, u, v).
- *Which pairs of variables can you solve for locally in terms of the other two at the point* $(1, 1, 1, 1)$? *What do you know about the resulting mapping G if it exists?*
- *Which pairs of variables can you solve for locally in terms of the other two at the point* $(0, 0, 0, 0)$? *What do you know about the resulting mapping G if it exists?*
- *Which pairs of variables can you solve for locally in terms of the other two at the point* $(1, 0, 1, 0)$? *What do you know about the resulting mapping G if it exists?*
- *Which pairs of variables can you solve for locally in terms of the other two at the point* $(0, 1, 0, 1)$? *What do you know about the resulting mapping G if it exists?*

Exercise 12.3.3 *Let* $F(x, y, u, v) = (u^2 + 3xu - v^2 - 7y, u^3 + 3u + 15x^2y^2 + v^3 + 10, 6xy^2u^3v^4)$. *This is a map from* $\Re^4$ *to* $\Re^3$. *So* $n = 1$ *and* $m = 3$ *in the Implicit Function Theorem.*

- *Find* $\boldsymbol{J_f}$. *use the order* (x, y, u, v).
- *Which three variables can you solve for locally in terms of the remaining one at the point* $(1, 1, 1, 1)$? *What do you know about the resulting mapping G if it exists?*
- *Which three variables can you solve for locally in terms of the remaining one at the point* $(0, 0, 0, 0)$? *What do you know about the resulting mapping G if it exists?*
- *Which three variables can you solve for locally in terms of the remaining one at the point* $(1, 0, 1, 0)$? *What do you know about the resulting mapping G if it exists?*
- *Which three variables can you solve for locally in terms of the remaining one at the point* $(0, 1, 0, 1)$? *What do you know about the resulting mapping G if it exists?*

Exercise 12.3.4 *Let* $F(x, y) = (\sin(x^2 + 3z), \cos(2xy) + 2z^2)$. *Here F maps* $\Re^3$ *to* $\Re^2$ *so* $n = 1$ *and* $m = 2$.

- *Find* $\boldsymbol{J_f}$. *use the order* (x, y, z).
- *Which pairs of variables can you solve for locally in terms of the other one at the point* $(1, 1, 2)$? *What do you know about the resulting mapping G if it exists?*
- *Which pairs of variables can you solve for locally in terms of the other one at the point* $(0, 0)$? *What do you know about the resulting mapping G if it exists?*
- *Which pairs of variables can you solve for locally in terms of the other one at the point* $(1, 0)$? *What do you know about the resulting mapping G if it exists?*
- *Which pairs of variables can you solve for locally in terms of the other one at the point* $(0, 1)$? *What do you know about the resulting mapping G if it exists?*

12.4 Constrained Optimization

Let's look at the problem of finding the minimum or maximum of a function $f(x, y)$ subject to the constraint $g(x, y) = c$ where c is a constant. Let's see how far we can get without explicitly using the implicit function theorem. Let's suppose we have a point (x_0, y_0) where an extremum value occurs and assume f and g are differentiable at that point. Then, at another point (x, y) which satisfies the constraint, we have

$$g(x, y) = g(x_0, y_0) + g_x(x_0, y_0)(x - x_0) + g_y(x_0, y_0)(y - y_0) + E_g(x, y, x_0, y_0)$$

But $g(x, y) = g(x_0, y_0) = c$, so we have

$$0 = g_x^0(x - x_0) + g_y^0(y - y_0) + E_g(x, y, x_0, y_0)$$

where we let $g_x^0 = g_x(x_0, y_0)$ and $g_y^0 = g_y(x_0, y_0)$.

Now, given an x, we assume we can find a y value so that $g(x, y) = c$ in $B_r(x)$ for some $r > 0$. Of course, this value need not be unique. Let $\phi(x)$ be the function defined by

$$\phi(x) = \{y | g(x, y) = c\}$$

For example, if $g(x, y) = c$ was the function $x^2 + y^2 = 25$, then $\phi(x) = \pm\sqrt{25 - x^2}$ for $-5 \leq x \leq 5$. Clearly, we can't find a full circle $B_r(x)$ when $x = -5$ or $x = 5$, so let's assume the point (x_0, y_0) where the extremum value occurs does have such a local circle around it where we can find corresponding y values for all x values in $B_r(x_0)$. So locally, we have a function $\phi(x)$ defined around x_0 so that $g(x, \phi(x)) = c$ for $x \in B_r(x_0)$. Then we have

$$0 = g_x^0(x - x_0) + g_y^0(\phi(x) - \phi(x_0)) + E_g(x, \phi(x), x_0, \phi(x_0))$$

Now divide through by $x - x_0$ to get

$$0 = g_x^0 + g_y^0\left(\frac{\phi(x) - \phi(x_0)}{x - x_0}\right) + \frac{E_g(x, \phi(x), x_0, \phi(x_0))}{x - x_0}$$

Thus,

$$g_y^0\left(\frac{\phi(x) - \phi(x_0)}{x - x_0}\right) = -g_x^0 - \frac{E_g(x, \phi(x), x_0, \phi(x_0))}{x - x_0}$$

and assuming $g_x^0 \neq 0$ and $g_y^0 \neq 0$, we can solve to find

$$\frac{\phi(x) - \phi(x_0)}{x - x_0} = -\frac{g_x^0}{g_y^0} - \frac{1}{g_y^0}\frac{E_g(x, \phi(x), x_0, \phi(x_0))}{x - x_0}$$

Since g is differentiable at $(x_0, \phi(x_0))$, as $x \to x_0$, $\frac{E_g(x,\phi(x),x_0,\phi(x_0))}{x-x_0} \to 0$. Thus, we have ϕ is differentiable and therefore also continuous with

$$\phi'(x_0) = \lim_{x \to x_0} \frac{\phi(x) - \phi(x_0)}{x - x_0} = -\frac{g_x^0}{g_y^0}$$

Now since both $g_x^0 \neq 0$ and $g_y^0 \neq 0$, the fraction $-\frac{g_x^0}{g_y^0} \neq 0$ too. This means locally $\phi(x)$ is either strictly increasing or strictly decreasing; i.e. is a strictly monotone function.

Let's assume $\phi'(x_0) > 0$ and so ϕ is increasing on some interval $(x_0 - R, x_0 + R)$. Let's also assume the extreme value is a minimum, so we know on this interval $f(x,y) \geq f(x_0,y_0)$ with $g(x,y)=c$. This means $f(x,\phi(x)) - f(x_0,\phi(x_0)) \geq 0$ on (x_0-R, x_0+R). Now do an expansion for f to get

$$f(x,\phi(x)) = f(x_0,\phi(x_0)) + f_x^0(x-x_0) + f_y^0(\phi(x)-\phi(x_0)) + E_f(x,\phi(x),x_0,\phi(x_0))$$

where $f_x^0 = f_x(x_0,\phi(x_0))$ and $f_y^0 = f_y(x_0,\phi(x_0))$. This implies

$$f(x,\phi(x)) - f(x_0,\phi(x_0)) = f_x^0(x-x_0) + f_y^0(\phi(x)-\phi(x_0)) + E_f(x,\phi(x),x_0,\phi(x_0))$$

Now we have assumed $f(x,\phi(x)) - f(x_0,\phi(x_0)) \geq 0$, so we have

$$f_x^0(x-x_0) + f_y^0(\phi(x)-\phi(x_0)) + E_f(x,\phi(x),x_0,\phi(x_0)) \quad \geq \quad 0$$

If $x - x_0 > 0$ we find

$$f_x^0 + f_y^0\left(\frac{\phi(x)-\phi(x_0)}{x-x_0}\right) + \frac{E_f(x,\phi(x),x_0,\phi(x_0))}{x-x_0} \quad \geq \quad 0$$

Now take the limit as $x \to x_0^+$ to find

$$f_x^0 + f_y^0\phi'(x_0) \quad \geq \quad 0$$

When $x - x_0 < 0$, we argue similarly to find

$$f_x^0 + f_y^0\phi'(x_0) \quad \leq \quad 0$$

Combining, we have

$$0 \leq f_x^0 + f_y^0\phi'(x_0) \quad \leq \quad 0$$

which tells us at this extreme value we have the equation $f_x^0 + f_y^0\phi'(x_0) = 0$. This implies $f_y^0 = -\frac{f_x^0}{\phi'(x_0)}$. Next note

$$\begin{bmatrix} f_x^0 \\ f_y^0 \end{bmatrix} = \begin{bmatrix} f_x^0 \\ -\frac{f_x^0}{\phi'(x_0)} \end{bmatrix} = \begin{bmatrix} f_x^0 \\ -f_x^0\left\{-\left(\frac{g_y^0}{g_x^0}\right)\right\} \end{bmatrix} = \frac{f_x^0}{g_x^0}\begin{bmatrix} g_x^0 \\ g_y^0 \end{bmatrix}$$

This says there is a scalar $\lambda = \frac{f_x^0}{g_x^0}$ so that

$$\nabla(f)(x_0,y_0) \quad = \quad \lambda\nabla(g)(x_0,y_0)$$

where $g(x_0,y_0) = c$. The value $\lambda = \frac{f_x^0}{g_x^0}$ is called the **Lagrange Multiplier** for this extremal problem. **This is the basis for the Lagrange Multiplier Technique for a constrained optimization problem.**

We can do a similar sort of analysis in the case the extremum is a maximum too. Our analysis assumes that the point (x_0,y_0) where the extremum occurs is like an interior point in the $\{(x,y)|g(x,y)=c\}$. That is, we assume for such an x_0 there is a interval $B_r(x_0)$ with any $x \in B_r(x_0)$ having a corresponding value $y=\phi(x)$ so that $g(x,\phi(x)) = c$. So the argument does not handle boundary points such as the ± 5 is our previous example.

To implement the Lagrange Multiplier technique, we define a new function

$$H(x, y, \lambda) = f(x, y) + \lambda(g(x, y) - c)$$

The critical points we see are the ones where the gradient of f is a multiple of the gradient of g for the reasons we discussed above. Note the λ we use here is the negative of the Lagrange Multiplier. To find them, we set the partials of H equal to zero.

$$\begin{aligned}
\frac{\partial H}{\partial x} &= 0 \Longrightarrow \frac{\partial f}{\partial x} = -\lambda \frac{\partial g}{\partial x} \\
\frac{\partial H}{\partial y} &= 0 \Longrightarrow \frac{\partial f}{\partial y} = -\lambda \frac{\partial g}{\partial x} \\
\frac{\partial H}{\partial \lambda} &= 0 \Longrightarrow g(x, y) = c
\end{aligned}$$

The first two lines are the statement that the gradient of f is a multiple of the gradient of g and the third line is the statement that the constraint must be satisfied. As usual, solving these three nonlinear equations gives the interior point solutions and we always must also include the boundary points that solve the constraints as well.

Example 12.4.1 *Extremize* $2x + 3y$ *subject to* $x^2 + y^2 = 25$.

Solution *Set* $H(x, y, \lambda) = (2x + 3y) + \lambda(x^2 + y^2 - 25)$. *The critical point equations are then*

$$\begin{aligned}
\frac{\partial H}{\partial x} &= 2 + \lambda(2x) = 0 \\
\frac{\partial H}{\partial y} &= 3 + \lambda(2y) = 0 \\
\frac{\partial H}{\partial \lambda} &= x^2 + y^2 - 25 = 0
\end{aligned}$$

Solving for λ, *we find* $\lambda = -1/x = -3/(2y)$. *Thus,* $x = 2y/3$.

Using this relationship in the constraint, we find $(2y/3)^2 + y^2 = 25$ *or* $13y^2/9 = 25$. *Thus,* $y^2 = 225/13$ *or* $y = \pm 15/\sqrt{13}$. *This means* $x = \pm 10/\sqrt{13}$.

We find $f(10/\sqrt{13}, 15/\sqrt{13}) = 45/\sqrt{13} = 12.48$, $f(-10/\sqrt{13}, -15/\sqrt{13}) = -45/\sqrt{13} = -12.48$, $f(-5, 0) = -10$, $f(5, 0) = 10$, $f(0, -5) = -15$, *and* $f(0, 5) = 15$. *So the extreme values here occur at the boundary points* $(0, \pm 5)$.

Example 12.4.2 *Extremize* $\sin(xy)$ *subject to* $x^2 + y^2 = 1$.

Solution *Set* $H(x, y, \lambda) = (\sin(xy)) + \lambda(x^2 + y^2 - 1)$. *The critical point equations are then*

$$\begin{aligned}
\frac{\partial H}{\partial x} &= y\cos(xy) + \lambda(2x) = 0 \\
\frac{\partial H}{\partial y} &= x\cos(xy) + \lambda(2y) = 0 \\
\frac{\partial H}{\partial \lambda} &= x^2 + y^2 - 1 = 0
\end{aligned}$$

Solving for λ *in the first equation, we find* $\lambda = -y\cos(xy)/2x$. *Now use this in the second equation to find* $x\cos(xy) - 2y^2\cos(xy)/2x = 0$. *Thus,* $\cos(xy)(x - y^2/x) = 0$.

The first case is $\cos(xy) = 0$. *This implies* $xy = \pi/2 + 2n\pi$. *These are rectangular hyperbolas and the closest one to the origin is* $xy = \pi/2 = 1.57$ *whose closest point to the origin is* $(\sqrt{\pi/2}, \sqrt{\pi/2}) = (1,25,1,25)$. *This point is outside of the constraint set, so this case does not matter.*

The second case is $x^2 = y^2$ *which, using the constraint, implies* $x^2 = 1/2$. *Thus there are four possibilities:* $(1/\sqrt{2}, 1/\sqrt{2})$, $(1/\sqrt{2}, -1/\sqrt{2})$, $(-1/\sqrt{2}, 1/\sqrt{2})$ *and* $(-1/\sqrt{2}, 1/\sqrt{2})$.
The other possibilities are the circle's boundary points $(0, \pm 1)$ *and* $(\pm 1, 0)$. *We find*
$\sin(1/\sqrt{2}, 1/\sqrt{2}) = \sin(-1/\sqrt{2}, -1/\sqrt{2}) = \sin(1/2) = .48$
$\sin(1/\sqrt{2}, -1/\sqrt{2}) = \sin(-1/\sqrt{2}, 1/\sqrt{2}) = -\sin(1/2) = -.48$
$\sin(\pm 1, 0) = 0$, $\sin(0, \pm 1) = 0$
So the extreme values occur here at the interior points of the constraint.

12.4.1 What Does the Lagrange Multiplier Mean?

Let's change our extremum problem by assuming the constraint constant is changing as a function of x; i.e. the problem is now find the extreme values of $f(x, y)$ subject to $g(x, y) = C(x)$. We can use our tangent plane analysis like before. We assume the extreme value for constraint value $C(x_0)$ occurs at an interior point of the constraint. We have

$$g(x, \phi(x)) - C(x) = 0$$

Then

$$0 = g_x^0 + g_y^0 \left(\frac{\phi(x) - \phi(x_0)}{x - x_0} \right) + \left(\frac{E_g(x, \phi(x), x_0, \phi(x_0))}{x - x_0} \right) - \left(\frac{C(x) - C(x_0)}{x - x_0} \right)$$

We assume the constraint bound function C is differentiable and so letting $x \to x_0$, we find

$$g_x^0 + g_y^0 \phi'(x_0) = C'(x_0)$$

Now assume $(x_1, \phi(x_1))$ extremizes f for the constraint bound value $C(x_1)$. We have

$$\begin{aligned} f(x_1, \phi(x_1)) - f(x_0, \phi(x_0)) &= f_x^0 (x_1 - x_0) + f_y^0 (\phi(x_1) - \phi(x_0)) \\ &+ E_f(x_1, \phi(x_1), x_0, \phi(x_0)) \end{aligned}$$

or

$$\begin{aligned} \left(\frac{f(x_1, \phi(x_1)) - f(x_0, \phi(x_0))}{x_1 - x_0} \right) &= f_x^0 + f_y^0 \left(\frac{\phi(x_1) - \phi(x_0)}{x_1 - x_0} \right) \\ &+ \left(\frac{E_f(x_1, \phi(x_1), x_0, \phi(x_0))}{x_1 - x_0} \right) \end{aligned}$$

Now assuming $C(x) \neq 0$ locally around x_0, we can write

$$\begin{aligned} &\left(\frac{f(x_1, \phi(x_1)) - f(x_0, \phi(x_0))}{C(x_1) - C(x_0)} \right) \left(\frac{C(x_1) - C(x_0)}{x_1 - x_0} \right) = \\ &f_x^0 + f_y^0 \left(\frac{\phi(x_1) - \phi(x_0)}{x_1 - x_0} \right) + \left(\frac{E_f(x_1, \phi(x_1), x_0, \phi(x_0))}{x_1 - x_0} \right) \end{aligned}$$

Now let $\Theta(x)$ be defined by $\Theta(x) = f(x, \phi(x))$ when $(x, \phi(x))$ is the extreme value for the problem of extremizing $f(x, y)$ subject to $g(x, y) = C(x)$. This is called the **Optimal Value Function** for this problem. Rewriting, we have

$$\left(\frac{\Theta(x_1)-\Theta(x_0)}{C(x_1)-C(x_0)}\right)\left(\frac{C(x_1)-C(x_0)}{x_1-x_0}\right) = f_x^0 + f_y^0\left(\frac{\phi(x_1)-\phi(x_0)}{x_1-x_0}\right) + \left(\frac{E_f(x_1,\phi(x_1),x_0,\phi(x_0))}{x_1-x_0}\right)$$

Now let $x_1 \to x_0$. We obtain

$$\lim_{x_1\to x_0}\left(\frac{\Theta(x_1)-\Theta(x_0)}{C(x_1)-C(x_0)}\right)C'(x_0) = f_x^0 + f_y^0\,\phi'(x_0)$$

Thus, the rate of change of the **optimal value** with respect to change in the constraint bound, $\frac{d\Theta}{dC}$ is well-defined and

$$\frac{d\Theta}{dC}(C_0)\,C'(x_0) = f_x^0 + f_y^0\,\phi'(x_0)$$

where C_0 is the value $C(x_0)$. Assuming $C'(x_0) \neq 0$, we have

$$\frac{d\Theta}{dC}(C_0) = \frac{f_x^0}{C'(x_0)} + \frac{f_y^0}{C'(x_0)}\,\phi'(x_0)$$

But we know at the extreme value for x_0 that $f_y^0 = \frac{g_y^0}{g_x^0} f_x^0$. Thus,

$$\begin{aligned}\frac{d\Theta}{dC}(C_0) &= \frac{f_x^0}{C'(x_0)} + \frac{g_y^0}{g_x^0}\frac{f_x^0}{C'(x_0)}\,\phi'(x_0) = \frac{f_x^0}{C'(x_0)}\left(1 + \frac{g_y^0}{g_x^0}\phi'(x_0)\right)\\ &= \frac{f_x^0}{g_x^0}\frac{1}{C'(x_0)}\left(g_x^0 + g_y^0\phi'(x_0)\right)\end{aligned}$$

But the Lagrange Multiplier λ_0 for extremal value for x_0 is $\lambda_0 = \frac{f_x^0}{g_x^0}$. So

$$\frac{d\Theta}{dC}(C_0) = \lambda_0\left(g_x^0 + g_y^0\phi'(x_0)\right)\frac{1}{C'(x_0)}$$

But $C'(x_0) = g_x^0 + g_y^0\phi'(x_0)$. Hence, we find $\frac{d\Theta}{dC}(C_0) = \lambda_0$.

We see from our argument that the Lagrange Multiplier λ_0 is the rate of change of optimal value with respect to the constraint bound. There are cases:

- $sign(\lambda_0) = sign(\frac{f_x^0}{g_x^0}) = \frac{+}{+} = +$ which implies the optimal value goes up if the constraint bound is perturbed.

- $sign(\lambda_0) = sign(\frac{f_x^0}{g_x^0}) = \frac{-}{-} = +$ which implies the optimal value goes up if the constrained bound is perturbed.

- $sign(\lambda_0) = sign(\frac{f_x^0}{g_x^0}) = \frac{+}{-} = -$ which implies the optimal value goes down if the constrained bound is perturbed.

- $sign(\lambda_0) = sign(\frac{f_x^0}{g_x^0}) = \frac{-}{+} = -$ which implies the optimal value goes down if the constrained bound is perturbed.

Hence, we can interpret the Lagrange Multiplier as a **price**: the change in optimal value due to a change in constraint is essentially the *loss of value* experienced due to the constraint change. For example, if $|\lambda_0|$ is very small, it says the rate of change of the optimal value with respect to constraint modification is small too. This implies the optimal value is insensitive to constraint modification.

12.4.2 Homework

Exercise 12.4.1 *Find the extreme values of* $\sin(xy)$ *subject to* $x^2 + 2y^2 = 1$ *using the Lagrange Multiplier Technique.*

Exercise 12.4.2 *Find the extreme values of* $\cos(xy)$ *subject to* $x^2 + 2y^2 = 1$ *using the Lagrange Multiplier Technique.*

Chapter 13

Linear Approximation Applications

Let's look at how we use tangent plane approximations in models and computations.

13.1 Linear Approximations to Nonlinear ODE

Here, we look at several nonlinear ODE applications.

13.1.1 An Insulin Model

This is a model we have discussed in (J. Peterson (10) 2016) which was first presented in (M. Braun (15) 1978) but we will go over it again here as it is very interesting and there are a lot of lessons to be learned from applying our ideas of tangent plane approximations to build practical models on nonlinear phenomena.

In diabetes there is too much sugar in the blood and the urine. This is a metabolic disease and if a person has it, they are not able to use up all the sugars, starches and various carbohydrates because they don't have enough **insulin**. Diabetes can be diagnosed by a **glucose tolerance test** (GTT). If you are given this test, you do an overnight fast and then you are given a large dose of sugar in a form that appears in the bloodstream. This sugar is called **glucose**. Measurements are made over about five hours or so of the concentration of glucose in the blood. These measurements are then used in the diagnosis of diabetes. It has always been difficult to interpret these results as a means of diagnosing whether a person has diabetes or not. Hence, different physicians interpreting the same data can come up with a different diagnosis, which is a pretty unacceptable state of affairs!

Let's discuss a criterion developed in the 1960s by doctors at the Mayo Clinic and the University of Minnesota that was fairly reliable. It showcases a lot of our modeling in this course and will give you another example of how we use our tools. We start with a simple model of the blood glucose regulatory system.

Glucose plays an important role in vertebrate metabolism because it is a source of energy. For each person, there is an optimal blood glucose concentration and large deviations from this leads to severe problems including death. Blood glucose levels are autoregulated via standard forward and backward interactions like we see in many biological systems. An example is the signal that is used to activate the creation of a protein which we discussed earlier. The signaling molecules are typically either bound to another molecule in the cell or are free. The equilibrium concentration of free signal is due to the fact that the rate at which signaling molecules bind equals the rate at which they split apart from their binding substrate. When an external message comes into the cell called a trigger, it induces a change in this careful balance which temporarily upgrades or degrades the equilibrium signal concentration. This then influences the protein concentration rate. Blood glucose concentrations work like this too, although the details differ. The blood glucose concentration is influenced by a variety of signaling molecules just like the protein creation rates can be. Here are some of them. The hormone that **decreases** blood glucose concentration is **insulin**. **Insulin** is a hormone secreted by the β cells of the pancreas. After we eat carbohydrates, our gastrointestinal tract sends a signal to the pancreas to secrete insulin. Also, the glucose in our blood directly stimulates the β cells to secrete insulin. We think insulin helps cells pull in the glucose needed for metabolic activity by attaching itself to membrane walls that are normally impenetrable. This attachment increases the ability of glucose to pass through to the inside of the cell where it can be used as *fuel*. So, if there is not enough insulin, cells don't have enough energy for their needs. The other hormones we will focus on all tend to **change** blood glucose concentrations also.

- **Glucagon** is a hormone secreted by the α cells of the pancreas. Excess glucose is stored in the liver in the form of **glycogen**. There is the usual equilibrium amount of storage caused by the rate of glycogen formation being equal to the rate of the reverse reaction that moves glycogen back to glucose. Hence the glycogen serves as a reservoir for glucose and when the body needs glucose, the rate balance is tipped towards conversion back to glucose to release needed glucose to the cells. The hormone **glucagon** increases the rate of the reaction that converts glycogen back to glucose and so serves an important regulatory function. **Hypoglycemia** (low blood sugar) and **fasting** tend to increase the secretion of the hormone glucagon. On the other hand, if the blood glucose levels increase, this tends to suppress glucagon secretion; i.e. we have another back and forth regulatory tool.

- **Epinephrine**, also called **adrenaline**, is a hormone secreted by the adrenal medulla. It is part of an emergency mechanism to quickly increase the blood glucose concentration in times of extremely low blood sugar levels. Hence, epinephrine also increases the rate at which glycogen converts to glucose. It also directly inhibits how much glucose is able to be pulled into muscle tissue because muscles use a lot of energy and this energy is needed elsewhere more urgently. It also acts on the pancreas directly to inhibit insulin production which keeps glucose in the blood. There is also another way to increase glucose by converting lactate into glucose in the liver. Epinephrine increases this rate also, so the liver can pump this extra glucose back into the blood stream.

- **Glucocorticoids** are hormones like **cortisol**, which are secreted by the adrenal cortex which influence how carbohydrates are metabolized which is turn increase glucose if the metabolic rate goes up.

- **Thyroxin** is a hormone secreted by the thyroid gland and it helps the liver form glucose from sources which are not carbohydrates such as glycerol, lactate and amino acids. So another way to upregulate glucose!

- **Somatotrophin** is called the growth hormone and it is secreted by the anterior pituitary gland. This hormone directly affects blood glucose levels (i.e. an increase in somatotrophin increases blood glucose levels and vice versa) but it also inhibits the effect of insulin on muscle and fat cell's permeability, which diminishes insulin's ability to help those cells pull glucose out of the blood stream. These actions can therefore increase blood glucose levels.

Now net hormone concentration is the sum of insulin plus the others. Let $\boldsymbol{H}$ denote this net hormone concentration. At normal conditions, call this concentration $\boldsymbol{H}_0$. There have been studies performed that show that under close to normal conditions, the interaction of the one hormone **insulin** with blood glucose completely dominates the net hormonal activity. That is normal blood sugar levels primarily depend on insulin-glucose interactions.

So if insulin increases from normal levels, it increases net hormonal concentration to $H_0 + \Delta H$ and decreases glucose blood concentration. On the other hand, if other hormones such as cortisol increased from base levels, this will make blood glucose levels go up. Since insulin dominates all activity at normal conditions, we can think of this increase in cortisol as a decrease in insulin with a resulting drop in blood glucose levels. A decrease in insulin from normal levels corresponds to a drop in net hormone concentration to $H_0 - \Delta H$. Now let G denote the blood glucose level. Hence, in our model an increase in H means a drop in G and a decrease in H means an increase in G!

13.1.1.1 Model Details

The idea of our model for diagnosing diabetes from the GTT is to find a simple dynamical model of this complicated blood glucose regulatory system in which the values of two parameters would give a nice criterion for distinguishing normal individuals from those with mild diabetes or those who are pre-diabetic. Here is what we will do. We describe the model as

$$\begin{aligned} G'(t) &= F_1(G, H) + J(t) \\ H'(t) &= F_2(G, H) \end{aligned}$$

where the function J is the external rate at which blood glucose concentration is being increased. There are two nonlinear interaction functions F_1 and F_2 because we know G and H have complicated interactions. Let's assume G and H have achieved optimal values G_0 and H_0 by the time the fasting patient has arrived at the hospital. Hence, we don't expect to have any contribution to $G'(0)$ and $H'(0)$; i.e. $F_1(G_0, H_0) = 0$ and $F_2(G_0, H_0) = 0$. We are interested in the deviation of G and H from their optimal values G_0 and H_0, so let $g = G - G_0$ and $h = H - H_0$. We can then write $G = G_0 + g$ and $H = H_0 + h$. The model can then be rewritten as

$$\begin{aligned} (G_0 + g)'(t) &= F_1(G_0 + g, H_0 + h) + J(t) \\ (H_0 + h)'(t) &= F_2(G_0 + g, H_0 + h) \end{aligned}$$

or

$$\begin{aligned} g'(t) &= F_1(G_0 + g, H_0 + h) + J(t) \\ h'(t) &= F_2(G_0 + g, H_0 + h) \end{aligned}$$

Now find the tangent plane approximations.

$$\begin{aligned} F_1(G_0 + g, H_0 + h) &= F_1(G_0, H_0) + \frac{\partial F_1}{\partial g}(G_0, H_0)\, g + \frac{\partial F_1}{\partial h}(G_0, H_0)\, h + E_{F_1} \\ F_2(G_0 + g, H_0 + h) &= F_2(G_0, H_0) + \frac{\partial F_2}{\partial g}(G_0, H_0)\, g + \frac{\partial F_2}{\partial h}(G_0, H_0)\, h + E_{F_2} \end{aligned}$$

but the terms $F_1(G_0, H_0) = 0$ and $F_1(G_0, H_0) = 0$, so we can simplify to

$$\begin{aligned} F_1(G_0 + g, H_0 + h) &= \frac{\partial F_1}{\partial g}(G_0, H_0)\, g + \frac{\partial F_1}{\partial h}(G_0, H_0)\, h + E_{F_1} \\ F_2(G_0 + g, H_0 + h) &= \frac{\partial F_2}{\partial g}(G_0, H_0)\, g + \frac{\partial F_2}{\partial h}(G_0, H_0)\, h + E_{F_2} \end{aligned}$$

It seems reasonable to assume that since we are so close to ordinary operating conditions, the errors E_{F_1} and E_{F_2} will be negligible. Thus our model approximation is

$$\begin{aligned} g'(t) &= \frac{\partial F_1}{\partial g}(\boldsymbol{G_0}, \boldsymbol{H_0})\, g + \frac{\partial F_1}{\partial h}(\boldsymbol{G_0}, \boldsymbol{H_0})\, h + \boldsymbol{J}(t) \\ h'(t) &= \frac{\partial F_2}{\partial g}(\boldsymbol{G_0}, \boldsymbol{H_0})\, g + \frac{\partial F_2}{\partial h}(\boldsymbol{G_0}, \boldsymbol{H_0})\, h \end{aligned}$$

The rest of the full analysis of this model can be found in (J. Peterson (10) 2016) as well as some curve fits to data using MATLAB.

In this next example, we assume we have a large population of cells $\boldsymbol{T}$ which consists of cells which are infected or altered in two distinct ways by a trigger $\boldsymbol{V}$ based on signals $\boldsymbol{I}$, $\boldsymbol{J}$ and $\boldsymbol{K}$. These two distinct populations of cells will be labeled $\boldsymbol{M}$ and $\boldsymbol{N}$. There are also non-infected cells, $\boldsymbol{H}$, and non-infected cells which will be removed due to auto immune action which we call $\boldsymbol{C}$, for collateral damage. We will be using the same approach to studying nonlinear interactions that was used in (J. Peterson and A. M. Kesson and N. J. C. King (12) June 21, 2017).

We assume the dynamics here are

$$\begin{aligned} \boldsymbol{C}'(t) &= F_1(\boldsymbol{C}, \boldsymbol{M}, \boldsymbol{N}) \\ \boldsymbol{M}'(t) &= F_2(\boldsymbol{C}, \boldsymbol{M}, \boldsymbol{N}) \\ \boldsymbol{N}'(t) &= F_3(\boldsymbol{C}, \boldsymbol{M}, \boldsymbol{N}) \end{aligned}$$

There are then three nonlinear interaction functions F_1, F_2 and F_3 because we know $\boldsymbol{C}$, $\boldsymbol{M}$ and $\boldsymbol{N}$ depend on each other's levels in very complicated ways. Usually, we assume the initial trigger dose $\boldsymbol{V_0}$ gives rise to some fraction of infected cells and the effect of the trigger will be different in the two cell populations $\boldsymbol{M}$ and $\boldsymbol{N}$.

Assumption 13.1.1 Number of infected cells

We assume the number of infected cells is $p_0\, \boldsymbol{V_0}$ which is split into $p_1\, p_0\, \boldsymbol{V_0}$ in population $\boldsymbol{N}$ and $p_2\, p_0\, \boldsymbol{V_0}$ in $\boldsymbol{M}$, where $p_1 + p_2 = 1$.

For example, a reasonable choice is $p_1 = 0.99$ and $p_2 = 0.01$. Thus, the total amount of trigger that goes into altered cells is $p_0\, \boldsymbol{V_0}$ and the amount of free trigger is therefore $(1 - p_0)\, \boldsymbol{V_0}$. Thus, we could expect $\boldsymbol{C_0} = 0$, $\boldsymbol{M_0} = p_2\, p_0\, \boldsymbol{V_0}$ and $\boldsymbol{N_0} = \boldsymbol{M_0} = p_1\, p_0\, \boldsymbol{V_0}$. However, we will explicitly assume we are starting from a point of equilibrium prior to the administration of the viral dose $\boldsymbol{V_0}$. We could assume there is always some level of collateral damage, $\boldsymbol{C_0}$ in a host, but we will not do that. We will therefore assume $\boldsymbol{C}$, $\boldsymbol{M}$ and $\boldsymbol{C}$ have achieved these values $\boldsymbol{C_0} = 0$, $\boldsymbol{M_0} = 0$ and $\boldsymbol{N_0} = 0$ right before the moment of alteration by the trigger. Hence, we don't expect there to be an initial contribution to $\boldsymbol{C}'(0)$, $\boldsymbol{M}'(0)$ and $\boldsymbol{N}'(0)$; i.e. $F_1(\boldsymbol{C_0}, \boldsymbol{M_0}, \boldsymbol{N_0}) = 0$, $F_2(\boldsymbol{C_0}, \boldsymbol{M_0}, \boldsymbol{N_0}) = 0$ and $F_3(\boldsymbol{C_0}, \boldsymbol{M_0}, \boldsymbol{N_0}) = 0$. We are interested in the deviation of $\boldsymbol{C}$, $\boldsymbol{M}$ and $\boldsymbol{N}$ from their optimal values $\boldsymbol{C_0}$, $\boldsymbol{M_0}$ and $\boldsymbol{N_0}$, so let $\boldsymbol{c} = \boldsymbol{C} - \boldsymbol{C_0}$, $\boldsymbol{m} = \boldsymbol{M} - \boldsymbol{M_0}$ and $\boldsymbol{n} = \boldsymbol{N} - \boldsymbol{N_0}$. We can then write $\boldsymbol{C} = \boldsymbol{C_0} + \boldsymbol{c}$, $\boldsymbol{M} = \boldsymbol{M_0} + \boldsymbol{m}$ and $\boldsymbol{N} = \boldsymbol{N_0} + \boldsymbol{n}$. The model can then be rewritten as

$$\begin{aligned} (\boldsymbol{C_0} + \boldsymbol{c})'(t) &= F_1(\boldsymbol{C_0} + \boldsymbol{c}, \boldsymbol{M_0} + \boldsymbol{m}, \boldsymbol{N_0} + \boldsymbol{n}) \\ (\boldsymbol{M_0} + \boldsymbol{m})'(t) &= F_2(\boldsymbol{C_0} + \boldsymbol{c}, \boldsymbol{M_0} + \boldsymbol{m}, \boldsymbol{N_0} + \boldsymbol{n}) \\ (\boldsymbol{M_0} + \boldsymbol{M})'(t) &= F_3(\boldsymbol{C_0} + \boldsymbol{c}, \boldsymbol{M_0} + \boldsymbol{m}, \boldsymbol{N_0} + \boldsymbol{n}) \end{aligned}$$

or

$$\begin{aligned} \boldsymbol{c}'(t) &= F_1(\boldsymbol{C_0} + \boldsymbol{c}, \boldsymbol{M_0} + \boldsymbol{m}, , \boldsymbol{N_0} + \boldsymbol{n}) \\ \boldsymbol{m}'(t) &= F_2(\boldsymbol{C_0} + \boldsymbol{c}, \boldsymbol{M_0} + \boldsymbol{m}, \boldsymbol{N_0} + \boldsymbol{n}) \\ \boldsymbol{n}'(t) &= F_3(\boldsymbol{C_0} + \boldsymbol{c}, \boldsymbol{M_0} + \boldsymbol{m}, \boldsymbol{N_0} + \boldsymbol{n}) \end{aligned}$$

Next, we do a standard tangent plane approximation on the nonlinear dynamics functions F_1, F_2 and F_3 to derive approximation dynamics. We find the approximate dynamics are

$$\begin{bmatrix} \boldsymbol{c}' \\ \boldsymbol{m}' \\ \boldsymbol{n}' \end{bmatrix} \approx \begin{bmatrix} F^o_{1c} & F^o_{1m} & F^o_{1n} \\ F^o_{2c} & F^o_{2m} & F^o_{2n} \\ F^o_{3c} & F^o_{3m} & F^o_{3n} \end{bmatrix} \begin{bmatrix} c \\ m \\ n \end{bmatrix}$$

where we now use a standard subscript scheme to indicate the partials. In (J. Peterson and A. M. Kesson and N. J. C. King (13) June 21, 2017), we show how to build additional models which include the signals IFN-γ ($\boldsymbol{I}$), $\boldsymbol{J}$ and $\boldsymbol{K}$.

Homework

Exercise 13.1.1 *Do the same sort of analysis for a system of two nonlinear differential equations. Assume the dynamics are sufficiently smooth to admit linearization around a given equilibrium point where the dynamics are*

$$\begin{aligned} x_1'(t) &= F_1(x_1, x_2) \\ x_2'(t) &= F_2(x_1, x_2) \end{aligned}$$

and the equilibrium point is (x_{10},x_{20}).

Letting $\boldsymbol{F} = \begin{bmatrix} F_1 \\ F_2 \end{bmatrix}$, $\boldsymbol{X} = \begin{bmatrix} x_1 \\ x_2 \end{bmatrix}$ *and* $\boldsymbol{X_0} = \begin{bmatrix} x_{10} \\ x_{20} \end{bmatrix}$ *show the linearization process here generates the linear system*

$$\boldsymbol{X'} = \boldsymbol{J_F}(\boldsymbol{X_0})\, \boldsymbol{X_0}, \quad \boldsymbol{X}(0) = \boldsymbol{X_0}$$

Finally, write down the explicit error estimates incurred in the linearization process.

Exercise 13.1.2 *Do the same sort of analysis for a system of three nonlinear differential equations. Assume the dynamics are sufficiently smooth to admit linearization around a given equilibrium point where the dynamics are*

$$\begin{aligned} x_1'(t) &= F_1(x_1, x_2, x_3) \\ x_2'(t) &= F_2(x_1, x_2, x_3) \\ x_3'(t) &= F_3(x_1, x_2, x_3) \end{aligned}$$

and the equilibrium point is (x_{10},x_{20},x_{30}).

Letting $\boldsymbol{F} = \begin{bmatrix} F_1 \\ F_2 \\ F_3 \end{bmatrix}$, $\boldsymbol{X} = \begin{bmatrix} x_1 \\ x_2 \\ x_3 \end{bmatrix}$ *and* $\boldsymbol{X_0} = \begin{bmatrix} x_{10} \\ x_{20} \\ x_{30} \end{bmatrix}$ *show the linearization process here generates the linear system*

$$\boldsymbol{X'} = \boldsymbol{J_F}(\boldsymbol{X_0})\, \boldsymbol{X_0}, \quad \boldsymbol{X}(0) = \boldsymbol{X_0}$$

Finally, write down the explicit error estimates incurred in the linearization process.

Exercise 13.1.3 *Do the same sort of analysis for a system of four nonlinear differential equations. Assume the dynamics are sufficiently smooth to admit linearization around a given equilibrium point where the dynamics are*

$$\begin{aligned} x_1'(t) &= F_1(x_1, x_2, x_3, x_4) \\ x_2'(t) &= F_2(x_1, x_2, x_3, x_4) \\ x_3'(t) &= F_3(x_1, x_2, x_3, x_4) \\ x_4'(t) &= F_4(x_1, x_2, x_3, x_4) \end{aligned}$$

and the equilibrium point is $(x_{10}, x_{20}, x_{30}, x_{40})$.

Letting $\boldsymbol{F} = \begin{bmatrix} F_1 \\ F_2 \\ F_3 \\ F_4 \end{bmatrix}$, $\boldsymbol{X} = \begin{bmatrix} x_1 \\ x_2 \\ x_3 \\ x_4 \end{bmatrix}$ *and* $\boldsymbol{X_0} = \begin{bmatrix} x_{10} \\ x_{20} \\ x_{30} \\ x_{40} \end{bmatrix}$ *show the linearization process here generates the linear system*

$$\boldsymbol{X}' = \boldsymbol{J_F}(\boldsymbol{X_0})\, \boldsymbol{X_0}, \quad \boldsymbol{X}(0) = \boldsymbol{X_0}$$

Finally, write down the explicit error estimates incurred in the linearization process.

Exercise 13.1.4 *This is just a general query about modeling of this sort. Look up nonlinear phenomena in whatever area interests you. The models here require that you have identified variables and interactions between the variables that can be modeled by differential equations. Often the variables of interest are based on lumped groups of other variables, as you saw in the diabetes and immune model.*

Now identify an equilibrium point. This is hard in general as this means you have to solve $\boldsymbol{F}(\boldsymbol{X_0}) = \boldsymbol{0}$ *which is a system of* n *nonlinear simultaneous equations. Often we can use the underlying physics, biology and so forth, to reason out something about this equilibrium point but not always.*

Then linearize about the equilibrium point and try to figure out how much error you are making in a local ball about $\boldsymbol{X_0}$ *in* $\Re^n$. *This is the key thing because if you can successfully argue this error is negligible for your purposes, perhaps the linearization model will provide insight into the actual dynamical system.*

So your exercise is to go and read about such applications!

13.2 Finite Difference Approximations in PDE

Another way to approximate the solution of linear partial differential equations is to use what are called *finite difference techniques*. Essentially, in these methods, the various partial derivatives are replaced by tangent line like approximations. We first must discuss how we handle the resulting approximation error carefully. Let's review how to approximate a function of two variables using Taylor series. Let's assume that $u(x,t)$ is a nice, smooth function of the two variables x (our spatial variable) and t (our time variable). For ease of discussion, we will focus on the space interval $[0, L]$ and the time interval $[0, M]$ where L and M are both positive numbers. We assume that u is continuous on the rectangle $\mathscr{D} = [0, L] \times [0, M]$. This means that $\lim_{(x,t)\to(x_0,t_0)} f(x,t) = f(x_0,t_0)$ for all pairs (x_0, t_0) in $\mathscr{D}$. Note that since the rectangle is two dimensional, there are an infinite number of ways the pairs (x,t) can approach (x_0, t_0). Continuity at the point (x_0, t_0) means that it does not matter how we do the approach; we always get the same answer. More precisely, if the positive tolerance ϵ is given, there is a positive number δ so that

$$\sqrt{(x-x_0)^2+(t-t_0)^2} < \delta \text{ and } (x,t) \in \mathscr{D} \implies |f(x,t)-f(x_0,t_0)| < \epsilon.$$

We also assume the partial derivatives up to order 4 are continuous on $\mathscr{D}$. Thus, we assume the continuity of many orders of partials. These terms get hard to write down as there are so many of them, so let's define $D^k_{ij}u$ to be the kth partial derivative of u with respect to x i times and t j times where $i+j$ must add up to k. More formally,

$$D^k_{ij}u = \frac{\partial^k u}{\partial x^i \partial t^j}, \quad \text{for all } i+j=k, \text{ with } i,j \geq 0.$$

So using this notation, $D^1_{10}u = \frac{\partial u}{\partial x}$, $D^2_{11}u = \frac{\partial^2 u}{\partial x \partial t}$, $D^4_{40}u = \frac{\partial^4 u}{\partial^4 x}$ and so forth. We will assume the partials $D^k_{ij}u$ are continuous on $\mathscr{D}$ up to and including $k=4$. It is known that continuous functions on a rectangle of the form $\mathscr{D}$ must be bounded. Hence, there are positive numbers B^k_{ij} so that for all (x,t) in $\mathscr{D}$, we have

$$|D^k_{ij}u(x,t)| < B^k_{ij}, \quad \text{for all } (x,t) \in \mathscr{D}.$$

Hence, the maximum of $|D^k_{ij}u(x,t)|$ over $\mathscr{D}$ exists and letting $||D^k_{ij}u|| = \max |D^k_{ij}u(x,t)|$ over all $(x,t) \in \mathscr{D}$, we have each $||D^k_{ij}u|| \leq D^k_{ij}$. This implies we can take the maximum of all these individual bounds and state that there is a single constant C so that

$$||D^k_{ij}u|| \leq C. \tag{13.1}$$

13.2.1 Approximating First Order Partials

Define $h(t) = u(x_0, t)$. The second order Taylor expansion of h is then

$$h(t) = h(t_0) + h'(t_0)(t-t_0) + \frac{1}{2}h''(c_t)(t-t_0)^2$$

where c_t is some point between t_0 and t. Using the chain rule for functions of two variables, it is easy to see $h(x_0) = u(x_0, y_0)$, $h'(t) = u_t(x_0, t)$ and $h''(t) = u_{tt}(x_0, t)$. Hence, we can rewrite the expansion as

$$u(x,t) = u(x_0,t_0) + u_t(x_0,t_0)(t-t_0) + \frac{1}{2}u_{tt}(x_0,c_t)(t-t_0)^2.$$

We can then write these another way by letting $\Delta t = t - t_0$. This gives

$$u(x_0, t_0+\Delta t) = u(x_0,t_0) + u_t(x_0,t_0)\Delta t + \frac{1}{2}u_{tt}(x_0,c_t)(\Delta t)^2.$$

We can use these expansions to estimate the first order partials in a variety of ways.

13.2.1.1 Central Differences

First, let's look at the central difference for the first order partial derivative with respect to t. We have

$$u(x_0, t_0-\Delta t) = u(x_0,t_0) - u_t(x_0,t_0)\Delta t + \frac{1}{2}u_{tt}(x_0,c_t^-)(\Delta t)^2$$

$$u(x_0, t_0 + \Delta t) = u(x_0, t_0) + u_t(x_0, t_0)\Delta t + \frac{1}{2}u_{tt}(x_0, c_t^+)(\Delta t)^2.$$

Subtracting, we find

$$u(x_0, t_0 + \Delta t) - u(x_0, t_0 - \Delta t,) = 2u_t(x_0, t_0)\Delta t + \frac{1}{2}\Big(u_{tt}(x_0, c_t^+) - u_{xx}(x_0, c_t^-)\Big)(\Delta t)^2.$$

Thus,

$$u_t(x_0, t_0) = \frac{u(x_0, t_0 + \Delta t) - u(x_0, , t_0 - \Delta t)}{2\Delta t} - \frac{1}{2}\Big(u_{tt}(x_0, c_t^+) - u_{tt}(x_0, c_t^-)\Big)\Delta t$$

Homework

Exercise 13.2.1 *For $f(x, y) = 2x^{10} + y^2$, approximate the first order partials using central differences.*

- *At $(0, 0)$ plot the approximate partials for h in the range 10^{-4} to 0.4 as well as the true values.*
- *At $(1, 1)$ plot the approximate partials for h in the range 10^{-4} to 0.4 as well as the true values.*
- *At $(2, 2)$ plot the approximate partials for h in the range 10^{-4} to 0.4 as well as the true values.*
- *At $(5, 5)$ plot the approximate partials for h in the range 10^{-4} to 0.4 as well as the true values.*

Exercise 13.2.2 *For $f(x, y) = 2x^{10} + y^2$, approximate the first order partials using central differences.*

- *At $(0, 0)$ plot the approximate partials for h in the range 10^{-4} to 0.4 as well as the true values.*
- *At $(1, 1)$ plot the approximate partials for h in the range 10^{-4} to 0.4 as well as the true values.*
- *At $(2, 2)$ plot the approximate partials for h in the range 10^{-4} to 0.4 as well as the true values.*
- *At $(5, 5)$ plot the approximate partials for h in the range 10^{-4} to 0.4 as well as the true values.*

Exercise 13.2.3 *For $f(x, y) = 2x^{10} + y^2\sin(x^4 + y^4)$, approximate the first order partials using central differences.*

- *At $(0, 0)$ plot the approximate partials for h in the range 10^{-4} to 0.4 as well as the true values.*
- *At $(1, 1)$ plot the approximate partials for h in the range 10^{-4} to 0.4 as well as the true values.*
- *At $(2, 2)$ plot the approximate partials for h in the range 10^{-4} to 0.4 as well as the true values.*
- *At $(5, 5)$ plot the approximate partials for h in the range 10^{-4} to 0.4 as well as the true values.*

Exercise 13.2.4 *For $f(x, y) = e^{2x^2+3y^2}$, approximate the first order partials using central differences.*

- *At $(0, 0)$ plot the approximate partials for h in the range 10^{-4} to 0.4 as well as the true values.*
- *At $(1, 1)$ plot the approximate partials for h in the range 10^{-4} to 0.4 as well as the true values.*

- *At* $(2, 2)$ *plot the approximate partials for* h *in the range* 10^{-4} *to* 0.4 *as well as the true values.*
- *At* $(5, 5)$ *plot the approximate partials for* h *in the range* 10^{-4} *to* 0.4 *as well as the true values.*

Exercise 13.2.5 *For* $f(x, y) = \sin(40(2x^3 + y^2))$*, approximate the first order partials using central differences.*

- *At* $(0, 0)$ *plot the approximate partials for* h *in the range* 10^{-4} *to* 0.4 *as well as the true values.*
- *At* $(1, 1)$ *plot the approximate partials for* h *in the range* 10^{-4} *to* 0.4 *as well as the true values.*
- *At* $(2, 2)$ *plot the approximate partials for* h *in the range* 10^{-4} *to* 0.4 *as well as the true values.*
- *At* $(5, 5)$ *plot the approximate partials for* h *in the range* 10^{-4} *to* 0.4 *as well as the true values.*

13.2.1.2 Forward Differences

Next, let's look at the forward difference. We have

$$u(x_0, t_0 + \Delta t) = u(x_0, t_0) + u_t(x_0, t_0)\Delta t + \frac{1}{2} u_{tt}(x_0, c_t^+)(\Delta t)^2.$$

Subtracting, we find

$$u(x_0, t_0 + \Delta t) - u(x_0, t_0) = u_t(x_0, t_0)\Delta t + \frac{1}{2} u_{tt}(x_0, c_t^+)(\Delta t)^2.$$

Thus,

$$u_t(x_0, t_0) = \frac{u(x_0, t_0 + \Delta t) - u(x_0, t_0)}{\Delta t} - \frac{1}{2} u_{tt}(x_0, c_t^+)\Delta t$$

Homework

Exercise 13.2.6 *For* $f(x, y) = 2x^{10} + y^2$*, approximate the first order partials using forward differences.*

- *At* $(0, 0)$ *plot the approximate partials for* h *in the range* 10^{-4} *to* 0.4 *as well as the true values.*
- *At* $(1, 1)$ *plot the approximate partials for* h *in the range* 10^{-4} *to* 0.4 *as well as the true values.*
- *At* $(2, 2)$ *plot the approximate partials for* h *in the range* 10^{-4} *to* 0.4 *as well as the true values.*
- *At* $(5, 5)$ *plot the approximate partials for* h *in the range* 10^{-4} *to* 0.4 *as well as the true values.*

Exercise 13.2.7 *For* $f(x, y) = 2x^{10} + y^2$*, approximate the first order partials using forward differences.*

- *At* $(0, 0)$ *plot the approximate partials for* h *in the range* 10^{-4} *to* 0.4 *as well as the true values.*
- *At* $(1, 1)$ *plot the approximate partials for* h *in the range* 10^{-4} *to* 0.4 *as well as the true values.*
- *At* $(2, 2)$ *plot the approximate partials for* h *in the range* 10^{-4} *to* 0.4 *as well as the true values.*

- *At* $(5, 5)$ *plot the approximate partials for* h *in the range* 10^{-4} *to* 0.4 *as well as the true values.*

Exercise 13.2.8 *For* $f(x, y) = 2x^{10} + y^2 \sin(x^4 + y^4)$, *approximate the first order partials using forward differences.*

- *At* $(0, 0)$ *plot the approximate partials for* h *in the range* 10^{-4} *to* 0.4 *as well as the true values.*
- *At* $(1, 1)$ *plot the approximate partials for* h *in the range* 10^{-4} *to* 0.4 *as well as the true values.*
- *At* $(2, 2)$ *plot the approximate partials for* h *in the range* 10^{-4} *to* 0.4 *as well as the true values.*
- *At* $(5, 5)$ *plot the approximate partials for* h *in the range* 10^{-4} *to* 0.4 *as well as the true values.*

Exercise 13.2.9 *For* $f(x, y) = e^{2x^2+3y^2}$, *approximate the first order partials using forward differences.*

- *At* $(0, 0)$ *plot the approximate partials for* h *in the range* 10^{-4} *to* 0.4 *as well as the true values.*
- *At* $(1, 1)$ *plot the approximate partials for* h *in the range* 10^{-4} *to* 0.4 *as well as the true values.*
- *At* $(2, 2)$ *plot the approximate partials for* h *in the range* 10^{-4} *to* 0.4 *as well as the true values.*
- *At* $(5, 5)$ *plot the approximate partials for* h *in the range* 10^{-4} *to* 0.4 *as well as the true values.*

Exercise 13.2.10 *For* $f(x, y) = \sin(40(2x^3 + y^2))$, *approximate the first order partials using forward differences.*

- *At* $(0, 0)$ *plot the approximate partials for* h *in the range* 10^{-4} *to* 0.4 *as well as the true values.*
- *At* $(1, 1)$ *plot the approximate partials for* h *in the range* 10^{-4} *to* 0.4 *as well as the true values.*
- *At* $(2, 2)$ *plot the approximate partials for* h *in the range* 10^{-4} *to* 0.4 *as well as the true values.*
- *At* $(5, 5)$ *plot the approximate partials for* h *in the range* 10^{-4} *to* 0.4 *as well as the true values.*

13.2.2 Approximating Second Order Partials

The approximation we wish to use for the second order partial u_{xx} is obtained like this. We fix the point (x_0, t_0) as usual and let $g(x) = u(x, t_0)$. The fourth order Taylor expansion of g is then

$$
\begin{aligned}
g(x) &= g(x_0) + g'(x_0)(x - x_0) + \frac{1}{2}g''(x_0)(x - x_0)^2 + \frac{1}{6}g'''(x_0)(x - x_0)^3 \\
&\quad + \frac{1}{24}g''''(c_x^+)(x - x_0)^4
\end{aligned}
$$

where c_x^+ is some point between x_0 and x. It is easy to see $g(x_0) = u(x_0, y_0)$, $g'(x) = u_x(x, t_0)$ and $g''(x) = u_{xx}(x, t_0)$ and so forth. Hence, letting $\Delta x = x - x_0$, we can rewrite the expansion as

$$
\begin{aligned}
u(x_0 + \Delta x, t_0) &= u(x_0, t_0) + u_x(x_0, t_0)\Delta x + \frac{1}{2}u_{xx}(x_0, t_0)(\Delta x)^2 + \frac{1}{6}u_{xxx}(x_0, t_0)(\Delta x)^3 \\
&\quad + \frac{1}{24}u_{xxxx}(c_x^+, t_0)(\Delta x)^4
\end{aligned}
$$

A similar expansion gives

$$\begin{aligned} u(x_0 - \Delta x, t_0) &= u(x_0, t_0) - u_x(x_0, t_0)\Delta x + \frac{1}{2}u_{xx}(x_0, t_0)(\Delta x)^2 - \frac{1}{6}u_{xxx}(x_0, t_0)(\Delta x)^3 \\ &+ \frac{1}{24}u_{xxxx}(c_x^+, t_0)(\Delta x)^4 \end{aligned}$$

Adding, we have

$$\begin{aligned} u(x_0 + \Delta x, t_0) + u(x_0 - \Delta x, t) &= 2u(x_0, t_0) + u_{xx}(x_0, t_0)(\Delta x)^2 + \frac{1}{24}(u_{xxxx}(c_x^+, t_0) \\ &+ u_{xxxx}(c_x^-, t_0))(\Delta x)^4 \end{aligned}$$

The intermediate value theorem tells us that a continuous function takes on every value between two points. Hence, there is a point c_x between c_x^+ and c_x^- so that

$$u_{xxxx}(c_x, t_0) = \frac{1}{2}(u_{xxxx}(c_x^+, t_0) + u_{xxxx}(c_x^-, t_0)).$$

We can then write the approximation as

$$u(x_0 + \Delta x, t) + u(x_0 - \Delta x, t) = 2u(x_0, t_0) + u_{xx}(x_0, t_0)(\Delta x)^2 + \frac{1}{12}u_{xxxx}(c_x, t_0)(\Delta x)^4$$

which tells us that

$$u_{xx}(x_0, t_0) = \frac{u(x_0 + \Delta x, t_0) + u(x_0 - \Delta x, t_0) - 2u(x_0, t_0)}{(\Delta x)^2} - \frac{1}{12}u_{xxxx}(c_x, t_0)(\Delta x)^2$$

13.2.3 Homework

Exercise 13.2.11 *For $f(x, y) = 2x^{10} + y^2$, approximate the second order partials using forward differences.*

- *At $(0, 0)$ plot the approximate partials for h in the range 10^{-4} to 0.4 as well as the true values.*
- *At $(1, 1)$ plot the approximate partials for h in the range 10^{-4} to 0.4 as well as the true values.*
- *At $(2, 2)$ plot the approximate partials for h in the range 10^{-4} to 0.4 as well as the true values.*
- *At $(5, 5)$ plot the approximate partials for h in the range 10^{-4} to 0.4 as well as the true values.*

Exercise 13.2.12 *For $f(x, y) = 2x^{10} + y^2$, approximate the second order partials using forward differences.*

- *At $(0, 0)$ plot the approximate partials for h in the range 10^{-4} to 0.4 as well as the true values.*
- *At $(1, 1)$ plot the approximate partials for h in the range 10^{-4} to 0.4 as well as the true values.*
- *At $(2, 2)$ plot the approximate partials for h in the range 10^{-4} to 0.4 as well as the true values.*
- *At $(5, 5)$ plot the approximate partials for h in the range 10^{-4} to 0.4 as well as the true values.*

Exercise 13.2.13 *For $f(x, y) = 2x^{10} + y^2 \sin(x^4 + y^4)$, approximate the second order partials using forward differences.*

- *At* $(0,0)$ *plot the approximate partials for* h *in the range* 10^{-4} *to* 0.4 *as well as the true values.*
- *At* $(1,1)$ *plot the approximate partials for* h *in the range* 10^{-4} *to* 0.4 *as well as the true values.*
- *At* $(2,2)$ *plot the approximate partials for* h *in the range* 10^{-4} *to* 0.4 *as well as the true values.*
- *At* $(5,5)$ *plot the approximate partials for* h *in the range* 10^{-4} *to* 0.4 *as well as the true values.*

Exercise 13.2.14 *For* $f(x,y) = e^{2x^2+3y^2}$*, approximate the second order partials using forward differences.*

- *At* $(0,0)$ *plot the approximate partials for* h *in the range* 10^{-4} *to* 0.4 *as well as the true values.*
- *At* $(1,1)$ *plot the approximate partials for* h *in the range* 10^{-4} *to* 0.4 *as well as the true values.*
- *At* $(2,2)$ *plot the approximate partials for* h *in the range* 10^{-4} *to* 0.4 *as well as the true values.*
- *At* $(5,5)$ *plot the approximate partials for* h *in the range* 10^{-4} *to* 0.4 *as well as the true values.*

Exercise 13.2.15 *For* $f(x,y) = \sin(40(2x^3 + y^2))$*, approximate the second order partials using forward differences.*

- *At* $(0,0)$ *plot the approximate partials for* h *in the range* 10^{-4} *to* 0.4 *as well as the true values.*
- *At* $(1,1)$ *plot the approximate partials for* h *in the range* 10^{-4} *to* 0.4 *as well as the true values.*
- *At* $(2,2)$ *plot the approximate partials for* h *in the range* 10^{-4} *to* 0.4 *as well as the true values.*
- *At* $(5,5)$ *plot the approximate partials for* h *in the range* 10^{-4} *to* 0.4 *as well as the true values.*

13.3 Approximating the Diffusion Equation

Using these approximations, we can approximate the diffusion equation $u_t = Du_{xx}$ using a forward difference for the first order partial with respect to time as

$$\begin{aligned} u_t(x_0,t_0) - Du_{xx}(x_0,t_0) &= \frac{u(x_0,t_0+\Delta t) - u(x_0,t_0)}{\Delta t} - \frac{1}{2}u_{tt}(x_0,c_t^+)\Delta t \\ -D\bigg(&\frac{u(x_0+\Delta x,t_0) + u(x_0-\Delta x,t_0) - 2u(x_0,t_0)}{(\Delta x)^2} - \frac{1}{12}u_{xxxx}(c_x,t_0)(\Delta x)^2\bigg) \end{aligned}$$

This can be reorganized as

$$\begin{aligned} u_t(x_0,t_0) - Du_{xx}(x_0,t_0) &= \frac{u(x_0,t_0+\Delta t) - u(x_0,t_0)}{\Delta t} \\ &-D\bigg(\frac{u(x_0+\Delta x,t_0) + u(x_0-\Delta x,t_0) - 2u(x_0,t_0)}{(\Delta x)^2}\bigg) \\ &-\frac{1}{2}u_{tt}(x_0,c_t^+)\Delta t + \frac{D}{12}u_{xxxx}(c_x,t_0)(\Delta x)^2 \end{aligned}$$

Now consider the error

$$\bigg| u_t(x_0,t_0) - Du_{xx}(x_0,t_0) - \frac{u(x_0,t_0+\Delta t) - u(x_0,t_0)}{\Delta t}$$

$$+D\left(\frac{u(x_0+\Delta x,t_0)+u(x_0-\Delta x,t_0)-2u(x_0,t_0)}{(\Delta x)^2}\right)\Bigg|$$
$$\leq \frac{1}{2}|u_{tt}(x_0,c_t^+)|\Delta t+\frac{D}{12}|u_{xxxx}(c_x,t_0)|(\Delta x)^2.$$

Then, using the estimates from Equation 13.1, we have

$$\Bigg|u_t(x_0,t_0)-Du_{xx}(x_0,t_0)-\frac{u(x_0,t_0+\Delta t)-u(x_0,t_0)}{\Delta t}$$
$$+D\left(\frac{u(x_0+\Delta x,t_0)+u(x_0-\Delta x,t_0)-2u(x_0,t_0)}{(\Delta x)^2}\right)\Bigg|$$
$$\leq \frac{1}{2}C\Delta t+\frac{D}{12}C(\Delta x)^2, \tag{13.2}$$

which clearly goes to zero as Δt and Δx go to zero. Hence, we know the replacement of the original partial derivatives in the diffusion equation by the finite difference approximations can be made as accurate as we wish by suitable choice of Δt and Δx. To see how we translate this into an approximate numerical method, divide the space interval $[0,L]$ into pieces of size Δx; then there will be $N\approx\frac{L}{\Delta x}$ such pieces within the interval. Divide the time interval $[0,T]$ in a similar way by Δt into $M\approx\frac{T}{\Delta t}$ subintervals.

The true solution at the point $(n\Delta x,m\Delta t)$ is then $u(n\Delta x,m\Delta t)$, which we will denote by u_{mn}. Hence, we know that the pair $(n\Delta x,m\Delta t)$ satisfies

$$\frac{u(n\Delta x,(m+1)\Delta t)-u(n\Delta x,m\Delta t)}{\Delta t}$$
$$-D\left(\frac{u((n+1)\Delta x,m\Delta t)+u((n-1)\Delta x,m\Delta t)-2u(n\Delta x,m\Delta t)}{(\Delta x)^2}\right)$$
$$=-\frac{1}{2}u_{tt}(x_0,c_t^+)\Delta t+\frac{D}{12}u_{xxxx}(c_x,t_0)(\Delta x)^2$$

Letting $E^1_{n,m}=-\frac{1}{2}u_{tt}(x_0,c_t^+)$ and $E^2_{n,m}=\frac{D}{12}u_{xxxx}(c_x,t_0)$, this is then rewritten as

$$\frac{u_{n,m+1}-u_{n,m}}{\Delta t}-D\left(\frac{u_{n+1,m}+u_{n-1,m}-2u_{n,m}}{(\Delta x)^2}\right) = E^1_{n,m}\Delta t+E^2_{n,m}(\Delta x)^2.$$

Now solve for the term $u_{n.m+1}$ to find

$$\begin{aligned}u_{n,m+1} &= u_{n,m}+D\frac{\Delta t}{(\Delta x)^2}\left(u_{n+1,m}+u_{n-1,m}-2u_{n,m}\right)\\ &+ E^1_{n,m}(\Delta t)^2+E^2_{n,m}\Delta t(\Delta x)^2.\end{aligned} \tag{13.3}$$

We therefore know the error we make in computing

$$u_{n,m+1} = u_{n,m}+D\frac{\Delta t}{(\Delta x)^2}\left(u_{n+1,m}+u_{n-1,m}-2u_{n,m}\right)$$

using the true solution values is reasonably small when Δx and Δt are also sufficiently small due to the error estimates in Equation 13.2. At this point, there is no relationship between Δx and Δt that is required for the approximation to work. However, when we take Equation 13.3 and solve it iteratively as a recursive equation, problems arise. Consider a full diffusion equation model on the domain $\mathscr{D}$:

$$\begin{aligned}
\frac{\partial u}{\partial t} - D\frac{\partial^2 u}{\partial x^2} &= 0 \\
u(0,t) &= 0, \text{ for } 0 \leq t \leq T \\
u(L,t) &= 0, \text{ for } 0 \leq t \leq T \\
u(x,0) &= f(x), \text{ for } 0 \leq x \leq L
\end{aligned}$$

The discrete boundary conditions at each *grid point* $(n\Delta x, m\Delta t)$ are then

$$u_{0,m} = u(0, m\Delta t) = 0, \ 1 \leq m \leq M \tag{13.4}$$

$$u_{N,m} = u(N\Delta x, m\Delta t) = 0, \ 1 \leq m \leq M \tag{13.5}$$

$$u_{n,0} = u(n\Delta x, 0) = f(n\Delta x) = f_n, \ 0 \leq n \leq N \tag{13.6}$$

as $N\Delta x = L$ and $M\Delta t = T$. Using these boundary conditions coupled with the discrete dynamics of Equation 13.3, we obtain a full recursive system to solve. First, recall the true values satisfy Equation 13.3. Letting $r = \Delta t/(\Delta x)^2$ and adding the boundary conditions, we obtain the full system

$$\begin{aligned}
u_{n,m+1} &= u_{n,m} + Dr(u_{n+1,m} + u_{n-1,m} - 2u_{n,m}) \\
&+ E^1_{n,m}(\Delta t)^2 + E^2_{n,m}\Delta t(\Delta x)^2
\end{aligned} \tag{13.7}$$

$$u_{0,m} = 0, \ 1 \leq m \leq M \tag{13.8}$$

$$u_{N,m} = 0, \ 1 \leq m \leq M \tag{13.9}$$

$$u_{n,0} = f_n, \ 0 \leq n \leq N \tag{13.10}$$

This gives us a recursive system we can solve of the form

$$v_{n,m+1} = v_{n,m} + Dr(v_{n+1,m} + v_{n-1,m} - 2v_{n,m}) \tag{13.11}$$

$$v_{0,m} = 0, \ 1 \leq m \leq M \tag{13.12}$$

$$v_{N,m} = 0, \ 1 \leq m \leq M \tag{13.13}$$

$$v_{n,0} = f_n, \ 0 \leq n \leq N \tag{13.14}$$

where $v_{n,m}$ is the solution to this discrete recursion system at the grid points. To solve this system, we start at time 0, time index value $m = 0$, and use Equation 13.8 to get

$$v_{1,1} = v_{1,0} + Dr(v_{2,0} + v_{0,0} - 2v_{1,0}).$$

We know $v_{0,0} = f_0$, $v_{1,0} = f_1$ and $v_{2,0} = f_2$. Hence, we have

$$v_{1,1} = f_1 + Dr(f_2 + f_0 - 2f_1).$$

We can do this for any n to find

$$\begin{aligned}
v_{n,1} &= v_{n,0} + Dr(v_{n+1,0} + v_{n-1,0} - 2v_{n,0}) \\
&= f_n + Dr(f_{n+1} + f_{n-1} - 2f_n)
\end{aligned}$$

Once the values at $m = 1$ have been found, we use the recursion to find the values at the next time step $m = 2$ and so on.

13.3.0.1 Homework

Exercise 13.3.1 *Write down the full set of discrete equations for a given* Δx *and* Δt *which must be solved in order to approximate this diffusion equation*

$$\begin{aligned}\frac{\partial u}{\partial t} - 4\,\frac{\partial^2 u}{\partial x^2} &= 0\\ u(0,t) &= 0,\ \textit{for}\ 0 \le t \le 10\\ u(5,t) &= 0,\ \textit{for}\ 0 \le t \le 10\\ u(x,0) &= f(x),\ \textit{for}\ 0 \le x \le 5\end{aligned}$$

for the data function

$$f(x) = \begin{cases} 300, & 3-0.4 < x < 3+0.4\\ 0, & \textit{else}\end{cases}$$

Exercise 13.3.2 *Write down the full set of discrete equations for a given* Δx *and* Δt *which must be solved in order to approximate this diffusion equation*

$$\begin{aligned}\frac{\partial u}{\partial t} - 8\,\frac{\partial^2 u}{\partial x^2} &= 0\\ u(0,t) &= 0,\ \textit{for}\ 0 \le t \le 12\\ u(6,t) &= 0,\ \textit{for}\ 0 \le t \le 12\\ u(x,0) &= f(x),\ \textit{for}\ 0 \le x \le 6\end{aligned}$$

for the data function

$$f(x) = \begin{cases} 50, & 1-0.2 < x < 1+0.2\\ 100, & 4-0.3 < x < 4+0.3\\ 0, & \textit{else}\end{cases}$$

Exercise 13.3.3 *Write down the full set of discrete equations for a given* Δx *and* Δt *which must be solved in order to approximate this diffusion equation*

$$\begin{aligned}\frac{\partial u}{\partial t} - 0.005\,\frac{\partial^2 u}{\partial x^2} &= 0\\ u(0,t) &= 0,\ \textit{for}\ 0 \le t \le 100\\ u(20,t) &= 0,\ \textit{for}\ 0 \le t \le 100\\ u(x,0) &= f(x),\ \textit{for}\ 0 \le x \le 20\end{aligned}$$

for the data function

$$f(x) = \begin{cases} 1000, & 2-0.2 < x < 2+0.2\\ 2000, & 6-0.2 < x < 6+0.2\\ 2000, & 10-0.2 < x < 10+0.2\\ 1000, & 16-0.4 < x < 16+0.4\\ 0, & \textit{else}\end{cases}$$

Exercise 13.3.4 *Write down the full set of discrete equations for a given* Δx *and* Δt *which must be solved in order to approximate this diffusion equation*

$$\frac{\partial u}{\partial t} - 0.00005\,\frac{\partial^2 u}{\partial x^2} = 0$$

$$\begin{aligned} u(0,t) &= 0, \textit{ for } 0 \leq t \leq 100 \\ u(12,t) &= 0, \textit{ for } 0 \leq t \leq 100 \\ u(x,0) &= f(x), \textit{ for } 0 \leq x \leq 12 \end{aligned}$$

for the data function

$$f(x) = \begin{cases} 50, & 2 - 0.2 < x < 2 + 0.2 \\ 500, & 5 - 0.2 < x < 5 + 0.2 \\ 1000, & 9 - 0.6 < x < 6 + 0.6 \\ 750, & 14 - 0.2 < x < 14 + 0.2 \\ 75, & 16 - 0.4 < x < 16 + 0.4 \\ 0, & \textit{else} \end{cases}$$

Exercise 13.3.5 *Write down the full set of discrete equations for a given Δx and Δt which must be solved in order to approximate this diffusion equation*

$$\begin{aligned} \frac{\partial u}{\partial t} - 2.6 \times 10^{-7} \frac{\partial^2 u}{\partial x^2} &= 0 \\ u(0,t) &= 0, \textit{ for } 0 \leq t \leq 150 \\ u(25,t) &= 0, \textit{ for } 0 \leq t \leq 150 \\ u(x,0) &= f(x), \textit{ for } 0 \leq x \leq 25 \end{aligned}$$

for the data function

$$f(x) = \begin{cases} 100, & 2 - 0.2 < x < 2 + 0.2 \\ 300, & 5 - 0.2 < x < 5 + 0.2 \\ 500, & 9 - 0.6 < x < 6 + 0.6 \\ 750, & 14 - 0.2 < x < 14 + 0.2 \\ 1500, & 16 - 0.4 < x < 16 + 0.4 \\ 0, & \textit{else} \end{cases}$$

13.3.1 Error Analysis

Let's denote the difference between the true grid point values and the discrete solution values by $w_{n,m} = u_{n,m} - v_{n,m}$. A simple subtraction shows that $w_{n,m}$ satisfies the following discrete system:

$$\begin{aligned} w_{n,m+1} &= w_{n,m} + Dr(w_{n+1,m} + w_{n-1,m} - 2w_{n,m}) \\ &\quad + E^1_{n,m}(\Delta t)^2 + E^2_{n,m}\Delta t(\Delta x)^2 && (13.15) \\ w_{0,m} &= 0, \ 1 \leq m \leq M && (13.16) \\ w_{N,m} &= 0, \ 1 \leq m \leq M && (13.17) \\ w_{n,0} &= 0, \ 0 \leq n \leq N && (13.18) \end{aligned}$$

From our previous estimates, we know $E^1_{n,m}(\Delta t)^2 + E^2_{n,m}\Delta t(\Delta x)^2 \leq C(\Delta t)^2 + \Delta t(\Delta x)^2$ and so we can overestimate these terms to obtain

$$|w_{n,m+1}| \leq |1 - 2Dr||w_{n,m}| + 2Dr(|w_{n+1,m}| + |w_{n-1,m}|) + C((\Delta t)^2 + \Delta t(\Delta x)^2)$$

Now since this equation holds for x grid positions, letting $||w^p|| = \max_{0 \leq N} |w_{n,p}|$, we see

$$|w_{n,m+1}| \leq |1 - 2Dr| \, ||w^m|| + 2Dr||w^m|| + E(\Delta t, \Delta x)$$

where $E(\Delta t, \Delta x)$ is the error term $C((\Delta t)^2 + \Delta t(\Delta x)^2)$. This equation holds for all indices n on the left-hand side which implies

$$||w^{m+1}| \leq (|1 - 2Dr| + 2Dr)||w^m|| + E(\Delta t, \Delta x)$$

Letting $g = |1 - 2Dr| + 2Dr$, we find the estimate

$$||w^1|| \leq g||w^0|| + E(\Delta t, \Delta x) = E(\Delta t, \Delta x)$$

since $||w^0|| = 0$ due to our error boundary conditions. The next steps are

$$\begin{aligned}
||w^2|| &\leq g||w^1|| + E(\Delta t, \Delta x) \leq (1 + g)E(\Delta t, \Delta x) \\
||w^3|| &\leq g||w^2|| + E(\Delta t, \Delta x) \leq (1 + g + g^2)E(\Delta t, \Delta x) \\
&\ldots \\
||w^N|| &\leq g||w^{N-1}|| + E(\Delta t, \Delta x) \leq (1 + g + g^2 + \ldots + g^{N-1})E(\Delta t, \Delta x).
\end{aligned}$$

If $g \leq 1$, we find $||w^N|| \leq NE(\Delta t, \Delta x)$. However, since $g \leq 1$, we know $|1-2Dr|+2Dr \leq 1$. This says $|1 - 2DR| \leq 1 - 2Dr$ which implies $Dr \leq 1/2$. Therefore, there is a relationship between Δt and Δx; $\Delta t \leq (\Delta x)^2/D$ Thus, the equation $g = |1-2Dr|+2Dr$ becomes $g = 1-2Dr+2Dr = 1$ and g is never actually less than 1. Hence, since $N \approx L/\Delta x$, we have

$$||w^N|| \leq \frac{L}{\Delta x}((\Delta t)^2 + \Delta t(\Delta x)^2) = \frac{LC}{\Delta x}\left(\frac{(\Delta x)^4}{D^2} + \frac{(\Delta x)^4}{D}\right) = \frac{LC(1 + D)}{D^2}(\Delta x)^3$$

which goes to zero as Δx goes to zero. Hence, in this case, the recursive equation has a well-behaved error and the solution to the recursion equation approaches the solution to the diffusion equation, which is finite. We will say our finite difference approximation is *stable* if the relationship $Dr \leq 1/2$ is satisfied.

Homework

Exercise 13.3.6 *Do the error analysis for the full set of discrete equations for a given Δx and Δt which must be solved in order to approximate this diffusion equation*

$$\begin{aligned}
\frac{\partial u}{\partial t} - 4\frac{\partial^2 u}{\partial x^2} &= 0 \\
u(0, t) &= 0, \textit{ for } 0 \leq t \leq 10 \\
u(5, t) &= 0, \textit{ for } 0 \leq t \leq 10 \\
u(x, 0) &= f(x), \textit{ for } 0 \leq x \leq 5
\end{aligned}$$

for the data function

$$f(x) = \begin{cases} 300, & 3 - 0.4 < x < 3 + 0.4 \\ 0, & \textit{else} \end{cases}$$

Exercise 13.3.7 *Do the error analysis for the full set of discrete equations for a given Δx and Δt which must be solved in order to approximate this diffusion equation*

$$\begin{aligned}
\frac{\partial u}{\partial t} - 8\,\frac{\partial^2 u}{\partial x^2} &= 0 \\
u(0,t) &= 0, \textit{ for } 0 \leq t \leq 12 \\
u(6,t) &= 0, \textit{ for } 0 \leq t \leq 12 \\
u(x,0) &= f(x), \textit{ for } 0 \leq x \leq 6
\end{aligned}$$

for the data function

$$f(x) = \begin{cases} 50, & 1 - 0.2 < x < 1 + 0.2 \\ 100, & 4 - 0.3 < x < 4 + 0.3 \\ 0, & \textit{else} \end{cases}$$

Exercise 13.3.8 *Do the error analysis for the full set of discrete equations for a given Δx and Δt which must be solved in order to approximate this diffusion equation*

$$\begin{aligned}
\frac{\partial u}{\partial t} - 0.005\,\frac{\partial^2 u}{\partial x^2} &= 0 \\
u(0,t) &= 0, \textit{ for } 0 \leq t \leq 100 \\
u(20,t) &= 0, \textit{ for } 0 \leq t \leq 100 \\
u(x,0) &= f(x), \textit{ for } 0 \leq x \leq 20
\end{aligned}$$

for the data function

$$f(x) = \begin{cases} 1000, & 2 - 0.2 < x < 2 + 0.2 \\ 2000, & 6 - 0.2 < x < 6 + 0.2 \\ 2000, & 10 - 0.2 < x < 10 + 0.2 \\ 1000, & 16 - 0.4 < x < 16 + 0.4 \\ 0, & \textit{else} \end{cases}$$

Exercise 13.3.9 *Do the error analysis for the full set of discrete equations for a given Δx and Δt which must be solved in order to approximate this diffusion equation*

$$\begin{aligned}
\frac{\partial u}{\partial t} - 0.00005\,\frac{\partial^2 u}{\partial x^2} &= 0 \\
u(0,t) &= 0, \textit{ for } 0 \leq t \leq 100 \\
u(12,t) &= 0, \textit{ for } 0 \leq t \leq 100 \\
u(x,0) &= f(x), \textit{ for } 0 \leq x \leq 12
\end{aligned}$$

for the data function

$$f(x) = \begin{cases} 50, & 2 - 0.2 < x < 2 + 0.2 \\ 500, & 5 - 0.2 < x < 5 + 0.2 \\ 1000, & 9 - 0.6 < x < 6 + 0.6 \\ 750, & 14 - 0.2 < x < 14 + 0.2 \\ 75, & 16 - 0.4 < x < 16 + 0.4 \\ 0, & \textit{else} \end{cases}$$

Exercise 13.3.10 *Do the error analysis for the full set of discrete equations for a given Δx and Δt which must be solved in order to approximate this diffusion equation*

$$\begin{aligned} \frac{\partial u}{\partial t} - 2.6 \times 10^{-7} \frac{\partial^2 u}{\partial x^2} &= 0 \\ u(0,t) &= 0, \textit{ for } 0 \leq t \leq 150 \\ u(25,t) &= 0, \textit{ for } 0 \leq t \leq 150 \\ u(x,0) &= f(x), \textit{ for } 0 \leq x \leq 25 \end{aligned}$$

for the data function

$$f(x) = \begin{cases} 100, & 2-0.2 < x < 2+0.2 \\ 300, & 5-0.2 < x < 5+0.2 \\ 500, & 9-0.6 < x < 6+0.6 \\ 750, & 14-0.2 < x < 14+0.2 \\ 1500, & 16-0.4 < x < 16+0.4 \\ 0, & \textit{else} \end{cases}$$

Note we can arrive at this result a little faster by assuming we can ignore the error $E(\Delta t, \Delta x)$ and looking at the solutions to the pure discrete recursion system only. We assume these have the form $w_{n,m} = g^m e^{in\beta}$ for a positive g, where recall $e^{i\theta} = \cos(\theta) + i \sin \theta$. Then, we find

$$\begin{aligned} g^{m+1} e^{in\beta} &= g^m e^{in\beta} + Dr\Big(g^m e^{i(n+1)\beta} + g^m e^{i(n-1)\beta} - 2g^m e^{in\beta}\Big) \\ &= g^m e^{in\beta}\Big(1 + Dr\Big(e^{i\beta} + e^{-i\beta} - 2\Big)\Big) \end{aligned}$$

But $2\cos(\beta) = e^{i\beta} + e^{-i\beta}$ and so we have

$$g^{m+1} e^{in\beta} = g^m e^{in\beta}\Big(1 + 2Dr(\cos(\beta) - 1)\Big)$$

Dividing, we have

$$g = 1 + 2Dr(\cos(\beta) - 1).$$

This is the *multiplier* that is applied at each time step. We can do the same analysis to show that solutions of the form $w_{n,m} = g^m e^{-in\beta}$ have the same multiplier. Then combinations of these two complex solutions lead to the usual two real solutions, $\phi_{n,m} = g^m \cos(n\beta)$ and $\psi_{n,m} = g^m \sin(n\beta)$. So the use of the complex form of the assumed solution is a useful way to obtain the multiplier g with minimal algebraic complications. To ensure the error does not go to infinity, we must have $0 < g < 1$. We want

$$0 < 1 - 2Dr(1 - \cos(\beta)) < 1$$

Since $1 - 2Dr(1 - \cos(\beta)) < 1$ always, the top inequality is satisfied. We also know $1 - 2Dr(1 - \cos(\beta)) > 1 - 2Dr$, so we can satisfy the bottom inequality if we choose r so that $0 < 1 - 2Dr < 1 - 2Dr(1 - \cos(\beta))$. This implies $Dr < 1/2$. Thus, stability is not guaranteed unless

$$D\Delta t < (\Delta x)^2.$$

13.3.2 Homework

Exercise 13.3.11 *Do the error analysis for the full set of discrete equations for a given Δx and Δt which must be solved in order to approximate this diffusion equation*

$$\begin{aligned} \frac{\partial u}{\partial t} - 4\frac{\partial^2 u}{\partial x^2} &= 0 \\ u(0,t) &= 0, \text{ for } 0 \le t \le 10 \\ u(5,t) &= 0, \text{ for } 0 \le t \le 10 \\ u(x,0) &= f(x), \text{ for } 0 \le x \le 5 \end{aligned}$$

for the data function

$$f(x) = \begin{cases} 300, & 3 - 0.4 < x < 3 + 0.4 \\ 0, & \text{else} \end{cases}$$

Exercise 13.3.12 *Do the simplified complex method error analysis for the full set of discrete equations for a given Δx and Δt which must be solved in order to approximate this diffusion equation*

$$\begin{aligned} \frac{\partial u}{\partial t} - 8\frac{\partial^2 u}{\partial x^2} &= 0 \\ u(0,t) &= 0, \text{ for } 0 \le t \le 12 \\ u(6,t) &= 0, \text{ for } 0 \le t \le 12 \\ u(x,0) &= f(x), \text{ for } 0 \le x \le 6 \end{aligned}$$

for the data function

$$f(x) = \begin{cases} 50, & 1 - 0.2 < x < 1 + 0.2 \\ 100, & 4 - 0.3 < x < 4 + 0.3 \\ 0, & \text{else} \end{cases}$$

Exercise 13.3.13 *Do the simplified complex method error analysis for the full set of discrete equations for a given Δx and Δt which must be solved in order to approximate this diffusion equation*

$$\begin{aligned} \frac{\partial u}{\partial t} - 0.005\frac{\partial^2 u}{\partial x^2} &= 0 \\ u(0,t) &= 0, \text{ for } 0 \le t \le 100 \\ u(20,t) &= 0, \text{ for } 0 \le t \le 100 \\ u(x,0) &= f(x), \text{ for } 0 \le x \le 20 \end{aligned}$$

for the data function

$$f(x) = \begin{cases} 1000, & 2 - 0.2 < x < 2 + 0.2 \\ 2000, & 6 - 0.2 < x < 6 + 0.2 \\ 2000, & 10 - 0.2 < x < 10 + 0.2 \\ 1000, & 16 - 0.4 < x < 16 + 0.4 \\ 0, & \text{else} \end{cases}$$

Exercise 13.3.14 *Do the simplified complex method error analysis for the full set of discrete equations for a given Δx and Δt which must be solved in order to approximate this diffusion equation*

$$\frac{\partial u}{\partial t} - 0.00005\frac{\partial^2 u}{\partial x^2} = 0$$

$$\begin{aligned} u(0,t) &= 0, \textit{ for } 0 \le t \le 100 \\ u(12,t) &= 0, \textit{ for } 0 \le t \le 100 \\ u(x,0) &= f(x), \textit{ for } 0 \le x \le 12 \end{aligned}$$

for the data function

$$f(x) = \begin{cases} 50, & 2-0.2 < x < 2+0.2 \\ 500, & 5-0.2 < x < 5+0.2 \\ 1000, & 9-0.6 < x < 6+0.6 \\ 750, & 14-0.2 < x < 14+0.2 \\ 75, & 16-0.4 < x < 16+0.4 \\ 0, & \textit{else} \end{cases}$$

Exercise 13.3.15 *Do the simplified complex method error analysis for the full set of discrete equations for a given Δx and Δt which must be solved in order to approximate this diffusion equation*

$$\begin{aligned} \frac{\partial u}{\partial t} - 2.6 \times 10^{-7}\frac{\partial^2 u}{\partial x^2} &= 0 \\ u(0,t) &= 0, \textit{ for } 0 \le t \le 150 \\ u(25,t) &= 0, \textit{ for } 0 \le t \le 150 \\ u(x,0) &= f(x), \textit{ for } 0 \le x \le 25 \end{aligned}$$

for the data function

$$f(x) = \begin{cases} 100, & 2-0.2 < x < 2+0.2 \\ 300, & 5-0.2 < x < 5+0.2 \\ 500, & 9-0.6 < x < 6+0.6 \\ 750, & 14-0.2 < x < 14+0.2 \\ 1500, & 16-0.4 < x < 16+0.4 \\ 0, & \textit{else} \end{cases}$$

13.4 Diffusion Equation Finite Difference Approximation Code

We now implement the finite difference scheme for a typical diffusion/ heat equation model. This is done in the function `NumericalHeatDirichlet` which uses the arguments `dataf`, the name of the data $f(x)$ on the bottom edge, `gamma`, the diffusion constant D, `L` and `T` and the dimension of the rectangle $\mathscr{D} = [0, L] \times [0, T]$. Stability implies that Δt can at most be $(1/2\gamma)\,(\delta x)^2$ and we allow the use of a fraction of that maximum by using the parameter `deltfrac` to determine Δt via the equation $\Delta t = $ deltfrac $1/(2\gamma)(\delta x)^2$. Finally, we set the desired Δx using the parameter `delx`. Here are the details. First we calculate the desired Δt.

Listing 13.1: **Find Δt**

```
% calculate delt
delt = .5*deltfrac*delx^2/gamma
```

We then find the number of time and space steps and set up the x and t data points for the grid.

Listing 13.2: **Setup Time and Space Grid Data Points**

```
% calculate N and M
N = round(L/delx);
M = round(T/delt);
% setup linspaces
x = linspace(0,L,N+1);
t = linspace(0,T,M+1);
```

We then implement the finite difference scheme using a double for loop construction.

Listing 13.3: **Implement Finite Difference Scheme**

```
% setup V
V = zeros(N+1,M+1);
% setup V(n,1) = data function (n)
for n = 1:N+1
  V(n,1) = dataf(x(n));
end
% setup r value
r = delt/(delx^2);
% find numerical solution to the heat equation
for m = 1:M
  for n = 2:N
    V(n,m+1) = gamma*r*V(n+1,m)+(1-2*gamma*r)*V(n,m) + gamma*r*V(n-1,m
        );
  end
end
```

Finally, we plot the solution surface.

Listing 13.4: **Plot Solution Surface**

```
% setup surface plot
[X,Time] = meshgrid(x,t);
mesh(X,Time,V','EdgeColor','blue');
xlabel('x axis');
ylabel('t axis');
zlabel('Solution');
title('Heat Equation on Square');
```

The full code is below.

Listing 13.5: **NumericalHeatDirichlet.m**

```
function NumericalHeatDirichlet(dataf,gamma,L,T,deltfrac,delx)
%
% dataf is our data function
% gamma is the diffusion coefficient
% L is the x interval
% T is the time interval
% deltfrac is how much of the maximum del t to use
```

```
%
% delx is our chosen delta x
% we calculate delt = .5*gamma*(delx^2);
% N = round(L/delx)
% M = round(T/delt)
%
% calculate delt
delt = .5*deltfrac*delx^2/gamma

% calculate N and M
N = round(L/delx);
M = round(T/delt);

% setup linspaces
x = linspace(0,L,N+1);
t = linspace(0,T,M+1);

% setup V
V = zeros(N+1,M+1);

% setup V(n,1) = data function (n)
for n = 1:N+1
  V(n,1) = dataf(x(n));
end

% setup r value
r = delt/(delx^2);
% find numerical solution to the heat equation
for m = 1:M
  for n = 2:N
    V(n,m+1) = gamma*r*V(n+1,m)+(1-2*gamma*r)*V(n,m) + gamma*r*V(n-1,m
        );
  end
end

% setup surface plot
[X,Time] = meshgrid(x,t);
mesh(X,Time,V','EdgeColor','blue');
xlabel('x axis');
ylabel('t axis');
zlabel('Solution');
title('Heat Equation on Square');
end
```

Here is a typical use for a heat equation model with diffusion coefficient 0.09 for a space interval of $[0,5]$ and a time interval of $[0,10]$ using a space step size of 0.2.

Listing 13.6: **Generating a Finite Difference Heat Equation Solution**

```
pulse = @(x) pulsefunc(x,2,.2,100);
NumericalHeatDirichlet(pulse,.09,5,10,.9,.2);
delt = 0.20000
```

This generates the plot we see in Figure 13.1.

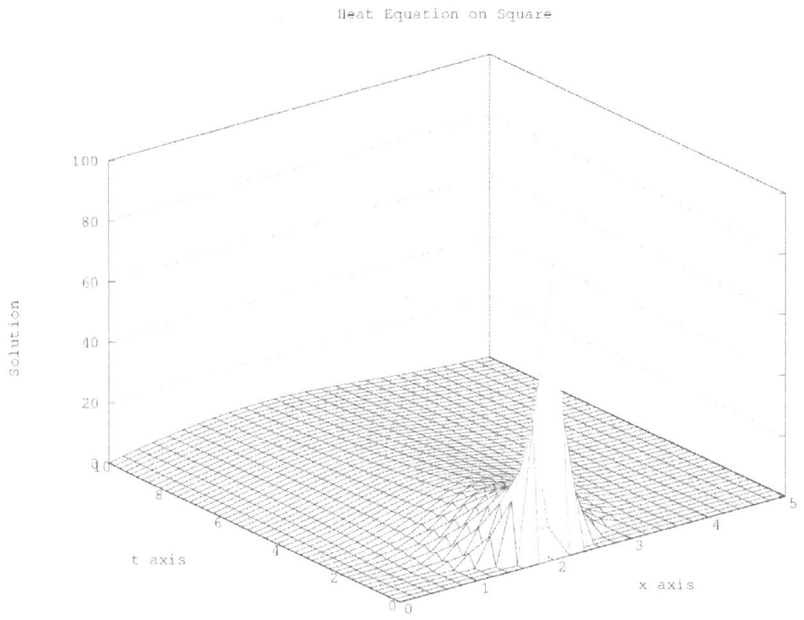

Figure 13.1: Finite difference solution to heat equation.

13.4.1 Homework

We can solve these diffusion models using Fourier Series approaches as we have discussed in (Peterson (21) 2020) (the more theoretical approach) and (J. Peterson (11) 2016) (more code oriented).

Exercise 13.4.1 *Find finite difference approximations to this diffusion equation*

$$\begin{aligned} \frac{\partial u}{\partial t} - 4\,\frac{\partial^2 u}{\partial x^2} &= 0 \\ u(0,t) &= 0, \textit{ for } 0 \le t \le 10 \\ u(5,t) &= 0, \textit{ for } 0 \le t \le 10 \\ u(x,0) &= f(x), \textit{ for } 0 \le x \le 5 \end{aligned}$$

for the data function

$$f(x) = \begin{cases} 300, & 3 - 0.4 < x < 3 + 0.4 \\ 0, & \textit{else} \end{cases}$$

for three appropriate choices of Δx and Δt. Show the surface plots of course.

Exercise 13.4.2 *Find finite difference approximations to this diffusion equation*

$$\begin{aligned} \frac{\partial u}{\partial t} - 8\,\frac{\partial^2 u}{\partial x^2} &= 0 \\ u(0,t) &= 0, \textit{ for } 0 \le t \le 12 \\ u(6,t) &= 0, \textit{ for } 0 \le t \le 12 \\ u(x,0) &= f(x), \textit{ for } 0 \le x \le 6 \end{aligned}$$

for the data function

$$f(x) = \begin{cases} 50, & 1 - 0.2 < x < 1 + 0.2 \\ 100, & 4 - 0.3 < x < 4 + 0.3 \\ 0, & \textit{else} \end{cases}$$

for three appropriate choices of $||\Delta x$ *and* Δt. *Show the surface plots of course.*

Exercise 13.4.3 *Find finite difference approximations to this diffusion equation*

$$\begin{aligned} \frac{\partial u}{\partial t} - 0.005 \frac{\partial^2 u}{\partial x^2} &= 0 \\ u(0,t) &= 0, \textit{ for } 0 \leq t \leq 100 \\ u(20,t) &= 0, \textit{ for } 0 \leq t \leq 100 \\ u(x,0) &= f(x), \textit{ for } 0 \leq x \leq 20 \end{aligned}$$

for the data function

$$f(x) = \begin{cases} 1000, & 2 - 0.2 < x < 2 + 0.2 \\ 2000, & 6 - 0.2 < x < 6 + 0.2 \\ 2000, & 10 - 0.2 < x < 10 + 0.2 \\ 1000, & 16 - 0.4 < x < 16 + 0.4 \\ 0, & \textit{else} \end{cases}$$

for three appropriate choices of $||\Delta x$ *and* Δt. *Show the surface plots of course.*

Exercise 13.4.4 *Find finite difference approximations to this diffusion equation*

$$\begin{aligned} \frac{\partial u}{\partial t} - 0.00005 \frac{\partial^2 u}{\partial x^2} &= 0 \\ u(0,t) &= 0, \textit{ for } 0 \leq t \leq 100 \\ u(12,t) &= 0, \textit{ for } 0 \leq t \leq 100 \\ u(x,0) &= f(x), \textit{ for } 0 \leq x \leq 12 \end{aligned}$$

for the data function

$$f(x) = \begin{cases} 50, & 2 - 0.2 < x < 2 + 0.2 \\ 500, & 5 - 0.2 < x < 5 + 0.2 \\ 1000, & 9 - 0.6 < x < 6 + 0.6 \\ 750, & 14 - 0.2 < x < 14 + 0.2 \\ 75, & 16 - 0.4 < x < 16 + 0.4 \\ 0, & \textit{else} \end{cases}$$

for three appropriate choices of $||\Delta x$ *and* Δt. *Show the surface plots of course.*

Exercise 13.4.5 *Find finite difference approximations to this diffusion equation*

$$\begin{aligned} \frac{\partial u}{\partial t} - 2.6 \times 10^{-7} \frac{\partial^2 u}{\partial x^2} &= 0 \\ u(0,t) &= 0, \textit{ for } 0 \leq t \leq 150 \\ u(25,t) &= 0, \textit{ for } 0 \leq t \leq 150 \\ u(x,0) &= f(x), \textit{ for } 0 \leq x \leq 25 \end{aligned}$$

for the data function

$$f(x) = \begin{cases} 100, & 2 - 0.2 < x < 2 + 0.2 \\ 300, & 5 - 0.2 < x < 5 + 0.2 \\ 500, & 9 - 0.6 < x < 6 + 0.6 \\ 750, & 14 - 0.2 < x < 14 + 0.2 \\ 1500, & 16 - 0.4 < x < 16 + 0.4 \\ 0, & \textit{else} \end{cases}$$

for three appropriate choices of $||\Delta x$ *and* Δt. *Show the surface plots of course.*

Part IV

Integration

Chapter 14

Integration in Multiple Dimensions

In this chapter, we are going to discuss Riemann Integration in $\Re^n$. This is a much more interesting thing than the development laid out in (Peterson (21) 2020) as the topology of $\Re$ is much simpler than the topology of $\Re^n$ for $n > 1$. We wish to do this from a more abstract perspective.

14.1 The Darboux Integral

Let $f : A \subset \Re^n \rightarrow \Re$ be a bounded function on the bounded set A. Enclose the set A inside a bounded hyper rectangle

$$R = [a_1, b_1] \times [a_2, b_2] \times \ldots \times [a_n, b_n] = \prod_{i=1}^{n} [a_i, b_i]$$

There is no need to try to find the tightest bounded box possible. We will show later the definition of integration we develop is independent of this choice. We extend f to R to $\hat{f}$ on R as follows:

$$\hat{f}(x) = \begin{cases} f(x), & x \in A \\ 0, & x \in R \setminus A \end{cases}$$

In $\Re^2$, we would have what we see in Figure 14.1(a) and in $\Re^3$, it could look like the image in Figure 14.1(b). The extension $\hat{f}$ allows us to define what we mean by integration without worrying about

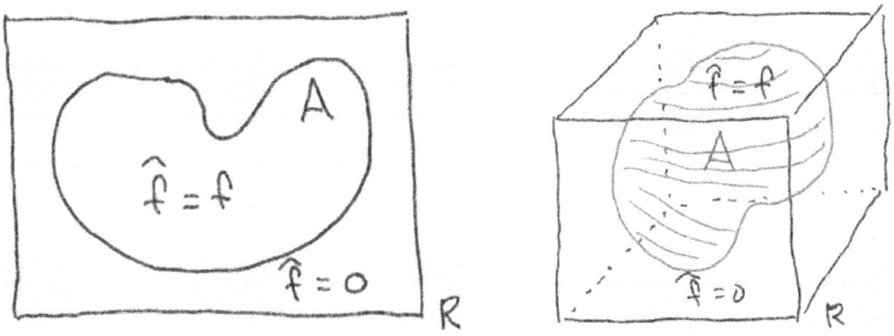

(a) A bounded set A in $\Re^2$. (b) A bounded set A in $\Re^3$.

Figure 14.1: Bounded integration domains.

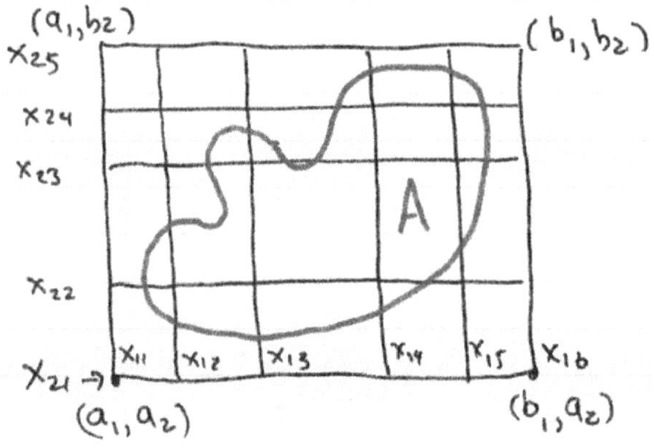

Figure 14.2: A bounded A and a partition P.

the boundary of A, ∂A. We begin by slicing R up into a *rectangular* grid to form what are called **Partitions**.

Definition 14.1.1 Partitions of the bounded set A

A partition P of A is defined as follows: For any bounded rectangle R containing A, subdivide each axis using a finite collection of points this way:

$$\begin{aligned} P_1 &= \{x_{11} = a_1 < x_{12} < x_{13} < \ldots < x_{1,p_1} = b_1\} \\ P_2 &= \{x_{21} = a_2 < x_{22} < x_{23} < \ldots < x_{2,p_2} = b_2\} \\ \vdots &= \vdots \\ P_k &= \{x_{k1} = a_k < x_{k2} < x_{k3} < \ldots < x_{k,p_k} = b_k\} \\ \vdots &= \vdots \\ P_n &= \{x_{n1} = a_n < x_{n2} < x_{n3} < \ldots < x_{n,p_n} = b_n\} \end{aligned}$$

The partition P is the collection of points in $\Re^n$ defined by

$$P = P_1 \times P_2 \times \ldots \times P_n$$

Note each P_k is a traditional partition as used in one dimensional integration theory. The sizes of each P_k are determined by the integers p_k and these need not be the same, of course.

Such a P defined a set of grid points in R and the lines of constant $x_{i,j}$ slice R and therefore also A into hyper rectangles. Figure 14.2 shows what this could look like in two dimensions. It is a lot more cluttered to try to draw this in three dimensions! And in higher than three dimensions, we cannot come up with visualizations at all, so it is best to be able to understand this abstractly. In the figure, we use the partition

$$P = P_1 \times P_2 = \{x_{11}, x12, x_{13}, x_{14}, x_{15}\} \times \{x_{21}, x_{22}, x_{23}, x_{24}, x_{25}\}$$

If you are familiar with code, you should see that if P_1 is represented by one `linspace` command and P_2, by another, the `meshgrid` command creates the partition P.

For simplicity, let's note a partition P determines a finite number of hyper rectangles of the form

$$S = [x_{1,j_1}; x_{1,j_1+1}] \times [x_{2,j_2}; x_{2,j_2+1}] \times \ldots \times [x_{n,j_n}; x_{n,j_n+1}]$$

where we are separating points in each $[\cdot, \cdot]$ pair by semicolons as using a comma becomes very confusing. Here the integers j_1 through j_n range over all the possible choices each P_i allows; i.e. $1 \leq j_i < p_i$. There are a large number of these rectangles, but still finitely many. The partition P will determine $p_1 \, p_2 \, \ldots p_n$ such rectangles and each rectangle will have a hyper volume

$$V(S) = \underbrace{(x_{1,j_1+1} - x_{1,j_1})}_{\text{length on axis 1}} \cdot \underbrace{(x_{2,j_1+1} - x_{2,j_1})}_{\text{length on axis 2}} \cdots \underbrace{(x_{n,j_1+1} - x_{n,j_1})}_{\text{length on axis } n}$$

In general, it is not important how we label these finitely many rectangles, so we will just say the partition P determines a finite number of rectangles N and label them as $S_1, \ldots, S_N$ or $(S_i)_{i=1}^n$. Note a rectangle S_j need not intersect A! For each rectangle S_i, define

$$m_i(f) = \inf_{(x_1,\ldots,x_n) \in S_i} \hat{f}(x_1, \ldots, x_n)$$

$$M_i(f) = \sup_{(x_1,\ldots,x_n) \in S_i} \hat{f}(x_1, \ldots, x_n)$$

These numbers are finite as we assume f is bounded on A. Since it is cumbersome to use the n - tuple notation for the points in S_i, we will use $\boldsymbol{x}$ to indicate this instead. Thus

$$m_i(f) = \inf_{\boldsymbol{x} \in S_i} \hat{f}(\boldsymbol{x}), \quad M_i(f) = \sup_{\boldsymbol{x} \in S_i} \hat{f}(\boldsymbol{x})$$

We can now define the Darboux lower and upper sums for f on A.

Definition 14.1.2 Darboux Lower and Upper Sums for f on A

Given a partition P of A with a bounding hyper rectangle R, P determines N hyper rectangles and the numbers $m_i(f)$ and $M_i(f)$ for $1 \leq i \leq N$. The Darboux Lower Sum associated with f, P, A and R is

$$L(f, P, A, R) = \sum_{i=1}^{N} m_i(f) V(S_i)$$

and the Darboux Upper Sum associated with f, P, A and R is

$$U(f, P, A, R) = \sum_{i=1}^{N} m_i(f) V(S_i)$$

The A is often understood from context and so we would usually write $L(f, P, R)$ and $U(F, P, R)$ to indicate these sums might depend on the choice of bounding hyper rectangle R. Since N depends on P, we generally write these as

$$L(f, P, R) = \sum_{i \in P} m_i(f) V(S_i), \quad U(f, P, R) = \sum_{i \in P} M_i(f) V(S_i)$$

Comment 14.1.1 *For a given partition P, it is always true that $L(f, P, R) \leq U(f, P, R)$ from the definition of m_i and M_i.*

Comment 14.1.2 *Since each partition P is dependent on the choice of R, if we had two bounding rectangles with partitions $P(R_1)$ and $P(R_2)$, we would know $R_1 = (R_1 \cap R_2) \cup (R_1 \cap R_2^C) \cup (R_2 \cap R_1^C)$. The partition $P(R_1)$ corresponds to possibly nonzero values of m_i and M_i only on $R_1 \cap R_2$. A similar argument shows the partition $Q(R_2)$ corresponds to possibly nonzero values of m_i and M_i only on $R_1 \cap R_2$. You can see $P(R_1)$ induces a partition of $R_1 \cap R_2$, $P(R_1 \cap R_2)$ and also $Q(R_2)$ induces a partition of $R_1 \cap R_2$, $Q(R_1 \cap R_2)$ Hence,*

$$\begin{aligned} L(f, P(R_1), R_1) &= L(f, P(R_1 \cap R_2), R_1 \cap R_2) \\ L(f, Q(R_2), R_2) &= L(f, Q(R_1 \cap R_2), R_1 \cap R_2) \end{aligned}$$

These sums could be different, of course.

However, by thinking about this just a little differently, we can establish the following result.

Theorem 14.1.1 Lower and Upper Darboux Sums are Independent of the Choice of Bounding Rectangle

$$L(f, P, R_1) = L(f, P, R_2), \quad U(f, P, R_1) = U(f, P, R_2)$$

Proof 14.1.1
First, note $R_1 = (R_1 \cap R_2) \cup (R_1 \cap R_2^C) \cup (R_2 \cap R_1^C)$. So a partition $P(R_1)$ induces a partition on $(R_1 \cap R_2)$ and there are infinitely many ways to extend $P(R_1)$ to the piece $R_2 \cap R_1^C$. Let $Q(R_2)$ be any such partition of R_2 which retains the partition of $P(R_1)$ on $R_1 \cap R_2$ and extends it on $R_2 \cap R_1^C$. Both of the partitions $P(R_1)$ and $Q(R_2)$ correspond to possibly nonzero values of m_i and M_i only on $R_1 \cap R_2$. Let $P(R_1 \cap R_2)$ denote the partition induced on $R_1 \cap R_2$ by $P(R_1)$. Then

$$L(f, P(R_1), R_1) = L(f, P(R_1 \cap R_2), R_1 \cap R_2) L(f, P(R_2), R_2)$$

Hence, the value of the lower sum doesn't depend on the choice of bounding rectangle. A similar argument shows the value of the upper sum are independent of the choice of bounding rectangle also.
■

Comment 14.1.3 *From Theorem 14.1.1, we know the lower and upper sums are independent of the choice of bounding rectangle and hence we no longer need to write $L(f, P, R)$ and $U(f, P, R)$. Hence, from now on we will simply write $L(f, P)$ and $U(f, P)$ where the partitions are determined from some choice of bounding rectangle R.*

If we add more points to any of the P_i collections that comprise P, we generate a new partition P' called a **refinement** of P.

Definition 14.1.3 A Refinement of a Partition

Given a partition P of A, a refinement P' of P is any partition of A which contains all the original points of P. Hence, P is a refinement of itself. Of course, the interesting refinements are the ones which add at least one point to P. These extra points will then introduce new rectangles.

In Figure 14.3, we see a refinement P' that adds one additional point. We also show the new rectangles generated. In this example, you can count the number of additional rectangles and see it is 18. It is very cluttered to draw what happens if you introduce two points and so on, so you have to learn

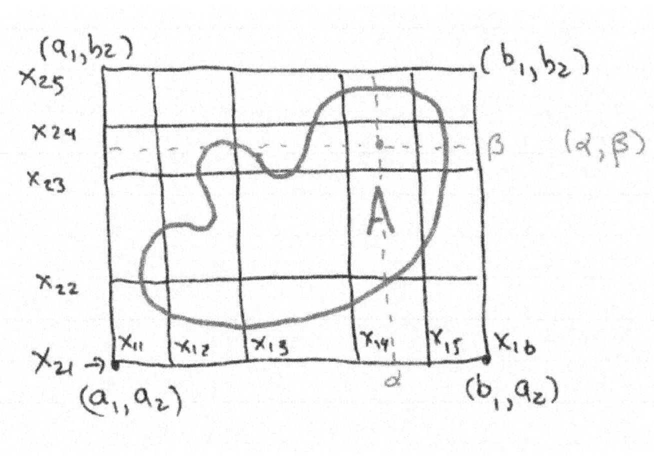

Figure 14.3: One point (α, β) is added to the partition P.

how to understand this from an abstract point of view. The new partition in the figure is

$$P' = [x_{11}, x_{12}, x_{13}, x_{14}, \alpha, x_{15}, x_{16}] \times [x_{21}, x_{22}, x_{23}, \beta, x_{24}, x_{25}]$$

Given two partitions P and Q of A, we often want to merge them into one partition by counting only counting points in the union of the collection of points that comprise P and the points that comprise Q once.

Definition 14.1.4 The Common Refinement of Two Partitions

Let P and Q be two partitions of A. Let U be the union of the collection of points from P and Q. On the ith axis, P and Q determine the collection of points

$$P_i = \{x_{i1}, \ldots, x_{i,p_i}\} \quad \text{and } Q_i = \{y_{i1}, \ldots, y_{i,q_j}\}$$

The union of P_i and Q_i gives the collection

$$P_i \cup Q_i = \{x_{i1}, \ldots, x_{i,p_i}, y_{i1}, \ldots, y_{i,q_j}\}$$

which determines a new ordering from low to high collection of points on the ith axis $G_i = \{z_{i1}, \ldots, z_{i,g_i}\}$ which only uses unique entries as all duplicates are removed. The collection of all $G_1 \times \ldots \times G_n$ gives the common refinement of P and Q denoted by $P \vee Q$.

Comment 14.1.4 *Note $P \vee Q$ is a refinement of both P and Q.*

Homework

Exercise 14.1.1 *Modify the Riemann Sum MATLAB code in (Peterson (21) 2020) for the two dimension case.*

Exercise 14.1.2 *Modify the Riemann Sum MATLAB code in (Peterson (21) 2020) for the two dimension case with uniform partitions and modify the code to draw the Riemann sums as surfaces.*

Exercise 14.1.3 *For*

$$P = \{-1, -0.7, 0, 0.5, 1.1, 1.8, 2.1, 2.7, 3.0\}$$

$$Q = \{4, 4, 3, 5, 1, 5, 6, 6, 1, 6, 8, 7.3, 8\}$$

draw the resulting partition $\pi = P \times Q$ *and find the maximum area that a rectangle* S_i *in* π *can have.*

Exercise 14.1.4 *For*

$$\begin{aligned} P &= \{-1, -0.7, 0, 0.5, 1.1, 1.8, 2.1, 2.7, 3.0\} \\ Q &= \{4, 4, 3, 5, 1, 5, 6, 6, 1, 6, 8, 7.3, 8\} \end{aligned}$$

add the point 1.5 *to* P *to create* P'. *Draw the resulting partition* $\pi' = P' \times Q$ *and find the maximum area that a rectangle* S_i *in* π' *can have.*

Exercise 14.1.5 *Let* A *be the circle of radius* 2 *centered at* $(2, 6)$. *For*

$$\begin{aligned} P &= \{-1, -0.7, 0, 0.5, 1.1, 1.8, 2.1, 2.7, 3.0\} \\ Q &= \{4, 4, 3, 5, 1, 5, 6, 6, 1, 6, 8, 7.3, 8\} \end{aligned}$$

$\pi = P \times Q$ *is a partition of a bounding rectangle containing this circle. Let* $f(x, y) = 2x^2 + 5y^2$. *Find the lower and upper sums associated with this partition. Draw the bounding box, the partition and the circle, and shade all the rectangles* S_i *that are outside the circle, one color; the rectangles inside the circle another color; and the rectangles that intersect the boundary of the circle another color.*

Exercise 14.1.6 *Let* A *be the circle of radius* 2 *centered at* $(2, 6)$. *For*

$$P = \{-1, -0.7, 0, 0.5, 1.1, 1.8, 2.1, 2.7, 3.0\} Q = \{4, 4, 3, 5, 1, 5, 6, 6, 1, 6, 8, 7.3, 8\}$$

$\pi = P \times Q$ *is a partition of a bounding rectangle containing this circle. Let* $f(x, y) = 2x^2 + 5y^2$. *Add the point* 1.5 *to* P *to create* P' *giving the* $\pi' = P' \times Q$. *Find the lower and upper sums associated with this partition. Draw the bounding box, the partition and the circle and shade all the rectangles* S_i *that are outside the circle, one color; the rectangles inside the circle another color; and the rectangles that intersect the boundary of the circle another color. What effect did adding the extra point have here?*

The next question is what happens to the Lower and Upper Darboux sums when you refine a partition?

Theorem 14.1.2 Lower and Upper Darboux Sums and Partition Refinements

Let P' *be a refinement of* P *for* A *with bounding hyper rectangle* R. *Then*

$$L(f, P, R) \leq L(f, P', R), \quad U(f, P, R) \geq U(f, P', R)$$

Of course, these results are independent of the bounding rectangle and could be written

$$L(f, P) \leq L(f, P'), \quad U(f, P) \geq U(f, P')$$

Proof 14.1.2
From now on, we will just call hyper rectangles by the simpler term, rectangles. We all know these are abstract things once $n > 3$*! Let's start by adding one point to* P. *Let* $\boldsymbol{z}$ *be the added point. Then* $\boldsymbol{z}$ *will be in some rectangle* S^* *determined by* P. *Note it could be on the boundary of* S^* *and not necessarily in the interior. Let's organize the indices of* P *and* P'. *Look at Figure 14.3 to see additional rectangles in a specific case.*

- *The indices of P can be put into three disjoint sets:*
  1. *The singleton index j_* which is the index corresponding to S^*.*
  2. *The indices which correspond to rectangles due to P which are not broken into two pieces by the addition of the extra point $\boldsymbol{z}$ to P. The rectangles here are exactly the same as the rectangles generated by P'. Let I denote this set of indices.*
  3. *The indices which correspond to rectangles which are broken into two pieces. Each of these rectangles S_j splits into two rectangles for P'. Let J denote this set of indices.*
- *The indices of P' can be put into disjoint sets also.*
  1. *For the index j_* from P, the additional point $\boldsymbol{z}$ causes the creation of 2^n new rectangles $T_{j*,i}$ for $1 \leq i \leq 2^n$. So $\sum_{i=1}^{2^n} V(T_{j_*,i}) = V(S^*)$. Also, since*

$$\inf_{\boldsymbol{x} \in T_{j_*,i}} \hat{f}(\boldsymbol{x}) \leq \inf_{\boldsymbol{x} \in S^*} \hat{f}(\boldsymbol{x})$$

  we have

$$m_{\mathit{for}\ T_{j_*,i}}(f) \quad \geq \quad m_{\mathit{for}\ S^*}$$

  which tells us

$$\sum_{i=1}^{2^n} m_{\mathit{for}\ T_{j_*,i}}(f)\, V(T_{j_*,i}) \geq m_{\mathit{for}\ S^*} \sum_{i=1}^{2^n} V(T_{j_*,i}) = m_{j_*} V(S^*).$$

  2. *The rectangles due to P' that are the same as the rectangles due to P that do not come from adding the extra point give a set of indices we call U. Let the infimum values here be called m_i' to distinguish them from infimum values m_j due to P and let the rectangles be called S_i' for the same reason. Then we have*

$$\sum_{i \in U} m_i'(f) V(S_i') \quad = \quad \sum_{j \in I} m_j(f) V(S_j)$$

  3. *The remaining indices correspond to the pairs of rectangles that come from the splitting due to the addition of the extra point for all rectangles from P except S^*. Call these indices V. For each rectangle, S_j from the index set J from P, we get two new rectangles: $S_j = H_{j,1} \cup H_{j,2}$ and $V(S_j) = V(H_{j,1}) + V(H_{j,2})$. Also,*

$$m_{\mathit{for}\ H_{j,i}}(f) \quad \geq \quad m_{\mathit{for}\ S_j}(f)$$

  and so

$$\begin{aligned} &\sum_{j \in J} m_{\mathit{for}\ H_{j,1}}(f)\, V(H_{j,1}) + m_{\mathit{for}\ H_{j,2}}(f)\, V(H_{j,2}) \\ &\geq \sum_{j \in J} m_{\mathit{for}\ S_j}(V(H_{j,1}) + V(H_{j,2})) = \sum_{j \in J} m_j V(S_j). \end{aligned}$$

So the lower sums are

$$\begin{aligned} L(f, P, R) &= \sum_{j \in P} m_j(f) V(S_j) = m_*(f) V(S^*) + \sum_{j \in I} m_j(f) V(S_j) + \sum_{j \in J} m_j(f) V(S_j) \\ L(f, P', R) &= \sum_{j \in P'} m_j'(f) V(S_j') \geq m_*(f) V(S^*) + \sum_{j \in I} m_j(f) V(S_j) + \sum_{j \in J} m_j(f) V(S_j) \end{aligned}$$

$$= L(f,P,R)$$

This shows the result is true for the addition of one extra point into one rectangle due to P.

The next step is an induction argument on the number of points added to one rectangle of P. Assume we have added k points to the rectangle S^* and that the result holds. Now add one more point to S^*. The partition we get by adding n points to S^* is what we call the partition Q and the partition we get by adding the $k+1$ point is the partition Q'. The argument above still works and so $L(f,Q',R) \geq L(f,Q,R)$.

Now we use induction again but this time in the case where we add a finite number of points to more than one rectangle. We assume we have added a finite number of points to k rectangles of P. Now add a finite number of points to a $k+1$ rectangle. Call the partition we get from adding a finite number of points to k rectangles S and the new partition we get by adding a finite number of points to the $k+1$ rectangle S'. Then the argument from the previous step works and we have $L(f,S',R) \geq L(f,S,R)$.

Of course, a very similar argument works for the other case and we prove $U(f,P',R) \leq U(f,P,R)$.
■

There are many consequences of this result and it allows us to define Darboux Integration.

Theorem 14.1.3 Lower Darboux Sums are Always less than Upper Darboux Sums

Let P and Q be partitions of A for any choice of bounding rectangle R. Then $L(f,P) \leq U(f,Q)$.

Proof 14.1.3
Let $P \vee Q$ be the common refinement of P and Q. Then, we have

$$L(f,P) \leq L(f,P \vee Q) \leq U(f,P \vee Q) \leq U(f,Q)$$

■

Theorem 14.1.3 immediately implies that for any fixed partition Q

$$\sup_P L(f,P) \leq U(f,Q)$$

Further, since $\sup_P L(f,P)$ is a lower bound for all $U(f,Q)$ no matter what Q is, we see

$$\sup_P L(f,P) \leq \inf_Q U(f,Q)$$

This leads to the lower and upper Darboux integrals.

Definition 14.1.5 The Lower and Upper Darboux Integral

For a bounded $f : A \subset \Re^n \rightarrow \Re$ where A is bounded the Lower Darboux Integral of f over A is

$$\underline{DI}(f, A) = \sup_P \; L(f, P)$$

and the Upper Darboux Integral of f over A is

$$\overline{DI}(f, A) = \inf_P \; U(f, P)$$

Comment 14.1.5 *From the definitions, it is clear $\underline{DI}(f, A) \leq \overline{DI}(f, A)$.*

When the lower and upper Darboux integral match, we obtain the Darboux integral.

Definition 14.1.6 The Darboux Integral

We say the bounded function $f : A \subset \Re^n \rightarrow \Re$ where A is bounded is Darboux Integrable on A if $\underline{DI}(f, A) = \overline{DI}(f, A)$. We denote this common value by $DI(f, A)$.

We eventually can tie this idea to our usual Riemann Integral but first we need to establish some results.

Theorem 14.1.4 The Riemann Criterion for Darboux Integrability

f is Darboux Integrable on A if and only if for all $\epsilon > 0$, there is a partition P_0 so that $U(f, P) - L(f, P) < \epsilon$ for all refinements P of P_0.

Proof 14.1.4

$\Longrightarrow$

Note the choice of bounding rectangle is not important here so we just assume there is one in the background that is used to determine the partitions. Assume f is Darboux Integrable on A. Then $\underline{DI}(f, A) = \overline{DI}(f, A)$. Using the Infimum and Supremum Tolerance Lemma, given $\epsilon > 0$, there are partitions P_ϵ and PQ_ϵ so that

$$\begin{aligned} \overline{DI}(f, A) &\leq U(f, Q_\epsilon) < \overline{DI}(f, A) + \epsilon/2 \\ \underline{DI}(f, A) - \epsilon/2 &< L(f, P_\epsilon) \leq \underline{DI}(f, A) \end{aligned}$$

Let $P_0 = P_\epsilon \vee Q_\epsilon$. Then,

$$U(f, P_0) - L(f, P_0) \leq U(f, Q_\epsilon) - L(f, P_\epsilon) < \overline{DI}(f, A) + \epsilon/2 - \underline{DI}(f, A) + \epsilon/2$$

But f is Darboux Integrable on A and so $\overline{DI}(f, A) - \underline{DI}(f, A) = 0$. Thus, $U(f, P_0) - L(f, P_0) < \epsilon$. Now if P is a refinement of P_0,

$$U(f, P) - L(f, P) \leq U(f, P_0) - L(f, P_0) < \epsilon$$

This completes the proof.

$\Longleftarrow$

Assume $\forall \, \epsilon > 0$, there is a partition P_0 of A so that $U(f, P) - L(f, P) < \epsilon$ for all refinements P of

P_0. Let $\epsilon > 0$ be given. Then there is a partition P_0^ϵ so that $U(f, P_0^\epsilon) < U(f, P_0^\epsilon) + \epsilon$. Hence,

$$\overline{DI}(f, A) = \inf_P U(f, P) \leq U(f, P_0^\epsilon) < L(f, P_0^\epsilon) + \epsilon$$

But then

$$\overline{DI}(f, A) - \epsilon < L(f, P_0^\epsilon) \leq \underline{DI}(f, A)$$

This implies $0 \leq \overline{DI}(f, A) - \underline{DI}(f, A) < \epsilon$. Since $\epsilon > 0$ is arbitrary, we have $\overline{DI}(f, A) = \underline{DI}(f, A)$. Hence f is Darboux Integrable on A. ■

Homework

Exercise 14.1.7 *Let A be the circle of radius 1.5 centered at $(2, 6)$. For*

$$\begin{aligned} P &= \{-1, -0.7, 0, 0.5, 1.1, 1.8, 2.1, 2.7, 3.0\} \\ Q &= \{4, 4, 3, 5, 1, 5, 6, 6, 1, 6, 8, 7.3, 8\} \end{aligned}$$

$\pi = P \times Q$ is a partition of a bounding rectangle containing this circle. Let $f(x, y) = 3x^2 + 2y^2$. Add the point 1.5 to P to create P' giving the refinement $\pi' = P' \times Q$. Find the lower and upper sums associated with this partition. Draw the bounding box, the partition and the circle, and shade all the rectangles S_i that are outside the circle one color; the rectangles inside the circle another color; and the rectangles that intersect the boundary of the circle another color. What effect did adding the extra point have here?

Exercise 14.1.8 *Let A be the circle of radius 1.5 centered at $(2, 6)$. For*

$$\begin{aligned} P &= \{-1, -0.7, 0, 0.5, 1.1, 1.8, 2.1, 2.7, 3.0\} \\ Q &= \{4, 4, 3, 5, 1, 5, 6, 6, 1, 6, 8, 7.3, 8\} \end{aligned}$$

$\pi = P \times Q$ is a partition of a bounding rectangle containing this circle. Let $f(x, y) = 3x^2 + 2y^2$. Add the point 1.5 to P to create P' and add the point 5.8 to Q to create Q' giving the refinement $\pi' = P' \times Q'$. Find the lower and upper sums associated with both the π and π' partition. Draw the bounding box, the partition. and the circle. and shade all the rectangles S_i that are outside the circle one color; the rectangles inside the circle another color; and the rectangles that intersect the boundary of the circle another color for both partitions. What effect did adding the extra two points have here?

Exercise 14.1.9 *Let $f(x, y, z) = c$ for some constant c. What are the lower and upper sums of f for any set A and partition P of a rectangle R containing A?*

Exercise 14.1.10 *Let A be the sphere of radius 1.5 centered at $(2, 6, 0.5)$. For*

$$\begin{aligned} P &= \{-1, -0.7, 0, 0.5, 1.1, 1.8, 2.1, 2.7, 3.0\} \\ Q &= \{4, 4, 3, 5, 1, 5, 6, 6, 1, 6, 8, 7.3, 8\} \\ S &= \{-2, -1.5, -1, 0, 0.6, 1.7, 2\} \end{aligned}$$

$\pi = P \times Q \times S$ is a partition of a bounding cube containing this circle. Let $f(x, y) = 3x^2 + 2y^2 + 4z^2$. Add the point 1.6 to P to create P' and add the point 5.9 to Q to create Q' and the point 0.3 to S to create S' giving the refinement $\pi' = P' \times Q' \times S'$. Find the lower and upper sums associated with both the π and π' partition. Draw the bounding cube, the partition, and the sphere, and shade all the cubes S_i that are outside the sphere one color; the cubes inside the sphere another color; and

the cubes that intersect the boundary of the sphere another color for both partitions. What effect did adding the extra three points have here?

Next, we prove an approximation result. But first a definition.

Definition 14.1.7 The Norm of a Partition

For the partition P of A, P determines N rectangles of form $S_k = \prod_{i=1}^n |a_i^k, b_i^k|$. For each rectangle, we can compute $d_k = \max_{1 \le i \le n} |b_i^k - a_i^k|$. The norm of P is

$$\|P\| = \max_k(\max_{1 \le i \le n} |b_i^k - a_i^k|) = \max_k(d_k)$$

Note, for a given rectangle S_k,

$$\begin{aligned} V(S) &= \prod_{1 \le i \le n} |b_i^k - a_i^k| \le \prod_{1 \le i \le n} d_k \\ &\le (\max_k) d_k^n = \|S\|^n \end{aligned}$$

Comment 14.1.6 *Suppose you were in 2D and you fixed the partition on the x_2 axis but let the partitioning of the x_1 axis get finer and finer. Then we would have $V(S) \to 0$ even though $\|S\|$ would not go to zero as the maximum axis distance is determined by the fixed partition of the x_2 axis.*

Theorem 14.1.5 Approximation of the Darboux Integral

If f is Darboux Integrable on A, then given a sequence of partitions (P_n) with $\|P_n\| \to 0$, we have

$$U(f, P_n) \downarrow DI(f, A), \quad L(f, P_n) \uparrow DI(f, A),$$

Proof 14.1.5
We will only do the upper sum case and leave the other one to you. Let $\epsilon > 0$ be given. Then there is a partition P_ϵ so that $U(f, P_\epsilon) \ge \overline{DI}(f, A)$ and $U(f, P_\epsilon) \le \overline{DI}(f, A) + \epsilon/2$. But f is Darboux integrable on A so we have

$$DI(f, A) \quad geq \quad U(f, P_\epsilon) < DI(f, A, P) + -\epsilon/2$$

Let $Q_n = P_n \vee P_\epsilon$. P_ϵ determines rectangles S_k of the form $\prod_{i=1}^n [a_i^k, b_i^k]$. Let

$$e_k = \min_{1 \le i \le n} (b_i^k - a_i^k), \quad e = \min_k(e_k)$$

For the tolerance $\xi = e/2$, there is an N so that if $n > N$, $\|P_n\| < e/2$. Pick any P_n with $n > N$. Any point in a rectangle S determined by P_ϵ is a corner point. If the rectangle S was given by $\prod[a_i^k, b_i^k]$, then the minimum distance between two points given by S is the minimum edge distance: i.e. $\min_{1 \le i \le n}(b_i^k - a_i^k) = e_k \ge e$. For a rectangle T given by P_n for $n > N$, if it is given by $\prod[c_i, d_i]$, the maximal distance between two points in T is $\max 1 \le i \le n(d_i - c_i) = \|P_n\| < e/2$. So if T contained two points $\boldsymbol{x}$ and $\boldsymbol{y}$ of some rectangle S of P_ϵ, letting $d(\boldsymbol{x}, \boldsymbol{y})$ denote the distance between the points, we would have

$$d(\boldsymbol{x}, \boldsymbol{y}) < e/2 \text{ and } d(\boldsymbol{x}, \boldsymbol{y}) > e$$

This is not possible. Hence, for $n > N$, the rectangles determined by $P_n \vee P_\epsilon$ contain at most one point of P_ϵ.

For a rectangle S containing one point of P_ϵ, we see this single point divides S into 2^n pieces T_k. Then we have a term of this form

$$\sup_{\boldsymbol{x}\in S} f(\boldsymbol{x})\, V(S) \quad - \quad \sum_{k=1}^{2^n} (\sup_{\boldsymbol{x}\in T_k} f(\boldsymbol{x}))\, V(T_j)$$

We can overestimate this term as

$$\begin{aligned}
&\sup_{\boldsymbol{x}\in S} f(\boldsymbol{x})\, V(S) - \sum_{k=1}^{2^n} (\sup_{\boldsymbol{x}\in T_k} f(\boldsymbol{x}))\, V(T_j) \\
&\leq \sup_{\boldsymbol{x}\in S} f(\boldsymbol{x})\, V(S) + \sup_{\boldsymbol{x}\in S} f(\boldsymbol{x}) \sum_{k=1}^{2^n} V(T_j) = 2 \sup_{\boldsymbol{x}\in S} f(\boldsymbol{x})\, V(S)
\end{aligned}$$

Let $B = \sup_{\boldsymbol{x}\in A} |f(\boldsymbol{x})|$. Then, for this rectangle

$$\underset{\boldsymbol{x}\in S}{f(\boldsymbol{x})}\, V(S) - \sum_{k=1}^{2^n} (\sup_{\boldsymbol{x}\in T_k} f(\boldsymbol{x}))\, V(T_j) \quad \leq \quad 2BV(S)$$

We can do this for each point in P_ϵ and its associated rectangle S. Let M be the number of points in P_ϵ. Overestimate each $V(S)$ by

$$V(S) \leq \max_{S \text{ determined by } P_n} V(S)$$

Now consider $U(f, P_n) - U(f, P_n \vee P_\epsilon)$. The rectangles from P_n that do not contain a point of P_ϵ occur in both sums and so these contributions zero out. What is left is the sum over the rectangles that contain points of P_ϵ which we have overestimated in the calculations above. Thus

$$U(f, P_n) - U(f, P_n \vee P_\epsilon) \quad \leq \quad BM \max_{S \text{ determined by } P_n} V(S)$$

We assume $\|P_n\| \to 0$ so we also know $\max_{S \text{ determined by } P_n} V(S) \to 0$. Hence, there is $\hat{N}$ so that

$$\max_{S \text{ determined by } P_n} V(S) \quad \leq \quad \frac{\epsilon}{4M(B+1)}$$

We conclude if $n > \max(N, \hat{N})$, then $U(f, P_n) - U(f, P_n \vee P_\epsilon) < 2BM \frac{\epsilon}{4M(B+1)} \leq \epsilon/2$. We know

$$DI(f, A) \quad \leq \quad U(f, P_n \vee P_\epsilon) \leq U(f, P_\epsilon) < DI(f, A) + \epsilon/2$$

and so

$$\begin{aligned}
U(f, P_n) - DI(f, A) &= U(f, P_n) - U(f, P_n \vee P_\epsilon) + U(f, P_n \vee P_\epsilon) - DI(f, A) \\
&< \epsilon/2 + \epsilon/2 < \epsilon
\end{aligned}$$

if $n > \max(N, \hat{N})$. This shows $U(f, P_n) \downarrow DI(f, A)$. A similar argument shows $L(f, P_n) \uparrow DI(f, A)$. ■

14.1.1 Homework

Exercise 14.1.11 *Let* $f(x,y) = e^{-x^2-y^2}$ *on* $[-1,1] \times [-1,1]$.

- *Write down the lower and upper Darboux sums for uniform partitions.*
- *Write MATLAB code to evaluate these lower and upper Darboux sums.*

Exercise 14.1.12 *Let* $f(x,y) = e^{-x^2-y^2-z^2}$ *on* $[-1,1] \times [-1,1] \times [-1,1]$.

- *Write down the lower and upper Darboux sums for uniform partitions.*
- *Write MATLAB code to evaluate these lower and upper Darboux sums.*

Exercise 14.1.13 *Let* $f(x,y) = e^{-x^2-y^2}$ *on* $x^2 + y^2 \leq 1$ *with bounding rectangle* $[-1,1] \times [-1,1]$.

- *Write down the lower and upper Darboux sums for uniform partitions.*
- *Write MATLAB code to evaluate these lower and upper Darboux sums.*

Exercise 14.1.14 *Let* $f(x,y) = e^{-x^2-y^2-z^2}$ *on* $x^2 + y^2 + z^2 \leq 1$ *with bounding rectangle* $[-1,1] \times [-1,1] \times [-1,1]$.

- *Write down the lower and upper Darboux sums for uniform partitions.*
- *Write MATLAB code to evaluate these lower and upper Darboux sums.*

14.2 The Riemann Integral in n Dimensions

There is another way to develop integration in $\Re^n$. We have gone through this carefully in $\Re$ in (Peterson (21) 2020) and what we do in $\Re^n$ is quite similar.

Let $f : A \subset \Re^n \rightarrow \Re$ be a bounded function on the bounded set A. Let R be any bounding rectangle that contains A and let P be any partition of A based on R. As before, this choice of bounding rectangle is immaterial and we simply choose one that is useful. Let $S_1, \ldots, S_N$ be the rectangles determined by P. An **in between** or **evaluation** set for P is any collection of points $\boldsymbol{z}_1, \ldots, \boldsymbol{z}_N$ where $\boldsymbol{z}_i \in S_i$ for the rectangle S_i determined by P. Such an **evaluation** set is denoted by $\boldsymbol{\sigma}$.

Definition 14.2.1 The Riemann Sum

The Riemann sum of f over A for partition P and evaluation set $\boldsymbol{\sigma}$ *is denoted by* $S(f, \sigma, P)$ *where*

$$S(f, \boldsymbol{\sigma}, P) = \sum_{i=1}^{N} \hat{f}(\boldsymbol{z}_i) V(S_i)$$

where $S_1, \ldots, S_N$ *are the rectangles determined by* P *and* $\boldsymbol{z}_i \in \boldsymbol{\sigma}$. *For convenience, we typically simply write* $S(f, \boldsymbol{\sigma}, P) = \sum_{S_i \in P} f(\boldsymbol{z}_i) V(S_i)$. *It is clear we have*

$$L(f, P) \leq S(f, \boldsymbol{\sigma}, P) \leq U(f, P)$$

The Riemann Integral is then defined like this:

Definition 14.2.2 The Riemann Integral

We say f is Riemann Integrable over A if there is a real number I so that for all $\epsilon > 0$, there is a partition P_ϵ so that

$$|S(f, \boldsymbol{\sigma}, P) - I| \quad < \quad \epsilon$$

for all partitions P that refine P_ϵ and any evaluation set $\boldsymbol{\sigma}$ for P. We typically use a relaxed notation for this and say $\boldsymbol{\sigma} \subset P$ for short. We denote the number I by the symbol $RI(f, A)$.

There is a wonderful connection between the Riemann integral and the Darboux Integral of f over A.

Theorem 14.2.1 The Equivalence of the Riemann and Darboux Integral

The following are true statements:

1. *f is Riemann Integrable on A implies f is Darboux Integrable on A and $RI(f, A) = DI(f, A)$.*
2. *f is Darboux Integrable on A implies f is Riemann Integrable on A and $DI(f, A) = RI(f, A)$.*

Proof 14.2.1
1: *We assume f is Riemann integrable on A. Then there is a number I so that given $\epsilon > 0$, there is a partition P_ϵ with*

$$I - \epsilon/6 \quad < \quad S(f, \boldsymbol{\sigma}, P) < I + \epsilon/6$$

for all refinements P of P_ϵ and $\boldsymbol{\sigma} \subset P$. Let R denote our arbitrary choice of bounding rectangle for A. Then given any rectangle S determined by P_ϵ, the Infimum and Supremum Tolerance lemmas tell us there are points $\boldsymbol{y_S}$ and $\boldsymbol{z_S}$ in S so that

$$\begin{aligned} m_S(f) \quad &\leq \quad f(\boldsymbol{y_S}) < m_S(f) + \frac{\epsilon}{6V(R)} \\ M_S(f) - \frac{\epsilon}{6V(R)} \quad &< \quad f(\boldsymbol{z_S}) \leq M_S(f) \end{aligned}$$

where m_S and M_S are the usual infimum and supremum over S of f.

$$L(f, P_\epsilon) \quad \leq \quad \sum_{S \in P_\epsilon} f(\boldsymbol{y_S}) V(S) < \sum_{S \in P_\epsilon} \left(m_S(f) + \frac{\epsilon}{6V(R)} \right) V(S)$$

or

$$\begin{aligned} L(f, P_\epsilon) \quad &\leq \quad \sum_{S \in P_\epsilon} f(\boldsymbol{y_S}) V(S) < L(f, P_\epsilon) + \frac{\epsilon}{6V(R)} \sum_{S \in P_\epsilon} V(S) \\ &= \quad L(f, P_\epsilon) + \frac{\epsilon}{6V(R)} V(R) = L(f, P_\epsilon) + \frac{\epsilon}{6} \end{aligned}$$

Let $\boldsymbol{\sigma}_{Lower} = \{\boldsymbol{y_S} | S \in P_\epsilon\}$. Then we have

$$L(f, P_\epsilon) \quad \leq \quad S(f, \boldsymbol{\sigma}_{Lower}, P_\epsilon) < L(f, P_\epsilon) + \frac{\epsilon}{6}$$

In a similar way, for $\boldsymbol{\sigma}_{Upper} = \{\boldsymbol{Z_S} | S \in P_\epsilon\}$, we find

$$U(f, P_\epsilon) - \frac{\epsilon}{6} \quad < \quad S(f, \boldsymbol{\sigma}_{Upper}, P_\epsilon) \leq U(f, P_\epsilon)$$

Hence,

$$U(f, P_\epsilon) - L(f, P_\epsilon) \quad < \quad S(f, \boldsymbol{\sigma}_{Upper}, P_\epsilon) - S(f, \boldsymbol{\sigma}_{Lower}, P_\epsilon) + \epsilon/6 + \epsilon/6$$

But we know how close the Riemann sums are to I. We find

$$S(f, \boldsymbol{\sigma}_{Upper}, P_\epsilon) - S(f, \boldsymbol{\sigma}_{Lower}, P_\epsilon) \quad < \quad I + \epsilon/6 - I + \epsilon/6 = \epsilon/3$$

Combining, we conclude

$$U(f, P_\epsilon) - L(f, P_\epsilon) \quad < \quad 2\epsilon/3 < \epsilon$$

This holds for any refinement of P_ϵ. Hence f satisfies the Riemann Criterion and so f is Darboux Integrable on A.

It remains to show $DI(f, A) = RI(f, A)$. Here are the details. Let $\epsilon > 0$ be given. From the definition of $\underline{DI}(f, A)$ and $\overline{DI}(f, A)$ there are partitions U_ϵ and V_ϵ so that

$$\begin{aligned} U(f, U_\epsilon) &< \overline{DI}(f, A) + \epsilon/4 = DI(f, A) + \epsilon/4 \\ L(f, V_\epsilon) &> \underline{DI}(f, A) - \epsilon/4 = DI(f, A) - \epsilon/4 \end{aligned}$$

Thus for the refinement $U_\epsilon \vee V_\epsilon$, we have

$$\begin{aligned} DI(f, A) - \epsilon/4 &< L(f, V_\epsilon) \leq L(f, U_\epsilon \vee V_\epsilon) \leq U(f, U_\epsilon \vee V_\epsilon) \\ &\leq U(f, U_\epsilon) < DI(f, A) + \epsilon/4 \end{aligned}$$

Also, since f is Riemann Integrable, there is a partition P_ϵ so that

$$|S(f, \boldsymbol{\sigma}, P) - RI(f, A)| < \epsilon/4$$

for all refinements P of P_ϵ and any $\boldsymbol{\sigma} \subset P$. Let $Q_\epsilon = U_\epsilon \vee V_\epsilon \vee P_\epsilon$. Then $|S(f, \boldsymbol{\sigma}, Q_\epsilon) - RI(f, A)| < \epsilon/4$ and

$$\begin{aligned} DI(f, A) - \epsilon/4 &< L(f, U_\epsilon \vee V_\epsilon) \leq L(f, Q_\epsilon) \leq S(f, \boldsymbol{\sigma}, Q_\epsilon) \\ &\leq U(f, Q_\epsilon) \leq U(f, U_\epsilon \vee V_\epsilon) < DI(f, A) + \epsilon/4 \end{aligned}$$

So

$$\begin{aligned} |RI(f, A) - DI(f, a)| &\leq |RI(f, A) - S(f, \boldsymbol{\sigma}, Q_\epsilon)| + |S(f, \boldsymbol{\sigma}, Q_\epsilon) - DI(f, a)| \\ &< \epsilon/4 + \epsilon/4 < \epsilon \end{aligned}$$

Since $\epsilon > 0$ is arbitrary, we see $RI(f, A) = DI(f, A)$.

2: *We now assume f is Darboux Integrable on A. Then, just as we argued in the last part of the proof of* **1**, *we see there is a partition $U_\epsilon \vee V_\epsilon$ so that*

$$\begin{aligned} DI(f, A) - \epsilon &< L(f, U_\epsilon \vee V_\epsilon) \leq L(f, Q) \\ &\leq S(f, \boldsymbol{\sigma}, Q) \leq U(f, Q) \\ &\leq U(f, U_\epsilon \vee V_\epsilon) < DI(f, A) + \epsilon \end{aligned}$$

for any refinement Q of $U_\epsilon \vee V_\epsilon$ and $\boldsymbol{\sigma} \subset Q$. We conclude $|S(f, \boldsymbol{\sigma}, Q) - DI(f, A)| < \epsilon$ for any refinement Q of $U_\epsilon \vee V_\epsilon$ and $\boldsymbol{\sigma} \subset Q$. This says f is Riemann Integrable on A with value $RI(f, A) = DI(f, A)$ ■

We can also prove the usual results about Riemann Integrability.

Theorem 14.2.2 The Riemann Integral is Linear

Let f and g be integrable on $\Omega \subset in\Re^n$. Then $\alpha f + \beta g$ is also integrable on Ω. Moreover $\int_\Omega (\alpha f + \beta g) dA = \alpha \int_\Omega f dA + \beta \int_| omegag dA$.

Proof 14.2.2
This proof is for you. ■

Next, we prove the approximation result for Riemann Integration.

Theorem 14.2.3 Approximation of the Riemann Integral

If f is Riemann Integrable on A, then given a sequence of partitions (P_n) with $\|P_n\| \to 0$, we have $S(f, \boldsymbol{\sigma_n}, P_n) \to RI(f, A)$ for any sequence $\boldsymbol{\sigma_n} \subset P_n$.

Proof 14.2.3
We know $L(f, P_n) \leq S(f, \boldsymbol{\sigma_n}, P_n) \leq U(f, P_n)$. Since f is also Darboux Integrable by the equivalence theorem, we know $DI(f, A) = RI(f, A)$ and $U(f, P_n) \downarrow DI(f, A)$ and $L(f, P_n) \uparrow DI(f, A)$. So $S(f, \boldsymbol{\sigma_n}, P_n) \to DI(f, A) = RI(f, A)$. ■

14.2.1 Homework

Exercise 14.2.1 *Prove Theorem 14.2.2.*

Exercise 14.2.2 *If f and g are integrable on $\Omega \subset \Re^n$, use an $\epsilon - \pi$ argument to prove $2f + 6g$ is also integrable on Ω.*

Exercise 14.2.3 *If f and g are integrable on $\Omega \subset \Re^n$, use an $\epsilon - \pi$ argument to prove $-3f + 5g$ is also integrable on Ω.*

Exercise 14.2.4 *We can prove $f(x, y) = e^{-x^2-y^2}$ on $\Omega = [-1, 1] \times [-1, 1]$ is Riemann Integrable. Write MATLAB code to approximate $\int_\Omega f$ using uniform partitions for a specified partition norm.*

Exercise 14.2.5 *We can prove $f(x, y) = e^{-x^2-y^2-z^2}$ on $\Omega = [-1, 1] \times [-1, 1] \times [-1, 1]$ is Riemann Integrable. Write MATLAB code to approximate $\int_\Omega f$ using uniform partitions for a specified partition norm.*

Exercise 14.2.6 *We can prove $f(x, y) = e^{-x^2-y^2}$ on $\Omega = \{(x, y) | x^2 + y^2 \leq 1\}$ with bounding rectangle on $[-1, 1] \times [-1, 1]$ is Riemann Integrable. Write MATLAB code to approximate $\int_\Omega f$ using uniform partitions for a specified partition norm.*

Exercise 14.2.7 *We can prove $f(x, y) = 2x^2 + 4y^2$ on $\Omega = [-1, 1] \times [-1, 1]$ is Riemann Integrable. Write MATLAB code to approximate $\int_\Omega f$ using uniform partitions for a specified partition norm.*

Exercise 14.2.8 *We can prove $f(x, y) = 2x^2 + 4y^2$ on $\Omega = \{(x, y) | x^2 + y^2 \leq 1\}$ with bounding rectangle $[-1, 1] \times [-1, 1]$ is Riemann Integrable. Write MATLAB code to approximate $\int_\Omega f$ using uniform partitions for a specified partition norm.*

14.3 Volume Zero and Measure Zero

Now that we have moved explicitly into $\Re^n$, you should see the idea of length, area and volume and so forth seem hazily defined. In (Peterson (21) 2020), one of the projects is the construction of Cantor type sets and the *size* of these sets can only be approached abstractly through the content of a set. We also developed a version of Lebesgue's Theorem which tells us a function is Riemann integrable if and only if its set of discontinuities is of content zero. Now we want to extend this discussion to more dimensions. Let's go back to the beginning first.

14.3.1 Measure Zero

What is the length of a single point? Note given any positive integer n, $x \in \overline{B(1/n, x)} = [x - 1/n, x + 1/n]$ and this interval has length $2/n$. Since n is arbitrary, it seems reasonable to define the length of $\{x\}$ to be

$$\text{length}\ \{x\} = \lim_{n \to \infty} 2/n = 0$$

What if we had a finite number of distinct points $\{x_1, \ldots, x_p\}$? Since the points are distinct, there is a N_0 so that

$$x_1 \in [x_1 - 1/n, x_1 + 1/n], \quad x_2 \in [x_2 - 1/n, x_2 + 1/n],$$

$$\vdots x_p \in [x_p - 1/n, x_p + 1/n]$$

with all these intervals disjoint when $n > N_0$. Hence,

$$\text{length}\ \{x_1, \ldots, x_p\} \leq 2/n + \ldots 2/n = 2p/n$$

and as $n \to 0$, we have length $\{x_1, \ldots, x_p\} = 0$. Let's try and make this more precise for sets in $\Re$ and for sets in $\Re^n$ also. Also, since the notion of *length* is not quite a normal length, let's start using the term **measure** instead.

Definition 14.3.1 Sets of Measure Zero

A set $A \subset \Re^n$ is said to have measure zero if for all $\epsilon > 0$, there is a collection of rectangles (S_i), countable or infinite, so that $A \subset \cup S_i$ with $\sum V(S_i) < \epsilon$.

Using this definition, we see measure $\{x\} = 0$ for any $x \in \Re$ and measure$\{x_1, \ldots, x_p\} = 0$ for any finite number of points in $\Re$. Here are some more examples.

Example 14.3.1 *$\mathbb{Q}$ is countable, so it can be enumerated as $\mathbb{Q} = (r_i)_{1 \leq i < \infty}$. Then, $r_i \in [r_1 - \epsilon/2^i, r_i + \epsilon/2^i]$ and $\mathbb{Q} \subset \cup[r_1 - \epsilon/2^i, r_i + \epsilon/2^i]$ with $\sum_{i=1}^{\infty} \epsilon/2^{i-1} = \epsilon/2 < \epsilon$. Hence measure $\mathbb{Q} = 0$.*

Example 14.3.2 *The line $y = mx + b$ has measure zero in $\Re^2$. We will show this a bit indirectly. First consider the segment of the line for $0 \leq x \leq 1$. Divide $[0, 1]$ into n pieces. Pick n boxes as shown in Figure 14.4 so that they overlap. The distance between $(i/n, m(i/n) + b)$ and $((i+1)/n, m((i+1)/n) + b)$ is*

$$d(i/n, m(i/n) + b), ((i+1)/n, m(i+1)/n + b) = \sqrt{(1/n)^2 + (m/n)^2} = \sqrt{(m^2+1)/n^2}$$

So if we use boxes as shown in Figure 14.5, the boxes overlap. We see the line segment of $y = mx+b$ *for* $0 \le x \le 1$ *is contained in* $\cup_{i=0}^{n}$ *Box*$_i$ *and*

$$\sum_{i=1}^{n} V(\,Box_i) \;=\; \sum_{i=0}^{n} \frac{4(1+m^2)}{n^2} = \frac{n+1}{n^2}\,4(1+m^2)$$

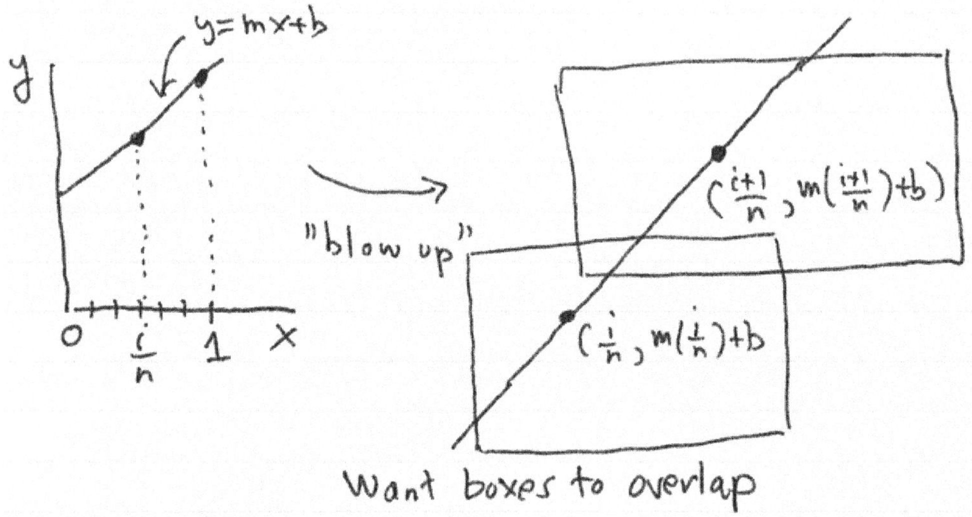

Figure 14.4: Covering boxes for $y = mx + b$, $0 \le x \le 1$.

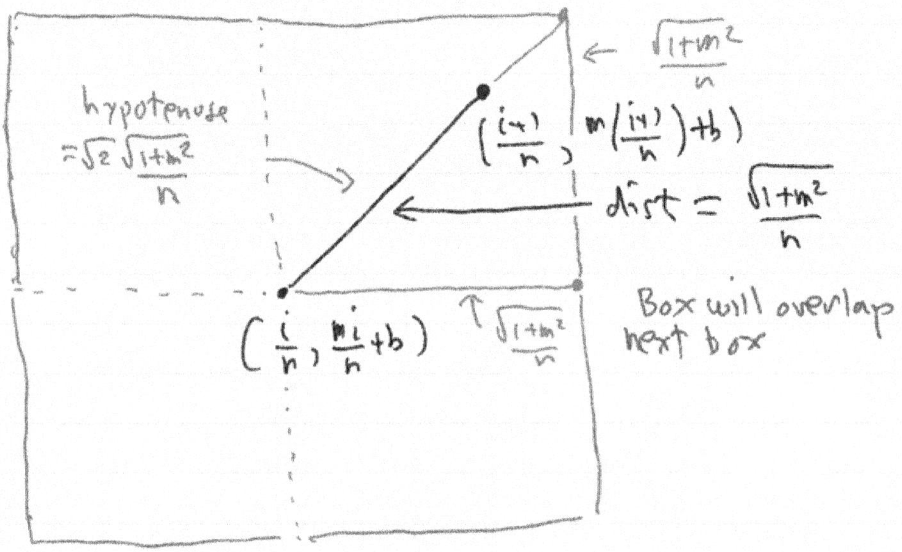

Figure 14.5: Overlapping covering boxes for $y = mx + b$, $0 \le x \le 1$.

This goes to zero as $n \to \infty$, *So the line segment of* $y = mx + b$ *for* $0 \leq x \leq 1$ *has measure zero.*

We can do this for $y = mx + b$ *for* $1 \leq x \leq 2$, $y = mx + b$ *for* $2 \leq x \leq 3$ *and so on and each of these line segments has measure zero. Now*

$$(\textit{line segment for } y = mx + b, x \geq 0) = \cup_{p=1}^{\infty}(\textit{line segment for } y = mx + b,\ p - 1 \leq x \leq p)$$

Let $E_p = (\textit{line segment for } y = mx + b,\ p - 1 \leq x \leq p)$. *Then*

$$(\textit{line segment for } y = mx + b, x \geq 0) = \cup_{p=1}^{\infty} E_p$$

Since each E_p *has measure zero, given* $\epsilon > 0$, *there is a collection of rectangles* $\{S_{i,p}\}$ *so that*

$$E_p \subset \cup S_{i,p}, \quad \sum V(S_{i,p}) < \epsilon/2^p$$

Hence,

$$(\textit{line segment for } y = mx + b, x \geq 0) \subset \cup_p \cup_i S_{ip}, \quad \sum_p \sum_i V(S_{ip}) < \sum_p \epsilon/2^p < \epsilon$$

This shows the measure $(\textit{line segment for } y = mx + b, x \geq 0) = 0$. *A similar argument show the same result for the case* $y = mx + b, x \leq 0$. *From this, it follows the measure*$(y = mx + b) = 0$.

The examples above indicate we can prove the following propositions.

Theorem 14.3.1 Finite Unions of Sets of Measure Zero Also Have Measure Zero

If $\{A_1, \ldots, A_p\}$ *is a finite collection of sets of measure zero, then* $\cup_{i=1}^{p} A_i$ *has measure zero also.*

Proof 14.3.1
This is left to you. ■

Theorem 14.3.2 Countable Unions of Sets of Measure Zero are Also Measure Zero

If $\{A_i | i \geq 1\}$ *is an infinite collection of sets of measure zero, then* $\cup_{i=1}^{\infty} A_i$ *has measure zero also.*

Proof 14.3.2
Again, this is left for you. ■

It is time for a harder example.

Example 14.3.3 *Consider the surface* $z = x^2 + y^2$ *in* $\Re^3$. *That is,* $f(x, y) = x^2 + y^2$. *We will show this is a set of measure zero in* $\Re^3$. *Let's consider only a finite piece of it in octant one. Note the projection of the surface to the* $x - y$ *plane is the circle* $x^2 + y^2 = 1$ *for* $z = 1$. *Hence, the square* $[-1, 1] \times [-1, 1]$ *contains the projection. Divide the square into* n^2 *uniform pieces of area* $1/n^2$. *These induce a corresponding subdivision of the disc* $\{(x, y)|x^2 + y^2 \leq 1\}$. *The corners of a patch* P_{ij} *in the* $x - y$ *plane are*

$$P_{i,j} = \begin{bmatrix} (i/n, (j+1)/m) & \cdots & ((i+1)/n, (j+1)/m) \\ \vdots & \vdots & \vdots \\ (i/n, j/n) & \cdots & ((i+1)/n, j/n) \end{bmatrix}$$

$$= \begin{bmatrix} (x_i, y_{j+1}) & \cdots & (x_{i+1}, y_{j+1}) \\ \vdots & \vdots & \vdots \\ (x_i, y_j) & \cdots & (x+i+1, y_j) \end{bmatrix}$$

and this square is mapped to a patch with corners

$$f(P_{i,j}) = \begin{bmatrix} z_{i,j+1} & \cdots & z_{i+1,j+1} \\ \vdots & \vdots & \vdots \\ z_{i,j} & \cdots & z_{i+1,j} \end{bmatrix}$$

where $z_{pq} = x_p^2 + y_q^2$*. Look at the square* $[0,1] \times [0,1]$*. Now* $z = x^2 + y^2$ *is a convex surface, the maximum occurs at the largest* i *and* j *for the patch, and the minimum is at the smallest* i *and* j*. So on the square* $P_{i,j}$*, the maximum and minimum surface values are*

$$\begin{aligned} z_{M,i,j} &= x_{i+1}^2 + y_{j+1}^2 = \left(\frac{i+1}{n}\right)^2 + \left(\frac{j+1}{n}\right)^2 \\ z_{m.i.j} &= x_i^2 + y_j^2 = \left(\frac{i}{n}\right)^2 + \left(\frac{j}{n}\right)^2 \end{aligned}$$

Thus,

$$Z_{M.i.j} - z_{m,i,j} = \frac{2i+1}{n^2} + \frac{2j+1}{n^2}$$

This patch of surface is contained in the box $B_{i,j} = P_{i,j} \times [z_{i,j} - 1/n, z_{M,i,j} + 1/n]$ *which has volume*

$$V(B_{i,j}) = \frac{1}{n^2}(z_{M,i,j} - z_{m,i,j} + \frac{2}{n}) = \frac{2}{n^3} + \frac{2i+1+2j+1}{n^4}$$

The surface above $[0,1] \times [0,1]$ *is contained in the union of the boxes* $B_{i,j}$ *and*

$$\begin{aligned} \sum_{i=1}^{n}\sum_{j=1}^{n} V(B_{i,j}) &= \sum_{i=1}^{n}\sum_{j=1}^{n} \frac{2}{n^3} + \sum_{i=1}^{n}\sum_{j=1}^{n} \frac{2i+2j+2}{n^4} \\ &= \frac{2}{n} + \sum_{i=1}^{n}\left(\frac{2i}{n^3} + \frac{2n(n+1)}{2n^4} + \frac{2n}{n^4}\right) \\ &= \frac{2}{n} + \frac{2n(n+1)}{2n^3} + \frac{2n^2(n+1)}{2n^4} + \frac{2n^2}{n^4} \end{aligned}$$

We see $\sum_{i=1}^{n}\sum_{j=1}^{n} V(B_{i,j}) \to 0$ *as* $n \to \infty$*. Thus, we conclude the measure of the surface above* $[0,1] \times [0,1]$ *is 0.*

A similar argument shows the measure of the surface above $[0,1] \times [-1,0]$*, above* $[-1,0] \times [0,1]$ *and above* $[-1,0] \times [-1,0]$ *are zero too. Hence the measure of the surface above* $[-1,1] \times [-1,1]$ *is zero. Let* Ω_n *be the surface above the annular region* $\{(x,y) | n^2 \leq x^2 + y^2 \leq (n+1)^2\}$*. Thus,* $n^2 \leq z \leq (n+1)^2$*. Using an argument similar to what we just did, we can show the measure of the surface above this annular region is zero. Note* Ω_0 *is contained in the surface above* $[-1,1] \times [-1,1]$ *and so the measure of* Ω_0 *is zero. The entire surface is the union of all the* Ω_n *for* $n \geq 0$ *and since each of these are measure zero, the entire surface in* $\Re^2$ *is measure zero.*

Homework

Exercise 14.3.1 *Prove Theorem 14.3.1*

Exercise 14.3.2 *Prove Theorem 14.3.2*

Exercise 14.3.3 *Prove* $y = x^2$ *has measure zero in* $\Re^2$.

Exercise 14.3.4 *Prove the surface* $z = 2x^2 + 3y^2$ *has measure zero in* $\Re^3$.

Exercise 14.3.5 *Prove* $y = 2x + 5$ *has measure zero in* $\Re^2$.

14.3.2 Volume Zero

Do all sets $A \in \Re^n$ have associated with them a generalization of volume for a box? Let's see if we can define a suitable notion. The volume of a rectangle in $\Re^n$, $R = \prod_{i=1}^n [a_i^k, b_i^k]$ is given by $V(R) = \prod_{i=1}^n (b_i^k - a_i^k)$. To generalize this idea, let's consider **characteristic functions**.

Definition 14.3.2 The Characteristic Function of a Set

If $A \subset \Re^n$, *the characterization of A is denoted by* $\mathbf{1_A}$ *and defined by*

$$\mathbf{1_A}(\boldsymbol{x}) = \begin{cases} 1, & \boldsymbol{x} \in A \\ 0, & \boldsymbol{x} \notin A \end{cases}$$

Let A be a rectangle in $\Re^n$. Then A is its own bounding rectangle. Let P be any partition of A with associated rectangles S_i. Then, $\inf_{\boldsymbol{x} \in S_i} \mathbf{1_A}(\boldsymbol{x}) = 1$ and $\sup_{\boldsymbol{x} \in S_i} \mathbf{1_A}(\boldsymbol{x}) = 1$ also. Thus,

$$\begin{aligned} L(\mathbf{1_A}, P) &= \sum_{S_i \in P} 1\, V(S_i) = V(A) \\ U(\mathbf{1_A}, P) &= \sum_{S_i \in P} 1\, V(S_i) = V(A) \end{aligned}$$

Thus, $\mathbf{1_A}$ is Darboux Integrable implying Riemann Integrable over A with value $V(A)$.

It is time to use a different notation for the value of our integrals.

Definition 14.3.3 The Integration Symbol

If $f : A \subset \Re^n \rightarrow \Re$ *for bounded A and f bounded is Riemann Integrable on A, the value of the Riemann Integral of f on A will be denoted by* $\int_A f$. *We sometimes want to remind ourselves explicitly that the integration is done over a subset of* $\Re^n$. *Since lower and upper sums involve volumes of rectangles in* $\Re^n$, *we will use the symbol* $\boldsymbol{dV_n}$ *to indicate this in the integration symbol. Hence, we can also write* $RI(f, A) = \int_A f \boldsymbol{dV_n}$.

Comment 14.3.1 *Our traditional one calculus integral would be* $\int_a^b f(x)dx$ *for the Riemann Integrable of the bounded function f on* $[a, b]$. *We could also write this as* $\int_{[a,b]} f\boldsymbol{dV_1}$ *which, of course, is a bit much. Just remember there are many ways to represent things. The new notation is better for an arbitrary number of dimensions n and not so great for* $n = 1$.

Using this definition, we see $V(A) = \int_A \mathbf{1_A}$ when A is a rectangle in $\Re^n$. This suggests a way to extend the idea of volume to some sets A of $\Re^n$.

Definition 14.3.4 The Volume of a Set

If A is a bounded set in $\Re^n$ and $\mathbf{1_A}$ is integrable, then the volume of A is defined to be $V(A) = \int_A \mathbf{1_A}$. As usual, the function $\mathbf{1_A}$ is extended to any bounded rectangle R by

$$\hat{\mathbf{1_a}} = \begin{cases} \mathbf{1_A}(\boldsymbol{x}), \boldsymbol{x} \in A \\ 0, \quad\quad \boldsymbol{x} \in R \setminus A \end{cases}$$

It is easy to show that if a set has volume zero, it is also has measure zero.

Theorem 14.3.3 Volume Zero Implies Measure Zero

If $A \subset \Re^n$ has volume zero, it is also measure zero.

Proof 14.3.3
Since A has volume zero, given $\epsilon > 0$, there is a partition P_ϵ so the $0 \leq U(\mathbf{1_A}, P_\epsilon) < \epsilon$ (remember all the lower sums here are zero!). But $U(\boldsymbol{A_A}, P_\epsilon) = \sum_{S_i \in P_\epsilon} V(S_i)$, so $A \subset \cup_{S_i \in P_\epsilon} S_i$ and $\sum_{S_i \in P_\epsilon} V(S_i) < \epsilon$. Since $\epsilon > 0$ is arbitrary, this shows A is measure zero. ■

The converse is not true. Let $A = \{(x, y, z) \in [0, 1] \times [0, 1] \times [0, 1] | x, y, z \in \mathbb{Q}\}$. Then

$$\mathbf{1_A}(\boldsymbol{x}) = \begin{cases} 1, & x, y, z \in \mathbb{Q} \cap [0, 1] \\ 0, & \text{otherwise} \end{cases}$$

Let the bounding rectangle be $R = [0, 1] \times [0, 1] \times [0, 1]$. Then for any partition P,

$$\begin{aligned} L(\mathbf{1_A}, P) &= \sum_{S_i \in P} 0\, V(S_i) = 0 \\ U(\mathbf{1_A}, P) &= \sum_{S_i \in P} 1\, V(S_i) = V(R) = 1 \end{aligned}$$

This is because any $S_i \in P$ contains rational triples. Thus, $\underline{DI}(\mathbf{1_A}) = 0$ and $\overline{DI}(\mathbf{1_A}) = 1$. We conclude $\mathbf{1_A}$ is not integrable and so its volume is not defined. However, the rational triples here are countable and so have measure zero.

Homework

Exercise 14.3.6 *Let $\Omega = [-1, 2] \times [2, 4]$. Compute $V(\Omega) = \int_\Omega \mathbf{1}_\Omega$. Note this is the usual area of this rectangle.*

Exercise 14.3.7 *Let $\Omega = [-1, 2] \times [2, 4] \times [-1, 2]$. Compute $V(\Omega) = \int_\Omega \mathbf{1}_\Omega$. Note this is the usual volume of this box.*

Exercise 14.3.8 *Let $\Omega = [-1, 2]$. Compute $V(\Omega) = \int_\Omega \mathbf{1}_\Omega$. Note this is the usual length of this interval.*

14.4 When is a Function Riemann Integrable?

We are now ready to consider the question of when a function is Riemann Integrable on A. The result we need is called Lebesgue's Theorem and we already have proven a version of this in (Peterson (21) 2020). But now we want to look at the situation in $\Re^n$ for $n > 1$ in general. First, a new way to look at continuity.

Definition 14.4.1 The Oscillation of a Function

Let $f : A \subset \Re^n \to \Re$ where A is an open set. The oscillation of f at $\boldsymbol{x_0} \in A$ is defined to be

$$\omega(f, \boldsymbol{x_0}) = \inf_{B(r,\boldsymbol{x_0}),r>0} \left(\sup_{\boldsymbol{x_1},\boldsymbol{x_2} \in B(r,\boldsymbol{x_0})} |f(\boldsymbol{x_1}) - f(\boldsymbol{x_2})| \right)$$

We leave it to you to prove this result.

Theorem 14.4.1 The Oscillation is Zero if and only f is Continuous

Let $f : A \subset \Re^n \to \Re$ where A is an open set. Then $\omega(f, \boldsymbol{x_0}) = 0$ if and only if f is continuous at $\boldsymbol{x_0}$.

Proof 14.4.1
This one's for you! ■

Now on to Lebesgue's Theorem.

Theorem 14.4.2 Lebesgue's Theorem

Let $f : A \subset \Re^n \to \Re$, A bounded and f bounded on A. Then f is Riemann Integrable on A if and only the set of points where $\hat{f}$ is not continuous is a set of measure zero.

Proof 14.4.2
$\Longrightarrow$
We assume the set of discontinuities of $\hat{f}$ has measure zero. Call this set D. For each $\epsilon > 0$, let $D_\epsilon = \{\boldsymbol{x} \mid \omega(\hat{f}, \boldsymbol{x}) \geq \epsilon\}$. Then $D = \cup_{\epsilon>0} D_\epsilon$ and each $D_\epsilon \subset D$ by Theorem 14.4.1. Assume $\boldsymbol{y}$ is a limit point of D_ϵ. If $\boldsymbol{y}$ is an isolated limit point, it is already in D_ϵ. If it is not isolated, there is a sequence $(\boldsymbol{y_n}) \subset D_\epsilon$ with $\boldsymbol{y_n} \neq \boldsymbol{y}$ so that $\boldsymbol{y_n} \to \boldsymbol{y}$. For any $r > 0$, since $\boldsymbol{y_n} \to \boldsymbol{y}$, there is N_r so that $\boldsymbol{y_n} \in B(r, \boldsymbol{y})$. Thus, there is an $s < r$ so that $B(s, \boldsymbol{y_n}) \subset B(r, \boldsymbol{y})$, i.e. $\boldsymbol{y} \in B(s, \boldsymbol{y_n})$ for any $n > N_r$. But then

$$\sup_{\boldsymbol{x_1},\boldsymbol{x_2} \in B(r,\boldsymbol{y})} |\hat{f}(\boldsymbol{x_1}) - \hat{f}(\boldsymbol{x_2})| \geq \sup_{\boldsymbol{x_1},\boldsymbol{x_2} \in B(s,\boldsymbol{y_n})} |\hat{f}(\boldsymbol{x_1}) - \hat{f}(\boldsymbol{x_2})| \geq \epsilon$$

because $\boldsymbol{y_n} \in D_\epsilon$. Since this is true for all r, we have

$$\inf_{B(r,\boldsymbol{x_0}),r>0} \left(\sup_{\boldsymbol{x_1},\boldsymbol{x_2} \in B(r,\boldsymbol{y})} |\hat{f}(\boldsymbol{x_1}) - \hat{f}(\boldsymbol{x_2})| \right) \geq \epsilon$$

which tells us $\omega(\hat{f}, \boldsymbol{y}) \geq \epsilon$ or $\boldsymbol{y} \in D_\epsilon$. Thus, D_ϵ is closed. Since it is inside the bounded set A, it is also bounded. We conclude D_ϵ is compact.

We know D has measure zero. thus, D_ϵ had measure zero as well. Then, there is a collection B_i of open rectangles (the definition uses closed rectangles, but it is easy to expand them a bit to make them open rectangles) so $D_\epsilon \subset \cup_i B_i$ and $\sum_i V(\overline{B_i}) < \epsilon$ where recall $\overline{B_i}$ is the closure of B_i. Since D_ϵ is topologically compact, the open cover (B_i) has a finite subcover $\{B_{i_1}, \ldots, B_{i_p}\}$ which we will call $\{B_1, \ldots, B_N\}$ by adding in any missing sets B_i as required with $N = i_p$. Hence, $D_\epsilon \subset \cup_{i=1}^N B_i$ and $\sum_{i=1}^N V(\overline{B_i}) < \epsilon$.

The collection $\{B_1, \ldots, B_N\}$ of possibly overlapping rectangles determines a partition P_0 of the bounding rectangle we choose for A. Pick a partition P which refines P_0. The new rectangles we form will either be inside the rectangles already made by P_0 or completely outside them. Thus, the rectangles S_i determined by P can be divided into two pieces with index sets I and J.

$$\begin{aligned} I &= \{i \in \{1, \ldots, N\} \mid \exists\, j \ni S_i \subset B_j\} \\ J &= \{i \in \{1, \ldots, N\} \mid S_i \subset (\cup_{j=1}^{n} B_j)^C\} \end{aligned}$$

Note, for each $j \in J$, S_j is disjoint from D_ϵ. Thus, if $\boldsymbol{z} \in S_j$, $j \in J$, $\omega(\hat{f}, \boldsymbol{z}) < \epsilon$. Since $\omega(\hat{f}, \boldsymbol{z})$ is defined using an infimum, there is $r_z > 0$ so that

$$\omega(\hat{f}, \boldsymbol{z}) \leq \sup_{\boldsymbol{x_1}, \boldsymbol{x_2} \in B(r_z, \boldsymbol{z})} |\hat{f}(\boldsymbol{x_1}) - \hat{f}(\boldsymbol{x_2})| < (1/2)(\omega(\hat{f}, \boldsymbol{z}) + \epsilon)$$

It is easier to use rectangles so let's switch to rectangles. Let $R^\circ(r_z, \boldsymbol{z})$ be an open rectangle contained in $B(r_z, \boldsymbol{z})$. Then we can say

$$\omega(\hat{f}, \boldsymbol{z}) \leq \sup_{\boldsymbol{x_1}, \boldsymbol{x_2} \in R^\circ(r_z, \boldsymbol{z})} |\hat{f}(\boldsymbol{x_1}) - \hat{f}(\boldsymbol{x_2})| < (1/2)(\omega(\hat{f}, \boldsymbol{z}) + \epsilon)$$

Let $\hat{\epsilon} = (1/2)(\omega(\hat{f}, \boldsymbol{z}) + \epsilon) < \epsilon$. Then, we have

$$-\hat{\epsilon} < \hat{f}(\boldsymbol{x_1}) - \hat{f}(\boldsymbol{x_2}) < \hat{\epsilon}, \quad \forall \boldsymbol{x_1}, \boldsymbol{x_2} \in R^\circ(r_z, \boldsymbol{z})$$

or

$$\hat{f}(\boldsymbol{x_2}) - \hat{\epsilon} < \sup_{\boldsymbol{x_1} \in R^\circ(r_z, \boldsymbol{z})} \hat{f}(\boldsymbol{x_1}) < \hat{f}(\boldsymbol{x_2}) + \hat{\epsilon}, \quad \forall \boldsymbol{x_2} \in R^\circ(r_z, \boldsymbol{z})$$

implying

$$\sup_{\boldsymbol{x_1} \in R^\circ(r_z, \boldsymbol{z})} \hat{f}(\boldsymbol{x_1}) \leq \inf_{\boldsymbol{x_2} \in R^\circ(r_z, \boldsymbol{z})} \hat{f}(\boldsymbol{x_2}) + \hat{\epsilon}$$

We conclude

$$\sup_{\boldsymbol{x_1} \in R^\circ(r_z, \boldsymbol{z})} \hat{f}(\boldsymbol{x_1}) - \inf_{\boldsymbol{x_2} \in R^\circ(r_z, \boldsymbol{z})} \hat{f}(\boldsymbol{x_2}) \leq \hat{\epsilon} < \epsilon$$

We know $S_j \subset \cup_{\boldsymbol{z} \in S_j} R^\circ(r_z, \boldsymbol{z})$ and since S_j is compact, there is a finite subcover

$$\{R^\circ(r_{z_1}, \boldsymbol{z_1}), \ldots, R^\circ(r_{z_q}, \boldsymbol{z_q})\}, \qquad S_j \subset \cup_{i=1}^{q} R^\circ(r_{z_i}, \boldsymbol{z_i})$$

The rectangles in the subcover define a partitioning scheme inside S_j. Choose any additional refinement of P so that the new rectangles are all inside some $R^\circ(r_{x_i}, \boldsymbol{z_i})$. Call this refinement P'. So for any rectangle E due to this new refinement of P, since it is contained in one of the sets in the subcover, we have

$$\sup_{\boldsymbol{x_1} \in E} \hat{f}(\boldsymbol{x_1}) - \inf_{\boldsymbol{x_2} \in E} \hat{f}(\boldsymbol{x_2}) \leq \hat{\epsilon} < \epsilon$$

Thus,

$$\sum_{E \in P'} \sup_{\boldsymbol{x} \in E} V(E) - \sum_{E \in P'} \inf_{\boldsymbol{x} \in E} V(E) \leq \hat{\epsilon} \sum_{E \in P'} V(E) = \hat{\epsilon} V(S_j) < \epsilon V(S_j)$$

We can carry out this process of refinement for each $j \in J$ to get a succession of refinements: $P \to P'_1 \to P'_2 \to \ldots \to P'_M$ where M is the number of indices in J. This gives a partition Q. The indices of Q still split into two pieces I and J as described, but the additional refinements we have introduced allow us to make some estimates for each $S_j \in J$. We can say

$$\begin{aligned} U(\hat{f}, Q) - L(\hat{f}, P) &= \sum_{E_i, i \in I} (\sup_{\boldsymbol{x} \in E_i} \hat{f}(\boldsymbol{x}))\, V(S_i) - \sum_{E_i, i \in I} (\inf_{\boldsymbol{x} \in E_i} \hat{f}(\boldsymbol{x}))\, V(S_i) \\ &+ \sum_{E_i, i \in J} (\sup_{\boldsymbol{x} \in E_i} \hat{f}(\boldsymbol{x}))\, V(S_i) - \sum_{E_J, i \in I} (\inf_{\boldsymbol{x} \in E_i} \hat{f}(\boldsymbol{x}))\, V(S_i) \\ &< \sum_{E_i, i \in I} (\sup_{\boldsymbol{x} \in E_i} \hat{f}(\boldsymbol{x}))\, V(S_i) - \sum_{E_i, i \in I} (\inf_{\boldsymbol{x} \in E_i} \hat{f}(\boldsymbol{x}))\, V(S_i) + \epsilon \sum_{E_i, i \in J} V(S_i) \\ &= \sum_{E_i, i \in I} (\sup_{\boldsymbol{x} \in E_i} \hat{f}(\boldsymbol{x}) - \inf_{\boldsymbol{x} \in E_i} \hat{f}(\boldsymbol{x}))\, V(S_i) + \epsilon V(R) \end{aligned}$$

where R is the bounding rectangle for A. Now f is bounded on A so there is a constant M so that $\hat{f}| < M$ on R. Thus,

$$U(\hat{f}, Q) - L(\hat{f}, Q) \quad < \quad 2M \sum_{E_i, i \in I} V(S_i) + \epsilon V(R)$$

But for each index $i \in I$, S_i is contained in some B_j and so $\cup_{i \in I} S_i \subset \cup_{j=1}^{N} B_i$ and $\sum_{i=1}^{N} V(B_j) < \epsilon$. We conclude

$$U(\hat{f}, Q) - L(\hat{f}, Q) \quad < \quad 2M\epsilon + \epsilon V(R)$$

This is enough to show f satisfies the Riemann Criterion and so f is integrable on A.

$\Longrightarrow$

Now we assume f is integrable on A. The set of discontinuities of $\hat{f}$ is the set

$$D \quad = \quad \{\boldsymbol{x} \in \Re^n | \omega(\hat{f}, \boldsymbol{x}) \neq 0\}$$

We also know $D = \cup_n D_{1/n}$ where $D_{1/n}$ is defined as in the first part of the proof. Since f is integrable, given $\xi > 0$, there is a partition P_ξ so that

$$U(\hat{f}, P_\xi) - L(\hat{f}, P_\xi) \quad = \quad \sum_{S_i \in P_\xi} (\sup_{\boldsymbol{x} \in S_i} \hat{f}(\boldsymbol{x}) - \inf_{\boldsymbol{x} \in S_i} \hat{f}(\boldsymbol{x}))\, V(S_i) < \xi$$

for all refinements P of P_ξ. Earlier, we showed $D_{1/n}$ is compact. Let's decompose $D_{1/n}$ in this way.

$$\begin{aligned} D_{1/n} \quad &= \quad \{\boldsymbol{x} \in D_{1/n} | \, \exists i \ni \boldsymbol{x} \in \partial S_i \in P_\xi\} \quad \Longleftarrow Q_1^n \\ &\quad \cup \{\boldsymbol{x} \in D_{1/n} | \, \exists i \ni \boldsymbol{x} \in Int(S_i^\circ) \in P_\xi\} \quad \Longleftarrow Q_2^n \end{aligned}$$

where ∂S_i is the usual notation for the boundary of S_i and S_i° denotes the interior of S_i. We know the measure of an edge of a rectangle is zero and we know countable unions of sets of measure zero are also zero. So we can conclude the measure of Q_1^n must be zero. It remains to see what the measure of Q_2^n is.

If S_i is a rectangle that intersects Q_2^n, then if $\boldsymbol{x} \in S_i$, there is a point $\boldsymbol{u} \in S$ and in $D_{1/n}$. Thus,

$$\omega(\hat{f}, \boldsymbol{u}) \quad = \quad \inf_{B(r, \boldsymbol{u}), r > 0} \; \sup_{\boldsymbol{x_1}, \boldsymbol{x_2} \in B(r, \boldsymbol{u})} |\hat{f}(\boldsymbol{x_1}) - \hat{f}(\boldsymbol{x_2})| \geq (1/n)$$

In particular, this can be rewritten in terms of open rectangles as

$$\omega(\hat{f}, \boldsymbol{u}) = \inf_{R^\circ(r,\boldsymbol{u}), r>0} \sup_{\boldsymbol{x_1}, \boldsymbol{x_2} \in R^\circ(r,\boldsymbol{u})} |\hat{f}(\boldsymbol{x_1}) - \hat{f}(\boldsymbol{x_2})| \geq (1/n)$$

where $R^\circ(r, \boldsymbol{u})$ is an open square rectangle with center $\boldsymbol{u}$ and axis dimensions r. Thus, for an open rectangle containing S, T°, we can say

$$\sup_{\boldsymbol{x_1}, \boldsymbol{x_2} \in T^\circ(r,\boldsymbol{u})} |\hat{f}(\boldsymbol{x_1}) - \hat{f}(\boldsymbol{x_2})| \geq (1/n)$$

and this implies

$$\sup_{\boldsymbol{x_1}, \boldsymbol{x_2} \in S} |\hat{f}(\boldsymbol{x_1}) - \hat{f}(\boldsymbol{x_2})| \geq (1/n)$$

Then, using the same (complicated!) arguments we used in the first part of this proof, we can say

$$\sup_{\boldsymbol{x} \in S} \hat{f}(\boldsymbol{x}) - \inf_{\boldsymbol{x} \in S} \hat{f}(\boldsymbol{x}) \geq (1/n)$$

Applying this argument to all such rectangles S that intersect Q_2^n, we find

$$\begin{aligned} (1/n) \sum_{S \cap Q_2^n \neq \emptyset} V(S) &\leq \sum_{S \cap Q_2^n \neq \emptyset} \left(\sup_{\boldsymbol{x} \in S} \hat{f}(\boldsymbol{x}) - \inf_{\boldsymbol{x} \in S} \hat{f}(\boldsymbol{x}) \right) V(S) \\ &\leq \sum_{S \in P_\xi} \left(\sup_{\boldsymbol{x} \in S} \hat{f}(\boldsymbol{x}) - \inf_{\boldsymbol{x} \in S} \hat{f}(\boldsymbol{x}) \right) V(S) < \xi \end{aligned}$$

Thus, the collection $V_\xi = \{S \cap Q_2^n \neq \emptyset\}$ is a set of rectangles that cover Q_2^n with $\sum_{S \cap Q_2^n \neq \emptyset} V(S) < n\xi$. Now since the measure of Q_1^n is zero, for $\xi > 0$, there is a collection of rectangles covering Q_1^n, $U_\xi = \{U_{1,\xi}, \ldots, U_{p,\xi}\}$ so that $\sum_{i=1}^p V(U_{i,\xi}) < \xi$. The collection $\{S \cap Q_2^n \neq \emptyset\} \cup U_\xi$ is cover for $D_{1/n}$ with $\sum_{S \in U_\xi \cup V_\xi} V(S) < (n+1)\xi$. Finally, letting $\xi = \epsilon/(n+1)$, we have found a collection whose union contains $D_{1/n}$ with $\sum_{S \in U_\xi \cup V_\xi} V(S) < \epsilon$. Since $\epsilon > 0$ is arbitrary because ξ was arbitrary, we see $D_{1/n}$ has measure zero. This tells us D also has measure zero. ■

Whew! That is an intense proof! Compare the approach we have used here to the one we used in $\Re$ in (Peterson (21) 2020). This one had to be able to handle rectangles in spaces of more than one dimension which is why it is much harder. Of course, this argument works just fine for $\Re$ although the rectangles degenerate to just closed intervals.

Example 14.4.1 *Consider the curve $x^2 + y^2 = 1$ in $\Re^2$. Let's show it has measure zero in $\Re^2$. We'll divide the curve into the four quadrants and just do the argument for quadrant one. In quadrant one, we have $y = \sqrt{1 - x^2}$ with $0 \leq x \leq 1$. Divide $[0, 1]$ into n uniform pieces and choose rectangles B_n centered at I/n as follows: Let $\boldsymbol{z_i} = (i/n, \sqrt{1 - (i/n)^2})$. Then the distance between $\boldsymbol{z_i}$ and $\boldsymbol{z_{i+1}}$ is*

$$d(\boldsymbol{z_i}, \boldsymbol{z_{i+1}}) = \sqrt{(1/n)^2 + (\sqrt{1 - ((i+1)/n)^2} - \sqrt{1 - (i/n)^2})^2}$$

Choose B_i to be the square centered at $\boldsymbol{z_i}$ whose distance to $\boldsymbol{z_{i+1}}$ is $\sqrt{2} d(\boldsymbol{z_i}, \boldsymbol{z_{i+1}})$. Then these squares overlap and their union covers the arc of $x^2 + y^2 = 1$ in quadrant one with summed area

$$\sum_{i=1}^n V(B_i) < \sum_{i=1}^n 4/n^2 = 4/n$$

Thus, for a given $\epsilon > 0$, there is N so that is $n > N$, the collection of rectangles covers the arc with summed volume less than ϵ. Hence the arc in quadrant one has measure zero in $\Re^2$. A similar argument works in the other quadrants and thus this curve has measure zero in $\Re^2$.

Example 14.4.2 *Now look at the integral of $f(x, y) = 1 - x^2 - y^2$ on A, the interior of the disc of radius one. Then let $R = [-1, 1] \times [-1, 1]$.*

$$\hat{f}(\boldsymbol{x}) = \begin{cases} 1 - x^2 - y^2, & \boldsymbol{x} \in D = \{(x, y) | x^2 + y^2 < 1\} \\ 0, & \boldsymbol{x} \in R \setminus D \end{cases}$$

$\hat{f}$ is clearly continuous on D and $\{(x, y) \in R | x^2 + y^2 > 1\}$. Also, the set of points where $x^2 + y^2 = 1$ is a set of measure zero in $\Re^2$. Hence, we know $\hat{f}$ is continuous everywhere except a set of measure zero and so f is integrable on D. So $\int_D f \boldsymbol{dV}$ exists and we can approximate it using any sequence of partitions whose norms go to zero. We usually would write $\int_D f \boldsymbol{dA}$ as we are using area ideas in $\Re^2$. Finally, we would normally write $\int_{x^2+y^2<1} (x^2 + y^2) dA_{xy}$ to be more specific.

14.4.1 Homework

Exercise 14.4.1 *Prove Theorem 14.4.1.*

Exercise 14.4.2 *Prove $x^2 + 4y^2$ is integrable on $[-1, 2] \times [3, 5]$.*

Exercise 14.4.3 *Prove $\sin(x^2 + 4xy + y^3)$ is integrable on the set A where $A = \{(x, y) | 0 \leq y \leq x^2\} \cap \{0 \leq y \leq 4\}$.*

Exercise 14.4.4 *Prove $\sin(x^2 + 4xy + y^3)$ is integrable on the set A where $A = \{(x, y) | x^2 \leq y \leq 6\}$.*

Exercise 14.4.5 *Prove $x^2 + y^2 + 4z^2$ is integrable on $D = \{(x, y, z) | z = x^2 + y^2 + z^2 \leq 25\}$.*

14.5 Integration and Sets of Measure Zero

Let $A \subset \Re^n$ be a set of measure zero. Then if A contains a rectangle $E = \prod_{i=1}^n [a_i, b_i]$, we would know $V(E) > 0$. But we also know A has measure zero, so given $\epsilon = V(E)/2$, there is a collection B_i with $A \subset \cup_i B_i$ and $\sum_i V(B_i) < V(E)/2$. However, $E \subset A \subset \cup_i B_i$ implies $V(E) \leq \sum_i B_i$ or $V(A) < V(A)/2$ which is not possible. We conclude if A is of measure zero, it cannot contain a rectangle such as E.

Let S be a rectangle containing A and now let's assume A is a bounded set. Let $f : A \to \Re$ be a bounded function which is integrable on A. Extend f to $\hat{f}$ on s as usual. Let P be a partition of $S = \{S_1, \ldots, S + M\}$. Since f is bounded, there are positives number m and M so that $m \leq f(\boldsymbol{x}) \leq M$ for $\boldsymbol{x} \in A$. Then

$$L(f, P) = \sum_{i \in P} \left(\inf_{\boldsymbol{x} \in S_i} \hat{f}(\boldsymbol{x}) \right) V(S_i)$$

Now

$$\inf_{\boldsymbol{x} \in S_i} \hat{f}(\boldsymbol{x}) = \inf \begin{cases} 0, & \boldsymbol{x} \in S_i \cap A^C \\ f(\boldsymbol{x}), & \boldsymbol{x} \in S_i \cap A \end{cases} \leq \inf \begin{cases} 0, & \boldsymbol{x} \in S_i \cap A^C \\ M, & \boldsymbol{x} \in S_i \cap A \end{cases} = M \inf_{\boldsymbol{x} \in S_i} \boldsymbol{1_A}$$

Thus,

$$L(f, P) \leq M \sum_{i \in P} \inf_{\boldsymbol{x} \in S_i} \boldsymbol{1_A} V(S_i)$$

If $\inf_{\boldsymbol{x} \in S_i} \mathbf{1_A} = 1$ for some index i, this would imply $S_i \cap A \neq \emptyset$. We know $A \subset \cup_i S_i$ though so this would force $S_i \subset A$. But by our argument at the start of this discussion, since A has measure zero, it is not possible for a nontrivial rectangle to be inside A. Thus, we must have $\inf_{\boldsymbol{x} \in S_i} \mathbf{1_A} = 0$ always. This tells us for all P

$$L(f, P) \leq M \sum_{i \in P} \inf_{\boldsymbol{x} \in S_i} \mathbf{1_A} \, V(S_i) \leq 0$$

We conclude $\underline{\int}_A f \leq 0$,

Next, we know $\sup_{\boldsymbol{x} \in S_i} \hat{f} = -\inf_{\boldsymbol{x} \in S_i} (-\hat{f})$, so

$$U(f, P) = \sum_{i \in P} \left(\sup_{\boldsymbol{x} \in S_i} \hat{f}(\boldsymbol{x}) \right) V(S_i) = -\sum_{i \in P} \left(\inf_{\boldsymbol{x} \in S_i} (-\hat{f}(\boldsymbol{x})) \right) V(S_i) = -L(-f, P)$$

Then, by the same arguments we just used, we see $-f \leq -m$ implies

$$L(-f, P) \leq (-m) \sum_{i \in P} \inf_{\boldsymbol{x} \in S_i} \mathbf{1_A} \, V(S_i)$$

implying $L(-f, P) \leq 0$. Thus, $-L(-f, P) \geq 0$ for all P. Therefore $U(f, P) \geq 0$ for all P and so $\overline{\int}_A f \geq 0$. We conclude

$$\underline{\int}_A f \leq 0 \leq \overline{\int}_A f$$

But f is integrable on A, so we have $0 \leq \int_A f \leq 0$ or $\int_A f = 0$.

We have proven an important theorem.

Theorem 14.5.1 If f is Integrable on A and A has Measure Zero, then the Integral of f on A is Zero

> *Let $A \subset \Re^n$ be bounded and have measure zero. If $f : A \to \Re$ is integrable on A, then $\int_A f = 0$.*

Proof 14.5.1
This is the argument we have just done. ■

Another interesting result is this:

Theorem 14.5.2 If the Non-negative Function f is Integrable on A with Value 0, then the Measure of the Set of Points where $f > 0$ is Measure Zero

> *Let $A \subset \Re^n$ be bounded and $f : A \to \Re$ be integrable on A and non-negative with integral value 0. Then the set of points where $f > 0$ is measure zero.*

Proof 14.5.2
Let $A_m = \{\boldsymbol{x} \in A | f(\boldsymbol{x}) > 1/m\}$ for any positive integer m. Pick $\epsilon > 0$. Since $\int_A f = 0$, there is a partition P_m so that

$$\int_A f \leq U(f, P_m) < \int_A f + \epsilon/m \Longrightarrow 0 \leq U(f, P_m) < \epsilon/m$$

as $\int_A f = 0$. *Then,*

$$\sum_{i \in P_m, S_i \cap A_m \neq \emptyset} \left(\sup_{\boldsymbol{x} \in S_i} \hat{f}(\boldsymbol{x}) \right) V(S_i) \leq \sum_{i \in P_m} \left(\sup_{\boldsymbol{x} \in S_i} \hat{f}(\boldsymbol{x}) \right) V(S_i) < \epsilon/m$$

If $S_i \cap A_m \neq \emptyset$, *then there is an* $\boldsymbol{x} \in S_i \cap A_m$, *implying there is an* $\boldsymbol{x}$ *with* $\hat{f}(\boldsymbol{x}) > 1/m$. *Hence,*

$$\sum_{i \in P_m : S_i \cap A_m \neq \emptyset} 1/mV(S_i) \leq \sum_{i \in P_m} \left(\sup_{\boldsymbol{x} \in S_i} \hat{f}(\boldsymbol{x}) \right) V(S_i) < \epsilon/m$$

or $\sum_{i \in P_m : S_i \cap A_m \neq \emptyset} V(S_i) < \epsilon$. *This tells us this collection of sets covers* A_m *with total summed volume smaller than* ϵ. *Since* ϵ *is arbitrary, we know* A_m *is measure zero which implies immediately the measure of the set of points where* $f > 0$ *is zero.* ■

And, a result that seems natural.

Theorem 14.5.3 If f is Integrable over A and f is Zero Except on a Set of Measure Zero, the Integral is Zero

Let $f : A \subset \Re^n \to \Re$ *where* A *is bounded be integrable. Assume* $B = \{\boldsymbol{x} \in A | f(\boldsymbol{x}) \neq 0\}$ *is measure zero. Then* $\int_A f = 0$.

Proof 14.5.3
Since f *is integrable, for any sequence of partitions* P_n *with* $\|P_n\| \to 0$, $S(f, \boldsymbol{\sigma_n}, P_n) \to \int_A f$ *where* $\boldsymbol{sigma_n}$ *is any evaluation set from* P_n. *Now if* S_{in} *is a rectangle in* P_n, *if all points* $\boldsymbol{z}$ *from* S_{in} *were in* B, *this would mean* B *contains a rectangle. But previous arguments tell us this is not possible since* B *has measure zero. Thus, each* S_{in} *contains a point not in* B. *Call this point* $\boldsymbol{z_{in}}$ *and let* $\boldsymbol{\sigma_n} = \{\boldsymbol{z_{in}} | i \in P_n\}$. *Then*

$$S(f, \boldsymbol{\sigma_n}, P_n) = \sum_{i \in P_n} f(\boldsymbol{z_{in}}) V(S_i) = 0$$

This shows $S(f, \boldsymbol{\sigma_n}, P_n) \to 0 = \int_A f$. ■

14.5.1 Homework

Exercise 14.5.1 *Why is* $\int_{x^2+y^2=1} e^{-x^2-y^2} dA = 0$?

Exercise 14.5.2 *Why is* $\int_{[-1,1]\times[-1,1]} e^{-x^2-y^2} dA = \int_{(-1,1]\times[-1,1)} e^{-x^2-y^2} dA$?

Exercise 14.5.3 *If* f *and* g *are integrable with* $f = g$ *on the interior of* $\Omega = [-1,1]\times[-1,1]\times[-2,5]$ *why is* $\int_\Omega (f-g) dA = 0$? *Why is* $\int_\Omega f dA = \int_\omega g dA$?

We all know the usual identity $\int_a^b f = \int_a^c f + \int_c^b f$ for any $c \in (a,b)$ for the one dimensional integral. It is a lot harder to make sense of this in higher dimensions. The next set of exercises will lead through how to do this for specific cases and then ask you to formulate an appropriate theorem. First, let's subdivide a rectangle in $\Re^2$ into two nice pieces.

Exercise 14.5.4 *Let* $\Omega = [-1,1] \times [-1,1]$, $\Omega_1 = [-1,0] \times [-1,1]$ *and* $\Omega_2 = [0,1] \times [-1,1]$. *Let* f *be integrable on* Ω.

- *Show* $f_1 = f I_{\Omega_1}$ *and* $f_2 = f I_{\Omega_2}$ *are also integrable.*

- *Show* $\int_\Omega f dA = \int_\Omega (f_1 + f_2) dA$.
- *Show* $\int_\Omega f_1 dA = \int_{\Omega_1} f dA$ *and* $\int_\Omega f_2 dA = \int_{\Omega_2} f dA$.
- *Thus, you can show* $\int_\Omega f dA = \int_{\Omega_1} f dA + \int_{\Omega_2} f dA$.

Exercise 14.5.5 *Now assume* Ω *has a well-defined volume (here actually area) with* $V(\Omega) = \int_\Omega \mathbf{1}_\Omega$. *We assume* $\Omega \subset [-1,1] \times [-1,1]$. *Let* $\Omega_1 = [-1,0] \times [-1,1] \cap \Omega$ *and* $\Omega_2 = [0,1] \times [-1,1] \cap \Omega$. *Let* f *be integrable on* Ω.

- *Show* $f_1 = f I_{\Omega_1}$ *and* $f_2 = f I_{\Omega_2}$ *are also integrable.*
- *Show* $\int_\Omega f dA = \int_\Omega (f_1 + f_2) dA$.
- *Show* $\int_\Omega f_1 dA = \int_{\Omega_1} f dA$ *and* $\int_\Omega f_2 dA = \int_{\Omega_2} f dA$.
- *Thus, you can show* $\int_\Omega f dA = \int_{\Omega_1} f dA + \int_{\Omega_2} f dA$.

Now, let's subdivide a rectangle in $\Re^2$ into four nice pieces.

Exercise 14.5.6 *Let* $\Omega = [-1,1] \times [-1,1]$, $\Omega_{11} = [-1,0] \times [-1,0]$, $\Omega_{12} = [0,1] \times [-1,0]$, $\Omega_{21} = [-1,0] \times [0,1]$, $\Omega_{12} = [0,1] \times [0,1]$ *and let* f *be integrable on* Ω.

- *Show* $f_{ij} = f I_{\Omega_{ij}}$ *is also integrable.*
- *Show* $\int_\Omega f dA = \int_\Omega (\sum_{1 \leq 2} \sum_{1 \leq j \leq 2} f_{ij} dA)$.
- *Show* $\int_\Omega f_{ij} dA = \int_{\Omega_{ij}} f dA$.
- *Thus, you can show* $\int_\Omega f dA = \sum_{1 \leq i \leq 2} \sum_{1 \leq j \leq 2} \int_{\Omega_{ij}} f dA$.

Exercise 14.5.7 *Now assume* Ω *has a well-defined volume (here actually area) with* $V(\Omega) = \int_\Omega \mathbf{1}_\Omega$. *We assume* $\Omega \subset [-1,1] \times [-1,1]$. *Let* $\Omega_{11} = [-1,0] \times [-1,0] \cap \Omega$, $\Omega_{12} = [0,1] \times [-1,0] \cap \Omega$, $\Omega_{21} = [-1,0] \times [0,1] \cap \Omega$, $\Omega_{12} = [0,1] \times [0,1] \cap \Omega$ *and let* f *be integrable on* Ω.

- *Show* $f_{ij} = f I_{\Omega_{ij}}$ *is also integrable.*
- *Show* $\int_\Omega f dA = \int_\Omega (\sum_{1 \leq 2} \sum_{1 \leq j \leq 2} f_{ij}) dA$.
- *Show* $\int_\Omega f_{ij} dA = \int_{\Omega_{ij}} f dA$.
- *Thus, you can show* $\int_\Omega f dA = \sum_{1 \leq i \leq 2} \sum_{1 \leq j \leq 2} \int_{\Omega_{ij}} f dA$.

Now let's subdivide into pieces using partitions.

Exercise 14.5.8 *Now let* P *and* Q *be partitions of* $[-1,1]$ *on the* x *axis and* $[-1,1]$ *on the* y *axis. Assume* Ω *has a well-defined volume (here actually area) with* $V(\Omega) = \int_\Omega \mathbf{1}_\Omega$. *We assume* $\Omega \subset [-1,1] \times [-1,1]$. *The partitions* P *and* Q *determine sub rectangles* R_{ij} *of* $[-1,1] \times [-1,1]$. *Let* $\Omega_{ij} = R_{ij} \cap \Omega$ *and assume* f *is integrable on* Ω.

- *Show* $f_{ij} = f I_{\Omega_{ij}}$ *is also integrable.*
- *Show* $\int_\Omega f dA = \int_\Omega (\sum_{i,j} f_{ij}) dA$.
- *Show* $\int_\Omega f_{ij} dA = \int_{\Omega_{ij}} f dA$.
- *Thus, you can show* $\int_\Omega f dA = \sum_{i,j} \int_{\Omega_{ij}} f dA$.

Exercise 14.5.9 *Now you should be able to see how to redo the exercises above for the rectangle* $[a,b]\times[c,d]$. *Assume* Ω *has a well-defined volume with* $V(\Omega) = \int_\Omega \mathbf{1}_\Omega$. *We assume* $\Omega \subset [a,b]\times[c,d]$. *Let* P *and* Q *be partitions of* $[a,b]$ *on the* x *axis and* $[c,d]$ *on the* y *axis. The partitions* P *and* Q *determine subrectangles* R_{ij} *like before. Let* $\Omega_{ij} = R_{ij} \cap \Omega$ *and assume* f *is integrable on* Ω.

- *Show* $f_{ij} = f I_{\Omega_{ij}}$ *is also integrable.*
- *Show* $\int_\Omega f dA = \int_\Omega (\sum_{i,j} f_{ij}) dA$.
- *Show* $\int_\Omega f_{ij} dA = \int_{\Omega_{ij}} f dA$.
- *Thus, you can show* $\int_\Omega f dA = \sum_{i,j} \int_{\Omega_{ij}} f dA$.

Now extend to a box in $\Re^3$.

Exercise 14.5.10 *Now redo the exercises above for the box* $[a,b]\times[c,d]\times[e,f]$. *Assume* Ω *has a well-defined volume with* $V(\Omega) = \int_\Omega \mathbf{1}_\Omega$. *We assume* $\Omega \subset [a,b]\times[c,d]\times[e,f]$. *Let* P, Q *and* S *be partitions of the three axes of the box. The partitions determine subboxes* R_{ijk}. *Let* $\Omega_{ijk} = R_{ijk} \cap |\Omega$ *and assume* f *is integrable on* Ω.

- *Show* $f_{ijk} = f I_{\Omega_{ijk}}$ *is also integrable.*
- *Show* $\int_\Omega f dV = \int_\Omega (\sum_{i,j,k} f_{ijk}) dV$.
- *Show* $\int_\Omega f_{ijk} dV = \int_{\Omega_{ijk}} f dV$.
- *Thus, you can show* $\int_\Omega f dV = \sum_{i,j,k} \int_{\Omega_{ijk}} f dV$.

Now think about how to generalize.

Exercise 14.5.11 *Extend these ideas to* $\Re^n$.

Exercise 14.5.12 *Formulate the theorem that lets you know* $\int_\Omega f dV = \sum_i \int_{\Omega_i} f dV$ *where* $\Omega = \cup_i \Omega_i$ *is a finite union. There have to be conditions on this union so you need to think about that.*

Chapter 15

Change of Variables and Fubini's Theorem

We now begin our discussion of how to make a change of variables in an integral. We start with linear maps.

15.1 Linear Maps

We start with the simplest case. We have this hazy notion of what it means for a set to have a volume: its indicator function must be integrable. But we really only know how to compute simple volumes such as the volumes of rectangles. The next proof starts the process of understanding change of variable theorems by looking at what happens to a set with volume when it is transformed with a linear map.

Theorem 15.1.1 Change of Variable for a Linear Map

If $L : \Re^n \rightarrow \Re^n$ is a linear map and $D \subset \Re^n$ is a set which has volume, i.e. $\boldsymbol{1_D}$ is integrable and $V(D) = \int_D \boldsymbol{1_D}$, then $L(D)$ also has volume as

$$V(L(D)) \quad = \quad \int_{L(D)} \boldsymbol{1_{L(D)}} = |detL| \int_D = \int_D |detL|\, \boldsymbol{1_D}$$

Proof 15.1.1

Note if L is not invertible, the range of L is at most a $n-1$ dimensional subspace of $\Re^n$ which has measure zero. Thus, $0 = \int_{L(D)} \boldsymbol{1_{L(D)}}$ and since $detL = 0$, this matches $detL \int_D \boldsymbol{1_D} = 0$. So this will work even if L is not invertible.

For any invertible matrix A there is a sequence of elementary matrices L_i so that

$$L_p L_{p-1} \cdots L_1 A \quad = \quad I$$

where each L_i is invertible. We discuss these things carefully in Section 9.2, but let's review a bit here. These matrices L_i come it two types: T_1 and T_2:

$$T_1 = \begin{bmatrix} 1 & 0 & \dots & 0 & 0 \\ 0 & 1 & \dots & 0 & 0 \\ 0 & 0 & 1 & 1, (i.j) \text{ position} & 0 \\ \vdots & \vdots & \vdots & \vdots & \vdots \\ 000 & 0 & 1 & & \end{bmatrix} = I + \Lambda_{ij}$$

where I is the usual identity matrix and Λ_{ij} is the matrix of all zeros except a 1 at the (i, j) position. Note $det(T_1) = 1$ and so $det(T_1^{-1}) = 1$ also. It is easy to see

$$T_1^{-1} = \begin{bmatrix} 1 & 0 & \dots & 0 & 0 \\ 0 & 1 & \dots & 0 & 0 \\ 0 & 0 & 1 & -1, (i.j) \text{ position} & 0 \\ \vdots & \vdots & \vdots & \vdots & \vdots \\ 000 & 0 & 1 & & \end{bmatrix} = I - \Lambda_{ij}$$

Next,

$$T_2 = \begin{bmatrix} 1 & 0 & \dots & 0 & 0 \\ 0 & 1 & \dots & 0 & 0 \\ 0 & 0 & 1 & & 0 \\ \vdots & \vdots & \vdots & c, (j, j) \text{ position} & \vdots \\ 000 & 0 & 1 & & \end{bmatrix}$$

It is easy to see

$$T_2 = \begin{bmatrix} 1 & 0 & \dots & 0 & 0 \\ 0 & 1 & \dots & 0 & 0 \\ 0 & 0 & 1 & & 0 \\ \vdots & \vdots & \vdots & 1/c, (j, j) \text{ position} & \vdots \\ 000 & 0 & 1 & & \end{bmatrix}$$

Note $det(T_2) = c$ and so $det(T_2^{-1}) = 1/c$ also. So T_2A creates a new matrix which scales row j by c and T_1A adds row j to row i. Thus, T_1T_2A adds c times row j to row i to create a new row i. The same is true for the inverses. So $T_2^{-1}A$ creates a new matrix which scales row j by $1/c$ and $T_1^{-1}A$ subtracts row j to row i. Thus, T_1T_2A subtracts $1/c$ times row j to row i to create a new row i.

The linear map L has a matrix representation A and so there are invertible linear maps λ_1 to λ_p so that $L = \lambda_1^{-1}\lambda_2^{-1}\cdots\lambda_p^{-1}$. Thus,

$$\begin{aligned} det(L) &= det(\lambda_1^{-1})\cdots det(\lambda_p^{-1}) \\ &= det(L_1^{-1})\cdots det(L_p^{-1}) = \prod_{i\in I}(1/c_i) \end{aligned}$$

where I is the set of indices where L_i corresponds to a Type 2 matrix. Note this is independent of the matrix representations of these maps.

If λ_i^{-1} is of type T_2, then $\lambda_i(x_1, x_2, \dots, x_n) = (x_1, x_2, \dots, 1/cx_j, \dots, x_n)$ for some j, and if S is a rectangle, we see $V(\lambda_i S) = |1/c|V(S)$.

If λ_i is of Type T_1, we have $\lambda_i(x_1, x_2, \ldots, x_n) = (x_1, x_2, \ldots, x_i + x_j, \ldots, x_n)$ for some i and j. Now apply λ_i to a rectangle S. The new rectangle $\lambda_i S$ preserves all the structure except in the $i-j$ plane. Then for $S = [a_1, b_1] \times \ldots \times [a_n, b_n]$, all of S is preserved except in the $2-3$ plane. There we have

$$\begin{bmatrix} (a_2, b_3) & \ldots & (b_2, b_3) \\ \vdots & \vdots & \vdots \\ (a_2, a_3) & \ldots & (b_2, a_3) \end{bmatrix} \longrightarrow \begin{bmatrix} (a_2 + b_4, b_3) & \ldots & (b_2 + b_4, b_3) \\ \vdots & \vdots & \vdots \\ (a_2 + a_4, a_3) & \ldots & (b_2 + a_4, a_3) \end{bmatrix}$$

We see the area of the original rectangle in the $2-3$ plane is $(b_2 - a_2)(b_3 - a_3)$ which is the same as the area of the parallelogram in the transformed rectangle. This experiment can be repeated for other choices. We have found the T_1 applied to a rectangle S does not change its volume. A similar analysis shows T_1^{-1} applied to a rectangle S does not change its volume.

Now pick a bounding rectangle R large enough to contain both D and $L(D)$. Since D has volume, by the integrable approximation theorem, given a sequence of partitions P_n with $\|P_n\| \to 0$, $U(\mathbf{1_D}, P_n) \downarrow \int_D \mathbf{1_D} = V(D)$ and $L(\mathbf{1_D}, P_n) \uparrow \int_D \mathbf{1_D} = V(D)$. Thus, for a given $\epsilon > 0$ there an N so that if $n > N$ $U(\mathbf{1_D}, P_n) < V(D) + \epsilon/|det(L)|$ and $L(\mathbf{1_D}, P_n) > V(D) - \epsilon/|det(L)|$. Hence, there are rectangles $S_1, \ldots, S_u$ in P_n, so that $\sum_{j=1}^{u} V(S_j) < V(D) + \epsilon/|det(L)|$ and rectangles (possibly different) $T_1, \ldots, T_v$ in P_n so that $\sum_{j=1}^{v} V(T_j) > V(D) - \epsilon//|det(L)|$. Then, from our arguments above

$$V(\lambda_1^{-1} \cdots \lambda_p^{-1}(S)) \;=\; = |det(L)| V(S)$$

Consider

$$U(\mathbf{1_{L(D)}}, P_n) \;=\; \sum_{S \in P_n} \sup_{\boldsymbol{x} \in S} (\mathbf{1_{L(D)}}(\boldsymbol{x})) V(S)$$

Now, if $\sup_{\boldsymbol{x} \in S}(\mathbf{1_{L(D)}}(\boldsymbol{x})) = 1$, this means there is $\boldsymbol{x} \in S$ so that $L_{-1}(\boldsymbol{x}) \in D$. This means there is a rectangle $T \in P_n$ so that $\sup_{\boldsymbol{x} \in T}(\mathbf{1_D}(\boldsymbol{x})) = 1$. Therefore, this 1 in the upper sum $U(\mathbf{1_{L(D)}}, P_n)$ also occurs in $U(\mathbf{1_D}, P_n)$. So every 1 in the upper sum $U(\mathbf{1_{L(D)}}, P_n)$ also occurs in $U(\mathbf{1_D}, P_n)$. We conclude

$$U(\mathbf{1_{L(D)}}, P_n) \;\leq\; U(\mathbf{1_D}, P_n)$$

Also,

$$L(\mathbf{1_{L(D)}}, P_n) \;=\; \sum_{S \in P_n} \inf_{\boldsymbol{x} \in S} (\mathbf{1_{L(D)}}(\boldsymbol{x})) V(S)$$

Now, if $\inf_{\boldsymbol{x} \in S}(\mathbf{1_{L(D)}}(\boldsymbol{x})) = 1$, this means there is $S \subset L(D)$ so that $L_{-1}(\boldsymbol{x}) \in D$ for all $\boldsymbol{x} \in S$; i.e. $L^{-1}(S) \subset D$. This does not mean we can be sure we can find a $T \in P_n$ which is inside D. Thus, this 1 in $L(\mathbf{1_{L(D)}}, P_n)$ need not be repeated in $L(\mathbf{1_D}, P_n)$. We conclude

$$U(\mathbf{1_{L(D)}}, P_n) \;\geq\; U(\mathbf{1_D}, P_n)$$

Therefore,

$$U(\mathbf{1_{L(D)}}, P_n) - L(\mathbf{1_{L(D)}}, P_n) \;\leq\; U(\mathbf{1_D}, P_n) - L(\mathbf{1_D}, P_n)$$

We also know

$$U(\mathbf{1}_{\boldsymbol{D}}, P_n) - L(\mathbf{1}_{\boldsymbol{D}}, P_n) \quad < \quad \epsilon$$

Combining, we see $\mathbf{1}_{\boldsymbol{L(D)}}$ *is integrable since it satisfies the Riemann Criterion. Also*

$$\begin{aligned} \sum_{j=1}^{p} (V(L(S_j))/|det(L)|) &< V(D) + \epsilon/|det(L)| \\ \sum_{j=1}^{q} (V(L(T_j))/|det(L)|) &< V(D) + \epsilon/|det(L)| \end{aligned}$$

or

$$\begin{aligned} U(\mathbf{1}_{\boldsymbol{L(D)}}, P_n) \leq U(\mathbf{1}_{\boldsymbol{D}}, P_n) &= \sum_{j=1}^{u} V(L(S_j)) < |det(L)|V(D) + \epsilon \\ L(\mathbf{1}_{\boldsymbol{L(D)}}, P_n) \geq L(\mathbf{1}_{\boldsymbol{D}}, P_n) &= \sum_{j=1}^{v} V(L(T_j)) < |det(L)|V(D) + \epsilon \end{aligned}$$

Since P_n *is a sequence of partitions with* $||P_n|| \to 0$*, we have*

$$\begin{aligned} \int_{L(D)} \mathbf{1}_{\boldsymbol{L(D)}} = \lim_{P_n} U(\mathbf{1}_{\boldsymbol{L(D)}}, P_n) &\leq |det(L)|V(D) + \epsilon \\ \int_{L(D)} \mathbf{1}_{\boldsymbol{L(D)}} = \lim_{P_n} L(\mathbf{1}_{\boldsymbol{L(D)}}, P_n) &\geq |det(L)|V(D) - \epsilon \end{aligned}$$

giving for all $\epsilon > 0$

$$\left| \int_{L(D)} \mathbf{1}_{\boldsymbol{L(D)}} - |det(L)|V(D) \right| < \epsilon$$

Thus, $L(D)$ *does have volume and* $\int_{L(D)} \mathbf{1}_{\boldsymbol{L(D)}} = |det(L)|V(D)|$. ■

It is easy to show that given two linearly independent vectors in $\Re^2$, say $\boldsymbol{V}$ and $\boldsymbol{W}$, the parallelogram formed by these two vectors has an area that can be computed by looking at the projection of $\boldsymbol{W}$ onto $\boldsymbol{V}$. If the angle between $\boldsymbol{V}$ and $\boldsymbol{W}$ is θ, then the area of the parallelogram is

$$\begin{aligned} A &= ||\boldsymbol{V}|| \; ||\boldsymbol{W}|| \; \sin(\theta) = ||\boldsymbol{V}|| \; ||\boldsymbol{W}|| \sqrt{\frac{||\boldsymbol{V}||^2 \; ||\boldsymbol{W}||^2 - < \boldsymbol{V}, \boldsymbol{W} >}{||\boldsymbol{V}||^2 \; ||\boldsymbol{W}||^2}} \\ &= \sqrt{|det(T)|^2} = |det(T)| \end{aligned}$$

where T is the matrix formed using the $\boldsymbol{V}$ and $\boldsymbol{W}$ as its columns. It is a lot harder to work this out in $\Re^3$ and trying to extend this idea to $\Re^n$ shows you there must be another way to reason this out. Theorem 15.1.1 is a reasonable way to attack this. It does require the large setup of a general integration theory in $\Re^n$ and the notion of extending volume in general, but it has the advantage of being quite clear logically. We suspect the **volume** of the parallelepiped in $\Re^n$ formed by n linearly independent vectors should be $det(T)$ where T is the matrix formed using the vectors as columns.

We can state this as follows:

Theorem 15.1.2 The Volume of a Rectangle in $\Re^n$

If $\boldsymbol{V}_1, \ldots, \boldsymbol{V}_n$ are linearly independent in $\Re^n$, the volume of the parallelepiped determined by them is $det(T)$

Proof 15.1.2
First, note if D is a rectangle in $\Re^n$, $\boldsymbol{1}_D(\boldsymbol{x}) = 1$ on all of D and it lacks continuity only on ∂D which is a set of measure zero. Thus, it is integrable. Theorem 15.1.1 applied with $L = I$ tells us $\int_D \boldsymbol{1}_D = V(D)$ which is the usual $\prod(b_i - a_i)$ where $[a_i, b_i]$ is the ith edge of the rectangle D. You can also prove this directly by looking at upper and lower sums without using this theorem and you should try that. In this case, $\boldsymbol{V}_1, \ldots, \boldsymbol{V}_n$ determine an invertible linear map L on $D = [0,1] \times \ldots \times [0,1]$ and so $\int_{L(D)} \boldsymbol{1}_{L(D)} = det(L)V(D) = det(L)$. ■

Example 15.1.1 *Here is a sample volume calculation in $\Re^3$. We pick three independent vectors, set up the corresponding matrix, and find its determinant.*

Listing 15.1: **Sample Volume Calculation**

```
V1 = [1;2;-3];
V2 = [-3;4;10];
V3 = [11;22;1];
A = [V1,V2,V3]
A =

     1    -3    11
     2     4    22
    -3    10     1

det(A)
ans =    340.00
```

We see $V(A) = 340$.

15.1.1 Homework

Exercise 15.1.1 *Prove if D is a rectangle in $\Re^n$, $V(D) = \prod(b_i - a_i)$ where $[a_i, b_i]$ is the ith edge of the rectangle D from a lower and upper sum argument without using Theorem 15.1.1.*

Exercise 15.1.2 *Here are the vectors. Find the volume of the parallelepiped determined by them.*

Listing 15.2: **Volume Calculation**

```
V1 = [-1;2;4;10];
V2 = [2;-3;5;1];
V3 = [8;-6;-7;1];
V4 = [9;1;2;4];
```

Exercise 15.1.3 *Show the volume of a parallelepiped in $\Re^n$ is independent of the choice of orthonormal basis.*

15.2 The Change of Variable Theorem

The Change of Variable result is a bit complicated to prove. So be prepared to follow some careful arguments! We start with an overestimate.

Theorem 15.2.1 Subsets of Sets with volume have volume

Let $D \subset \Re^n$ be open and bounded. Assume $g(D)$ has volume. Then if $V \subset D$, $g(V)$ also have volume.

Proof 15.2.1

Since $g(V)$ has volume, for given $\epsilon > 0$, there is a partition P_ϵ of $g(D)$ so that

$$U(\hat{g}, P) - L(\hat{g}, P) \quad < \quad \epsilon$$

Since any refinement P of P_ϵ is also a partition of $g(V)$ and if $h = g\mathbf{1}_{\boldsymbol{V}}$, it is easy to see

$$U(\hat{h}, P) - L(\hat{h}, P) \quad \leq \quad U(\hat{g}, P) - L(\hat{g}, P) < \epsilon$$

which implies h satisfies the Riemann Criterion and so h is integrable and thus $g(G)$ has volume. ■

Our first change of variable theorem involves the integration of a simple constant.

Theorem 15.2.2 The Change of Variable Theorem for a Constant Map

Let $D \subset \Re^n$ be open and bounded. Let $g : D \rightarrow \Re^n$ have continuous partial derivatives on D and assume g is $1-1$ on D with $\boldsymbol{J_g}(\boldsymbol{x}) \neq 0$ on D. If $g(D)$ has volume, $\int_{g(D)} 1 = \int_D |\det \boldsymbol{J_G}|$.

Proof 15.2.2

Let $C(s, \boldsymbol{x_0})$ be a cube of edge size s inside the open set D. The $C(s, \boldsymbol{x_0})$ has the form

$$C(s, \boldsymbol{x_0}) \quad = \quad \{\boldsymbol{x} \in \Re^n | \, \|\boldsymbol{x} - \boldsymbol{x_0}\|_\infty < s\}$$

and $V(C(s, \boldsymbol{x_0})) = (2s)^n$. Since g has continuous partial derivatives, using the Mean Value Theorem, we can say there are points $\boldsymbol{u_i}$ on $[\boldsymbol{x_0}, \boldsymbol{x}]$ so that

$$g_i(\boldsymbol{x}) - g_i(\boldsymbol{x_0}) \quad = \quad \sum_{k=1}^n g_{i,x_k}(\boldsymbol{u_i})\,(x_k - x_{0,k})$$

Note each $\boldsymbol{u_i} \in C(s, \boldsymbol{x_0})$. We can then overestimate

$$\begin{aligned} |g_i(\boldsymbol{x}) - g_i(\boldsymbol{x_0})| \quad &\leq \quad \sum_{k=1}^n |g_{i,x_k}(\boldsymbol{u_i})| \, \|\boldsymbol{x} - \boldsymbol{x_0}\|_\infty \leq \sum_{k=1}^n |g_{i,x_k}(\boldsymbol{u_i})| \, s \\ &\leq \quad n \, \|\boldsymbol{J_g}\|_{Fr}(\boldsymbol{u_i}) \, s \leq n \max_{\boldsymbol{y} \in C(s,\boldsymbol{x_0})} \|\boldsymbol{J_g}\|_{Fr}(\boldsymbol{y}) \, s \end{aligned}$$

as $\sqrt{n} < n$. Hence,

$$\|g(\boldsymbol{x}) - g(\boldsymbol{x_0})\|_\infty \quad \leq \quad \max_{\boldsymbol{y} \in C(s,\boldsymbol{x_0})} \|\boldsymbol{J_g}\|_{Fr}(\boldsymbol{y}) \, ns$$

implying

$$V(g(C(s,\boldsymbol{x_0}))) \leq (\max_{\boldsymbol{y}\in C(s,\boldsymbol{x_0})} \|\boldsymbol{J_g}\|_{Fr}(\boldsymbol{y}))^n \, n^n \, V(C(s,\boldsymbol{x_0}))$$

Thus, $g(C(s,\boldsymbol{x_0}))$ *is contained in a cube centered at* $g(\boldsymbol{x_0})$ *defined by*

$$E(s,\boldsymbol{x_0}) = \left\{\boldsymbol{z} \mid \|\boldsymbol{z} - g(\boldsymbol{x_0})\|_\infty \leq n \max_{\boldsymbol{y}\in C(s,\boldsymbol{x_0})} \|\boldsymbol{J_g}(\boldsymbol{y})\|_{Fr}\right\}$$

Now let L be an invertible linear map. Then, if S has volume, $L^{-1}(S)$ *also has volume and*

$$V(L^{-1}(S)) = \int_{L^{-1}(S)} 1 = \int_S (\det(L^{-1}))\, 1 = \det(L^{-1})V(S)$$

Let $S = g(C(s,\boldsymbol{x_0}))$. *We know S has volume because of Theorem 15.2.1.*

Thus, for any invertible map L,

$$\begin{aligned} det(L^{-1})\, V(g(C(s,\boldsymbol{x_0}))) &= V((L^{-1}g)(C(s,\boldsymbol{x_0}))) \\ &= \left(\max_{\boldsymbol{y}\in C(s,\boldsymbol{x_0})} \|(L^{-1}\boldsymbol{J_g})(\boldsymbol{y})\|_1\right)^n n^n \, V(C(s,\boldsymbol{x_0})) \end{aligned}$$

implying

$$\begin{aligned} V(g(C(s,\boldsymbol{x_0}))) &= V((L^{-1}g)(C(s,\boldsymbol{x_0}))) \\ &= (detL)\left(\max_{\boldsymbol{y}\in C(s,\boldsymbol{x_0})} \|(L^{-1}\boldsymbol{J_g})(\boldsymbol{y})\|_1\right)^n n^n \, V(C(s,\boldsymbol{x_0})) \end{aligned}$$

Now divide $C(s,\boldsymbol{x_0})$ *into* M *cubes* $C_1,\ldots,C_M$ *with centers* $\boldsymbol{x_1},\ldots,\boldsymbol{x_M}$ *and apply* $L_i = \boldsymbol{J_g}(\boldsymbol{x_i})$ *to each cube* C_k. *We find*

$$V(g(C_k)) = (detL_k)\left(\max_{\boldsymbol{y}\in C(s,\boldsymbol{x_0})} \|(L_k^{-1}\boldsymbol{J_g})(\boldsymbol{y})\|_1\right)^n n^n \, V(C_k)$$

and thus

$$V(g(C(s,\boldsymbol{x_0}))) = \sum_{k=1}^{M} V(g(C_k)) = \sum_{k=1}^{M} (detL_k)\left(\max_{\boldsymbol{y}\in C(s,\boldsymbol{x_0})} \|(L_k^{-1}\boldsymbol{J_g})(\boldsymbol{y})\|_1\right)^n V(C_k)$$

The function $(\boldsymbol{J_g}(\boldsymbol{y}))^{-1}$ *is uniformly continuous on* $C(s,\boldsymbol{x_0})$ *with respect to the Frobenius (i.e. sup) norm on matrices (you should be able to check this easily) and so*

$$\lim_{\boldsymbol{z}\to\boldsymbol{y}} (\boldsymbol{J_g}(\boldsymbol{z}))^{-1} = (\boldsymbol{J_g}(\boldsymbol{y}))^{-1}$$

which tells us

$$\lim_{\boldsymbol{z}\to\boldsymbol{y}} \boldsymbol{J_g}(\boldsymbol{z})^{-1}\boldsymbol{J_g}(\boldsymbol{y}) = \boldsymbol{J_g}(\boldsymbol{y})^{-1}\,\boldsymbol{J_g}(\boldsymbol{y}) = I$$

Thus, given $\epsilon > 0$, *there is a* $\delta > 0$ *so that*

$$\|\boldsymbol{z} - \boldsymbol{y}\| < \delta \implies \|\boldsymbol{J_g}(\boldsymbol{z})^{-1}\boldsymbol{J_g}(\boldsymbol{y}) - I\|_\infty < \epsilon$$

But this says

$$\|\boldsymbol{z} - \boldsymbol{y}\| < \delta \implies \|I\|_\infty - \epsilon < \|\boldsymbol{J_g}(\boldsymbol{z})^{-1}\boldsymbol{J_g}(\boldsymbol{y})\|_\infty < \|I\|_\infty + \epsilon$$

or

$$\|\boldsymbol{z} - \boldsymbol{y}\| < \delta \implies 1 - \epsilon < \|\boldsymbol{J_g}(\boldsymbol{z})^{-1}\boldsymbol{J_g}(\boldsymbol{y})\|_\infty < 1 + \epsilon$$

Now we find the subdivision of cubes we need. Let $\epsilon = 1/p$ *and for the* δ *associated with* $\epsilon = 1/p$, *choose it so that* $\delta_p < 1/p$. *Choose the cubes* $C_1, \ldots, C_{M_p}$ *and centers* $\boldsymbol{x_1} < \ldots, \boldsymbol{x_{M_p}}$ *so that* $\|\boldsymbol{x_i} - \boldsymbol{y}\| < \delta_p$. *Then, we must have*

$$\|\boldsymbol{z} - \boldsymbol{x_i}\| < \delta_p \implies 1 - 1/p < \|\boldsymbol{J_g}(\boldsymbol{x_i})^{-1}\boldsymbol{J_g}(\boldsymbol{y})\|_\infty < 1 + 1/p$$

We now have the estimate

$$\begin{aligned} V(g(C(s, \boldsymbol{x_0}))) &\leq \sum_{i=1}^{M_p} \left(\max_{\boldsymbol{y} \in C_i}(1 + 1/p)\right)^n |det(\boldsymbol{J_g}(\boldsymbol{x_i}))|\, n^n\, V(C_i) \\ &\leq \sum_{i=1}^{M_p} (1 + 1/p)^n\, det(\boldsymbol{J_g}(\boldsymbol{x_i}))\, n^n\, V(C_i) \end{aligned}$$

Since $(1 + 1/p) \to 1$ *as* $p \to \infty$, *for a given* ϵ, *there is a* p_0 *so that* $(1 + 1/p)^n\, n^n < 1 + \epsilon$. *We conclude if* $p > p_0$,

$$V(g(C_i)) < (1 + \epsilon)\, |det(\boldsymbol{J_g}(\boldsymbol{x_i}))|\, V(C_i)$$

and so

$$V(g(C(s, \boldsymbol{x_0}))) < \sum_{i=1}^{M_p} (1 + \epsilon)\, |det(\boldsymbol{J_g}(\boldsymbol{x_i}))|\, V(C_i)$$

Now, let's connect this to Riemann sums. Let P_p *be the partition of* $g(C(s, \boldsymbol{x_0}))$ *determined by these cubes. Then for the evaluation set* $\boldsymbol{\sigma_p} = \{\boldsymbol{x_1}, \ldots, \boldsymbol{x_{M_P}}\}$, *we have*

$$\sum_{i \in P_p} det(\boldsymbol{J_g}(\boldsymbol{x_i}))\, V(C_i) = S(|det(\boldsymbol{J_g})|, \boldsymbol{\sigma_p}, P_p)$$

implying for $p > p_0$,

$$V(g(C(s, \boldsymbol{x_0}))) \leq (1 + \epsilon) S(|det(\boldsymbol{J_g})|, \boldsymbol{\sigma_p}, P_p)$$

We know $|det(\boldsymbol{J_g})|$ *is integrable on* $C(s, \boldsymbol{x_0})$ *and since* $\|P_p\| \to 0$, *we know* $S(|det(\boldsymbol{J_g})||, \boldsymbol{\sigma_p}, P_p) \to \int_{g(C(s,\boldsymbol{x_0}))} |det(\boldsymbol{J_g})|$. *Since* $\epsilon > 0$ *is arbitrary, we know*

$$V(g(C(s, \boldsymbol{x_0}))) \leq \int_{g(C(s,\boldsymbol{x_0}))} |det(\boldsymbol{J_g})|$$

To finish, we let P *be a partition of* $g(D)$ *into cubes* C_i. *Then each* C_i *is contained in a cube* W_i *and* $d(D) \subset \cup W_i$. *Since* $g(D)$ *has volume, so does* $g(C_i)$ *and*

$$\int_{g(C_i)} \boldsymbol{1_{g(C_i)}} = V(g(C_i))$$

Hence,

$$\begin{aligned}\int_{g(D)} \mathbf{1}_{g(D)} &= V(g(D)) = \sum \int_{g(C_i)} \mathbf{1}_{g(C_i)} = \sum V(g(C_i)) \\ &\leq \sum \int_{C_i} |det(\boldsymbol{J_g})| = \int_D |det(\boldsymbol{J_g})|\end{aligned}$$

We have shown one part of the inequality we need: $\int_{g(D)} \mathbf{1}_{g(D)} \leq \int_D |det(\boldsymbol{J_g})|$.

We can also apply this argument to the composition $f \circ g$. *We have for* $p > p_0$

$$\begin{aligned}S(f, \boldsymbol{\sigma_p}, P_p) &\leq \sum_{P_p} f(g(\boldsymbol{x_i}))V(g(C_i)) \\ &< \sum_{P_p} f(g(\boldsymbol{x_i}))(1+\epsilon)|det(\boldsymbol{J_g}(\boldsymbol{x_i}))|\, V(C_i)\end{aligned}$$

Hence,

$$\int_{g(C(s,\boldsymbol{x_0}))} f \leq (1+\epsilon) \int_{C(s,\boldsymbol{x_0})} f \circ g |det(\boldsymbol{J_g})|$$

and since $\epsilon > 0$ *is arbitrary, we have* $\int_{g(C(s,\boldsymbol{x_0}))} f \leq \int_{C(s,\boldsymbol{x_0})} f \circ g|det(\boldsymbol{J_g})|$. *It then follows that* $\int_{(f\circ g)(D)} \mathbf{1}_{(f\circ g)(D)} = \int_{g(D)} f \leq \int_D f \circ g|det(\boldsymbol{J_g})|$. *Let's apply this result to* $\phi = (f \circ g)|det(\boldsymbol{J_g})|$ *and* $\psi = g^{-1}$. *Then, we have*

$$\int_{\psi(g(D))} \phi \leq \int_{g(D)} (\phi \circ \psi)|det(\boldsymbol{J_\psi})|$$

or

$$\begin{aligned}\int_D (f \circ g)|det(\boldsymbol{J_g})| &\leq \int_{g(D)} (((f \circ g)|det(\boldsymbol{J_g})|) \circ g^{-1})|det(\boldsymbol{J_{g^{-1}}})| \\ &\leq \int_{g(D)} f|det(\boldsymbol{J_g}(g^{-1}))|\,|det(\boldsymbol{J_{g^{-1}}})| = \int_{g(D)} f\end{aligned}$$

In particular, for the constant map $f = \mathbf{1}$, $f \circ g = \mathbf{1} \circ g = \mathbf{1}$ *and we have*

$$\int_D (\mathbf{1} \circ g)|det(\boldsymbol{J_g})| = \int_D |det(\boldsymbol{J_g})| \leq \int_{g(D)} \mathbf{1}$$

This provides the other half of the inequality we need. Combining, we have shown $\int_D det(\boldsymbol{J_g})| = \int_{g(D)} \mathbf{1}_{g(D)}$. *So we have shown the Change of Variable Theorem works for constant maps* f. ■

Next, we extend from a constant map to a general map.

Theorem 15.2.3 The Change of Variable Theorem for a General Map

Let $D \subset \Re^n$ *be open and bounded. Let* $g : D \rightarrow \Re^n$ *have continuous partial derivatives on* D *and assume* g *is* $1-1$ *on* D *with* $\boldsymbol{J_g}(\boldsymbol{x}) \neq 0$ *on* D. *If* $g(D)$ *has volume, then for* $f : g(D) \rightarrow \Re$ *integrable,* $f \circ g|det\, \boldsymbol{J_g}|$ *is also integrable and* $\int_{g(D)} f = \int_D f \circ g|det\, \boldsymbol{J_g}|$.

Proof 15.2.3

We assume f is integrable on $g(D)$. Let R be a rectangle containing $g(D)$ and P be a partition of $g(D)$ using this rectangle. For any rectangle S determined by P, define the function h_S by $h_S(\boldsymbol{x}) = \inf_{\boldsymbol{x} \in S} f(\boldsymbol{x})$. Then, h_S is a constant on S. Then

$$L(f,P) = \sum_{S \in P} (\inf_{\boldsymbol{X} \in S} f(\boldsymbol{x}))\, V(S) = \sum_{S \in P} h_S V(S) = \sum_{S \in P} \int_{S \in P} h_S$$

Now apply the Change of Variable Theorem to the constant function h_S. We have

$$\int_S h_S = \int_{g^{-1}(S)} h_S \circ g |det\, \boldsymbol{J_g}|$$

So

$$\int_{g^{-1}(S)} h_S \circ g |det\, \boldsymbol{J_g}| = \int_{g^{-1}(S)} \left(\inf_{\boldsymbol{x} \in g^{-1}(S)} f \circ g(\boldsymbol{x}) \right) |det\, \boldsymbol{J_g}|$$

Combining,

$$\begin{aligned} L(f,P) &= \sum_{S \in P} \int_{S \in P} h_S = \sum_{S \in P} \int_{g^{-1}(S)} \left(\inf_{\boldsymbol{x} \in g^{-1}(S)} f \circ g(\boldsymbol{x}) \right) |det\, \boldsymbol{J_g}| \\ &\leq \sum_{S \in P} \int_{g^{-1}(S)} f \circ g(\boldsymbol{x})\, |det\, \boldsymbol{J_g}| \end{aligned}$$

But $\sum_{S \in P} \int_{g^{-1}(S)} = \int_D$. Hence, we have shown $L(f,P) \leq \int_D f \circ g |det \boldsymbol{J_g}|$. This implies

$$\underline{\int}_{g(D)} f = \sup_P L(f,P) \leq \int_D f \circ g |det \boldsymbol{J_g}|$$

Since we assume f is integrable over $g(D)$, we conclude $\int_{g(D)} f \leq \int_D f \circ g |det \boldsymbol{J_g}|$.

We can use a similar argument to show

$$\inf_P U(f,P) \geq \int_D f \circ g |det \boldsymbol{J_g}|$$

Thus, $\overline{\int}_{g(D)} f \circ g \geq= \int_D f \circ g |det \boldsymbol{J_g}$. But we assume f is integrable on $g(D)$ and therefore, we have shown

$$\int_D f \circ g |det \boldsymbol{J_g}| \leq \int_{g(D)} f \leq \int_D f \circ g |det \boldsymbol{J_g}|$$

which shows $\int_{g(D)} f = \int_D f \circ g |det \boldsymbol{J_g}|$. We have shown the Change of Variable Theorem hold for a general f. ■

15.2.1 Homework

Exercise 15.2.1 *Work out the Change of Variable results for polar coordinate transformations.*

Exercise 15.2.2 *Work out the Change of Variable results for cylindrical coordinate transformations.*

Exercise 15.2.3 *Work out the Change of Variable results for spherical coordinate transformations.*

Exercise 15.2.4 *Work out the Change of Variable results for ellipsoidal coordinate transformations.*

Exercise 15.2.5 *Work out the Change of Variable results for paraboloidal coordinate transformations.*

15.3 Fubini Type Results

We all remember how to evaluate double integrals for the area of a region D by rewriting them like this

$$\int_D f = \int_a^b \int_{f(x)}^{g(x)} F(x,y)dy\, dx$$

where $\int_D f$ is the two dimensional integral we have been developing. Expressing this integral in terms of successive single variable integrals leads to what are called **iterated** integrals which is how we commonly evaluate them in our calculus classes. Thus, using the notation $\int dx$, $\int dy$ etc. to indicate these one dimensional integrals along one axis only, we want to know when

$$\begin{aligned}
\int_D F &= \int \left(\int F(x,y)\, dy \right) dx = \int \int F(x,y)\, dy\, dx \\
\int_D F &= \int \left(\int F(x,y)\, dx \right) dy = \int \int F(x,y)\, dx\, dy \\
\int_\Omega G &= \int \left(\int \left\{ \int G(x,y,z)\, dy \right\} dx \right) dz = \int \int \int G(x,y,z)\, dy\, dx\, dz \\
\int_\Omega G &= \int \left(\int \left\{ \int G(x,y,z)\, dy \right\} dz \right) dx = \int \int \int G(x,y,z)\, dy\, dz\, dx \\
\int_\Omega G &= \int \left(\int \left\{ \int G(x,y,z)\, dx \right\} dy \right) dz = \int \int \int G(x,y,z)\, dx\, dy\, dz \\
\int_\Omega G &= \int \left(\int \left\{ \int G(x,y,z)\, dx \right\} dz \right) dy = \int \int \int G(x,y,z)\, d\,xd\,zdy \\
\int_\Omega G &= \int \left(\int \left\{ \int G(x,y,z)\, dz \right\} dx \right) dy = \int \int \int G(x,y,z)\, dz\, dx\, dy \\
\int_\Omega G &= \int \left(\int \left\{ \int G(x,y,z)\, dz \right\} dy \right) dx = \int \int \int G(x,y,z)\, dz\, dy\, dx
\end{aligned}$$

where D is a bounded open subset of $\Re^2$ and Ω is a bounded open subset of $\Re^3$. Note in the Ω integration there are many different ways to organize the integration. If we were doing an integration over $\Re^n$ there would in general be a lot of choices! The conditions under which we can do this give rise to what are called Fubini type theorems. If general, you think of the bounded open set V as being in $\Re^{n+m} = \Re^n \times \Re^m$, label the volume elements as $\boldsymbol{dV}_n$ and $\boldsymbol{dV}_m$ and want to know when

$$\begin{aligned}
\int_V F &= \int \left(\int F(x_1, \ldots, x_n, x_{n+1}, \ldots, x_{n+m})\, \boldsymbol{dV}_n \right) \boldsymbol{dV}_m \\
&= \int \left(\int F \boldsymbol{dV}_n \right) \boldsymbol{dV}_m = \int \int F\, \boldsymbol{dV}_n\, \boldsymbol{dV}_m
\end{aligned}$$

We are only going to work out a few of these type theorems so that you can get a taste for the ways we prove these results.

15.3.1 Fubini on a Rectangle

Let's specialize to the bounded set $D = [a, b] \times [c, d] \subset \Re^2$. Assume $f : D \to \Re$ is continuous. There are three integrals here:

- The integral over the two dimensional set D, $\int_D f \boldsymbol{dV}_2$, where the subscript 2 reminds us that this is the two dimensional situation.

- The integral of functions like $g : [a, b] \to \Re$. Assume g is continuous. Then the one dimensional integral would be represented as $\int_a^b g(x)dx$ in the integration treatment (Peterson (21) 2020) but here we will use $\int_{[a,b]} g\boldsymbol{dV}_1$.

- The integral of functions like $h : [c, d] \to \Re$. Assume h is continuous. Then the one dimensional integral $\int_c^d h(y)dy$ is called $\int_{[c,d]} h\boldsymbol{dV}_2$. Note although the choice of integration variable y in $\int_c^d g(y)dy$ is completely arbitrary, it makes a lot of sense to use this notation instead of $\int_c^d g(x)dx$ because we want to cleanly separate what we are doing on each axis from the other axis. So we have to exercise a bit of discipline here to set up the variable names and notation to help us understand.

Now any partition P of D divides the bounding box R we choose to enclose D into rectangles. If $P = P_1 \times P_2$ where $P_1 = \{a = x_0, x_1, \ldots, x_{n-1}, x_n = b\}$ and $P_2 = \{c = y_0, y_1, \ldots, y_{m-1}, y_m = d\}$, the rectangles have the form

$$S_{ij} = [x_i, x_{i+1}] \times [y_j, y_{j+1}] = \Delta x_i \times \Delta y_j$$

where $\Delta x_i = x_{i+1} - x_i$ and $\Delta y_j = y_{j+1} - y_j$. Thus, for an evaluation set $\boldsymbol{\sigma}$ with points $(z_i, w_j) \in S_{ij}$, the Riemann sum is

$$S(f, \boldsymbol{\sigma}, P) = \sum_{(i,j)\in P} f(z_i, w_j)\, \Delta x_i\, \Delta y_j = \sum_{(i,j)\in P} f(z_i, w_j)\, \Delta y_j\, \Delta x_i$$

Note the order of the terms in the area calculation for each rectangle do not matter. Also, we could call $\Delta x_i = \Delta V_{1i}$ (for the first axis) and $\Delta y_j = \Delta V_{2j}$ (for the second axis) and rewrite as

$$S(f, \boldsymbol{\sigma}, P) = \sum_{(i,j)\in P} f(z_i, w_j)\, \Delta V_{1i}\, \Delta V_{2j} = \sum_{(i,j)\in P} f(z_i, w_j)\, \Delta V_{2j}\, \Delta V_{1i}$$

We can also revert to writing $\sum_{(i,j)\in P}$ as a double sum to get

$$S(f, \boldsymbol{\sigma}, P) = \sum_{i=1}^{n} \left(\sum_{j=1}^{m} f(z_i, w_j)\, \Delta V_{2j} \right) \Delta V_{1i} = \sum_{j=1}^{m} \left(\sum_{i=1}^{n} f(z_i, w_j)\, \Delta V_{1i} \right) \Delta V_{2j}$$

Then the idea is this:

- **y first then x**: For $\sum_{i=1}^n \left(\sum_{j=1}^m f(z_i, w_j)\, \Delta V_{2j} \right) \Delta V_{1i}$, fix the variables on axis one and suppose

$$\lim_{\|P_2\|\to 0} \sum_{j=1}^{m} f(z_i, w_j)\, \Delta V_{2j} = \int_{z_i \times [c,d]} f(z_i, y)dy$$

To avoid using specific sizes like m for P_2 here, we would usually just say

$$\lim_{\|P_2\|\to 0} \sum_{j\in P_2} f(z_i, w_j)\, \Delta V_{2j} \quad = \quad \int_{z_i\times[c,d]} f(z_i, y)dy$$

Now, for this to happen we need the function $f(\cdot, y)$ to be integrable for each choice of x for slot one. An easy way to do this is to assume f is continuous in $(x, y) \in D$, which will force $f(\cdot, y)$ to be a continuous function of y for each x. Let's look at this integral more carefully. We have defined a new function $G(x)$ by

$$G(x) \quad = \quad \int_{[c,d]} f(x, y)dy = \int_c^d f(x, y)dy$$

If $G(x)$ is integrable on $[a, b]$, we can say

$$\lim_{\|P_1\|\to 0} \sum_{i=1}^{n} \left(\lim_{\|P_2\|\to 0} \sum_{j=1}^{m} f(z_i, w_j)\, \Delta V_{2j} \right) \Delta V_{1i} =$$

$$\lim_{\|P_1\|\to 0} \sum_{i=1}^{n} \left(\int_{z_i\times[c,d]} f(z_i, y)dy \right) \Delta V_{1i} = \lim_{\|P_1\|\to 0} \sum_{i=1}^{n} G(z_i)\Delta V_{1i} = \int_a^b G(x)dx$$

We can resay this again. Let ΔV_{ij} be the area $\Delta x_i\, \Delta y_j$. Then

$$\int_D f\boldsymbol{dV}_2 \quad = \quad \lim_{\|P1\times P_2\|\to 0} \sum_{(i,j)\in P_1\times P_2} f(z_i, w_j)\Delta V_{ij}$$

This can be rewritten as

$$\lim_{\|P_1\|\to 0} \sum_{i\in P_1} \left(\lim_{\|P_2\|\to 0} \sum_{j\in P_2} f(z_i, w_j)\, \Delta V_{2j} \right) \Delta V_{1i} = \lim_{\|P_1\|\to 0} \sum_{i\in P_1} \left(\int_c^d f(z_i, y)\, dy \right) \Delta V_{1i}$$

$$= \int_a^b \left(\int_c^d f(x, y)\, dy \right) dx = \int_a^b \int_c^d f(x, y)\, dy\, dx$$

- **x first then y**: For $\sum_{j=1}^{m} \left(\sum_{i=1}^{n} f(z_i, w_j)\, \Delta V_{1i} \right) \Delta V_{2j}$, fix the variables on axis two and suppose

$$\lim_{\|P_1\|\to 0} \sum_{i=1}^{n} f(z_i, w_j)\, \Delta V_{1i} \quad = \quad \int_{[a,b]\times w_j} f(x, w_j)dx$$

Of course, this is the same as

$$\lim_{\|P_1\|\to 0} \sum_{i\in P_1} f(z_i, w_j)\, \Delta V_{1i} \quad = \quad \int_{[a,b]\times w_j} f(x, w_j)dx$$

Now, for this to happen we need the function $f(x, \cdot)$ to be integrable for each choice of y for slot two. If we assume f is continuous in $(x, y) \in D$, this will force $f(x, \cdot)$ to be a continuous function of x for each y. Let's look at this integral more carefully. We have defined a new function $H(y)$ by

$$H(y) = \int_{[a,b]} f(x,y)dx = \int_a^b f(x,y)dx$$

If $H(y)$ is integrable on $[c,d]$, we can say

$$\begin{aligned}
\lim_{\|P_2\|\to 0} \sum_{j=1}^{m} \left(\lim_{\|P_1\|\to 0} \sum_{i=1}^{n} f(z_i, w_j)\, \Delta V_{1i} \right) \Delta V_{2j} &= \lim_{\|P_2\|\to 0} \sum_{j=1}^{m} \left(\int_{[a,b]\times w_j} f(x, w_j)dx \right) \Delta V_{2j} \\
&= \lim_{\|P_2\|\to 0} \sum_{j=1}^{m} H(w_j)\Delta V_{2j} = \int_c^d H(y)dy
\end{aligned}$$

or

$$\begin{aligned}
\int_D f\boldsymbol{dV}_2 &= \lim_{\|P1\times P_2\|\to 0} \sum_{(i,j)\in P_1\times P_2} f(z_i, w_j)\Delta V_{ij} \\
&= \lim_{\|P_2\|\to 0} \sum_{j\in P_2} \left(\lim_{\|P_1\|\to 0} \sum_{i\in P_1} f(z_i, w_j)\, \Delta V_{1i} \right) \Delta V_{2j} \\
&= \lim_{\|P_2\|\to 0} \sum_{j\in P_2} \left(\int_a^b f(x, w_j)\, dx \right) \Delta V_{2j} \\
&= \int_c^d \left(\int_a^b f(x,y)\, dx \right) dy = \int_c^d \int_a^b f(x,y)\, dx\, dy
\end{aligned}$$

We have laid out an approximate chain of reasoning to prove our first Fubini result. Let's get to it.

Theorem 15.3.1 Fubini's Theorem on a Rectangle: 2D

Let $f : [a,b] \times [c,d] \to \Re$ be continuous. Then

$$\begin{aligned}
\int_D f\boldsymbol{dV}_2 &= \int_a^b \left(\int_c^d f(x,y)\, dy \right) dx = \int_a^b \int_c^d f(x,y)\, dy\, dx \\
\int_D f\boldsymbol{dV}_2 &= \int_c^d \left(\int_a^b f(x,y)\, dx \right) dy = \int_c^d \int_a^b f(x,y)\, dx\, dy
\end{aligned}$$

Proof 15.3.1

For $x \in [a,b]$, let G be defined by $G(x) = \int_c^d f(x,y)dy$. We need to show G is continuous for each x. Note since f is continuous on $[a,b] \times [c,d]$, given $\epsilon > 0$, there is a $\delta > 0$ so that

$$\sqrt{(x'-x)^2 + (y'-y)^2} < \delta \implies |f(x,y) - f(x',y')| < \epsilon$$

In particular, for fixed $y' = y$, there is a $\delta > 0$ so that

$$|x' - x| < \delta \implies |f(x',y) - f(x,y)| < \epsilon/(d-c)$$

Thus, if $|x' - x| < \delta$,

$$|G(x') - G(x)| = \left| \int_c^d (f(x',y) - f(x,y))dy \right| \le \int_c^d |f(x',y) - f(x,y)|dy$$

$$\leq \quad (d-c)\,\epsilon/(d-c) = \epsilon$$

and so we know G is continuous in x for each y. A similar argument shows the function H defined at each $y \in [c,d]$ by $H(y) = \int_a^b f(x,y)dy$ is also continuous in y for each x.

Since f is continuous on $[a,b] \times [c,d]$, and the boundary of D has measure zero, we see the set of discontinuities of f on D is a set of measure zero and so f is integrable on D. Let $P_n = P_{1n} \times P_{2n}$ be a sequence of partitions with $\|P_n\| \to 0$. Then, we know for any evaluation set σ of P, that

$$\int_D f \quad = \quad \lim_{\|P_n\| \to 0} \sum_{i \in P_{1n}} \sum_{j \in P_{2n}} f(z_i, w_j)\, \Delta x_i\, \Delta y_j$$

and following the reasoning we sketch out earlier, we can choose to organize this as **x first then y** *or* **y first then x**. *We will do the case of* **x first then y** *only and leave the details of the other case to you. fix $\epsilon > 0$, Then there is N, so that $n > N$ implies*

$$\left| \sum_{j \in P_{2n}} \left(\sum_{i \in P_{1n}} f(z_i, w_j)\, \Delta x_i \right) \Delta y_j - \int_D f \right| < \epsilon/2$$

Since $f(x, w_j)$ is continuous in x for each w_j, we know $\int_a^b f(x, w_j)dx$ exists and so for any sequence of partitions Q_n of $[a,b]$ with $\|Q_n\| \to 0$, using as the evaluation set the points z_i, we have

$$\sum_{i \in Q_n} f(z_i, w_j)\, \Delta x_i \quad \to \quad \int_a^b f(x, w_j)dx.$$

Since $\|P_n\| \to 0$, so does the sequence P_{1n}, there is $N_1 < N$ so that $n > N_1$ implies

$$\left| \sum_{i \in P_{1n}} f(z_i, w_j)\, \Delta x_i - \int_a^b f(x, w_j)dx \right| = \left| \sum_{i \in P_{1n}} f(z_i, w_j)\, \Delta x_i - H(w_j) \right| \quad < \quad \epsilon/(4(d-c))$$

Similarly, since $\|P_{2n}\| \to 0$ and H is integrable, there is $N_2 > N_1 > N$ so that

$$\left| \sum_{j \in P_{2n}} H(w_j)\, \Delta y_j - \int_c^d H(y)dy \right| < \epsilon/4$$

where the points w_j form the evaluation set we use in each P_{2n}. Thus,

$$\begin{aligned}
&\left| \sum_{j \in P_{2n}} \left(\sum_{i \in P_{1n}} f(z_i, w_j)\, \Delta x_i \right) \Delta y_j - \int_c^d H(y)dy \right| \\
&= \left| \sum_{j \in P_{2n}} \left(\sum_{i \in P_{1n}} f(z_i, w_j)\, \Delta x_i - H(w_j) + H(w_j) \right) \Delta y_j - \int_c^d H(y)dy \right| \\
&\leq \left| \sum_{j \in P_{2n}} \left(\sum_{i \in P_{1n}} f(z_i, w_j)\, \Delta x_i - H(w_j) \right) \Delta y_j \right| + \left| \sum_{j \in P_{2n}} H(w_j)\Delta y_j - \int_c^d H(y)dy \right| \\
&\leq \sum_{j \in P_{2n}} \left| \sum_{i \in P_{1n}} f(z_i, w_j)\, \Delta x_i - H(w_j) \right| \Delta y_j + \left| \sum_{j \in P_{2n}} H(w_j)\Delta y_j - \int_c^d H(y)dy \right| \\
&< \sum_{j \in P_{2n}} \epsilon/(4(d-c))\Delta y_j + \epsilon/4 = \epsilon/2
\end{aligned}$$

We can now complete the argument. For $n > N_2$,

$$\begin{aligned}\left|\int_c^d H(y)dy - \int_d f\right| &\leq \left|\sum_{j\in P_{2n}} \left(\sum_{i\in P_{1n}} f(z_i, w_j)\, \Delta x_i\right) \Delta y_j - \int_D f\right| \\ &+ \left|\sum_{j\in P_{2n}} \left(\sum_{i\in P_{1n}} f(z_i, w_j)\, \Delta x_i\right) \Delta y_j - \int_c^d H(y)dy\right| \\ &< \epsilon/2 + \epsilon/2 = \epsilon\end{aligned}$$

Thus

$$\int_D f = \int_c^d H(y)dy = \int_c^d \left(\int_a^b f(x,y)\, dx\right) dy = \int_c^d \int_a^b f(x,y)\, dx\, dy$$

A very similar argument handles the case **y first then x** *giving*

$$\int_D f = \int_a^b G(x)dx = \int_a^b \left(\int_c^c f(x,y)\, dy\right) dx = \int_a^b \int_c^d f(x,y)\, dy\, dx$$

■

As you can see, the argument to prove Theorem 15.3.1 is somewhat straightforward but you have to be careful to organize the inequality estimates. A similar approach can prove the more general theorem.

Theorem 15.3.2 Fubini's Theorem for a Rectangle: n dimensional

Let $f : R_1 \times R_2 \to \Re$ be continuous where R_1 is a rectangle in $\Re^n$ and $R_2 \in \Re^m$. Then

$$\begin{aligned}\int_D f\boldsymbol{dV}_{n+m} &= \int_{R_1} \left(\int_{R_2} f(\boldsymbol{x},\boldsymbol{y})\, \boldsymbol{dV}_m\right) \boldsymbol{dV}_n = \int_{R_1} \int_{R_2} f(\boldsymbol{x},\boldsymbol{y})\, \boldsymbol{dV}_m\, \boldsymbol{dV}_n \\ \int_D f\boldsymbol{dV}_{n+m} &= \int_{R_2} \left(\int_{R_1} f(\boldsymbol{x},\boldsymbol{y})\, \boldsymbol{dV}_n\right) \boldsymbol{dV}_m = \int_{R_2} \int_{R_1} f(\boldsymbol{x},\boldsymbol{y})\, \boldsymbol{dV}_n\, \boldsymbol{dV}_m\end{aligned}$$

where $\boldsymbol{x}$ are the variables $x_1, \ldots, x_n$ from $\Re^n$ and $\boldsymbol{y}$ are the variables $x_{n+1}, \ldots, x_{n+m}$.

Proof 15.3.2

For example, $f : [a_1, b_1] \times [a_2, b_2] \times [a_3, b_3] \times [a_4, b_4] \to \Re$ would use $R_1 = [a_1, b_1] \times [a_2, b_2]$ and $R_2 = [a_3, b_3] \times [a_4, b_4]$. The proof here is quite similar. We prove $G(\boldsymbol{x})$ and $H(\boldsymbol{y})$ are continuous since f is continuous. Then, since f is integrable, given any sequence of partitions P_n with $\|P_n\| \to 0$, there is N so that $n > N$ implies

$$\left|\sum_{j\in P_{2n}} \left(\sum_{i\in P_{1n}} f(\boldsymbol{z}_i, \boldsymbol{w}_j)\, \Delta \boldsymbol{x}_i\right) \Delta \boldsymbol{y}_j - \int_D f\right| < \epsilon/2$$

where $P_n = P_{1n} \times P_{2n}$ like usual except P_{1n} is a partition of R_1 and P_{2n} is a partition of R_2. The summations like $\sum_{j\in P_{2n}}$ are still interpreted in the same way. The rest of the argument is straightforward. ■

Now let's specialize to some useful two dimensional situations. Consider the situation shown in Figure 15.1.

Theorem 15.3.3 Fubini's Theorem: 2D: a Top and Bottom Curve

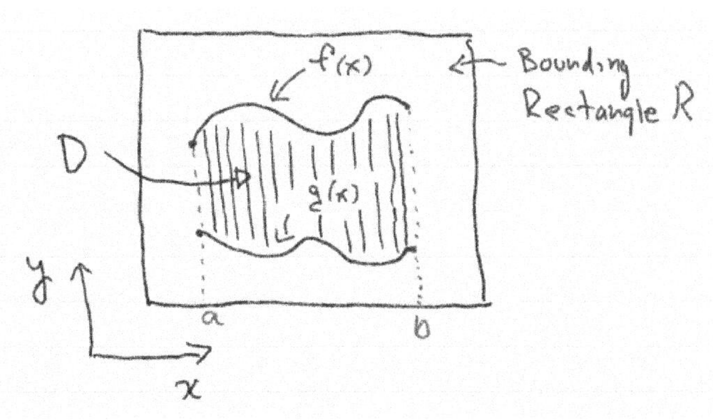

Figure 15.1: Fubini's Theorem in 2D: a top and a bottom curve.

Let f and g be continuous real-valued functions on $[a,b]$ and $F : D \rightarrow \Re$ be continuous on D where $D = \{(x,y) : a \leq x \leq b,\ g(x) \leq y \leq f(x)\}$. Then

$$\int_D F d\boldsymbol{V}_2 = \int_a^b \int_{g(x)}^{f(x)} dy\, dx$$

Proof 15.3.3

Let's apply Theorem 15.3.2 on the rectangle $[a,b] \times [g_m, f_M]$ where g_m is the minimum value of g on $[a,b]$ which we know exists because g is continuous on a compact domain. Similarly f_M is the maximum value of f on $[a,b]$ which also exists. Then the D showing in Figure 15.1 is contained in $[a,b] \times [g_m, f_M]$. We see $\hat{F}$ is continuous on this rectangle except possibly on the curves given by f and g. We also know the set $S_f = \{(x, f(x)) | a \leq x \leq b\}$ and the set $S_g = \{(x, g(x)) | a \leq x \leq b\}$ both have measure zero in $\Re^2$. Thus $\hat{F}$ has a discontinuity set of measure zero and so F is integrable on $[a,b] \times [g_m, f_M]$ which matches the integral of F on D.

Next, since F is continuous on the interior of D, for fixed x in $[a,b]$, $G(y) = F(x,y)$ is continuous on $[g(x), f(x)]$. Of course, this function need not be continuous at the endpoints but that is a set of measure zero in $\Re^1$ anyway. Hence, $\hat{G}$ is integrable on $[g_m, f_M]$ and hence G is integrable on $[g(x), f(x)]$. By Theorem 15.3.2, we conclude

$$\begin{aligned} \int_{[a,b]\times[g_m,f_M]} F d\boldsymbol{V}_2 &= \int_D F d\boldsymbol{V}_2 = \int_a^b \int_{g_m}^{f_M} G(y)\, dy\, dx \\ &= \int_a^b \int_{g(x)}^{f(x)} F(x,y)\, dy\, dx \end{aligned}$$

■

Next, look at a typical closed curve situation.

Theorem 15.3.4 Fubini's Theorem: 2D: a Top and Bottom Closed Curve

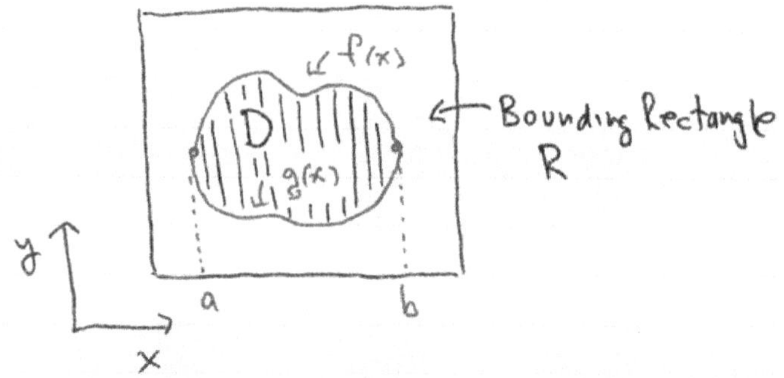

Figure 15.2: Fubini's Theorem in 2D: a top and bottom closed curve.

Let f and g be continuous real-valued functions on $[a,b]$ which form the bottom and top halves of a closed curve and $F : D \rightarrow \Re$ be continuous on D where $D = \{(x,y) : a \leq x \leq b,\ g(x) \leq y \leq f(x)\}$. Then

$$\int_D F d\boldsymbol{V}_2 = \int_a^b \int_{g(x)}^{f(x)} dy\, dx$$

Proof 15.3.4
From Figure 15.2, we see the closed curve defines a box $[a,b] \times [g_m, f_M]$ where g_m is the minimum value of g and f_M is the maximum value of f on $[a,b]$. These exist because f and g are continuous on the compact domain $[a,b]$. The rest of the argument is exactly the same as the proof of Theorem 15.3.3. We have

$$\begin{aligned} \int_{[a,b]\times[g_m,f_M]} F d\boldsymbol{V}_2 &= \int_D F d\boldsymbol{V}_2 = \int_a^b \int_{g_m}^{f_M} G(y)\, dy\, dx \\ &= \int_a^b \int_{g(x)}^{f(x)} F(x,y)\, dy\, dx \end{aligned}$$

■

It is easy to see we could prove similar theorems for a region $D = \{(x,y) | c \leq y \leq d,\ g(y) \leq x \leq f(y)\}$ for functions f and g continuous on $[c,d]$ and F continuous on D. We would find

$$\int_{[g_m,f_M]\times[c,d]} F d\boldsymbol{V}_2 = \int_D F d\boldsymbol{V}_2 = \int_c^d \int_{g(y)}^{f(y)} F(x,y)\, dx\, dy$$

Call the top-bottom region a TB region and the left-right region a LR region. Then you should be able to see how you can break a lot of regions into finite combinations of TB and LR regions and apply Fubini's Theorem to each piece to complete the integration.

Comment 15.3.1 *We can work out similar theorems in Re^3, $\Re^4$ and so forth. Naturally, this process gets complicated as the dimensions increase. Still, the principles of the arguments are the same. The boundary surfaces are all of measure zero as they are a strict subspace of $\Re^n$. You should try a few to see how the arguments are constructed. This is how you learn how things work: by doing.*

15.3.2 Homework

Exercise 15.3.1 *Prove Fubini's Theorem for rectangles in* $\Re^3$*. It is enough to prove it for one of the six cases.*

Exercise 15.3.2 *Since we can approximate these integrals using Riemann sums, can you estimate how computationally expensive this gets using uniform partitions? Would it be easy to compute a Riemann approximation to a 10 dimensional integral over a rectangle?*

Exercise 15.3.3 *This problem shows you how we can use these ideas to do something interesting!*

1. *Compute the two dimensional* $I_R = \int_{D_R} e^{-x^2-y^2}$ *where* $D_R = B(R,(0,0))$ *for all* R *by using the polar coordinate transformation. Can you compute* $\lim_{R\to\infty}$ *for this result?*
2. *Let* $J_R = \int_{S_R} e^{-x^2-y^2}dxdy$*. Prove* $\lim_{R\to\infty} J_R = \lim_{R\to\infty} I_R$.
3. *What is the value of* $\int_{-\infty}^{\infty} e^{-x^2}dx$*?*

Exercise 15.3.4 *Draw a bounded 2D region in* $\Re^2$ *which is constructed from two TB's and one LR and convince yourself how Fubini's theorem is applied to compute the integral.*

Exercise 15.3.5 *Draw an arbitrary bounded region in* $\Re^2$ *and decompose it into appropriate TB and LR regions on which Fubini's Theorem can be applied and convince yourself how the total integral is computed.*

Chapter 16

Line Integrals

It is now time to set the stage for the connection between a type of one dimensional integral and an equivalent two dimensional integral. This is challenging material and it has deep meaning. However, we will start out slow in this chapter and show you a low-level version of what it all means. Let's start with the idea of a **path** or **curve** in $\Re^2$.

16.1 Paths

Let's begin with curves in two dimensions.

Definition 16.1.1 Two Dimensional Curves

Let $[a,b]$ be a finite interval of real numbers. Let f and g be two continuously differentiable functions on $[a,b]$. The path defined by these functions is the set of ordered pairs

$$\mathscr{C} = \{(x,y) | x = f(t), y = g(t), \quad a \leq t \leq b\}$$

If the starting and ending point of the curve are the same, we say the curve is **closed**.

Comment 16.1.1 *We often abuse this notation and talk about the path $\mathscr{C}$ being given by the pairs (x,y) with*

$$x = x(t), \quad y = y(t), \quad a \leq t \leq b$$

even though the letters x and y are being used in two different ways. In practice, it is not that hard to get the two uses separate.

The curve $\mathscr{C}$ has a tangent line that is defined at each point t_0 given by

$$\begin{bmatrix} x \\ y \end{bmatrix} = \begin{bmatrix} x(t_0) \\ y(t_0) \end{bmatrix} + c \begin{bmatrix} x'(t_0) \\ y'(t_0) \end{bmatrix}$$

and has a tangent vector $\boldsymbol{T}(t_0)$ given by

$$\boldsymbol{T}(t_0) = \begin{bmatrix} x'(t_0) \\ y'(t_0) \end{bmatrix}$$

and an associated vector perpendicular to $\boldsymbol{T}(t_0)$ given

$$\boldsymbol{N}(t_0) = \begin{bmatrix} -y'(t_0) \\ x'(t_0) \end{bmatrix}$$

You can see $< \boldsymbol{T}(t_0), \boldsymbol{N}(t_0) >= 0$. Also note

$$\boldsymbol{N}^*(t_0) = \begin{bmatrix} y'(t_0) \\ -x'(t_0) \end{bmatrix}$$

is also perpendicular to $\boldsymbol{T}(t_0)$. Both choices for this perpendicular vector are called the **Normal** vector to the curve $\mathscr{C}$ at t_0 Using the right-hand rule, we can calculate $\boldsymbol{T}(t_0) \times \boldsymbol{N}(t_0)$. This will either be $\boldsymbol{k}$ or $-\boldsymbol{k}$ depending on whether to get to $\boldsymbol{N}(t_0)$ from $\boldsymbol{T}(t_0)$, we need to move our right hand counterclockwise (ccw) (this gives $+\boldsymbol{k}$) or clockwise (this gives $-\boldsymbol{k}$). Look at Figure 16.1 to get a feel for this. We are deliberately showing you hand drawn figures as that is what we would do in a lecture and what you should learn to do on your scratch paper as you read this.

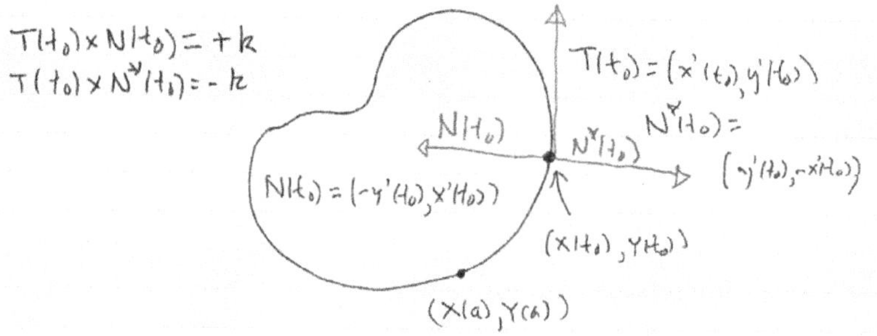

Figure 16.1: A simple closed curve with tangent and normal vectors.

In Figure 16.2, we show a self-intersecting curve like a figure eight. Note in the right half, the tangent vector moves ccw and in the left half, the tangent vector moves cw. We indicate the direction of both $\boldsymbol{N}$ and $\boldsymbol{N}^* = -\boldsymbol{N}$ also. Note, if $\boldsymbol{T}$ is chosen, we cannot decide which of $\boldsymbol{N}$ or $-\boldsymbol{N}$. The point $\boldsymbol{Q}$ here indicates where the tangent vector switches from moving ccw to moving cw. Then in Figure 16.3, we show the same figure eight curve with what we would clearly decide was the inside and outside of the curve. Now the curve $\mathscr{C}$ is defined by the two functions $x(t)$ and $y(t)$ which are the components of a vector field we can call γ. Thus, we can identify the curve $\mathscr{C}$ and the range of the vector function γ. Hence, in Figure 16.3, instead of labeling the curve as $\mathscr{C}$ we call it γ. We use the γ notation with some changes in definition in Chapter 17 where we discuss another way of looking at all of this using the language of differential forms. And, thinking more carefully about insides and outsides of curves will lead us into ideas from homology. So there is much to do! But for right now, we can already see it is not so easy to decide how to pick the normal – should it be $\boldsymbol{N}$ or $-\boldsymbol{N}$? It appears it should be $\boldsymbol{N}$ is $\boldsymbol{T}$ is moving ccw and $-\boldsymbol{N}$ if $\boldsymbol{T}$ is moving cw. Now, scribble out a picture of a curve with two self intersections or four self intersections with concomitant changes in ccw and cw movement of $\boldsymbol{T}$ as the parameter t increases. We want to believe, based on simple examples, that each closed curve γ divides $\Re^2$ into three disjoint pieces: the **interior**, **exterior** and the curve itself.

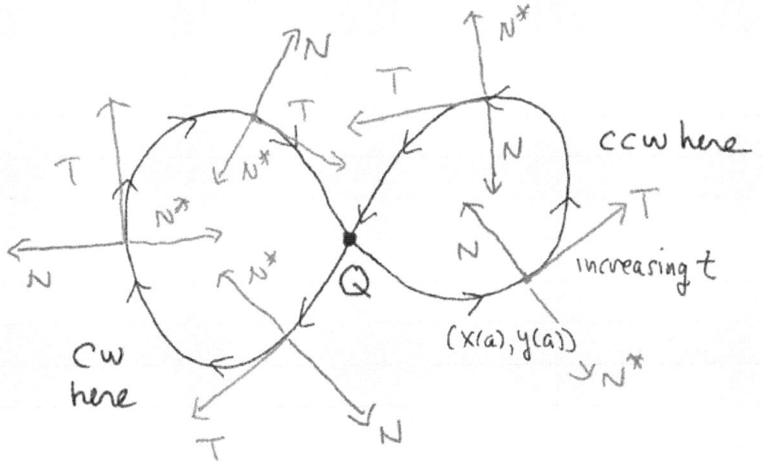

Figure 16.2: A figure eight curve with tangent and normal vectors.

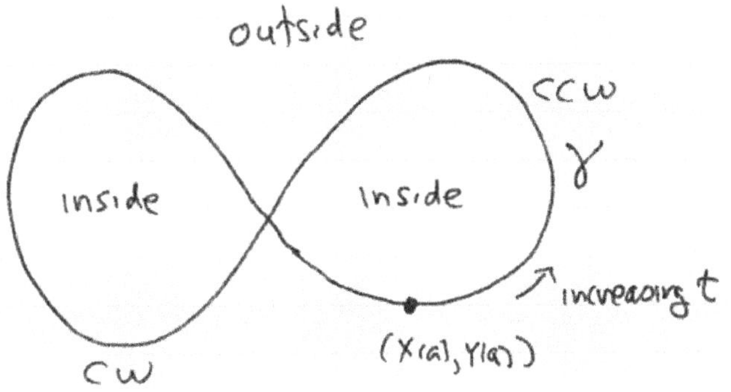

Figure 16.3: The inside and outside of a curve.

Hence, if we want to point towards the interior of the closed curve, we should use $\boldsymbol{N}(t)$ if a point on the curve is moving ccw and otherwise, use $-\boldsymbol{N}(t)$. Also, note we can choose $\pm\boldsymbol{N}$ and $\boldsymbol{T}$ to be unit vectors by simply dividing them by their length $||\boldsymbol{T}(t)||$ and $||\boldsymbol{N}(t)||$, respectively. The only time this becomes impossible is when these norms are zero. This only occurs at a point t where both $x'(t)$ and $y'(t) = 0$. We can usually figure out a way to define a nice tangent and normal anyway, but we won't complicate the discussion here by that. Now, as long as the functions $x(t)$ and $y(t)$ are continuously differentiable on $[a, b]$, consider a vector field

$$\boldsymbol{F} = \begin{bmatrix} A(x, y) \\ B(x, y) \end{bmatrix}$$

defined on a open set U of $\Re^2$ which contains the curve $\mathscr{C}$. We define the action of $\boldsymbol{F}$ along $\mathscr{C}$ to be the **line integral** of $\boldsymbol{F}$ along $\mathscr{C}$ from the starting point of the curve to its ending point using a very specific Riemann integral.

Definition 16.1.2 The Line Integral of a Force Field along a Curve

Let the vector field $\boldsymbol{F}$ on an open set U in $\Re^2$ be defined by

$$\boldsymbol{F} = \begin{bmatrix} A(x,y) \\ B(x,y) \end{bmatrix}$$

where A and B are at least continuous at each point on $\mathscr{C}$. Then the line integral of F along $\mathscr{C}$ from $\boldsymbol{P} = (x(a), y(a))$ to $\boldsymbol{q} = (x(b), y(b))$ is defined by

$$\begin{aligned} \int_{\mathscr{C}} \left\langle \boldsymbol{F}, \begin{bmatrix} x' \\ y' \end{bmatrix} \right\rangle dt &= \int_{\mathscr{C}} \left(\left\langle \begin{bmatrix} A(x,y) \\ B(x,y) \end{bmatrix}, \begin{bmatrix} x'(t) \\ y'(t) \end{bmatrix} \right\rangle \right) dt \\ &= \int_a^b (A(x(t), y(t))x'(t) + B(x(t), y(t))y'(t))\, dt \end{aligned}$$

Since $\Phi(t) = A(x(t), y(t))x'(t) + B(x(t), y(t))y'(t)$ is continuous on $[a, b]$, this Riemann integral exists and so the line integral is well-defined.

Example 16.1.1 *Suppose each point on $\mathscr{C}$ represents the electric field at the point t. Then how do we compute the total work done in moving a test charge from the beginning point of the curve to the ending point? First, note the function*

$$\Phi(t) = \boldsymbol{A}(x(t), y(t))\, x'(t) + \boldsymbol{B}(x(t), y(t))\, y'(t)$$

is a continuous function on $[a, b]$ and so is Riemann Integrable. Hence, the value of this integral is given by

$$\lim_{n \to \infty} S(\Phi, \pi_n, \sigma_n)$$

for any sequence of partitions $\{\pi_n\}$ whose norms go to zero and σ_n is any evaluation set in π_n. Then, for a given partition, the charge in the portion of the path between partition point t_j and t_{j+1} can be approximated by

$$\begin{aligned} &\boldsymbol{A}(x(z_j), y(z_j))(x(t_{j+1}) - x(t_j)) + \boldsymbol{B}(x(z_j), y(z_j))(y(t_{j+1}) - y(t_j)) = \\ &\left(\boldsymbol{A}(x(z_j), y(z_j)) \frac{x(t_{j+1}) - x(t_j)}{t_{j+1} - t_j} + \boldsymbol{B}(x(z_j), y(z_j)) \frac{y(t_{j+1}) - y(t_j)}{t_{j+1} - t_j} \right) (t_{j+1} - t_j) \end{aligned}$$

for any choice of z_y in $[t_j, t_{j+1}]$. Now apply the Mean Value Theorem to see

$$\begin{aligned} &\left(\boldsymbol{A}(x(z_j), y(z_j)) \frac{x(t_{j+1}) - x(t_j)}{t_{j+1} - t_j} + \boldsymbol{B}(x(z_j), y(z_j)) \frac{y(t_{j+1}) - y(t_j)}{t_{j+1} - t_j} \right) (t_{j+1} - t_j) = \\ &\left(\boldsymbol{A}(x(z_j), y(z_j))x'(s_j) + \boldsymbol{B}(x(z_j), y(z_j))y'(s_j) \right) (t_{j+1} - t_j) \end{aligned}$$

The approximation to the mass on the portion $[t_j, t_{j+1}]$ can just as well be done setting $z_j = s_j$ in the $\boldsymbol{A}$ and $\boldsymbol{B}$ terms. Thus, the approximation can be

$$\Phi(s_j)\,(t_{j+1} - t_j) = \left(\boldsymbol{A}(x(s_j), y(s_j))x'(s_j) + \boldsymbol{B}(x(s_j), y(s_j))y'(s_j) \right) (t_{j+1} - t_j)$$

The Riemann Approximation Theorem then tells us

$$\lim_{n \to \infty} \sum_{\pi_n} \Phi(s_j) \Delta t_j = \int_a^b \Phi(t) dt$$

which is how we defined the line integral.

Comment 16.1.2 *In the same setting as the previous example, we could find the work done along a closed path* $\mathscr{C}$. *The same line integral would represent this work.*

Example 16.1.2 *What if the mass at a point on the string represented by the curve* $\mathscr{C}$ *was represented by the scalar* $m(x(t), y(t))$? *The mass of the piece of the string corresponding to the subinterval* $[t_j, t_{j+1}]$ *for a given partition* π *of* $[a, b]$ *can be approximated by*

$$\begin{aligned} M_j &= m(x(s_j), y(s_j)) \sqrt{(x(t_{j+1}) - x(t_j))^2 + (y(t_{j+1}) - y(t_j))^2} \\ &= m(x(s_j), y(s_j)) \sqrt{\left(\frac{x(t_{j+1}) - x(t_j)}{t_{j+1} - t_j}\right)^2 + \left(\frac{y(t_{j+1}) - y(t_j)}{t_{j+1} - t_j}\right)^2} \, (t_{j+1} - t_j) \end{aligned}$$

From the Mean Value Theorem, we know there is an z_j *between* t_j *and* t_{j+1} *so that*

$$\begin{bmatrix} x(t_{j+1}) - x(t_j) \\ y(t_{j+1}) - y(t_j) \end{bmatrix} = (t_{j+1} - t_j) \begin{bmatrix} x'(z_j) \\ y'(z_j) \end{bmatrix}$$

Thus, we have

$$M_j = m(x(s_j), y(s_j)) \sqrt{(x'(z_j))^2 + (y'(z_j))^2} \, (t_{j+1} - t_j)$$

Choosing $s_j = z_j$ *always, we find the Riemann sums here are*

$$S(\Phi, \pi, \sigma) = \sum_{\pi} m(x(z_j), y(z_j)) \sqrt{(x'(z_j))^2 + (y'(z_j))^2} \, (t_{j+1} - t_j)$$

where $\Phi(t) = m(x(t), y(t))\sqrt{x'(t)^2 + y'(t)^2}$ *which is a nice continuous function on* $[a, b]$. *Hence, the Riemann Integral* $\int_a^b \Phi(t)\, dt$ *exists and can be used to define the mass on the curve. Note, this argument, while similar to the one we used for the line integrals, is not the same as* $m(x, y)$ *is not a vector function.*

16.1.1 Homework

Exercise 16.1.1 *Let the vector field be defined by* $\boldsymbol{F} = \begin{bmatrix} x^2 + 2y^2 \\ 4xy \end{bmatrix}$ *and the path* $\mathscr{C}$ *by* $\begin{bmatrix} x(t) = 2t \\ y(t) = t^2 \end{bmatrix}$.

- *Set up the line integral for the work done in moving a particle from* $\boldsymbol{P} = \begin{bmatrix} 2 \\ 1 \end{bmatrix}$ *to* $\boldsymbol{Q} = \begin{bmatrix} 6 \\ 9 \end{bmatrix}$.
- *Compute the value of the line integral using a uniform partition in MATLAB.*
- *Reverse the path and move from* $\boldsymbol{Q}$ *to* $\boldsymbol{P}$ *instead. Set up the line integral again.*
- *Compute the value of the reversed work done line integral.*
- *Compute the value of the work done in moving from* $\boldsymbol{P}$ *to* $\boldsymbol{Q}$ *and then back to* $\boldsymbol{Q}$ *using the reversed path.*

Exercise 16.1.2 *Let the vector field be defined by* $\boldsymbol{F} = \begin{bmatrix} x^2y^2 \\ 4sin(xy) \end{bmatrix}$ *and the path* $\mathscr{C}$ *by* $\begin{bmatrix} x(t) = \cos(3t) \\ y(t) = 2\sin(4t) \end{bmatrix}$.

- *Set up the line integral for the work done in moving a particle from* $t = 0$ *to* $t = \pi/4$, *i.e.* $\boldsymbol{P} = \begin{bmatrix} x(0) \\ y(0) \end{bmatrix}$ *to* $\boldsymbol{Q} = \begin{bmatrix} x(\pi/4) \\ y(\pi/4) \end{bmatrix}$.
- *Compute the value of the line integral using a uniform partition in MATLAB.*
- *Reverse the path and move from* $\boldsymbol{Q}$ *to* $\boldsymbol{P}$ *instead. Set up the line integral again.*
- *Compute the value of the reversed work done line integral.*
- *Compute the value of the work done in moving from* $\boldsymbol{P}$ *to* $\boldsymbol{Q}$ *and then back to* $\boldsymbol{Q}$ *using the reversed path.*

Exercise 16.1.3 *Let the vector field be defined by* $\boldsymbol{F} = \begin{bmatrix} \sqrt{x^2 + 2y^2} \\ 2 + x^2 + y^2 \end{bmatrix}$ *and the path* $\mathscr{C}$ *which is the union of the paths* $\mathscr{C}_1$ *given by* $\begin{bmatrix} x(t) = 1 \\ y(t) = t \end{bmatrix}$ *for* $0 \leq t \leq 1$ *and* $\mathscr{C}_2$ *given by* $\begin{bmatrix} x(t) = 1 - t \\ y(t) = 1 \end{bmatrix}$ *for* $0 \leq t \leq 1$.

- *Set up the line integral for the work done in moving a particle from* $\boldsymbol{P} = \begin{bmatrix} 1 \\ 0 \end{bmatrix}$ *to* $\boldsymbol{Q} = \begin{bmatrix} 0 \\ 1 \end{bmatrix}$ *along this path.*
- *Compute the value of the line integral using a uniform partition in MATLAB.*
- *Reverse the path and move from* $\boldsymbol{Q}$ *to* $\boldsymbol{P}$ *instead. Set up the line integral again.*
- *Compute the value of the reversed work done line integral.*
- *Compute the value of the work done in moving from* $\boldsymbol{P}$ *to* $\boldsymbol{Q}$ *and then back to* $\boldsymbol{Q}$ *using the reversed path.*

Exercise 16.1.4 *Let the vector field be defined by* $\boldsymbol{F} = \begin{bmatrix} x^2 + 2y^2 \\ 2xy \end{bmatrix}$ *and the path* $\mathscr{C}$ *by* $\begin{bmatrix} x(t) = t^2 \\ y(t) = 3t \end{bmatrix}$.

- *Set up the line integral for the work done in moving a particle from* $\boldsymbol{P} = \begin{bmatrix} x(0) \\ y(0) \end{bmatrix}$ *to* $\boldsymbol{Q} = \begin{bmatrix} x(2) \\ y(2) \end{bmatrix}$.
- *Compute the value of the line integral using a uniform partition in MATLAB.*
- *Reverse the path and move from* $\boldsymbol{Q}$ *to* $\boldsymbol{P}$ *instead. Set up the line integral again.*
- *Compute the value of the reversed work done line integral.*
- *Compute the value of the work done in moving from* $\boldsymbol{P}$ *to* $\boldsymbol{Q}$ *and then back to* $\boldsymbol{Q}$ *using the reversed path.*

Exercise 16.1.5 *Let the vector field be defined by* $\boldsymbol{F} = \begin{bmatrix} 1 - x^2 - y^2 \\ -2xy \end{bmatrix}$ *and the path* $\mathscr{C}$ *by* $\begin{bmatrix} x(t) = 2t \\ y(t) = t^2 \end{bmatrix}$.

- *Set up the line integral for the work done in moving a particle from* $\boldsymbol{P} = \begin{bmatrix} 2 \\ 1 \end{bmatrix}$ *to* $\boldsymbol{Q} = \begin{bmatrix} 6 \\ 9 \end{bmatrix}$.
- *Compute the value of the line integral using a sequence of uniform partitions in MATLAB.*
- *Reverse the path and move from* $\boldsymbol{Q}$ *to* $\boldsymbol{P}$ *instead. Set up the line integral again.*
- *Compute the value of the reversed work done line integral.*
- *Compute the value of the work done in moving from* $\boldsymbol{P}$ *to* $\boldsymbol{Q}$ *and then back to* $\boldsymbol{Q}$ *using the reversed path.*

16.2 Conservative Force Fields

Let's restrict our attention to nice **closed** curves. We want them to be **simple** which means they do not have self intersections so the figure eight type curves are not allowed. We also want them to have finite length as we can calculate from the usual arc length formulae. Recall the length of the curve $\mathscr{C}$ fro a to b is

$$L_a^b = \int_a^b \sqrt{x'(t)^2 + y'(t)^2}\, dt$$

Since we assume $x(t)$ and $y(t)$ are continuously differentiable on $[a, b]$, we know the integrand is continuous on $[a, b]$ and hence it is bounded by some $B > 0$. Thus, $L_a^b \leq B\,(b - a)$. So our curves do have finite arc length. Such curves are called **rectifiable**. We also want the normal vector to be uniquely defined so it points towards the interior of the closed curve. We usually think of the tangent vector as moving ccw here, so the normal we pick is always the $\boldsymbol{N}$ instead of the $-\boldsymbol{N}$. This kind of curve is called **orientable**. Together, using the first letters of these properties, we want to consider **SCROC**s: i.e. **simple, closed, rectifiable** and **orientable** curves. So what is a **conservative** force field $\boldsymbol{F}$?

Definition 16.2.1 Conservative Force Field

Let the vector field $\boldsymbol{F}$ be defined on an open set U in $\Re^2$ by

$$\boldsymbol{F} = \begin{bmatrix} A(x,y) \\ B(x,y) \end{bmatrix}$$

where A and B are at least continuous at each point in U. Let $\mathscr{C}$ be a SCROC corresponding to the vector function on $[a, b]$ (this domain is dependent on the choice of $\mathscr{C}$).

$$\boldsymbol{\gamma} = \begin{bmatrix} x(t) \\ y(t) \end{bmatrix}$$

We say $\boldsymbol{F}$ is a Conservative Force Field if $\int_{\mathscr{C}} < \boldsymbol{F}, \boldsymbol{\gamma}' > \,dt = 0$ for all SCROCs $\mathscr{C}$.

Comment 16.2.1 *Let the start point be $\boldsymbol{P}$ and the endpoint be $\boldsymbol{Q}$. Let $\mathscr{C}_1$ be a path that goes from $\boldsymbol{P}$ to $\boldsymbol{Q}$ and $\mathscr{C}_2$ be a path that moves in the reverse. Let's also assume the path $\boldsymbol{C}$ obtained by traversing $\mathscr{C}_1$ followed by $\mathscr{C}_2$. The paths are associated with vector functions $\boldsymbol{\gamma_1}$ and $\boldsymbol{\gamma_2}$. We would write this as $\mathscr{C} = \mathscr{C}_1 + \mathscr{C}_2$ for convenience. Then, if $\boldsymbol{F}$ is conservative, we have*

$$\int_{\mathscr{C}} < \boldsymbol{F}, \boldsymbol{\gamma}' > \,dt = \int_{\mathscr{C}_1} < \boldsymbol{F}, \boldsymbol{\gamma_1}' > + \int_{\mathscr{C}_2} < \boldsymbol{F}, \boldsymbol{\gamma_2}' > = 0$$

This implies

$$\int_{\mathscr{C}_1} < \boldsymbol{F}, \boldsymbol{\gamma_1}' > = \int_{-\mathscr{C}_2} < \boldsymbol{F}, \boldsymbol{\gamma_2}' >$$

where the integration over $-\mathscr{C}_2$ simply indicates we are moving on $\mathscr{C}_2$ backwards. This says the path we take from $\boldsymbol{P}$ to $\boldsymbol{Q}$ doesn't matter. In other words, the value we obtain by the line integral depends only on the start and final point. We call this property **path independence** *of the line integral.*

Let's assume $\boldsymbol{F}$ is a conservative force field. Let the start point be $\boldsymbol{P} = (x_s, y_s)$ and the endpoint be some $\boldsymbol{Q} = (x_e = x_s + h, y_e = y_s + h)$. Since $\boldsymbol{P}$ is an interior point of U, there is a ball of radius $r > 0$ about $\boldsymbol{P}$ which is in U. So we can pick an endpoint $\boldsymbol{Q}$ like this in this ball. You can look at Figure 16.4 to see the kind of paths we are going to use in the argument below. Since the value of

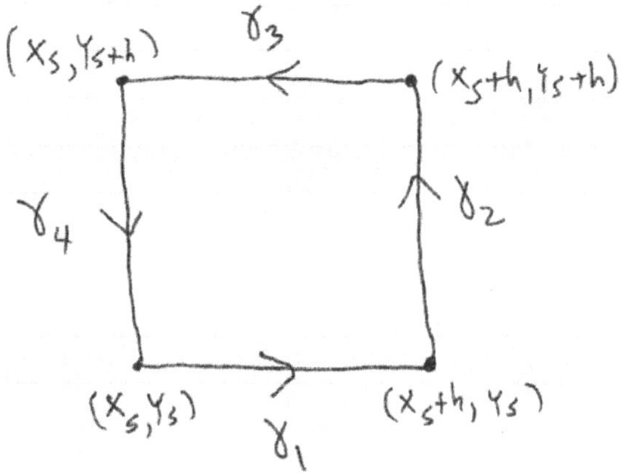

Figure 16.4: A closed path to prove $A_y = B_x$.

the line integral is independent of path, look at the path

$$\begin{aligned} \mathscr{C}_1 : \boldsymbol{\gamma_1} &= \begin{bmatrix} x_1(t) = x_s + th \\ y_1(t) = y_s \end{bmatrix}, \quad 0 \leq t \leq 1 \\ \mathscr{C}_2 : \boldsymbol{\gamma_2} &= \begin{bmatrix} x_2(t) = x_s + h \\ y_2(t) = y_s + th \end{bmatrix}, \quad 0 \leq t \leq 1 \end{aligned}$$

Then, we can calculate the line integrals $\int_{\mathscr{C}_1}$ and $\int_{\mathscr{C}_2}$ to find

$$\begin{aligned} &\int_{\mathscr{C}_1} < \boldsymbol{F}, \boldsymbol{\gamma_1}' > + \int_{\mathscr{C}_2} < \boldsymbol{F}, \boldsymbol{\gamma_2}' > \\ &= \int_0^1 A(x_s + th, y_s)\, hdt + \int_0^1 B(x_e, y_s + th)\, hdt \end{aligned}$$

Now we can go back to $\boldsymbol{P}$ along the path

$$\begin{aligned} \mathscr{C}_3 : \boldsymbol{\gamma_3} &= \begin{bmatrix} x_3(t) = x_e - th = x_s + (1-t)h \\ y_3(t) = y_e \end{bmatrix}, \quad 0 \leq t \leq 1 \\ \mathscr{C}_4 : \boldsymbol{\gamma_4} &= \begin{bmatrix} x_4(t) = x_s \\ y_4(t) = y_e - th = y_s + (1-t)h \end{bmatrix}, \quad 0 \leq t \leq 1 \end{aligned}$$

Then, we can calculate the line integrals $\int_{\mathscr{C}_3}$ and $\int_{\mathscr{C}_4}$ to find

$$\begin{aligned} &\int_{\mathscr{C}_3} < \boldsymbol{F}, \boldsymbol{\gamma_3}' > + \int_{\mathscr{C}_4} < \boldsymbol{F}, \boldsymbol{\gamma_4}' > \\ &= \int_0^1 A(x_e - th, y_e)\, (-h)dt + \int_0^1 B(x_s, y_e - th)\, (-h)dt \end{aligned}$$

Because of path independence. we have $\int_{\mathscr{C}_1} + \int_{\mathscr{C}_2} = -\int_{\mathscr{C}_3} - \int_{\mathscr{C}_4}$ and so

$$\int_0^1 A(x_s + th, y_s)\, hdt + \int_0^1 B(x_e, y_s + th)\, hdt$$

$$= -\int_0^1 A(x_e - th, y_e)\,(-h)dt - \int_0^1 B(x_s, y_e - th)\,(-h)dt$$

Thus,

$$h \int_0^1 \Big(A(x_s + th, y_s) - A(x_e - th, y_e) \Big) dt = h \int_0^1 \Big(B(x_s, y_e - th) - B(x_e, y_s + th) \Big) dt$$

Hence, canceling the common h,

$$\int_0^1 \Big(A(x_s + th, y_s) - A(x_e - th, y_e) \Big) dt = \int_0^1 \Big(B(x_s, y_e - th) - B(x_e, y_s + th) \Big) dt$$

Next, let $h \to 0$:

$$\lim_{h \to 0} \int_0^1 \Big(A(x_s + th, y_s) - A(x_e - th, y_e) \Big) dt = \lim_{h \to 0} \int_0^1 \Big(B(x_s, y_e - th) - B(x_e, y_s + th) \Big) dt$$

The integrals are continuous with respect to h and so

$$\int_0^1 \Big(A(x_s, y_s) - A(x_s, y_e) \Big) dt \quad = \quad \int_0^1 \Big(B(x_s, y_s) - B(x_e, y_s) \Big) dt$$

If we assume A and B have first order partials, then

$$\int_0^1 \Big(-A_y(x_s, y_s)h - E_A(x_s, y_s, h) \Big) dt \quad = \quad \int_0^1 \Big(-B_x(x_s, y_s)h - E_B(x_s, y_s, h) \Big) dt$$

Thus,

$$A_y(x_s, y_s) + \int_0^1 \frac{E_A(x_s, y_s, h)}{h} dt \quad = \quad B_x(x_s, y_s) + \int_0^1 \frac{E_B(x_s, y_s, h)}{h} dt$$

Now let $h \to 0$ again to obtain the final result: $A_y(x_s, y_s) = B_x(x_s, y_s)$. We can state this as an important result.

Theorem 16.2.1 Conservative Force Fields Imply $A_y = B_x$

> *Let the conservative vector field* $\boldsymbol{F}$ *be defined on an open set* U *in* $\Re^2$ *by*
>
> $$\boldsymbol{F} \quad = \quad \begin{bmatrix} A(x, y) \\ B(x, y) \end{bmatrix}$$
>
> *where* A *and* B *have first order partials at each point in* U*. Then* $A_y = B_x$ *in* U*.*

Proof 16.2.1
We have just gone through this argument. ■

16.2.1 Homework

Exercise 16.2.1 *Let the vector field be defined by* $\boldsymbol{F} = \begin{bmatrix} x^2 + 2y^2 \\ 4xy \end{bmatrix}$. *Is* $\boldsymbol{F}$ *a conservative force field?*

Exercise 16.2.2 *Let the vector field be defined by* $\boldsymbol{F} = \begin{bmatrix} x^2y^2 \\ 4sin(xy) \end{bmatrix}$. *Is* $\boldsymbol{F}$ *a conservative force field?*

Exercise 16.2.3 *Let the vector field be defined by* $\boldsymbol{F} = \begin{bmatrix} \sqrt{x^2+2y^2} \\ 2+x^2+y^2 \end{bmatrix}$. *Is* $\boldsymbol{F}$ *a conservative force field?*

Exercise 16.2.4 *Let the vector field be defined by* $\boldsymbol{F} = \begin{bmatrix} x^2+2y^2 \\ 2xy \end{bmatrix}$. *Is* $\boldsymbol{F}$ *a conservative force field?*

Exercise 16.2.5 *Let the vector field be defined by* $\boldsymbol{F} = \begin{bmatrix} 1-x^2-y^2 \\ -2xy \end{bmatrix}$. *Is* $\boldsymbol{F}$ *a conservative force field?*

16.3 Potential Functions

To understand this at a deeper level, we need to look at connectedness more. Let's introduce the idea of a set U being **path connected**.

Definition 16.3.1 Path Connected Sets

The set V *in* $\Re^2$ *is path connected if given* $\boldsymbol{P}$ *and* $\boldsymbol{Q}$ *is* V, *there is a path* $\mathscr{C}$ *connecting* $\boldsymbol{P}$ *to* $\boldsymbol{Q}$.

Comment 16.3.1 *It is not true that a connected open set has to be path connected. There is a standard counterexample of this which you should see. We need to extend connectedness to closed sets.*

Definition 16.3.2 Connected Closed Sets

The closed set C *in* $\Re^2$ *is connected if we cannot find two open sets* A *and* B *so that* $C = A' \cup B'$ *with* $A' \cap B = \emptyset$ *and* $B' \cap A = \emptyset$ *where, recall,* A' *is the closure of* A.

To find our counterexample, consider

$$S = \{(x,y) \in \Re^2 | y = \sin(1/x), x > 0\} \cup (\{0\} \times [-1,1])$$

Let $S_+ = \{(x,y) \in \Re^2 | y = \sin(1/x), x > 0\}$ *and* $S_0 = \{0\} \times [-1,1]$. *It is easy to see any two points in* S_+ *can be connected by a path. Just parameterize the portion of the curve* $y = \sin(1/x)$ *that connects the two points:* $x(t) = t$ *and* $y(t) = \sin(1/t)$ *for suitable* $[a,b]$. *Now take a point in* S_+ *and a point in* S_0 *and try to find a path that connects the two points. There is no way a path that begins in* S_+ *can* jump *to* S_0. *Hence, the set* S *is not path connected. If you think about it though,* S *itself is connected.*
Since the set of cluster points of $\sin(1/x)$ *at* 0 *in* $[-1,1]$, *the closure of* S_+ *must be all of* S. *Now if* S *was not connected, we could write* $S = A' \cup B'$ *where* A *and* B *are open sets with* $A' \cap B = \emptyset$ *and* $B' \cap A = \emptyset$. *But then* $A \cap S$ *and* $B \cap S$ *is a decomposition of* S *and since* S_+ *is connected, we can assume without loss of generality* $B \cap S_+ = \emptyset$. *Hence,* $S_+ = A \cap S_+ \subset A$ *which implies* $S \subset A'$. *That tells us* S *does not intersect* B' *and so* $S = A'$ *implying* S *is connected. This argument works in the other case,* $A \cap S = \emptyset$ *as well. Thus, as* S_+ *is connected, its closure* S *must be connected also.*

So S *is a subset of* $\Re^2$ *which is connected but not path connected.*

We can now say much more about conservative force fields. Let's consider a force field $\boldsymbol{F}$ in U and now assume $\boldsymbol{A}$ and $\boldsymbol{A}$ to have continuous first order partials with $A_y = -B_x$ in U. Let's also assume the open set U is path connected. Pick a point (x_0, y_0) in U and any other point (x, y) in U. Since $\boldsymbol{A}_y = \boldsymbol{B}_x$ in U, we know the line integrals over paths connecting these two points are independent of path. The point (x, y) is an interior point in U and so there is a radius $r > 0$ with $B_r(x, y)$ in U. For h sufficiently small, the lines connecting (x, y) to $(x + h, y)$ and $(x + h, y)$ to $(x, h, y + h)$ are in U too. Let a curve connecting (x_0, y_0) to (x, y) be chosen and call it $\mathscr{C}$ with corresponding vector function $\boldsymbol{\gamma}$. Define the function $\phi(x, y) = \int_{\mathscr{C}} < \boldsymbol{F}, \boldsymbol{\gamma}' dt >$ as usual. Then since we are free to choose any path to get from (x, y) to $(x + h, y + h)$ we wish, we can consider the new paths

$$\begin{aligned} \mathscr{C}_1 : \boldsymbol{\gamma}_1 &= \begin{bmatrix} x_1(t) = x + th \\ y_1(t) = y \end{bmatrix}, \quad 0 \le t \le 1 \\ \mathscr{C}_2 : \boldsymbol{\gamma}_2 &= \begin{bmatrix} x_2(t) = x + h \\ y_2(t) = y + th \end{bmatrix}, \quad 0 \le t \le 1 \end{aligned}$$

We illustrate this in Figure 16.5. Then

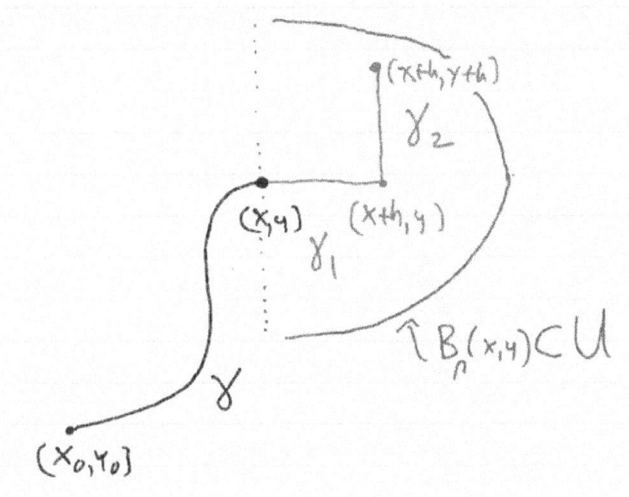

Figure 16.5: Defining the function ϕ on U.

$$\lim_{h \to 0^+} \frac{\phi(x + h, y) - \phi(x, y)}{h} = \lim_{h \to 0^+} \frac{1}{h} \int_0^1 \boldsymbol{A}(x + th, y) h dt = \lim_{h \to 0^+} \int_0^1 \boldsymbol{A}(x + th, y) dt$$

Now apply the Mean Value Theorem for integrals to find

$$\lim_{h \to 0^+} \frac{\phi(x + h, y) - \phi(x, y)}{h} = \lim_{h \to 0^+} \boldsymbol{A}(x + ch, y)$$

where c is between 0 and h. Hence, as $h \to 0^+$, we find $(\phi_x(x, y))^+ = \boldsymbol{A}(x, y)$. A similar argument, with a slightly different picture of course, shows $(\phi_x(x, y))^- = \boldsymbol{A}(x, y)$. Hence, $\phi_x = \boldsymbol{A}$.

We can argue in a very similar fashion to show $\phi_y = \boldsymbol{B}$. We have proven another result. This function ϕ is called a **potential** function corresponding to $\boldsymbol{F}$.

Theorem 16.3.1 The Potential Function for $\boldsymbol{F}$

Let the vector field $\boldsymbol{F}$ be defined on an path connected open set U in $\Re^2$ by

$$\boldsymbol{F} = \begin{bmatrix} A(x,y) \\ B(x,y) \end{bmatrix}$$

where A and B have continuous first order partials in U. Then $\boldsymbol{F}$ is a conservative force field if and only if there is a function ϕ on U so that $\nabla\phi = \boldsymbol{F}$. Moreover, for any path $\mathscr{C}$ with start point $\boldsymbol{P} = (x_s, y_s)$ and end point $\boldsymbol{Q} = (x_e, y_e)$,

$$\int_{\mathscr{C}} < \boldsymbol{F}, \gamma' > dt = \phi(x_e, y_e) - \phi(x_s, y_s)$$

Proof 16.3.1

($\Longrightarrow$):

If $\boldsymbol{F}$ is a conservative force field, we know the line integrations are path independent and by the argument above, we can find the desired ϕ function.

($\Longleftarrow$):

If such a function ϕ exists, then $\boldsymbol{A}_x = \boldsymbol{B}_y$ and $\boldsymbol{F}$ is a conservative force field.

Finally, if $\boldsymbol{F}$ is conservative

$$\begin{aligned} \int_{\mathscr{C}} < \boldsymbol{F}, \gamma' > dt &= \int_a^b \Big(\boldsymbol{A}(x(t), y(t))x'(t) + \boldsymbol{B}(x(t), y(t))y'(t) \Big) dt \\ &= \int_a^b \Big(\phi_x(x(t), y(t))x'(t) + \phi_y(x(t), y(t))y'(t) \Big) dt \\ &= \int_a^b \phi'(t)\, dt = \phi(x_e, y_e) - \phi(x_s, y_s) \end{aligned}$$

■

How do you find a potential function for a conservative force field? Well, it is easy. Consider these examples.

Example 16.3.1 *Let the vector field be defined by $\boldsymbol{F} = \begin{bmatrix} x^2 + y^2 \\ 2xy \end{bmatrix}$ We see $\boldsymbol{F}$ is a conservative force field as $\partial_y(x^2+y^2) = 2y = \partial_x(2xy)$. To construct the potential functions $\phi(x,y)$, follow this script.*

- *Integrate $A(x,y) = x^2 + y^2$ with respect to x only thinking of y as a constant. We'll represent this* partial integration *using the symbol $\int_x$. Then*

$$\phi(x,y) = \int_x (x^2 + y^2) dx + g(y) = \frac{x^3}{3} + xy^2 + g(y)$$

where we add the arbitrary function of y, $g(y)$, as the constant of integration.

- *Since the field is conservative, we must have $B(x,y) = \phi_y = \partial_u$ of the answer above. Thus,*

$$\begin{aligned} 2xy = \phi_y &= \partial_y(\frac{x^3}{2} + xy^2 + g(y)) \\ &= 2xy + g'(y) \end{aligned}$$

Hence, $g'(y) = 0$ and $g(y)$ is any constant c.

- *The potential functions are thus $\phi(x,y) = \frac{x^3}{3} + xy^2 + c$.*

Of course, you can do the other order: $\int_y$ first instead of what we do here.

Example 16.3.2 *Let the vector field be defined by $\boldsymbol{F} = \begin{bmatrix} 1 - x^2 - y^2 \\ -2xy \end{bmatrix}$ We see $\boldsymbol{F}$ is a conservative force field as $\partial_y(1 - x^2 - y^2) = -2y = \partial_x(-2xy)$. To construct the potential functions $\phi(x,y)$:*

- *Integrate $A(x,y) = x^2 + y^2$ with respect to x only thinking of y as a constant.*

$$\begin{aligned} \phi(x,y) &= \int_x (1 - x^2 - y^2)dx + g(y) \\ &= \frac{x - x^3}{3} - xy^2 + g(y) \end{aligned}$$

where we add the arbitrary function of y, $g(y)$, as the constant of integration.

- *Since the field is conservative, we must have $B(x,y) = \phi_y = \partial_u$ of the answer above. Thus,*

$$\begin{aligned} -2xy = \phi_y &= \partial_y(x - \frac{x^3}{2} - xy^2 + g(y)) \\ &= -2xy + g'(y) \end{aligned}$$

Hence, $g'(y) = 0$ and $g(y)$ is any constant c.

- *The potential functions are thus $\phi(x,y) = x - \frac{x^3}{3} - xy^2 + c$.*

Again, you could use the $\int_y$ order instead.

16.3.1 Homework

Exercise 16.3.1 *Let the vector field be defined by $\boldsymbol{F} = \begin{bmatrix} \frac{x}{\sqrt{x^2+y^2}} \\ \frac{x}{\sqrt{x^2+y^2}} \end{bmatrix}$ and the path $\mathscr{C}$ by $\begin{bmatrix} x(t) = 2t \\ y(t) = t^2 \end{bmatrix}$.*

- *Show $\boldsymbol{F}$ is a conservative force field.*
- *Find the potential functions $\phi(x,y)$.*
- *Use the potential function to find the work done in moving from $\boldsymbol{P} = \begin{bmatrix} x(1) \\ y(1) \end{bmatrix}$ to $\boldsymbol{Q} = \begin{bmatrix} x(4) \\ y(4) \end{bmatrix}$.*
- *Compute the value of the line integral using the potential function.*
- *Reverse the path and move from $\boldsymbol{Q}$ to $\boldsymbol{P}$ instead. Again, find the work done using the potential function.*
- *Compute the value of the work done in moving from $\boldsymbol{P}$ to $\boldsymbol{Q}$ and then back to $\boldsymbol{Q}$ using the reversed path.*

Exercise 16.3.2 *Let the vector field be defined by $\boldsymbol{F} = \begin{bmatrix} \frac{x}{\sqrt{1+x^2+y^2}} \\ \frac{x}{\sqrt{1+x^2+y^2}} \end{bmatrix}$ and the path $\mathscr{C}$ by $\begin{bmatrix} x(t) = 2\cos(t) \\ y(t) = 3t \end{bmatrix}$.*

- *Show $\boldsymbol{F}$ is a conservative force field.*

- *Find the potential functions* $\phi(x, y)$.
- *Use the potential function to find the work done in moving from* $\boldsymbol{P} = \begin{bmatrix} x(0) \\ y(0) \end{bmatrix}$ *to* $\boldsymbol{Q} = \begin{bmatrix} x(\pi/2) \\ y(\pi/2) \end{bmatrix}$.
- *Compute the value of the line integral using the potential function.*
- *Reverse the path and move from* $\boldsymbol{Q}$ *to* $\boldsymbol{P}$ *instead. Again, find the work done using the potential function.*
- *Compute the value of the work done in moving from* $\boldsymbol{P}$ *to* $\boldsymbol{Q}$ *and then back to* $\boldsymbol{Q}$ *using the reversed path.*

Exercise 16.3.3 *Let the vector field be defined by* $\boldsymbol{F} = \begin{bmatrix} \frac{2x}{1+x^2+y^2} \\ \frac{2y}{1+x^2+y^2} \end{bmatrix}$ *and the path* $\mathscr{C}$ *by* $\begin{bmatrix} x(t) = 3 + 2t \\ y(t) = 7 - \sin(t) \end{bmatrix}$.

- *Show* $\boldsymbol{F}$ *is a conservative force field.*
- *Find the potential functions* $\phi(x, y)$.
- *Use the potential function to find the work done in moving from* $\boldsymbol{P} = \begin{bmatrix} 1 \\ 0 \end{bmatrix}$ *to* $\boldsymbol{Q} = \begin{bmatrix} 0 \\ 1 \end{bmatrix}$ *along this path.*
- *Compute the value of the line integral using the potential function.*
- *Reverse the path and move from* $\boldsymbol{Q}$ *to* $\boldsymbol{P}$ *instead. Again, find the work done using the potential function.*
- *Compute the value of the work done in moving from* $\boldsymbol{P}$ *to* $\boldsymbol{Q}$ *and then back to* $\boldsymbol{Q}$ *using the reversed path.*

Exercise 16.3.4 *Let the vector field be defined by* $\boldsymbol{F} = \begin{bmatrix} 2xe^{-x^2-y^2} \\ 2ye^{-x^2-y^2} \end{bmatrix}$ *and the path* $\mathscr{C}$ *is the union of the paths* $\mathscr{C}_1$ *and* $\mathscr{C}_2$ *where* $\mathscr{C}_1$ *is* $\begin{bmatrix} x(t) = 1 \\ y(t) = t \end{bmatrix}$ *for* $0 \leq t \leq 1$ *and* $\mathscr{C}_2$ *is* $\begin{bmatrix} x(t) = 1 - t \\ y(t) = 1 \end{bmatrix}$ *for* $0 \leq t \leq 1$.

- *Show* $\boldsymbol{F}$ *is a conservative force field.*
- *Find the potential functions* $\phi(x, y)$.
- *Use the potential function to find the work done in moving from* $\boldsymbol{P} = \begin{bmatrix} 1 \\ 0 \end{bmatrix}$ *to* $\boldsymbol{Q} = \begin{bmatrix} 0 \\ 1 \end{bmatrix}$ *along this path.*
- *Compute the value of the line integral using the potential function.*
- *Reverse the path and move from* $\boldsymbol{Q}$ *to* $\boldsymbol{P}$ *instead. Again, find the work done using the potential function.*
- *Compute the value of the work done in moving from* $\boldsymbol{P}$ *to* $\boldsymbol{Q}$ *and then back to* $\boldsymbol{Q}$ *using the reversed path.*

Exercise 16.3.5 *Let the vector field be defined by* $\boldsymbol{F} = \begin{bmatrix} -\frac{5x}{(1+x^2+y^2)^{7/2}} \\ -\frac{5y}{(1+x^2+y^2)^{7/2}} \end{bmatrix}$ *and the path* $\mathscr{C}$ *is the union of the paths* $\mathscr{C}_1$, $\mathscr{C}_2$ *and* $\mathscr{C}_3$ *where* $\mathscr{C}_1$ *is* $\begin{bmatrix} x(t) = 0 \\ y(t) = 3t \end{bmatrix}$ *for* $0 \le t \le 1$, $\mathscr{C}_2$ *is* $\begin{bmatrix} x(t) = 4t \\ y(t) = 3 \end{bmatrix}$ *for* $0 \le t \le 1$ *and* $\mathscr{C}_3$ *is* $\begin{bmatrix} x(t) = 4+t \\ y(t) = 3+t \end{bmatrix}$ *for* $0 \le t \le 1$.

- *Show* $\boldsymbol{F}$ *is a conservative force field.*
- *Find the potential functions* $\phi(x, y)$.
- *Use the potential function to find the work done in moving from* $\boldsymbol{P} = \begin{bmatrix} 1 \\ 0 \end{bmatrix}$ *to* $\boldsymbol{Q} = \begin{bmatrix} 5 \\ 4 \end{bmatrix}$ *along this path.*
- *Compute the value of the line integral using the potential function.*
- *Reverse the path and move from* $\boldsymbol{Q}$ *to* $\boldsymbol{P}$ *instead. Again, find the work done using the potential function.*
- *Compute the value of the work done in moving from* $\boldsymbol{P}$ *to* $\boldsymbol{Q}$ *and then back to* $\boldsymbol{Q}$ *using the reversed path.*

16.4 Green's Theorem

There is a connection between the line integral over a SCROC and a double integral over the area enclosed by the curve. From our earlier discussions, since curves can be very complicated with their interiors and exteriors not so clear, by restricting our interest to a SCROC, we can look carefully at the meat of the idea and not get lost in extraneous details. Remember, this is what we always must do: look for the **core** of the thing. We start with a very simple SCROC as shown in Figure 16.6. The curve is the finite rectangle $[a, b] \times [c, d]$, the open set $U = (a, b) \times (c, d)$, and we let ∂U, the boundary of U, be given by the curve which traverses the rectangle ccw. Thus, the normal vector always points in.

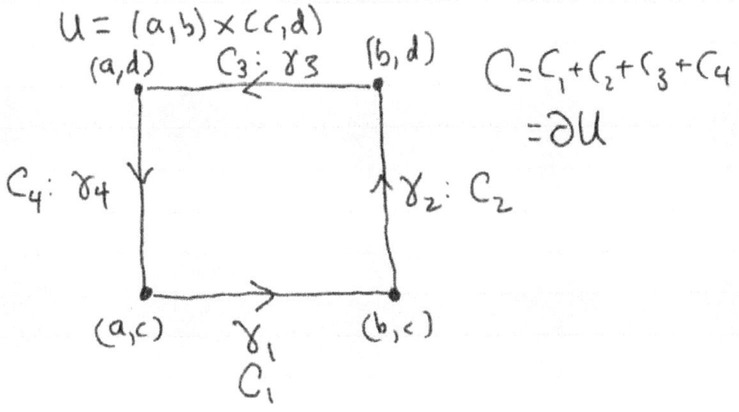

Figure 16.6: Green's Theorem for a rectangle.

The curve $\mathscr{C} = \mathscr{C}_1 + \mathscr{C}_2 + \mathscr{C}_3 + \mathscr{C}_4$, as shown in Figure 16.6. We use very simple definitions for $\boldsymbol{\gamma_1}$ through $\boldsymbol{\gamma_4}$ here.

$$\begin{aligned} \mathscr{C}_1 : \boldsymbol{\gamma_1} &= \begin{bmatrix} x,\ a \leq x \leq b \\ c \end{bmatrix}, \quad \mathscr{C}_2 : \boldsymbol{\gamma_2} = \begin{bmatrix} x+h \\ y,\ c \leq y \leq d \end{bmatrix} \\ \mathscr{C}_3 : \boldsymbol{\gamma_3} &= \begin{bmatrix} x,\ b \rightarrow a \\ d \end{bmatrix}, \quad \mathscr{C}_4 : \boldsymbol{\gamma_4} = \begin{bmatrix} a \\ y,\ d \rightarrow c \end{bmatrix}, \end{aligned}$$

The line integral over $\mathscr{C}$ is then

$$\begin{aligned} &\int_a^b \boldsymbol{A}(x,c)dx + \int_c^d \boldsymbol{B}(b,y)dy + \int_b^a \boldsymbol{A}(x,d)dx + \int_d^c \boldsymbol{B}(a,y)dy = \\ &\int_a^b \Big(\boldsymbol{A}(x,c) - \boldsymbol{A}(x,d)\Big)dx + \int_c^d \Big(\boldsymbol{B}(b,y) - \boldsymbol{B}(a,y)\Big)dy \\ &\int_a^b \int_c^d -\boldsymbol{A}_y(x,y)\ dydx + \int_c^d \int_a^b \boldsymbol{B}_x(x,y)\ dxdy \end{aligned}$$

This can be rewritten as

$$\int_{\mathscr{C}} < \boldsymbol{F}, \boldsymbol{\gamma}' > dt \quad = \quad \int_a^b \int_c^d (\boldsymbol{B}_x - \boldsymbol{A}_y)dxdy$$

This is our first version of this result

$$\int_{\partial U} < \boldsymbol{F}, \boldsymbol{\gamma}' > dt \quad = \quad \int\int_U (\boldsymbol{B}_x - \boldsymbol{A}_y)dA$$

where dA is the usual area element notation in a double integral. The open set U here is particularly simple. Next, let's look at a more interesting ∂U. This is a SCROC which encloses an area which can be described either with a left and right curve (so the area is $\int\int dxdy$) or with a bottom and top curve (so the area is $\int\int dydx$).

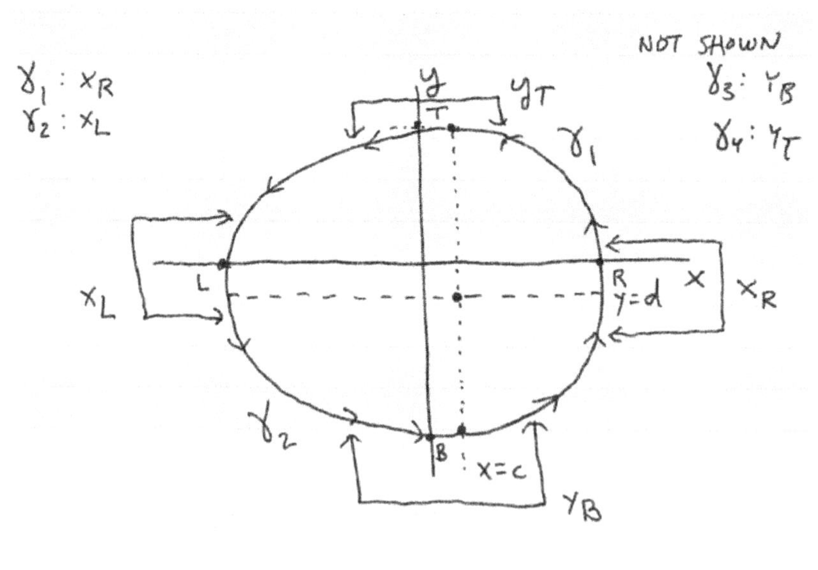

Figure 16.7: Green's Theorem for a SCROC: Right/ Left and Bottom/ Top.

The curve $\mathscr{C} = \mathscr{C}_1 + \mathscr{C}_2$ as shown in Figure 16.7. We use

$$\mathscr{C}_1 : \gamma_1 \quad = \quad \begin{bmatrix} x = x_R(y) \\ y,\ B \le y \le T \end{bmatrix}, \quad \mathscr{C}_2 : \gamma_2 = \begin{bmatrix} x = x_L(y) \\ y,\ T \to B \end{bmatrix}$$

The line integral over $\mathscr{C}$ is then

$$\int_B^T \boldsymbol{A}(x_R(y), y) x'_R(y) dy + \int_B^T \boldsymbol{B}(x_R(y), y) dy - \int_B^T \boldsymbol{A}(x_L(y, y)) x'_L(y) dy$$
$$- \int_B^T \boldsymbol{B}(x_L(y), y) dy = \int_B^T \boldsymbol{A}(x_R(y), y)\, x'_R(y)\, dy - \int_B^T \boldsymbol{A}(x_L(y), y)\, x'_L(y)\, dy$$
$$+ \int_B^T (\boldsymbol{B}(x_R(y), y) - \boldsymbol{B}(x_L(y), y))\, dy$$

First, note

$$\int_B^T (\boldsymbol{B}(x_R(y), y) - \boldsymbol{B}(x_L(y), y))\, dy \quad = \quad \int_B^T \int_{x_L(y)}^{x_R(y)} \boldsymbol{B}_x(x, y)\, dx\, dy$$

Next, consider

$$\int_B^T \boldsymbol{A}(x_R(y), y) x'_R(y) dy - \int_B^T \boldsymbol{A}(x_L(y), y) x'_L(y) dy = \int_B^d \boldsymbol{A}(x_R(y), y) x'_R(y) dy$$
$$+ \int_d^T \boldsymbol{A}(x_R(y), y) x'_R(y) dy - \int_B^d \boldsymbol{A}(x_L(y), y) x'_L(y) dy - \int_d^T \boldsymbol{A}(x_L(y), y) x'_L(y) dy$$

We can regroup this as follows:

$$\int_B^d \Big(\boldsymbol{A}(x_R(y), y) x'_R(y) - \boldsymbol{A}(x_L(y), y) x'_L(y) \Big) dy$$
$$+ \int_d^T \Big(\boldsymbol{A}(x_R(y), y) x'_R(y) - \boldsymbol{A}(x_L(y), y) x'_L(y) \Big) dy$$

Now when $x = x_R(y)$ on $[B, d]$, $y = y_B(x)$ on $[c, R]$; on $[d, T]$, $y = y_T(x)$ on $[R, c]$. Also, when $x = x_L(y)$ on $[B, d]$, $y = y_B(x)$ on $[c, L]$; on $[d, T]$, $y = y_T(x)$ on $[L, c]$. Making these changes of variables, we now have

$$\int_B^T \boldsymbol{A}(x_R(y), y) x'_R(y) dy - \int_B^T \boldsymbol{A}(x_L(y, y)) x'_L(y) dy =$$
$$\int_c^R (\boldsymbol{A}(x, y_B(x)) - \boldsymbol{A}(x, y_T(x)))\, dx + \int_c^L (\boldsymbol{A}(x, y_B(x)) - \boldsymbol{A}(x, y_T(x)))\, dy$$

or

$$\int_B^T \boldsymbol{A}(x_R(y), y) x'_R(y) dy - \int_B^T \boldsymbol{A}(x_L(y), y) x'_L(y) dy =$$
$$\int_L^R (\boldsymbol{A}(x, y_B(x)) - \boldsymbol{A}(x, y_T(x))) dx = \int_L^R \int_{y_B(x)}^{y_T(x)} -\boldsymbol{A}_y(x, y)\, dx\, dy$$

Combining our results, we see we have shown

$$\int_{\mathscr{C}} < \boldsymbol{F}, \gamma' > dt \quad = \quad \int_a^b \int_c^d (\boldsymbol{B}_x - \boldsymbol{A}_y) dx dy$$

This is our second version of this result

$$\int\limits_{\partial U} < \boldsymbol{F}, \gamma' > dt \quad = \quad \int\int\limits_{U} (\boldsymbol{B}_x - \boldsymbol{A}_y) dA$$

Also, we would get the same result if we had chosen to move around the boundary of our region using the y_B and y_T curves, which would have determined γ_3 and γ_4. For our arguments to work, it was essential that we could invert the equations $x = x_R(y)$ and $x = x_L(y)$ on suitable intervals. If the region determined by the SCROC, doesn't allow us to do this, we have difficulties as we can see in the next example. We now examine a SCROC laid out bottom to top. We show this in Figure 16.8. In this figure, for a change of pace, we label the bottom and top curves f_B and f_T, respectively. You should get used to using a variety of notations for these kinds of arguments.

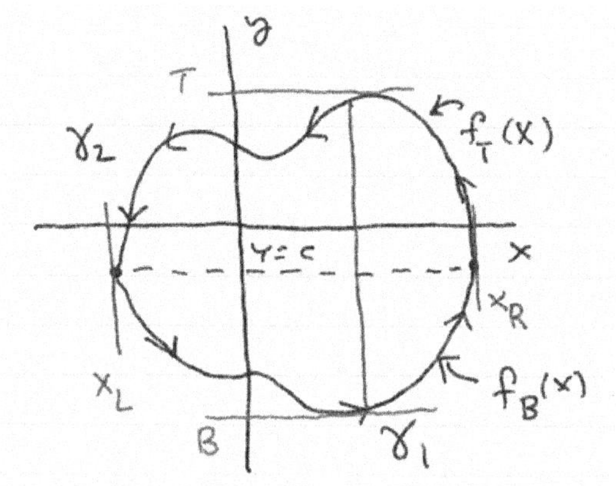

Figure 16.8: Green's Theorem for a SCROC: Bottom Top.

The curve $\mathscr{C} = \mathscr{C}_1 + \mathscr{C}_2$ as shown in Figure 16.8. We use

$$\mathscr{C}_1 : \gamma_1 \quad = \quad \begin{bmatrix} x,\; x_L \leq x \leq x_R \\ y = f_B(x), \end{bmatrix}, \quad \mathscr{C}_2 : \gamma_2 = \begin{bmatrix} x, x_R \rightarrow x_L \\ y = f_T(x), \end{bmatrix}$$

The line integral over $\mathscr{C}$ is then

$$\int_{x_L}^{x_R} \boldsymbol{A}(x, f_B(x)) dx + \int_{x_L}^{x_R} \boldsymbol{B}(x, f_B(x)) f_B'(x) dx + \int_{x_R}^{x_L} \boldsymbol{A}(x, f_T(x)) dx$$
$$+ \int_{x_R}^{x_L} \boldsymbol{B}(x, f_T(x)) f_T'(x) dx =$$
$$\int_{x_L}^{x_R} \Big(A(x, f_B(x)) - A(x, f_T(x)) \Big) dx + \int_{x_L}^{x_R} \boldsymbol{B}(x, f_B(x)) f_B'(x) dx$$
$$- \int_{x_L}^{x_R} \boldsymbol{B}(x, f_T(x)) f_T'(x) dx$$

The first integral can be rewritten giving

$$\int_{x_L}^{x_R} \boldsymbol{A}(x, f_B(x))dx + \int_{x_L}^{x_R} \boldsymbol{B}(x, f_B(x))f_B'(x)dx + \int_{x_R}^{x_L} \boldsymbol{A}(x, f_T(x))dx$$
$$+ \int_{x_R}^{x_L} \boldsymbol{B}(x, f_T(x))f_T'(x)dx = \int_{x_L}^{x_R} \int_{f_B(x)}^{f_T(x)} -\boldsymbol{A}_y(x,y)\, dy\, dx$$
$$+ \int_{x_L}^{x_R} \boldsymbol{B}(x, f_B(x))f_B'(x)dx - \int_{x_L}^{x_R} \boldsymbol{B}(x, f_T(x))f_T'(x)dx$$

However, in the last two integrals above, we cannot use the arguments from the last example as we don't know how to invert y_B and y_T on appropriate intervals. This is why the last example was structured the way it was. We need the inversion information. But, all is not lost. Consider the region shown in Figure 16.9.

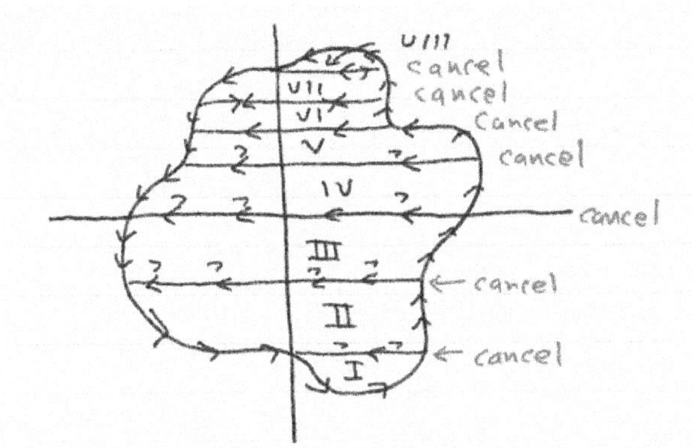

Figure 16.9: Green's Theorem for a more general SCROC.

In this more general case, we divide the region into pieces which have the right structure (i.e. a nice x_L, x_R, y_B and y_T) so that the inversions can be done. In the figure, you see we define many closed paths which are set up so that the horizontal movements across the figure always cancel out. The lower path and the path above it move in opposite directions and so the net contribution to the line integral is zero. The part that is nonzero is the line integral we want. The second version of Green's Theorem applies to each piece. So in general, we can prove Green's Theorem for any SCROC although it does get quite involved!

So let's state Green's Theorem now.

Theorem 16.4.1 Green's Theorem for a SCROC

Let $\mathscr{C}$ be a SCROC enclosing the open set U in $\Re^2$. Let γ be the vector function used in $\mathscr{C}$ and let ∂U denote the boundary of U which is traversed ccw by γ. Then, for any vector field $\boldsymbol{F}$

$$\int_{\partial U} < \boldsymbol{F}, \gamma' > dt \quad = \quad \int\int_{U} (\boldsymbol{B}_x - \boldsymbol{A}_y)\, dA$$

where $\boldsymbol{A}$ and $\boldsymbol{B}$ are the components of $\boldsymbol{F}$.

16.4.1 Homework

Exercise 16.4.1 *Let the vector field be defined by* $\boldsymbol{F} = \begin{bmatrix} \frac{x}{\sqrt{x^2+y^2}} \\ \frac{x}{\sqrt{x^2+y^2}} \end{bmatrix}$ *and the simple closed path* $\mathscr{C}$ *is the unit square starting at* $(0,0)$ *and moving counterclockwise to* $(1,0)$*, then* $(1,1)$*, then back to* $(0,1)$ *and finally back to* $(0,0)$*. This is the path* $\mathscr{C}$ *defined by four simple straight paths:* $\mathscr{C}_1$ *is* $\begin{bmatrix} x(t) = t \\ y(t) = 0 \end{bmatrix}$ *for* $0 \leq t \leq 1$*,* $\mathscr{C}_2$ *is* $\begin{bmatrix} x(t) = 1 \\ y(t) = t \end{bmatrix}$ *for* $0 \leq t \leq 1$*,* $\mathscr{C}_3$ *is* $\begin{bmatrix} x(t) = 1-t \\ y(t) = 1 \end{bmatrix}$ *for* $0 \leq t \leq 1$ *and* $\mathscr{C}_4$ *is* $\begin{bmatrix} x(t) = 0 \\ y(t) = 1-t \end{bmatrix}$ *for* $0 \leq t \leq 1$*.*

- *Determine if this is a conservative force field.*
- *Draw the curve and indicate the enclosed area.*
- *Use Green's Theorem to write the line integral* $\int < \boldsymbol{F}, \gamma' >$ *as a double integral over the region enclosed by* $\mathscr{C}$ *in two ways: integrating in the order* $dsdy$ *and then integrating in the order* $dydx$*. The value of this double integral will be zero as this is a conservative force field.*

Exercise 16.4.2 *Let the vector field be defined by* $\boldsymbol{F} = \begin{bmatrix} \frac{x}{\sqrt{1+x^2+y^2}} \\ \frac{x}{\sqrt{1+x^2+y^2}} \end{bmatrix}$ *and the simple closed path* $\mathscr{C}$ *by* $\begin{bmatrix} x(t) = 2\cos(t) \\ y(t) = 3\sin(t) \end{bmatrix}$ *for* $0 \leq t \leq 2\pi$*.*

- *Determine if* $\boldsymbol{F}$ *is a conservative force field.*
- *Draw the curve and indicate the enclosed area.*
- *Use Green's Theorem to write the line integral* $\int < \boldsymbol{F}, \gamma' >$ *as a double integral over the region enclosed by* $\mathscr{C}$ *in two ways: integrating in the order* $dsdy$ *and then integrating in the order* $dydx$*. The value of this double integral will be zero as this is a conservative force field. Note this double integral would be impossible to do by hand as the region enclosed by the curve is an ellipse.*

Exercise 16.4.3 *Let the vector field be defined by* $\boldsymbol{F} = \begin{bmatrix} \frac{2x}{1+x^2+y^2} \\ \frac{2y}{1+x^2+y^2} \end{bmatrix}$ *and the simple closed path* $\mathscr{C}$ *is defined as the union of the following four curves:* $\mathscr{C}_1$ *is* $\begin{bmatrix} x(t) = 1-t \\ y(t) = 1-t^2 \end{bmatrix}$ *for* $0 \leq t \leq 1$*,* $\mathscr{C}_2$ *is* $\begin{bmatrix} x(t) = t \\ y(t) = 1-t^2 \end{bmatrix}$ *for* $0 \leq t \leq 1$*,* $\mathscr{C}_3$ *is* $\begin{bmatrix} x(t) = -1+t \\ y(t) = -1+(-1+t)^2 \end{bmatrix}$ *for* $0 \leq t \leq 1$ *and* $\mathscr{C}_4$ *is* $\begin{bmatrix} x(t) = t \\ y(t) = 1-t^2 \end{bmatrix}$ *for* $0 \leq t \leq 1$*.*

- *Determine if* $\boldsymbol{F}$ *is a conservative force field.*
- *Draw the curve and indicate the enclosed area.*
- *Use Green's Theorem to write the line integral* $\int < \boldsymbol{F}, \gamma' >$ *as a double integral over the region enclosed by* $\mathscr{C}$ *in two ways: integrating in the order* $dsdy$ *and then integrating in the order* $dydx$*. The value of this double integral will be zero as this is a conservative force field.*

Exercise 16.4.4 *Let the vector field be defined by* $\boldsymbol{F} = \begin{bmatrix} 2xe^{-x^2-y^2} \\ 2ye^{-x^2-y^2} \end{bmatrix}$ *and the simple closed path* $\mathscr{C}$ *is defined as the union of the following four curves:* $\mathscr{C}_1$ *is* $\begin{bmatrix} x(t) = 1-t \\ y(t) = 1-t^2 \end{bmatrix}$ *for* $0 \leq t \leq 1$*,*

$\mathscr{C}_2$ *is* $\begin{bmatrix} x(t) = t \\ y(t) = 1 - t^2 \end{bmatrix}$ *for* $0 \leq t \leq 1$, $\mathscr{C}_3$ *is* $\begin{bmatrix} x(t) = -1 + t \\ y(t) = -1 + (-1 + t)^2 \end{bmatrix}$ *for* $0 \leq t \leq 1$ *and* $\mathscr{C}_4$ *is* $\begin{bmatrix} x(t) = t \\ y(t) = 1 - t^2 \end{bmatrix}$ *for* $0 \leq t \leq 1$.

- *Determine if* $\boldsymbol{F}$ *is a conservative force field.*
- *Draw the curve and indicate the enclosed area.*
- *Use Green's Theorem to write the line integral* $\int < \boldsymbol{F}, \gamma' >$ *as a double integral over the region enclosed by* $\mathscr{C}$ *in two ways: integrating in the order* $dsdy$ *and then integrating in the order* $dydx$*. The value of this double integral will be zero as this is a conservative force field.*

Exercise 16.4.5 *Let the vector field be defined by* $\boldsymbol{F} = \begin{bmatrix} -\frac{5x}{(1+x^2+y^2)^{7/2}} \\ -\frac{5y}{(1+x^2+y^2)^{7/2}} \end{bmatrix}$ *and the simple closed path* $\mathscr{C}$ *is defined as the union of the following curves:* $\mathscr{C}_1$ *is* $\begin{bmatrix} x(t) = t \\ y(t) = 0 \end{bmatrix}$ *for* $0 \leq t \leq 1$, $\mathscr{C}_2$ *is* $\begin{bmatrix} x(t) = 1 \\ y(t) = t \end{bmatrix}$ *for* $0 \leq t \leq 1$ *and* $\mathscr{C}_3$ *is* $\begin{bmatrix} x(t) = 1 - t \\ y(t) = 1 - t \end{bmatrix}$ *for* $0 \leq t \leq 1$.

- *Determine if* $\boldsymbol{F}$ *is a conservative force field.*
- *Draw the curve and indicate the enclosed area.*
- *Use Green's Theorem to write the line integral* $\int < \boldsymbol{F}, \gamma' >$ *as a double integral over the region enclosed by* $\mathscr{C}$ *in two ways: integrating in the order* $dsdy$ *and then integrating in the order* $dydx$*. The value of this double integral will be zero as this is a conservative force field.*

Exercise 16.4.6 *Let the vector field be defined by* $\boldsymbol{F} = \begin{bmatrix} x^2 + 2y^2 \\ 4xy \end{bmatrix}$ *and the simple closed path* $\mathscr{C}$ *is defined as the union of the following curves:* $\mathscr{C}_1$ *is* $\begin{bmatrix} x(t) = t \\ y(t) = 0 \end{bmatrix}$ *for* $0 \leq t \leq 2$, $\mathscr{C}_2$ *is* $\begin{bmatrix} x(t) = 2 \\ y(t) = t \end{bmatrix}$ *for* $0 \leq t \leq 2$ *and* $\mathscr{C}_3$ *is* $\begin{bmatrix} x(t) = 2 - t \\ y(t) = 2 - t \end{bmatrix}$ *for* $0 \leq t \leq 2$.

- *Determine if* $\boldsymbol{F}$ *is a conservative force field.*
- *Draw the curve and indicate the enclosed area.*
- *Use Green's Theorem to write the line integral* $\int < \boldsymbol{F}, \gamma' >$ *as a double integral over the region enclosed by* $\mathscr{C}$ *in two ways: integrating in the order* $dsdy$ *and then integrating in the order* $dydx$*. The value of this double integral will be zero as this is a conservative force field.*

Exercise 16.4.7 *Let the vector field be defined by* $\boldsymbol{F} = \begin{bmatrix} x^2y^2 \\ 4 \\ sin(xy) \end{bmatrix}$ *and the simple closed path* $\mathscr{C}$ *is defined as the union of the following curves:* $\mathscr{C}_1$ *is* $\begin{bmatrix} x(t) = t \\ y(t) = 0 \end{bmatrix}$ *for* $0 \leq t \leq 1$, $\mathscr{C}_2$ *is* $\begin{bmatrix} x(t) = 1 \\ y(t) = t \end{bmatrix}$ *for* $0 \leq t \leq 1$ *and* $\mathscr{C}_3$ *is* $\begin{bmatrix} x(t) = 1 - t \\ y(t) = 1 - t \end{bmatrix}$ *for* $0 \leq t \leq 1$.

- *Determine if* $\boldsymbol{F}$ *is a conservative force field.*
- *Draw the curve and indicate the enclosed area.*

- *Use Green's Theorem to write the line integral $\int < \boldsymbol{F}, \gamma' >$ as a double integral over the region enclosed by $\mathscr{C}$ in two ways: integrating in the order $dsdy$ and then integrating in the order $dydx$.*

- *Use a uniform Riemann sum to evaluate the double integral and a uniform Riemann sum to evaluate the line integral. Use MATLAB to do the computations. They should match.*

Exercise 16.4.8 *Let the vector field be defined by* $\boldsymbol{F} = \begin{bmatrix} \sqrt{x^2 + 2y^2} \\ 2 + x^2 + y^2 \end{bmatrix}$ *and the simple closed path* $\mathscr{C}$ *is defined as the union of the following curves:* $\mathscr{C}_1$ *is* $\begin{bmatrix} x(t) = t \\ y(t) = 0 \end{bmatrix}$ *for* $0 \leq t \leq 3$, $\mathscr{C}_2$ *is* $\begin{bmatrix} x(t) = 3 \\ y(t) = t \end{bmatrix}$ *for* $0 \leq t \leq 3$ *and* $\mathscr{C}_3$ *is* $\begin{bmatrix} x(t) = 3 - t \\ y(t) = 3 - t \end{bmatrix}$ *for* $0 \leq t \leq 3$.

- *Determine if* $\boldsymbol{F}$ *is a conservative force field.*

- *Draw the curve and indicate the enclosed area.*

- *Use Green's Theorem to write the line integral $\int < \boldsymbol{F}, \gamma' >$ as a double integral over the region enclosed by $\mathscr{C}$ in two ways: integrating in the order $dsdy$ and then integrating in the order $dydx$.*

- *Use a uniform Riemann sum to evaluate the double integral and a uniform Riemann sum to evaluate the line integral. Use MATLAB to do the computations. They should match.*

Exercise 16.4.9 *Let the vector field be defined by* $\boldsymbol{F} = \begin{bmatrix} x^2 + 2y^2 \\ 2x^3 y \end{bmatrix}$ *and the simple closed path* $\mathscr{C}$ *is defined as the union of the following four curves:* $\mathscr{C}_1$ *is* $\begin{bmatrix} x(t) = 1 - t \\ y(t) = 1 - t^2 \end{bmatrix}$ *for* $0 \leq t \leq 1$, $\mathscr{C}_2$ *is* $\begin{bmatrix} x(t) = t \\ y(t) = 1 - t^2 \end{bmatrix}$ *for* $0 \leq t \leq 1$, $\mathscr{C}_3$ *is* $\begin{bmatrix} x(t) = -1 + t \\ y(t) = -1 + (-1 + t)^2 \end{bmatrix}$ *for* $0 \leq t \leq 1$ *and* $\mathscr{C}_4$ *is* $\begin{bmatrix} x(t) = t \\ y(t) = 1 - t^2 \end{bmatrix}$ *for* $0 \leq t \leq 1$.

- *Determine if* $\boldsymbol{F}$ *is a conservative force field.*

- *Draw the curve and indicate the enclosed area.*

- *Use Green's Theorem to write the line integral $\int < \boldsymbol{F}, \gamma' >$ as a double integral over the region enclosed by $\mathscr{C}$ in two ways: integrating in the order $dsdy$ and then integrating in the order $dydx$.*

- *Use a uniform Riemann sum to evaluate the double integral and a uniform Riemann sum to evaluate the line integral. Use MATLAB to do the computations. They should match.*

Exercise 16.4.10 *Let the vector field be defined by* $\boldsymbol{F} = \begin{bmatrix} 1 - x^2 - 5y^2 \\ -2xy \end{bmatrix}$ *and the simple closed path path* $\mathscr{C}$ *by* $\begin{bmatrix} x(t) = 2\cos(t) \\ y(t) = 3\sin(t) \end{bmatrix}$ *for* $0 \leq t \leq 2\pi$.

- *Determine if* $\boldsymbol{F}$ *is a conservative force field.*

- *Draw the curve and indicate the enclosed area.*

- *Use Green's Theorem to write the line integral $\int < \boldsymbol{F}, \gamma' >$ as a double integral over the region enclosed by $\mathscr{C}$ in two ways: integrating in the order $dsdy$ and then integrating in the order $dydx$.*

- *Use a uniform Riemann sum to evaluate the double integral and a uniform Riemann sum to evaluate the line integral. Use MATLAB to do the computations. They should match.*

16.5 Green's Theorem for Images of the Unit Square

We can also prove a version of Green's Theorem for mappings that smoothly take $[0,1] \times [0,1]$ to $\Re^2$ in a one-to-one way.

Definition 16.5.1 Oriented 2-Cells

We assume the coordinates of $\mathbb{I}^2$ are (u,v) and the coordinates of the image $\boldsymbol{F}(\mathbb{I}^2) = D$ are (x,y). We say D in $\Re^2$ is an oriented 2-cell if there is a 1-1 continuously differentiable map $\boldsymbol{G} : [0,1] \times [0,1] \to D$ where $\boldsymbol{G}$ is actually defined on an open set U containing $[0,1] \times [0,1]$ satisfying $\boldsymbol{G}([0,1] \times [0,1]) = D$. In addition, we assume $\det \boldsymbol{J_G}(u,v) > 0$ on U. For convenience, we let $\mathbb{I}^2 = [0,1] \times [0,1]$.

Consider the picture shown in Figure 16.10. We use $\boldsymbol{G_1}$ and $\boldsymbol{G_2}$ as the component functions for $\boldsymbol{F}$.

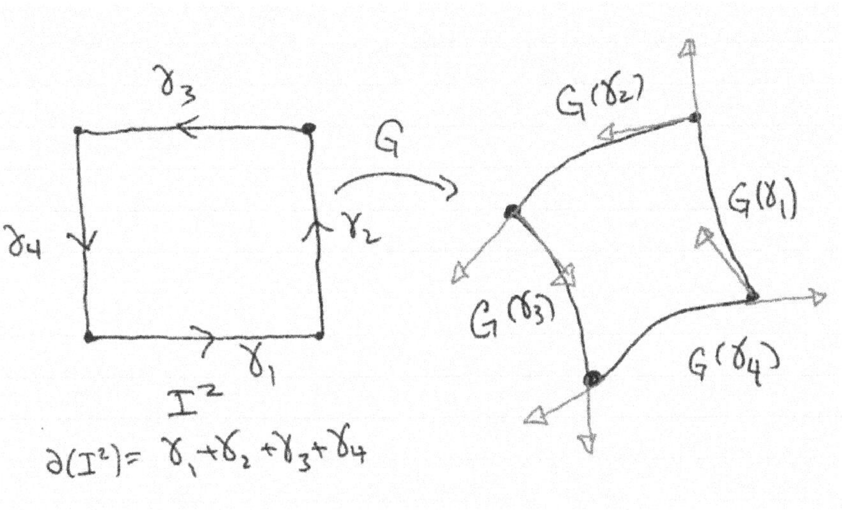

Figure 16.10: An oriented 2-cell.

Thus,

$$\boldsymbol{G} = \begin{bmatrix} \boldsymbol{G_1} \\ \boldsymbol{G_2} \end{bmatrix} \Longrightarrow \boldsymbol{J_G}(u,v) = \begin{bmatrix} \boldsymbol{G_1}_u(u,v) & \boldsymbol{G_1}_v(u,v) \\ \boldsymbol{G_2}_u(u,v) & \boldsymbol{G_2}_v(u,v) \end{bmatrix}$$

The edges of $\mathbb{I}^2$ are mapped by $\boldsymbol{G}$ into curves in $\Re^2$ as shown in the figure. The edge corresponding to γ_1 is the trace $\boldsymbol{G}(u,0)$ and the tangent along this trace is

$$\boldsymbol{T}(u,0) = \begin{bmatrix} \boldsymbol{G_1}_u(u,0) & \boldsymbol{G_1}_v(u,0) \\ \boldsymbol{G_2}_u(u,0) & \boldsymbol{G_2}_v(u,0) \end{bmatrix} \begin{bmatrix} 1 \\ 0 \end{bmatrix} = \begin{bmatrix} \boldsymbol{G_1}_u(u,0) & \boldsymbol{G_2}_u(u,0) \end{bmatrix}$$

Hence, the limiting tangent vector as we approach $(1,0)$ along $\boldsymbol{F}(\gamma_1)$ is

$$\boldsymbol{T}_{\gamma_1}(1,0) = \begin{bmatrix} \boldsymbol{G_1}_u(1,0) & \boldsymbol{G_2}_u(1,0) \end{bmatrix}$$

On the other hand, if we look at the trace for the edge given by γ_2, we find

$$\boldsymbol{T}(1,y) = \begin{bmatrix} \boldsymbol{G_1}_u(1,v) & \boldsymbol{G_2}_v(1,v) \\ \boldsymbol{G_2}_u(1,v) & \boldsymbol{G_2}_v(1,v) \end{bmatrix} \begin{bmatrix} 0 \\ 1 \end{bmatrix} = \begin{bmatrix} \boldsymbol{G_1}_v(1,v)\boldsymbol{G} - \boldsymbol{2}_v(1,v) \end{bmatrix}$$

At the point $(1,0)$, on the curve $\boldsymbol{F}(\gamma_2)$, we find the limiting tangent vector

$$\boldsymbol{T}_{\gamma_2}(1,0) = \begin{bmatrix} \boldsymbol{G_1}_v(1,0) & \boldsymbol{G_2}_v(1,0) \end{bmatrix}$$

Note

$$\boldsymbol{T}_{\gamma_1}(1,0) \times \boldsymbol{T}_{\gamma_2} = det\boldsymbol{J_g}(1,0)\boldsymbol{k} = (\boldsymbol{G_1}_u(1,0)\boldsymbol{G_2}_v(1,0) - \boldsymbol{G_1}_v(1,0)\boldsymbol{G_2}_u(1,0)) > 0$$

by assumption. Hence as we rotate $\boldsymbol{T}_{\gamma_1}(1,0)$ into $\boldsymbol{T}_{\gamma_2}(1,0)$, by the right-hand rule the cross product points up. Hence, the curve is moving ccw because of the assumption on the positivity of the determinant of the Jacobian. We can do this sort of analysis at all the vertices we get when the edges in $\Re^2$ under the map $\boldsymbol{F}$ are joined together. Since $det\boldsymbol{J_G}(1,0)$, it is not possible for the tangent vectors at the vertices to align as if they did, the determinant would be zero there. Thus, the mapping $\boldsymbol{G}$ will always generate four curves with four vertices where the tangent vectors do not line up. We often call the angles between the limiting value tangent vectors at the vertices the **interior** angles of D. From what we have just said, since the curve must be moving ccw, these interior angles must be less than π. That rules out some D's we could draw.

To connect the line integrals around $\partial\mathbb{I}^2$ to the line integrals around $\boldsymbol{G}(\partial\mathbb{I}^2)$, we define what is called the **pullback** of $\boldsymbol{F}$ where $\boldsymbol{F}$ is the vector field in the line integral. This *pulls* $\boldsymbol{F}$ which acts on $\boldsymbol{G}(\partial\mathbb{I}^2)$ in (x,y) space *back* to a function defined on (u,v) space. The pullback of $\boldsymbol{F}$ is denoted by $\boldsymbol{F}^*$ and is defined by

$$\boldsymbol{F}^*(u,v) = \begin{bmatrix} \boldsymbol{A}(\boldsymbol{G_1}(u,v), \boldsymbol{G_2}(u,v)) & \boldsymbol{B}(\boldsymbol{G_1}(u,v), \boldsymbol{G_2}(u,v)) \end{bmatrix} \begin{bmatrix} \boldsymbol{G_1}_u & \boldsymbol{G_1}_v \\ \boldsymbol{G_2}_u & \boldsymbol{G_2}_v \end{bmatrix}$$

where, for convenience, we leave out the (u,v) in the Jacobian. So the line integral is

$$\int_{\partial\mathbb{I}^2} < \boldsymbol{F}^*, \gamma' > dt = \int_{\partial\mathbb{I}^2} \left\langle \begin{bmatrix} A(G_1, G_2) \\ B(G_1, G_2) \end{bmatrix}, \begin{bmatrix} \boldsymbol{G_1}_u & \boldsymbol{G_1}_v \\ \boldsymbol{G_2}_u & \boldsymbol{G_2}_v \end{bmatrix} \gamma' \right\rangle dt$$

Now make the change of variables $x = \boldsymbol{G_1}(u,v)$ and $y = \boldsymbol{G_2}(u,v)$. Then

$$\begin{bmatrix} dx \\ dy \end{bmatrix} = \begin{bmatrix} \boldsymbol{G_1}_u(u,v) & \boldsymbol{G_1}_v(u,v) \\ \boldsymbol{G_2}_u(u,v) & \boldsymbol{G_2}_v(u,v) \end{bmatrix} \begin{bmatrix} du \\ dv \end{bmatrix}$$

Thus, after the change of variables, we have

$$\int_{\partial\mathbb{I}^2} < \boldsymbol{F}^*, \gamma' > dt = \int_{\boldsymbol{G}(\partial\mathbb{I}^2)} \left\langle \begin{bmatrix} \boldsymbol{A}(x,y) \\ \boldsymbol{B}(x,y) \end{bmatrix}, (\,\boldsymbol{G}(\gamma)\,)' \right\rangle dt$$

Thus, we know how to define the line integral around the boundary of the image of the unit square under a nice mapping. Green's Theorem applies to the left side, so we have

$$\int\int_{\mathbb{I}^2} \Big((\boldsymbol{A}\boldsymbol{G_1}_v + \boldsymbol{B}\boldsymbol{G_2}_v)_u - (\boldsymbol{A}\boldsymbol{G_1}_u + \boldsymbol{B}\boldsymbol{G_2}_u)_v \Big)\, dA_{uv} =$$

$$\int_{\boldsymbol{G}(\partial\mathbb{I}^2)} \left\langle \begin{bmatrix} \boldsymbol{A}(x,y) & \boldsymbol{B}(x,y) \end{bmatrix}, (\,\boldsymbol{G}(\gamma)\,)' \right\rangle dt$$

Let's simplify the integrand for $\int\int$ over $\mathbb{I}_2$. We use $\boldsymbol{A}_1$, $\boldsymbol{A}_2$ etc. to indicate the partials of $\boldsymbol{A}$ with respect to the first and second argument. Also, we must now assume more smoothness of $\boldsymbol{G}$ as we want to take second order partials and we want the mixed order partials to match. So now $\boldsymbol{G}$ has continuous second order partials.

$$\begin{aligned}&(\boldsymbol{A}\boldsymbol{G_1}_v + \boldsymbol{B}\boldsymbol{G_2}_v)_u - (\boldsymbol{A}\boldsymbol{G_1}_u + \boldsymbol{B}\boldsymbol{G_2}_u)_v = \\ &(\boldsymbol{A}_1\boldsymbol{G}_{1u} + \boldsymbol{A}_2\boldsymbol{G}_{2u})\boldsymbol{G}_{1v} + \boldsymbol{A}\boldsymbol{G}_{1vu} + (\boldsymbol{B}_1\boldsymbol{G}_{1u} + \boldsymbol{B}_2\boldsymbol{G}_{2u})\boldsymbol{G}_{2v} + \boldsymbol{B}\boldsymbol{G}_{2vu} \\ &-(\boldsymbol{A}_1\boldsymbol{G}_{1v} + \boldsymbol{A}_2\boldsymbol{G}_{2v})\boldsymbol{G}_{1u} - \boldsymbol{A}\boldsymbol{G}_{1uv} - (\boldsymbol{B}_1\boldsymbol{G}_{1v} + \boldsymbol{B}_2\boldsymbol{G}_{2v})\boldsymbol{G}_{2u} - \boldsymbol{B}\boldsymbol{G}_{2uv}\end{aligned}$$

Now cancel some terms to get

$$\begin{aligned}&(\boldsymbol{A}\boldsymbol{G_1}_v + \boldsymbol{B}\boldsymbol{G_2}_v)_u - (\boldsymbol{A}\boldsymbol{G_1}_u + \boldsymbol{B}\boldsymbol{G_2}_u)_v = (\boldsymbol{A}_1\boldsymbol{G}_{1u} + \boldsymbol{A}_2\boldsymbol{G}_{2u})\boldsymbol{G}_{1v} \\ &+(\boldsymbol{B}_1\boldsymbol{G}_{1u} + \boldsymbol{B}_2\boldsymbol{G}_{2u})\boldsymbol{G}_{2v} - (\boldsymbol{A}_1\boldsymbol{G}_{1v} + \boldsymbol{A}_2\boldsymbol{G}_{2v})\boldsymbol{G}_{1u} - (\boldsymbol{B}_1\boldsymbol{G}_{1v} + \boldsymbol{B}_2\boldsymbol{G}_{2v})\boldsymbol{G}_{2u}\end{aligned}$$

The terms with $\boldsymbol{A}_1$ and $\boldsymbol{B}_2$ also cancel giving

$$\begin{aligned}(\boldsymbol{A}\boldsymbol{G_1}_v + \boldsymbol{B}\boldsymbol{G_2}_v)_u - (\boldsymbol{A}\boldsymbol{G_1}_u + \boldsymbol{B}\boldsymbol{G_2}_u)_v &= (\boldsymbol{B}_1 - \boldsymbol{A}_2)(\boldsymbol{G}_{2u}\boldsymbol{G}_{1v} - \boldsymbol{G}_{2v}\boldsymbol{G}_{1u}) \\ &= (\boldsymbol{B}_1 - \boldsymbol{A}_2)\ det\boldsymbol{J_G}\end{aligned}$$

What have we shown? We have this chain of results:

$$\begin{aligned}\int\int_{\boldsymbol{G}(\mathbb{I}^2)} (\boldsymbol{B}_1 - \boldsymbol{A}_2)\, dA_{xy} &= \int\int_{\mathbb{I}^2} (\boldsymbol{B}_1 - \boldsymbol{A}_2)\ det\boldsymbol{J_G}\, dA_{uv} \\ &= \int\int_{\mathbb{I}^2} \Big((\boldsymbol{A}\boldsymbol{G_1}_v + \boldsymbol{B}\boldsymbol{G_2}_v)_u - (\boldsymbol{A}\boldsymbol{G_1}_u + \boldsymbol{B}\boldsymbol{G_2}_u)_v\Big)\, dA_{uv} \\ &= \int_{\partial\mathbb{I}^2} < \boldsymbol{F}^*, \gamma' > dt \\ &= \int_{\boldsymbol{G}(\partial\mathbb{I}^2)} \left\langle \begin{bmatrix}\boldsymbol{A}(x,y)\\ \boldsymbol{B}(x,y)\end{bmatrix}, (\,\boldsymbol{G}(\gamma)\,)' \right\rangle dt\end{aligned}$$

This is the proof for Green's Theorem for Oriented 2-Cells. Let's state it formally.

Theorem 16.5.1 Green's Theorem for Oriented 2-Cells

Assume $\boldsymbol{G}$ is a 1-1 map from the oriented 2-cell $\mathbb{I}^2$ to $\boldsymbol{G}(\mathbb{I}^2) = D$ in $\Re^2$. Assume $\boldsymbol{G}$ has continuous second order partials on an open set U which contains $[0,1] \times [0,1]$. Let $\mathscr{C}$ be a path around $\partial\mathbb{I}^2$ with associated vector function γ which traverses the boundary ccw. Then,

$$\int_{\boldsymbol{G}(\partial\mathbb{I}^2)} \left\langle \begin{bmatrix}\boldsymbol{A}(x,y)\\ \boldsymbol{B}(x,y)\end{bmatrix}, (\,\boldsymbol{G}(\gamma)\,)' \right\rangle dt = \int\int_{\boldsymbol{G}(\mathbb{I}^2)} (\boldsymbol{B}_1 - \boldsymbol{A}_2)\, dA_{xy}$$

Proof 16.5.1
We have just gone through this argument. ■

Comment 16.5.1 *You probably noticed how awkward these arguments sometimes were and there was a considerable amount of manipulation involved. Everything is considerably simplified when we begin using the language of differential forms which is discussed in Chapter 17. If ω is a 1-form* *defined on $\boldsymbol{G}(\mathbb{I}^2)$ and $\boldsymbol{d\omega}$ is the 2-form obtained from $\boldsymbol{\omega}$, the above theorem looks like this:*

$$\int_{\boldsymbol{G}(\partial\mathbb{I}^2)} \omega = \int\int_{\boldsymbol{G}(\mathbb{I}^2)} d\omega$$

Further, the integral symbols are simplified a bit: it is understood $\int$ over a two dimensional subset can be handled by $\int\int$. So context tells what to do and we just say

$$\int_{\boldsymbol{G}(\partial\mathbb{I}^2)} \omega = \int_{\boldsymbol{G}(\mathbb{I}^2)} d\omega$$

Comment 16.5.2 *When you study one dimensional things, the work is much simpler. You can see how much more complicated our arguments become when we pass to differentiation and integration over portions of $\Re^2$. Make sure you think about this as it helps you begin to see how to free yourself from one dimensional bias and prepares you for higher dimensions than two. Some of the things we have begun to wrestle with are:*

- *What exactly is a curve in $\Re^2$? What do we mean by the interior and exterior of a closed curve?*
- *What do we mean by the smoothness of functions on subsets of $\Re^2$?*
- *Clearly, some of what we do with line integrals is a way to generalize the Fundamental Theorem of Calculus. You should see this is both an interesting problem and a difficult one as well.*

16.5.1 Homework

Exercise 16.5.1 *Let the vector field be defined by* $\boldsymbol{G} = \begin{bmatrix} \frac{2}{3}u + \frac{1}{3}v \\ \frac{2}{5}u + \frac{3}{5}v \end{bmatrix}$.

The simple closed path $\mathscr{C}$ is defined as the union of the following four curves:

$\mathscr{C}_1$ *is* $\begin{bmatrix} x(t) = t \\ y(t) = 0 \end{bmatrix}$ *for* $0 \le t \le 1$, $\mathscr{C}_2$ *is* $\begin{bmatrix} x(t) = 1 \\ y(t) = t \end{bmatrix}$ *for* $0 \le t \le 1$,
$\mathscr{C}_3$ *is* $\begin{bmatrix} x(t) = 1 - t \\ y(t) = 1 \end{bmatrix}$ *for* $0 \le t \le 1$, $\mathscr{C}_4$ *is* $\begin{bmatrix} x(t) = 0 \\ y(t) = 1 - t \end{bmatrix}$ *for* $0 \le t \le 1$.

- *Draw the boundary curves $\boldsymbol{G}$ determined in the target plane with coordinates (x, y). The easiest way to do this is to use MATLAB and the function* `UnitSquareImage` *which is*

Listing 16.1: **UnitSquareImage.m**

```
function UnitSquareImage(f,g)
%
F = @(U) [f(U);g(U)];
Bottom = @(t) [t;0];
Right =@(t) [1;t];
Top =@(t) [1-t;1];
Left =@(t) [0;1-t];
T = linspace(0,1,31);
for i=1:31
  YB(:,i) = F(Bottom(T(i)));
```

```
    YR(:,i) = F(Right(T(i)));
    YT(:,i) = F(Top(T(i)));
    YL(:,i) = F(Left(T(i)));
  end
  clf;
  hold on
  plot(YB(1,:),YB(2,:),'r');
  plot(YR(1,:),YR(2,:),'g');
  plot(YT(1,:),YT(2,:),'b');
  plot(YL(1,:),YL(2,:),'b');
  xlabel('x');
  ylabel('y');
  title('Unit Square Image under G');
  hold off
  end
```

Using this code is fairly simple:

Listing 16.2: **Unit Square Image under G One**

```
>> f = @(U) (2/3)*U(1)+(1/3)*U(2);
>> g = @(U)(2/5)*U(1) + (3/5)*U(2);
>> UnitSquareImage(f,g);
```

This generates the plot seen in Figure 16.11.

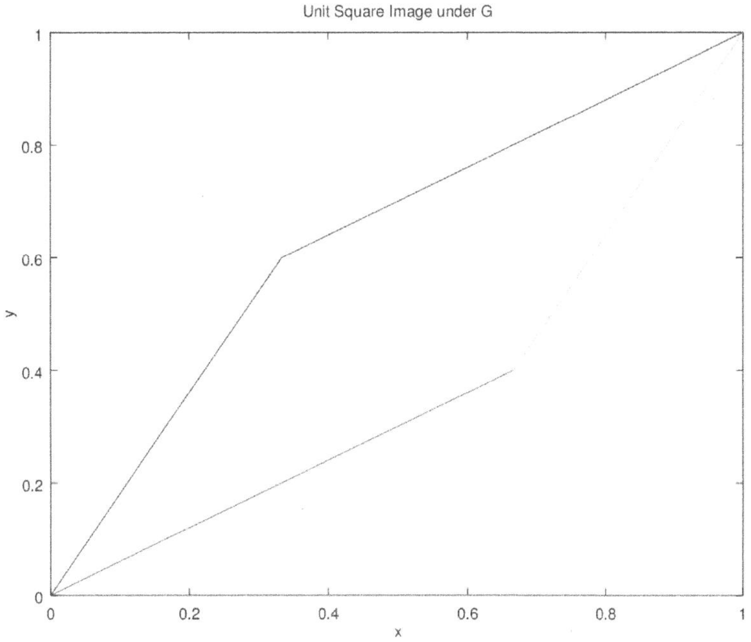

Figure 16.11: Image of the unit square under $((2/3)u = (1/3)v, (2/5)u + (3/5)v)$.

- *Show G is* $1-1$.
- *Find* $\boldsymbol{J_G}$ *and show* $\det(\boldsymbol{J_G}) > 0$ *on an open set containing* $\mathbb{I}^2$.
- *Write down the line integral and the area integral and evaluate both using a uniform Riemann sum. They should match.*

Exercise 16.5.2 *Let the vector field be defined by* $\boldsymbol{G} = \begin{bmatrix} \frac{5}{6}u + \frac{1}{6}v \\ \frac{1}{3}u + \frac{2}{3}v + \frac{1}{15}u^2 - \frac{1}{15}v^2 \end{bmatrix}$.

The simple closed path $\mathscr{C}$ *is defined just as in the previous exercise.*

- *Draw the boundary curves* $\boldsymbol{G}$ *determines in the target plane with coordinates* (x, y) *using MATLAB. Simple code is this:*

Listing 16.3: **Unit Square Image under G Two**

```
>> f =@(U) (5/6)*U(1)+(1/6)*U(2);
>> g = @(U) (1/3)*U(1) + (2/3)*U(2) + (1/15)*U(1).^2 -(1/15)*U(2)
    .^2;
>> UnitSquareImage(f,g);
```

where we use the function `UnitSquareImage` *again. This generates the plot seen in Figure 16.12.*

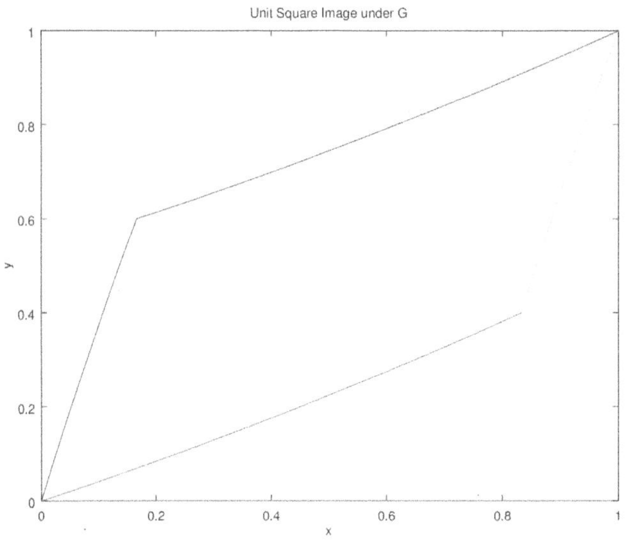

Figure 16.12: Image of the unit square under $(5/6)u + (1/6)v, (1/3)u + (2/3)v + (1/15)u^2 - (1/15)v^2$.

- *Show G is* $1-1$.
- *Find* $\boldsymbol{J_G}$ *and show* $\det(\boldsymbol{J_G}) > 0$ *on an open set containing* $\mathbb{I}^2$.
- *Write down the line integral and the area integral and evaluate both using a uniform Riemann sum. They should match.*

Exercise 16.5.3 *Let the vector field be defined by* $\boldsymbol{G} = \begin{bmatrix} u^2 + v^2 + 4u + 3v \\ -u^2 + v^2 + 2u + 2v \end{bmatrix}$.

The simple closed path $\mathscr{C}$ *is defined just as in the previous exercise.*

- *Draw the boundary curves* $\boldsymbol{G}$ *determines in the target plane with coordinates* (x, y) *using MATLAB using code similar to the other exercises.*
- *Show* G *is* $1-1$.
- *Find* $\boldsymbol{J_G}$ *and show* $\det(\boldsymbol{J_G}) > 0$ *on an open set containing* $\mathbb{I}^2$.
- *Write down the line integral and the area integral and evaluate both using a uniform Riemann sum. They should match.*

Exercise 16.5.4 *Rethink the analysis we did in this section of maps of the unit square assuming the* $\det(\boldsymbol{J_G}) < 0$. *What changes, if anything?*

16.6 Motivational Notation

Recall, we often use the notation $dx = x'(t)dt$ as a convenient way to handle substitution in an integration. We can use this idea to repackage the idea of a line integral of a force field along a curve. We know

$$\begin{aligned} \int_{\mathscr{C}} \left\langle \boldsymbol{F}, \begin{bmatrix} x' \\ y' \end{bmatrix} \right\rangle dt &= \int_{\mathscr{C}} \left(\left\langle \begin{bmatrix} A(x,y) \\ B(x,y) \end{bmatrix}, \begin{bmatrix} x'(t) \\ y'(t) \end{bmatrix} \right\rangle \right) dt \\ &= \int_a^b (A(x(t), y(t))x'(t) + B(x(t), y(t))y'(t))\, dt \end{aligned}$$

Let's rewrite using differential notation. We get

$$\int_{\mathscr{C}} \left\langle \boldsymbol{F}, \begin{bmatrix} x' \\ y' \end{bmatrix} \right\rangle dt = \int_a^b (A(x(t), y(t))dx + B(x(t), y(t))dy)\, dt$$

This suggests we combine the vector field and differential notation and define something we will call a 1-form which has the look

$$\boldsymbol{\omega} = \boldsymbol{A dx + B dy}$$

and define

$$\int_{\mathscr{C}} \omega = \int_a^b (A(x(t), y(t))dx + B(x(t), y(t))dy)\, dt$$

We define the **exterior derivative** (more on this later) of $\boldsymbol{\omega}$ to be

$$\boldsymbol{d\omega} = (\boldsymbol{B}_1 - \boldsymbol{A}_2)\, \boldsymbol{dxdy} = (\boldsymbol{B}_1 - \boldsymbol{A}_2)\, dA_{xy}$$

Now go back to the oriented 2-cell problem. Then,

$$\begin{aligned} \boldsymbol{\omega} &= \boldsymbol{A du + B dv} \\ \int_{\boldsymbol{G}(\mathbb{I}^2)} \boldsymbol{\omega} &= \int_{\boldsymbol{G}(\gamma)} (A(u(t), v(t))du + B(u(t), v(t))dv)\, dt \\ \boldsymbol{d\omega} &= (\boldsymbol{B}_1(u, v) - \boldsymbol{A}_2(u, v))\boldsymbol{dudv} = (\boldsymbol{B}_1 - \boldsymbol{A}_2)\, dA_{uv} \end{aligned}$$

The **pullback** of $\boldsymbol{G}$ is then

$$\begin{aligned}\boldsymbol{G}^*(\boldsymbol{\omega}) &= \boldsymbol{A}(\boldsymbol{G_1},\boldsymbol{G_2})(\boldsymbol{G}_{1u}\boldsymbol{du}+\boldsymbol{G}_{1v}\boldsymbol{dv})+\boldsymbol{B}(\boldsymbol{G_1},\boldsymbol{G_2})(\boldsymbol{G}_{2u}\boldsymbol{du}+\boldsymbol{G}_{2v}\boldsymbol{dv})\\ &= (\boldsymbol{A}(\boldsymbol{G_1},\boldsymbol{G_2})\boldsymbol{G}_{1u}+\boldsymbol{B}(\boldsymbol{G_1},\boldsymbol{G_2})\boldsymbol{G}_{2u})\boldsymbol{du}+(\boldsymbol{A}(\boldsymbol{G_1},\boldsymbol{G_2})\boldsymbol{G}_{1v}+\boldsymbol{B}(\boldsymbol{G_1},\boldsymbol{G_2})\boldsymbol{G}_{2v})\boldsymbol{dv}\end{aligned}$$

and so the exterior derivative gives

$$\begin{aligned}\boldsymbol{d}(\boldsymbol{G}^*(\boldsymbol{\omega})) &= \Big(-(\boldsymbol{A}(\boldsymbol{G_1},\boldsymbol{G_2})\boldsymbol{G}_{1u}+\boldsymbol{B}(\boldsymbol{G_1},\boldsymbol{G_2})\boldsymbol{G}_{2u})_2+\\ &\quad (\boldsymbol{A}(\boldsymbol{G_1},\boldsymbol{G_2})\boldsymbol{G}_{1v}+\boldsymbol{B}(\boldsymbol{G_1},\boldsymbol{G_2})\boldsymbol{G}_{2v})_1\Big)\,\boldsymbol{dudv}\end{aligned}$$

We also define the **pullback** of the exterior derivative

$$\boldsymbol{G}^*(\boldsymbol{d\omega}) = (\boldsymbol{B}_1-\boldsymbol{A}_2)\,det\boldsymbol{J_G}$$

Using the same calculations as before, we then find

$$\boldsymbol{d}(\boldsymbol{G}^*(\boldsymbol{\omega})) = (\boldsymbol{B}_1-\boldsymbol{A}_2)\,det\boldsymbol{J_G}\,\boldsymbol{dudv}=\boldsymbol{G}^*(\boldsymbol{d\omega})$$

Now go back to our arguments for Green's Theorem for the oriented 2-cells. Using this new notation, our proof now goes like this:

$$\int\!\!\int_{\boldsymbol{G}(\mathbb{I}^2)} \boldsymbol{d\omega} = \int\!\!\int_{\mathbb{I}^2} \boldsymbol{G}^*(\boldsymbol{d\omega}) = \int\!\!\int_{\mathbb{I}^2} \boldsymbol{d}(\boldsymbol{G}^*(\boldsymbol{\omega})) = \int_{\partial\mathbb{I}^2} \boldsymbol{G}^*(\boldsymbol{\omega}) = \int_{\boldsymbol{G}(\partial\mathbb{I}^2)} \boldsymbol{\omega}$$

We hide the details of the differential area $\boldsymbol{dxdy}$ and so forth. We can also rewrite using context for whether it is a single or double integral and just write

$$\int_{\boldsymbol{G}(\mathbb{I}^2)} \boldsymbol{d\omega} = \int_{\mathbb{I}^2} \boldsymbol{G}^*(\boldsymbol{d\omega}) = \int_{\mathbb{I}^2} \boldsymbol{d}(\boldsymbol{G}^*(\boldsymbol{\omega})) = \int_{\partial\mathbb{I}^2} \boldsymbol{G}^*(\boldsymbol{\omega}) = \int_{\boldsymbol{G}(\partial\mathbb{I}^2)} \boldsymbol{\omega}$$

In the next chapter we will explore differential forms more carefully.

16.6.1 Homework

Exercise 16.6.1 *Let the vector fields be defined by*

$$\begin{aligned}\boldsymbol{F} &= \begin{bmatrix} -\frac{5x}{(1+x^2+y^2)^{7/2}} \\ -\frac{5y}{(1+x^2+y^2)^{7/2}} \end{bmatrix}\\ \boldsymbol{G} &= \begin{bmatrix} \frac{2}{3}u+\frac{1}{3}v \\ \frac{2}{5}u+\frac{3}{5}v \end{bmatrix}\end{aligned}$$

- *Find the corresponding differential form* $\boldsymbol{\omega}$ *for* $\boldsymbol{F}$.
- *Find* $\boldsymbol{d\omega}$.
- *Find the pullback* $\boldsymbol{G}^*(\boldsymbol{\omega})$.
- *Find the exterior derivative* $\boldsymbol{d}(\boldsymbol{G}^*(\boldsymbol{\omega}))$.
- *Find the pullback of the exterior derivative* $\boldsymbol{G}^*(\boldsymbol{d\omega})$. *Note* $\boldsymbol{G}^*(\boldsymbol{d\omega})$ *is the same as* $\boldsymbol{d}(\boldsymbol{G}^*(\boldsymbol{\omega}))$.

- *Write out the details of the chain*

$$\int_{G(\mathbb{I}^2)} d\omega = \int_{\mathbb{I}^2} G^*(d\omega) = \int_{\mathbb{I}^2} d(G^*(\omega)) = \int_{\partial\mathbb{I}^2} G^*(\omega) = \int_{G(\partial\mathbb{I}^2)} \omega$$

Exercise 16.6.2 *Let the vector fields be defined by*

$$\begin{aligned} F &= \begin{bmatrix} -\frac{5x}{(1+x^2+y^2)^{7/2}} \\ -\frac{5y}{(1+x^2+y^2)^{7/2}} \end{bmatrix} \\ G &= \begin{bmatrix} \frac{5}{6}u + \frac{1}{6}v \\ \frac{1}{3}u + \frac{2}{3}v + \frac{1}{15}u^2 - \frac{1}{15}v^2 \end{bmatrix} \end{aligned}$$

- *Find the corresponding differential form* ω *for* F.
- *Find* $d\omega$.
- *Find the pullback* $G^*(\omega)$.
- *Find the exterior derivative* $d(G^*(\omega))$.
- *Find the pullback of the exterior derivative* $G^*(d\omega)$. *Note* $G^*(d\omega)$ *is the same as* $d(G^*(\omega))$.
- *Write out the details of the chain*

$$\int_{G(\mathbb{I}^2)} d\omega = \int_{\mathbb{I}^2} G^*(d\omega) = \int_{\mathbb{I}^2} d(G^*(\omega)) = \int_{\partial\mathbb{I}^2} G^*(\omega) = \int_{G(\partial\mathbb{I}^2)} \omega$$

Exercise 16.6.3 *Let the vector fields be defined by*

$$\begin{aligned} F &= \begin{bmatrix} -\frac{5x}{(1+x^2+y^2)^{7/2}} \\ -\frac{5y}{(1+x^2+y^2)^{7/2}} \end{bmatrix} \\ G &= \begin{bmatrix} u^2 + v^2 + 4u + 3v \\ -u^2 + v^2 + 2u + 2v \end{bmatrix} \end{aligned}$$

- *Find the corresponding differential form* ω *for* F.
- *Find* $d\omega$.
- *Find the pullback* $G^*(\omega)$.
- *Find the exterior derivative* $d(G^*(\omega))$.
- *Find the pullback of the exterior derivative* $G^*(d\omega)$. *Note* $G^*(d\omega)$ *is the same as* $d(G^*(\omega))$.
- *Write out the details of the chain*

$$\int_{G(\mathbb{I}^2)} d\omega = \int_{\mathbb{I}^2} G^*(d\omega) = \int_{\mathbb{I}^2} d(G^*(\omega)) = \int_{\partial\mathbb{I}^2} G^*(\omega) = \int_{G(\partial\mathbb{I}^2)} \omega$$

Exercise 16.6.4 *Let the vector fields be defined by*

$$\begin{aligned} \boldsymbol{F} &= \begin{bmatrix} -\frac{5x}{(1+x^2+y^2)^{7/2}} \\ -\frac{5y}{(1+x^2+y^2)^{7/2}} \end{bmatrix} \\ \boldsymbol{G} &= \begin{bmatrix} u^2+v^2+4u+3v \\ -u^2+v^2+2u+2v \end{bmatrix} \end{aligned}$$

- *Find the corresponding differential form* $\boldsymbol{\omega}$ *for* $\boldsymbol{F}$.
- *Find* $\boldsymbol{d\omega}$.
- *Find the pullback* $\boldsymbol{G}^*(\boldsymbol{\omega})$.
- *Find the exterior derivative* $\boldsymbol{d}(\boldsymbol{G}^*(\boldsymbol{\omega}))$.
- *Find the pullback of the exterior derivative* $\boldsymbol{G}^*(\boldsymbol{d\omega})$. *Note* $\boldsymbol{G}^*(\boldsymbol{d\omega})$ *is the same as* $\boldsymbol{d}(\boldsymbol{G}^*(\boldsymbol{\omega}))$.
- *Write out the details of the chain*

$$\int_{\boldsymbol{G}(\mathbb{I}^2)} \boldsymbol{d\omega} = \int_{\mathbb{I}^2} \boldsymbol{G}^*(\boldsymbol{d\omega}) = \int_{\mathbb{I}^2} \boldsymbol{d}(\boldsymbol{G}^*(\boldsymbol{\omega})) = \int_{\partial\mathbb{I}^2} \boldsymbol{G}^*(\boldsymbol{\omega}) = \int_{\boldsymbol{G}(\partial\mathbb{I}^2)} \boldsymbol{\omega}$$

Exercise 16.6.5 *Let the vector fields be defined by*

$$\begin{aligned} \boldsymbol{F} &= \begin{bmatrix} 1-x^2-5y^2 \\ -2xy \end{bmatrix} \\ \boldsymbol{G} &= \begin{bmatrix} \frac{2}{3}u+\frac{1}{3}v \\ \frac{2}{5}u+\frac{3}{5}v \end{bmatrix} \end{aligned}$$

- *Find the corresponding differential form* $\boldsymbol{\omega}$ *for* $\boldsymbol{F}$.
- *Find* $\boldsymbol{d\omega}$.
- *Find the pullback* $\boldsymbol{G}^*(\boldsymbol{\omega})$.
- *Find the exterior derivative* $\boldsymbol{d}(\boldsymbol{G}^*(\boldsymbol{\omega}))$.
- *Find the pullback of the exterior derivative* $\boldsymbol{G}^*(\boldsymbol{d\omega})$. *Note* $\boldsymbol{G}^*(\boldsymbol{d\omega})$ *is the same as* $\boldsymbol{d}(\boldsymbol{G}^*(\boldsymbol{\omega}))$.
- *Write out the details of the chain*

$$\int_{\boldsymbol{G}(\mathbb{I}^2)} \boldsymbol{d\omega} = \int_{\mathbb{I}^2} \boldsymbol{G}^*(\boldsymbol{d\omega}) = \int_{\mathbb{I}^2} \boldsymbol{d}(\boldsymbol{G}^*(\boldsymbol{\omega})) = \int_{\partial\mathbb{I}^2} \boldsymbol{G}^*(\boldsymbol{\omega}) = \int_{\boldsymbol{G}(\partial\mathbb{I}^2)} \boldsymbol{\omega}$$

Exercise 16.6.6 *Let the vector fields be defined by*

$$\begin{aligned} \boldsymbol{F} &= \begin{bmatrix} 1-x^2-5y^2 \\ -2xy \end{bmatrix} \\ \boldsymbol{G} &= \begin{bmatrix} \frac{5}{6}u+\frac{1}{6}v \\ \frac{1}{3}u+\frac{2}{3}v+\frac{1}{15}u^2-\frac{1}{15}v^2 \end{bmatrix} \end{aligned}$$

- *Find the corresponding differential form* $\boldsymbol{\omega}$ *for* $\boldsymbol{F}$.

- *Find* $\boldsymbol{d}\omega$.
- *Find the pullback* $\boldsymbol{G}^*(\omega)$.
- *Find the exterior derivative* $\boldsymbol{d}(\boldsymbol{G}^*(\omega))$.
- *Find the pullback of the exterior derivative* $\boldsymbol{G}^*(\boldsymbol{d}\omega)$. *Note* $\boldsymbol{G}^*(\boldsymbol{d}\omega)$ *is the same as* $\boldsymbol{d}(\boldsymbol{G}^*(\omega))$.
- *Write out the details of the chain*

$$\int_{\boldsymbol{G}(\mathbb{I}^2)} \boldsymbol{d}\omega = \int_{\mathbb{I}^2} \boldsymbol{G}^*(\boldsymbol{d}\omega) = \int_{\mathbb{I}^2} \boldsymbol{d}(\boldsymbol{G}^*(\omega)) = \int_{\partial\mathbb{I}^2} \boldsymbol{G}^*(\omega) = \int_{\boldsymbol{G}(\partial\mathbb{I}^2)} \omega$$

Exercise 16.6.7 *Let the vector fields be defined by*

$$\begin{aligned} \boldsymbol{F} &= \begin{bmatrix} 1 - x^2 - 5y^2 \\ -2xy \end{bmatrix} \\ \boldsymbol{G} &= \begin{bmatrix} u^2 + v^2 + 4u + 3v \\ -u^2 + v^2 + 2u + 2v \end{bmatrix} \end{aligned}$$

- *Find the corresponding differential form* ω *for* $\boldsymbol{F}$.
- *Find* $\boldsymbol{d}\omega$.
- *Find the pullback* $\boldsymbol{G}^*(\omega)$.
- *Find the exterior derivative* $\boldsymbol{d}(\boldsymbol{G}^*(\omega))$.
- *Find the pullback of the exterior derivative* $\boldsymbol{G}^*(\boldsymbol{d}\omega)$. *Note* $\boldsymbol{G}^*(\boldsymbol{d}\omega)$ *is the same as* $\boldsymbol{d}(\boldsymbol{G}^*(\omega))$.
- *Write out the details of the chain*

$$\int_{\boldsymbol{G}(\mathbb{I}^2)} \boldsymbol{d}\omega = \int_{\mathbb{I}^2} \boldsymbol{G}^*(\boldsymbol{d}\omega) = \int_{\mathbb{I}^2} \boldsymbol{d}(\boldsymbol{G}^*(\omega)) = \int_{\partial\mathbb{I}^2} \boldsymbol{G}^*(\omega) = \int_{\boldsymbol{G}(\partial\mathbb{I}^2)} \omega$$

Exercise 16.6.8 *Let the vector fields be defined by*

$$\begin{aligned} \boldsymbol{F} &= \begin{bmatrix} 1 - x^2 - 5y^2 \\ -2xy \end{bmatrix} \\ \boldsymbol{G} &= \begin{bmatrix} u^2 + v^2 + 4u + 3v \\ -u^2 + v^2 + 2u + 2v \end{bmatrix} \end{aligned}$$

- *Find the corresponding differential form* ω *for* $\boldsymbol{F}$.
- *Find* $\boldsymbol{d}\omega$.
- *Find the pullback* $\boldsymbol{G}^*(\omega)$.
- *Find the exterior derivative* $\boldsymbol{d}(\boldsymbol{G}^*(\omega))$.
- *Find the pullback of the exterior derivative* $\boldsymbol{G}^*(\boldsymbol{d}\omega)$. *Note* $\boldsymbol{G}^*(\boldsymbol{d}\omega)$ *is the same as* $\boldsymbol{d}(\boldsymbol{G}^*(\omega))$.

- *Write out the details of the chain*

$$\int_{\boldsymbol{G}(\mathbb{I}^2)} \boldsymbol{d\omega} = \int_{\mathbb{I}^2} \boldsymbol{G}^*(\boldsymbol{d\omega}) = \int_{\mathbb{I}^2} \boldsymbol{d}(\boldsymbol{G}^*(\boldsymbol{\omega})) = \int_{\partial\mathbb{I}^2} \boldsymbol{G}^*(\boldsymbol{\omega}) = \int_{\boldsymbol{G}(\partial\mathbb{I}^2)} \boldsymbol{\omega}$$

Chapter 17

Differential Forms

The idea of a **differential form** is very powerful and one that you are not exposed to in the first courses in analysis. So it is time to rectify that and introduce you to a new and useful way of looking at functions of many variables. We started you thinking about the concept at the end of the last chapter as part of a motivational argument to show you how the formalism of differential forms enables us to write down line integrals and their equivalent surface integrals is a nice compact notation. Let's look at this much more closely now.

17.1 One Forms

Let's begin with the idea of fields of vectors.

Definition 17.1.1 Tangent Vector Fields

A tangent vector field $\boldsymbol{V}$ on $\Re^n$ is a function $\boldsymbol{V}$ which assigns to each point $\boldsymbol{P} \in \Re^n$ a vector $\boldsymbol{v_P}$. In addition, we let $\Re^n_{\boldsymbol{P}} = \boldsymbol{T_P}(\Re^n)$ denote the set of all vectors $\boldsymbol{w}$ in $\Re^n$ of the form $\boldsymbol{w} = \boldsymbol{u} + \boldsymbol{P}$ where $\boldsymbol{u}$ in $\Re^n$ is arbitrary.

Comment 17.1.1 *The set $\Re^n_{\boldsymbol{P}} = \boldsymbol{T_P}(\Re^n)$ is clearly an n dimensional vector space over $\Re$ which consists of all the translations of vectors in $\Re^n$ by a fixed vector $\boldsymbol{P}$. We also call this the vector space* **rooted at P**.

We also want to look at **smooth** functions on an open set U in $\Re^n$.

Definition 17.1.2 Smooth Functions on $\Re^n$

A smooth function or C^∞ function on the open set U in $\Re^n$ is a mapping $\boldsymbol{f} : U \rightarrow \Re$ whose partial derivatives of all orders exist and are continuous.

We define a differential 1-form on U or simply a 1-form on U as follows:

Definition 17.1.3 One Forms on an Open Set U

A differential 1-form or just 1-form ω on U is a mapping $\omega : U \subset \Re^2 \rightarrow \Re$ whose value is determined by a pair of smooth functions $\boldsymbol{A}$ and $\boldsymbol{B}$ on U. The one form ω is therefore associated with a vector field $\boldsymbol{H} = \begin{bmatrix} \boldsymbol{A} \\ \boldsymbol{B} \end{bmatrix}$ which means $\boldsymbol{H} : U \rightarrow \Re^2$ with smooth component functions. We also write this as $\boldsymbol{H} = \boldsymbol{A}\boldsymbol{i} + \boldsymbol{B}\boldsymbol{j}$. where $\boldsymbol{i}$ and $\boldsymbol{j}$ are the standard orthonormal basis for $\Re^2$. We define the action of ω on a vector $\boldsymbol{V}$ rooted at the point $\boldsymbol{P}$ with components V_1 and V_2 and P_1 and P_2, respectively to be

$$\omega(\boldsymbol{V_P}) = \boldsymbol{A}(P_1, P_2)V_1 + \boldsymbol{B}(P_1, P_2)V_2$$

and it is easy to see ω is linear.

Homework

Exercise 17.1.1 *Prove the action of the 1-form ω on rooted vectors is linear.*

Exercise 17.1.2 *Let $\boldsymbol{H}(x, y) = \begin{bmatrix} x^2 + 3y^2 \\ 5xy \end{bmatrix}$ define the 1-form ω.*

- *Find the action of ω on the vector $\boldsymbol{V} = \begin{bmatrix} -3 \\ 5 \end{bmatrix}$ rooted at $\boldsymbol{P} = \begin{bmatrix} -1 \\ 1 \end{bmatrix}$.*
- *Find the action of ω on the vector $\boldsymbol{V} = \begin{bmatrix} 6 \\ 2 \end{bmatrix}$ rooted at $\boldsymbol{P} = \begin{bmatrix} 0 \\ 0 \end{bmatrix}$.*

Exercise 17.1.3 *Let $\boldsymbol{H}(x, y) = \begin{bmatrix} 5x^2y^4 + 3y^2 \\ 5x + y^2 \end{bmatrix}$ define the 1-form ω.*

- *Find the action of ω on the vector $\boldsymbol{V} = \begin{bmatrix} -13 \\ 15 \end{bmatrix}$ rooted at $\boldsymbol{P} = \begin{bmatrix} -11 \\ 1 \end{bmatrix}$.*
- *Find the action of ω on the vector $\boldsymbol{V} = \begin{bmatrix} -6 \\ -2 \end{bmatrix}$ rooted at $\boldsymbol{P} = \begin{bmatrix} 1 \\ 0 \end{bmatrix}$.*

Exercise 17.1.4 *Let $\boldsymbol{H}(x, y) = \begin{bmatrix} 5 + x + 2y - xy \\ 2 + 6x - 7y + 2y^2 \end{bmatrix}$ define the 1-form ω.*

- *Find the action of ω on the vector $\boldsymbol{V} = \begin{bmatrix} -3 \\ 1 \end{bmatrix}$ rooted at $\boldsymbol{P} = \begin{bmatrix} -1 \\ 1 \end{bmatrix}$.*
- *Find the action of ω on the vector $\boldsymbol{V} = \begin{bmatrix} -3 \\ -1 \end{bmatrix}$ rooted at $\boldsymbol{P} = \begin{bmatrix} 1 \\ 1 \end{bmatrix}$.*

To understand this better, we need to look at the **dual vector space** to a vector space.

Definition 17.1.4 Algebraic Dual of a Vector Space

Let $\mathscr{V}$ be a vector space over $\Re$. The algebraic dual to $\mathscr{V}$ is denoted by $\mathscr{V}^*$ and is defined by

$$\mathscr{V}^* = \{f : \mathscr{V} \rightarrow \Re | f \text{ is linear}\}$$

Such a function f is called a linear functional on $\mathscr{V}$.

If the vector space $\mathscr{V}$ was also an inner product space and hence had an induced norm or if $\mathscr{V}$ had a norm without an inner product, then we could ask if a linear functional was continuous. Let $\|\cdot\|$ be this norm. Recall f is continuous at $\boldsymbol{p} \in \mathscr{V}$ if for any $\epsilon > 0$, there is a ball $B_\delta(p) = \{\boldsymbol{x} \in \mathscr{V} : \|x - p\| < \delta\}$ in $\mathscr{V}$ so $|f(\boldsymbol{x}) - f(\boldsymbol{pp})| < \epsilon$. We cannot prove this here but we will show you where the proof breaks down. We discuss this properly in (Peterson (22) 2020). Let $\boldsymbol{E}$ be a basis for $\mathscr{V}$ with norm $\|\cdot\|$. and we can assume each $\boldsymbol{E_j}$ has norm one. If f is a linear functional on $\mathscr{V}$, let $\xi_j = f(\boldsymbol{E_j})$. Then for any $\boldsymbol{x} \in \mathscr{V}$, there is a unique representation with respect to this basis

$$\boldsymbol{x} \;=\; c_i\boldsymbol{E_1} + \ldots + c_n\boldsymbol{E_n}$$

and so by the linearity of f

$$f(\boldsymbol{x}) \;=\; x_i f(\boldsymbol{E_1}) + \ldots + x_n(\boldsymbol{E_n}) = \sum_{i=1}^{n} x_i\xi_i$$

Thus

$$|f(\boldsymbol{x}) - f(\boldsymbol{p})| \;\leq\; = \sum_{i=1}^{n} |x_i - p_i|\,|\xi_i| \leq f_E \sum_{i=1}^{n} |x_i - p_i|$$

where $f_E = \max_{1 \leq i \leq n} |\xi_i|$ and p_i are the components of $\boldsymbol{p}$ with respect to the basis $\boldsymbol{E}$. So given an $\epsilon > 0$, we can make $|f(x) - f(p)| < \epsilon$ if we choose $\sum_{i=1}^{n} |x_i - p_i| < \epsilon/f_E$. However, this does not help as

$$\|\boldsymbol{x} - \boldsymbol{p}\| \;=\; \|\sum_{i=1}^{n}(c_i - p_i)\boldsymbol{E_i}\| \leq \sum_{i=1}^{n} |c_i - p_i|\,\|\boldsymbol{E_i}\| = \sum_{i=1}^{n} |c_i - p_i|$$

But choosing $\|\boldsymbol{x} - \boldsymbol{p}\| < \delta = \epsilon/f_E$, does not guarantee that $\sum_{i=1}^{n} |c_i - p_i| < \epsilon/f_E$ which we need. So we are missing a way to handle the $\|\boldsymbol{x} - \boldsymbol{p}\|$ correctly here. To fill this in, we need more details about how normed linear spaces work which we go over in (Peterson (22) 2020). The set of continuous linear functionals on a vector space $\mathscr{V}$ is denoted by $\mathscr{V}'$ and it turns out $\mathscr{V}' = \mathscr{V}^*$ when $\mathscr{V}$ is finite dimensional normed linear space.

The vector space $\boldsymbol{X} = (\Re^2, \{\boldsymbol{i}, \boldsymbol{j}\})$ is traditionally thought of as column vectors and the dual space to $(\Re^2, \{\boldsymbol{i}, \boldsymbol{j}\})$ is then $\boldsymbol{X}^*$ the set of linear functionals on $\boldsymbol{X}$ which is the same as $\boldsymbol{X}'$ the set of continuous linear functionals on $\boldsymbol{X}$. The dual to $(\Re^2, \{\boldsymbol{i}, \boldsymbol{j}\})$ is interpreted as the set of row vectors and a basis dual to any basis $\boldsymbol{E} = \{\boldsymbol{E_1}, \boldsymbol{E_2}\}$ is $\boldsymbol{F} = \{\boldsymbol{F_1}, \boldsymbol{F_2}\}$ where

$$\begin{aligned} \boldsymbol{F_1}(\boldsymbol{E_1}) &= 1, \quad \boldsymbol{F_1}(\boldsymbol{E_2}) = 0 \\ \boldsymbol{F_2}(\boldsymbol{E_1}) &= 0, \quad \boldsymbol{F_2}(\boldsymbol{E_2}) = 1 \end{aligned}$$

Hence,

$$\boldsymbol{F_1}\left(\begin{bmatrix} V_1 \\ V_2 \end{bmatrix}\right) = V_1, \quad \boldsymbol{F_2}\left(\begin{bmatrix} V_1 \\ V_2 \end{bmatrix}\right) = V_2$$

Hence, we can identify $\boldsymbol{F_1}$ and $\boldsymbol{F_2}$ as follows

$$\begin{aligned} \boldsymbol{F_1} &= \begin{bmatrix} 1 & 0 \end{bmatrix} \Longrightarrow \boldsymbol{F_1}\left(\begin{bmatrix} V_1 \\ V_2 \end{bmatrix}\right) = \begin{bmatrix} 1 & 0 \end{bmatrix} \begin{bmatrix} V_1 \\ V_2 \end{bmatrix} = V_1 \\ \boldsymbol{F_2} &= \begin{bmatrix} 0 & 1 \end{bmatrix} \Longrightarrow \boldsymbol{F_2}\left(\begin{bmatrix} V_1 \\ V_2 \end{bmatrix}\right) = \begin{bmatrix} 0 & 1 \end{bmatrix} \begin{bmatrix} V_1 \\ V_2 \end{bmatrix} = V_2 \end{aligned}$$

The usual notation for the dual basis to $\{\boldsymbol{i}, \boldsymbol{j}\}$ is not $\boldsymbol{F_1}$ and $\boldsymbol{F_2}$. Instead, it is $\boldsymbol{dx} = \boldsymbol{F_1}$ and $\boldsymbol{dy} = \boldsymbol{F_2}$. Hence, we define the action of a 1-form $\boldsymbol{\omega}$ on a vector $\boldsymbol{V}$ rooted at $\boldsymbol{P}$ as follows:

$$\begin{aligned}
\boldsymbol{\omega}\left(\begin{bmatrix} V_1 \\ V_2 \end{bmatrix}\right) &= \left(\begin{bmatrix} \boldsymbol{F_1} & \boldsymbol{F_2} \end{bmatrix} \boldsymbol{H}\right)\left(\begin{bmatrix} V_1 \\ V_2 \end{bmatrix}\right) = \left(\begin{bmatrix} \boldsymbol{F_1} & \boldsymbol{F_2} \end{bmatrix} \begin{bmatrix} \boldsymbol{A} \\ \boldsymbol{B} \end{bmatrix}\right)\left(\begin{bmatrix} V_1 \\ V_2 \end{bmatrix}\right) \\
&= (\boldsymbol{A}\boldsymbol{F_1} + \boldsymbol{B}\boldsymbol{F_2})\left(\begin{bmatrix} V_1 \\ V_2 \end{bmatrix}\right) = \boldsymbol{A}(P_1, P_2)\boldsymbol{F_1}\left(\begin{bmatrix} V_1 \\ V_2 \end{bmatrix}\right) + \boldsymbol{B}(P_1, P_2)\boldsymbol{F_2}\left(\begin{bmatrix} V_1 \\ V_2 \end{bmatrix}\right) \\
&= \boldsymbol{A}(P_1, P_2)V_1 + \boldsymbol{Q}(P_1, P_2)V_2
\end{aligned}$$

This is quite a mouthful. It is easier to follow if we use the $\boldsymbol{dx}$ and $\boldsymbol{dy}$ notation as define

$$\begin{aligned}
\boldsymbol{\omega}\left(\begin{bmatrix} V_1 \\ V_2 \end{bmatrix}\right) &= (\boldsymbol{A}\boldsymbol{dx} + \boldsymbol{B}\boldsymbol{dy})\left(\begin{bmatrix} V_1 \\ V_2 \end{bmatrix}\right) = \boldsymbol{A}(P_1, P_2)\boldsymbol{dx}\left(\begin{bmatrix} V_1 \\ V_2 \end{bmatrix}\right) + \boldsymbol{B}(P_1, P_2)\boldsymbol{dy}\left(\begin{bmatrix} V_1 \\ V_2 \end{bmatrix}\right) \\
&= \boldsymbol{A}(P_1, P_2)V_1 + \boldsymbol{B}(P_1, P_2)V_2
\end{aligned}$$

To be even more succinct, for a vector $\boldsymbol{V}$ with the usual components V_1 and V_2 rooted at $\boldsymbol{P}$:

$$\begin{aligned}
\boldsymbol{\omega}(\boldsymbol{V}) &= (\boldsymbol{A}\boldsymbol{dx} + \boldsymbol{B}\boldsymbol{dy})(\boldsymbol{V}) = \boldsymbol{A}(P_1, P_2)\boldsymbol{dx}(\boldsymbol{V}) + \boldsymbol{B}(P_1, P_2)\boldsymbol{dy}(\boldsymbol{V}) \\
&= \boldsymbol{A}(P_1, P_2)V_1 + \boldsymbol{B}(P_1, P_2)V_2
\end{aligned}$$

We can interpret these results a bit differently by thinking of the dual space to $\Re^2_{\boldsymbol{P}}$ where we now think of this space as the space of all tangent vectors attached to a point $\boldsymbol{P}$ where $\boldsymbol{P}$ has coordinates (x_0, y_0). Consider a curve $\mathscr{C}$ in $\Re^2$ given by the smooth curve $f(x, y) = c$ for some constant c. We can think of this as the level curve of the surface $z = f(x, y)$ for $z = c$. Then, the tangent plane at (x_0, y_0) has the normal vector $\boldsymbol{N} = f_x^0\boldsymbol{i} \times f_y^0\boldsymbol{j}$ which is rooted at the point (x_0, y_0). This determines $\boldsymbol{P} = \begin{bmatrix} x_0 \\ y_0 \end{bmatrix}$. The tangent vector line is thus $\boldsymbol{L_P} = f_x^0(x - x_0)\boldsymbol{i} + f_y^0(y - y_0)\boldsymbol{j}$ and the tangent vector itself is $\boldsymbol{V_P} = f_x^0\boldsymbol{i} + f_y^0\boldsymbol{j} + \boldsymbol{P}$. The simplest case is the one determined by $z = f(x, y) = x + y = c$. The tangent vector line is then $\boldsymbol{V_P} = \boldsymbol{1}(x - x_0) + \boldsymbol{1}(y - y_0)$. Any point in $\Re^2_{\boldsymbol{P}}$ can thus be interpreted as the vector $\boldsymbol{Q} = f_x^0\boldsymbol{i} + f_y^0\boldsymbol{j}$ for some smooth function $f(x, y) = c$ where $\boldsymbol{i}$ and $\boldsymbol{j}$ are rooted at $\boldsymbol{P}$. Another way to look at it is the $\boldsymbol{i}$ is a unit vector in the direction of x rooted at x_0 and $\boldsymbol{j}$ is a unit vector in the direction of y rooted at y_0. A linear functional ϕ acting on $\Re^2_{\boldsymbol{P}}$ thus gives

$$\phi(c_1(x - x_0)\boldsymbol{i} + c_2(y - y_0)\boldsymbol{j}) = c_1\phi((x - x_0)\boldsymbol{i}) + c_2\phi((y - y_0)\boldsymbol{j})$$

Now the simplest function, $x + y = c$, gives the tangent plane $z = \boldsymbol{1}(x - x_0) + \boldsymbol{1}(y - y_0)$ and we have

$$\phi(\boldsymbol{1}(x - x_0)\boldsymbol{i} + \boldsymbol{1}(y - x_0)\boldsymbol{j}) = \boldsymbol{1}\phi((x - x_0)\boldsymbol{i}) + \boldsymbol{1}\phi((y - y_0)\boldsymbol{j})$$

This suggests we rethink the dual basis. We know

$$\begin{aligned}
\boldsymbol{F_1}\left(\begin{bmatrix} x - x_0 \\ 0 \end{bmatrix}\right) &= 1, \quad \text{and } \frac{\partial}{\partial x}(x - x_0) = \frac{d}{dx}(x - x_0) = \boldsymbol{1} \\
\boldsymbol{F_2}\left(\begin{bmatrix} 0 \\ (y - y_0) \end{bmatrix}\right) &= 1, \quad \text{and } \frac{\partial}{\partial y}(y - y_0) = \frac{d}{dy}(y - y_0) = \boldsymbol{1}
\end{aligned}$$

This suggests we define

$$\boldsymbol{F_1}\left(\begin{bmatrix} x - x_0 \\ 0 \end{bmatrix}\right) = \frac{\partial}{\partial x}(x - x_0) \Longrightarrow \boldsymbol{F_1} \equiv \frac{\partial}{\partial x}$$

$$F_2\left(\begin{bmatrix} 0 \\ (y-y_0) \end{bmatrix}\right) = \frac{\partial}{\partial y}(y-y_0) \Longrightarrow F_2 \equiv \frac{\partial}{\partial y}$$

With this reinterpretation, we have

$$\phi(c_1(x-x_0)\boldsymbol{i} + c_2(y-y_0)\boldsymbol{j}) = c_1\frac{\partial}{\partial x} + c_2\frac{\partial}{\partial x}$$

and so

$$\phi(f_x^0(x-x_0)\boldsymbol{i} + f_y^0(y-y_0)\boldsymbol{j}) = f_x^0\frac{\partial}{\partial x} + f_y^0\frac{\partial}{\partial x}$$

Thus, in terms of the dual basis, we would write $\phi = c_1\frac{\partial}{\partial x} + c_2\frac{\partial}{\partial y}$ and

$$\begin{aligned}\phi(f_x^0(x-x_0)\boldsymbol{i} + f_y^0(y-y_0)\boldsymbol{j}) &= \left(c_1\frac{\partial}{\partial x} + c_2\frac{\partial}{\partial y}\right)\left(f_x^0(x-x_0)\boldsymbol{i} + f_y^0(y-y_0)\boldsymbol{j}\right) \\ &= c_1 f_x^0 + c_2 f_y^0\end{aligned}$$

The common symbols for this dual basis are $\boldsymbol{dx} = \frac{\partial}{\partial x}$ and $\boldsymbol{dy} = \frac{\partial}{\partial y}$. Thus, a linear functional is written $\phi = c_1\boldsymbol{dx} + c_2\boldsymbol{dy}$.

Homework

Exercise 17.1.5

- *What is the dual to $\Re^3$ with the standard orthonormal basis? What is the dual basis?*
- *What is the dual to $\Re^4$ with the standard orthonormal basis? What is the dual basis?*
- *What is the dual to $\Re^n$ with the standard orthonormal basis? What is the dual basis?*

Exercise 17.1.6 *This exercise is about thinking through all these definitions. We now think of $\Re^2_{\boldsymbol{P}}$ as the set of all tangent vectors attached to $\boldsymbol{P}$. This means the tangent vectors are associated with a curve in $\Re^2$ whose tangent vector at $\boldsymbol{P}$ gives a vector rooted at $\boldsymbol{P}$. We know an element ϕ in the dual of $\Re^2_{\boldsymbol{P}}$ is a linear function that acts on tangent vectors and returns scalars. Further, we know a basis for the dual space is $\frac{\partial}{\partial x}$ and $\frac{\partial}{\partial y}$ so that $\phi = c_1\frac{\partial}{\partial x} + c_2\frac{\partial}{\partial y}$ or $\phi = c_1\boldsymbol{dx} + c_2\boldsymbol{dy}$.*

- *Explain what the dual to $\Re^3_{\boldsymbol{P}}$ is following the discussion above.*
- *Explain what the dual to $\Re^4_{\boldsymbol{P}}$ is following the discussion above. At this point, switch to the variables x_1, x_2, x_3, x_4.*
- *Explain what the dual to $\Re^n_{\boldsymbol{P}}$ is following the discussion above. Use the variables $x_1, \ldots, x_n$.*

Exercise 17.1.7 *Consider the curve $f(x,y) = x^2 + 3y^2 = 2$.*

- *Find the tangent plane $x_0 = .2$ and $y_0 = \sqrt{(2-x_0^2)/3}$.*
- *Find the tangent vector $\boldsymbol{V_P}$ at to this curve at (x_0, y_0), $\boldsymbol{V_P}$ where $P = \begin{bmatrix} x_0 \\ y_0 \end{bmatrix}$.*
- *Let $\phi = 2\boldsymbol{dx} + 4\boldsymbol{dy}$. Find $\phi(\boldsymbol{V_P})$.*

Exercise 17.1.8 *Consider the curve $f(x,y) = 3x^2 + 6y^2 = 2$.*

- *Find the tangent plane at $x_0 = .3$ and $y_0 = \sqrt{(2-3x_0^2)/6}$.*

- *Find the tangent vector* $\boldsymbol{V_P}$ *to this curve at* (x_0, y_0), $\boldsymbol{V_P}$ *where* $P = \begin{bmatrix} x_0 \\ y_0 \end{bmatrix}$. *This is* $\boldsymbol{V_P} = f_x^0(x - x_0)\boldsymbol{i} + f_y^0(y - y_0)\boldsymbol{j}$

- *Let* $\phi = 12\boldsymbol{dx} - 3\boldsymbol{dy}$. *Find* $\phi(\boldsymbol{V_P})$.

Exercise 17.1.9 *Consider the curve* $f(x, y) = x^2y^4 = 8$.

- *Find the tangent plane at* $x_0 = .4$ *and* $y_0 = -(8/x_0^2)^{1/4}$.

- *Find the tangent vector* $\boldsymbol{V_P}$ *to this curve at* (x_0, y_0), $\boldsymbol{V_P}$ *where* $P = \begin{bmatrix} x_0 \\ y_0 \end{bmatrix}$.

- *Let* $\phi = 8\boldsymbol{dx} + 45\boldsymbol{dy}$. *Find* $\phi(\boldsymbol{V_P})$.

17.1.1 Smooth Paths

Next, we need to define a smooth path in U.

Definition 17.1.5 Smooth Paths

Let U *be an open set in* $\Re^2$. *A smooth path in* U *is a mapping* $\gamma : [a, b] \rightarrow U$ *for some finite interval* $[a, b]$ *in* $\Re$ *which is continuous on* $[a, b]$ *and differentiable on* (a, b). *We assume* γ *can be extended to a smooth* C^∞ *map in a neighborhood of* $[a, b]$. *Hence*

$$\gamma(t) = \begin{bmatrix} x(t) \\ y(t) \end{bmatrix}, \quad a \leq t \leq b$$

where x *and* y *are* C^∞ *on* $(a - \epsilon, b + \epsilon)$ *for some positive* ϵ.

Comment 17.1.2 *The initial point of the path is* $\gamma(a)$ *and the final point is* $\gamma(b)$. *We say* γ *is a path from* $\gamma(a)$ *to* $\gamma(b)$.

If $\gamma(t_0) \neq \boldsymbol{0}$, then at the point $\gamma(t_0)$, there is a two dimensional vector space rooted at $\gamma(t_0)$ we can call $\boldsymbol{T}(t_0)$ defined by the span of the vectors

$$\boldsymbol{E_1}(t_0) = \boldsymbol{i}, \quad \boldsymbol{E_2}(t_0) = \boldsymbol{j}$$

Note $\boldsymbol{T}(t_0)$ is determined by the tangent vector to the curve $\gamma(t)$ at the point t_0. Any vector in this vector space has the representation rooted at $\gamma(t_0)$ given by

$$\begin{bmatrix} p \\ q \end{bmatrix} = (t - t_0) \begin{bmatrix} \alpha\, x'(t_0) \\ \beta\, y'(t_0) \end{bmatrix} = \alpha\, x'(t_0)\,(t - t_0)\boldsymbol{E_1}(t_0) + \beta\, y'(t_0)(t - t_0)\boldsymbol{E_2}(t_0)$$

This shows $\{\boldsymbol{E_1}(t_0), \boldsymbol{E_2}(t_0)\}$ is a basis for $\boldsymbol{T}(t_0)$. We define a dual basis $\{\boldsymbol{dx}(t_0), \boldsymbol{dy}(t_0)\}$ by

$$\boldsymbol{dx}(t_0) = \boldsymbol{E_1}^T(t_0), \quad \boldsymbol{dy}(t_0) = \boldsymbol{E_2}^T(t_0)$$

Then at each t, we can represent $\gamma(t)$ in $\boldsymbol{T}(t_0)$ by

$$\begin{bmatrix} x(t) \\ y(t) \end{bmatrix} = \begin{bmatrix} \alpha(t)\, x'(t_0) \\ \beta(t)\, y'(t_0) \end{bmatrix} = \gamma(t_0) + \alpha(t)\, x'(t_0)\, \boldsymbol{E_1}(t_0) + \beta(t)\, y'(t_0)\, \boldsymbol{E_2}(t_0)$$

Now if $\boldsymbol{f}$ is a linear functional on $\boldsymbol{T}(t_0)$, then

$$\begin{aligned} &\boldsymbol{f}\Big(\alpha(t)\, x'(t_0)\, \boldsymbol{E_1}(t_0) + \beta(t)\, y'(t_0)\, \boldsymbol{E_2}(t_0)\Big) \\ &= \alpha(t)\, x'(t_0) \boldsymbol{f}\Big(\boldsymbol{E_1}(t_0)\Big) + \beta(t)\, y'(t_0)\, \boldsymbol{f}\Big(\boldsymbol{E_2}(t_0)\Big) \\ &= \alpha(t)\, x'(t_0)\, c_f + \beta(t)\, y'(t_0)\, d_f \end{aligned}$$

as the action of $\boldsymbol{f}$ is completely determined by how it acts on the basis for $\boldsymbol{T}(t_0)$. Now consider

$$\begin{aligned} &(c_1 \boldsymbol{dx} + c_2 \boldsymbol{dy})\Big(\gamma(t_0) + \alpha(t)\, x'(t_0)\, \boldsymbol{E_1}(t_0) + \beta(t)\, y'(t_0)\, \boldsymbol{E_2}(t_0)\Big) \\ &= c_1\, \alpha(t)\, x'(t_0) + c_2\, \beta(t)\, y'(t_0) \end{aligned}$$

For equality, we need

$$\begin{aligned} \alpha(t)\, x'(t_0)\, c_f &= c_1\, \alpha(t)\, x'(t_0) \Longrightarrow c_1 = c_f \\ c_2\, \beta(t)\, y'(t_0) &= \beta(t)\, y'(t_0)\, d_f \Longrightarrow c_2 = d_f \end{aligned}$$

We conclude the span of $\{\boldsymbol{dx}, \boldsymbol{dy}\}$ is the dual of $\boldsymbol{T}(t_0)$. Thus, the action of $\boldsymbol{dx}$ and $\boldsymbol{dy}$ on the path γ should be defined to be

$$\boldsymbol{dx}(\gamma(t)) = x'(t), \quad \boldsymbol{dy}(\gamma(t)) = y'(t)$$

Note this is same as what we had before except now we have the chain rule involved. We have a linear functional

$$\phi = c_1 \boldsymbol{dx} + c_2 \boldsymbol{dy} = c_1 \frac{\partial}{\partial x}\frac{d}{dt} + c_2 \frac{\partial}{\partial y}\frac{d}{dt}$$

With all this done, we have an understanding of how to interpret the action of the 1-form $\boldsymbol{\omega}$ on the path γ. At each t, we define

$$\begin{aligned} \boldsymbol{\omega}(\gamma) &= \boldsymbol{P}(x(t_0), y(t_0))\, \boldsymbol{dx}(t_0)(\gamma(t)) + \boldsymbol{Q}(x(t_0), y(t_0))\, \boldsymbol{dy}(t_0)(\gamma(t)) \\ &= \boldsymbol{P}(x(t_0), y(t_0))\, x'(t_0) + \boldsymbol{Q}(x(t_0), y(t_0))\, y'(t_0) \end{aligned}$$

This is quite a notational morass, so we usually say this sloppier and let context be our guide. The action of the 1-form $\boldsymbol{\omega}$ on the smooth path γ is

$$\begin{aligned} \boldsymbol{\omega}(\gamma(t)) &= \boldsymbol{P}(\gamma(t))\boldsymbol{dx}(\gamma(t)) + \boldsymbol{Q}(\gamma(t))\boldsymbol{dy}(\gamma(t)) \\ &= \boldsymbol{P}(x(t), y(t))x'(t) + \boldsymbol{Q}(x(t), y(t))y'(t) = \begin{bmatrix} \boldsymbol{P} \\ \boldsymbol{Q} \end{bmatrix} (\gamma(t))\, \gamma'(t) \end{aligned}$$

Homework

Exercise 17.1.10 *Let* $\boldsymbol{H}(x, y) = \begin{bmatrix} x^2 + 3y^2 \\ 5xy \end{bmatrix}$ *define the 1-form* $\boldsymbol{\omega}$ *and let the smooth path be* $\gamma(t) = \begin{bmatrix} 2\cos(t) \\ 2\sin(t) \end{bmatrix}$ *for* $0 \leq t \leq 2\pi$. *Find* $\boldsymbol{\omega}(\gamma(t))$.

Exercise 17.1.11 *Let* $\boldsymbol{H}(x,y) = \begin{bmatrix} 5x^2y^4 + 3y^2 \\ 5x + y^2 \end{bmatrix}$ *define the 1-form* ω *and let the smooth path be* $\gamma(t) = \begin{bmatrix} 2\cos(2t) \\ 3\sin(2t) \end{bmatrix}$ *for* $0 \leq t \leq \pi$. *Find* $\boldsymbol{\omega}(\gamma(t))$.

Exercise 17.1.12 *Let* $\boldsymbol{H}(x,y) = \begin{bmatrix} 5 + x + 2y - xy \\ 2 + 6x - 7y + 2y^2 \end{bmatrix}$ *define the 1-form* ω *and let the smooth path be* $\gamma(t) = \begin{bmatrix} 5\cos(2t) \\ 3\sin(2t) \end{bmatrix}$ *for* $0 \leq t \leq \pi$. *Find* $\boldsymbol{\omega}(\gamma(t))$.

We can then define the integral $\int_\gamma \omega$.

Definition 17.1.6 Integration of a 1-Form over a Smooth Path

Let γ *be a smooth path in the open set* U *in* $\Re^2$ *and let* $\boldsymbol{\omega} = \boldsymbol{P}d\boldsymbol{x} + \boldsymbol{Q}d\boldsymbol{y}$ *be a 1-form. Then*

$$\int_\gamma \omega = \int_a^b \left(\boldsymbol{P}(x(t),y(t))\, x'(t) + \boldsymbol{Q}(x(t),y(t))\, y'(t) \right) dt$$

Note this is a well-defined Riemann Integrable as the integrand is continuous.

Recall, in Chapter 16, these types of integrals are called **Line Integrals**. Hence, what we went through above can be interpreted from another point of view. Suppose each point on the smooth path $\gamma(t)$ represents the charge of the path at the point t. Then how do we compute the total charge of the path? First, note the function

$$U(t) = \boldsymbol{P}(x(t),y(t))\, x'(t) + \boldsymbol{Q}(x(t),y(t))\, y'(t)$$

is a continuous function on $[a,b]$ and hence, it is Riemann Integrable. Hence, the value of this integral is given by

$$\lim_{n \to \infty} S(U, \pi_n, \sigma_n)$$

for any sequence of partitions $\{\pi_n\}$ whose norms go to zero and σ_n is any evaluation set in π_n. Then, for a given partition, the mass in the portion of the path between partition point t_j and t_{j+1} can be approximated by

$$\boldsymbol{P}(x(t_j),y(t_j))(x(t_{j+1}) - x(t_j)) + \boldsymbol{Q}(x(t_j),y(t_j))(y(t_{j+1}) - y(t_j)) =$$
$$\left(\boldsymbol{P}(x(t_j),y(t_j)) \frac{x(t_{j+1}) - x(t_j)}{t_{j+1} - t_j} + \boldsymbol{Q}(x(t_j),y(t_j)) \frac{y(t_{j+1}) - y(t_j)}{t_{j+1} - t_j} \right)(t_{j+1} - t_j)$$

Now apply the Mean Value Theorem to see

$$\left(\boldsymbol{P}(x(t_j),y(t_j)) \frac{x(t_{j+1}) - x(t_j)}{t_{j+1} - t_j} + \boldsymbol{Q}(x(t_j),y(t_j)) \frac{y(t_{j+1}) - y(t_j)}{t_{j+1} - t_j} \right)(t_{j+1} - t_j) =$$
$$\left(\boldsymbol{P}(x(t_j),y(t_j))x'(s_j) + \boldsymbol{Q}(x(t_j),y(t_j))y'(s_j) \right)(t_{j+1} - t_j)$$

The approximation to the mass on the portion $[t_j, t_{j+1}]$ can just as well be done setting $t_j = s_j$ in the $\boldsymbol{P}$ and $\boldsymbol{Q}$ terms. Thus, the approximation can be

$$U(s_j)\,(t_{j+1} - t_j) = \left(\boldsymbol{P}(x(s_j),y(s_j))x'(s_j) + \boldsymbol{Q}(x(s_j),y(s_j))y'(s_j) \right)(t_{j+1} - t_j)$$

The Riemann Approximation Theorem then tells us

$$\lim_{n\to\infty} \sum_{\pi_n} U(s_j)\Delta t_j = \int_a^b U(t)dt$$

which is the same result as before. We see our interpretation of the integration of the 1-form ω over the smooth path γ is the same as what we call the line integral of the vector field $\boldsymbol{P}\boldsymbol{i} + \boldsymbol{Q}\boldsymbol{j}$ on the path given by the smooth curve γ.

Homework

Exercise 17.1.13 *Let* $\boldsymbol{H}(x,y) = \begin{bmatrix} x^2 + 3y^2 \\ 5xy \end{bmatrix}$ *define the 1-form* ω *and let the smooth path be* $\gamma(t) = \begin{bmatrix} 2\cos(t) \\ 2\sin(t) \end{bmatrix}$ *for* $0 \leq t \leq 2\pi$. *Set up* $\int_\gamma \omega$ *and then evaluate it using an appropriate uniform Riemann Sum.*

Exercise 17.1.14 *Let* $\boldsymbol{H}(x,y) = \begin{bmatrix} 5x^2y^4 + 3y^2 \\ 5x + y^2 \end{bmatrix}$ *define the 1-form* ω *and let the smooth path be* $\gamma(t) = \begin{bmatrix} 2\cos(2t) \\ 3\sin(2t) \end{bmatrix}$ *for* $0 \leq t \leq \pi$. *Set up* $\int_\gamma \omega$ *and then evaluate it using an appropriate uniform Riemann Sum.*

Exercise 17.1.15 *Let* $\boldsymbol{H}(x,y) = \begin{bmatrix} 5 + x + 2y - xy \\ 2 + 6x - 7y + 2y^2 \end{bmatrix}$ *define the 1-form* ω *and let the smooth path be* $\gamma(t) = \begin{bmatrix} 5\cos(2t) \\ 3\sin(2t) \end{bmatrix}$ *for* $0 \leq t \leq \pi$. *Set up* $\int_\gamma \omega$ *and then evaluate it using an appropriate uniform Riemann Sum.*

17.1.2 What is a Winding Number?

Consider a simple circle in the plane parameterized by $x(t) = \cos(t)$, $y(t) = \sin(t)$ for $0 \leq t \leq 2\pi$. Hence, the smooth path here is

$$\gamma(t) = \begin{bmatrix} \cos(t) \\ \sin(t) \end{bmatrix}$$

for $0 \leq t \leq 2\pi$ where we start at the point $(1,0)$ at $t = 0$ and end there as well at $t = 2\pi$. This is a circle in the plane of radius 1 centered at the origin. From the center of the circle, if you draw a vector to a point outside the circle, the variable t represents the angle from the positive x axis measured counterclockwise (ccw) to this line. You can imagine that the picture we draw would be very similar even if we chose an anchor point different from the center. We would still have a well-defined angle t. If we choose an arbitrary starting angle t_0, the angle we measure as we move ccw around the circle would start at t_0 and would end right back at the start point with the new angle $t_0 + 2\pi$. Hence, there is no way around the fact that as we move ccw around the circle, the angle we measure has a discontinuity. To allow for more generality later, let this angle be denoted by $\theta(t)$ which in our simple example is just $\theta(t) = t$. Since $\tan(\theta(t)) = y(t)/x(t)$, we have the general equation for the rate of change of the angle:

$$\theta'(t) = \frac{-y(t)x'(t) + x(t)y'(t)}{x^2(t) + y^2(t)}$$

Of course, here this reduces to $\theta'(t) = 1$ as we really just have $\theta(t) = t$ which has a very simple derivative! Note we can use this to define a 1-form ω by

$$\omega = \frac{-y}{x^2+y^2}\,\boldsymbol{dx} + \frac{x}{x^2+y^2}\,\boldsymbol{dy}$$

Then, consider this scaled integral of ω on the path γ:

$$\begin{aligned} \frac{1}{2\pi}\int_\gamma \omega &= \frac{1}{2\pi}\int_0^{2\pi} \frac{-y(t)x'(t) + x(t)y'(t)}{x^2(t)+y^2(t)}\,dt \\ &= \frac{1}{2\pi}\int_0^{2\pi} 1\,dt = 1. \end{aligned}$$

On the other hand, suppose we put the anchor point of our angle measurement system outside the circle. So imagine the reference line for the angle $\theta(t)$ to be moved from a vector starting at the origin of the circle to a point outside the circle to a new vector starting outside the circle. For convenience, assume this new vector is in quadrant one. At the point this vector is rooted, it determines a local coordinate system for the plane whose positive x axis direction is given by the bottom of the usual reference triangle we draw at this point. In Figure 17.1, we show a typical setup. Note the angles measured start at θ_1, increase to θ_2 and then decrease back to θ_1. Then, since the angles are now measured from the base point (x_0, y_0) outside the circle, we set up the line integral a bit differently. We find the change in angle is now zero and there is no discontinuous jump in the angle.

$$\begin{aligned} &\frac{1}{2\pi}\int_{\theta_1}^{\theta_2} \frac{-(y(t)-y_0)x'(t) + (x(t)-x_0)y'(t)}{(x(t)-x_0)^2 + (y(t)-y_0)^2}\,dt + \\ &\frac{1}{2\pi}\int_{\theta_2}^{\theta_1} \frac{-(y(t)-y_0)x'(t) + (x(t)-x_0)y'(t)}{(x(t)-x_0)^2 + (y(t)-y_0)^2}\,dt = 0. \end{aligned}$$

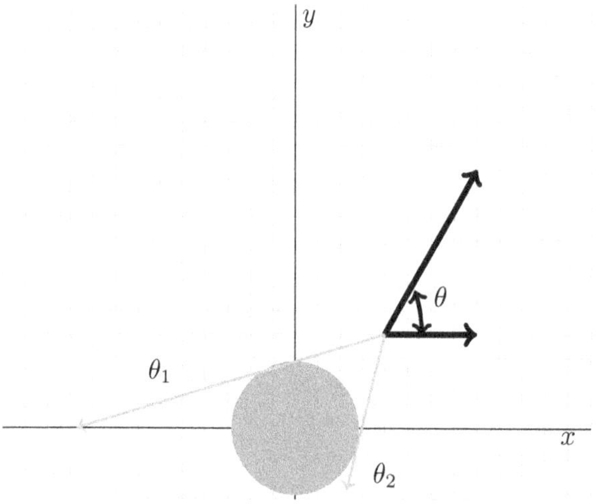

The angle reference system is outside the circle. The angle θ_1 is the first angle needed for the circle and the angle θ_2 is the last.

Figure 17.1: Angle measurements from outside the circle.

In general, if we try to measure the angle using a point inside the circle, we always get $+1$ for this integral and if we try to measure the angle from a reference point outside the circle we get 0. This integer we are calculating is called the *winding number* associated with the particular curve given by our circle and we will return to it later in more detail. Now, if we had a parameterized curve in the plane given by the pair of functions $(x(t), y(t))$, the line integrals we have just used still are a good way to define the change in angle as we move around the curve. We will define the *winding number* of the smooth path γ relative to a point $\boldsymbol{P}$ to mean we are measuring the change in angle as we move around the curve γ from its start point to its finish point where we assume the start and finish point are the same. Such smooth paths are called **closed**. So if functions that determine γ are defined on $[a, b]$, a closed curve means $(x(a), y(a)) = (x(b), y(b))$. Hence, closed curves are nice analogues of the simplest possible closed curve, the circle at the origin. The winding number, $W(\gamma, \boldsymbol{P})$, is defined to be

$$W(\gamma, \boldsymbol{P}) = \frac{1}{2\pi} \int_\gamma \omega$$

where we use as 1-form

$$\omega = \frac{-(y - y_0)}{(x - x_0)^2 + (y - y_0)^2}\, \boldsymbol{dx} + \frac{(x - x_0)}{(x - x_0)^2 + (y - y_0)^2}\, \boldsymbol{dy}$$

where the coordinates of P are (x_0, y_0).
Now let's back up a bit. In Figure 17.1, we found the change in angle as we moved around the circle was zero giving us a winding number of zero because we had placed the angle measurement systems reference outside the circle. Another way of saying this is that $W(\gamma, \boldsymbol{P}) = 0$ for γ being the circle centered at the origin when $\boldsymbol{P}$ is outside the circle. A little thought tells you that this should be true for more arbitrary paths γ although, of course, there is lots of detail we are leaving out as well as subtleties. Now look at Figure 17.1 again and think of the circle and its interior as a collection of points we cannot probe with our paths γ. We can form closed loops around any point P in the plane in general and we can *deform* these closed paths into new closed paths that let $\boldsymbol{P}$ pop out of the inside of γ as long as we are free to alter the path. Altering the path means we find new parameterizations $(x(t), y(t))$ which pass through new points in the plane. We are free to do this kind of alteration as freely as we want as long as we don't pick a $\boldsymbol{P}$ inside the gray circle. Think about it. If we had a closed path γ that enclosed a point $\boldsymbol{P}$ inside the gray circle, then we cannot deform this γ into a new path that excludes $\boldsymbol{P}$ from its interior because we are not allowed to let the points $(x(t), y(t))$ on the new path enter the gray circle's interior. So we can probe for the existence of this collection of points which have been excluded by looking at the winding numbers of paths γ. We look for paths γ, whose winding numbers relative to a point $\boldsymbol{P}$ inside the gray circle, all have nonzero winding numbers. Another way of saying this is that if $\boldsymbol{P}$ is **not** inside the gray circle, there are paths γ with $\boldsymbol{P}$ **not** in their interior and so their winding number is zero. But for $\boldsymbol{P}$ inside the gray circle, we cannot find any closed paths γ whose interior excludes $\boldsymbol{P}$ and hence their winding numbers are always not zero (in our simple thought experiment, they are $+1$). Also, please note we keep talking about the **interior** of a smooth path γ as if it is always easy to figure out what that is. In general, it is not! So we still have much to discuss. The first book you can read in this area is the book on *Algebraic Topology* (W. Fulton (29) 1995), which requires you to be focused and ready to learn new ways to think. But these are sophisticated ideas and you should not be disheartened by their difficulty!

Homework
In these exercises use the 1-form

$$\omega^0 = \frac{-(y - y_0)}{(x - x_0)^2 + (y - y_0)^2}\, \boldsymbol{dx} + \frac{(x - x_0)}{(x - x_0)^2 + (y - y_0)^2}\, \boldsymbol{dy}$$

Exercise 17.1.16 *For the 1-form ω^0, let the smooth path be $\gamma(t) = \begin{bmatrix} 2\cos(t) \\ 2\sin(t) \end{bmatrix}$ for $0 \leq t \leq 2\pi$.*

- *For $(x_0, y_0) = (0, 0)$, set up $\int_\gamma \omega^0$ and then evaluate it using an appropriate uniform Riemann Sum.*

- *For* $(x_0, y_0) = (0.5, 0.5)$*, set up* $\int_\gamma \omega^0$ *and then evaluate it using an appropriate uniform Riemann Sum.*
- *For* $(x_0, y_0) = (1, 1)$*, set up* $\int_\gamma \omega^0$ *and then evaluate it using an appropriate uniform Riemann Sum.*
- *For* $(x_0, y_0) = (2, 2)$*, set up* $\int_\gamma \omega^0$ *and then evaluate it using an appropriate uniform Riemann Sum.*

Exercise 17.1.17 *For the 1-form* ω^0*, let the smooth path be* $\gamma(t) = \begin{bmatrix} 2\cos(t) \\ 2\sin(t) \end{bmatrix}$ *for* $0 \leq t \leq 4\pi$*.*

- *For* $(x_0, y_0) = (0, 0)$*, set up* $\int_\gamma \omega^0$ *and then evaluate it using an appropriate uniform Riemann Sum.*
- *For* $(x_0, y_0) = (0.5, 0.5)$*, set up* $\int_\gamma \omega^0$ *and then evaluate it using an appropriate uniform Riemann Sum.*
- *For* $(x_0, y_0) = (1, 1)$*, set up* $\int_\gamma \omega^0$ *and then evaluate it using an appropriate uniform Riemann Sum.*
- *For* $(x_0, y_0) = (2, 2)$*, set up* $\int_\gamma \omega^0$ *and then evaluate it using an appropriate uniform Riemann Sum.*

Exercise 17.1.18 *For the 1-form* ω^0*, let the smooth path be* $\gamma(t) = \begin{bmatrix} 2\cos(2t) \\ 3\sin(2t) \end{bmatrix}$ *for* $0 \leq t \leq \pi$*.*

- *For* $(x_0, y_0) = (0, 0)$*, set up* $\int_\gamma \omega^0$ *and then evaluate it using an appropriate uniform Riemann Sum.*
- *For* $(x_0, y_0) = (0.5, 0.5)$*, set up* $\int_\gamma \omega^0$ *and then evaluate it using an appropriate uniform Riemann Sum.*
- *For* $(x_0, y_0) = (3, 2)$*, set up* $\int_\gamma \omega^0$ *and then evaluate it using an appropriate uniform Riemann Sum.*
- *For* $(x_0, y_0) = (4, 4)$*, set up* $\int_\gamma \omega^0$ *and then evaluate it using an appropriate uniform Riemann Sum.*

Exercise 17.1.19 *For the 1-form* ω^0*, let the smooth path be* $\gamma(t) = \begin{bmatrix} 2\cos(2t) \\ 3\sin(2t) \end{bmatrix}$ *for* $0 \leq t \leq 2\pi$*.*

- *For* $(x_0, y_0) = (0, 0)$*, set up* $\int_\gamma \omega^0$ *and then evaluate it using an appropriate uniform Riemann Sum.*
- *For* $(x_0, y_0) = (0.5, 0.5)$*, set up* $\int_\gamma \omega^0$ *and then evaluate it using an appropriate uniform Riemann Sum.*
- *For* $(x_0, y_0) = (3, 2)$*, set up* $\int_\gamma \omega^0$ *and then evaluate it using an appropriate uniform Riemann Sum.*
- *For* $(x_0, y_0) = (2, 4)$*, set up* $\int_\gamma \omega^0$ *and then evaluate it using an appropriate uniform Riemann Sum.*

Exercise 17.1.20 *For the 1-form* ω^0*, let the smooth path be* $\gamma(t) = \begin{bmatrix} 5\cos(2t) \\ 2\sin(2t) \end{bmatrix}$ *for* $0 \leq t \leq 2\pi$*.*

- *For* $(x_0, y_0) = (0, 0)$*, set up* $\int_\gamma \omega^0$ *and then evaluate it using an appropriate uniform Riemann Sum.*

- *For* $(x_0, y_0) = (0.5, 0.5)$*, set up* $\int_\gamma \omega^0$ *and then evaluate it using an appropriate uniform Riemann Sum.*
- *For* $(x_0, y_0) = (-3, -2)$*, set up* $\int_\gamma \omega^0$ *and then evaluate it using an appropriate uniform Riemann Sum.*
- *For* $(x_0, y_0) = (2, 4)$*, set up* $\int_\gamma \omega^0$ *and then evaluate it using an appropriate uniform Riemann Sum.*

Exercise 17.1.21 *For the 1-form* ω^0*, let the smooth path be* $\gamma(t) = \begin{bmatrix} 5\cos(2t) \\ 2\sin(2t) \end{bmatrix}$ *for* $0 \leq t \leq 6\pi$.

- *For* $(x_0, y_0) = (0, 0)$*, set up* $\int_\gamma \omega^0$ *and then evaluate it using an appropriate uniform Riemann Sum.*
- *For* $(x_0, y_0) = (0.5, 0.5)$*, set up* $\int_\gamma \omega^0$ *and then evaluate it using an appropriate uniform Riemann Sum.*
- *For* $(x_0, y_0) = (-3, -2)$*, set up* $\int_\gamma \omega^0$ *and then evaluate it using an appropriate uniform Riemann Sum.*
- *For* $(x_0, y_0) = (2, 4)$*, set up* $\int_\gamma \omega^0$ *and then evaluate it using an appropriate uniform Riemann Sum.*

17.2 Exact and Closed 1-Forms

Let U be an open set in $\Re^2$. We have already defined what a smooth function on U is, but now we need to know more about the topology of the open set U itself. We need the idea of a connected set. There are several ways to do this. We will start with the simplest.

Definition 17.2.1 Connected Sets

An open set U *in* $\Re^n$ *is connected if it is not possible to find two disjoint open sets* A *and* B *so that* $U = A \cup B$*. If* U *is not connected, the sets* A *and* B *are called the components of* U*.*

An important fact about the relationship between connected open sets and smooth functions is this:

Theorem 17.2.1 All Smooth Functions with Zero Derivatives are Constant if and only if the Domain is Connected

Let $\boldsymbol{f} : U \subset \Re^n \to \Re$ *be a smooth function on the open set* U*. Then*

$$\text{all smooth } \boldsymbol{f} \ni \boldsymbol{\nabla f} = \boldsymbol{0} \text{ are constant} \quad \Longleftrightarrow \quad U \text{ is connected}$$

Proof 17.2.1

$\Longrightarrow$:

If $\boldsymbol{\nabla f} = \boldsymbol{0}$ *in* U*, then it is easy to see the second order partials of* $\boldsymbol{f}$ *are identically* 0*. Given* $\boldsymbol{x_0}$ *in* U*, it is an interior point and so there is a neighborhood of it,* $B_r(\boldsymbol{x_0})$ *in* U*. Hence, for any* $\boldsymbol{x} \in B_r(\boldsymbol{x_0})$*, we know*

$$\boldsymbol{f}(\boldsymbol{x}) = \boldsymbol{f}(\boldsymbol{x_0}) + \boldsymbol{\nabla f}(\boldsymbol{x_0})(\boldsymbol{x} - \boldsymbol{x_0}) + \frac{1}{2}\left[\boldsymbol{x} - \boldsymbol{x_0}\right]^T \begin{bmatrix} \boldsymbol{f}_{11}(\boldsymbol{c}) & \ldots & \boldsymbol{f}_{1n}(\boldsymbol{c}) \\ \vdots & & \vdots \\ \boldsymbol{f}_{n1}(\boldsymbol{c}) & \ldots & \boldsymbol{f}_{nn}(\boldsymbol{c}) \end{bmatrix} \left[\boldsymbol{x} - \boldsymbol{x_0}\right]$$

where $\mathbf{c}$ is a point on the line between $\boldsymbol{x}_0$ and $\boldsymbol{x}$. This point is also in $B_r(\boldsymbol{x}_0)$. However, since all the second order partials are zero, the Hessian matrix here is always zero. Thus, we have $\boldsymbol{f}(\boldsymbol{x}) = \boldsymbol{f}(\boldsymbol{x}_0)$. Since these points were arbitrarily chosen, we see $\boldsymbol{f}$ is constant on $B_r(\boldsymbol{x}_0)$; i.e. we have shown such functions $\boldsymbol{f}$ must be locally constant on U.

For the given $\boldsymbol{x}_0$ and a given $\boldsymbol{f}$, let $A = \boldsymbol{f}^{-1}(\mathbf{d})$ where $\mathbf{d} = \boldsymbol{f}(\boldsymbol{x}_0)$. If $\boldsymbol{y} \in A$, then because $\boldsymbol{y}$ is an interior point, there is a neighborhood $B_s(\boldsymbol{y})$ contained in U. Hence, $B_s(\boldsymbol{y}) \cap A$ is an open set with $\boldsymbol{z} in A$ for all $\boldsymbol{z}$ in $B_s(\boldsymbol{y}) \cap A$. This shows $\boldsymbol{y}$ is an interior point of A and so A is open. Now let $B = U \setminus A$. Pick any $\boldsymbol{y}$ in B. Then $\boldsymbol{f}(\boldsymbol{y}) \neq \mathbf{d}$ and since $\boldsymbol{f}$ is locally constant at $\boldsymbol{y}$, we know there is a neighborhood about $\boldsymbol{y}$ with corresponding function values not in A. Hence, they are in B which shows $\boldsymbol{y}$ is an interior point of B. We conclude B is an open set and we have written $U = A \cup B$ with A and B disjoint and open. This shows U is not connected. This is a contradiction and so we must conclude B is empty and $U = \boldsymbol{f}^{-1}(\mathbf{d})$. Thus any such $\boldsymbol{f}$ must be constant on all of U.

$\Longleftarrow$:

If U were not connected, we can see using the argument in the first part that $\boldsymbol{f}$ is a constant on each component of U. This would mean $\boldsymbol{f}$ might not be constant on U. Thus, U must be connected in this case. ■

Note we could use any orthonormal basis $\boldsymbol{E}$ here, but we might as well just do the internal mapping from $\boldsymbol{E}$ to $\{\boldsymbol{i}, \boldsymbol{j}\}$ mentally and just think of everything in terms of a starting basis which is the traditional standard one.

Let's look at a few results in this area.

Theorem 17.2.2 A Simple Recapture Theorem in $\Re^2$

Let f be a C^∞ function in the open set U in $\Re^2$. Let the 1-form $\omega = f_x \boldsymbol{dx} + f_y \boldsymbol{dy}$. Then $\int_\gamma \omega = f(\gamma(b)) - f(\gamma(a))$ where the notation $f(\gamma(t))$ means $f(x(t), y(t))$ where $x(t)$ and $y(t)$ are the component functions of the smooth path γ.

Proof 17.2.2

This is a simple application of the Fundamental Theorem of Calculus.

$$\begin{aligned} \left(f(\gamma(t))\right)' &= f_x(x(t), y(t))\, x'(t) + f_y(x(t), y(t))\, y'(t) \\ &\Longrightarrow \int_\gamma \omega = \int_a^b \left(f(\gamma(t))\right)' dt = f(\gamma(b)) - f(\gamma(a)) \end{aligned}$$

■

Comment 17.2.1 *The 1-form $\omega = f_x \boldsymbol{dx} + f_y \boldsymbol{dy}$ occurs very frequently and we use a special notation for it: $\boldsymbol{df} = f_x \boldsymbol{dx} + f_y \boldsymbol{dy}$.*

An important idea is that of **exactness**.

Definition 17.2.2 Exact 1-Forms

*Let f be a C^∞ function in the open set U in $\Re^2$ with corresponding 1-form $\boldsymbol{df} = f_x \boldsymbol{dx} + f_y \boldsymbol{dy}$. A 1-form ω is a **differential** of f if $\omega = \boldsymbol{df}$. A 1-form ω is called **exact** when $\omega = \boldsymbol{df}$ for some smooth f.*

Theorem 17.2.3 The Angle Function is not Exact on $\Re^2 \setminus \{0.0\}$

For $f(x,y) = \tan^{-1}(y/x)$,

$$\boldsymbol{df} = \frac{-y}{x^2+y^2}\,\boldsymbol{dx} + \frac{x}{x^2+y^2}\,\boldsymbol{dy}$$

Note f is not a smooth function on $\Re^2 \setminus \{0,0\}$ but $\boldsymbol{df}$ is a smooth 1-form on $\Re^2 \setminus \{0,0\}$. In fact, there is **no** *smooth function g on $\Re^2 \setminus \{0,0\}$ so that $\boldsymbol{df} = \boldsymbol{dg}$.*

Proof 17.2.3

By direct calculation, for the smooth path

$$\gamma(t) = \begin{bmatrix} \cos(t) \\ \sin(t) \end{bmatrix}$$

for $0 \le t \le 2\pi$, we have $\int_\gamma \boldsymbol{df} = 2\pi$ and $\gamma(0) = \gamma(2\pi)$. Note, f_x and f_y fail to exist at some points on the path, so the argument we used in Theorem 17.2.2 does not apply. However, using the argument we used in Theorem 17.2.2 does work for a smooth g; we have

$$\int_\gamma \omega = \int_0^{2\pi} \Big(f(\gamma(t))\Big)' \, dt = f(\gamma(2\pi)) - f(\gamma(0)) = 0$$

Hence, no such g can exist. ■

Homework

We are going to redo previous exercises using our new language of exactness.

Exercise 17.2.1 *Let the vector field be defined by $\boldsymbol{F} = \begin{bmatrix} \frac{x}{\sqrt{x^2+y^2}} \\ \frac{x}{\sqrt{x^2+y^2}} \end{bmatrix}$.*

- *Show $\boldsymbol{F}$ is a conservative force field.*
- *Find the potential functions $\phi(x,y)$.*
- *The differential form here is $\boldsymbol{\omega} = \boldsymbol{F_1 dx} + \boldsymbol{F_2 dy}$. Show $\boldsymbol{\omega}$ is exact and hence $\boldsymbol{\omega} = \boldsymbol{d\phi}$.*
- *Where is ω defined?*
- *Compute $\int_{\gamma,(1,1)}^{(4,7)} \boldsymbol{\omega}$ for any smooth curve in the domain of $\boldsymbol{\omega}$ connecting $(1,1)$ to $(4,7)$.*

Exercise 17.2.2 *Let the vector field be defined by $\boldsymbol{F} = \begin{bmatrix} \frac{x}{\sqrt{1+x^2+y^2}} \\ \frac{x}{\sqrt{1+x^2+y^2}} \end{bmatrix}$.*

- *Show $\boldsymbol{F}$ is a conservative force field.*
- *Find the potential functions $\phi(x,y)$.*
- *The differential form here is $\boldsymbol{\omega} = \boldsymbol{F_1 dx} + \boldsymbol{F_2 dy}$. Show $\boldsymbol{\omega}$ is exact and hence $\boldsymbol{\omega} = \boldsymbol{d\phi}$.*
- *Where is ω defined?*
- *Compute $\int_{\gamma,(1,1)}^{(4,7)} \boldsymbol{\omega}$ for any smooth curve in the domain of $\boldsymbol{\omega}$ connecting $(1,1)$ to $(4,7)$.*

Exercise 17.2.3 *Let the vector field be defined by* $\boldsymbol{F} = \begin{bmatrix} \frac{2x}{1+x^2+y^2} \\ \frac{2y}{1+x^2+y^2} \end{bmatrix}$.

- *Show* $\boldsymbol{F}$ *is a conservative force field.*
- *Find the potential functions* $\phi(x, y)$.
- *The differential form here is* $\boldsymbol{\omega} = \boldsymbol{F_1 dx} + \boldsymbol{F_2 dy}$. *Show* $\boldsymbol{\omega}$ *is exact and hence* $\boldsymbol{\omega} = \boldsymbol{d\phi}$.
- *Where is* ω *defined?*
- *Compute* $\int_{\gamma,(1,1)}^{(4,7)} \boldsymbol{\omega}$ *for any smooth curve in the domain of* $\boldsymbol{\omega}$ *connecting* $(1, 1)$ *to* $(4, 7)$.

Exercise 17.2.4 *Let the vector field be defined by* $\boldsymbol{F} = \begin{bmatrix} 2xe^{-x^2-y^2} \\ 2ye^{-x^2-y^2} \end{bmatrix}$.

- *Show* $\boldsymbol{F}$ *is a conservative force field.*
- *Find the potential functions* $\phi(x, y)$.
- *The differential form here is* $\boldsymbol{\omega} = \boldsymbol{F_1 dx} + \boldsymbol{F_2 dy}$. *Show* $\boldsymbol{\omega}$ *is exact and hence* $\boldsymbol{\omega} = \boldsymbol{d\phi}$.
- *Where is* ω *defined?*
- *Compute* $\int_{\gamma,(1,1)}^{(4,7)} \boldsymbol{\omega}$ *for any smooth curve in the domain of* $\boldsymbol{\omega}$ *connecting* $(1, 1)$ *to* $(4, 7)$.

Exercise 17.2.5 *Let the vector field be defined by* $\boldsymbol{F} = \begin{bmatrix} -\frac{5x}{(1+x^2+y^2)^{7/2}} \\ -\frac{5y}{(1+x^2+y^2)^{7/2}} \end{bmatrix}$.

- *Show* $\boldsymbol{F}$ *is a conservative force field.*
- *Find the potential functions* $\phi(x, y)$.
- *The differential form here is* $\boldsymbol{\omega} = \boldsymbol{F_1 dx} + \boldsymbol{F_2 dy}$. *Show* $\boldsymbol{\omega}$ *is exact and hence* $\boldsymbol{\omega} = \boldsymbol{d\phi}$.
- *Where is* ω *defined?*
- *Compute* $\int_{\gamma,(1,1)}^{(4,7)} \boldsymbol{\omega}$ *for any smooth curve in the domain of* $\boldsymbol{\omega}$ *connecting* $(1, 1)$ *to* $(4, 7)$.

Exercise 17.2.6 *Let the vector field be defined by* $\boldsymbol{F} = \begin{bmatrix} 5x^2 + 7y^2 \\ 3xy \end{bmatrix}$.

- *Show* $\boldsymbol{F}$ *is a not conservative force field.*
- *Why is* $\boldsymbol{\omega}$ *not exact?*
- *Where is* ω *defined?*
- *How do you compute* $\int_{\gamma,(1,1)}^{(4,7)} \boldsymbol{\omega}$ *for any smooth curve in the domain of* $\boldsymbol{\omega}$ *connecting* $(1, 1)$ *to* $(4, 7)$.

Exercise 17.2.7 *Let the vector field be defined by* $\boldsymbol{F} = \begin{bmatrix} 5x^2 - 7y^2 \\ x + 2y + 3xy \end{bmatrix}$.

- *Show* $\boldsymbol{F}$ *is a not conservative force field.*
- *Why is* $\boldsymbol{\omega}$ *not exact?*
- *Where is* ω *defined?*
- *How do you compute* $\int_{\gamma,(1,1)}^{(4,7)} \boldsymbol{\omega}$ *for any smooth curve in the domain of* $\boldsymbol{\omega}$ *connecting* $(1, 1)$ *to* $(4, 7)$.

17.2.1 Smooth Segmented Paths

We can also glue smooth paths together to build **segmented paths**.

Definition 17.2.3 Smooth Segmented Paths

A smooth segmented path γ is a finite sequence of smooth paths $\{\gamma_1, \ldots, \gamma_n\}$ where each γ_i is a smooth path and the final point of γ_i is the initial point of γ_{i+1}. We write

$$\gamma = \gamma_1 + \ldots + \gamma_n$$

and for any 1 - form ω, we define

$$\int_\gamma \omega = \int_{\gamma_1} \omega + \ldots + \int_{\gamma_n} \omega$$

The first result is to extend Theorem 17.2.2 to segmented paths.

Theorem 17.2.4 The Recapture Theorem for Segmented Paths

Let γ be a smooth segmented path in the open set U in $\Re^2$. Then if $\omega = \boldsymbol{df}$ in U

$$\int_\gamma \omega = f(\boldsymbol{Q}) - f(\boldsymbol{P})$$

where $\boldsymbol{P} = \gamma_1(a)$ and $\boldsymbol{Q} = \gamma_n(b)$. $\boldsymbol{P}$ is the initial and $\boldsymbol{Q}$ is the final point of the path.

Proof 17.2.4
We can apply Theorem 17.2.2 to each piece of the segmented path.

$$\int_\gamma \omega = \sum_{i=1}^{n} \Big(f(\gamma_i(b)) - f(\gamma_i(a)) \Big)$$

But $\gamma_i(b) = \gamma_{i+1}(a)$. Hence,

$$\int_\gamma \omega = f(\gamma_n(b)) + \sum_{i=1}^{n-1} \Big(f(\gamma_{i+1}(a)) - f(\gamma_i(a)) \Big)$$

But the summation here is telescoping and gives

$$\int_\gamma \omega = f(\gamma_n(b)) - f(\gamma_1(a))$$

which is the result. ∎

What about the converse?

Theorem 17.2.5 Equivalent Characterizations of Exactness

Let ω be a 1-form on the open set U in $\Re^2$. Then the following are equivalent:

1. $\int_\gamma \omega = \int_\delta \omega$ *for any smooth segmented paths* γ *and* $\boldsymbol{\delta}$ *with the same domain,* $[a, b]$*, and the same initial and final points.*
2. $\int_\gamma \omega = 0$ *for smooth segmented paths* γ *that are closed; i.e. their initial and final points are the same.*
3. $\omega = \boldsymbol{df}$ *for a smooth function* f *on* U*; i.e.* ω *is* **exact**.

Proof 17.2.5
First, the inverse of the path γ *will be denoted by* γ^{-1} *and is defined by* $\gamma^{-1} = \gamma(b + a - t)$.

(2 $\Longrightarrow$ **1:)**
Note $\int_{\gamma^{-1}} \omega = -\int_\gamma \omega$*. So given paths* γ *and* $\boldsymbol{\delta}$ *in* U *with the same initial and final points, let* $\boldsymbol{\tau}$ *be the closed path*

$$\boldsymbol{\tau} = \gamma_1 + \ldots + \gamma_n - \boldsymbol{\delta}_1 + \ldots + \boldsymbol{\delta}_m$$

Then by (2), $\int_{\boldsymbol{\tau}} \omega = 0$ *which implies* $\int_\gamma \omega = \int_\delta \omega$.

(1 $\Longrightarrow$ **2:)**
Let γ *be a closed path and let* $\boldsymbol{\delta}$ *be a constant path, i.e. it never leaves the initial point. By (1),* $\int_\gamma \omega = \int_\delta \omega = 0$.

(1 $\Longrightarrow$ **3:)**
Let $\omega = \boldsymbol{U dx} + \boldsymbol{V dy}$ *for concreteness. Assume* U *is connected. If* U *has more than one component, you can do the argument that follows on each component. Pick a fixed point* $\boldsymbol{P_0}$ *in* U*. Let the smooth segmented path* γ *have initial point* $\boldsymbol{P_0}$ *and final point* $\boldsymbol{P} = (x, y)$ *in* U*. For the point* $\boldsymbol{P'} = (x + h, y)$*, define the path* $\boldsymbol{\sigma}$ *by* $\boldsymbol{\sigma}(t) = (x + t, y)$ *for* $0 \leq t \leq h$*. Then* $\gamma + \boldsymbol{\sigma}$ *is a segmented path from* $\boldsymbol{P_0}$ *to* $\boldsymbol{P'}$*. For any* $\boldsymbol{Q}$ *in* U*, define* f *in* U *by* $f(\boldsymbol{Q}) = \int_{\boldsymbol{\tau}} \omega$ *for any smooth segmented path* $\boldsymbol{\tau}$ *from* $\boldsymbol{P_0}$ *to* $\boldsymbol{Q}$*. Since* $\boldsymbol{P}$ *is in* U*,* $\boldsymbol{P}$ *is an interior point and so for sufficiently small* h*,* $\boldsymbol{P'}$ *is in* U *also. Thus* $f(\boldsymbol{P'})$ *is well-defined and*

$$\frac{f(x+h,y) - f(x,y)}{h} = \frac{1}{h}\left(\int_{\gamma+\sigma} \omega - \int_\gamma \omega\right) = \frac{1}{h}\int_\sigma \omega$$

$$= \frac{1}{h}\int_0^h \boldsymbol{U}(x+t,y)dt$$

Now apply the Mean Value Theorem to $g(h) = \int_0^h \boldsymbol{U}(x+t,y)dt$ *whose derivative is* $g'(h) = \boldsymbol{U}(x+h,y)$ *by the Fundamental Theorem of Calculus. Thus,* $g(h) - g(0) = g'(c)\,h$ *for some* c *between* 0 *and* h*. Hence,*

$$\frac{1}{h}\int_0^h \boldsymbol{U}(x+t,y)dt = \frac{1}{h}\Big(\boldsymbol{U}(x+c,y)\,h\Big) = \boldsymbol{U}(x+c,y)$$

But $\boldsymbol{U}$ *is continuous, so as* $h \to 0$,

$$\frac{\partial f}{\partial x}(\boldsymbol{P}) = \lim_{h\to 0}\frac{f(x+h,y) - f(x,y)}{h} = \lim_{h\to 0}\boldsymbol{U}(x+c,y) = \boldsymbol{U}(x,y) = \boldsymbol{U}(\boldsymbol{P})$$

A similar argument shows $\frac{\partial f}{\partial x}(\boldsymbol{P}) = \boldsymbol{V}(\boldsymbol{P})$. This shows $\boldsymbol{\omega} = \boldsymbol{df}$.

*(**3** $\Longrightarrow$ **1**:)*
If $\boldsymbol{\omega} = \boldsymbol{df}$, then $\int_\gamma \omega = f(\boldsymbol{Q}) - f(\boldsymbol{P})$ where $\boldsymbol{P}$ and $\boldsymbol{Q}$ are the initial and final point of γ. This is also true for another path $\boldsymbol{\delta}$ with the same initial and final points. This shows (1) holds. ■

Homework

Exercise 17.2.8 *Let the vector field be defined by* $\boldsymbol{F} = \begin{bmatrix} \frac{x}{\sqrt{x^2+y^2}} \\ \frac{x}{\sqrt{x^2+y^2}} \end{bmatrix}$.

- *Consider the smooth segmented path $\mathscr{C}$ which is the unit square starting at $(0,0)$ and moving counterclockwise to $(1,0)$, then $(1,1)$, then back to $(0,1)$ and finally back to $(0,0)$. This is the path $\mathscr{C}$ defined by four simple straight paths: $\mathscr{C}_1$ is $\begin{bmatrix} x(t) = t \\ y(t) = 0 \end{bmatrix}$ for $0 \leq t \leq 1$, $\mathscr{C}_2$ is $\begin{bmatrix} x(t) = 1 \\ y(t) = t \end{bmatrix}$ for $0 \leq t \leq 1$, $\mathscr{C}_3$ is $\begin{bmatrix} x(t) = 1-t \\ y(t) = 1 \end{bmatrix}$ for $0 \leq t \leq 1$ and $\mathscr{C}_4$ is $\begin{bmatrix} x(t) = 0 \\ y(t) = 1-t \end{bmatrix}$ for $0 \leq t \leq 1$. Why is $\int_{\mathscr{C}} \omega = 0$?*
- *Consider the smooth segmented path $\mathscr{C}$ and the simple closed path $\mathscr{C}$ is defined as the union of the following curves: $\mathscr{C}_1$ is $\begin{bmatrix} x(t) = t \\ y(t) = 0 \end{bmatrix}$ for $0 \leq t \leq 3$, $\mathscr{C}_2$ is $\begin{bmatrix} x(t) = 3 \\ y(t) = t \end{bmatrix}$ for $0 \leq t \leq 3$ and $\mathscr{C}_3$ is $\begin{bmatrix} x(t) = 3-t \\ y(t) = 3-t \end{bmatrix}$ for $0 \leq t \leq 3$. Why is $\int_{\mathscr{C}} \omega = 0$?*
- *Consider the smooth segmented path $\mathscr{C}$ and the simple closed path $\mathscr{C}$ is defined as the union of the following four curves: $\mathscr{C}_1$ is $\begin{bmatrix} x(t) = 1-t \\ y(t) = 1-t^2 \end{bmatrix}$ for $0 \leq t \leq 1$, $\mathscr{C}_2$ is $\begin{bmatrix} x(t) = t \\ y(t) = 1-t^2 \end{bmatrix}$ for $0 \leq t \leq 1$, $\mathscr{C}_3$ is $\begin{bmatrix} x(t) = -1+t \\ y(t) = -1+(-1+t)^2 \end{bmatrix}$ for $0 \leq t \leq 1$ and $\mathscr{C}_4$ is $\begin{bmatrix} x(t) = t \\ y(t) = 1-t^2 \end{bmatrix}$ for $0 \leq t \leq 1$. Why is $\int_{\mathscr{C}} \omega = 0$?*

Exercise 17.2.9 *Let the vector field be defined by* $\boldsymbol{F} = \begin{bmatrix} \frac{x}{\sqrt{1+x^2+y^2}} \\ \frac{x}{\sqrt{1+x^2+y^2}} \end{bmatrix}$.

- *Consider the smooth segmented path $\mathscr{C}$ which is the unit square starting at $(0,0)$ and moving counterclockwise to $(1,0)$, then $(1,1)$, then back to $(0,1)$ and finally back to $(0,0)$. This is the path $\mathscr{C}$ defined by four simple straight paths: $\mathscr{C}_1$ is $\begin{bmatrix} x(t) = t \\ y(t) = 0 \end{bmatrix}$ for $0 \leq t \leq 1$, $\mathscr{C}_2$ is $\begin{bmatrix} x(t) = 1 \\ y(t) = t \end{bmatrix}$ for $0 \leq t \leq 1$, $\mathscr{C}_3$ is $\begin{bmatrix} x(t) = 1-t \\ y(t) = 1 \end{bmatrix}$ for $0 \leq t \leq 1$ and $\mathscr{C}_4$ is $\begin{bmatrix} x(t) = 0 \\ y(t) = 1-t \end{bmatrix}$ for $0 \leq t \leq 1$. Why is $\int_{\mathscr{C}} \omega = 0$?*
- *Consider the smooth segmented path $\mathscr{C}$ and the simple closed path $\mathscr{C}$ is defined as the union of the following curves: $\mathscr{C}_1$ is $\begin{bmatrix} x(t) = t \\ y(t) = 0 \end{bmatrix}$ for $0 \leq t \leq 3$, $\mathscr{C}_2$ is $\begin{bmatrix} x(t) = 3 \\ y(t) = t \end{bmatrix}$ for $0 \leq t \leq 3$ and $\mathscr{C}_3$ is $\begin{bmatrix} x(t) = 3-t \\ y(t) = 3-t \end{bmatrix}$ for $0 \leq t \leq 3$. Why is $\int_{\mathscr{C}} \omega = 0$?*
- *Consider the smooth segmented path $\mathscr{C}$ and the simple closed path $\mathscr{C}$ is defined as the union of the following four curves: $\mathscr{C}_1$ is $\begin{bmatrix} x(t) = 1-t \\ y(t) = 1-t^2 \end{bmatrix}$ for $0 \leq t \leq 1$, $\mathscr{C}_2$ is $\begin{bmatrix} x(t) = t \\ y(t) = 1-t^2 \end{bmatrix}$ for*

$0 \leq t \leq 1$, $\mathscr{C}_3$ *is* $\begin{bmatrix} x(t) = -1+t \\ y(t) = -1+(-1+t)^2 \end{bmatrix}$ *for* $0 \leq t \leq 1$ *and* $\mathscr{C}_4$ *is* $\begin{bmatrix} x(t) = t \\ y(t) = 1-t^2 \end{bmatrix}$ *for* $0 \leq t \leq 1$. *Why is* $\int_{\mathscr{C}} \boldsymbol{\omega} = 0$?

Exercise 17.2.10 *Let the vector field be defined by* $\boldsymbol{F} = \begin{bmatrix} \frac{2x}{1+x^2+y^2} \\ \frac{2y}{1+x^2+y^2} \end{bmatrix}$.

- *Consider the smooth segmented path* $\mathscr{C}$ *which is the unit square starting at* $(0,0)$ *and moving counterclockwise to* $(1,0)$, *then* $(1,1)$, *then back to* $(0,1)$ *and finally back to* $(0,0)$. *This is the path* $\mathscr{C}$ *defined by four simple straight paths:* $\mathscr{C}_1$ *is* $\begin{bmatrix} x(t) = t \\ y(t) = 0 \end{bmatrix}$ *for* $0 \leq t \leq 1$, $\mathscr{C}_2$ *is* $\begin{bmatrix} x(t) = 1 \\ y(t) = t \end{bmatrix}$ *for* $0 \leq t \leq 1$, $\mathscr{C}_3$ *is* $\begin{bmatrix} x(t) = 1-t \\ y(t) = 1 \end{bmatrix}$ *for* $0 \leq t \leq 1$ *and* $\mathscr{C}_4$ *is* $\begin{bmatrix} x(t) = 0 \\ y(t) = 1-t \end{bmatrix}$ *for* $0 \leq t \leq 1$. *Why is* $\int_{\mathscr{C}} \boldsymbol{\omega} = 0$?
- *Consider the smooth segmented path* $\mathscr{C}$ *and the simple closed path* $\mathscr{C}$ *is defined as the union of the following curves:* $\mathscr{C}_1$ *is* $\begin{bmatrix} x(t) = t \\ y(t) = 0 \end{bmatrix}$ *for* $0 \leq t \leq 3$, $\mathscr{C}_2$ *is* $\begin{bmatrix} x(t) = 3 \\ y(t) = t \end{bmatrix}$ *for* $0 \leq t \leq 3$ *and* $\mathscr{C}_3$ *is* $\begin{bmatrix} x(t) = 3-t \\ y(t) = 3-t \end{bmatrix}$ *for* $0 \leq t \leq 3$. *Why is* $\int_{\mathscr{C}} \boldsymbol{\omega} = 0$?
- *Consider the smooth segmented path* $\mathscr{C}$ *and the simple closed path* $\mathscr{C}$ *is defined as the union of the following four curves:* $\mathscr{C}_1$ *is* $\begin{bmatrix} x(t) = 1-t \\ y(t) = 1-t^2 \end{bmatrix}$ *for* $0 \leq t \leq 1$, $\mathscr{C}_2$ *is* $\begin{bmatrix} x(t) = t \\ y(t) = 1-t^2 \end{bmatrix}$ *for* $0 \leq t \leq 1$, $\mathscr{C}_3$ *is* $\begin{bmatrix} x(t) = -1+t \\ y(t) = -1+(-1+t)^2 \end{bmatrix}$ *for* $0 \leq t \leq 1$ *and* $\mathscr{C}_4$ *is* $\begin{bmatrix} x(t) = t \\ y(t) = 1-t^2 \end{bmatrix}$ *for* $0 \leq t \leq 1$. *Why is* $\int_{\mathscr{C}} \boldsymbol{\omega} = 0$?

Exercise 17.2.11 *Let the vector field be defined by* $\boldsymbol{F} = \begin{bmatrix} 2xe^{-x^2-y^2} \\ 2ye^{-x^2-y^2} \end{bmatrix}$.

- *Consider the smooth segmented path* $\mathscr{C}$ *which is the unit square starting at* $(0,0)$ *and moving counterclockwise to* $(1,0)$, *then* $(1,1)$, *then back to* $(0,1)$ *and finally back to* $(0,0)$. *This is the path* $\mathscr{C}$ *defined by four simple straight paths:* $\mathscr{C}_1$ *is* $\begin{bmatrix} x(t) = t \\ y(t) = 0 \end{bmatrix}$ *for* $0 \leq t \leq 1$, $\mathscr{C}_2$ *is* $\begin{bmatrix} x(t) = 1 \\ y(t) = t \end{bmatrix}$ *for* $0 \leq t \leq 1$, $\mathscr{C}_3$ *is* $\begin{bmatrix} x(t) = 1-t \\ y(t) = 1 \end{bmatrix}$ *for* $0 \leq t \leq 1$ *and* $\mathscr{C}_4$ *is* $\begin{bmatrix} x(t) = 0 \\ y(t) = 1-t \end{bmatrix}$ *for* $0 \leq t \leq 1$. *Why is* $\int_{\mathscr{C}} \boldsymbol{\omega} = 0$?
- *Consider the smooth segmented path* $\mathscr{C}$ *and the simple closed path* $\mathscr{C}$ *is defined as the union of the following curves:* $\mathscr{C}_1$ *is* $\begin{bmatrix} x(t) = t \\ y(t) = 0 \end{bmatrix}$ *for* $0 \leq t \leq 3$, $\mathscr{C}_2$ *is* $\begin{bmatrix} x(t) = 3 \\ y(t) = t \end{bmatrix}$ *for* $0 \leq t \leq 3$ *and* $\mathscr{C}_3$ *is* $\begin{bmatrix} x(t) = 3-t \\ y(t) = 3-t \end{bmatrix}$ *for* $0 \leq t \leq 3$. *Why is* $\int_{\mathscr{C}} \boldsymbol{\omega} = 0$?
- *Consider the smooth segmented path* $\mathscr{C}$ *and the simple closed path* $\mathscr{C}$ *is defined as the union of the following four curves:* $\mathscr{C}_1$ *is* $\begin{bmatrix} x(t) = 1-t \\ y(t) = 1-t^2 \end{bmatrix}$ *for* $0 \leq t \leq 1$, $\mathscr{C}_2$ *is* $\begin{bmatrix} x(t) = t \\ y(t) = 1-t^2 \end{bmatrix}$ *for* $0 \leq t \leq 1$, $\mathscr{C}_3$ *is* $\begin{bmatrix} x(t) = -1+t \\ y(t) = -1+(-1+t)^2 \end{bmatrix}$ *for* $0 \leq t \leq 1$ *and* $\mathscr{C}_4$ *is* $\begin{bmatrix} x(t) = t \\ y(t) = 1-t^2 \end{bmatrix}$ *for* $0 \leq t \leq 1$. *Why is* $\int_{\mathscr{C}} \boldsymbol{\omega} = 0$?

Exercise 17.2.12 *Let the vector field be defined by* $\boldsymbol{F} = \begin{bmatrix} -\frac{5x}{(1+x^2+y^2)^{7/2}} \\ -\frac{5y}{(1+x^2+y^2)^{7/2}} \end{bmatrix}$.

- *Consider the smooth segmented path* $\mathscr{C}$ *which is the unit square starting at* $(0,0)$ *and moving counterclockwise to* $(1,0)$*, then* $(1,1)$*, then back to* $(0,1)$ *and finally back to* $(0,0)$*. This is the path* $\mathscr{C}$ *defined by four simple straight paths:* $\mathscr{C}_1$ *is* $\begin{bmatrix} x(t) = t \\ y(t) = 0 \end{bmatrix}$ *for* $0 \leq t \leq 1$*,* $\mathscr{C}_2$ *is* $\begin{bmatrix} x(t) = 1 \\ y(t) = t \end{bmatrix}$ *for* $0 \leq t \leq 1$*,* $\mathscr{C}_3$ *is* $\begin{bmatrix} x(t) = 1-t \\ y(t) = 1 \end{bmatrix}$ *for* $0 \leq t \leq 1$ *and* $\mathscr{C}_4$ *is* $\begin{bmatrix} x(t) = 0 \\ y(t) = 1-t \end{bmatrix}$ *for* $0 \leq t \leq 1$*. Why is* $\int_{\mathscr{C}} \omega = 0$*?*

- *Consider the smooth segmented path* $\mathscr{C}$ *and the simple closed path* $\mathscr{C}$ *is defined as the union of the following curves:* $\mathscr{C}_1$ *is* $\begin{bmatrix} x(t) = t \\ y(t) = 0 \end{bmatrix}$ *for* $0 \leq t \leq 3$*,* $\mathscr{C}_2$ *is* $\begin{bmatrix} x(t) = 3 \\ y(t) = t \end{bmatrix}$ *for* $0 \leq t \leq 3$ *and* $\mathscr{C}_3$ *is* $\begin{bmatrix} x(t) = 3-t \\ y(t) = 3-t \end{bmatrix}$ *for* $0 \leq t \leq 3$*. Why is* $\int_{\mathscr{C}} \omega = 0$*?*

- *Consider the smooth segmented path* $\mathscr{C}$ *and the simple closed path* $\mathscr{C}$ *is defined as the union of the following four curves:* $\mathscr{C}_1$ *is* $\begin{bmatrix} x(t) = 1-t \\ y(t) = 1-t^2 \end{bmatrix}$ *for* $0 \leq t \leq 1$*,* $\mathscr{C}_2$ *is* $\begin{bmatrix} x(t) = t \\ y(t) = 1-t^2 \end{bmatrix}$ *for* $0 \leq t \leq 1$*,* $\mathscr{C}_3$ *is* $\begin{bmatrix} x(t) = -1+t \\ y(t) = -1+(-1+t)^2 \end{bmatrix}$ *for* $0 \leq t \leq 1$ *and* $\mathscr{C}_4$ *is* $\begin{bmatrix} x(t) = t \\ y(t) = 1-t^2 \end{bmatrix}$ *for* $0 \leq t \leq 1$*. Why is* $\int_{\mathscr{C}} \omega = 0$*?*

Exercise 17.2.13 *Let the vector field be defined by* $\boldsymbol{F} = \begin{bmatrix} 2x^2 + 5y^2 \\ 7xy \end{bmatrix}$.

- *Consider the smooth segmented path* $\mathscr{C}$ *which is the unit square starting at* $(0,0)$ *and moving counterclockwise to* $(1,0)$*, then* $(1,1)$*, then back to* $(0,1)$ *and finally back to* $(0,0)$*. This is the path* $\mathscr{C}$ *defined by four simple straight paths:* $\mathscr{C}_1$ *is* $\begin{bmatrix} x(t) = t \\ y(t) = 0 \end{bmatrix}$ *for* $0 \leq t \leq 1$*,* $\mathscr{C}_2$ *is* $\begin{bmatrix} x(t) = 1 \\ y(t) = t \end{bmatrix}$ *for* $0 \leq t \leq 1$*,* $\mathscr{C}_3$ *is* $\begin{bmatrix} x(t) = 1-t \\ y(t) = 1 \end{bmatrix}$ *for* $0 \leq t \leq 1$ *and* $\mathscr{C}_4$ *is* $\begin{bmatrix} x(t) = 0 \\ y(t) = 1-t \end{bmatrix}$ *for* $0 \leq t \leq 1$*. Why don't we know* $\int_{\mathscr{C}} \omega = 0$ *without computing it?*

- *Consider the smooth segmented path* $\mathscr{C}$ *and the simple closed path* $\mathscr{C}$ *is defined as the union of the following curves:* $\mathscr{C}_1$ *is* $\begin{bmatrix} x(t) = t \\ y(t) = 0 \end{bmatrix}$ *for* $0 \leq t \leq 3$*,* $\mathscr{C}_2$ *is* $\begin{bmatrix} x(t) = 3 \\ y(t) = t \end{bmatrix}$ *for* $0 \leq t \leq 3$ *and* $\mathscr{C}_3$ *is* $\begin{bmatrix} x(t) = 3-t \\ y(t) = 3-t \end{bmatrix}$ *for* $0 \leq t \leq 3$*. Why don't we know* $\int_{\mathscr{C}} \omega = 0$ *without computing it?*

- *Consider the smooth segmented path* $\mathscr{C}$ *and the simple closed path* $\mathscr{C}$ *is defined as the union of the following four curves:* $\mathscr{C}_1$ *is* $\begin{bmatrix} x(t) = 1-t \\ y(t) = 1-t^2 \end{bmatrix}$ *for* $0 \leq t \leq 1$*,* $\mathscr{C}_2$ *is* $\begin{bmatrix} x(t) = t \\ y(t) = 1-t^2 \end{bmatrix}$ *for* $0 \leq t \leq 1$*,* $\mathscr{C}_3$ *is* $\begin{bmatrix} x(t) = -1+t \\ y(t) = -1+(-1+t)^2 \end{bmatrix}$ *for* $0 \leq t \leq 1$ *and* $\mathscr{C}_4$ *is* $\begin{bmatrix} x(t) = t \\ y(t) = 1-t^2 \end{bmatrix}$ *for* $0 \leq t \leq 1$*. Why don't we know* $\int_{\mathscr{C}} \omega = 0$ *without computing it?*

17.3 Two and Three Forms

A 1 -form $\omega = \boldsymbol{A}dx + \boldsymbol{B}dy$ is therefore a mapping $\omega : \Re^2_{\boldsymbol{P}} = \boldsymbol{T_P}(\Re^2) \rightarrow \Re$ defined by

$$\omega(\boldsymbol{V_P}) = \boldsymbol{A}(P_1, P_2)V_1 + \boldsymbol{B}(P_1, P_2)V_2$$

Now consider mappings of the form $\psi = \boldsymbol{A}dxdy$. Then, we define the action of ψ as follows:

Definition 17.3.1 2-Forms

Let U be an open set in $\Re^2$. A 2-form $\psi : \Re^2_{\boldsymbol{P}} \times \Re^2_{\boldsymbol{P}} \rightarrow \Re$ of the form $\boldsymbol{A}\ \boldsymbol{dx}\ \boldsymbol{dy}$ for a smooth function $\boldsymbol{A}$ on U has action defined by

$$\begin{aligned} \psi(\boldsymbol{V_P}, \boldsymbol{W_P}) &= \boldsymbol{A}\ \boldsymbol{dx}\ \boldsymbol{dy}(\boldsymbol{V_P}, \boldsymbol{W_P}) \\ &= \boldsymbol{A}(P_1, P_2)\Big(\boldsymbol{dx}(V_P)\boldsymbol{dy}(\boldsymbol{W_P}) - \boldsymbol{dx}(\boldsymbol{W_P})\boldsymbol{dy}(\boldsymbol{V_P})\Big) \\ &= \boldsymbol{A}(P_1, P_2)(V_1\, W_2 - W_1\, V_2) \end{aligned}$$

Note it is clear from this definition that $\boldsymbol{dx}\ \boldsymbol{dy} = -\boldsymbol{dy}\ \boldsymbol{dx}$. Further, $\boldsymbol{dx}\ \boldsymbol{dx} = \boldsymbol{dy}\ \boldsymbol{dy} = 0$.

Next, we can look at 3-forms. To do this, we have to figure out a way to alternate the algebraic sign of the terms in the expansion of $\boldsymbol{dxdydz}$ as well as move the setting to open sets U in $\Re^3$. You probably have seen the permutation groups S_n on n symbols before. Let's refresh that memory. First, let's look at two symbols:

Definition 17.3.2 The Symmetric Group S_2

For concreteness, the symmetric groups S_2 will be phrased in terms of positive integers. S_2 is the group consisting of the two ways the symbols 1 and 2 can be shuffled. These are the ordering 12 and the ordering 21. The ordering 21 can be obtained from 12 by flipping the entries: i.e. $12 \rightarrow 21$. Since one flip is involved, we define the algebraic sign of the ordering 21 to be -1 while since there are an even number of flips (zero to be exact) the algebraic sign of 12 is $+$. We let $\sigma(i_1\, i_2)$ be the sign of the ordering.

Next, three symbols:

Definition 17.3.3 The Symmetric Group S_3

For concreteness, the symmetric groups S_3 will be phrased in terms of positive integers. S_3 is the group consisting of the ways the symbols 1, 2 and 3 can be shuffled. We let $\sigma(i_1\, i_2\, i_3)$ be the sign of the ordering.

- 123 *does not need reordering so* $\sigma = +1$.
- 213*:* $123 \rightarrow 213$*; so one flip implying* $\sigma = -1$.
- 321*:* $123 \rightarrow 132 \rightarrow 312 \rightarrow 321$*: three flips implying* $\sigma = (-1)^3 = -1$.
- 132*:* $123 \rightarrow 132$*: so one flip implying* $\sigma = -1$
- 312*:* $123 \rightarrow 132 \rightarrow 312$*: so two flips implying* $\sigma = (-1)^2 + +1$.
- 231*:* $123 \rightarrow 213 \rightarrow 231$*: so two flips implying* $\sigma = (-1)^2 = +1$.

Using the symmetric group S_3, we can define what we mean by $\boldsymbol{dxdydz}$.

Definition 17.3.4 3-Forms

> *Let U be an open subset of $\Re^3$. Then a 3-form $\psi : \Re^3_{\boldsymbol{P}} \times \Re^3_{\boldsymbol{P}} \times \Re^3_{\boldsymbol{P}} \rightarrow \Re$ of the form $\boldsymbol{A}\, \boldsymbol{dx}^1\, \boldsymbol{dx}^2\, \boldsymbol{dx}^3$ for a smooth function $\boldsymbol{A}$ on U has action defined by*
>
> $$\begin{aligned} \psi(\boldsymbol{X}^1_{\boldsymbol{P}}, \boldsymbol{X}^2_{\boldsymbol{P}}, \boldsymbol{X}^3_{\boldsymbol{P}}) &= \boldsymbol{A}\, \boldsymbol{dx}^1\, \boldsymbol{dx}^2\, \boldsymbol{dx}^3(\boldsymbol{X}^1_{\boldsymbol{P}}, \boldsymbol{X}^2_{\boldsymbol{P}}, \boldsymbol{X}^3_{\boldsymbol{P}}) \\ &= \boldsymbol{A}(P_1, P_2, P_3) \sum_{(i_1 i_2 i_3) \in S_3} \sigma(i_1 i_2 i_3) \boldsymbol{dx}^1(X^{i_1}_{\boldsymbol{P}}) \boldsymbol{dx}^2(\boldsymbol{X}^{i_2}_{\boldsymbol{P}}) \boldsymbol{dx}^3(\boldsymbol{X}^{i_3}_{\boldsymbol{P}}) \end{aligned}$$

Comment 17.3.1 *It is easier to remember this using determinants. For a 2-form*

$$\begin{aligned} \boldsymbol{dx}\, \boldsymbol{dy}(\boldsymbol{V_P}, \boldsymbol{W_P}) &= \boldsymbol{dx}(V_P) \boldsymbol{dy}(\boldsymbol{W_P}) - \boldsymbol{dx}(\boldsymbol{W_P}) \boldsymbol{dy}(\boldsymbol{V_P}) \\ &= (V_1\, W_2 - W_1\, V_2) = det \begin{bmatrix} V_1 & V_2 \\ W_1 & W_2 \end{bmatrix} \\ &= \left\langle det \begin{bmatrix} \boldsymbol{i} & \boldsymbol{j} & \boldsymbol{k} \\ V_1 & V_2 & 0 \\ W_1 & W_2 & 0 \end{bmatrix}, \boldsymbol{k} \right\rangle \end{aligned}$$

Comment 17.3.2 *Hence, for a 3-form, we can use determinants also:*

$$\begin{aligned} &\boldsymbol{dx}^1\, \boldsymbol{dx}^2\, \boldsymbol{dx}^3(\boldsymbol{X}^1_{\boldsymbol{P}}, \boldsymbol{X}^2_{\boldsymbol{P}}, \boldsymbol{X}^3_{\boldsymbol{P}}) \\ &(+)\boldsymbol{dx}^1\, \boldsymbol{dx}^2\, \boldsymbol{dx}^3(\boldsymbol{X}^1_{\boldsymbol{P}}, \boldsymbol{X}^2_{\boldsymbol{P}}, \boldsymbol{X}^3_{\boldsymbol{P}})(-)\boldsymbol{dx}^1\, \boldsymbol{dx}^2\, \boldsymbol{dx}^3(\boldsymbol{X}^2_{\boldsymbol{P}}, \boldsymbol{X}^1_{\boldsymbol{P}}, \boldsymbol{X}^3_{\boldsymbol{P}}) \\ &(-)\boldsymbol{dx}^1\, \boldsymbol{dx}^2\, \boldsymbol{dx}^3(\boldsymbol{X}^3_{\boldsymbol{P}}, \boldsymbol{X}^2_{\boldsymbol{P}}, \boldsymbol{X}^1_{\boldsymbol{P}})(-)\boldsymbol{dx}^1\, \boldsymbol{dx}^2\, \boldsymbol{dx}^3(\boldsymbol{X}^1_{\boldsymbol{P}}, \boldsymbol{X}^3_{\boldsymbol{P}}, \boldsymbol{X}^2_{\boldsymbol{P}}) \\ &(+)\boldsymbol{dx}^1\, \boldsymbol{dx}^2\, \boldsymbol{dx}^3(\boldsymbol{X}^3_{\boldsymbol{P}}, \boldsymbol{X}^1_{\boldsymbol{P}}, \boldsymbol{X}^2_{\boldsymbol{P}})(+)\boldsymbol{dx}^1\, \boldsymbol{dx}^2\, \boldsymbol{dx}^3(\boldsymbol{X}^2_{\boldsymbol{P}}, \boldsymbol{X}^3_{\boldsymbol{P}}, \boldsymbol{X}^1_{\boldsymbol{P}}) \end{aligned}$$

Thus, we have

$$\begin{aligned} \boldsymbol{dx}^1\, \boldsymbol{dx}^2\, \boldsymbol{dx}^3(\boldsymbol{X}^1_{\boldsymbol{P}}, \boldsymbol{X}^2_{\boldsymbol{P}}, \boldsymbol{X}^3_{\boldsymbol{P}}) &= X^{11}\, X^{22}\, X^{33} - X^{21}\, X^{12}\, X^{33} - X^{31}\, X^{22}\, X^{13} \\ &- X^{11}\, X^{32}\, X^{23} + X^{31}\, X^{12}\, X^{23} + X^{21}\, X^{32}\, X^{13} \end{aligned}$$

We can rewrite this as

$$\begin{aligned} \boldsymbol{dx}^1\, \boldsymbol{dx}^2\, \boldsymbol{dx}^3(\boldsymbol{X}^1_{\boldsymbol{P}}, \boldsymbol{X}^2_{\boldsymbol{P}}, \boldsymbol{X}^3_{\boldsymbol{P}}) &= X^{11}(X^{22}X^{33} - X^{23}X^{32}) - X^{12}(X^{31}X^{23} - X^{21}X^{33}) \\ &+ X^{13}(X^{21}X^{32} - X^{22}X^{31}) \end{aligned}$$

This can be reorganized like follows:

$$\begin{aligned} \boldsymbol{dx}^1\, \boldsymbol{dx}^2\, \boldsymbol{dx}^3(\boldsymbol{X}^1_{\boldsymbol{P}}, \boldsymbol{X}^2_{\boldsymbol{P}}, \boldsymbol{X}^3_{\boldsymbol{P}}) &= X^{11}\, det \begin{bmatrix} X^{22} & X^{23} \\ X^{32} & X^{33} \end{bmatrix} - X^{12}\, \det \begin{bmatrix} X^{21} & X^{23} \\ X^{31} & X^{33} \end{bmatrix} \\ &+ X^{13} det \begin{bmatrix} X^{21} & X^{22} \\ X^{31} & X^{32} \end{bmatrix} \\ &- det \begin{bmatrix} X^{11} & X^{12} & X^{13} \\ X^{21} & X^{22} & X^{23} \\ X^{31} & X^{32} & X^{33} \end{bmatrix} \end{aligned}$$

Comment 17.3.3 *From the 3-form expansion, if you look at $\boldsymbol{dx}^2 \boldsymbol{dx}^2 \boldsymbol{dx}^3$, we would find*

$$\boldsymbol{dx}^2\, \boldsymbol{dx}^2\, \boldsymbol{dx}^3(\boldsymbol{X}^1_{\boldsymbol{P}}, \boldsymbol{X}^2_{\boldsymbol{P}}, \boldsymbol{X}^3_{\boldsymbol{P}}) = det \begin{bmatrix} X^{21} & X^{22} & X^{23} \\ X^{21} & X^{22} & X^{23} \\ X^{31} & X^{32} & X^{33} \end{bmatrix}$$

Since two rows in the determinant are the same, the determinant is zero. A similar argument shows $\boldsymbol{dx^i dx^j dx^3}$ *is zero if any two superscripts are the same. Thus,*

$$\boldsymbol{dxdxdz} = 0, \quad \boldsymbol{dxdydy} = 0, \quad \boldsymbol{dxdzdz} = 0$$

and so forth. Finally, changing order gives

$$\boldsymbol{dx^2\, dx^1\, dx^3}(\boldsymbol{X^1_P}, \boldsymbol{X^2_P}, \boldsymbol{X^3_P}) \quad = \quad det \begin{bmatrix} X^{21} & X^{22} & X^{23} \\ X^{11} & X^{12} & X^{13} \\ X^{31} & X^{32} & X^{33} \end{bmatrix}$$

which interchanges two rows in the matrix and hence the determinant changes sign. Thus, we know

$$\boldsymbol{dxdydz} = -\boldsymbol{dydxdz}, \quad \boldsymbol{dxdydy} = -\boldsymbol{dxdzdy}, \quad \boldsymbol{dxdydz} = -\boldsymbol{dzdydx}$$

and so forth.

Homework

Exercise 17.3.1 *Let*

$$\begin{aligned} \boldsymbol{H}(x,y) &= \begin{bmatrix} 2xy \\ 4x^2 - y^2 \end{bmatrix}, \quad \boldsymbol{F}(x,y) = 5x^2y^2z^3, \quad \boldsymbol{G}(x,y,z) = 2x^2 + y^2 + 4yz^3 \\ \boldsymbol{W_1} &= \begin{bmatrix} 2 \\ 3 \end{bmatrix}, \quad \boldsymbol{W_2} = \begin{bmatrix} -1 \\ 5 \end{bmatrix} \\ \boldsymbol{V_1} &= \begin{bmatrix} 2 \\ 3 \\ -4 \end{bmatrix}, \quad \boldsymbol{V_2} = \begin{bmatrix} -1 \\ 5 \\ 1 \end{bmatrix}, \quad \boldsymbol{V_3} = \begin{bmatrix} -2 \\ 1 \\ 8 \end{bmatrix} \end{aligned}$$

- *Compute the action of the 1-form* $\omega = \boldsymbol{H_1 dx} + \boldsymbol{H_2 dy}$ *on* $\boldsymbol{W_1}$*:* $\boldsymbol{H_1}(W_{11}, W_{12})W_{11} + \boldsymbol{H_2}(W_{11}, W_{12})W_{12}$.

- *Compute the two form* $\Theta = \boldsymbol{F dxdy}$*'s action on* $(\boldsymbol{W_1}, \boldsymbol{W_2})$*:* $\boldsymbol{\Theta}(\boldsymbol{W_1}, \boldsymbol{W_2})$.

- *Compute the three form* $\Psi = \boldsymbol{G dxdydz}$ *action on* $(\boldsymbol{V_1}, \boldsymbol{V_2}, \boldsymbol{V_3})$*:* $\boldsymbol{\Psi}(\boldsymbol{V_1}, \boldsymbol{V_2}, \boldsymbol{V_3})$.

Exercise 17.3.2 *Let*

$$\begin{aligned} \boldsymbol{H}(x,y) &= \begin{bmatrix} 2x + y^2 \\ 4 + x^2 - +4y^2 \end{bmatrix}, \quad \boldsymbol{F}(x,y) = 5x^5y^3 + x^2 + y^2 + 2, \quad \boldsymbol{G}(x,y,z) = 2x^2 + 3y^2yz^3 \\ \boldsymbol{W_1} &= \begin{bmatrix} -1 \\ 4 \end{bmatrix}, \quad \boldsymbol{W_2} = \begin{bmatrix} 2 \\ 5 \end{bmatrix} \\ \boldsymbol{V_1} &= \begin{bmatrix} 6 \\ 1 \\ 2 \end{bmatrix}, \quad \boldsymbol{V_2} = \begin{bmatrix} 1 \\ 4 \\ -2 \end{bmatrix}, \quad \boldsymbol{V_3} = \begin{bmatrix} -8 \\ 3 \\ -4 \end{bmatrix} \end{aligned}$$

- *Compute the action of the 1-form* $\omega = \boldsymbol{H_1 dx} + \boldsymbol{H_2 dy}$ *on* $\boldsymbol{W_1}$*:* $\boldsymbol{H_1}(W_{11}, W_{12})W_{11} + \boldsymbol{H_2}(W_{11}, W_{12})W_{12}$.

- *Compute the two form* $\Theta = \boldsymbol{F dxdy}$*'s action on* $(\boldsymbol{W_1}, \boldsymbol{W_2})$*:* $\boldsymbol{\Theta}(\boldsymbol{W_1}, \boldsymbol{W_2})$.

- *Compute the three form* $\Psi = \boldsymbol{G dxdydz}$ *action on* $(\boldsymbol{V_1}, \boldsymbol{V_2}, \boldsymbol{V_3})$*:* $\boldsymbol{\Psi}(\boldsymbol{V_1}, \boldsymbol{V_2}, \boldsymbol{V_3})$.

Exercise 17.3.3 *Let*

$$\begin{aligned} \boldsymbol{H}(x,y) &= \begin{bmatrix} 2+x+y \\ 4+x^2+3xy-2y^2 \end{bmatrix}, \quad \boldsymbol{F}(x,y) = 15\sqrt{x^2+y^2+z^4}, \\ \boldsymbol{G}(x,y,z) &= 12x^2+2y^2+9yz^6 \\ \boldsymbol{W_1} &= \begin{bmatrix} 12 \\ 1 \end{bmatrix}, \quad \boldsymbol{W_2} = \begin{bmatrix} 2 \\ 7 \end{bmatrix} \\ \boldsymbol{V_1} &= \begin{bmatrix} 4 \\ 3 \\ -4 \end{bmatrix}, \quad \boldsymbol{V_2} = \begin{bmatrix} -1 \\ 5 \\ 7 \end{bmatrix}, \quad \boldsymbol{V_3} = \begin{bmatrix} 5 \\ 1 \\ 8 \end{bmatrix} \end{aligned}$$

- *Compute the action of the 1-form* $\boldsymbol{\omega} = \boldsymbol{H_1 dx} + \boldsymbol{H_2 dy}$ *on* $\boldsymbol{W_1}$*:* $\boldsymbol{H_1}(W_{11}, W_{12})W_{11} + \boldsymbol{H_2}(W_{11}, W_{12})W_{12}$.
- *Compute the two form* $\Theta = \boldsymbol{F dx dy}$*'s action on* $(\boldsymbol{W_1}, \boldsymbol{W_2})$*:* $\boldsymbol{\Theta}(\boldsymbol{W_1}, \boldsymbol{W_2})$.
- *Compute the three form* $\Psi = \boldsymbol{G dx dy dz}$ *action on* $(\boldsymbol{V_1}, \boldsymbol{V_2}, \boldsymbol{V_3})$*:* $\boldsymbol{\Psi}(\boldsymbol{V_1}, \boldsymbol{V_2}, \boldsymbol{V_3})$.

Exercise 17.3.4 *Let*

$$\begin{aligned} \boldsymbol{H}(x,y) &= \begin{bmatrix} 2xy^2+4x=10 \\ 4x^2y^2 \end{bmatrix}, \quad \boldsymbol{F}(x,y) = 5x^2y^2+z^2, \\ \boldsymbol{G}(x,y,z) &= 2+4x^2+6y^2+4z+2z^2 \\ \boldsymbol{W_1} &= \begin{bmatrix} 2 \\ 5 \end{bmatrix}, \quad \boldsymbol{W_2} = \begin{bmatrix} 9 \\ 1 \end{bmatrix} \\ \boldsymbol{V_1} &= \begin{bmatrix} 2 \\ 7 \\ -4 \end{bmatrix}, \quad \boldsymbol{V_2} = \begin{bmatrix} 6 \\ 5 \\ 1 \end{bmatrix}, \quad \boldsymbol{V_3} = \begin{bmatrix} -2 \\ 14 \\ 8 \end{bmatrix} \end{aligned}$$

- *Compute the action of the 1-form* $\boldsymbol{\omega} = \boldsymbol{H_1 dx} + \boldsymbol{H_2 dy}$ *on* $\boldsymbol{W_1}$*:* $\boldsymbol{H_1}(W_{11}, W_{12})W_{11} + \boldsymbol{H_2}(W_{11}, W_{12})W_{12}$.
- *Compute the two form* $\Theta = \boldsymbol{F dx dy}$*'s action on* $(\boldsymbol{W_1}, \boldsymbol{W_2})$*:* $\boldsymbol{\Theta}(\boldsymbol{W_1}, \boldsymbol{W_2})$.
- *Compute the three form* $\Psi = \boldsymbol{G dx dy dz}$ *action on* $(\boldsymbol{V_1}, \boldsymbol{V_2}, \boldsymbol{V_3})$*:* $\boldsymbol{\Psi}(\boldsymbol{V_1}, \boldsymbol{V_2}, \boldsymbol{V_3})$.

Part V

Applications

Chapter 18

The Exponential Matrix

Let A be a $n \times n$ matrix. We will let $\|A\|_{Fr}$ be the Frobenius norm of A; hence, we know $\|A\boldsymbol{x}\| \leq \|A\|_{Fr}\ \|\boldsymbol{x}\|$. For convenience, we will simply let $\|A\|_{Fr}$ be denoted by $\|A\|$. Before we begin, let's consider the product of two $n \times n$ matrices A and B. We have

$$C_{ij} \;=\; (AB)_{ij} = \sum_{k=1}^{n} A_{ik}B_{kj} \leq \sqrt{\sum_{k=1}^{n} A_{ik}^2}\ \sqrt{\sum_{k=1}^{n} B_{kj}^2}$$

Thus

$$\begin{aligned}
\sum_{i=1}^{n}\sum_{j=1}^{n} C_{ij}^2 &\leq \sum_{i=1}^{n}\sum_{j=1}^{n}\sum_{k=1}^{n} A_{ik}^2 \sum_{k=1}^{n} B_{kj}^2 \\
&= \left(\sum_{i=1}^{n}\sum_{k=1}^{n} A_{ik}^2\right)\left(\sum_{j=1}^{n}\sum_{k=1}^{n} B_{kj}^2\right) = \|A\|_{Fr}^2\ \|B\|_{Fr}^2
\end{aligned}$$

Therefore $\|AB\|_{Fr} \leq \|A\|_{Fr}\ \|B\|_F\|$. We will use this in a bit.

Homework

Exercise 18.0.1 *Let*

$$A \;=\; \begin{bmatrix} 1 & 3 & -2 \\ -4 & 11 & 8 \\ 2 & -3 & 9 \end{bmatrix}, \quad B = \begin{bmatrix} 10 & -3 & -22 \\ 4 & 7 & 1 \\ 9 & 2 & -8 \end{bmatrix}$$

Compute $\|A\|_{Fr}$, $\|B\|_{Fr}$ and $\|AB\|_{Fr}$ and verify the proposed inequality does indeed work.

Exercise 18.0.2 *Let*

$$A \;=\; \begin{bmatrix} 10 & 3 \\ 5 & 12 \end{bmatrix}, \quad B = \begin{bmatrix} -9 & 1 \\ 14 & 5 \end{bmatrix}$$

Compute $\|A\|_{Fr}$, $\|B\|_{Fr}$ and $\|AB\|_{Fr}$ and verify the proposed inequality does indeed work.

18.1 The Exponential Matrix

Now consider the finite sum

$$P_n = I + A + \frac{A^2}{2!} + \frac{A^3}{3!} + \cdots + \frac{A^n}{n!}$$

Does $\lim_{n \to \infty} P_n$ exist? i.e. is there a matrix P such that $P = lim_{n \to \infty} P_n$? Thus, we want to show

$$\forall \epsilon > 0, \ \exists N \ni \ \|P - P_n\| < \epsilon, \quad \text{if } n > N$$

Formally, letting $A^0 = I$, we suspect the matrix we seek is expressed as the following infinite series.

$$P = \sum_{j=0}^{\infty} \frac{A^j}{j!}$$

It is straightforward to show the set of all $n \times n$ real matrices form a complete normed vector space under the Frobenius norm. This is not surprising as if $M_{n,n}$ denotes this set of matrices, it can be identified with $\Re^{n^2}$ and this space is complete using the usual $\|\cdot\|_2$ norm. The Frobenius norm is essentially just the $\|\cdot\|_2$ norm applied to objects requiring a double summation. So if we can show the sequence of partial sums is a Cauchy sequence with respect to the Frobenius norm, we know there is a unique matrix P which can be denoted by the series. For $n > m$, $P_n - P_m = \sum_{j=m+1}^{n} \frac{A^j}{j!}$ and

$$\|P_n - P_m\| \leq \sum_{j=m+1}^{n} \frac{\|A^j\|}{j!} \leq \sum_{j=m+1}^{n} \frac{\|A\|^j}{j!}$$

where we have used induction to show $\|A^n\| \leq \|A\|^n$. But $\|A\| = \alpha$ is a constant, so

$$\|P_n - P_m\| \leq \sum_{j=m+1}^{n} \frac{(\alpha)^j}{j!}$$

We know that

$$e^{\alpha} = \sum_{j=0}^{\infty} \frac{(\alpha)^j}{j!}$$

converges and so it is a Cauchy sequence of real numbers. Therefore, given $\epsilon > 0$, there is an N so that $\sum_{j=m+1}^{n} \alpha^j/(j!) < \epsilon$ when $n > m > N$. We conclude if $n > m > N$,

$$\|P_n - P_m\| \leq \sum_{j=m+1}^{n} \frac{(\alpha)^j}{j!} < \epsilon$$

Thus the sequence of partial sums is a Cauchy sequence and by the completeness of $M_{n,n}$ with the Frobenius norm, there is a unique matrix P so that $P = \lim_{n \to \infty} P_n$ in the Frobenius norm. The partial sums P_n are exactly the partial sums we would expect for a function like e^A even though A is a matrix. Hence, we use this sequence of partial sums to define what we mean by e^A which is called the **matrix exponential function**. Since this analysis works for any matrix A, in particular, given a value of t, it works for the matrix At to give the matrix e^{At}.

Homework

Exercise 18.1.1 *Prove the set of square matrices over $\Re$ with the Frobenius norm is a complete normed vector space over $\Re$.*

Exercise 18.1.2 *Let*

$$A = \begin{bmatrix} 2 & 1 \\ 0 & 2 \end{bmatrix}$$

Find a formula for A^n for all n and compute the nth partial sum of e^A. What do you think happens as $n \to \infty$?

Exercise 18.1.3 *Let*

$$A = \begin{bmatrix} 3 & 1 \\ 0 & 3 \end{bmatrix}$$

Find a formula for A^n for all n and compute the nth partial sum of e^A. What do you think happens as $n \to \infty$?

Exercise 18.1.4 *Let*

$$A = \begin{bmatrix} 2 & 0 \\ 1 & 2 \end{bmatrix}$$

Find a formula for A^n for all n and compute the nth partial sum of e^A. What do you think happens as $n \to \infty$?

18.2 The Jordan Canonical Form

In an advanced course in linear algebra, you will learn about the **Jordan Canonical Form** of a matrix. We can prove every matrix A has one and it is similar to the decomposition we find for symmetric matrices in a way. A crucial part of the Jordan Canonical form are what are called **Jordan Blocks**. Consider the three 3×3 matrices below where λ is an eigenvalue (these could possibly be submatrices of a larger matrix).

$$\underset{(1)}{\begin{bmatrix} \lambda & 1 & 0 \\ 0 & \lambda & 1 \\ 0 & 0 & \lambda \end{bmatrix}} \quad \underset{(2)}{\begin{bmatrix} \lambda & 0 & 0 \\ 0 & \lambda & 1 \\ 0 & 0 & \lambda \end{bmatrix}} \quad \underset{(3)}{\begin{bmatrix} \lambda & 0 & 0 \\ 0 & \lambda & 0 \\ 0 & 0 & \lambda \end{bmatrix}}$$

How many eigenvectors do we get for each of these? We know that for (3) we will get 3 distinct eigenvectors, but what about for (1) and (2)? Recall, to get the eigenvalues of a matrix we look at

$$\det(A - \mu I) = 0 \qquad (\det(\mu I - A) = 0)$$

and then to find the eigenvectors we solve

$$(A - \mu I)(\boldsymbol{v}) = \boldsymbol{0}$$

(Here μ is used to denote the eigenvalues because λ is being used in the matrices.) So, for matrix (1) we get

$$\det(A - \mu I) = \begin{vmatrix} \lambda - \mu & 1 & 0 \\ 0 & \lambda - \mu & 1 \\ 0 & 0 & \lambda - \mu \end{vmatrix} = 0$$

$$(\lambda - \mu)^3 = 0$$

So $\mu = \lambda$ and λ is our eigenvalue. To get the eigenvectors of (1) we solve

$$\left[\begin{pmatrix} \lambda & 1 & 0 \\ 0 & \lambda & 1 \\ 0 & 0 & \lambda \end{pmatrix} - \begin{pmatrix} \lambda & 0 & 0 \\ 0 & \lambda & 0 \\ 0 & 0 & \lambda \end{pmatrix} \right] \begin{pmatrix} v_1 \\ v_2 \\ v_3 \end{pmatrix} = \mathbf{0}$$

or

$$\begin{pmatrix} 0 & 1 & 0 \\ 0 & 0 & 1 \\ 0 & 0 & 0 \end{pmatrix} \begin{pmatrix} v_1 \\ v_2 \\ v_3 \end{pmatrix} = \begin{pmatrix} 0 \\ 0 \\ 0 \end{pmatrix}$$

We see that $v_2 = v_3 = 0$ and v_1 is arbitrary. Thus from (1) we only get one distinct eigenvector. The eigenvalue λ has algebraic multiplicity 3 as it occurs three times in the characteristic equation. It has geometric multiplicity 1 as it has only one eigenvector. Hence, the eigenspace for this eigenvalue of algebraic multiplicity three is one dimensional. It is clear this means, in this case, the eigenvectors associated with the eigenvalue do not give a subspace of the same dimension as the algebraic multiplicity. Note the eigenvalues are on the diagonal and the superdiagonal right above that is all 1's. This is the first of the Jordan Canonical Blocks possible for this eigenvalue of multiplicity 3. Note there are 2 1s in the superdiagonal and there are $(3 - 2)$ eigenvectors here.

In a similar fashion we can show that matrix (2) has 2 distinct eigenvectors. Note it can be written like

$$\left[\begin{array}{c|cc} \lambda & 0 & 0 \\ \hline 0 & \lambda & 1 \\ 0 & 0 & \lambda \end{array} \right]$$

and the eigenvector equation is now

$$\begin{pmatrix} 0 & 0 & 0 \\ 0 & 0 & 1 \\ 0 & 0 & 0 \end{pmatrix} \begin{pmatrix} v_1 \\ v_2 \\ v_3 \end{pmatrix} = \begin{pmatrix} 0 \\ 0 \\ 0 \end{pmatrix}$$

Hence, $v3 = 0$ and the other two are arbitrary. This tells us there are two independent eigenvectors. Note the algebraic multiplicity is still 3 but now the geometric multiplicity is now 2. Note the structure of this matrix is different from (1). The $(1, 1)$ entry is just λ and this top position corresponds to a one dimensional eigenspace. The bottom 2×2 submatrix has 1's on the diagonal and a 1 on the superdiagonal about that. The number of 1's on the superdiagonal is 1 and there are $3 - 1$ or 2 eigenvectors now. This is the second type of Jordan Canonical block associated with an eigenvalue of algebraic multiplicity 3.

In case (3), the eigenvalues form the diagonal and the superdiagonal about it is all 0's. Note there are 3 eigenvectors now which is the same as $(3 - 0)$ where the 0 is the number of 1s on the superdiagonal.

Here the eigenvalue has algebraic multiplicity 3 and geometric multiplicity 3 also. This is the third type of Jordan Canonical block associated with an eigenvalue of algebraic multiplicity 3.

The Jordan Canonical Blocks for an eigenvalue λ of algebraic multiplicity 4 are then

$$\begin{array}{cccc}
\begin{bmatrix} \lambda & 1 & 0 & 0 \\ 0 & \lambda & 1 & 0 \\ 0 & 0 & \lambda & 1 \\ 0 & 0 & 0 & \lambda \end{bmatrix} &
\begin{bmatrix} \lambda & 0 & 0 & 0 \\ 0 & \lambda & 1 & 0 \\ 0 & 0 & \lambda & 1 \\ 0 & 0 & 0 & \lambda \end{bmatrix} &
\begin{bmatrix} \lambda & 0 & 0 & 0 \\ 0 & \lambda & 0 & 0 \\ 0 & 0 & \lambda & 1 \\ 0 & 0 & 0 & \lambda \end{bmatrix} &
\begin{bmatrix} \lambda & 0 & 0 & 0 \\ 0 & \lambda & 0 & 0 \\ 0 & 0 & \lambda & 0 \\ 0 & 0 & 0 & \lambda \end{bmatrix} \\
(1) & (2) & (3) & (4) \\
AM = 4 & AM = 4 & AM = 4 & AM = 4 \\
GM = 1 & GM = 2 & GM = 3 & GM = 4
\end{array}$$

where AM and GM are the algebraic and geometric multiplicity of the eigenvalue in each Jordan Canonical Block. From now on, let a Jordan Canonical Block be denoted by JCB for convenience.

Homework

Exercise 18.2.1 *Let*

$$A = \begin{pmatrix} -2 & 0 & 0 \\ 0 & -2 & 0 \\ 0 & 0 & -2 \end{pmatrix}$$

- *Compute* A^0 *to* A^5 *and compute*
- *Compute* $A^0 t^0,\ A^1 \frac{t^5}{1!},\ A^2 \frac{t^5}{2!},\ \ldots,\ A^5 \frac{t^5}{5!}$.
- *Compute* $\sum_{i=0}^{5} A^i \frac{t^I}{i!}$. *Does this result remind you of anything?*

Exercise 18.2.2 *Let*

$$A = \begin{pmatrix} -2 & 1 & 0 \\ 0 & -2 & 1 \\ 0 & 0 & -2 \end{pmatrix}$$

- *Compute* A^0 *to* A^5 *and compute*
- *Compute* $A^0 t^0,\ A^1 \frac{t^5}{1!},\ A^2 \frac{t^5}{2!},\ \ldots,\ A^5 \frac{t^5}{5!}$.
- *Compute* $\sum_{i=0}^{5} A^i \frac{t^I}{i!}$. *Does this result remind you of anything?*

Exercise 18.2.3 *Let*

$$A = \begin{pmatrix} -2 & 1 & 0 \\ 0 & -2 & 1 \\ 0 & 0 & -2 \end{pmatrix}$$

- *Compute* A^0 *to* A^5 *and compute*
- *Compute* $A^0 t^0,\ A^1 \frac{t^5}{1!},\ A^2 \frac{t^5}{2!},\ \ldots,\ A^5 \frac{t^5}{5!}$.
- *Compute* $\sum_{i=0}^{5} A^i \frac{t^I}{i!}$. *Does this result remind you of anything?*

Exercise 18.2.4 *Let*

$$A = \begin{pmatrix} -1 & 0 & 0 \\ 0 & -1 & 0 \\ 0 & 0 & -1 \end{pmatrix}$$

- *Compute A^0 to A^5 and compute*
- *Compute $A^0 t^0$, $A^1 \frac{t^5}{1!}$, $A^2 \frac{t^5}{2!}$, ..., $A^5 \frac{t^5}{5!}$.*
- *Compute $\sum_{i=0}^{5} A^i \frac{t^I}{i!}$. Does this result remind you of anything?*

Exercise 18.2.5 *Let*

$$A = \begin{pmatrix} -1 & 1 & 0 \\ 0 & -1 & 1 \\ 0 & 0 & -1 \end{pmatrix}$$

- *Compute A^0 to A^5 and compute*
- *Compute $A^0 t^0$, $A^1 \frac{t^5}{1!}$, $A^2 \frac{t^5}{2!}$, ..., $A^5 \frac{t^5}{5!}$.*
- *Compute $\sum_{i=0}^{5} A^i \frac{t^I}{i!}$. Does this result remind you of anything?*

Exercise 18.2.6 *Let*

$$A = \begin{pmatrix} -1 & 1 & 0 \\ 0 & -1 & 1 \\ 0 & 0 & -1 \end{pmatrix}$$

- *Compute A^0 to A^5 and compute*
- *Compute $A^0 t^0$, $A^1 \frac{t^5}{1!}$, $A^2 \frac{t^5}{2!}$, ..., $A^5 \frac{t^5}{5!}$.*
- *Compute $\sum_{i=0}^{5} A^i \frac{t^I}{i!}$. Does this result remind you of anything?*

From the exercise above, we can see a pattern when we multiply a JCB by itself many times. Let's work that out. Here is one particular JCB.

$$A = \begin{pmatrix} \lambda & 1 & 0 \\ 0 & \lambda & 1 \\ 0 & 0 & \lambda \end{pmatrix}$$

$$A^2 = \begin{pmatrix} \lambda & 1 & 0 \\ 0 & \lambda & 1 \\ 0 & 0 & \lambda \end{pmatrix} \begin{pmatrix} \lambda & 1 & 0 \\ 0 & \lambda & 1 \\ 0 & 0 & \lambda \end{pmatrix} = \begin{pmatrix} \lambda^2 & 2\lambda & 1 \\ 0 & \lambda^2 & 2\lambda \\ 0 & 0 & \lambda^2 \end{pmatrix}$$

$$A^3 = \begin{pmatrix} \lambda^2 & 2\lambda & 1 \\ 0 & \lambda^2 & 2\lambda \\ 0 & 0 & \lambda^2 \end{pmatrix} \begin{pmatrix} \lambda & 1 & 0 \\ 0 & \lambda & 1 \\ 0 & 0 & \lambda \end{pmatrix} = \begin{pmatrix} \lambda^3 & 3\lambda^2 & 3\lambda \\ 0 & \lambda^3 & 3\lambda \\ 0 & 0 & \lambda^3 \end{pmatrix}$$

We will get the same sort of formulas for A^4, A^5, etc. You can see if we could compute e^{At} for a JCB then we could compute it for a matrix written in a form containing only JCBs.

The general Jordan Canonical Block is thus

$$J = \underbrace{\begin{bmatrix} \lambda & 1 & & 0 \\ & \ddots & \ddots & \\ & & \ddots & 1 \\ 0 & & & \lambda \end{bmatrix}}_{n} \left.\vphantom{\begin{bmatrix} \lambda \\ \\ \\ \lambda \end{bmatrix}}\right\} n$$

We can rewrite this as

$$J = \lambda I + Z, \quad Z = \begin{pmatrix} 0 & 1 & & 0 \\ & \ddots & \ddots & \\ & & \ddots & 1 \\ 0 & & & 0 \end{pmatrix}$$

with ones on the super diagonal. It turns out that the matrix Z is nilpotent, i.e. $Z^n = 0$ for some n. Let's show this:

$$Z = \begin{bmatrix} 0 & 1 & & 0 \\ & \ddots & \ddots & \\ & & \ddots & 1 \\ 0 & & & 0 \end{bmatrix}, \quad Z^2 = \begin{bmatrix} 0 & 1 & & 0 \\ & \ddots & \ddots & \\ & & \ddots & 1 \\ 0 & & & 0 \end{bmatrix} \begin{bmatrix} 0 & 1 & & 0 \\ & \ddots & \ddots & \\ & & \ddots & 1 \\ 0 & & & 0 \end{bmatrix}$$

where the components are

$$z_{ij}^2 = \sum_{l=1}^{n} z_{il} z_{lj} = z_{i,i+1} z_{i+1,j} = \begin{cases} 1 & \text{if } j = i+2 \\ 0 & \text{otherwise} \end{cases}$$

So

$$Z^2 = \begin{bmatrix} 0 & 0 & 1 & & 0 \\ & \ddots & \ddots & \ddots & \\ & & \ddots & \ddots & 1 \\ & & & \ddots & 0 \\ 0 & & & & 0 \end{bmatrix}, \quad Z^3 = Z^2 Z = Z Z^2$$

with components

$$z_{ij}^3 = \sum_{l=1}^{n} z_{il}^2 z_{lj} = z_{i,i+2} z_{i+2,j} = \begin{cases} 1 & \text{if } j = i+3 \\ 0 & \text{otherwise} \end{cases}$$

We find

$$Z^3 = \begin{bmatrix} 0 & 0 & 0 & 1 & & 0 \\ & \ddots & \ddots & \ddots & \ddots & \\ & & \ddots & \ddots & \ddots & 1 \\ & & & \ddots & \ddots & 0 \\ & & & & \ddots & 0 \\ 0 & & & & & 0 \end{bmatrix} \begin{matrix} \\ \\ \leftarrow \text{main-3} \ (i, i+3) \\ \leftarrow \text{main-2} \ (i, i+2) \\ \leftarrow \text{main-1} \ (i, i+1) \\ \leftarrow \text{main} \end{matrix}$$

We can only go so far though. After $n-1$ iterations, we have

$$Z^{n-1} = \begin{bmatrix} 0 & & 1 \\ & \ddots & \\ 0 & & 0 \end{bmatrix}, \quad Z^n = 0$$

Therefore Z is a nilpotent matrix.

18.2.1 Homework

Exercise 18.2.7 *Let*

$$A = \begin{pmatrix} -2 & 1 & 0 \\ 0 & -2 & 0 \\ 0 & 0 & 3 \end{pmatrix}$$

- *Compute A^0 to A^5 and compute*
- *Compute $A^0 t^0$, $A^1 \frac{t^5}{1!}$, $A^2 \frac{t^5}{2!}$, ..., $A^5 \frac{t^5}{5!}$.*
- *Compute $\sum_{i=0}^{5} A^i \frac{t^I}{i!}$. Does this result remind you of anything?*

Exercise 18.2.8 *Let*

$$A = \begin{pmatrix} 3 & 1 & 0 & 0 \\ 0 & 3 & 1 & 0 \\ 0 & 0 & 3 & 0 \\ 0 & 0 & 0 & 4 \end{pmatrix}$$

- *Compute A^0 to A^5 and compute*
- *Compute $A^0 t^0$, $A^1 \frac{t^5}{1!}$, $A^2 \frac{t^5}{2!}$, ..., $A^5 \frac{t^5}{5!}$.*
- *Compute $\sum_{i=0}^{5} A^i \frac{t^I}{i!}$. Does this result remind you of anything?*

Exercise 18.2.9 *Let*

$$A = \begin{pmatrix} 3 & 1 & 0 & 0 \\ 0 & 3 & 1 & 0 \\ 0 & 0 & 3 & 0 \\ 0 & 0 & 0 & 4 \end{pmatrix}$$

- *Compute* A^0 *to* A^5 *and compute*
- *Compute* $A^0 t^0$, $A^1 \frac{t^5}{1!}$, $A^2 \frac{t^5}{2!}$, …, $A^5 \frac{t^5}{5!}$.
- *Compute* $\sum_{i=0}^{5} A^i \frac{t^I}{i!}$. *Does this result remind you of anything?*

Exercise 18.2.10 *Let*

$$A = \begin{pmatrix} 3 & 1 & 0 & 0 \\ 0 & 3 & 0 & 0 \\ 0 & 0 & 4 & 1 \\ 0 & 0 & 0 & 4 \end{pmatrix}$$

- *Compute* A^0 *to* A^5 *and compute*
- *Compute* $A^0 t^0$, $A^1 \frac{t^5}{1!}$, $A^2 \frac{t^5}{2!}$, …, $A^5 \frac{t^5}{5!}$.
- *Compute* $\sum_{i=0}^{5} A^i \frac{t^I}{i!}$. *Does this result remind you of anything?*

Exercise 18.2.11 *Let*

$$A = \begin{pmatrix} 2 & 0 & 0 & 0 \\ 0 & 3 & 0 & 0 \\ 0 & 0 & 4 & 1 \\ 0 & 0 & 0 & 4 \end{pmatrix}$$

- *Compute* A^0 *to* A^5 *and compute*
- *Compute* $A^0 t^0$, $A^1 \frac{t^5}{1!}$, $A^2 \frac{t^5}{2!}$, …, $A^5 \frac{t^5}{5!}$.
- *Compute* $\sum_{i=0}^{5} A^i \frac{t^I}{i!}$. *Does this result remind you of anything?*

Exercise 18.2.12 *Let*

$$A = \begin{pmatrix} 3 & 1 & 0 & 0 & 0 \\ 0 & 3 & 0 & 0 & 0 \\ 0 & 0 & 4 & 1 & 0 \\ 0 & 0 & 0 & 4 & 0 \\ 0 & 0 & 0 & 0 & 6 \end{pmatrix}$$

- *Compute* A^0 *to* A^5 *and compute*
- *Compute* $A^0 t^0$, $A^1 \frac{t^5}{1!}$, $A^2 \frac{t^5}{2!}$, …, $A^5 \frac{t^5}{5!}$.
- *Compute* $\sum_{i=0}^{5} A^i \frac{t^I}{i!}$. *Does this result remind you of anything?*

18.3 Exponential Matrix Calculations

Theorem 18.3.1 If A **and** B **Commute,** $e^{A+B} = e^A e^B$

Let A *and* B *be* $n \times n$ *matrices which commute; i.e.* $AB = BA$. *Then* $e^{A+B} = e^A e^B$.

Proof 18.3.1
We claim

$$\sum_{i=0}^{n} (A^i/(i!)) \sum_{j=0}^{n} (B^j/(j!)) = \sum_{k=0}^{n} \sum_{j=0}^{k} \frac{A^j B^{k-j}}{j!(k-j)!}$$

To see this, consider the array

$$\begin{bmatrix} (0,0) & (0,1) & (0,2) & \dots & (0,n) \\ (1,0) & (1,1) & (1,2) & \dots & (1,n) \\ \vdots & \vdots & \vdots & \vdots & \vdots \\ (n,0) & (n,1) & (n,2) & \dots & (n,n) \end{bmatrix}$$

Summing over all $(n+1)^2$ entries indexed by this array is the same as summing over the diagonals here

$$\begin{bmatrix} (i+j)=0 & (i+j)=1 & (i+j)=2 & \dots & (i+j)=n \\ (i+j)=1 & (i+j)=2 & \vdots & \vdots & (i+j)=n+1 \\ (i+j)=2 & \vdots & \vdots & \vdots & \vdots \\ \vdots & (i+j)=n & \vdots & \vdots & \vdots \\ (i+j)=n & (i+j)=n+1 & \dots & \dots & (i+j)=2n \end{bmatrix}$$

Let Z_0 be the set of non-negative integers. The scheme shown above defines a 1-1 and onto mapping g from $Z_0 \times Z_0$ to Z_0. If we let $a_{ij} = A^i B^j/(i!j!)$, we are wondering if the $\sum_i \sum_j a_{ij}$ is the same as the $\sum a_{g(i,j)}$ where $a_{g(i,j)}$ is the particular index in the scheme above. Thus, we are wondering if

$$\begin{aligned} \sum_{i=0}^{n} \sum_{j=0}^{n} (A^i/(i!)) (B^j/(j!)) &= a_{g(0,0)} + (a_{g(1,0)} + a_{g(0,1)}) + (a_{g(2,0)} + a_{g(1,1)} + a_{g(0,2)}) + \dots \\ &= \sum_{k=0}^{n} \sum_{j=0}^{k} \frac{A^j B^{k-j}}{j!(k-j)!} \end{aligned}$$

The $\sum_{i=0}^{n} \sum_{j=0}^{n} (A^i/(i!)) (B^j/(j!))$ converges to $e^A e^B$. Hence, it is a Cauchy sequence in the Frobenius norm. If you think about the indexing scheme, we can see

$$\left(\sum_{i+j=2n} + \sum_{i+j=3n} + \dots + \sum_{i+j=pn} \right) A^i B^j/(i!j!)$$

does not contain the terms $\sum_{i=0}^{n} \sum_{j=0}^{n} A^i B^j/(i!j!)$. Further all of these terms are contained in the difference

$$\left(\sum_{i=0}^{pn} \sum_{j=0}^{pn} - \sum_{i=0}^{n} \sum_{j=0}^{n} \right) A^i/(i!) B^j/(j!)$$

$$\left(\sum_{i=0}^{n} \sum_{j=n+1}^{pn} + \sum_{i=n+1}^{pn} \sum_{j=0}^{n} + \sum_{i=n+1}^{pn} \sum_{j=n+1}^{pn} \right) A^i/(i!) B^j/(j!)$$

Thus,

$$\begin{aligned}
&\left\| \left(\sum_{i+j=2n} + \sum_{i+j=3n} + \ldots + \sum_{i+j=pn} \right) A^i B^j/(i!j!) \right\| \\
&\leq \left(\sum_{i+j=2n} + \sum_{i+j=3n} + \ldots + \sum_{i+j=pn} \right) \|A\|^i \, \|B\|^j/(i!j!) \\
&\leq \left(\sum_{i=0}^{n} \sum_{j=n+1}^{pn} + \sum_{i=n+1}^{pn} \sum_{j=0}^{n} + \sum_{i=n+1}^{pn} \sum_{j=n+1}^{pn} \right) \|A\|^i \, \|B\|^j/(i!j!)
\end{aligned}$$

Now we can simplify. We have

$$\begin{aligned}
&\sum_{i=0}^{n} \|A\|^i/(i!) \sum_{j=n+1}^{pn} \|B\|^j/(j!) + \sum_{i=n+1}^{pn} \|A\|^i/(i!) \sum_{j=0}^{n} \|B\|^j/(j!) \\
&+ \sum_{i=n+1}^{pn} \|A\|^i/(i!) \sum_{j=n+1}^{pn} \|B\|^j/(j!) \\
&\leq e^{\|A\|} \sum_{j=n+1}^{pn} \|B\|^j/(j!) + e^{\|B\|} \sum_{i=n+1}^{pn} \|A\|^i/(i!) + \sum_{i=n+1}^{pn} \sum_{j=n+1}^{pn} \|A\|^i \, \|B\|^j/(i!j!)
\end{aligned}$$

Hence, given $\epsilon > 0$*, there are integers* Q_1*,* Q_2 *and* Q_3 *so that*

$$\begin{aligned}
pn > n+1 > Q_1 \quad &\Longrightarrow \quad e^{\|A\|} \sum_{j=n+1}^{pn} \|B\|^j/(j!) < \epsilon/3 \\
pn > n+1 > Q_2 \quad &\Longrightarrow \quad e^{\|B\|} \sum_{i=n+1}^{pn} \|A\|^j/(j!) < \epsilon/3 \\
pn > n+1 > Q_3 \quad &\Longrightarrow \quad \sum_{i=n+1}^{pn} \sum_{j=n+1}^{pn} \|A\|^i \, \|B\|^j/(i!j!) < \epsilon/3
\end{aligned}$$

because $\sum_{i=0}^{n} \|A\|^i/(i!) \rightarrow e^{\|A\|}$ *and* $\sum_{j=0}^{n} \|B\|^j/(j!) \rightarrow e^{\|B\|}$*. Thus, for* $pn > n+1 > \max\{Q_1, Q_2, Q_3\}$*, we have*

$$\left\| \left(\sum_{i+j=2n} + \sum_{i+j=3n} + \ldots + \sum_{i+j=pn} \right) A^i B^j/(i!j!) \right\| < \epsilon$$

which tells us the sequence $\sum_{g(i,j)=0}^{g(i,j)=n}$ *is a Cauchy sequence and so converges to matrix* C*. Finally, consider*

$$\begin{aligned}
\|e^A e^B - C\| \quad \leq \quad & \left\| e^A e^B - \sum_{i=0}^{n} \sum_{j=0}^{n} A^i B^j/(i!j!) \right\| + \left\| C - \sum_{g(i,j)=0}^{g(i,j)=N} A^i B^j/(i!j!) \right\| \\
+ \quad & \left\| \sum_{i=0}^{n} \sum_{j=0}^{n} A^i B^j/(i!j!) - \sum_{g(i,j)=0}^{g(i,j)=N} A^i B^j/(i!j!) \right\|
\end{aligned}$$

For a given $\epsilon > 0$, the first and second terms can be made less than $\epsilon/3$ because of the convergence of the sums to $E^A e^B$ and C respectively. Thus, there is an Q_1 so that if $n > Q_1$,

$$\|e^A e^B - C\| < 2\epsilon/3 + \left\| \sum_{i=0}^{n} \sum_{j=0}^{n} A^i B^j/(i!j!) - \sum_{g(i,j)=0}^{g(i,j)=N} A^i B^j/(i!j!) \right\|$$

The last term is analyzed just like we did in the argument to show the sequence defining C was a Cauchy sequence. We find there is a Q_2 so that is $n > N > Q_2$

$$\left\| \sum_{i=0}^{n} \sum_{j=0}^{M} A^i B^j/(i!j!) - \sum_{g(i,j)=0}^{g(i,j)=N} A^i B^j/(i!j!) \right\| < \epsilon/3$$

Thus, for large enough choices of n and N, we have $\|e^A e^B - C\| < \epsilon$. Since ϵ is arbitrary, this shows $C = e^A e^B$.

Thus,

$$\begin{aligned} e^A e^B &= \left(I + A + \frac{A^2}{2!} + \frac{A^3}{3!} + \cdots + \frac{A^k}{k!} + \cdots \right) \\ &\quad \left(I + B + \frac{B^2}{2!} + \frac{B^3}{3!} + \cdots + \frac{B^k}{k!} + \cdots \right) \\ &= I + \frac{B}{1!} + \frac{A}{1!} + \\ &\quad \frac{B^2}{2!} + \frac{AB}{1!1!} + \frac{A^2}{2!} + \frac{B^3}{3!} + \frac{AB^2}{1!2!} + \frac{A^2B}{2!1!} + \frac{A^3}{3!} + \ldots \\ &\quad \frac{B^k}{k!} + \frac{AB^{k-1}}{1!(k-1)!} + \frac{A^2B^{k-2}}{2!(k-2)!} + \ldots \end{aligned}$$

Then, using the fact that A and B commute,

$$\begin{aligned} e^A e^B &= \sum_{n=0}^{\infty} \sum_{j=0}^{n} \frac{A^j B^{n-j}}{j!(n-j)!} = \sum_{n=0}^{\infty} \frac{1}{n!} \sum_{j=0}^{n} \frac{n!}{j!(n-j)!} A^j B^{n-j} \\ &= \sum_{n=0}^{\infty} \frac{1}{n!} \sum_{j=0}^{n} \binom{n}{j} A^j B^{n-j} = \sum_{n=0}^{\infty} \frac{(A+B)^n}{n!} \end{aligned}$$

since $\sum_{j=0}^{n} \binom{n}{j} A^j B^{n-j} = (A+B)^n$. This shows $e^A e^B e^{A+B}$. ∎

Homework

Exercise 18.3.1 *Prove $e^A e^B e^C = e^{A+B+C}$ for $n \times n$ matrices A, B and C.*

18.3.1 Jordan Block Matrices

In applying this to a Jordan Block we get

$$J = \lambda I + Z$$

$$e^{Jt} = e^{\lambda It+Zt} = e^{\lambda It}e^{Zt}$$

where

$$\begin{aligned} e^{\lambda It} &= I + t\lambda I + \frac{(t\lambda I)^2}{2!} + \cdots + \frac{(t\lambda I)^n}{n!} + \cdots = I + t\lambda I + \frac{(t\lambda)^2 I}{2!} + \cdots + \frac{(t\lambda)^n I}{n!} + \cdots \\ &= \mathbf{diag}[1 + \lambda t + \frac{(\lambda t)^2}{2!} + \cdots + \frac{(\lambda t)^n}{n!} + \cdots] = \mathbf{diag}[e^{\lambda t}] = e^{\lambda t} I \end{aligned}$$

where **diag** denotes a diagonal matrix. We also find

$$\begin{aligned} e^{Zt} &= I + tZ + \frac{t^2 Z^2}{2!} + \cdots + \frac{t^{n-1} Z^{n-1}}{(n-1)!} + \cdots \\ &= \begin{bmatrix} 1 & & & \\ & \ddots & & \\ & & \ddots & \\ & & & 1 \end{bmatrix} + t \begin{bmatrix} 0 & 1 & & \\ & \ddots & \ddots & \\ & & \ddots & 1 \\ & & & 0 \end{bmatrix} + \cdots + \frac{t^{n-1}}{(n-1)!} \begin{bmatrix} 0 & 0 & & 1 \\ & \ddots & \ddots & \\ & & \ddots & 0 \\ & & & 0 \end{bmatrix} \\ &= \begin{bmatrix} 1 & t & \frac{t^2}{2!} & & \frac{t^{n-1}}{(n-1)!} \\ & \ddots & \ddots & \ddots & \\ & & \ddots & \ddots & \frac{t^2}{2!} \\ & & & \ddots & t \\ & & & & 1 \end{bmatrix} \begin{matrix} \\ \\ \leftarrow \text{main-2} \\ \leftarrow \text{main-1} \\ \leftarrow \text{main} \end{matrix} \end{aligned}$$

So then

$$e^{Jt} = e^{\lambda It}e^{Zt} = e^{\lambda t} \begin{bmatrix} 1 & t & \frac{t^2}{2} & & \frac{t^{n-1}}{(n-1)!} \\ & \ddots & \ddots & \ddots & \\ & & \ddots & \ddots & \frac{t^2}{2} \\ & & & \ddots & t \\ & & & & 1 \end{bmatrix}$$

So if the Jordan form of the matrix A was

$$J = \begin{bmatrix} \boxed{2} & & & & & & & \\ & \boxed{\begin{matrix} 2 & 1 \\ & 2 \end{matrix}} & & & & \\ & & \boxed{\begin{matrix} -3 & 0 \\ & -3 \end{matrix}} & \\ & & & \boxed{\begin{matrix} -3 & 1 \\ & -3 \end{matrix}} \end{bmatrix}$$

then

$$e^{Jt} = \begin{bmatrix} \boxed{e^{2t}} & & & & & & \\ & e^{2t} & te^{2t} & & & & \\ & & e^{2t} & & & & \\ & & & e^{-3t} & & & \\ & & & & e^{-3t} & & \\ & & & & & e^{-3t} & te^{-3t} \\ & & & & & & e^{-3t} \end{bmatrix}$$

Homework

Exercise 18.3.2 *Let*

$$A = \begin{pmatrix} -2 & 1 \\ 0 & -2 \end{pmatrix}, \quad B = \begin{pmatrix} 4 & 1 \\ 0 & 4 \end{pmatrix}$$

Is $e^{A+B} = e^A e^B$*?*

Exercise 18.3.3 *Let*

$$A = \begin{pmatrix} -2 & 1 \\ 0 & -2 \end{pmatrix}, \quad B = \begin{pmatrix} 4 & 1 \\ 0 & 8 \end{pmatrix}$$

Is $e^{A+B} = e^A e^B$*?*

Exercise 18.3.4 *Let*

$$A = \begin{pmatrix} 3 & 1 \\ 0 & 4 \end{pmatrix}, \quad B = \begin{pmatrix} -2 & 6 \\ 0 & 4 \end{pmatrix}$$

Is $e^{A+B} = e^A e^B$*?*

18.3.2 General Matrices

In general, for an $n \times n$ matrix A there is an invertible matrix P such that

$$A = PJP^{-1}$$

where J is in Jordan canonical form. So then,

$$e^{At} = I + PJP^{-1}t + \overbrace{(PJP^{-1}PJP^{-1})}^{PJ^2P^{-1}}\frac{t^2}{2!} + \cdots + (PJ^nP^{-1})\frac{t^n}{n!}$$

and we can easily prove we can factor out the P and P^{-1} to get

$$\begin{aligned} e^{At} &= P(I + J + \frac{J^2}{2!} + \cdots + \frac{J^n}{n!})P^{-1} \\ &= Pe^{Jt}P^{-1} \end{aligned}$$

But how easy are P and P^{-1} to find? Look at a 2×2 matrix A having eigenvalues λ_1 and λ_2 and eigenvectors v_1 and v_2 such that $v_1 \perp v_2$ and each has unit length. Then, as we know

$$\begin{aligned}
\underbrace{\begin{pmatrix} v_1^T \\ v_2^T \end{pmatrix}}_{Q} A \underbrace{\begin{pmatrix} v_1 & v_2 \end{pmatrix}}_{Q^T} &= \begin{pmatrix} v_1^T \\ v_2^T \end{pmatrix} \begin{pmatrix} Av_1 & Av_2 \end{pmatrix} = \begin{pmatrix} v_1^T \\ v_2^T \end{pmatrix} \begin{pmatrix} \lambda_1 v_1 & \lambda_2 v_2 \end{pmatrix} \\
&= \begin{pmatrix} \lambda_1 v_1^T v_1 & \lambda_2 v_1^T v_2 \\ \lambda_1 v_2^T v_1 & \lambda_2 v_2^T v_2 \end{pmatrix} = \begin{pmatrix} \lambda_1 & 0 \\ 0 & \lambda_2 \end{pmatrix}
\end{aligned}$$

So $QAQ^T = J$ where J is the above matrix in Jordan canonical form. Then $AQ^T = Q^T J \Longrightarrow A = Q^T JQ$ and set $P = Q^T = (v_1, v_2) \Longrightarrow P^{-1} = Q$. We can therefore do this fairly easily for symmetric matrices. From this we can see that certain matrices have some nice forms and things just fall out nicely.
If the matrix is not necessarily symmetric, we can still find these matrices easily.

Assume A is a 3×3 matrix whose JCB is $J = \begin{bmatrix} \lambda & 1 & 0 \\ 0 & \lambda & 1 \\ 0 & 0 & \lambda \end{bmatrix}$ for some nonzero λ. If we want to find a matrix P so that $A = PJP^{-1}$, this is equivalent to finding P so that $AP = PJ$. Let $U = \begin{bmatrix} 0 & 1 & 0 \\ 0 & 0 & 1 \\ 0 & 0 & 0 \end{bmatrix}$ and then note since $PI = IP$, we have

$$AP = PJ = P(\lambda I + U) \Longrightarrow AP - \lambda IP + PU$$

Let P_1, P_2 and P_3 be the columns of P. Then we have

$$\begin{bmatrix}(A - \lambda I)P_1, & (A - \lambda I)P_2 & (A - \lambda I)P_3\end{bmatrix} = \begin{bmatrix}0 & P_1, P_2\end{bmatrix}$$

Thus we must find solutions to

$$\begin{aligned}
(A - \lambda I)P_1 &= 0 \Longrightarrow P_1 = E \\
(A - \lambda I)P_2 &= P_1 \Longrightarrow (A - \lambda I)P_2 = E, \text{ let } P_2 = F. \\
(A - \lambda I)P_3 &= P_2 \Longrightarrow (A - \lambda I)P_3 = F \\
&\Longrightarrow (A - \lambda I)^2 P_3 = E, \text{ let } P_3 = G
\end{aligned}$$

where E is the eigenvector for λ. For this example, the geometric multiplicity of the eigenvalue is 1 because of the structure of the JCB. It is straightforward to see $\{E, F, G\}$ is a basis. Thus $P = \begin{bmatrix} E & F & G \end{bmatrix}$ and since the columns are linearly independent, we must have P^{-1} exists. So this is the matrix we seek.

If we had a bit more complicated multiple JCB situation, we could have A is a 7×7 matrix with JCB block structure given by

$$J = \begin{bmatrix} \begin{bmatrix} \lambda & 1 & 0 & 0 \\ 0 & \lambda & 1 & 0 \\ 0 & 0 & \lambda & 1 \\ 0 & 0 & 0 & \lambda \end{bmatrix} & \begin{bmatrix} 0 & 0 & 0 \\ 0 & 0 & 0 \\ 0 & 0 & 0 \\ 0 & 0 & 0 \end{bmatrix} \\ \begin{bmatrix} 0 & 0 & 0 & 0 \\ 0 & 0 & 0 & 0 \\ 0 & 0 & 0 & 0 \end{bmatrix} & \begin{bmatrix} \mu & 1 & 0 \\ 0 & \mu & 1 \\ 0 & 0 & \mu \end{bmatrix} \end{bmatrix}$$

Again we want to find a matrix P so that so that $AP = PJ$. The eigenvalue λ has algebraic multiplicity 4 and geometric multiplicity 1 and the eigenvalue μ has algebraic multiplicity 3 and geometric multiplicity 1. Following the procedure we just used, we find vectors solving

$$\begin{aligned}
(A-\lambda I)P_1^\lambda &= 0 \Longrightarrow P_1^\lambda = E^\lambda \\
(A-\lambda I)P_2^\lambda &= P_1^\lambda \Longrightarrow (A-\lambda I)P_2^\lambda = E^\lambda, \text{ Let } P_2^\lambda = F^\lambda. \\
(A-\lambda I)P_3^\lambda &= P_1^\lambda + P_2^\lambda \Longrightarrow (A-\lambda I)P_3^\lambda = F^\lambda \\
&\Longrightarrow (A-\lambda I)^2 P_3^\lambda = E^\lambda, \text{ Let } P_3^\lambda = G^\lambda. \\
(A-\lambda I)P_4^\lambda &= P_3^\lambda \Longrightarrow (A-\lambda I)P_3^\lambda = G^\lambda \\
&\Longrightarrow (A-\lambda I)^3 P_4^\lambda = E^\lambda, \text{ Let } P_4^\lambda = H^\lambda.
\end{aligned}$$

Thus, $P^\lambda = \left[E^\lambda, F^\lambda, G^\lambda, H^\lambda\right]$ and $\{E^\lambda, F^\lambda, G^\lambda, H^\lambda\}$ is a linearly independent set. Then we find the vectors solving

$$\begin{aligned}
(A-\lambda I)P_1^\mu &= 0 \Longrightarrow P_1^\mu = E^\mu \\
(A-\lambda I)P_2^\mu &= P_1^\mu \Longrightarrow (A-\mu I)P_2^\mu = E, \text{ Let } P_2^\mu = F^\mu. \\
(A-\lambda I)P_3^\mu &= P_1^\mu + P_2^\mu \Longrightarrow (A-\mu I)P_3^\mu = F^\mu \\
&\Longrightarrow (A-\mu I)^2 P_3^\mu = E^\mu, \text{ Let } P_3^\mu = G^\mu.
\end{aligned}$$

So, $P^\mu = \left[E^\mu, F^\mu, G^\mu\right]$ and $\{E^\mu, F^\mu, G^\mu\}$ is a linearly independent set. Hence,

$$\{E^\lambda, F^\lambda, G^\lambda, H^\lambda, E^\mu, F^\mu, G^\mu\}$$

is a basis. The matrix $P = \left[P^\lambda, P^\mu\right]$ is then invertible and is the matrix we need.

Of course, it is not so easy to find the JCB blocks due to A. If we found the eigenvalues using, say MATLAB, we have a reasonable probability of understanding which eigenvalues are repeated, although floating point errors could cause some issues. Also, we have not addressed what we do if the eigenvalues are complex conjugate pairs that are repeated. That is a discussion for another time but we encourage you to look it up!

18.3.3 Homework

Exercise 18.3.5 *Follow our discussion to find P for the case of an eigenvalue λ of multiplicity 3 with geometric multiplicity 2.*

Exercise 18.3.6 *Follow our discussion to find P for the case of an eigenvalue λ of multiplicity 4 with geometric multiplicity 2 and 3.*

Exercise 18.3.7 *Follow our discussion to find P for the case of an eigenvalue λ of multiplicity 5 with geometric multiplicity 2, 3 and 4.*

Exercise 18.3.8 *In the case of an eigenvalue λ of multiplicity 3 with geometric multiplicity 1, we found*

$$\begin{aligned}
(A-\lambda I)P_1 &= 0 \Longrightarrow P_1 = E \\
(A-\lambda I)P_2 &= P_1 \Longrightarrow (A-\lambda I)P_2 = E, \textit{ Let } P_2 = F. \\
(A-\lambda I)P_3 &= P_1 + P_2 \Longrightarrow (A-\lambda I)P_3 = F \\
&\Longrightarrow (A-\lambda I)^2 P_3 = E, \textit{ Let } P_3 = G.
\end{aligned}$$

- *Prove E, F and G are a basis.*
- *Prove* $(A - \lambda I)^2 P_3 = E$ *implies* $(A - \lambda I)P_3 = F$.

Exercise 18.3.9 *Let*

$$A = \begin{pmatrix} -2 & 1 \\ 1 & -2 \end{pmatrix}$$

- *Find the eigenvalues and eigenvectors of A using MATLAB. This gives you the decomposition* $A = PDP^T$ *as usual where D is the diagonal matrix of the eigenvalues of A.*
- *Compute* A^n *by hand showing all the steps.*
- *Compute* e^A *by hand and by theory.*

Exercise 18.3.10 *Let*

$$A = \begin{pmatrix} -2 & 1 & 3 \\ 1 & 4 & -1 \\ 3 & -1 & 5 \end{pmatrix}$$

- *Find the eigenvalues and eigenvectors of A using MATLAB. This gives you the decomposition* $A = PDP^T$ *as usual where D is the diagonal matrix of the eigenvalues of A.*
- *Compute* A^n *by hand showing all the steps.*
- *Compute* e^A *by hand and by theory.*

Exercise 18.3.11 *Let*

$$A = \begin{pmatrix} 3 & 1 \\ 1 & 3 \end{pmatrix}, \quad B = \begin{pmatrix} -2 & 6 \\ 6 & 4 \end{pmatrix}$$

- *Find the eigenvalues and eigenvectors of A and B using MATLAB. This gives you the decompositions* $A = PD_1P^T$ *as usual where* D_1 *is the diagonal matrix of the eigenvalues of A and* $B = QD_2Q^T$ *as usual, where* D_2 *is the diagonal matrix of the eigenvalues of B.*
- *Compute* e^A *and* e^B *by hand and by theory.*
- *Compute* e^{A+B} *by hand and by theory.*
- *Compute* $e^A e^B$ *by hand and by theory. This should match* e^{A+B}.

Exercise 18.3.12 *Let*

$$A = \begin{pmatrix} -2 & 1 & 3 \\ 1 & 4 & -1 \\ 3 & -1 & 5 \end{pmatrix}, \quad B = \begin{pmatrix} -5 & -1 & 4 \\ -1 & 2 & -3 \\ 4 & -3 & 7 \end{pmatrix}$$

- *Find the eigenvalues and eigenvectors of A and B using MATLAB. This gives you the decompositions* $A = PD_1P^T$ *as usual where* D_1 *is the diagonal matrix of the eigenvalues of A and* $B = QD_2Q^T$ *as usual, where* D_2 *is the diagonal matrix of the eigenvalues of B.*
- *Compute* e^A *and* e^B *by hand and by theory.*
- *Compute* e^{A+B} *by hand and by theory.*
- *Compute* $e^A e^B$ *by hand and by theory. This should match* e^{A+B}.

18.4 Applications to Linear ODE

Consider the following system:

$$x' = = \begin{bmatrix} x_1' \\ x_2' \\ \vdots \\ x_n' \end{bmatrix} = \begin{bmatrix} a_{11} & \cdots & \cdots & a_{1n} \\ \vdots & \cdots & \cdots & \vdots \\ \vdots & \cdots & \cdots & \vdots \\ a_{n1} & \cdots & \cdots & a_{nn} \end{bmatrix} \begin{bmatrix} x_1 \\ x_2 \\ \vdots \\ x_n \end{bmatrix} = Ax, \quad x(t_0) = x_0 = \begin{bmatrix} x_{01} \\ x_{02} \\ \vdots \\ x_{0n} \end{bmatrix}$$

Theorem 18.4.1 The Derivative of the Exponential Matrix

Let A be an $n \times n$ matrix. Then $\frac{d}{dt}\left(e^{At}\right) = Ae^{At}$.

Proof 18.4.1
Since $e^{At} = \sum_{j=0}^{\infty} \frac{A^j}{j!} t^j$, to find its derivative, we look at

$$\frac{d}{dt}\left(e^{At}\right) = \frac{d}{dt}\left(\sum_{j=0}^{\infty} \frac{A^j}{j!} t^j\right)$$

We want to use an analogue of the derivative interchange theorem we proved in (Peterson (21) 2020) to show

$$\frac{d}{dt}\left(\sum_{j=0}^{\infty} \frac{A^j}{j!} t^j\right) = \sum_{j=0}^{\infty} \frac{d}{dt}\left(\frac{A^j}{j!} t^j\right)$$

The series we have here is a series of matrices and convergence is with respect to the Frobenius norm so it is not immediately clear the earlier results apply. For a series of real numbers, we would perform the following checks applied to the sequence of partial sums S_n. First, we fix a T. We need to check the following:

1. *S_n is differentiable on $[0, T]$. It is easy to see*

$$\left(\sum_{j=0}^{n} \frac{A^j}{j!} t^j\right)' = \sum_{j=0}^{n} j \frac{A^j}{j!} t^{j-1}$$

   as the derivative $Bf(t)$ for a matrix B and a differentiable function f is easily seen to be $Bf'(t)$ by looking at the derivative of each component.

2. *S_n' is Riemann Integrable on $[0, t]$: This is true as the integral of $Bf(t)$ for any integrable f is $B \int_0^t f(s)ds$ for any matrix B by looking at components.*

3. *There is at least one point $t_0 \in [0, t]$ such that the sequence $(S_n(t_0))$ converges. This is true as the series here converges at all t.*

4. *$S_n' \stackrel{unif}{\longrightarrow} y$ on $[0, T]$ and the limit function y is continuous. This one is tougher. What we have shown in our work so far is that S_n is a Cauchy sequence with respect to the Frobenius norm. In fact, we know that if $T_n(t) = \sum_{i=0}^{n} \|A\|^i/(i!)$ is a partial sum of $e^{\|A\|t}$ which we know converges uniformly on any finite interval $[0, T]$, we use it to bound the partial sums S_n. Hence, for any $\epsilon > 0$, there is N so that $n > m > N$ implies $\|S_n(t) - S_m(t)\|_{Fr} \leq$*

$|T_n(t) - T_m(t)| < \epsilon$ *on* $[0, T]$. *If you look at the arguments we use in proving these sorts of theorems in (Peterson (21) 2020), you can see how they can be modified to show* S_n *converges uniformly on* $[0, T]$ *to a continuous function* e^{At}. *Everything just said can be modified a bit to prove* S_n' *converges uniformly to some matrix function* D *on* $[0, T]$ *as well.*

You can then modify the proof of the derivative interchange theorem to show there is a function W *on* $[0, T]$ *so that* $S_n \xrightarrow{unif} W$ *on* $[0, T]$ *and* $W' = D$. *Since limits are unique, we then have* $W = e^{At}$ *with* $(e^{At})' = D$. *This is the same as saying the interchange works and*

$$\frac{d}{dt}\left(\sum_{j=0}^{\infty} \frac{A^j}{j!} t^j\right) = \sum_{j=0}^{\infty} \frac{d}{dt}\left(\frac{A^j}{j!} t^j\right)$$

Now

$$\begin{aligned}
\sum_{j=0}^{\infty} \frac{d}{dt}\left(\frac{A^j t^J}{j!}\right) &= \sum_{j=0}^{\infty} \frac{A^j j t^{j-1}}{j!} = \sum_{j=1}^{\infty} \frac{A^j j t^{j-1}}{j!} = \sum_{j=1}^{\infty} \frac{A^j t^{j-1}}{(j-1)!} \\
&= A(I + tA + \frac{t^2}{2} A^2 + \cdots + \frac{t^n}{n!} A^n + \cdots) = Ae^{At}
\end{aligned}$$

where it is straightforward to prove we can factor the A *out of the series by a partial sum argument which we leave to you. So we get that* $\frac{d}{dt}(e^{At}) = Ae^{At}$. ■

Homework

Exercise 18.4.1 *If*

$$A(t) = \begin{pmatrix} A_{11}(t) & \ldots & A_{1,n}(t) \\ \vdots & \vdots & \vdots \\ A_{n1}(t) & \ldots & A_{nn}(t) \end{pmatrix}$$

where each A_{ij} *is differentiable, then*

$$A'(t) = \begin{pmatrix} A'_{11}(t) & \ldots & A'_{1,n}(t) \\ \vdots & \vdots & \vdots \\ A'_{n1}(t) & \ldots & A'_{nn}(t) \end{pmatrix}$$

Exercise 18.4.2 *Let*

$$A = \begin{pmatrix} -2 & 1 & 3 \\ 1 & 4 & -1 \\ 3 & -1 & 5 \end{pmatrix}$$

- *Find* e^{At}.
- *Find* $(e^{At})'$ *directly as it is a matrix of differentiable functions.*
- *Find* Ae^{At}. *This should match* $(e^{At})'$.

Exercise 18.4.3 *Let*

$$A = \begin{pmatrix} 2 & 3 & -1 \\ 3 & 4 & 5 \\ -1 & 5 & 6 \end{pmatrix}$$

- *Find* e^{At}.
- *Find* $(e^{At})'$ *directly as it is a matrix of differentiable functions.*
- *Find* Ae^{At}. *This should match* $(e^{At})'$.

Exercise 18.4.4 *Let*

$$A = \begin{pmatrix} 1 & 2 & -1 \\ 2 & 2 & 4 \\ -1 & 4 & -5 \end{pmatrix}$$

- *Find* e^{At}.
- *Find* $(e^{At})'$ *directly as it is a matrix of differentiable functions.*
- *Find* Ae^{At}. *This should match* $(e^{At})'$.

Exercise 18.4.5 *Let*

$$A = \begin{pmatrix} 8 & 5 & 6 \\ 5 & 4 & -7 \\ 6 & -7 & 15 \end{pmatrix}$$

- *Find* e^{At}.
- *Find* $(e^{At})'$ *directly as it is a matrix of differentiable functions.*
- *Find* Ae^{At}. *This should match* $(e^{At})'$.

Exercise 18.4.6 *Let*

$$A = \begin{pmatrix} -5 & 1 & 13 \\ 1 & -3 & -11 \\ 13 & -11 & -2 \end{pmatrix}$$

- *Find* e^{At}.
- *Find* $(e^{At})'$ *directly as it is a matrix of differentiable functions.*
- *Find* Ae^{At}. *This should match* $(e^{At})'$.

18.4.1 The Homogeneous Solution

Theorem 18.4.2 The General Solution to a Linear ODE System

Let A *be a* $n \times n$ *matrix. Then* $\Phi(t) = e^{At} \cdot C$ *is a solution to* $\boldsymbol{x}' = A\boldsymbol{x}$.

Proof 18.4.2
We note

$$\Phi'(t) = \frac{d}{dt}\left(e^{At} \cdot C\right)$$

and if we write e^{At} *using its column vectors*

$$e^{At} = [\Phi_1(t), \cdots, \Phi_n(t)]$$

then

$$\begin{aligned}
\frac{d}{dt}\left(e^{At}\cdot C\right) &= \frac{d}{dt}\left(\begin{bmatrix} \Phi_1(t) & \cdots & \Phi_n(t) \end{bmatrix}\begin{bmatrix} C_1 \\ \vdots \\ C_n \end{bmatrix}\right) \\
&= \frac{d}{dt}\left(\sum_{i=1}^{n} \Phi_i(t) C_i\right) = \sum_{i=1}^{n} \frac{d}{dt}\left(\Phi_i(t) C_i\right) \\
&= \sum_{i=1}^{n} C_i \Phi_i'(t) = \begin{bmatrix} \Phi_1'(t) & \cdots & \Phi_n'(t) \end{bmatrix}\begin{bmatrix} C_1 \\ \vdots \\ C_n \end{bmatrix} \\
&= \frac{d}{dt}\left(e^{At}\right)\cdot C = Ae^{At}\cdot C = A\Phi(t)
\end{aligned}$$

This shows Φ is a solution to the system. ■

We also know that $e^{At} = Pe^{Jt}P^{-1}C$ where J is the Jordan canonical form of A. For the structure of the Jordan Canonical Form of A, you can see its columns are linearly independent functions. Thus, letting the columns of e^{Jt} be $\Psi(t)$, we see

$$\begin{bmatrix} \Phi_1 & \ldots & \Phi_n \end{bmatrix} = P\begin{bmatrix} \Psi_1 & \ldots & \Psi_n \end{bmatrix} P^{-1}$$

To see the columns of e^{At} are linearly independent, consider the usual linear dependence equation

$$\alpha_1\Phi_1 + \ldots + \alpha_n\Phi_n = \mathbf{0} \implies \begin{bmatrix} \Phi_1 & \ldots & \Phi_n \end{bmatrix}\begin{bmatrix} \alpha_1 \\ \vdots \\ \alpha_n \end{bmatrix} = \mathbf{0}$$

This implies

$$P\begin{bmatrix} \Psi_1 & \ldots & \Psi_n \end{bmatrix} P^{-1}\begin{bmatrix} \alpha_1 \\ \vdots \\ \alpha_n \end{bmatrix} = \mathbf{0} \text{ or } \begin{bmatrix} \Psi_1 & \ldots & \Psi_n \end{bmatrix} P^{-1}\begin{bmatrix} \alpha_1 \\ \vdots \\ \alpha_n \end{bmatrix} = P^{-1}\mathbf{0} = \mathbf{0}$$

Since e^{Jt} has independent columns, the only solution here is

$$P^{-1}\begin{bmatrix} \alpha_1 \\ \vdots \\ \alpha_n \end{bmatrix} = \mathbf{0} \implies \begin{bmatrix} \alpha_1 \\ \vdots \\ \alpha_n \end{bmatrix} = \mathbf{0}$$

We conclude the functions Φ are independent. Since $\Phi_i' = A\Phi_i$, we also know each Φ is a solution to $\boldsymbol{x}' = A\boldsymbol{x}$. The solutions $\Phi = e^{At}C$ for arbitrary $C \in \Re^n$ are the span of the linearly independent functions Φ_1 to Φ_n. So the functions Φ_1 to Φ_n form a basis for the set of solutions to $\boldsymbol{x}' = A\boldsymbol{x}$.

When we search for solutions to $\boldsymbol{x}' = A\boldsymbol{x}$, we look for solutions of the form $\boldsymbol{x} = Ve^{rt}$ and we find the only scalars r that work are the eigenvalues of A. Of course, if a root is repeated the eigenspace for that eigenvalue need not have the same geometric multiplicity as its algebraic multiplicity. The structure of the Jordan Canonical block here shows us what happens. Any solution satisfies $\boldsymbol{x}' = P\,J\,P^{-1}\boldsymbol{x}$ and letting $\boldsymbol{y} = P^{-1}\boldsymbol{x}$, we see we are searching for solutions to $\boldsymbol{y}' = J\boldsymbol{y}$. Now focus on one JCB, say J_i. Then setting all variables in $\boldsymbol{y_i}$ to be zero except for the ones that concern

J_i, we look for solutions $\boldsymbol{y_i}' = J_i \boldsymbol{y_i}$ and we know we either get e^{λ_i} solutions with corresponding eigenvectors, or if the eigenspace is too small, we also get solutions such as $t^k e^{\lambda_i}$ for appropriate k. So the only solutions we can get to this system are linear combinations of the columns of e^{Jt}. Hence, any solution is a linear combination of the columns of e^{At}. We conclude the solution space of $\boldsymbol{x}' = A\boldsymbol{x}$ is the n dimensional vector space with basis $\{\Phi_1, \ldots, \Phi_n\}$ and any function in the span of this basis can be written as $e^{At}C$ for some $C \in \Re^n$. Let's summarize:

Theorem 18.4.3 The Solution Space to a Linear System of ODE

Let $S = \{\boldsymbol{y} \mid \boldsymbol{y}' = A\boldsymbol{y}\}$ Then

1. *S is an n-dimensional vector space and any basis for S consists of n linearly independent solutions to $\boldsymbol{x}' = A\boldsymbol{x}$.*
2. *The columns of e^{At} give a basis for S.*
3. *e^{At} is invertible and we write its inverse as e^{-At}.*
4. *$\Phi(t) = e^{A(t-t_0)}\boldsymbol{x_0}$ is the solution to $\boldsymbol{x}' = A\boldsymbol{x}$, $\boldsymbol{x}(t_0) = \boldsymbol{x_0}$.*

Proof 18.4.3

Most of this has already been proven. Note the columns of e^{At} are linearly independent and so it does have an inverse. Thus

$$e^{(A-A)t} = I \implies e^A\, e^{-A} = I \implies e^{-A} = (e^A)^{-1}$$

Next, note $\boldsymbol{x}' = A\boldsymbol{x}$, $\boldsymbol{x}(t_0) = \boldsymbol{x_0}$ has the general solution $\Phi(t) = e^{At}C$. Using the initial condition we can find

$$\Phi(t_0) = \boldsymbol{x_0} \quad = \quad e^{A\,t_0}C \implies C = e^{-A\,t_0}\boldsymbol{x_0}$$

So $\Phi(t) = e^{At}\left(e^{-At_0}\right)\boldsymbol{x_0} = e^{A(t-t_0)}\boldsymbol{x_0}$. ■

In general, if you can solve a linear ODE, the set of linearly independent solutions to the problem forms a finite dimensional vector space. One way to find all of these solutions is to find e^{At} but if you can find them in another way, you can form the matrix whose columns are the linearly independent solutions to the problem. We call this the **fundamental matrix** for the system. To help set this in place, let's look at second order systems.

Example 18.4.1

$$\begin{aligned} x'(t) &= -3\,x(t) + 4\,y(t) \\ y'(t) &= -1\,x(t) + 2\,y(t) \\ x(0) &= 2 \\ y(0) &= -4 \end{aligned}$$

First, note the matrix-vector form is

$$\begin{bmatrix} x'(t) \\ y'(t) \end{bmatrix} = \begin{bmatrix} -3 & 4 \\ -1 & 2 \end{bmatrix} \begin{bmatrix} x(t) \\ y(t) \end{bmatrix}.$$
$$\begin{bmatrix} x(0) \\ y(0) \end{bmatrix} = \begin{bmatrix} 2 \\ -4 \end{bmatrix}.$$

The coefficient matrix A is thus

$$A = \begin{bmatrix} -3 & 4 \\ -1 & 2 \end{bmatrix}.$$

The characteristic equation is thus

$$\det\left(r\,I - \begin{bmatrix} -3 & 4 \\ -1 & 2 \end{bmatrix}\right) = (r+2)(r-1) = 0.$$

The eigenvalues of the coefficient matrix A are $r_1 = -2$ *and* $r = 1$. *The general solution will then be of the form*

$$\begin{bmatrix} x(t) \\ y(t) \end{bmatrix} = a\,\boldsymbol{E}_1\,e^{-2t} + b\,\boldsymbol{E}_2\,e^{t}$$

The non-normalized **eigenvectors** *associated with these eigenvalues are*

$$\boldsymbol{E}_1 = \begin{bmatrix} 1 \\ 1/4 \end{bmatrix}, \quad \boldsymbol{E}_2 = \begin{bmatrix} 1 \\ 1 \end{bmatrix}$$

The general solution is therefore

$$\begin{aligned} \begin{bmatrix} x(t) \\ y(t) \end{bmatrix} &= a \begin{bmatrix} x_1(t) \\ y_1(t) \end{bmatrix} + b \begin{bmatrix} x_2(t) \\ y_2(t) \end{bmatrix} \\ &= a \begin{bmatrix} 1 \\ 1/4 \end{bmatrix} e^{-2t} + b \begin{bmatrix} 1 \\ 1 \end{bmatrix} e^{t}. \end{aligned}$$

and the fundamental matrix for this system is

$$\boldsymbol{\Theta}(t) = \left[\boldsymbol{E_1}e^{-2t} \quad \boldsymbol{E_2}e^{t}\right]$$

Of course $\boldsymbol{\Phi}(t) = e^{At}$ *is also a fundamental matrix.*

Let's do one more.

Example 18.4.2 *For the system below*

$$\begin{aligned} \begin{bmatrix} x'(t) \\ y'(t) \end{bmatrix} &= \begin{bmatrix} -20 & 12 \\ -13 & 5 \end{bmatrix} \begin{bmatrix} x(t) \\ y(t) \end{bmatrix} \\ \begin{bmatrix} x(0) \\ y(0) \end{bmatrix} &= \begin{bmatrix} -1 \\ 2 \end{bmatrix} \end{aligned}$$

Find the fundamental matrix $\boldsymbol{\Theta}(t)$

Solution *The characteristic equation is*

$$det\left(r \begin{bmatrix} 1 & 0 \\ 0 & 1 \end{bmatrix} - \begin{bmatrix} -20 & 12 \\ -13 & 5 \end{bmatrix}\right) = (r+8)(r+7) = 0$$

Hence, the **eigenvalues**, *of the characteristic equation are* $r_1 = -8$ *and* $r_2 = -7$. *The associated* **eigenvectors** *for these eigenvalues are*

$$\boldsymbol{E}_1 = \begin{bmatrix} 1 \\ 1 \end{bmatrix}, \quad \boldsymbol{E}_2 = \begin{bmatrix} 1 \\ 13/12 \end{bmatrix}$$

The general solution to our system is thus

$$\begin{bmatrix} x(t) \\ y(t) \end{bmatrix} = a \begin{bmatrix} 1 \\ 1 \end{bmatrix} e^{-8t} + b \begin{bmatrix} 1 \\ 13/12 \end{bmatrix} e^{-7t}$$

and the fundamental matrix for this system is

$$\boldsymbol{\Theta}(t) = \begin{bmatrix} \boldsymbol{E}_1 e^{-8t} & \boldsymbol{E}_2 e^{-7t} \end{bmatrix}$$

Also, $\boldsymbol{\Phi}(t) = e^{At}$ *is a fundamental matrix.*

18.4.2 Homework

Exercise 18.4.7 *Find the fundamental matrix* $\boldsymbol{\theta}(t)$ *for this system*

$$\begin{aligned} x' &= 3\,x + y \\ y' &= 5\,x - y \\ x(0) &= 4 \\ y(0) &= -6 \end{aligned}$$

Exercise 18.4.8 *Find the fundamental matrix* $\boldsymbol{\theta}(t)$ *for this system.*

$$\begin{aligned} x' &= x + 4\,y \\ y' &= 5\,x + 2\,y \\ x(0) &= 4 \\ y(0) &= -5 \end{aligned}$$

Exercise 18.4.9 *Find the fundamental matrix* $\boldsymbol{\theta}(t)$ *for this system.*

$$\begin{aligned} x' &= -3\,x + y \\ y' &= -4\,x + 2\,y \\ x(0) &= 1 \\ y(0) &= 6 \end{aligned}$$

Exercise 18.4.10 *Let*

$$A = \begin{pmatrix} -2 & 1 & 3 \\ 1 & 4 & -1 \\ 3 & -1 & 5 \end{pmatrix}$$

Consider the differential equation $\boldsymbol{x}' = A\boldsymbol{x}$ *where* $\boldsymbol{x} = \begin{bmatrix} x_1 \\ x_2 \\ x_3 \end{bmatrix}$.

- *Find* $\Phi(t) = e^{At}$.
- *Verify each column of* $\Phi(t)$ *solves* $\boldsymbol{x}' = A\boldsymbol{x}$.
- *Verify the columns of* $\Phi(t)$ *are linearly independent.*
- *Find* $(e^{At})'$ *directly as it is a matrix of differentiable functions.*
- *Find* Ae^{At}. *This should match* $(e^{At})'$.

- *If* $\boldsymbol{x}(0) = \begin{bmatrix} 1 \\ -2 \\ 4 \end{bmatrix}$, *verify* $\Phi(t)\boldsymbol{x}(0)$ *solves the initial value problem* $\boldsymbol{x}' = A\boldsymbol{x}$, $\boldsymbol{x}(0) = \begin{bmatrix} 1 \\ -2 \\ 4 \end{bmatrix}$ *by direct calculation.*

Exercise 18.4.11 *Let*

$$A = \begin{pmatrix} -2 & 1 \\ 1 & 4 \end{pmatrix}$$

Consider the differential equation $\boldsymbol{x}' = A\boldsymbol{x}$ *where* $\boldsymbol{x} = \begin{bmatrix} x_1 \\ x_2 \end{bmatrix}$.

- *Find* $\Phi(t) = e^{At}$.
- *Verify each column of* $\Phi(t)$ *solves* $\boldsymbol{x}' = A\boldsymbol{x}$.
- *Verify the columns of* $\Phi(t)$ *are linearly independent.*
- *Find* $(e^{At})'$ *directly as it is a matrix of differentiable functions.*
- *Find* Ae^{At}. *This should match* $(e^{At})'$.
- *If* $\boldsymbol{x}(0) = \begin{bmatrix} 3 \\ -2 \end{bmatrix}$, *verify* $\Phi(t)\boldsymbol{x}(0)$ *solves the initial value problem* $\boldsymbol{x}' = A\boldsymbol{x}$, $\boldsymbol{x}(0) = \begin{bmatrix} 3 \\ -2 \end{bmatrix}$ *by direct calculation.*

Exercise 18.4.12 *Let*

$$A = \begin{pmatrix} 1 & 1 & -3 \\ 1 & 2 & -1 \\ -3 & -1 & 3 \end{pmatrix}$$

Consider the differential equation $\boldsymbol{x}' = A\boldsymbol{x}$ *where* $\boldsymbol{x} = \begin{bmatrix} x_1 \\ x_2 \\ x_3 \end{bmatrix}$.

- *Find* $\Phi(t) = e^{At}$.
- *Verify each column of* $\Phi(t)$ *solves* $\boldsymbol{x}' = A\boldsymbol{x}$.
- *Verify the columns of* $\Phi(t)$ *are linearly independent.*
- *Find* $(e^{At})'$ *directly as it is a matrix of differentiable functions.*
- *Find* Ae^{At}. *This should match* $(e^{At})'$.
- *If* $\boldsymbol{x}(0) = \begin{bmatrix} 10 \\ -2 \\ 5 \end{bmatrix}$, *verify* $\Phi(t)\boldsymbol{x}(0)$ *solves the initial value problem* $\boldsymbol{x}' = A\boldsymbol{x}$, $\boldsymbol{x}(0) = \begin{bmatrix} 10 \\ -2 \\ 5 \end{bmatrix}$ *by direct calculation.*

Exercise 18.4.13 *Let*

$$A = \begin{pmatrix} -2 & 1 & 3 & -6 \\ 1 & 4 & 8 & 2 \\ 3 & 8 & 5 & 1 \\ -6 & 2 & 1 & 7 \end{pmatrix}$$

Consider the differential equation $\boldsymbol{x}' = A\boldsymbol{x}$ *where* $\boldsymbol{x} = \begin{bmatrix} x_1 \\ x_2 \\ x_3 \\ x_4 \end{bmatrix}$.

- *Find* $\Phi(t) = e^{At}$.
- *Verify each column of* $\Phi(t)$ *solves* $\boldsymbol{x}' = A\boldsymbol{x}$.
- *Verify the columns of* $\Phi(t)$ *are linearly independent.*
- *Find* $(e^{At})'$ *directly as it is a matrix of differentiable functions.*
- *Find* Ae^{At}. *This should match* $(e^{At})'$.
- *If* $\boldsymbol{x}(0) = \begin{bmatrix} 1 \\ -2 \\ 4 \\ -3 \end{bmatrix}$, *verify* $\Phi(t)\boldsymbol{x}(0)$ *solves the initial value problem* $\boldsymbol{x}' = A\boldsymbol{x}$, $\boldsymbol{x}(0) = \begin{bmatrix} 1 \\ -2 \\ 4 \\ -3 \end{bmatrix}$ *by direct calculation.*

18.5 The Non-Homogeneous Solution

Now let's look at what are called non-homogeneous systems. Now consider the non-homogeneous system

$$\boldsymbol{x}' = A\boldsymbol{x} + b(t), \quad \boldsymbol{x}(t_0) = \boldsymbol{x_0}$$

We say Ψ is a particular solution if we have $\boldsymbol{\Psi}' = A\boldsymbol{\Psi} + b(t)$. These are hard to find in practice, but we can find a general formula that is a good start. If we have a particular solution in hand, the general solution is then $\boldsymbol{\Phi}(t) = E^{A(t-t_0)}C + \Psi(t)$.

Theorem 18.5.1 A Particular Solution to a Linear ODE System

Let b be a continuous function. Then $\Psi(t) = e^{At}\int_{t_0}^{t} e^{-As}\, b(s)ds$ *is a particular solution to*

$$\boldsymbol{x}' = A\boldsymbol{x} + b(t)$$

Here, integration is defined on each component in the usual way. Hence, the solution to the system is

$$\Phi(t) = e^{A(t-t_0)}x_0 + \int_{t_0}^{t} e^{-A(t-s)}b(s)ds$$

Proof 18.5.1

This is a straightforward calculation. The general solution to the non-homogeneous system should be

$$\Phi(t) = e^{At}C + e^{At}\int_{t_0}^{t} e^{-As}b(s)ds$$

Using the initial condition we get that the actual solution to the above non-homogeneous system is

$$\Phi(t) = e^{A(t-t_0)}\boldsymbol{x_0} + \int_{t_0}^{t} e^{-A(t-s)} b(s)ds$$

where the second part holds since

$$e^{At}\int_{t_0}^{t} e^{-As} b(s)ds \quad = \quad \int_{t_0}^{t} e^{At}e^{-As} b(s)ds = \int_{t_0}^{t} e^{A(t-s)} b(s)ds$$

Note,

$$\begin{aligned}\Psi' &= Ae^{At}\int_{t_0}^{t} e^{-As} b(s)ds + e^{At}\left(\int_{t_0}^{t} e^{-As} b(s)ds\right)' \\ &= Ae^{At}\int_{t_0}^{t} e^{-As} b(s)ds + e^{At}\left(e^{-At} b(t)\right)\end{aligned}$$

as the Fundamental Theorem of Calculus holds for these vectors functions which is easily seen by looking at components. Therefore

$$\Psi' \quad = \quad Ae^{At}\int_{t_0}^{t} e^{-As} b(s)ds + b(t) = A\Psi + b(t)$$

The solution to the system is thus

$$\Phi(t) = e^{A(t-t_0)}x_0 + \int_{t_0}^{t} e^{-A(t-s)} b(s)ds$$

■

Comment 18.5.1 *We see then that if we were able to find e^{At} we would then be able to solve many of these linear systems!*

Finally, we can also find the solution using the fundamental matrix $\boldsymbol{\Theta}(t)$ if we can find the linearly independent solutions in some way. We find the particular solution can be written as follows:

Theorem 18.5.2 The Particular Solution to a Linear ODE System Using the Matrix of Solutions

Let b be a continuous function. Then $\boldsymbol{F}(t) = \boldsymbol{\Theta}(t)\int_{t_0}^{t}\boldsymbol{\Theta}^{-1}(s)\, b(s)ds$ is a particular solution to

$$\boldsymbol{x}' \quad = \quad A\boldsymbol{x} + b(t)$$

Here, integration is defined on each component in the usual way. The solution to the system is

$$x(t) = \boldsymbol{\theta}(t)\boldsymbol{\Theta}^{-1}(t_0)x_0 + \boldsymbol{\Theta}(t)\int_{t_0}^{t}\boldsymbol{\Theta}^{-1}(s)\, b(s)ds$$

Of course, the problem here is that finding the inverse $\boldsymbol{\Theta}^{-1}(t)$ in general is very hard! This is a good reason to favor the e^{At} approach as it has an easy inverse, $e{-}At$!

18.5.1 Homework

Exercise 18.5.1 *Find the fundamental matrix $\boldsymbol{\theta}(t)$ for this system*

$$\begin{aligned} x' &= 3\,x + y \\ y' &= 5\,x - y \\ x(0) &= 4 \\ y(0) &= -6. \end{aligned}$$

and then find the solution to the non-homogeneous problem with $f(t) = \begin{bmatrix} 2 \\ 5 \end{bmatrix}$ in integral form.

Exercise 18.5.2 *Find the fundamental matrix $\boldsymbol{\theta}(t)$ for this system.*

$$\begin{aligned} x' &= x + 4\,y \\ y' &= 5\,x + 2\,y \\ x(0) &= 4 \\ y(0) &= -5 \end{aligned}$$

and then find the solution to the non-homogeneous problem with $f(t) = \begin{bmatrix} 2 \\ 5 \end{bmatrix}$ in integral form.

Exercise 18.5.3 *Find the fundamental matrix $\boldsymbol{\theta}(t)$ for this system.*

$$\begin{aligned} x' &= -3\,x + y \\ y' &= -4\,x + 2\,y \\ x(0) &= 1 \\ y(0) &= 6 \end{aligned}$$

and then find the solution to the non-homogeneous problem with $f(t) = \begin{bmatrix} 2 \\ 5 \end{bmatrix}$ in integral form.

Exercise 18.5.4 *Let*

$$A = \begin{pmatrix} -2 & 1 & 3 \\ 1 & 4 & -1 \\ 3 & -1 & 5 \end{pmatrix}, \quad f(t) = \begin{pmatrix} 2t \\ -4t^2 \\ 2+4t \end{pmatrix}$$

Consider the differential equation $\boldsymbol{x}' = A\boldsymbol{x} + f$ with initial condition $\boldsymbol{x}(0) = \begin{bmatrix} 1 \\ -2 \\ 4 \end{bmatrix}$.

- *Find $\Phi(t) = e^{At}$.*
- *Write down the formula for the particular solution $\Psi(t)$ in integral form.*
- *Verify directly the particular solution satisfies the differential equation.*
- *Verify $e^{At}\boldsymbol{x}(\mathbf{0}) + \psi(t)$ solves the initial value problem.*

18.6 A Diagonalizable Test Problem

We can find e^{At} a variety of ways. Here is one that is different from what we have done before. Take

$$A = \begin{bmatrix} 0 & 1 \\ -2 & -3 \end{bmatrix}$$

Let's find e^{At}.

1. Find eigenvalues:

$$\begin{vmatrix} -r & 1 \\ -2 & 3-r \end{vmatrix} = 3r + r^2 - 2 = (r+2)(r+1) - 0$$

So eigenvalues are $r_1 = -2, r_2 = -1$.

2. The Cayley - Hamilton Theorem (which we have not discussed and which you can look up) tells us A satisfies the characteristic equation. Hence, we know $A^2 + 3A + 2I = \mathbf{0}$. So

$$\begin{aligned} A^2 &= -3A - 2I, \quad A^3 = A^2 A = -3A^2 - 2A = -3(-3A - 2I) - 2A = 7A - 6I \\ &\vdots \\ A^n &= a_n A + b_n\, I \end{aligned}$$

Now

$$\begin{aligned} e^A &= I + A + \frac{A^2}{2!} + \frac{A^3}{3!} + \cdots + \frac{A^n}{n!} + \cdots \\ &= I + A + \left(\frac{a_2}{2!}A + \frac{b_2}{2!}I\right) + \cdots + \left(\frac{a_n}{n!}A + \frac{b_n}{n!}I\right) + \cdots \\ &= \left(1 + \frac{b_2}{2!} + \frac{b_3}{3!} + \cdots\right) I + \left(1 + \frac{a_2}{2!} + \frac{a_3}{3!} + \cdots\right) A \end{aligned}$$

and we know that both of these sequences in parenthesis converge because e^A converges. So $e^A = \alpha I + \beta A$ for some α and β. Then e^{At} can be written as $e^{At} = \alpha(t) I + \beta(t) A$. Next,

$$\begin{bmatrix} 0 & 1 \\ -2 & -3 \end{bmatrix} = P \overbrace{\begin{bmatrix} -2 & 0 \\ 0 & -1 \end{bmatrix}}^{J} P^{-1}$$

From before we know that $e^{At} = Pe^{Jt}P^{-1}$, so

$$\begin{aligned} Pe^{Jt}P^{-1} &= P\alpha(t)P^{-1} + \beta(t)PJP^{-1} \\ P\begin{bmatrix} e^{-2t} & 0 \\ 0 & e^{-t} \end{bmatrix}P^{-1} &= P\left[\alpha(t)I + \beta(t)\begin{bmatrix} -2 & 0 \\ 0 & -1 \end{bmatrix}\right]P^{-1} \end{aligned}$$

i.e.

$$e^{-2t} = \alpha(t) - 2\beta(t) \qquad (1), \quad e^{-t} = \alpha(t) - \beta(t) \qquad (2)$$

Taking the negative of (2) we get

$$e^{-2t} = \alpha(t) - 2\beta(t), \quad -e^{-t} = -\alpha(t) + \beta(t)$$

Adding these we get

$$e^{-2t} - e^{-t} = -\beta(t) \qquad \text{or} \qquad \beta(t) = e^{-t} - e^{-2t}$$

So

$$\alpha(t) = \beta(t) + e^{-t} = 2e^{-t} - e^{-2t}$$

Therefore

$$\begin{aligned} e^{At} &= \left(2e^{-t} - e^{-2t}\right)\begin{pmatrix} 1 & 0 \\ 0 & 1 \end{pmatrix} + \left(e^{-t} - e^{-2t}\right)\begin{pmatrix} 0 & 1 \\ -2 & -3 \end{pmatrix} \\ &= \begin{pmatrix} 2e^{-t} - e^{-2t} & e^{-t} - e^{-2t} \\ -2e^{-t} + 2e^{-2t} & -e^{-t} + 2e^{-2t} \end{pmatrix} \end{aligned}$$

and we know that each of the columns in e^{At} should be solutions to the linear system of equations associated with the matrix A.

Now solve the system $\boldsymbol{x}' = A\boldsymbol{x}$. As mentioned before, both columns of e^{At} should be solutions to the above system. Let's check it for

$$\Phi_1 = \begin{pmatrix} 2e^{-t} - e^{-2t} \\ -2e^{-t} + 2e^{-2t} \end{pmatrix}$$

First,

$$\Phi_1' = \begin{pmatrix} -2e^{-t} + 2e^{-2t} \\ 2e^{-t} - 4e^{-2t} \end{pmatrix}$$

and second,

$$A\Phi_1 = \begin{pmatrix} 0 & 1 \\ -2 & -3 \end{pmatrix}\begin{pmatrix} 2e^{-t} - e^{-2t} \\ -2e^{-t} + 2e^{-2t} \end{pmatrix} = \begin{pmatrix} -2e^{-t} + 2e^{-2t} \\ 2e^{-t} - 4e^{-2t} \end{pmatrix}$$

So $\Phi_1' = A\Phi_1$ and Φ_1 is a solution to the system. In a similar fashion it can be shown that Φ_2 is also a solution. Thus e^{At} contains the fundamental solutions to the system.

18.6.1 Homework

Exercise 18.6.1 *Find e^{At} for this system using the Cayley - Hamilton technique. Compare to the matrix $\mathbf{theta}(t)$ we found earlier. What do you conclude?*

$$\begin{aligned} x' &= 3\,x + y \\ y' &= 5\,x - y \\ x(0) &= 4 \\ y(0) &= -6. \end{aligned}$$

Exercise 18.6.2 *Find e^{At} for this system using the Cayley - Hamilton technique. Compare to the matrix $\mathbf{theta}(t)$ we found earlier. What do you conclude?*

$$\begin{aligned} x' &= x + 4\,y \\ y' &= 5\,x + 2\,y \\ x(0) &= 4 \end{aligned}$$

$$y(0) = -5.$$

Exercise 18.6.3 *Find e^{At} for this system using the Cayley - Hamilton technique. Compare to the matrix* $\mathbf{theta}(t)$ *we found earlier. What do you conclude?*

$$\begin{aligned} x' &= -3\,x + y \\ y' &= -4\,x + 2\,y \\ x(0) &= 1 \\ y(0) &= 6. \end{aligned}$$

18.7 Simple Jordan Blocks

Now try finding e^{At} using this method for a matrix A which is similar to a Jordan Block. The matrix

$$A = \begin{pmatrix} 3 & 2 \\ -2 & -1 \end{pmatrix}$$

has Jordan Block

$$J = \begin{pmatrix} 1 & 1 \\ 0 & 1 \end{pmatrix}$$

This is a Jordan block and we know that for this matrix

$$e^{Jt} = \begin{pmatrix} e^t & te^t \\ 0 & e^t \end{pmatrix}$$

Now let's use the method above to find e^{At}.

1. Find eigenvalues: We know they are $r_1 = 1, r_2 = 1$.
2. Use the Cayley - Hamilton Theorem:

$$(A - I)(A - I) = 0 \Longrightarrow A^2 - 2A + I = 0 \Longrightarrow A^2 = 2A - I$$

3. $e^{At} = \alpha(t)I + \beta(t)A$

$$\begin{aligned} Pe^{J(t)}P^{-1} &= P[\alpha(t)I + \beta(t)J]P^{-1} \\ Pe^{\begin{bmatrix} 1 & 1 \\ 0 & 1 \end{bmatrix} t} P^{-1} &= P\left[\ \alpha(t)I + \beta(t)\begin{bmatrix} 1 & 1 \\ 0 & 1 \end{bmatrix}\right]P^{-1} \\ &= P\begin{bmatrix} \alpha(t) & 0 \\ 0 & \alpha(t) \end{bmatrix} + \begin{bmatrix} \beta(t) & \beta(t) \\ 0 & \beta(t) \end{bmatrix} P^{-1} \end{aligned}$$

So,

$$\begin{aligned} e^t &= \alpha(t) + \beta(t) && (1) \\ te^t &= \beta(t) && (2) \\ e^t &= \alpha(t) + \beta(t) && (3) \end{aligned}$$

From (2) we get $\beta(t) = te^t$, and using (2), (1) gives us $\alpha(t) = e^t - te^t$. Therefore,

$$e^{At} = (1 - t)e^t I + te^t A$$

$$\begin{aligned}
&= (1-t)e^t \begin{pmatrix} 1 & 0 \\ 0 & 1 \end{pmatrix} + te^t \begin{pmatrix} 1 & 1 \\ 0 & 1 \end{pmatrix} \\
&= \begin{pmatrix} (1-t)e^t & 0 \\ 0 & (1-t)e^t \end{pmatrix} + \begin{pmatrix} te^t & te^t \\ 0 & te^t \end{pmatrix} \\
&= \begin{pmatrix} e^t & te^t \\ 0 & e^t \end{pmatrix}
\end{aligned}$$

and this is what we expected. So this method works for matrices having repeated eigenvalues and for matrices that are not diagonalizable. Again, note that we can find e^{At} and we never have to find P or P^{-1}.

18.7.1 Homework

Exercise 18.7.1 *For the system*

$$\begin{aligned}
x'(t) &= 3x(t) + y(t) \\
y'(t) &= -x(t) + y(t) \\
x(0) &= 2 \\
y(0) &= 3
\end{aligned}$$

find e^{At} *by finding the Jordan block.*

Exercise 18.7.2 *For the system*

$$\begin{aligned}
x'(t) &= 5\,x(t) + 4\,y(t) \\
y'(t) &= -16\,x(t) - 11\,y(t) \\
x(0) &= 5 \\
y(0) &= -6
\end{aligned}$$

find e^{At} *by finding the Jordan block.*

Exercise 18.7.3 *For the system*

$$\begin{aligned}
x'(t) &= 5\,x(t) - 9\,y(t) \\
y'(t) &= 4\,x(t) - 7\,y(t) \\
x(0) &= 10 \\
y(0) &= -20
\end{aligned}$$

find e^{At} *by finding the Jordan block.*

Exercise 18.7.4 *For the system*

$$\begin{aligned}
x'(t) &= 2\,x(t) + y(t) \\
y'(t) &= -4\,x(t) + 6\,y(t) \\
x(0) &= -4 \\
y(0) &= -12
\end{aligned}$$

find e^{At} *by finding the Jordan block.*

Chapter 19

Nonlinear Parametric Optimization Theory

Let's look at some optimization problems to see how the material we have been working through applies. We will not have much homework here as this is mostly a long conversation. But we hope to make you think a bit more deeply about optimization using our tools. To do this better, you need to take a nice theoretical course in differential equations and optimization!

First, here is a very relaxed way to think about a classical optimization problem. Consider the task of

$$\begin{aligned}
&\min \int_0^t f_o(s, x(s), x'(s), \mu(s)) \text{ subject to} \\
&x'(t) = f(t, x(t), \mu(t)), \quad 0 \leq t \leq T \\
&x(0) = x_0
\end{aligned}$$

As stated, it is not very clear what is going on. We are supposed to find a minimum value of an integral, so there should be some conditions of the integrand f_o to ensure $f_o(s, x(s), \dot{x}(s), \mu(s))$ is an integrable function. This problem probably makes sense even if the arguments of f_0 are vectors too. We clearly need to restrict our attention to some class of interesting functions. For example, we could rephrase the problem as

$$\begin{aligned}
&\min_{x, x', \mu \in C^1([0,T])} \int_0^t f_o(s, x(s), x'(s), \mu(s)) \text{ subject to} \\
&x'(t) = f(t, x(t), \mu(t)), \quad 0 \leq t \leq T \\
&x(0) = x_0
\end{aligned}$$

This still does not say anything about f_0. A more careful version would be as follows.

Let's assume $f_0 : [0, t] \times D \subset \Re^n \times \Re^n \times \Re^n \rightarrow \Re$ is continuous on D with continuous partials on D. The minimization problem is

$$\begin{aligned}
&\min_{\boldsymbol{x}, \boldsymbol{x}', \boldsymbol{\mu} \in C^1([0,T])} \int_0^t f_o(s, \boldsymbol{x}(s), \boldsymbol{x}'(s), \boldsymbol{\mu}(s)) \text{ subject to} \\
&\boldsymbol{x}'(t) = f(t, \boldsymbol{x}(t), \boldsymbol{\mu}(t)), \quad 0 \leq t \leq T \\
&\boldsymbol{x}(0) = \boldsymbol{x_0}
\end{aligned}$$

Here is a specific one.

Example 19.0.1

$$\min \int_0^1 \overbrace{[\underbrace{x^2(s) + (x'(s))^2}_{\text{like energy}} + \underbrace{\mu^2(s)}_{\text{control cost}}]}^{f_0-\text{not dependent on time}}\, ds$$

subject to

$$\begin{aligned} x'(t) &= x(t) + \mu(t) + \underbrace{e^{-t}}_{\text{damping term}} \qquad 0 \leq t \leq T \\ x(0) &= 1 \end{aligned}$$

We call x the state variable and μ the control variable. It's possible that we could have a control constraint which bounds the control. So we could have something like $-1 \leq \mu(t) \leq 1$. We could also have a state constraint which says that $x(t)$ lives in a certain set; i.e. $x(t) \in C$ for some set C. Hence, these problems can be pretty hard.

19.1 The More Precise and Careful Way

Let T be a fixed real number and define some underlying spaces.

- The **control space** is $\mathcal{U} = \{\boldsymbol{\mu} : [0,T] \rightarrow \Re^m | \mu$ has some properties$\}$. These properties might be the control functions are continuous, piecewise continuous (meaning they can have a finite number of discontinuities), piecewise differentiable (meaning they can have a finite number of points where the right- and left-hand derivatives exist but don't match) and so forth. Each function in the control space is called a **control**.

- The **State Space** is $X = \{\boldsymbol{x} : [0,T] \rightarrow \Re^n \mid \boldsymbol{x}'$ continuous on $[0,T]\}$. This is the same as $C^1([0,T])$. Each function in the state space is called a **state**.

Let $f_0 : [0,T] \times \Re^n \times \Re^n \times \Re^m \rightarrow \Re$ and assume f_0 is continuous on $[0,T] \times \Re^n \times \Re^n \times \Re^m$ i.e., $\forall\ \epsilon > 0,\ \exists\ \delta > 0$ so that

$$\|(t, \boldsymbol{u}, \boldsymbol{v}, \boldsymbol{w}) - (t_0, \boldsymbol{u}_0, \boldsymbol{v}_0, \boldsymbol{w}_0)\| < \delta \implies |f_0(t, \boldsymbol{u}, \boldsymbol{v}, \boldsymbol{w}) - f_0(t_0, \boldsymbol{u}_0, \boldsymbol{v}_0, \boldsymbol{w}_0)| < \epsilon$$

The norm $\|\cdot\|$ here is

$$\|(t, \boldsymbol{u}, \boldsymbol{v}, \boldsymbol{w}) - (t_0, \boldsymbol{u}_0, \boldsymbol{v}_0, \boldsymbol{w}_0)\| = \sqrt{|t - t_0|^2 + \|\boldsymbol{u} - \boldsymbol{u}_0\|^2_{\Re^n} + \|\boldsymbol{v} - \boldsymbol{v}_0\|^2_{\Re^n} + \|\boldsymbol{w} - \boldsymbol{w}_0\|^2_{\Re^m}}$$

This means that if $\boldsymbol{x}(t)$, $\boldsymbol{x}'(t)$ and $\boldsymbol{\mu}(t)$ are continuous functions on $[0,T]$, then since the composition of continuous functions is continuous, $f(t, \boldsymbol{x}(t), \boldsymbol{x}'(t), \boldsymbol{\mu}(t))$ will be continuous on $[0,T]$. Let's choose $\mathcal{U} = PC([0,T])$, the set of piecewise continuous functions on $[0,T]$. We know that even if $\mu(t)$ is a piecewise continuous function the above will still hold except at the places where the control is not continuous because there is only a finite number of discontinuities. Hence, f_0 will also be piecewise continuous. Now the ODE is

$$\begin{bmatrix} x_1' \\ \vdots \\ x_n' \end{bmatrix} = \begin{bmatrix} f_1(t, \boldsymbol{x}(t), \boldsymbol{\mu}(t)) \\ \vdots \\ f_n(t, \boldsymbol{x}(t), \boldsymbol{\mu}(t)) \end{bmatrix}, \quad 0 \leq t \leq T, \quad \begin{bmatrix} x_1(0) \\ \vdots \\ x_n(0) \end{bmatrix} = \begin{bmatrix} x_{01} \\ \vdots \\ x_{0n} \end{bmatrix}$$

or $\boldsymbol{x}' = f(t, \boldsymbol{x}(t), \boldsymbol{\mu}(t))$. $0 \leq t \leq T$, $\boldsymbol{x}(0) = \boldsymbol{x_0}$ where $f : [0, T] \times \Re^n \times \Re^m \rightarrow \Re$ and f is continuous on $[0, T] \times \Re^n \times \Re^m$.

Now, how do we describe the minimization? Let

$$\hat{J}(\boldsymbol{x_0}, \boldsymbol{\mu}) = \int_0^T f_0(s, \boldsymbol{x}(s), \boldsymbol{x}'(s), \boldsymbol{\mu}(s))ds$$

where μ is a specific control function and $\boldsymbol{x}$ is the solution to the ODE for that control function choice. Hence, if there is such a solution and the solution is at least piecewise continuous, then for the piecewise continuous control $\boldsymbol{\mu}$, the integrand is piecewise continuous and so the integral exists. But perhaps the ODE does not have a continuous solution for a given control $\boldsymbol{\mu}$ which in other courses we would find would mean the solution $\boldsymbol{x}$ would have $\|\boldsymbol{x}\| \rightarrow \infty$. In that case, the integral could go unbounded. Hence, in general, $\hat{J} : \Re^n \times \mathcal{U} \rightarrow \Re \bigcup \{\pm\infty\}$. We want to find

$$J(x_0) = \inf_{\boldsymbol{\mu} \in \mathcal{U}} \hat{J}(\boldsymbol{x_0}, \boldsymbol{\mu})$$

This function is called the **Optimal Value Function** and understanding its properties is very important in many application areas. For example, a really good question would be **how does the optimal value function vary when the initial data x_0 changes?** So our full optimization problem is

$$\begin{aligned} &\inf_{\boldsymbol{\mu} \in \mathcal{U}} \hat{J}(\boldsymbol{x_0}, \boldsymbol{\mu}), \text{ subject to} \\ &\boldsymbol{x}' = f(t, \boldsymbol{x}(t), \boldsymbol{\mu}(t)), \quad 0 \leq t \leq T \\ &\boldsymbol{x}(0) = \boldsymbol{x_0} \end{aligned}$$

or

$$\begin{aligned} &\inf_{\boldsymbol{\mu} \in \mathcal{U}} \int_0^T f_0(s, \boldsymbol{x}(s), \boldsymbol{x}'(s), \boldsymbol{\mu}(s))ds, \text{ subject to} \\ &\boldsymbol{x}' = f(t, \boldsymbol{x}(t), \boldsymbol{\mu}(t)), \quad 0 \leq t \leq T \\ &\boldsymbol{x}(0) = \boldsymbol{x_0} \end{aligned}$$

The most general way to write this is: Find

$$J(t_0, x_0) = \inf_{\boldsymbol{\mu} \in \mathcal{U}} \hat{J}(t_0, x_0, \boldsymbol{\mu})$$

where

$$\hat{J}(t_0, x_0, \mu) = \ell(t_0, \boldsymbol{x_0}) + \int_0^T f_0(s, \boldsymbol{x}(s), \boldsymbol{x}'(s), \boldsymbol{\mu}(s))ds$$

subject to

$$\begin{aligned} \boldsymbol{x}' &= f(t, \boldsymbol{x}(t), \boldsymbol{\mu}(t)), \quad 0 \leq t \leq T \\ \boldsymbol{x}(t_0) &= \boldsymbol{x_0} \\ \boldsymbol{x}(t) &\in \Omega(t, \boldsymbol{\mu}(t)) \\ \boldsymbol{\mu}(t) &\in \Lambda(t, \boldsymbol{x}(t)) \end{aligned}$$

with $\ell(t_0, \boldsymbol{x_0})$ a penalty on initial conditions. For example, this could be some random noise added from some probability distribution. The set Ω is a constraint set on the values the state can take at time t given the value of the control at the time. The set Λ is a constraint set on the values the control is permitted to have at time t given the state value at that time. As you may imagine, adding state and control constraints makes this problem much harder to find solutions to. We would like to see that $J(t_0, \boldsymbol{x_0})$ is a finite number. So we want to find a control which actually allows us to achieve the **infimum**. This is a very hard problem both theoretically and practically.

Example 19.1.1 A Constrained Optimal Control Problem
Consider this problem which has constraints on the control.

$$inf \int_0^1 \left[\frac{1}{2}\boldsymbol{x}^T Q\boldsymbol{x} + \frac{1}{2}(\boldsymbol{x}')^T R\boldsymbol{x}' + \frac{1}{2}\boldsymbol{\mu}^T S\boldsymbol{\mu}\right] ds$$

Subject to:

$$\begin{aligned} \boldsymbol{x}' &= A\boldsymbol{x} + B\boldsymbol{\mu} \\ -1 &\leq \mu \leq 1 \\ \boldsymbol{x}(t_0) &= \boldsymbol{x_0} \end{aligned}$$

The idea here is to take a control from any space that we want. Then we solve the differential equation, hopefully obtaining a solution, and then use this solution along with the control to find a value for the integral. Our hope is that there is a control which gives us a smallest value for the integral. For the above problem, the solution to the differential equation is given by

$$\begin{aligned} \boldsymbol{x}(t) &= e^{A(t-t_0)}\boldsymbol{x_0} + e^{At}\int_{t_0}^t e^{-As}\, B\,\boldsymbol{\mu}(s)ds \\ &= e^{A(t-t_0)}x_0 + \int_{t_0}^t e^{A(t-s)}B\mu(s)ds. \end{aligned}$$

Now how big is $\boldsymbol{x}(t)$?

$$\|\boldsymbol{x}(t)\| \leq \|e^{A(t-t_0)}\|\,\|\boldsymbol{x_0}\| + \int_{t_0}^t \|e^{A(t-s)}\|\,\|B\|\,\|\boldsymbol{\mu}(s)\|ds$$

We get a nice bound on the first term if all of the eigenvalues have a negative real part. We know $e^{At} = Pe^{Jt}P^{-1}$, so the norm of the exponential matrix is determined by the structure of the Jordan Canonical Blocks. Any eigenvalue which is positive or has positive real part will give us a norm which is unbounded over long time scales. We encourage you to work this out. Hence, we value problems where we can guarantee the eigenvalues of A have negative real parts. And, provided our control is bounded, we can bound the second term. So we can get a bound on $\boldsymbol{x}(t)$ in these cases. We get a bound even if the eigenvalues have positive real parts but then the norm of the solution can get arbitrarily large, which is not what we usually want.

Now, in general, we can write the solution as

$$\boldsymbol{x}(t) = \Phi(t)\Phi^{-1}(t_0)\boldsymbol{x_0} + \Phi(t)\int_{t_0}^t \Phi^{-1}(s)B\mu(s)ds$$

where $\Phi(t)$ is the fundamental matrix composed of linearly independent solutions. For nonlinear problems this is very difficult to find and it is here that the problem lies.

Comment 19.1.1 *This control problem is called an* **LQR** *problem, where* **LQR** *stands for* **Linear Quadratic regulator**. *Often there are state variables $\boldsymbol{x}$ and variables $\boldsymbol{y}$ which are actually observed, called* **observed variables**. *This simply adds an extra constraint of the form $\boldsymbol{y} = C\boldsymbol{x}$.*

19.2 Unconstrained Parametric Optimization

Given $L : \Re^n \to \Re$, the problem is to find $\inf_{x \in \Re^n} L(x)$. We have already discussed such extremum problems, so we will be brief here. Under sufficient smoothness conditions, if x_0 is a local minimum, then

$$\nabla L(\boldsymbol{x_0}) = \begin{bmatrix} \frac{\partial L}{\partial x_1}(\boldsymbol{x_0}) \\ \vdots \\ \frac{\partial L}{\partial x_n}(\boldsymbol{x_0}) \end{bmatrix} = \mathbf{0}$$

The directional derivative of L in the direction of the unit vector $\boldsymbol{u}$ from a point $\boldsymbol{x}$ is

$$D_u L(x) = \left\langle \nabla L(\boldsymbol{x}), \boldsymbol{u} \right\rangle = \left(\nabla L(\boldsymbol{x}) \right)^T \boldsymbol{u} = \begin{bmatrix} \frac{\partial L}{\partial x_1}(\boldsymbol{x}) \\ \vdots \\ \frac{\partial L}{\partial x_n}(\boldsymbol{x}) \end{bmatrix}^T \begin{bmatrix} u_1 \\ \vdots \\ u_n \end{bmatrix}$$

where $u_1^2 + \cdots + u_n^2 = 1$.

In two dimensions, $\Re^2$, assume L has a local minimum at $\vec{x}_0 = \begin{bmatrix} x_{01} \\ x_{02} \end{bmatrix}$, i.e. $\exists\ r > 0$ such that if $x \in B(x_0; r)$, then $L(x) \geq L(x_0)$; a generic illustration of this is shown in Figure 19.1. This

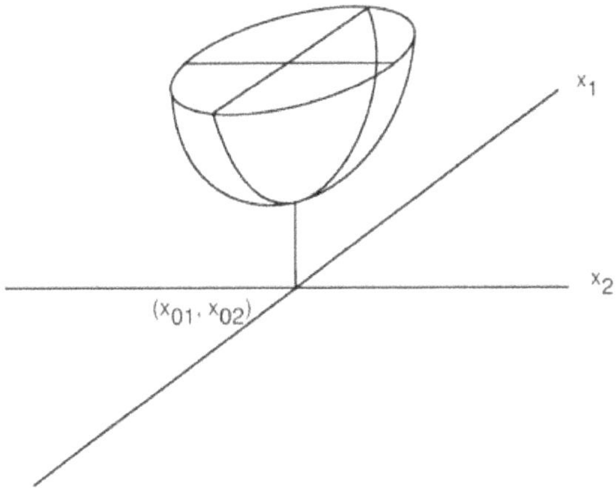

Figure 19.1: Two dimensional local minimum.

looks like a paraboloid locally. In fact, if we look at the cross section of this surface through the line L_θ, $0 \leq \theta \leq 2\pi$, we obtain the two dimensional slice shown in Figure 19.2. Now restrict your attention to the trace in the L surface above this line lying inside the cylinder of radius $\frac{r}{2}$ about x_0 as shown in Figure 19.3. Then for a given θ, we obtain the parameterized curve given by $x_3(t) = L(x_1, x_2) = L(x_{01} + t\cos(\theta), x_{02} + t\sin(\theta))$ for $-\frac{r}{2} < t < \frac{r}{2}$. This is illustrated in Figure 19.4. Now $\boldsymbol{x} = \begin{bmatrix} x_1 \\ x_2 \end{bmatrix}$, so

$$\frac{dx_3}{dt} = \left\langle \nabla L(\boldsymbol{x}), \begin{bmatrix} \cos\theta \\ \sin\theta \end{bmatrix} \right\rangle = \frac{\partial L}{\partial x_1}(\boldsymbol{x})\cos\theta + \frac{\partial L}{\partial x_2}(\boldsymbol{x})\sin\theta.$$

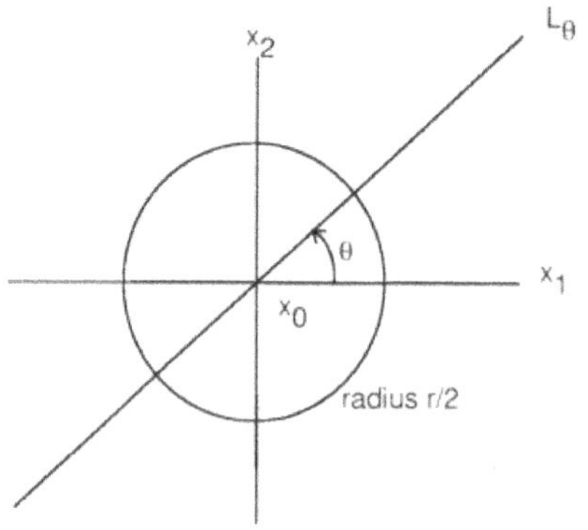

Figure 19.2: L_θ slice.

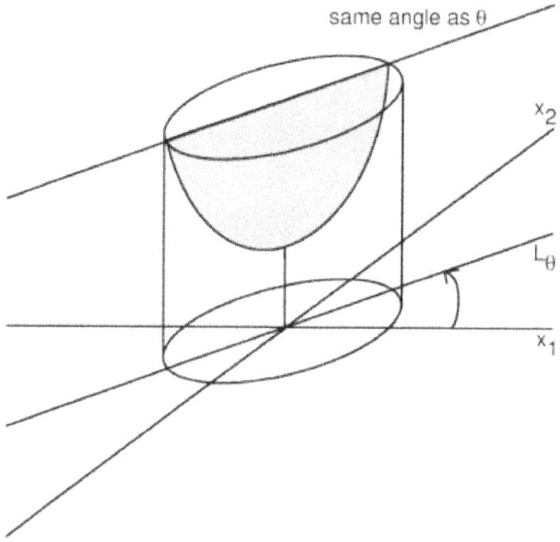

Figure 19.3: Local cylinder.

Then

$$\begin{aligned}\frac{d^2x_3}{dt^2} &= \cos\theta L_{x_1,x_1}(\boldsymbol{x})\,\frac{\partial x_1}{\partial t} + \cos\theta L_{x_1,x_2}(\boldsymbol{x})\,\frac{\partial x_2}{\partial t}\\ &\quad + \sin\theta L_{x_2,x_1}(\boldsymbol{x})\frac{\partial x_1}{\partial t} + \sin\theta L_{x_2,x_2}(\boldsymbol{x})\,\frac{\partial x_2}{\partial t}\\ &= \cos^2\theta L_{x_1,x_1}(\boldsymbol{x}) + 2\sin\theta\cos\theta L_{x_1,x_2}(\boldsymbol{x}) + \sin^2\theta L_{x_2,x_2}(\boldsymbol{x})\end{aligned}$$

where we assume the mixed order partials match. Of course, if we assume the partials are continuous locally this is true. From this, we get the general quadratic form

$$\frac{d^2x_3}{dt^2} = \begin{bmatrix}\cos\theta\\ \sin\theta\end{bmatrix}^T \boldsymbol{H_L}(\boldsymbol{x})\begin{bmatrix}\cos\theta\\ \sin\theta\end{bmatrix}.$$

Hence, at the local minimum $\boldsymbol{x_0}$,

$$\frac{d^2x_3}{dt^2} = \begin{bmatrix}\cos\theta\\ \sin\theta\end{bmatrix}^T \boldsymbol{H_L}(\boldsymbol{x_0})\begin{bmatrix}\cos\theta\\ \sin\theta\end{bmatrix}.$$

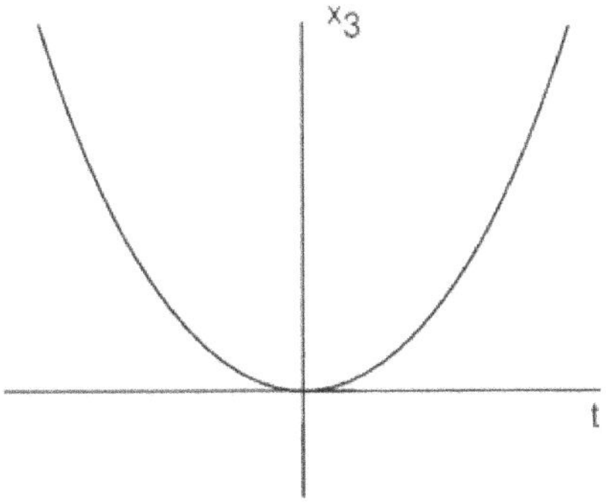

Figure 19.4: Local cylinder cross section.

Since $\boldsymbol{x_0}$ is a local minimum

$$\frac{d^2 x_3}{dt^2} = \boldsymbol{E}_\theta^T \boldsymbol{H_L}(\boldsymbol{x_0}) \boldsymbol{E}_\theta > 0 \qquad \forall\ 0 < \theta < 2\pi$$

where

$$\boldsymbol{E}_\theta = \begin{bmatrix} \cos\theta \\ \sin\theta \end{bmatrix}$$

Any vector in $\Re^2$ can be written as $r\boldsymbol{E_\theta}$, so we have shown

$$\boldsymbol{x}^T \boldsymbol{H_L}(\boldsymbol{x_0}) \boldsymbol{x} > 0 \qquad \forall\ \boldsymbol{x} \in \Re^2$$

i.e. $\boldsymbol{H_L}(\boldsymbol{x_0})$ is a positive definite matrix implying it is diagonalizable and has determinant greater than zero. We note that at $\theta = 0$ we get $L_{x_1,x_1}(\boldsymbol{x_0}) > 0$. So for the matrix $\boldsymbol{H_L}(\boldsymbol{x_0})$ we know at a minimum:

1. The first principle minor is $L_{x_1,x_1}(\boldsymbol{x_0})$ and its determinant is positive.
2. The next principle minor is $\boldsymbol{H_L}(\boldsymbol{x_0})$ itself and since it is positive definite symmetric we know its determinant is positive.

So at the local minimum $\boldsymbol{x_0}$, we expect the following behavior:

- $\boldsymbol{\nabla} L(\boldsymbol{x_0}) = \boldsymbol{0}$.
- $\boldsymbol{H_L}(\boldsymbol{x_0})$ is symmetric positive definite.
- The first principle minor has positive determinant.
- The second principle minor has positive determinant.

These are all results we have already worked out, but here we used a different method of attack. For this $\Re^2$ case, we have

$$\frac{d^2 x_3}{dt^2} = \cos^2\theta f_{x_1,x_1}(\boldsymbol{x_0}) + 2\sin\theta\cos\theta f_{x_1,x_2}(\boldsymbol{x_0}) + \sin^2\theta f_{x_2,x_2}(\boldsymbol{x_0}) > 0 \qquad \forall\ 0 < \theta < 2\pi$$

Using $\alpha = \cos\theta$ and $\beta = \sin\theta$, rewrite this as

$$\frac{\alpha^2}{\alpha^2+\beta^2} f_{x_1,x_1}(\boldsymbol{x_0}) + \frac{2\alpha\beta}{\alpha^2+\beta^2} f_{x_1,x_2}(\boldsymbol{x_0}) + \frac{\beta^2}{\alpha^2+\beta^2} f_{x_2,x_2}(\boldsymbol{x_0}) > 0 \qquad \forall\ \alpha, \beta$$

So

$$\alpha^2 f_{x_1,x_1}(\boldsymbol{x_0}) + 2\alpha\beta f_{x_1,x_2}(\boldsymbol{x_0}) + \beta^2 f_{x_1,x_1}(\boldsymbol{x_0}) > 0 \qquad \forall\ \alpha, \beta$$

Now factoring out $f_{x_1,x_1}(\boldsymbol{x_0})$ and completing the square we get the following:

$$f_{x_1,x_1}(\boldsymbol{x_0}) \left[\alpha^2 + \frac{2\beta f_{x_1,x_2}(\boldsymbol{x_0})}{f_{x_1,x_1}(\boldsymbol{x_0})} + \frac{\beta^2 (f_{x_1,x_1}(\boldsymbol{x_0}))^2}{(f_{x_1,x_1}(\boldsymbol{x_0}))^2}\right] - \frac{\beta^2 (f_{x_1,x_2}(\boldsymbol{x_0}))^2}{f_{x_1,x_1}(\boldsymbol{x_0})} + \beta^2 f_{x_2,x_2}(\boldsymbol{x_0}) \ > \ 0$$

Simplifying,

$$f_{x_1,x_1}(\boldsymbol{x_0}) \left[\alpha + \frac{\beta f_{x_1,x_2}(\boldsymbol{x_0})}{f_{x_1,x_1}(\boldsymbol{x_0})}\right]^2 + \beta^2 \left[f_{x_2,x_2}(\boldsymbol{x_0}) - \frac{(f_{x_1,x_2}(\boldsymbol{x_0}))^2}{f_{x_1,x_1}(\boldsymbol{x_0})}\right] \ > \ 0$$

Thus

$$f_{x_1,x_1}(\boldsymbol{x_0}) \left[\alpha + \frac{\beta f_{x_1,x_2}(\boldsymbol{x_0})}{f_{x_1,x_1}(\boldsymbol{x_0})}\right]^2 + \beta^2 \left[\frac{f_{x_1,x_1}(\boldsymbol{x_0})\, f_{x_2,x_2}(\boldsymbol{x_0}) - (f_{x_1,x_2}(\boldsymbol{x_0}))^2}{f_{x_1,x_1}(\boldsymbol{x_0})}\right] \ > \ 0$$

Since we know $f_{x_1,x_1}(\boldsymbol{x_0}) > 0$, this forces

$$f_{x_1,x_1}(\boldsymbol{x_0})\, f_{x_2,x_2}(\boldsymbol{x_0}) - (f_{x_1,x_2}(\boldsymbol{x_0}))^2 \ > \ 0$$

This is precisely the determinant of the second minor which we expected from our theory to be positive at a minimum. You should also go back and look at the argument for the $\Re^2$ case and $\Re^n$ cases we have already presented to help you see the big picture.

19.3 Constrained Parametric Optimization

Consider the problem:

$$\inf_{\boldsymbol{\mu} \in \Re^m} L(\boldsymbol{x}, \boldsymbol{\mu})$$

subject to

$$f(\boldsymbol{x}, \boldsymbol{\mu}) = \boldsymbol{0}$$

where $f(\boldsymbol{x}, \boldsymbol{\mu}) = \boldsymbol{0}$ is the same as saying

$$\begin{aligned} f_1(x, \mu) &= 0 \\ &\vdots \\ f_n(x, \mu) &= 0 \end{aligned}$$

where $\boldsymbol{x} \in \Re^n$ is the **state**, $\boldsymbol{\mu} \in \Re^m$ is the **control** and $L : \Re^n \times \Re^m \rightarrow \Re$ is the **performance index**. The function $f : \Re^n \times \Re^m \rightarrow \Re^n$ is the **constraint**. We assume the functions f and L have sufficient smoothness.

Example 19.3.1

$$\inf_{\mu \in \Re} x^2 + \mu^2$$

subject to

$$x^2 = 2\mu$$

The Figures 19.5 and 19.6 show what is happening.

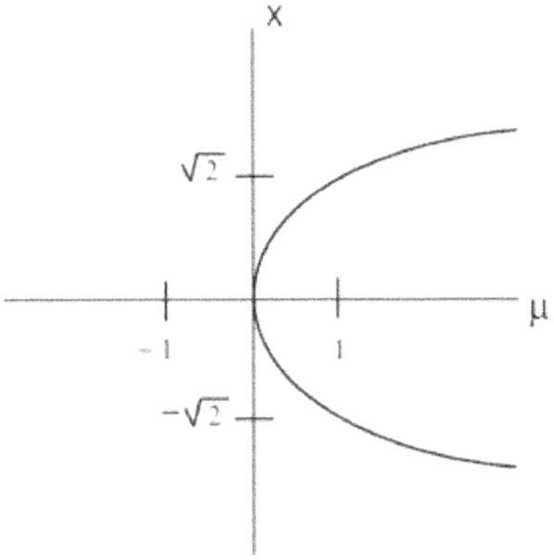

Figure 19.5: Minimize $x^2 + \mu^2$ subject to $x^2 = 2\mu$.

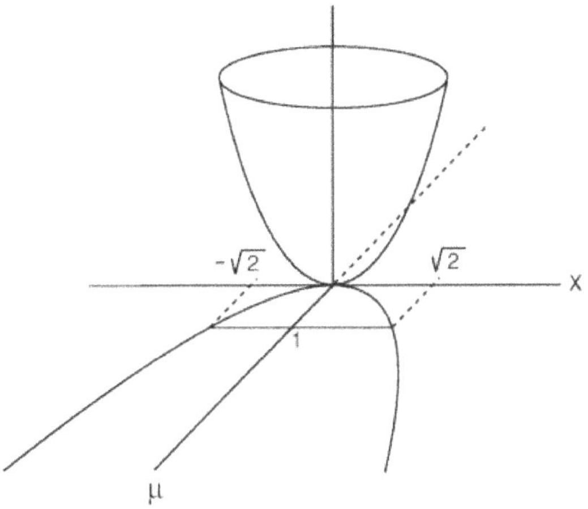

Figure 19.6: Solution to minimize $x^2 + \mu^2$ Subject to $x^2 = 2\mu$.

Here the minimum is clearly at $(0, 0)$. *The idea here is to pick* μ *and then find an* x *which satisfies the given constraints. This is done until the minimum of* $L(x, \mu)$ *over all* μ *is found. Note: Here* x *and* μ *are* **not** *independent of one another. They may be linked in the constraints somehow.*

Let's look at the general problem in detail. We start by looking at Hessian approximations in detail. We want to learn how to use them to derive estimates.

19.3.1 Hessian Error Estimates

To save some typing, we will start using the superscript $()^o$ to indicate an expression evaluated at the point $(\boldsymbol{x}_0, \boldsymbol{\mu}_0)$. The superscript $()^\theta$ refers to the expression evaluated at the point on the line segment $[(\boldsymbol{x}_0, \boldsymbol{\mu}_0), (\boldsymbol{x}_0 + \Delta\boldsymbol{x}, \boldsymbol{\mu}_0 + \Delta\boldsymbol{\mu})]$. In general, given a matrix A, we know that

$$\|A\boldsymbol{x}\| \leq \|A\|_{Fr}\,\|\boldsymbol{x}\|$$

and so

$$\|\boldsymbol{x}^T A\boldsymbol{x}\| = |<\boldsymbol{x}, A\boldsymbol{x}>| \leq \|\boldsymbol{x}\|\,\|A\boldsymbol{x}\| \leq \|A\|_{Fr}\,\|\boldsymbol{x}\|^2$$

Now consider the Hessian approximations of the functions L and f_i.

$$\begin{aligned} L(\boldsymbol{x_0} + \Delta\boldsymbol{x}, \boldsymbol{\mu_0} + \Delta\boldsymbol{\mu}) &= L^0 + (\boldsymbol{\nabla}_x L^0)^T \Delta x + (\boldsymbol{\nabla}_\mu L^0)^T \Delta\mu \\ &\quad + \frac{1}{2} \begin{bmatrix} \Delta\boldsymbol{x} \\ \Delta\boldsymbol{\mu} \end{bmatrix}^T \begin{bmatrix} \boldsymbol{H_L}^{\theta}_{x,x} & \boldsymbol{H_L}^{\theta}_{x,\mu} \\ \boldsymbol{H_L}^{\theta}_{\mu,x} & \boldsymbol{H_L}^{\theta}_{\mu,\mu} \end{bmatrix} \begin{bmatrix} \Delta\boldsymbol{x} \\ \Delta\boldsymbol{\mu} \end{bmatrix} \end{aligned}$$

where

$$\boldsymbol{H_L}_{x,x} = \begin{bmatrix} L_{x_1,x_1} & \dots & L_{x_1,x_n} \\ \vdots & & \vdots \\ L_{x_n,x_1} & \dots & L_{x_n,x_n} \end{bmatrix}, \quad \boldsymbol{H_L}_{x,\mu} = \begin{bmatrix} L_{x_1,\mu_1} & \dots & L_{x_1,\mu_m} \\ \vdots & & \vdots \\ L_{x_n,\mu_1} & \dots & L_{x_n,\mu_m} \end{bmatrix}$$

$$\boldsymbol{H_L}_{\mu,x} = \begin{bmatrix} L_{\mu_1,x_1} & \dots & L_{\mu_1,x_n} \\ \vdots & & \vdots \\ L_{\mu_m,x_1} & \dots & L_{\mu_m,x_n} \end{bmatrix}, \quad \boldsymbol{H_L}_{\mu,\mu} = \begin{bmatrix} L_{\mu_1,\mu_1} & \dots & L_{\mu_1,\mu_m} \\ \vdots & & \vdots \\ L_{\mu_m,\mu_1} & \dots & L_{\mu_m,\mu_m} \end{bmatrix}$$

Expand the Hessian term:

$$\begin{aligned} &\frac{1}{2} \begin{bmatrix} \Delta\boldsymbol{x} \\ \Delta\boldsymbol{\mu} \end{bmatrix}^T \begin{bmatrix} \boldsymbol{H_L}^{\theta}_{x,x} & \boldsymbol{H_L}^{\theta}_{x,\mu} \\ \boldsymbol{H_L}^{\theta}_{\mu,x} & \boldsymbol{H_L}^{\theta}_{\mu,\mu} \end{bmatrix} \begin{bmatrix} \Delta\boldsymbol{x} \\ \Delta\boldsymbol{\mu} \end{bmatrix} \\ &= \frac{1}{2} \begin{bmatrix} \Delta\boldsymbol{x} \\ \Delta\boldsymbol{\mu} \end{bmatrix}^T \begin{bmatrix} \boldsymbol{H_L}^{\theta}_{x,x}\Delta\boldsymbol{x} + \boldsymbol{H_L}^{\theta}_{x,\mu}\Delta\boldsymbol{\mu} & \\ \boldsymbol{H_L}^{\theta}_{\mu,x}\Delta\boldsymbol{x} & \boldsymbol{H_L}^{\theta}_{\mu,\mu}\Delta\boldsymbol{\mu} \end{bmatrix} \\ &= \frac{1}{2}\Big(\boldsymbol{H_L}^{\theta}_{x,x}(\Delta\boldsymbol{x})^T\Delta\boldsymbol{x} + \boldsymbol{H_L}^{\theta}_{x,\mu}(\Delta\boldsymbol{x})^T\Delta\boldsymbol{\mu} + \boldsymbol{H_L}^{\theta}_{\mu,x}(\Delta\boldsymbol{\mu})^T\Delta\boldsymbol{x} + \boldsymbol{H_L}^{\theta}_{\mu,\mu}(\Delta\boldsymbol{\mu})^T\Delta\boldsymbol{\mu}\Big) \end{aligned}$$

Assuming sufficient smoothness for L, in any closed and bounded domain $\Omega \subset \Re^n \times \Re^m$, there exists a bound $B_L(\bar{\Omega})$ such that

$$\max_{\bar{\Omega}} \left(\left|\frac{\partial^2 L}{\partial x_i \partial \mu_j}\right|, \left|\frac{\partial^2 L}{\partial x_i \partial x_j}\right|, \left|\frac{\partial^2 L}{\partial \mu_i \partial \mu_j}\right| \right) < B_L^2(\bar{\Omega})$$

Thus,

$$\begin{aligned} \|\boldsymbol{H_L}_{x,x}\|_{Fr} &\leq nM_L(\bar{\Omega}), \quad \|\boldsymbol{H_L}_{x,\mu}\|_{Fr} \leq \sqrt{nm}\, M_L(\bar{\Omega}) \\ \|\boldsymbol{H_L}_{\mu,x}\|_{Fr} &\leq \sqrt{nm}\, M_L(\bar{\Omega}), \quad \|\boldsymbol{H_L}_{\mu,\mu}\|_{Fr} \leq mM_L(\bar{\Omega}) \end{aligned}$$

Let $M_L(\bar{\Omega}) = \max\{\sqrt{nm}, n, m\} M_L(\bar{\Omega})$. Then we have

$$\begin{aligned} \|\boldsymbol{H_L}_{x,x}\|_{Fr} &< M_L(\bar{\Omega}), \quad \|\boldsymbol{H_L}_{x,\mu}\|_{Fr} < M_L(\bar{\Omega}) \\ \|\boldsymbol{H_L}_{\mu,x}\|_{Fr} &< M_L(\bar{\Omega}), \quad \|\boldsymbol{H_L}_{\mu,\mu}\|_{Fr} < M_L(\bar{\Omega}) \end{aligned}$$

for any $\bar{\Omega}$ containing $(\boldsymbol{x}, \boldsymbol{\mu})$. So then

$$\frac{1}{2} \begin{bmatrix} \Delta\boldsymbol{x} \\ \Delta\boldsymbol{\mu} \end{bmatrix}^T \begin{bmatrix} \boldsymbol{H_L}^{\theta}_{x,x} & \boldsymbol{H_L}^{\theta}_{x,\mu} \\ \boldsymbol{H_L}^{\theta}_{\mu,x} & \boldsymbol{H_L}^{\theta}_{\mu,\mu} \end{bmatrix} \begin{bmatrix} \Delta\boldsymbol{x} \\ \Delta\boldsymbol{\mu} \end{bmatrix} \leq \frac{M_L(\bar{\Omega})}{2} \left[\|\Delta\boldsymbol{x}\|^2 + 2\|\Delta\boldsymbol{x}\|\|\Delta\boldsymbol{\mu}\| + \|\Delta\boldsymbol{\mu}\|^2 \right]$$

$$= \frac{M_L(\bar{\Omega})}{2}(\|\Delta \boldsymbol{x}\| + \|\Delta \boldsymbol{\mu}\|)^2$$

We know for two non-negative numbers a and b, $(a+b)^2 \leq 2(a^2+b^2)$. Thus,

$$\frac{1}{2}\begin{bmatrix}\Delta \boldsymbol{x} \\ \Delta \boldsymbol{\mu}\end{bmatrix}^T \begin{bmatrix}\boldsymbol{H_L}^{\theta}_{x,x} & \boldsymbol{H_L}^{\theta}_{x,\mu} \\ \boldsymbol{H_L}^{\theta}_{\mu,x} & \boldsymbol{H_L}^{\theta}_{\mu,\mu}\end{bmatrix}\begin{bmatrix}\Delta \boldsymbol{x} \\ \Delta \boldsymbol{\mu}\end{bmatrix} \leq M_L(\bar{\Omega})(\|\Delta \boldsymbol{x}\|^2 + \|\Delta \boldsymbol{\mu}\|^2)$$

Now

$$L(\boldsymbol{x_0}+\Delta \boldsymbol{x}, \boldsymbol{\mu_0}+\Delta \boldsymbol{\mu}) - L^0 - \boldsymbol{\nabla}_x L^0 \Delta x - \boldsymbol{\nabla}_\mu L^0 \Delta \mu$$
$$= \frac{1}{2}\begin{bmatrix}\Delta \boldsymbol{x} \\ \Delta \boldsymbol{\mu}\end{bmatrix}^T \begin{bmatrix}\boldsymbol{H_L}^{\theta}_{x,x} & \boldsymbol{H_L}^{\theta}_{x,\mu} \\ \boldsymbol{H_L}^{\theta}_{\mu,x} & \boldsymbol{H_L}^{\theta}_{\mu,\mu}\end{bmatrix}\begin{bmatrix}\Delta \boldsymbol{x} \\ \Delta \boldsymbol{\mu}\end{bmatrix}$$

Let

$$\eta_L(\Delta \boldsymbol{x}, \Delta \boldsymbol{\mu}) = \frac{1}{2}\begin{bmatrix}\Delta \boldsymbol{x} \\ \Delta \boldsymbol{\mu}\end{bmatrix}^T \begin{bmatrix}\boldsymbol{H_L}^{\theta}_{x,x} & \boldsymbol{H_L}^{\theta}_{x,\mu} \\ \boldsymbol{H_L}^{\theta}_{\mu,x} & \boldsymbol{H_L}^{\theta}_{\mu,\mu}\end{bmatrix}\begin{bmatrix}\Delta \boldsymbol{x} \\ \Delta \boldsymbol{\mu}\end{bmatrix}$$

Now for any fixed R we can set $\bar{\Omega} = \bar{B}(R, (x, \boldsymbol{\mu}))$. Then in this ball

$$\Delta L = (\boldsymbol{\nabla}_x L^0)^T \Delta \boldsymbol{x} + (\boldsymbol{\nabla}_\mu L^0)^T \Delta \boldsymbol{\mu} + \eta_L(\Delta \boldsymbol{x}, \Delta \boldsymbol{\mu})$$

where $\Delta L = L(\boldsymbol{x_0}+\Delta \boldsymbol{x}, \boldsymbol{\mu_0}+\Delta \boldsymbol{\mu}) - L^0$. Since all the partials of L are continuous locally, for sufficiently small R, L is differentiable and thus we know $\eta_L \to 0$ as $\Delta x, \Delta \mu \to 0$ and $\|\eta_L\|/\|(\Delta \boldsymbol{x}, \Delta \boldsymbol{\mu})\| \to 0$ as $\|(\Delta \boldsymbol{x}, \Delta \boldsymbol{\mu})\| \to 0$. Of course, we have already shown this ourselves as the estimate

$$\eta_L(\Delta \boldsymbol{x}, \Delta \boldsymbol{\mu}) \leq M_L(\bar{\Omega})(\|\Delta \boldsymbol{x}\|^2 + \|\Delta \boldsymbol{\mu}\|^2)$$

clearly shows this. Also, in using the same argument as above on each f_i we get

$$\Delta f_i = \boldsymbol{\nabla}_x f_i^o \Delta x + \boldsymbol{\nabla}_\mu f_i^0 \Delta \mu + \eta_{f_i}(\Delta x, \Delta \mu)$$

and $\eta_{f_i} \to 0$ as $\Delta x, \Delta \mu \to 0$. Here

$$\eta_{f_i}(\Delta \boldsymbol{x}, \Delta \boldsymbol{\mu}) = \frac{1}{2}\begin{bmatrix}\Delta \boldsymbol{x} \\ \Delta \boldsymbol{\mu}\end{bmatrix}^T \begin{bmatrix}\boldsymbol{H_{f_i}}^{\phi}_{x,x} & \boldsymbol{H_{f_i}}^{\phi}_{x,\mu} \\ \boldsymbol{H_{f_i}}^{\phi}_{\mu,x} & \boldsymbol{H_{f_i}}^{\phi}_{\mu,\mu}\end{bmatrix}\begin{bmatrix}\Delta \boldsymbol{x} \\ \Delta \boldsymbol{\mu}\end{bmatrix}$$

with

$$|\eta_{f_i}(\Delta x, \Delta \mu)| \leq M_{f_i}(\bar{\Omega})\left(\|\Delta x\|^2 + \|\Delta \mu\|^2\right)$$

for all i, $1 \leq i \leq n$. So in putting all of the Δf_i into one vector we see

$$\underbrace{\begin{bmatrix}\Delta f_1 \\ \Delta f_2 \\ \vdots \\ \Delta f_n\end{bmatrix}}_{n\times 1} = \underbrace{\begin{bmatrix}\frac{\partial f_1^0}{\partial x_1} & \cdots & \frac{\partial f_1^0}{\partial x_n} \\ \frac{\partial f_2^0}{\partial x_1} & \cdots & \frac{\partial f_2^0}{\partial x_n} \\ \vdots & \vdots & \vdots \\ \frac{\partial f_n^0}{\partial x_1} & \cdots & \frac{\partial f_n^0}{\partial x_n}\end{bmatrix}}_{n\times n}\underbrace{\begin{bmatrix}\Delta x_1 \\ \Delta x_2 \\ \vdots \\ \Delta x_n\end{bmatrix}}_{n\times 1} + \underbrace{\begin{bmatrix}\frac{\partial f_1^0}{\partial \mu_1} & \cdots & \frac{\partial f_1^0}{\partial \mu_m} \\ \frac{\partial f_2^0}{\partial \mu_1} & \cdots & \frac{\partial f_2^0}{\partial \mu_m} \\ \vdots & \vdots & \vdots \\ \frac{\partial f_n^0}{\partial \mu_1} & \cdots & \frac{\partial f_n^0}{\partial \mu_m}\end{bmatrix}}_{n\times m}\underbrace{\begin{bmatrix}\Delta \mu_1 \\ \Delta \mu_2 \\ \vdots \\ \Delta \mu_m\end{bmatrix}}_{m\times 1}$$

$$+ \underbrace{\begin{bmatrix} \eta_{f_1}(\Delta x, \Delta \mu) \\ \eta_{f_2}(\Delta x, \Delta \mu) \\ \vdots \\ \eta_{f_n}(\Delta x, \Delta \mu) \end{bmatrix}}_{n \times 1}$$

We can rewrite this in terms of gradients:

$$\underbrace{\begin{bmatrix} \Delta f_1 \\ \vdots \\ \Delta f_n \end{bmatrix}}_{n\times 1} = \underbrace{\begin{bmatrix} (\boldsymbol{\nabla}^0_{f_1,x})^T \\ \vdots \\ (\boldsymbol{\nabla}^0_{f_n,x})^T \end{bmatrix}}_{n\times n} \underbrace{\begin{bmatrix} \Delta x_1 \\ \vdots \\ \Delta x_n \end{bmatrix}}_{n\times 1} + \underbrace{\begin{bmatrix} (\boldsymbol{\nabla}^0_{f_1,\mu})^T \\ \vdots \\ (\boldsymbol{\nabla}^0_{f_n,\mu})^T \end{bmatrix}}_{n\times m} \underbrace{\begin{bmatrix} \Delta \mu_1 \\ \vdots \\ \Delta \mu_m \end{bmatrix}}_{m\times 1} + \underbrace{\begin{bmatrix} \eta_{f_1}(\Delta x, \Delta \mu) \\ \vdots \\ \eta_{f_n}(\Delta x, \Delta \mu) \end{bmatrix}}_{n\times 1}$$

Now let

$$\boldsymbol{J_f}^0_x = \begin{bmatrix} (\boldsymbol{\nabla}^0_{f_1,x})^T \\ \vdots \\ (\boldsymbol{\nabla}^0_{f_n,x})^T \end{bmatrix}, \quad \boldsymbol{J_f}^0_\mu = \begin{bmatrix} (\boldsymbol{\nabla}^0_{f_1,\mu})^T \\ \vdots \\ (\boldsymbol{\nabla}^0_{f_n,\mu})^T \end{bmatrix}, \quad \eta_f = \begin{bmatrix} \eta_{f_1}(\Delta x, \Delta \mu) \\ \vdots \\ \eta_{f_n}(\Delta x, \Delta \mu) \end{bmatrix}$$

Then, we have

$$\Delta f = \boldsymbol{J_f}^0_x \Delta x + \boldsymbol{J_f}^0_\mu \Delta \mu + \eta_f(\Delta x, \Delta \mu)$$

We also have the estimate

$$\|\eta_f(\Delta x, \Delta \mu)\| \leq M_f(\bar{\Omega})(\|\Delta x\|^2 + \|\Delta \mu\|^2)n$$

where

$$M_f(\bar{\Omega}) = \max(M_{f_1}(\bar{\Omega}), \cdots, M_{f_n}(\bar{\Omega})).$$

To summarize, we can write

$$\begin{aligned} \Delta L &= (\boldsymbol{\nabla}_x L^0)^T \Delta x + (\boldsymbol{\nabla}_\mu L^0)^T \Delta \mu + \eta_L(\Delta x, \Delta \mu) \\ \Delta f &= \boldsymbol{J_f}^0_x \Delta x + \boldsymbol{J_f}^0_\mu \Delta \mu + \eta_f(\Delta x, \Delta \mu) \end{aligned}$$

and we know that the error terms go to zero as Δx and $\Delta \mu$ go to zero.

19.3.2 A First Look at Constraint Satisfaction

Now we also know that constraint satisfaction, i.e. satisfying $f = \boldsymbol{0}$, implies that $\Delta f = \boldsymbol{0}$. So then

$$\boldsymbol{0} = \boldsymbol{J_f}^0_x \Delta x + \boldsymbol{J_f}^0_\mu \Delta \mu + \eta_f(\Delta x, \Delta \mu)$$

Assuming $(\boldsymbol{J_f}^0_x)^{-1}$ exists and is continuous, we then get

$$\Delta x = -(\boldsymbol{J_f}^0_x)^{-1} \boldsymbol{J_f}^0_\mu \Delta \mu - (\boldsymbol{J_f}^0_x)^{-1} \eta_f(\Delta x, \Delta \mu)$$

Note the Δx on both sides of this equation. We cannot really solve for Δx of course, but ignoring that we see the above expression yields

$$\|(\boldsymbol{J_f}^0_x)^{-1} \eta_f(\Delta x, \Delta \mu)\| \leq \|(\boldsymbol{J_f}^0_x)^{-1} \boldsymbol{J_f}^0_\mu \Delta \mu\| + \|\Delta x\| \leq \|(\boldsymbol{J_f}^0_x)^{-1}\| \, \|\boldsymbol{J_f}^0_\mu \Delta \mu\| + \|\Delta x\|$$

$$\leq \quad \|(\boldsymbol{J_f}_x^0)^{-1}\| \, \|\boldsymbol{J_f}_\mu^0\| \, \|\Delta\mu\| + \|\Delta x\|$$

From our discussions of the bounds due to the smoothness of the functions here, there is an overestimate $B(\bar{\Omega})$ such that

$$\|(\boldsymbol{J_f}_x^0)^{-1}\| \, \|\boldsymbol{J_f}_\mu^0\| \quad \leq \quad B(\bar{\Omega})$$

and so

$$\|(\boldsymbol{J_f}_x^0)^{-1}\eta_f(\Delta x, \Delta\mu)\| \quad \leq \quad B(\bar{\Omega}) \, \|\Delta\mu\| + \|\Delta x\|$$

Using this, we get that the change in performance ΔL is given by

$$\begin{aligned}
\Delta L &= (\boldsymbol{\nabla}_x L^0)^T\Big(-(\boldsymbol{J_f}_x^0)^{-1}\boldsymbol{J_f}_\mu^0\Delta\mu - (\boldsymbol{J_f}_x^0)^{-1}\eta_f(\Delta x, \Delta\mu)\Big) + (\boldsymbol{\nabla}_\mu L^0)^T\Delta\mu + \eta_L(\Delta x, \Delta\mu)\\
&= -(\boldsymbol{\nabla}_x L^0)^T(\boldsymbol{J_f}_x^0)^{-1}\boldsymbol{J_f}_\mu^0\Delta\mu - (\boldsymbol{\nabla}_x L^0)^T(\boldsymbol{J_f}_x^0)^{-1}\eta_f(\Delta x, \Delta\mu)\\
&+ (\boldsymbol{\nabla}_\mu L^0)^T\Delta\mu + \eta_L(\Delta x, \Delta\mu)
\end{aligned}$$

Letting $\alpha = (\boldsymbol{\nabla}_\mu L^0)^T - (\boldsymbol{\nabla}_x L^0)^T(\boldsymbol{J_f}_x^0)^{-1}\boldsymbol{J_f}_\mu^0$ we see for a minimum, $\Delta L > 0$. Thus,

$$0 \quad \leq \quad \alpha\Delta\mu - (\boldsymbol{\nabla}_x L^0)^T(\boldsymbol{J_f}_x^0)^{-1}\eta_f(\Delta x, \Delta\mu) + \eta_L(\Delta x, \Delta\mu)$$

19.3.3 Constraint Satisfaction and the Implicit Function Theorem

We want to show the $\alpha = 0$ at the minimum of L subject to the constraints, but as written Δx and $\Delta\mu$ are dependent of one another. What we need is to find how Δx depends on $\Delta\mu$ so we can write an inequality equation only in $\Delta\mu$. We will do this by stepping back, rewriting a few things, and invoking the Implicit Function Theorem. It is the Implicit Function Theorem that will allow us to write Δx in terms of $\Delta\mu$ only, rather than in terms of itself and Δx. At the local minimum $(\boldsymbol{x_0}, \boldsymbol{\mu_0})$, we have constraint satisfaction and so

$$\begin{aligned}
0 &\leq \Delta L = (\boldsymbol{\nabla}_x L^0)^T\Delta x + (\boldsymbol{\nabla}_\mu L^0)^T\Delta\mu + \eta_L(\Delta x, \Delta\mu)\\
\boldsymbol{0} &= \Delta f = \boldsymbol{J_f}_x^0\,\Delta x + \boldsymbol{J_f}_\mu^0\,\Delta\mu + \eta_f(\Delta x, \Delta\mu)
\end{aligned}$$

where the Hessian terms in η_L and η_f are evaluated at in between points

$$(\boldsymbol{x}^\theta, \boldsymbol{\mu}^\theta) \quad = \quad (1-\theta)(\boldsymbol{x_0}, \boldsymbol{\mu_0}) + \theta(\boldsymbol{x_0} + \Delta x, \boldsymbol{\mu_0} + \Delta\mu)$$

Here we could assume that $(\boldsymbol{J_f}_x^0)^{-1}$ exists and solve for Δx like we did last time, but we must realize that there is a relationship between Δx and $\Delta\mu$ and that we need to do something else to be precise. We can use the Implicit Function Theorem to write Δx in terms of $\Delta\mu$ to provide this precision. The Implicit Function Theorem, Theorem 12.3.1, has been carefully proven and discussed in Chapter 12. If we translate it to our situation, it would be stated as follows:

Implicit Function Theorem
Let $U \subset \Re^{n+'m}$ be an open set. Let $\boldsymbol{u} \in U$ be written as $\boldsymbol{u} = (\boldsymbol{x}, \boldsymbol{\mu})$ where $\boldsymbol{x} \in \Re^n$ and $\boldsymbol{\mu} \in \Re^m$. Assume $f : U \to \Re^m$ has continuous first order partials in U and there is a point $(\boldsymbol{x_0}, \boldsymbol{\mu_0}) \in U$ satisfying $f(\boldsymbol{x_0}, \boldsymbol{\mu_0}) = 0$. This is the constraint satisfaction condition. Also assume $det\boldsymbol{J_f}_x^0 \neq 0$ Then there is an open set V_0 containing $\boldsymbol{\mu_0} \in \Re^m$ and $g : V_0 \to \Re^n$ with continuous partials so that $g(\boldsymbol{\mu_0}) = \boldsymbol{x_0}$, $f(g(\boldsymbol{\mu}), \boldsymbol{\mu}) = \boldsymbol{0}$ on V_0.

This tells us we have constraint satisfaction: $f(g(\boldsymbol{\mu_0} + \Delta\mu), \boldsymbol{\mu_0} + \Delta\mu) = \boldsymbol{0}$ for all $\boldsymbol{\mu_0} + \Delta\mu \in V_0$. Also, since $g(\boldsymbol{\mu_0}) = \boldsymbol{x_0}$, letting $g(\boldsymbol{\mu_0} + \Delta\mu) = \boldsymbol{x}$, then $\Delta x = g(\boldsymbol{\mu_0} + \Delta\mu) - g(\boldsymbol{\mu_0}) = \boldsymbol{x} - \boldsymbol{x_0}$ which is our usual notation.

Recall that we had from before

$$0 \leq \Delta L = (\boldsymbol{\nabla}_x L^0)^T \Delta x + (\boldsymbol{\nabla}_\mu L^0)^T \Delta\mu + \frac{1}{2}\begin{bmatrix} \Delta \boldsymbol{x} \\ \Delta \boldsymbol{\mu} \end{bmatrix}^T \begin{bmatrix} \boldsymbol{H_L}^{\theta}_{x,x} & \boldsymbol{H_L}^{\theta}_{x,\mu} \\ \boldsymbol{H_L}^{\theta}_{\mu,x} & \boldsymbol{H_L}^{\theta}_{\mu,\mu} \end{bmatrix} \begin{bmatrix} \Delta \boldsymbol{x} \\ \Delta \boldsymbol{\mu} \end{bmatrix}$$

Let $\Omega = B(r, \boldsymbol{x_0}, \boldsymbol{\mu_0})$ and apply the Implicit Function Theorem. Associated with Ω will be an open set V_0. Choose a ball $B(\rho, \boldsymbol{x_0})$ such that $B(\rho, \boldsymbol{x_0}) \subset V_0$. Then for all $\Delta x \in B(\rho, \boldsymbol{\mu_0})$ we know

$$\begin{aligned} &0 \leq \Delta L = (\boldsymbol{\nabla}_x L^0)^T \Delta x + (\boldsymbol{\nabla}_\mu L^0)^T \Delta\mu + \\ &\frac{1}{2}\begin{bmatrix} g(\boldsymbol{\mu_0} + \Delta\mu) - g(\boldsymbol{\mu_0}) \\ \Delta\mu \end{bmatrix}^T \begin{bmatrix} \boldsymbol{H_L}^{\theta}_{x,x} & \boldsymbol{H_L}^{\theta}_{x,\mu} \\ \boldsymbol{H_L}^{\theta}_{\mu,x} & \boldsymbol{H_L}^{\theta}_{\mu,\mu} \end{bmatrix} \begin{bmatrix} g(\boldsymbol{\mu_0} + \Delta\mu) - g(\boldsymbol{\mu_0}) \\ \Delta\mu \end{bmatrix} \end{aligned}$$

Now consider

$$\boldsymbol{H_L}^{\theta}(\Delta\mu) = \frac{1}{2}\begin{bmatrix} g(\boldsymbol{\mu_0} + \Delta\mu) - g(\boldsymbol{\mu_0}) \\ \Delta\mu \end{bmatrix}^T \begin{bmatrix} \boldsymbol{H_L}^{\theta}_{x,x} & \boldsymbol{H_L}^{\theta}_{x,\mu} \\ \boldsymbol{H_L}^{\theta}_{\mu,x} & \boldsymbol{H_L}^{\theta}_{\mu,\mu} \end{bmatrix} \begin{bmatrix} g(\boldsymbol{\mu_0} + \Delta\mu) - g(\boldsymbol{\mu_0}) \\ \Delta\mu \end{bmatrix}$$

The matrix in the middle is evaluated at the point

$$\begin{aligned} (\boldsymbol{x}^\theta, \boldsymbol{\mu}^\theta) &= (1-\theta)(\boldsymbol{x_0}, \boldsymbol{\mu_0}) + \theta(\boldsymbol{x_0} + \Delta x, \boldsymbol{\mu_0} + \Delta\mu) = (\boldsymbol{x_0}, \boldsymbol{\mu_0}) + \theta(\Delta x, \Delta\mu) \\ &= (g(\boldsymbol{\mu_0}), \boldsymbol{\mu_0}) + \theta(g(\boldsymbol{\mu_0} + \Delta\mu) - g(\boldsymbol{\mu_0}), \Delta\mu) \end{aligned}$$

Note here that

$$(\boldsymbol{x}^\theta, \boldsymbol{\mu}^\theta) - (\boldsymbol{x_0}, \boldsymbol{\mu_0}) = \theta(\Delta x, \Delta\mu) = \theta(g(\boldsymbol{\mu_0} + \Delta\mu) - g(\boldsymbol{\mu_0}), \Delta\mu)$$

is in $B(r, (\boldsymbol{0}, \boldsymbol{0}))$ as $g(\boldsymbol{\mu_0} + \Delta\mu, \boldsymbol{\mu_0} + \Delta\mu)$ is in $B(r, (\boldsymbol{x_0}, \boldsymbol{\mu_0}))$. Note also that by our assumptions $\boldsymbol{H_L}^{\theta}(\Delta\mu)$ is continuous. So now

$$0 \leq \Delta L = (\boldsymbol{\nabla}_x L^0)^T \Delta x + (\boldsymbol{\nabla}_\mu L^0)^T \Delta\mu + H_L^\theta(\Delta\mu)$$

and because constraints are satisfied

$$0 = \boldsymbol{J_f}^0_x \Delta x + \boldsymbol{J_f}^0_\mu \Delta\mu + \boldsymbol{H_f}^{\phi}(\Delta\mu)$$

where

$$\begin{aligned} \boldsymbol{H_{f_i}}^{\phi}(\Delta\mu) &= \frac{1}{2}\begin{bmatrix} g(\boldsymbol{\mu_0} + \Delta\mu) - g(\boldsymbol{\mu_0}) \\ \Delta\mu \end{bmatrix}^T \begin{bmatrix} \boldsymbol{H_{f_i}}^{\theta}_{x,x} & \boldsymbol{H_{f_i}}^{\theta}_{x,\mu} \\ \boldsymbol{H_{f_i}}^{\theta}_{\mu,x} & \boldsymbol{H_{f_i}}^{\theta}_{\mu,\mu} \end{bmatrix} \begin{bmatrix} g(\boldsymbol{\mu_0} + \Delta\mu) - g(\boldsymbol{\mu_0}) \\ \Delta\mu \end{bmatrix} \\ \boldsymbol{H_f}^{\phi}(\Delta\mu) &= \begin{bmatrix} \boldsymbol{H_{f_1}}^{\phi}(\Delta\mu) \\ \vdots \\ \boldsymbol{H_{f_n}}^{\phi}(\Delta\mu) \end{bmatrix} \end{aligned}$$

Hence, $H_f^\phi(\Delta\mu)$ is constructed in the same way as $H_L^\theta(\Delta\mu)$ but the intermediate point is now $(\boldsymbol{x}^\phi, \boldsymbol{\mu}^\phi)$. Then, since we know $\boldsymbol{J_f}^0_x$ is invertible, we have

$$\Delta x = -(\boldsymbol{J_f}^0_x)^{-1} \boldsymbol{J_f}^0_\mu \Delta\mu - (\boldsymbol{J_f}^0_x)^{-1} \boldsymbol{H_f}^{\phi}(\Delta\mu)$$

which is an expression for Δx written in terms of $\Delta\mu$ only. This is what we wanted all along. Now, plugging in for Δx in ΔL we get

$$\begin{aligned} 0 \leq \Delta L &= (\nabla_x L^0)^T \Big(-(J_{f_x}^0)^{-1} J_{f_\mu}^0 \Delta\mu - (J_{f_x}^0)^{-1} H_f^{\phi}(\Delta\mu) \Big) + (\nabla_\mu L^0)^T \Delta\mu + H_L^{\theta}(\Delta\mu) \\ &= \Big((\nabla_\mu L^0)^T - (\nabla_x L^0)^T (J_{f_x}^0)^{-1} J_{f_\mu}^0 \Big) \Delta\mu - (\nabla_x L^0)^T (J_{f_x}^0)^{-1} H_f^{\phi}(\Delta\mu) + H_L^{\theta}(\Delta\mu) \end{aligned}$$

Now let $\alpha = (\nabla_\mu L^0)^T - (\nabla_x L^0)^T (J_{f_x}^0)^{-1} J_{f_\mu}^0$ and $\beta(\Delta\mu) = -(\nabla_x L^0)^T (J_{f_x}^0)^{-1} H_f^{\phi}(\Delta\mu) + H_L^{\theta}(\Delta\mu)$. We then have

$$0 \leq \alpha \, \Delta\mu + \beta(\Delta\mu)$$

for $\Delta\mu \in B(\rho, \mathbf{0})$ and we see β is continuous on $B(\rho, \mathbf{0})$. Our goal is to show that $\alpha = 0$, so that we will have the known necessary conditions for constrained optimization. In trying to show this we will attempt to get a bound on $\beta(\Delta\mu)$ and then try to trap α in between two arbitrarily small quantities, thus showing that α must be 0. There are two approaches that we will take. Our first attempt will be an incorrect approach and will lead us to a dead end. Our second approach will back off a little, summarizing some of this material once again, and proceed on to the result that we are trying to get to: namely, that $\alpha = 0$.

19.3.3.1 An Incorrect Approach

Picking up where we left off, since β is continuous on $B(\rho\boldsymbol{\mu_0})$ we know that given $\epsilon > 0$, there is a $\delta > 0$ so that

$$\Delta\mu \in B(\delta, \mathbf{0}) \implies |\beta(\Delta\mu) - \beta(0)| < \epsilon$$

But $\beta(0) = 0$ so

$$\Delta\mu \in B(\delta, \mathbf{0}) \implies |\beta(\Delta\mu)| < \epsilon$$

Choose $\Delta\mu = c\boldsymbol{E_i}$ for c sufficiently small. Then we have if $|c| < \delta$,

$$\Delta\mu = \begin{bmatrix} 0 \\ \vdots \\ c \\ \vdots \\ 0 \end{bmatrix} \leftarrow j^{th}\text{slot} \;, \qquad \text{and } 0 \leq \begin{bmatrix} \alpha_1 \\ \vdots \\ \alpha_j \\ \vdots \\ \alpha_m \end{bmatrix}^T \begin{bmatrix} 0 \\ \vdots \\ c \\ \vdots \\ 0 \end{bmatrix} + \epsilon = \alpha_j c + \epsilon$$

So $0 \leq \alpha_j c + \epsilon$ if $|c| < \delta$.

Choose $c = \frac{\delta}{2}$	Choose $c = -\frac{\delta}{2}$
$0 \leq \alpha_j \frac{\delta}{2} + \epsilon$	$0 \leq \alpha_j \left(-\frac{\delta}{2}\right) + \epsilon$
$-\epsilon \leq \alpha_j \frac{\delta}{2}$	$-\epsilon \leq \alpha_j \left(-\frac{\delta}{2}\right)$
$-\frac{2\epsilon}{\delta} \leq \alpha_j$	$\frac{-2\epsilon}{-\delta} \geq \alpha_j$

In combining these two inequalities we get

$$-\frac{2\epsilon}{\delta} \leq \alpha_j \leq \frac{2\epsilon}{\delta}$$

We would like to say here that both sides of this inequality go to zero as $\epsilon \to 0$, but since we do not know anything about how ϵ depends on δ, we have no idea what the ratio $\frac{\epsilon}{\delta}$ does as $\epsilon \to 0$. So we can see that this approach is not going to work! Arghh!

19.3.3.2 The Correct Approach

To back off a little bit, we know that there is always a linkage between the controls and the states. Look at $B(\frac{\rho}{2}, \boldsymbol{\mu_0})$. Then

$$\begin{aligned}
\boldsymbol{x_0} &= g(\boldsymbol{\mu_0}) \\
\boldsymbol{x_0} + \Delta x &= g(\boldsymbol{\mu_0} + \Delta\mu) \\
f(g(\boldsymbol{\mu_0} + \Delta\mu), \boldsymbol{\mu_0} + \Delta\mu) &\in B(r, \boldsymbol{x_0}, \boldsymbol{\mu_0}) \\
f(g(\boldsymbol{\mu_0} + \Delta\mu), \boldsymbol{\mu_0} + \Delta\mu) &= \boldsymbol{0},\ \Delta\mu \in B(\frac{\rho}{2}, \boldsymbol{\mu_0})
\end{aligned}$$

Constraint satisfaction then gives us

$$\boldsymbol{0} = \boldsymbol{J_f}{}_x^0 \Delta x + \boldsymbol{J_f}{}_\mu^0 \Delta\mu + \boldsymbol{H_f}^\phi(\Delta\mu)$$

for $\Delta\mu \in B(\frac{\rho}{2}, \boldsymbol{\mu_0})$. Now examine the last term $\boldsymbol{H_f}^\phi(\Delta\mu)$. Writing $H_f(\Delta\mu)$ out we see the ith component is

$$\begin{aligned}
(\, H_f(\Delta\mu)\,)_i &= \frac{1}{2}(\Delta x)^T\, \boldsymbol{H_{f_i}}{}_{xx}^\phi \Delta x + \frac{1}{2}(\Delta\mu)^T\, \boldsymbol{H_{f_i}}{}_{x\mu}^\phi \Delta x \\
&+ \frac{1}{2}(\Delta x)^T\, \boldsymbol{H_{f_i}}{}_{\mu x}^\phi \Delta\mu + \frac{1}{2}(\Delta\mu)^T\, \boldsymbol{H_{f_i}}{}_{\mu\mu}^\phi \Delta\mu
\end{aligned}$$

and so

$$\begin{aligned}
2\,\|(\, H_f(\Delta\mu)\,)_i\| &\leq \|\boldsymbol{H_{f_i}}{}_{xx}^\phi\|\, \|\Delta x\|^2 + \|\boldsymbol{H_{f_i}}{}_{x\mu}^\phi\|\, \|\Delta x\|\, \|\Delta\mu\| \\
&+ \|\boldsymbol{H_{f_i}}{}_{\mu x}^\phi\|\, \|\delta x\|\, \|\Delta\mu\| + \|\boldsymbol{H_{f_i}}{}_{\mu\mu}^\phi\|\, \|\Delta\mu\|^2
\end{aligned}$$

But,

$$g(\boldsymbol{\mu_0} + \Delta\mu) - g(\boldsymbol{\mu_0}) = \boldsymbol{J_g}{}_\mu^0 \Delta\mu + \frac{1}{2}(\Delta\mu)^T \boldsymbol{H_g}{}_\mu^\xi \Delta\mu$$

for an intermediate point ξ. Therefore

$$\|\Delta x\| \leq \|\boldsymbol{J_g}{}_\mu^0\|\, \|\Delta\mu\| + \frac{1}{2}\|\boldsymbol{H_g}{}_\mu^\xi\|\, \|\Delta\mu\|^2$$

Now for any fixed R we can set $\bar{\Omega} = \bar{B}(R, (x, \boldsymbol{\mu}))$, which is a compact set. Since all the partials here are continuous, this means all the gradient and Hessian terms have maximum values on $\bar{\Omega}$. Thus (this is a bit messy as there are a lot of upper bounds!)

$$\begin{aligned}
\|\boldsymbol{H_{f_i}}{}_{xx}^\phi\| &\leq B_{1i}(\bar{\Omega}), \quad \|\boldsymbol{H_{f_i}}{}_{x\mu}^\phi\| \leq B_{2i}(\bar{\Omega}) \\
\|\boldsymbol{H_{f_i}}{}_{\mu\mu}^\phi\| &\leq B_{3i}(\bar{\Omega}), \quad \|\boldsymbol{H_{f_i}}{}_{\mu\mu}^\phi\| \leq B_{4i}(\bar{\Omega}) \\
\|\boldsymbol{H_g}{}_\mu^\xi\| &\leq C(\bar{\Omega})
\end{aligned}$$

Thus,

$$\begin{aligned} 2\,\|(\,H_f(\Delta\mu)\,)_i\| &\le B_{1i}(\bar{\Omega})\,\|\Delta x\|^2 + B_{2i}(\bar{\Omega})\|\Delta x\|\,\|\Delta\mu\| + B_{3i}(\bar{\Omega})\,\|\Delta x\|\,\|\Delta\mu\| \\ &\quad + B_{4i}(\bar{\Omega})\,\|\Delta\mu\|^2 \end{aligned}$$

Let

$$\begin{aligned} B(\bar{\Omega}) &= \max_{1\le i\le n} \{B_{1i}(\bar{\Omega}), B_{2i}(\bar{\Omega}), B_{3i}(\bar{\Omega}), B4i(\bar{\Omega})\} \\ D(\bar{\Omega}) &= \max\{\|\boldsymbol{J_g}_{\mu}^{0}\|, \frac{1}{2}C(\bar{\Omega})\} \end{aligned}$$

$$\|\Delta x\| \le D(\bar{\Omega})\|\Delta\mu\| + D(\bar{\Omega})\|\Delta\mu\|^2$$

Now assume $\|\Delta\mu\| < 1$, then $\|\Delta x\| \le 2D(\bar{\Omega})\|\Delta\mu\|$. We can now do our final estimate:

$$\begin{aligned} 2\max_{1\le i\le n}\|(\,H_f(\Delta\mu)\,)_i\| &\le B(\bar{\Omega})\Big(\|\Delta x\|^2 + 2\|\Delta x\|\|\Delta\mu\| + \|\Delta\mu\|^2\Big) \\ &= B(\bar{\Omega})((\|\Delta x\| + \|\Delta\mu\|)^2) \le 2B(\bar{\Omega})(\|\Delta x\|^2 + \|\Delta\mu\|^2) \end{aligned}$$

Thus

$$\max_{1\le i\le n}\|(\,H_f(\Delta\mu)\,)_i\| \le B(\bar{\Omega})(\|\Delta x\|^2 + \|\Delta\mu\|^2)$$

Next, using the estimate for $\|\Delta x\|$, we find

$$\max_{1\le i\le n}\|(\,H_f(\Delta\mu)\,)_i\| \le B(\bar{\Omega})(4D(\bar{\Omega})^2 + 1)\,\|\Delta\mu\|^2$$

Let

$$\xi(\bar{\Omega}) = B(\bar{\Omega})(4D(\bar{\Omega})^2 + 1)$$

and we have shown

$$\max_{1\le i\le n}\|(\,H_f(\Delta\mu)\,)_i\| \le \xi(\bar{\Omega})\,\|\Delta\mu\|^2$$

Thus, to repeat, we have the following constraint equations

$$0 = \boldsymbol{J_f}_{x}^{0}\Delta x + \boldsymbol{J_f}_{\mu}^{0}\Delta\mu + \boldsymbol{H_f}^{\phi}(\Delta\mu)$$

where

$$\|\boldsymbol{H_f}^{\phi}(\Delta\mu)\| = \max_{1\le i\le n}\|(\,H_f(\Delta\mu)\,)_i\| \le \xi(\bar{\Omega})\|\Delta\mu\|^2$$

Using the same argument we used above we will also get that

$$0 \le \Delta L = (\boldsymbol{\nabla}_x L^0)^T\Delta x + (\boldsymbol{\nabla}_\mu L^0)^T\Delta\mu + H_L^{\theta}(\Delta\mu)$$

$$\|\boldsymbol{H_L}^{\theta}(\Delta\mu)\| \le \zeta(\bar{\Omega})\|\Delta\mu\|^2$$

some constant $\zeta(\bar{\Omega})$. And now since we know $\boldsymbol{J_f}^0_{xx}$ is invertible we know from the constraint equation that

$$\Delta x = -(\boldsymbol{J_f}^0_x)^{-1}\, \boldsymbol{J_f}^0_\mu \Delta\mu - (\boldsymbol{J_f}^0_x)^{-1}\, \boldsymbol{H_f}^\phi(\Delta\mu)$$

So putting this into the expression for ΔL we will get an expression that we have seen before. Namely,

$$\begin{aligned} 0 \le \Delta L &= (\boldsymbol{\nabla}^0_x)^T\left(-(\boldsymbol{J_f}^0_x)^{-1}\, \boldsymbol{J_f}^0_\mu \Delta\mu - (\boldsymbol{J_f}^0_x)^{-1}\, \boldsymbol{H_f}^\phi\, \Delta\mu\right) + (\boldsymbol{\nabla}_\mu L^0)^T\Delta\mu + H^\theta_L\Delta\mu \\ &= \left((\boldsymbol{\nabla}_\mu L^0)^T - (\boldsymbol{\nabla}_x L^0)^T\, (\boldsymbol{J_f}^0_x)^{-1}\, \boldsymbol{J_f}^0_\mu\right)\Delta\mu - (\boldsymbol{\nabla}_x L^0)^T\, (\boldsymbol{J_f}^0_x)^{-1}\, \boldsymbol{H_f}^\phi(\Delta\mu) + H^\theta_L(\Delta\mu) \end{aligned}$$

Letting

$$\alpha \;=\; (\boldsymbol{\nabla}_\mu L^0)^T - (\boldsymbol{\nabla}_x L^0)^T\, (\boldsymbol{J_f}^0_x)^{-1}\, \boldsymbol{J_f}^0_\mu$$

and

$$Q(\Delta\mu) \;=\; -(\boldsymbol{\nabla}_x L^0)^T\, (\boldsymbol{J_f}^0_x)^{-1}\, \boldsymbol{H_f}^\phi(\Delta\mu) + H^\theta_L(\Delta\mu)$$

we get

$$0 \;\le\; \Delta L = \alpha\, \Delta\mu + Q(\Delta\mu).$$

From all of the above work and the fact the $(\boldsymbol{J_f}^0_x)^{-1}$ exists, we have

$$\begin{aligned} |Q(\Delta\mu) &\le \|\boldsymbol{\nabla}_x L^0\|\, \|(\boldsymbol{J_f}^0_x)^{-1}\|\, \xi(\bar{\Omega})\|\Delta\mu\|^2 + C(\bar{\Omega})\|\Delta\mu\|^2 \\ &= \left(\|\boldsymbol{\nabla}_x L^0\|\, \|(\boldsymbol{J_f}^0_x)^{-1}\|\, \xi(\bar{\Omega})\| + C(\bar{\Omega})\right)\|\Delta\mu\|^2 \end{aligned}$$

Now let

$$K(\bar{\Omega}) \;=\; \|\boldsymbol{\nabla}_x L^0\|\, \|(\boldsymbol{J_f}^0_x)^{-1}\|\, \xi(\bar{\Omega})\| + C(\bar{\Omega})$$

and we have shown $|Q(\Delta\mu) \le K(\bar{\Omega})\|\Delta\mu\|^2$. So now we have the inequality

$$0 \le \alpha\Delta\mu + K\|\Delta\mu\|^2.$$

Notice that this is similar to the inequality that we had at the end of the other approach. This time though we will be able to get the result that we want, i.e. $\alpha = 0$. Now choose all $\Delta\mu_i = 0$ except for $\Delta\mu_j$, and let $\Delta\mu_j = \frac{1}{\sqrt{K}p}$. For p large enough, we know that $\Delta\mu \in \bar{B}(\frac{\rho}{2}, \boldsymbol{\mu_0})$. Looking at the jth component of the inequality we see

$$0 \;\le\; \alpha_j\frac{1}{\sqrt{K}p} + K\frac{1}{Kp^2} \Longrightarrow -\frac{1}{p^2} \le \frac{1}{p\sqrt{K}}\alpha_j \Longrightarrow -\frac{p\sqrt{K}}{p^2} \le \alpha_j \Longrightarrow -\frac{\sqrt{K}}{p} \le \alpha_j$$

Now choose $\Delta\mu_j = -\frac{1}{\sqrt{K}p}$. Then the jth component of the inequality becomes

$$0 \;\le\; \alpha_j\left(\frac{-1}{\sqrt{K}p}\right) + K\frac{1}{Kp^2} \Longrightarrow \alpha_j\left(\frac{1}{\sqrt{K}p}\right) \le \frac{1}{p^2} \Longrightarrow \alpha_j \le \frac{\sqrt{K}}{p}$$

Combining these we get that

$$-\frac{\sqrt{K}}{p} \leq \alpha_j \leq \frac{\sqrt{K}}{p}$$

Since p is arbitrarily large, this implies $\alpha_j = 0$. This same argument works for any of the components $\alpha_1, \alpha_2, \cdots, \alpha_n$. So we are finally able to conclude that $\alpha = 0$, and we have found the necessary conditions for constrained optimization.

Theorem 19.3.1 Constrained Optimization

Consider the problem

$$\min_{\boldsymbol{\mu} \in \Re^m} L(\boldsymbol{x}, \boldsymbol{\mu})$$

subject to

$$\left.\begin{aligned} f_1(x, \mu) &= 0 \\ &\vdots \\ f_n(x, \mu) &= 0 \end{aligned}\right\} \boldsymbol{f}(\boldsymbol{x}, \boldsymbol{\mu}) = 0$$

where $\boldsymbol{x} \in \Re^n$ *is the* **state**, $\boldsymbol{\mu} \in \Re^m$ *is the* **control** *and* $L : \Re^n \times \Re^m \rightarrow \Re$ *is the* **performance index**. *The function* $f : \Re^n \times \Re^m \rightarrow \Re^n$ *is the* **constraint**. *We assume the functions f and L have continuous partials locally about* $(\boldsymbol{x}_0, \boldsymbol{\mu}_0)$. *Then if* $(\boldsymbol{x}_0, \boldsymbol{\mu}_0)$ *is a point where L is minimized and* $(\boldsymbol{J}_{\boldsymbol{f}_x}(\boldsymbol{x}_0, \boldsymbol{\mu}_0))^{-1}$ *exists, then the following equation must be satisfied:*

$$(\boldsymbol{\nabla}_\mu L^0)^T - (\boldsymbol{\nabla}_x L^0)^T \, (\boldsymbol{J}_{\boldsymbol{f}_x^0})^{-1} \, \boldsymbol{J}_{\boldsymbol{f}_\mu^0} \;=\; \boldsymbol{0}$$

where $()^0$ *indicates terms that are evaluated at* $(\boldsymbol{x}_0, \boldsymbol{\mu}_0)$. *Hence, finding the points* $(\boldsymbol{x}_0, \boldsymbol{\mu}_0)$ *that satisfy this equation are possible candidates for the minima of L subject to* $f = \boldsymbol{0}$.

Proof 19.3.1
We have just finished a very long-winded explanation! ■

Homework

Exercise 19.3.1 *Write down the equations that must be satisfied for the problem*

$$\min_{\mu \in \Re} x^2 + \mu^2$$

subject to

$$2x^2 + \mu^2 \;=\; 4$$

Exercise 19.3.2 *Write down the equations that must be satisfied for the problem*

$$\min_{(y,z) \in \Re^2} x^2 + y^2 + z^2$$

subject to

$$\begin{aligned} x^2 y^2 + z^2 &= 10 \\ 2x + 4y + 8z &= 5 \end{aligned}$$

Exercise 19.3.3 *Write down the equations that must be satisfied for the problem*

$$\min_{(y,z)\in\Re^2} 4x^2 + 20y^2z^2$$

$$\textit{subject to}$$

$$2x + 4y + 5z = 8$$

Exercise 19.3.4 *Write down the equations that must be satisfied for the problem*

$$\min_{\mu\in\Re^3} x^2 + 4y + \mu_1^2 + 5\mu_2^2 + 6\mu_3^4$$

$$\textit{subject to}$$

$$\begin{aligned} 2x^2 + 3y^2\mu_1^2 &= 4 \\ 3x^2 + 5y^2\mu_2^2 + \mu_3^2 &= 5 \end{aligned}$$

Exercise 19.3.5 *Write down the equations that must be satisfied for the problem*

$$\min_{\mu\in\Re^3} x^2 + 5y^2 + 4z + 2\mu_1^2 + 4\mu_2^2 + 8\mu_3^2$$

$$\textit{subject to}$$

$$\begin{bmatrix} 2 & 4 & 5 & -2 & 3 & 1 \\ -1 & 5 & 8 & 3 & 1 & -5 \\ 6 & 2 & -5 & 10 & 2 & 7 \end{bmatrix} \begin{bmatrix} x \\ y \\ z \\ \mu_1 \\ \mu_2 \\ \mu_3 \end{bmatrix} = \begin{bmatrix} 1 \\ 2 \\ -4 \end{bmatrix}$$

19.4 Lagrange Multipliers

We could also approach this problem using the Lagrange Multiplier technique. If we let

$$\Phi(x, \mu, \lambda) = L(x, \mu) + \lambda^T f(x, \mu)$$

then Φ is minimized when

$$\begin{aligned} \boldsymbol{\nabla}_x \Phi &= \mathbf{0} \Longrightarrow \boldsymbol{\nabla}_x L^0 + \lambda^T \boldsymbol{J}_{\boldsymbol{f}_x} f^0 = \mathbf{0}, \quad * \\ \boldsymbol{\nabla}_\mu \Phi &= 0 \Longrightarrow \boldsymbol{\nabla}_\mu L^0 + \lambda^T \boldsymbol{J}_{\boldsymbol{f}_\mu} f^0, \quad ** \\ \boldsymbol{\nabla}_\lambda \Phi &= 0 \Longrightarrow f(\boldsymbol{x}, \boldsymbol{\mu}), \quad \text{constraint equations} \end{aligned}$$

If $\boldsymbol{J}_{\boldsymbol{f}_x} f^0$ is invertible, we get from (*) that

$$\lambda^T = -\boldsymbol{\nabla}_x L^0 \, (\boldsymbol{J}_{\boldsymbol{f}_x} f^0)^{-1}$$

and so using (**) we see

$$\mathbf{0} = \boldsymbol{\nabla}_\mu L^0 - \boldsymbol{\nabla}_x L^0 (\boldsymbol{J}_{\boldsymbol{f}_x} f^0)^{-1} \boldsymbol{J}_{\boldsymbol{f}_\mu} f^0$$

which is the transpose of the same equation that we had before. To get some sort of idea what the λ_i's are we know that to first order

$$\Delta L \approx (\boldsymbol{\nabla}_x L^0)^T \Delta x + (\boldsymbol{\nabla}_\mu L^0)^T \Delta \mu$$

$$\Delta f \approx \boldsymbol{J_f}_x^0 \Delta x + \boldsymbol{J_f}_\mu^0 \Delta \mu.$$

Now suppose we set $\Delta\mu = 0$ and we let Δx vary. Then

$$\begin{aligned} \Delta L &\approx (\boldsymbol{\nabla}_x L^0)^T \Delta x \\ \Delta f &\approx \boldsymbol{J_f}_x^0 \Delta x. \end{aligned}$$

So

$$\Delta L \approx (\boldsymbol{\nabla}_x L^0)^T \, (\boldsymbol{J_f}_x^0)^{-1} \Delta f \Longrightarrow \Delta L \approx -\lambda^T \Delta f \Longrightarrow \Delta L \approx -\sum_i \lambda_i \Delta f_i$$

and thus

$$\lambda_i \approx -\frac{\Delta L}{\Delta f_i}.$$

Here the Lagrange Multipliers λ_i have an interpretation of the cost for violating the constraint f_i, i.e. they are the change in performance with respect to the change in a constraint. Of course, the argument above is very loose! But suggestive arguments can often be proven later with precision. The hardest part of all in some ways is trying to discover conjectures. Think of the pricing interpretation as a challenge! How would you prove it carefully?

Homework

Exercise 19.4.1 *Write down the equations that must be satisfied for the problem*

$$\begin{aligned} &\min_{\mu \in \Re} x^2 + \mu^2 \\ &\textit{subject to} \\ &2x^2 + \mu^2 = 4 \end{aligned}$$

using Lagrange multipliers.

Exercise 19.4.2 *Write down the equations that must be satisfied for the problem*

$$\begin{aligned} \min_{(y,z) \in \Re^2} \; & x^2 + y^2 + z^2 \\ & \textit{subject to} \\ x^2 y^2 + z^2 &= 10 \\ 2x + 4y + 8z &= 5 \end{aligned}$$

using Lagrange multipliers.

Exercise 19.4.3 *Write down the equations that must be satisfied for the problem*

$$\begin{aligned} \min_{(y,z) \in \Re^2} \; & 4x^2 + 20y^2 z^2 \\ & \textit{subject to} \\ 2x + 4y + 5z &= 8 \end{aligned}$$

using Lagrange multipliers.

Exercise 19.4.4 *Write down the equations that must be satisfied for the problem*

$$\min_{\mu \in \Re^3} x^2 + 4y + \mu_1^2 + 5\mu_2^2 + 6\mu_3^4$$

$$\text{subject to}$$

$$\begin{aligned} 2x^2 + 3y^2\mu_1^2 &= 4 \\ 3x^2 + 5y^2\mu_2^2 + \mu_3^2 &= 5 \end{aligned}$$

using Lagrange multipliers.

Exercise 19.4.5 *Write down the equations that must be satisfied for the problem*

$$\min_{\mu \in \Re^3} x^2 + 5y^2 + 4z + 8\mu_1^2 + 4\mu_2^2 + 8\mu_3^2$$

$$\text{subject to}$$

$$\begin{bmatrix} 2 & 4 & 5 & -2 & 3 & 1 \\ -1 & 5 & 8 & 3 & 1 & -5 \\ 6 & 2 & -5 & 10 & 2 & 7 \end{bmatrix} \begin{bmatrix} x \\ y \\ z \\ \mu_1 \\ \mu_2 \\ \mu_3 \end{bmatrix} = \begin{bmatrix} 1 \\ 2 \\ -4 \end{bmatrix}$$

using Lagrange multipliers.

Part VI

Summing It All Up

Chapter 20

Summing It All Up

We have now come to the end of these notes, which means you have covered the material in the first two basic analysis texts. We have not covered all of the things we wanted to, but we view that as a plus: there is more to look forward to! In particular, we hope we have encouraged your interest in these extensions of the ideas from the first volume on analysis to a bit more of calculus on $\Re^n$. Our aims are modest here. We know from experience how little we actually understood these ideas when we were taking these courses. We dutifully, like you, read the notes, memorized the material and did some problems. We were never really exposed to the inverse and implicit function theorems, as courses in that haven't really been taught since the 1980s or so. We also had not seen a reasonably careful development of calculus in $\Re^2$ and $\Re^3$ other than the traditional third semester engineering calculus which covered Green's and Stokes's Theorems somewhat loosely. There was much yet to understand. Our understanding of those things came from the course we took in physics which covered basic electricity and magnetism in which Gauss's law was used to derive voltage across a capacitor in a clear way. But there was a lot of hand waving. Integration in 2D involving surface integrals was connected to integration in 3D over volumes contained in a closed surface in a *just calculate and don't worry too much about it* point of view. In this volume, we decided to do a careful exposition of integration in $\Re^n$ and to build the connection between line integrals and integrals in 2D over the area of a region encoded by a SCROC very carefully. We even introduced some unifying notions involving differential forms. We therefore delay a proper introduction to the theorems connecting $\int_{\partial U} \omega$ and $\int_U d\omega$ until we can develop the subject of manifolds and manifolds with a boundary more carefully. We will do some of that in (Peterson (20) 2020).

We have learned most of our analysis, algebra, topology, computation, and much more after we finished our formal education. So we have learned to read a lot, think even more, and stare at the wall trying to convince our family we are thinking deep thoughts. There are many books on these things and you just have to find the ones that fit your style of self-learning. We hope you like our approach of course! Old books are often useful as they are cheap and so you don't have to invest a lot. We still like holding a book in our hands and writing in it, as editing pdfs on a display is not quite right. In addition to the books we mentioned in the summary for (Peterson (21) 2020), let's add some more which will enable you to grow mathematically. But to each their own thing. You should start learning more about topology and algebra and how that connects to differential geometry. Lots of modern physics such as condensed matter physics, loop quantum gravity and even immunology uses these ideas.

- **Calculus on Manifolds** (M. Spivak (17) 1965) tries to introduce you to $\Re^n$ integration and it is a classic but we have always found it hard to read. Lots of details are left out but it has great but succinct coverage of interesting ideas. We discuss some of this in (Peterson (23) 2020) but not in as much abstract detail.

- **Topology Illustrated** (Savelier (28) 2016) is a great introduction to topology which we wished we had had available to read back in the day!

- **An Illustrated Introduction to Topology and Homotopy** (Kalajdzievski (14) 2015) is probably the first book on homotopy that really made sense. Highly recommended!

- **A Visual Introduction to Differential Forms and Calculus on Manifolds** (Fortney (5) 2018) is a good introduction which is very useful as it helps explain the other text on differential forms listed below.

- **Geometry of Differential Forms** (S. Morita (27) 1998) is about differential forms. This one is kind of hard to read, but with hard work opens, more doors for you. You are almost ready to read this one based on this second volume.

Let's summarize what we have done in the first two volumes. The first volume (Peterson (21) 2020) is a critical course in the use of abstraction and it is just one in a sequence of courses which prepare students to become practicing scientists. It is important to balance the theory and abstraction with clear explanation and argument so that students who are from many areas can follow this text and use it profitably for self-study even if they cannot take this material as a course. Many professionals need to add this sort of training to their toolkit later and they will do that if the text is accessible. This is a primary text for a typical junior or senior year course in basic analysis. It can also be used as a supplementary text for anyone whose work requires that they begin to assimilate more abstract mathematical concepts as part of their professional growth after graduation or as a supplement to deficiencies in undergraduate preparation that leaves them unprepared for the jump to the first graduate level analysis course. Students in other disciplines, such as biology and physics, also need a more theoretical discussion of these ideas, and the writing is designed to help such students use this book for self-study as their individual degree programs do not have enough leeway in them to allow them to take this course. This text is for a two semester sequence.

First Semester: Sequences, Continuity and Differentiation: This semester is designed to cover through the consequences of differentiation for functions of one variable and to cover the basics leading up to extremal theory for functions of two variables. Along the way, a lot of attention is paid to developing the theory underlying these ideas properly. The usual introductory chapters on the real line, basic logic and set theory are not covered as students have seen that before and frankly are bored to see it again. The style here is to explain very carefully with many examples and to always leave pointers to higher level concepts that would be covered in the other text. The study of convex functions and lower semicontinuity are also introduced so that students can see there are alternate types of smoothness that are useful.

The number e is developed and all of its properties from the sequence definition $e = \lim_n (1+1/n)^n$. Since ideas from sequences of functions and series are not yet known, all of the properties of the exponential and logarithm function must be explained using limit approaches, which helps tie together all of the topics that have been discussed. The pointer to the future here is that the exponential and logarithm functions are developed also in the second semester by defining the logarithm as a Riemann integral and all the properties are proven using other sorts of tools. It is good for the students to see alternate pathways.

Note that this text eschews the development of these ideas in a metric space setting, although we do talk about metrics and norms as appropriate. It is very important to develop the derivative and its consequences on the real line, and while there is a simplicity and economy of expression if convergence and so on is handled using a general metric, the proper study of differentiation in $\Re^n$ is not as amenable to the metric space choice of exposition. That sort of discussion is done in a later text.

Chapters 1 through 14 are the first semester of undergraduate analysis.

Second Semester: Riemann Integration, Sequences and Series of Functions The second half of this text is about developing the theory of Riemann Integration and sequences and series of functions carefully. Also, student horizons are expanded a bit by showing them some basic topology in $\Re^2$ and $\Re^3$ and revisiting sequential and topological compactness in these higher dimensions. This ties in well with the last chapter of semester one. These ideas are used at the end to prove the pointwise convergence of the Fourier Series of a function which is a great application of these ideas. In the text, Chapters 15 through 28 are for the second semester.

Once the first year of training is finished, there is a lot more to do. The first basic analysis course essentially discusses the abstract concepts underlying the study of calculus on the real line. A few higher dimensional concepts are touched on such as the development of rudimentary topology in $\Re^2$ and $\Re^3$, compactness and the tests for extrema for functions of two variables, but that is not a proper study of calculus concepts in two or more variables. A full discussion of the $\Re^n$ based calculus is quite complex and even this second basic analysis text cannot cover all the important things. The chosen focus here is on differentiation in $\Re^n$ and important concepts about mappings from $\Re^n$ to $\Re^m$ such as the inverse and implicit function theorem and change of variable formulae for multidimensional integration. These topics alone require much discussion and setup. These topics intersect nicely with many other important applied and theoretical areas which are no longer covered in mathematical science curricula. The knowledge here allows a quantitatively inclined person to more properly develop multivariable nonlinear ODE models for themselves and to learn the proper background to study differential geometry and manifolds among many other applications.

However, this course is just not taught at all anymore. It is material that students at all levels must figure out on their own. Most of the textbooks here are extremely terse and hard to follow as they assume a lot of abstract sophistication from their readers. This text is designed to be a self-study guide to this material. It is also designed to be taught from, but at my institution, it would be very difficult to find the requisite 10 students to register so that the course could be taught. Students who are coming in as master's students generally do not have the ability to allocate a semester course like this to learn this material. Instead, even if they have a solid introduction from the first basic analysis text on analysis, they typically jump to third basic course which is an introduction to very abstract concepts such as metric spaces, normed linear spaces and inner products spaces along with many other needed deeper ideas. Such a transition is always problematic and the student is always trying to catch up on the holes in their background.

Also, many multivariable concepts and the associated theory are used in probability, operations research and optimization to name a few. In those courses, $\Re^n$ based ideas must be introduced and used despite the students not having a solid foundational course in such things. Hence, good students are reading about this themselves. This text is intended to give them a reasonable book to read and study from. Hence, this second volume would be a primary text for a senior year course that follows the first basic analysis course in basic analysis. However, this course is not taught very often now, so it is designed for self-study and recommended as the appropriate self-study text for anyone whose work requires that they begin to assimilate more abstract mathematical concepts as part of their professional growth after graduation or as a supplement to deficiencies in undergraduate preparation that leaves them unprepared for the jump to the first graduate level analysis course, (Peterson (22) 2020).

Now that you have learned the first two sets of material here as laid out (Peterson (21) 2020) and (Peterson (23) 2020), you are ready to learn additional abstraction which is taught in three more basis courses (Peterson (22) 2020) (analysis in abstract spaces), (Peterson (19) 2019) (measure theory and abstract integration theory) and (Peterson (20) 2020) (topology, differential geometry and advanced functional analysis). Understanding this material will let you will easily pass the preliminary exams that are used to determine your entry into graduate programs. However, more importantly, they prepare you for research and what might be called the *life of the mind*. You should not simply try to guess or memorize how to answer questions from the past samples of such preliminary exams. You should simply learn the material and then success in the examination is guaranteed. Note, to become a properly trained mathematical scientist, you should learn more and the five volumes we have written provide a reasonable core set of things to add to your toolkit. This is our take on the right mix of topics based on how we teach, the things we use frequently in our thinking, and so forth. So we admit it is probably biased. But we have worked hard over the last 30 years to make decisions on the ideas that we think are essential or core and we wanted to let you read about them because we think it is helpful to think about how to learn this kind of abstraction.

You and your instructors have a pact actually: you must learn the core concepts and those of us who teach these courses must provide satisfying educational experiences that let you learn these ideas efficiently. Of course, we always expect you to work hard, but if you want to be a mathematical scientist or use advanced mathematics to make a living, you have all the motivation you need! And, as we said in the first volume, it goes without saying we always assume that you are eager and interested in the material!

We think it is very important that in a mathematics department, there is a sequence of courses covering **analysis** which trains our students to think carefully in a new way. An undergraduate mathematics major is no longer exposed to adequate levels of abstraction in their early years. For many reasons, the mathematics curriculum now includes less mathematics, more operations research (OR), more statistics and probability and more computational training. Departments no longer train the students in powerful computer languages such as **C** and **C++** and instead encourage them to use integrated development environments such as **MATLAB** or **Octave**, **Sage**, **Maple** and **Mathematica** which is supposed to free the students from many of the details of using a tool to solve a problem. One also increasingly uses scripting languages such as **Python** as a way to glue together complicated computational procedures and further erode the student's ability to reason from first principles themselves.

Also, over the last thirty years or so, as pressures to include courses outside of mathematics proper such as OR etc. have grown, the ability of the incoming students who come into college to handle abstraction in the first three semester calculus courses has changed dramatically. For example, it used to be routine to teach perhaps $20 - 30$ increasingly complicated ways to do *integration by substitution*, which taught the students how to see *patterns* within complexity. As more substitution techniques were mastered, students became increasingly better at *seeing* the hidden core of the integral which was inside the complexity of the initial formulation. Make no mistake about what this means: students were being trained to learn how to **abstract** from a complicated thing, the *essence* hidden inside. This is a very difficult thing to teach and the wonderful thing about humans is that most of us can jump from a set of practice problems in a substitution technique to a gestalt understanding of "Oh, I see that now" and once that leap is made, it is never really lost. So over the years, teachers of this material have slowly eroded the students exposure to such exercises in gaining facility at using abstraction.

In addition, many colleges and universities do not have enough mathematics majors to field a theoretical course in linear algebra which exposes the students at a critical post calculus time period to new levels of abstraction. For example, most such courses now are very computational in nature using **MATLAB** or **Sage** to illustrate the ideas with little theoretical background. The first course in differential equations can either be taught before taking linear algebra, while taking linear algebra or after taking linear algebra. The net effect of this is that one cannot, in general, use too much abstraction in the first differential equations course. Also, in calculus three, proper coverage of vector analysis is lacking as there is not enough time nor enough theoretical background to do Stokes theorem and surface integrals properly. Hence, many students finish the calculus sequence, a course in differential equations and a course in linear algebra without much theoretical understanding of many important concepts. These are then the students who will take the first semester of real analysis and also the first semester of group/ ring theory.

The issues faced in teaching abstract material are difficult but not insurmountable. The *Nuts and Bolts Foundation* is concerned that our young people are losing the ability to connect practical aspects of diagnosing problems and building solutions to the theoretical knowledge they learn. They think (B. Bergeron (2, page 7) 2010)

> more has to be done in [our] educational system. For example, shop classes - once popular - are now rare. As a result, most students who finish high school are functionally illiterate when it comes to basic mechanical skills such as the ability to read a ruler. [Also, many employers have a complaint about] new engineering graduates. Apparently, they often can't build anything because they don't know how things are made. Theoretical knowledge alone just doesn't cut it on the shop floor.

They also believe (B. Bergeron (2, page 7) 2010)

> resources [must be] devoted to the pleasures of 'tinkering' – getting away from ... video games and TV sets and into the backyard building things. In that way, we will create the next generation of artisans, inventors, engineers, repairman, and skilled workers – in short, a self-sufficient, self-sustaining society.

This idea of *tinkering* and creative play can also be applied to what Resnick (M. Resnick and J. Maloney and A. Monroy-Hernandez and N. Rusk and E. Eastmond and K. Brennan and A. Millner and E. Rosenbaum and J. Silver and B. Silverman and Y. Kafai (16) 2009) refers to as *digital literacy*.

> It has become commonplace to refer to young people as "digital natives" due to their apparent fluency with digital technologies. Indeed, many young people are very comfortable sending text messages, playing online games, and browsing the web. But does that really make them fluent with new technologies? Though they interact with digital media all the time, few are able to create their own games, animations, or simulations. It's as if they can "read" but not "write."
>
> As we see it, digital fluency requires not just the ability to chat, browse, and interact but also the ability to design, create, and invent with new media. To do [this], you need to learn some type of programming. The ability to program provides important benefits. For example, it greatly expands the range of what you can create (and how you can express yourself) with the computer. It also expands the range of what you can learn. In particular, programming supports "computational thinking," helping you learn important problem-solving and design strategies (such as modularization and iterative design) that carry over to non programming domains. And since programming involves the creation of external representations of your problem-solving processes, programming provides you with opportunities to reflect on your own thinking, even to think about thinking itself (M. Resnick and J. Maloney and A. Monroy-Hernandez and N. Rusk and E. Eastmond and K. Brennan and A. Millner and E. Rosenbaum and J. Silver and B. Silverman and Y. Kafai (16, page 62) 2009).

So it would be nice, in our opinion, to have undergraduate *and* graduate courses designed to develop such a sense of general literacy in the students. This would include the ability to use computation and simulation to develop insight, to build appropriate blends of mathematically enabled science for the purpose of exploration and utilize creative play in generalized learning. These attitudes would, of course, enhance the student's ability to problem solve and integrate theoretical and practical knowledge. A general philosophy might be one we often use in our own classes. One should be developing *tool builders* rather than *tool users*. No matter how careful the training, all classwork is eventually obsolete in the light of rapidly advancing technology and science. Hence, what is of paramount importance, is to teach students how to think for themselves; to learn how to rapidly scan lots of information in books and manuals to gather the gist and to know when to assemble a new tool from scratch to efficiently solve their current problem. In many of our engineering jobs, we were routinely simply handed a mountain of information in the form of poorly written manuals and handwritten documentation and told to make something happen in a few days. Well, you just can't read all of the stuff in a few days, so you have to be good at sifting through the pile for the useful nugget. This also means you have to be good at taking the task you are given and breaking it down into smaller problems more amenable to attack and then use tools to develop more insight. This definitely involves what Resnick said above. You have to create new representations of the process you are using to solve the problems given to you and reflect on how you are thinking about your tasks. All of these comments are true also in the case of teaching courses in abstract mathematics. You have to know how to take the knowledge you are taught in the classroom experience or from your own self-taught learning experiences and apply it to something novel.

So how does a teacher inculcate this kind of attitude in a class? For us, we always try to get the students engaged intellectually by asking questions verbally, by applying the things discussed to real problems and reiterating that it is important to carry a lot of material locally in your head. "Look," we would say, "if you are working for a company and are at the water cooler or lunch when you get into conversations with your boss, the boss will appreciate your ability to think on your feet. If you always say you have to go look it up, your boss will slowly but surely learn to rely on others who can have an informed opinion using their skills on the fly." So the bottom line is that a student does need to master their coursework in such a way that they can draw on it in real-life situations without access to a book or notes. They become leaders then and that is what is needed. This is why it is important to memorize definitions, theorems and build a large list of examples and counterexamples in theoretical mathematics and applied work. Internalizing such information is the first step in making it part of our problem solving process. What all of us who teach this stuff are after is to somehow teach students how to pull out of complicated situations and data, the bones of the underlying ideas buried inside the mess.

Currently, virtually all mathematics programs eschew the use of programming and instead use integrated programming environments such as MATLAB/Octave or Sage. In general, these environments teach poor programming skills and students who only know how to solve computational problems within that framework do not know enough about how to solve problems outside of that paradigm. This is part of the being a *tool user* versus a *tool builder* tension seen in modern society. We believe we need to rethink our approaches to teaching courses at the university to stress the ideas mentioned above: learn literacy in problem solving, tool building not just tool using, and the ability to skim lots of information quickly to get the basics assimilated. The problem with all of this, is that the teacher's role is huge! All students are essentially unique and helping a student with their journey towards becoming their best requires a lot of effort *tailored for each student.* In (D. Hecht (3, page 206) 2004), a comment is made about how a particularly ineffective psychologist looks at their patient load:

> He was ... one of that breed of psychologists who looked for a tidy, encompassing theory that wrapped the human psyche into a neat diagnostic bundle. The trailing ends, the parts that didn't fit, were to be ignored or cut to size. It was the outlook of a man accustomed to dealing with human problems in quantity: to treating an unending flow of short-term patients, managing their acute stages and referring them on, but never having to dig in for the long haul and the messy, irregular, and highly individual process of healing.

Of course, if you simply replace *psychologist* by *teacher*, it is clear this comment is equally relevant to the teaching and research profession. Many teachers want to automate their assignments with standardized testing, large class sizes so economies of scale can be brought to bear on resource allocation issues, and evaluation procedures that minimize subjectivity. The problem is that in all of our years in teaching, there are always students who don't fit into any grading scheme we devise. Our life is easier if they do, but *it seems like our job is to help them find their path to greatness, if possi-*

ble. The students have many problems of their own to work out and that will always be so; however, in the classes one teaches, one needs to offer them tools to help with that process rather than hinder it.

Therefore, our philosophy is that all parts of the analysis courses must be integrated, so we didn't want to use mathematics, science or computer approaches for their own intrinsic value. We want the students to learn to be generalists and always look for connective approaches. Also, models should be carefully chosen to illustrate the basic idea that we know far too much detail about virtually any complex system we can think of. Hence, we must learn to throw away information in the search of the appropriate abstraction. The resulting ideas can then be phrased in terms of mathematics and simulated or solved with computer based tools. However, the results are not useful, and must be discarded and the model changed, if the predictions and illuminating insights we gain from the model are incorrect. We must always remember that throwing away information allows for the possibility of mistakes. This is a hard lesson to learn, but important. This is a needed skill in learning abstract analysis as well. There is an overwhelming amount of detail and we have to find ways to train the students to decide what parts of that detail are relevant to their needs.

In general, in abstract mathematics courses such as analysis, a big problem is how should a student study so they can figure out this complicated material? Many teachers have faced this challenge. A good set of guiding principles comes from another text that tries to combine mathematics and biology **Mathematical Models of Social Evolution: A Guide for the Perplexed** by Richard McElreath and Robert Boyd published by the University of Chicago Press in 2007 (R. McElreath and R. Boyd (25) 2007). They are teaching evolutionary models to biologists, anthropologists and psychologists who, at best, have had a calculus course a long time in the past. You might think this is a very different situation than what all of us face in teaching analysis, but it is not really. When we teach the first undergraduate analysis course, our audience have all had three courses in calculus, one course in differential equations and a course in proof and perhaps a computational linear algebra course. But they hardly *know* this material. Most of them have dutifully passed the tests in these courses but do not really *know* the ideas. So in many ways it is like we are teaching people who have had their mathematics a long time in the past! McElreath and Boyd state the problem very nicely:

> Imagine a field in which nearly all important theory is written in Latin, but most researchers can barely read Latin and certainly cannot speak it. Everyone cites the Latin papers, and a few even struggle through them for implications. However, most rely upon a tiny cabal of fluent Latin-speakers to develop and vet important theory.
>
> Things aren't quite so bad as this in evolutionary biology [think a combination of mathematics, science, engineering and computation here instead!]. Like all good caricature, it exaggerates but also contains a grain of truth. Most of the important theory developed in the last 50 years is at least partly in the form of mathematics, the "Latin" of the field, yet few animal behaviorists and behavioral ecologists understand formal evolutionary models. The problem is more acute among those who take an evolutionary approach to human behavior, as students in anthropology and psychology are customarily even less numerate than their biology colleagues.
>
> We would like to see the average student and researcher in animal behavior, behavioral ecology, and evolutionary approaches to human behavior [mathematical scientists] become more fluent in the "Latin" of our fields. Increased fluency will help empiricists better appreciate the power and limits of the theory, and more sophisticated consumers will encourage theorists to address issues of empirical importance. Both of us teach courses to this end, and this book arose from our experiences teaching anxious, eager, and hard-working students the basic tools and results of theoretical evolutionary ecology.

They note the most common problem is that textbooks are too mathematical at the expense of the science. The same thing occurs in the analysis texts. They are too abstract and short on examples and insight. The *key* thing is to always include many examples and worked out steps. McElreath and Boyd have suggestions to help with learning complicated material:

> While existing theory texts are generally excellent, they often assume too much mathematical background. Typical students in animal behavior, behavioral ecology, anthropology or psychology had one semester of calculus long ago. It was hard and didn't seem to have much to do with their interests, and, as a result, they have forgotten most of it. Their algebra skills have atrophied from disuse, and even factoring a polynomial is only an ancient memory, as if from a past life. They have never had proper training in probability or dynamical systems. In our experience, these students need more hand holding to get them started.
>
> This book does more hand-holding than most, but ultimately learning mathematical theory is just as hard as learning a foreign language. To this end, we have some suggestions. In order to get the full use of this book, the reader should
>
> 1. Read the book in order. Each chapter builds on the last. ...
> 2. Work through the examples within each chapter. Math is like karate: it must be learned by doing. If you really want to understand this stuff, you should work through every derivation and understand every step. [So] read the chapters at two levels, for general understanding and for skill development.
> 3. Work all the problems. ...
> 4. Be patient, both with the material and yourself. Learning to understand and build formal evolutionary models is much like learning a language. This book is a first course, and it can help the reader understand the grammar and essential vocabulary. It is even a passable pocket dictionary. However, only practice and use of the language will build fluency. Students sometimes jump to the conclusion that they are stupid because they do not immediately understand a theoretical paper of interest. This is not justified. No one quickly learns to comprehend and produce mathematical arguments, any more than one can quickly become fluent in Latin.

They let you know that the student needs a lot of practice at working through all the steps in a derivation so that they really understand.

> This book has lots of algebra....The reason is that beginning students need it. If you really want to learn this material, you have to follow the derivations one mathematical step at a time. Students that haven't done much math in a while often have a shaky grasp on how to do the algebra; they make lots of mistakes, and even more important, don't have much information about how to get from point A to point B in a mathematical argument. All this means that they can find themselves stuck, unable to derive the next result, and unsure of whether they don't understand something fundamental or the obstacle is just a mathematical trick or algebraic error. Many years of teaching this material convinces us that the best remedy is to show lots of intermediate steps in derivations.

We follow this plan too, but we also expect that the students work hard to understand all the steps so that their growth path is on target to helping them become a modern analyst. Finally, they talk about how we use computation in our modeling work. Their comments are very true. We know many mathematicians who study biology, chemistry, ecology or physics by having the equations that some other group has decided provide a model handed to them without understanding how the model is derived. And they don't want to understand that! They simply put a thin veneer of science into their mathematics. Their work is usually impossible for scientists to use as it is divorced from the science. The same is true about computational approaches. All of these things need to be fully integrated with the science. As they say:

> There is a growing number of modelers who know very little about analytic methods. Instead, these researchers focus on computer simulations of complex systems. When computers were slow and memory was tight, simulation was not a realistic option. Without analytic methods, it would have taken years to simulate even moderately complex systems. With the rocketing ascent of computer speed and plummeting price of hardware, it has become increasingly easy to simulate very complex systems. This makes it tempting to give up on analytic methods, since most people find them difficult to learn and understand.
>
> There are several reasons why simulations are poor substitutes for analytic models.

Equations Talk Equations – given the proper training – really do speak to you. They provide intuitions about the reasons an evolutionary [think more general biological models!] system behaves the way it does, and these reasons can be read from the expressions that define the dynamics and resting states of the system. Analytic models therefore tell us things that we must infer, often with great difficulty, from simulation results. Analytic models can provide proofs, while simulations only provide a collection of examples.

Sensitivity Analysis It is difficult to explore the sensitivity of simulations to changes in parameter values. Parameters are quantities that specify assumptions of the model for a given run of the simulation - things like population size, mutation rate, and the value of a resource. In analytic models, the effects of these changes can be read directly from equations or by using various analytic techniques. In simulations, there are no analogous results. Instead the analyst has to run large numbers of simulations, varying the parameters in all combinations. For a small number of parameters, this may not be so bad. But let's assume a model has four parameters of interest, each of which has only 10 interesting values. Then we require 10^4 simulations. If there are any stochastic effects in the model, we will need maybe 100 or 1000 or 10,000 times as many....[M]anaging and interpreting the large amounts of data generated from the rest of the combinations can be a giant project, and this data-management problem will remain no matter how fast computers become in the future. Technology cannot save us here. When simple analytic methods can produce the same results, simulation should be avoided, both for economy and sanity.

Computer Programs are hard to communicate and verify There is as yet no standard way to communicate the structure of a simulation, especially a complicated "agent-based" simulation, in which each organism or other entity is kept track of independently. Often, key aspects of the model are never mentioned at all. Subtle and important details of how organisms reproduce or interact have benefited from generations of notational standardization, and even unmentioned assumptions can be read from expressions in the text. Thus it is much easier for other researchers to verify and reproduce modeling results in the analytic case. Bugs are all too common, and simulations are rarely replicated, so this is not a minor virtue.

Overspecification The apparent ease of simulation often tempts the modeler to put in every variable which might matter, leading to complicated and uninterpretable models of an already complicated world. Surprising results can emerge from simulations, effects we cannot explain. In these cases, it is hard to tell what exactly the models have taught us. We had a world we didn't understand and now we have added a model we don't understand.

If the temptation to overspecify is resisted, however, simulation and analytic methods complement each other. Each is probably most useful when practiced along side each other. There are plenty of important problems for which it is simply impossible to derive analytic results. In these cases, simulation is the only solution. And many important analytic expressions can be specified entirely as mathematical expressions but cannot be solved, except numerically. For these reasons, we would prefer formal and simulation models be learned side by side.

Of course, in the analysis courses, one wants to convince the students that not only do equations talk but the underlying abstractions speak to them too! In general, the transition to the courses that are more specifically targeted to abstract content is difficult for the students and difficult for the teacher. As outlined above, the methods used to teach material that is by necessity abstract in nature requires us to move slowly and to very carefully motivate all that is done. However, students, whether in biology, engineering or mathematics, come unprepared for such abstraction now and one must learn how to gently lead them into both an appreciation of this way of thinking and also greater and greater skill in applying these skills.

We think of the pathway to analysis competence as achievable using the sequence of five courses we have discussed earlier. In the summary to the first book, we explained our choices and now that you have mastered the first two books, it is appropriate to go over this outline again. Recall the core ideas are as follows:

- **Basic Analysis One**: This covers the fundamental ideas of calculus on the real line. Hence, sequences, function limits, continuity and differentiation, compactness and all the usual consequences as well as Riemann integration, sequences and series of functions and so forth. It is important to add pointers to extensions of these ideas to more general things regularly even though they are not gone over in detail. The problem with analysis on the real line is that the real line is everything: it is a metric space, a vector space, a normed linear space and an inner product space and so on. So many ideas that are actually separate things are conflated because of the special nature of $\Re$. **Basic Analysis One** is a two semester sequence which is different from the other texts to be discussed.

- **Basic Analysis Two**: A proper study of calculus in $\Re^n$ is no longer covered in most undergraduate and graduate curricula. Most students are learning this on their own or not learning it properly at all even though they get a master's or Ph.D. in mathematics. This course covers differentiation in $\Re^n$, integration in $\Re^2$ and $\Re^3$, the inverse and implicit function theorem, and connections of these things to optimization.

- **Basic Analysis Three**: This covers the basics of three new kinds of spaces: metric spaces, normed linear spaces and inner product spaces. Since the students do not know measure theory at this point, many of the examples come from the sequence spaces ℓ^p. However, one can introduce many of the needed ideas on operators, compactness and completeness even with that restriction. This includes what is called **Linear Functional Analysis**. In general, the standards are discussed: the Hahn - Banach Theorem, the open and closed mapping theorems and some spectral theory for linear operators of various kinds.

- **Basic Analysis Four**: This covers the extension of what is meant by the length of an interval to more general ideas on the *length* or *measure* of a set. This is done abstractly first and then specialized to what is called **Lebesgue Measure**. In addition, more general notions are usually covered along with many ideas on the convergence of functions with respect to various notions.

- **Basic Analysis Five**: Here, we connect topology and analysis more clearly. There is a full discussion of topological and linear topological spaces, differential geometry, some degree theory and distribution theory.

You can see this sequence laid out in Figure 20.1. Since what we discuss in the first four volumes is still essentially what is a *primer* for the start of learning even more analysis and mathematics at this level, in Figure 20.1 we have explicitly referred to our texts using that label. The first volume is the one in the figure we call **A Primer On Analysis**, the second (this text) is **Primer Two: Escaping The Real Line**, the third is **Primer Three: Basic Abstract Spaces**, the fourth is **Primer Four: Measure Theory** and the fifth is **Primer Five: Functional Analysis**. Keep in mind these new labels as they show up in Figure 20.3 also. The typical way a graduate student takes this sequence is shown in Figure 20.2. There are lots of problems with how mathematically interested students learn this material. An undergraduate major here would be required to take a two semester sequence in core real analysis, however, given the competition for graduate students in a mathematical sciences graduate program, it is often true that incoming master's degree students are admitted with deficiencies. One often has people come into the program with just one semester of core real analysis. Such students should take the second semester when they come into the program so that they can be exposed to Riemann integration theory and sequences of functions among other important things, but not all want to, as it means taking a deficiency course. However, from Figure 20.2, you can see that jumping straight to **Basic Analysis Three** on Linear Analysis without proper training will almost always lead to inadequate understanding. Such students will have a poorly developed core set of mathematics, which is not what is wanted. Hence, one needs to do as much as one can to train all undergraduate students in two full semesters of core real analysis. You are now nicely trained and fit well into the journey into abstraction laid out in Figure 20.2.

Even if students take the full two semesters of core real analysis at the graduate level, the undergraduate curriculum today does not in general offer courses in many areas:

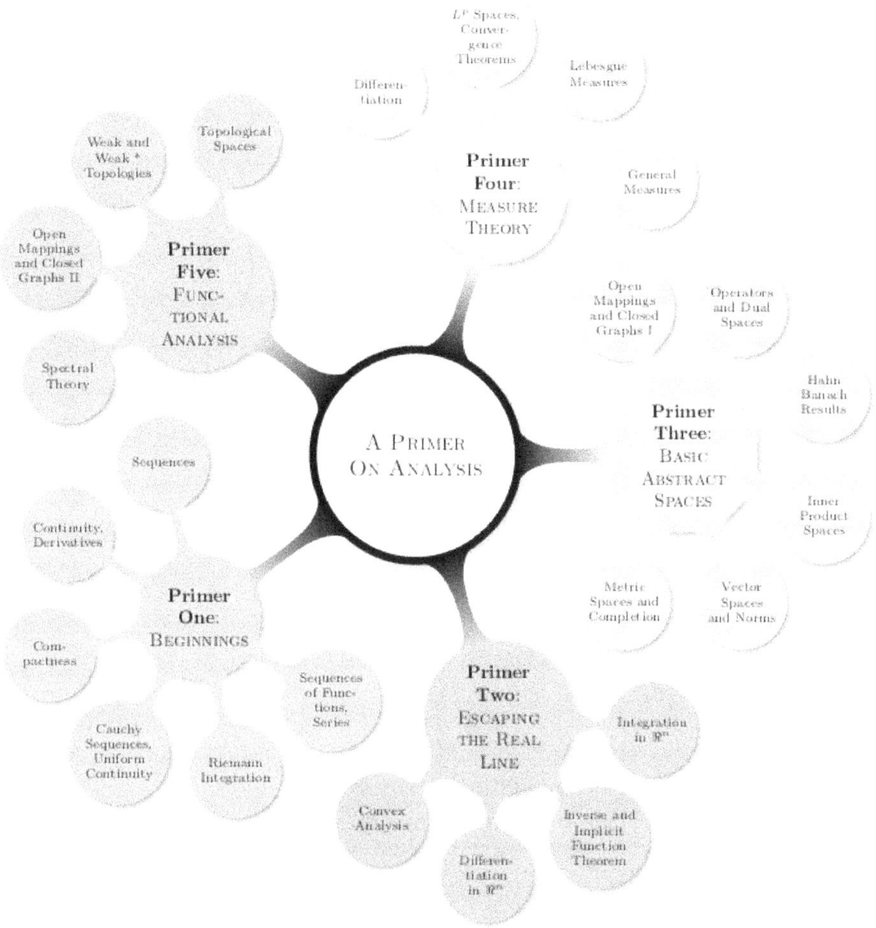

Figure 20.1: The general structure of the five core basic analysis courses.

- Basic Ordinary Differential Equation (ODE) Theory: existence of solutions, continuous dependence of the solution on the data and the structure of the solutions to linear ODEs. Since most students do not take **Basic Analysis Two**, they have not been exposed to differential calculus in $\Re^n$, the inverse and implicit function theorem and so forth. Many universities have replaced this type of course with a course in Dynamical Systems which cannot assume a rigorous background in the theory of ODE and so the necessary background must be discussed in a brief fashion. Again a text such as **Basic Analysis Two** designed for both self-study and as a regular class is an important tool for the mathematical science major to have available to them.

- Basic Theoretical Numerical Analysis: the discussion of the theory behind the numerical solution of ODEs, linear algebra factorizations such as QR and optimization algorithms. This is the course where one used to discuss operator norms for matrices as part of trying to understand our code implementations.

- Basic Manifold Theory: the discussion of the extension of $\Re^n$ topology and analysis to the

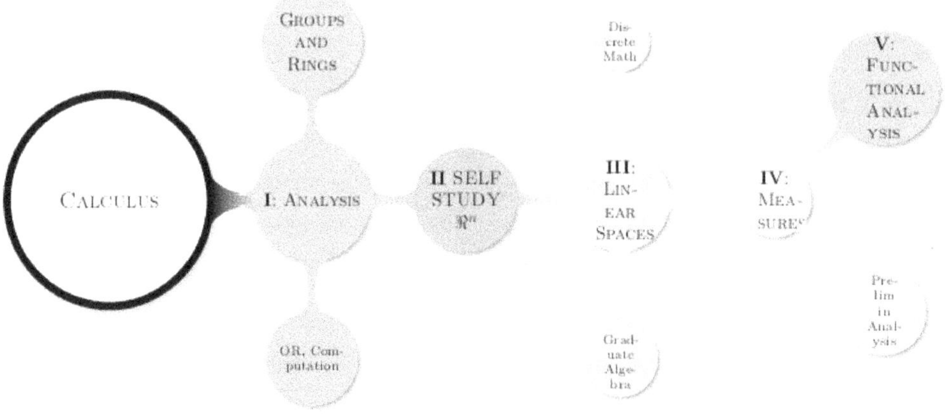

Figure 20.2: The analysis education pathway.

more general case of sets of objects locally like $\Re^n$. This material is pertinent to many courses in physics and to numerical Partial Differential Equations (PDE) as manifolds with a boundary are essential. Note this type of course needs **Basic Analysis Two** material, more topology and a more general theory of integration on chains.

Of course, no student can take all of these courses and so the question is how to prepare them to either study them on their own or to study them later as part of their further graduate education. We argue that the ideas covered in **Basic Analysis Two** are essential to mathematical growth and even to the expanded use of mathematics in models seen in biology. For example, in our own work, we used differential models in three or four variables in (J. Peterson and A. M. Kesson and N. J. C. King (12) June 21, 2017) and (J. Peterson and A. M. Kesson and N. J. C. King (13) June 21, 2017). These papers appear in an immunological journal, not a mathematical biology journal, and since the mathematical background of the typical immunologist at best includes one semester of calculus and this model needs **Basic Analysis Two** material or at least the course (J. Peterson (9) 2016), we are asking the immunology community to grow mathematically and add these kinds of ideas into their toolkits. But also notice that our mathematical students have not been exposed to these ideas either even if they take the full sequence of **Basic Analysis One**. Ideas from algebraic topology can also be used to model signals into a complex biological or engineering system and both mathematical sciences and physical sciences students have not been exposed to those ideas. Again, it is clear one must inculcate into the students the ability to read and think critically on their own, *outside of class and instructors* so they can become the scientists of the future. The way these proposed analysis courses fit into the grand scheme of things is shown in Figure 20.3.

So here's to the future! Now that you have a much needed familiarity with basic analysis in $\Re^n$ and a better understanding the integration, you are ready to move into deeper levels of abstraction. As discussed, this next journey starts in the next book (Peterson (22) 2020) which focuses on the study of metric, normed and inner product spaces in more depth. You are welcome to get started on that one next! In most graduate level mathematical sciences departments, you are now at the stage where you are preparing for the preliminary examinations in analysis. These typically cover the material in the books one, three and four of our series. Just remember to master the material in the courses you take at your university, supplement as needed from books such as ours and learn to think carefully and deeply. You are preparing yourself for a life of research and the better you learn this material the easier it will be for you to read journal papers and think about solving problems that have not been solved before. Study all the proofs carefully as they tell you how to approach new problems.

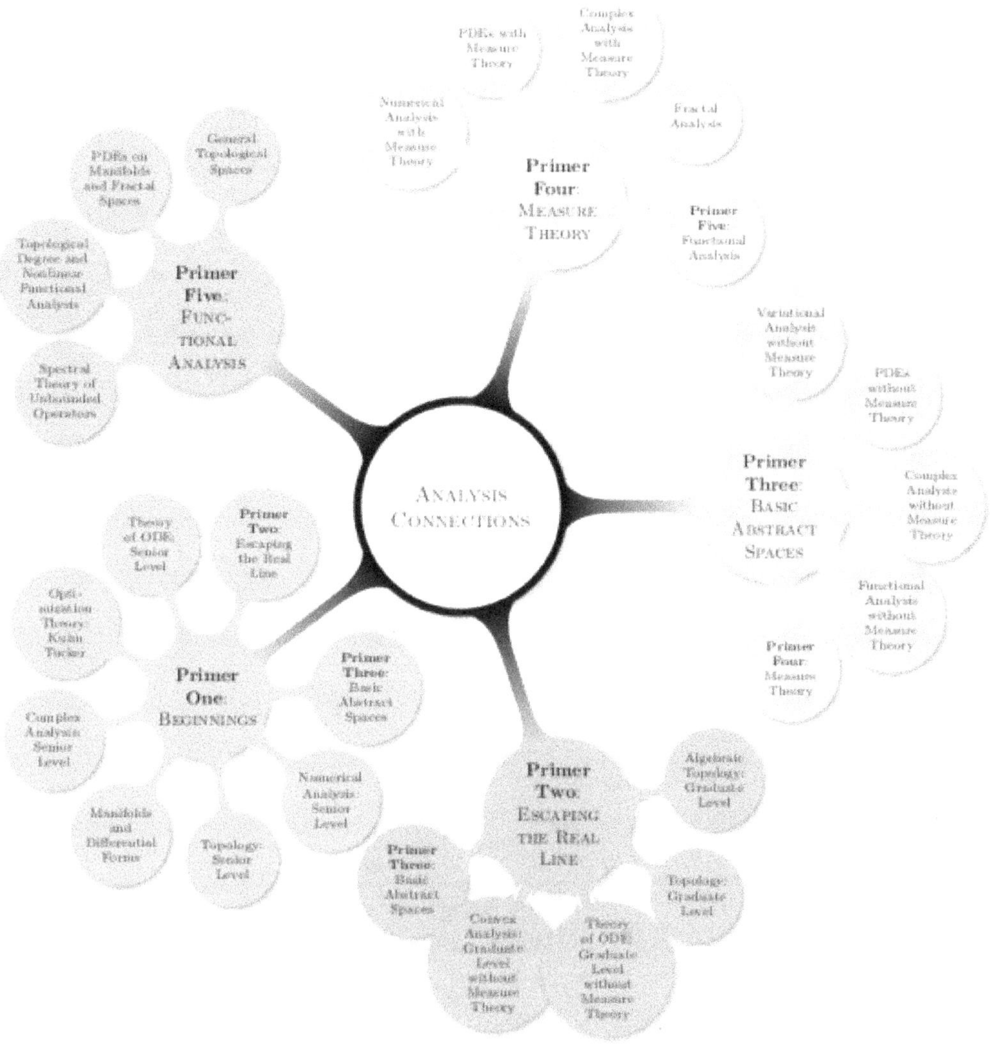

Figure 20.3: Analysis connections.

In the proof of each theorem, ask yourself what role each assumption plays in the proof so that you understand why the proof fails if you don't make that assumption. And remember, a new conjecture you have later in your own work always follows this pattern: it is true and you just haven't found the proof yet or it is false and there is a counterexample. This is a tricky thing to navigate and the best advice we can give you is to really, really learn the background material such as is in the books we have written, and know the examples well. The final piece of advice is to learn to redefine your problem: when you are stuck, a lot of times your failure leads to a better formulation of the problem you want to solve. It is the equivalent of walking around a wall instead of just butting into it over and over!

So enjoy your journey and get started!

Part VII

References

References

[1] A. E. Roy. *Orbital Motion*. Adam Hilger, Ltd, Bristol, 1982.

[2] B. Bergeron. Mind/ Iron: Little Hands Build Big Dreams. *Servo*, 8(1):6–7, 2010.

[3] D. Hecht. *Land of Echoes*. Bloomsbury Publishing, 2004.

[4] John W. Eaton, David Bateman, Søren Hauberg, and Rik Wehbring. *GNU Octave version 5.2.0 manual: a high-level interactive language for numerical computations*, 2020. URL https://www.gnu.org/software/octave/doc/v5.2.0/.

[5] J. Fortney. *A Visual Introduction to Differential Forms and Calculus on Manifolds*. Birkhäuser, 2018.

[6] Free Software Foundation. *GNU General Public License Version 3*, 2020. URL http://www.gnu.org/licenses/gpl.html.

[7] G. Sutton and O. Biblarz. *Rocket Propulsion Elements*. John Wiley & Sons, Inc, New York, 2001.

[8] H. Curtis. *Orbital Mechanics for Engineering Students, Third Edition*. Butterworth-Heinemann, 2014.

[9] J. Peterson. *Calculus for Cognitive Scientists: Derivatives, Integration and Modeling*. Springer Series on Cognitive Science and Technology, Springer Science+Business Media Singapore Pte Ltd. 152 Beach Road, #22-06/08 Gateway East Singapore 189721, Singapore, 2016. URL http://dx.doi.org/10.1007/978-981-287-874-8.

[10] J. Peterson. *Calculus for Cognitive Scientists: Higher Order Models and Their Analysis*. Springer Series on Cognitive Science and Technology, Springer Science+Business Media Singapore Pte Ltd. 152 Beach Road, #22-06/08 Gateway East Singapore 189721, Singapore, 2016. URL http://dx.doi.org/10.1007/978-981-287-877-9.

[11] J. Peterson. *Calculus for Cognitive Scientists: Partial Differential Equation Models*. Springer Series on Cognitive Science and Technology, Springer Science+Business Media Singapore Pte Ltd. 152 Beach Road, #22-06/08 Gateway East Singapore 189721, Singapore, 2016. URL http://dx.doi.org/10.1007/978-981-287-880-9.

[12] J. Peterson and A. M. Kesson and N. J. C. King. A Theoretical Model of the West Nile Virus Survival Data. *BMC Immunology*, 18(Suppl 1)(22):24–38, June 21, 2017. URL http://dx.doi.org/10.1186/s12865-017-0206-z.

[13] J. Peterson and A. M. Kesson and N. J. C. King. A Model of Auto Immune Response. *BMC Immunology*, 18(Suppl 1)(24):48–65, June 21, 2017. URL http://dx.doi.org/10.1186/s12865-017-0208-x.

[14] S. Kalajdzievski. *An Illustrated Introduction to Topology and Homotopy*. Chapman and Hall/CRC, 2015.

[15] M. Braun. *Differential Equations and Their Applications*. Springer-Verlag, 1978.

[16] M. Resnick and J. Maloney and A. Monroy-Hernandez and N. Rusk and E. Eastmond and K. Brennan and A. Millner and E. Rosenbaum and J. Silver and B. Silverman and Y. Kafai. Scratch Programming for All. *Communications of the ACM*, 52(11):60–67, 2009.

[17] M. Spivak. *Calculus on Manifolds*. Perseus Press, 1965.

[18] MATLAB. *Version Various (R2010a) - (R2019b)*, 2018 - 2020. URL `https://www.mathworks.com/products/matlab.html`.

[19] J. Peterson. *Basic Analysis IV: Measure Theory and Integration*. CRC Press, Boca Raton, Florida 33487, 2019.

[20] J. Peterson. *Basic Analysis V: Functional Analysis and Topology*. CRC Press, Boca Raton, Florida 33487, 2020.

[21] J. Peterson. *Basic Analysis I: Functions of a Real Variable*. CRC Press, Boca Raton, Florida 33487, 2020.

[22] J. Peterson. *Basic Analysis III: Mappings on Infinite Dimensional Spaces*. CRC Press, Boca Raton, Florida 33487, 2020.

[23] J. Peterson. *Basic Analysis II: A Modern Calculus in Many Variables*. CRC Press, Boca Raton, Florida 33487, 2020.

[24] R. Bate and D. Mueller and J. White. *Fundamentals of Astrodynamics*. Dover, 1971.

[25] R. McElreath and R. Boyd. *Mathematical Models of Social Evolution: A Guide for the Perplexed*. University of Chicago Press, 2007.

[26] R. Wrede and M. Spiegel. *Schaum's Outline of Advanced Calculus*. McGraw Hill, 2010.

[27] S. Morita. *Geometry of Differential Forms*. American Mathematical Society, 1998.

[28] P. Savelier. *Topology Illustrated*. Peter Savelier, 2016.

[29] W. Fulton. *Algebraic Topology: A First Course*. Graduate Texts in Mathematics 153, Springer NY, 1995.

[30] W. Thompson. *Introduction to Space Dynamics*. Dover, 1961.

Part VIII

Detailed Index

Index